Exploration
of the Universe

Exploration of the Universe
Second Edition

George Abell
University of California, Los Angeles

Holt, Rinehart and Winston
New York   Chicago   San Francisco   Atlanta
Dallas   Montreal   Toronto   London   Sydney

Acknowledgments

Cover photograph: Veil nebula.

    (Courtesy Mount Wilson and Palomar Observatories.)

Half-title photograph: M8, Lagoon Nebula in Sagittarius.

    (Courtesy Mount Wilson and Palomar Observatories.)

Frontispiece photograph: The "Whirlpool" Galaxy, M51.

    (U.S. Naval Observatory; Flagstaff Station.)

Library of Congress Catalog Card Number: 69-13567

SBN: 03-075955-2

Printed in the United States of America

 0123 41 987654

Preface
to Second Edition

"Nature," wrote Emerson, "is always consistent, though she feigns to contravene her own laws." It is, of course, the business of Science to discover the consistency of Nature — the laws or rules that describe her behavior. It is difficult to imagine a more appropriate field than astronomy — the oldest science — with which to demonstrate this consistency, nor with which to exhibit the apparent contravention of Nature's own order. Thus we apply well-understood laws of celestial mechanics to travel to the moon, but are dumbfounded in our attempts to understand the mysterious quasars and pulsars.

Astronomy, then, is an old and precise, as well as contemporary and perplexing, science. How can one be educated and yet ignorant of what man has learned and is striving to learn of the universe about us? How, in this day when our students are calling on us to teach them what is "relevant," can they consider irrelevant the very forces and circumstances that brought our world into existence, and later us into being, and which, in the broad sense, guide our destiny, and which will ultimately decide our fate?

In the older tradition, science in general, and astronomy in particular, were part of the humanities — as *natural philosophy*. An introduction to science is still often included among the requirements for a degree at colleges and universities. It is for this purpose that *Exploration of the Universe* was written. Although it is hoped that the book will be of interest and value to prospective science students, it is written mostly for that great captive audience of liberal-arts students. Further, it is recognized that it may be the text for the only good course in science they ever will take, so it is first a text in science, and second a text on the application of the scientific method to our exploration of the universe.

I have attempted here to make astronomy meaningful and understandable to a person without special training in science or mathematics. Often, I have adopted a semihistorical approach to show how man's ideas develop, and something of the human side of science. I hope that in reading through a vast maze of facts and new and difficult ideas, the student can somehow escape becoming bogged down in detail and trivia, and instead sense some of the excitement, drama, and grandeur of astronomy. Of course, a science text is not a text if it is merely descriptive and presents the "gee-whizes" of its field. Frankly, such a book is not even very interesting to the curious reader. Rather, a science text must show how we have learned what we know, or at least what we think we know. Science, in a sense, is a great structure of logic, based on relatively few postulates. It must be shown how each logical step, or component of the structure, follows from and builds upon the previous steps, or components.

It is the latter that makes science so difficult for the nonscientist. Astronomy is mathematical, of course, but in this elementary introduction I have avoided mathematics beyond that which is learned in the first few weeks of high-school courses in algebra and geometry. I think that it is not mathematics in the computational sense that bothers students; it is that they must learn new thought processes — how to put simple ideas together and to follow logical thought processes that may be more complex than those they are used to. A student cannot *memorize* his way through a good course in science — he must learn to *think*. The exercises at the ends of the chapters are, for the most part, not of the type that may be answered by thumbing back through the text. They are intended to encourage the student to think about the subject matter, to try to understand it, and to apply new ideas. The perplexed student should be guided by the maxim: a complexity is nothing more than a disarrangement of simplicities.

Students occasionally become so interested in a particular subject that they wish to learn more or to delve deeper. Usually they must resort to more advanced literature, but sometimes they can be given additional information and/or insight in a short space. For these students, a considerable amount of extra material is included in *Exploration of the Universe,* and is set in somewhat smaller print. The headings of sections containing supplementary material are marked with an asterisk, and end-of-chapter exercises pertaining to these sections are also marked with an asterisk. These additional sections are either somewhat difficult or rather specialized, and it is neither intended nor expected that they will be read by the typical nonscience student; they can be omitted entirely without loss of continuity.

Exploration of the Universe is designed for a one- or two-semester course. There is, however, too much material to be learned thoroughly in a one-semester or quarter course. Instructors will vary in their judgment on which chapters or sections are most expendable. I recommend that the entire book (other than the fine-print parts) be read, at least quickly, even in the shorter course. Some

parts, however, should be emphasized. My choice for the chapters that can be glossed over slightly, or even omitted, is: 6, 8, 15, 16, 17, 19, 25, 28, 29, and 31. I have also prepared a brief edition of *Exploration of the Universe* specifically for the one-quarter or one-semester course.

I have made every effort to make this revised edition as up-to-date as possible as it goes to press, even to the point of incorporating new data into the proof (to the chagrin of my publisher). But in a field that evolves as rapidly as astronomy no text can be up-to-date — even after the first few months. If all goes as planned, before this book is in the hands of many students, Americans will have landed on the moon and returned to earth. The Apollo Program and unmanned planetary explorations over the next few years are certain to add to, and probably change, some of our ideas about the solar system. If the Orbiting Astronomical Observatory (just launched at the time of writing) continues to function properly, we shall have a new view of the universe in ultraviolet radiation. Pulsars are being discovered so rapidly (in late 1968) that rather than trying to give the latest figure for the number known, I debated giving a telephone number the reader could dial to learn the most recent total. I cannot even guess when we will understand the natures of the pulsars and of the quasars. In short, the burden is on the instructor of an astronomy course to keep himself abreast of new developments by following the scientific periodicals, no matter what textbook he may be using.

I have made every effort to weed out errors from the First Edition — both printing errors and errors of fact. Yet, both kinds of errors unquestionably persist. I am most grateful to the many correspondents who informed me of such slips in the First Edition; these good people are too numerous to list here, but I would like to mention a few who were especially helpful: Sally H. Dieke, H. Herbert Howe, Charles Hyder, Clyde Tombaugh, and Peter Wehinger. I shall continue to appreciate having mistakes and suggestions of all kinds called to my attention, so that future printings of this book may be improved.

In the preparation of the First Edition, I am most indebted to Mr. Paul Wylie and Drs. Daniel Popper, Paul Routly, and John Schopp, who carefully read the manuscript and helped me greatly with many fine suggestions. Additional reviewers who have read sections of this new edition include: Drs. Grant Athay, William Bidelman, Owen Gingerich, Kenneth Greisen, Louis Henyey, William Kaula, Dimitri Mihalas, Thornton Page, Stanton Peale, Hyron Spinrad, Victor Thoren, Harold Weaver, George Wetherill, and Harold Zirin. All have been most helpful, but none has seen the final revision, and I alone must take responsibility for the remaining goofs.

Most of all, I am grateful to my long-suffering family. They had to wait through four years of weekends for me to finish writing the First Edition; then I added insult to injury by ignoring them for another year while I revised it! It is to them that any merit from this work is heartily dedicated.

Los Angeles, California   G.O.A.
March 1969

Contents

Exploration
of the Universe

CHAPTER 1

The Scope
of Astronomy

To the ancient observer the earth seemed vast and immobile — the center of the universe. Indeed, the entire realm of the universe, the whole scope of his knowledge and imagination, was the *realm of the earth*. He regarded the sky as a huge inverted bowl overhead, or actually, as a great hollow sphere completely surrounding the earth. This celestial sphere turned daily, carrying around the sky, the sun, moon, stars, and planets that seemed to be affixed to its inner surface. It was natural, of course, for early man to attach such central importance to his earth. His experience had not prepared him to accept the sky as virtually empty space extending to such unimaginable — perhaps infinite — depths that the earth, in comparison, becomes a mere point in the cosmos.

1.1 DEVELOPMENT OF ASTRONOMICAL THOUGHT

Certain original and imaginative individuals, however, began to make systematic observations of the positions and motions of the celestial objects, to formulate more precise theories to explain their behavior, and to subject such theories to critical tests to ascertain their validity. Thus, with the development of the scientific procedure, man's concept of the universe began to change until, a few centuries ago, he had come to the realization that the earth was one of several planets that revolve in precise paths and in regular periods about the sun.

The sun, man learned, is a great, luminous, gaseous sphere with a diameter more than a hundred times that of the earth. The earth, on the other hand, is more or less like the rest of the planets — cold and solid. Further, man learned that there are many lesser objects as well in the solar system. Some are minor planets, tiny planets, most of them not over a few miles across, circling the sun by the thousands. Others, in unknown numbers, are comets, loose swarms of solid particles embedded in tenuous gases, revolving about the sun in elongated orbits that carry them billions of miles into space and eventually back to the vicinity of the sun in long, but regular, periods. Still others are meteoroids, cluttering space by the countless trillions, particles that range in size down to microscopic dimensions.

FIG. 1-1   The Sun. *(Mount Wilson and Palomar Observatories.)*

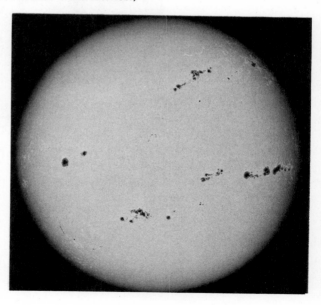

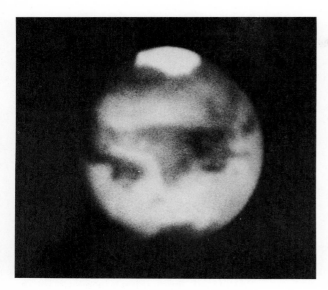

FIG. 1-2   The planet Mars, photographed in orange light in August, 1956. *(Mount Wilson and Palomar Observatories.)*

The earth, in other words, is only one member of a large family of objects in the solar system, all of which derive their light and heat from that central and by far most massive body, the sun. Until recent times, however, it was the planets, not the sun, that commanded the principal attention of astronomers. These were other worlds, in some respects like the earth. It was natural to investigate them, to wonder and speculate about them. Was it possible that some of them might also maintain life? Man's concept of space had moved out to encompass the solar system — the *realm of the planets.*

Human curiosity, however, does not long remain inattentive to what lies beyond. As early as 1600, the monk Giordano Bruno was burned alive in Rome for writing such heresies as that the stars themselves are suns. But in 1838, Bessel's detection of the parallax of the star 61 Cygni (a slight angular displacement due to the earth's orbital motion) made possible the first direct measurement of a stellar distance. The investigations that followed proved conclusively that the stars are suns, more or less like our own sun, but at such immense distances that they appear as mere points. Light, which travels with a speed of 186,000 mi/sec, requires minutes or hours to reach us from the planets but years to travel from and between the stars. Once more man's concept of space had enlarged, now to encompass the *realm of the stars.*

It had been suspected in the middle eighteenth century, and demonstrated convincingly by William Herschel in 1787, that the stars are not strewn over the sky at random but rather make up a huge assemblage, or *galaxy.* Herschel described the assemblage as shaped like a "grindstone." The Milky Way, a luminous band completely circling the sky, is the light from the myriads of stars that are projected into our line of sight as we look edge on through the grindstone. In addition to individual stars, it was found that the Galaxy contains clusters of stars and diffuse luminous clouds, or *nebulae.* Although the disklike character of the Galaxy was understood throughout the nineteenth century, astronomers had no clear idea of its extent until well into the twentieth century. It is a system of about 100 billion stars — suns like our own — stretching across an expanse of space that light may require 100,000 years to traverse. The scope of astronomy had now extended to include the *realm of the Galaxy.*

Immanuel Kant had speculated in 1755 that the sun was a part of a system or great cluster of stars, and that there were other such clusters, or "island universes," distributed through space far beyond the borders of our own system. From time to time in the decades to follow, the *island universe* theory gained the attention of astronomers. It seemed possible that many of the nebulae were composed of stars and might actually lie out beyond the stars of our Galaxy — "candidates" for such outposts of space. By the close of the nineteenth century, however, some of the nebulae had

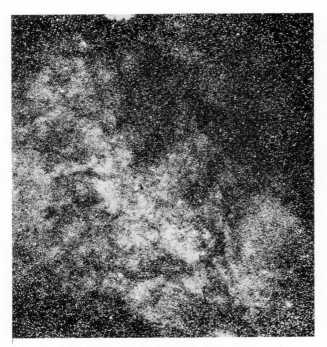

FIG. 1-3 Star clouds and dark obscuring matter in the Milky Way in the region of Sagittarius. *(Yerkes Observatory.)*

been shown to be of a definitely gaseous nature, and even as late as 1920 the weight of astronomical opinion did not favor the notion that any of the nebulae could be *island universes* — galaxies in their own right.

The controversy over the theory continued, however, until 1924, when it was finally settled. Kant had been right. Cepheid variables were discovered in some of the "nebulae." These variables are stars that can be recognized because of the peculiar and characteristic manner in which they vary in their brightness. The Cepheid variables in our own galaxy are known to be very luminous stars; yet in the nebulae they appear very faint, which indicates that these Cepheids, and the nebulae in which they reside, must lie at distances far greater than the farthest stars in our own galaxy.

Subsequent investigations have shown that most of the other nebulae — in fact, all but the comparatively few that are glowing gases between the stars of the Milky Way — *are* galaxies. They speckle space to the depths of the observable universe — to distances requiring light billions of years to cross. The number of observable galaxies must be close to a billion. Most of these appear as insignificant, slightly irregular specks, barely visible on photographs made over long time exposures with the world's greatest telescopes. The number of remote galaxies too faint to observe cannot even be estimated with any degree of confidence. Yet each galaxy is a system of billions of stars, each star a sun like our own. Even in our galaxy, which is probably only one of many billions, the nearest other suns are so remote as to appear as points of light. Man's concept of the universe has enlarged to encompass the *realm of the galaxies.*

Since World War II, new techniques have been developed which make it possible to receive and record radio radiation emitted by celestial objects. Still more recently, instruments carried by rockets, artificial satellites, and space probes have permitted the detection of energy in the far ultraviolet, x-rays, and even gamma rays — adding new data to that obtained from ground-based observatories. This new information has raised the inevitable new problems — for example, the source of x-rays from certain peculiar stars and of radio waves from peculiar galaxies. Especially perplexing is the nature of the quasi-stellar objects or "quasars" — stellar-appearing objects that emit much of their energy at radio wavelengths and which may be the most remote things yet observed. Many other sources of radio waves may be galaxies too remote to be visible through even our greatest optical telescopes. Some of the feeble radio radiation coming from all directions in space is now thought to be energy generated billions of years ago from a primeval state of the universe, and may be an immensely

FIG. 1-4   The spiral galaxy M81 in Ursa Major.
(Lick Observatory.)

important clue to our investigations in the *realm of cosmology.*

Today, in the new age of space technology, man stands at the "threshold of space." Yet, he is not really at the threshold of the astronomical universe. Even if man should go beyond the moon, to the planets, or even if, in some unforeseen, unimaginable manner, he should succeed in traveling at nearly the speed of light and reaching the nearest stars — even then, he would scarcely have begun on a journey through *space* in the astronomical sense of the word.

1.2 ASTRONOMY AS A SCIENCE

Astronomy is the science of the universe. The astronomer, of necessity, must examine his subjects from afar; the whole of the cosmos is his laboratory.

Except for his limited exploration of the moon and neighboring planets, made possible by the new space technology, and for his observations of cosmic rays and of fallen meteorites, he has no direct contact with the objects of his investigations. In this respect astronomy differs from the other sciences.

The astronomer must call upon all his resources of ingenuity and employ more clever devices than the most astute detective to solve the problems of his business. In this respect astronomy is representative of the other sciences.

Man, by nature, is curious about his surroundings, particularly about unexplained phenomena. To satisfy his curiosity, he may employ one of two entirely dissimilar approaches. The first is to invent, to make up an explanation out of the imagination. The procedure is extremely "efficient" — it answers all the big questions. Here, of course, is the root of myth, superstition, and primitive religion. One such "religion" that grew rather naturally from a very early and invented idea of the nature of the universe is astrology.

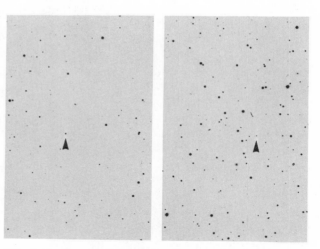

FIG. 1-5   Quasistellar objects BSO-1 and 3C-9; 200-inch telescope photograph. *(Mount Wilson and Palomar Observatories.)*

The other approach is the *scientific method.* One gathers clues by making observations or performing experiments. With these clues at hand, one formulates tentative hypotheses to explain the phenomena or the experimental results. Then — the crucial step — one tests these hypotheses by predicting from them new phenomena or the results of new experiments. If the tests fail, the hypotheses must be discarded in favor of better ones.

The purpose of science is to find order in the chaos of natural phenomena. Science attempts to *represent* nature as simply and accurately as possible with *natural laws* — descriptions of how nature *behaves.* Note that science *describes the how* but does not *explain the why* of nature; it makes no attempt to establish the true and absolute "nature of things." This latter activity belongs in the province of religion. Science, then, is a *method,* not a *subject.* It is a *method* for the organized investigation of nature.

The scientific method is the most successful means we have found for achieving the aims of science. It has made possible our modern technology with its skyscrapers, airplanes, television, and rockets. It does not, however, yield absolutes of knowledge. Rather it allows us to describe how nature has behaved in the past and how we might expect it to behave in the future. The *laws* of science are the rules of the game that nature seems to play. Some of these rules are extremely well established and appear never to be violated. We cannot, of course, consider them absolute and binding on nature; yet they are so consistent that we believe in them almost religiously, and predic-

FIG. 1-6   Stonehenge, an assemblage of stones in Western England, believed to have been an astronomical shrine of early Druids. *(Rapho Guillumette Pictures.)*

tions based on them are, to all intents and purposes, "established facts." For example, if we raise a stone and drop it, we can be quite certain that it will fall to the ground. Similarly, all known phenomena are so consistent with the hypothesis that the earth goes around the sun, and not vice versa, that we regard the earth's motion as "fact," not "theory." Science, however, deals with other postulates that are by no means so firmly established. We shall come across many of these in the chapters to follow.

Although science itself does not ascribe an absolutism to its laws, it does not follow that one theory is as good as another. We hear much from cranks and charlatans proposing new (and not so new) theories in the name of science. They dwell on the scientific errors of the past, on "academic prejudice" (against new ideas that conflict with established theory), and on their inability, against such persecution, to obtain general acceptance of their new revelations, or even to be heard. They are not, however, always so unheard. The public invests millions of dollars a year in the purchase of crank literature. In 1950 a best-selling book, from a highly reputable publisher, described the eruption from Jupiter of a comet, which stopped or slowed the rotation of the earth, and eventually became the planet Venus. The most elementary knowledge of the natures of comets and planets, and of the simplest and best-established physical law, is sufficient to reveal the absolute absurdity of the idea. The author, in common with nearly all authors of such hypotheses, acknowledged that his theory is counter to established laws of science, and that if it is to be accepted those laws must be rejected. To the general public, uneducated in even elementary science, and used to many technical miracles, the scientific-sounding jargon and forceful arguments of cranks are often convincing.

Of course, scientists *do* make mistakes — and frequently. Even the most eminent of them often disagree. But wrong tracks and conflicting views are almost always at the frontiers of knowledge, where progress is made by the trial and error process of the scientific method. Moreover, "established" ideas are under continual reevaluation and, if necessary, revision; but they are not likely to be rejected entirely. Newton's laws have not been abandoned because they fail to apply to the motions of electrons in atoms or to bodies moving with speeds near that of light (for which the quantum theory and relativity, respectively, are superior descriptions); they are as valid as ever within the domain in which they have been shown to apply. If our most basic rules of science are wrong, then it is hard to understand how repeated tests continue to verify them, let alone how our modern technology ever came to be. In any case, the burden rests on new theories to predict the results of tests or observations more accurately than the old ones.

One of the most important, and perhaps most difficult, tasks for a student taking a course in science is to learn to evaluate what he hears and reads in terms of basic and well-tested natural laws and to tell the crank from the legitimate investigator. The task is not always easy for the scientist himself, and there exists a middle ground where the distinction itself is at best vague. The student should learn to develop at least a critical and objective attitude and to distinguish open-mindedness from gullibility.

Astronomy is probably the oldest of the sciences; it is also one of the most exciting and rapidly developing fields of knowledge today. What we have learned of the universe may well stagger the imagination, but no less impressive is the list of unsolved problems. A text in astronomy must therefore be somewhat of a "progress report" to the student; in no other field do textbooks go out of date so rapidly. Astronomy is an excellent subject with which to demonstrate the methods of the detectives known as scientists. In scientific investigation, however, unlike detective fiction, the case is never closed. Certain facets may be "cleared up" from time to time, but new "developments" continually complicate the picture.

1.3 CONTRIBUTIONS OF ASTRONOMY

Astronomy, as well as serving as an excellent example of the scientific approach, is important for the effect its contributions have had on our everyday lives. Observations of the motions of the planets led Newton to his great discovery of the laws of motion and gravitation — the basis of much of our modern technology. Further, many mathematical techniques used daily by our engineers were first developed to solve astronomical problems. The science of optics was initiated and developed largely by astronomers for the investigation of astronomical problems. Yet it has led to the invention not only of the telescope, but also of the microscope, so vital to medicine and other fields unrelated to astronomy. The field of spectros-copy, which had an astronomical origin and is still widely applied to astronomical problems, has proved invaluable to physics, chemistry, metallurgy, biology, and other areas of knowledge that bear directly on our daily living. Of course, time-keeping, navigation, and geodesy are closely related to astronomy. The vastly important field of physics, in fact, developed largely as an offshoot of astronomy.

Perhaps the most spectacular contribution of astronomy in the twentieth century has been the realization that thermonuclear energy is the source of energy of the sun and other stars. This realization has led to experiments in nuclear physics, and to man's effort to duplicate the source of stellar energy. It is, perhaps, a dubious honor to astronomy that man first succeeded in this venture with the production of nuclear bombs.

Now let us begin the story of man's exploration of the universe. We shall see how astronomers dabble in their laboratory — the universe itself.

CHAPTER 2

Ancient Astronomy

2.1 EARLIEST ASTRONOMERS

Primitive speculations on the nature of the universe must date from prehistoric times. It is difficult to state definitely when the earliest observations of a more or less quantitative sort were made, or when astronomy as a science began. Certainly, in many of the ancient civilizations the regularity of the motions of celestial bodies was recognized and attempts were made to keep track of and predict celestial phenomena. In particular, the invention of and keeping of a calendar requires at least some knowledge of astronomy — the basic units of the calendar being the day, the month (originally, the 29- and 30-day cycling of the moon's phases), and the year of seasons (the *tropical year*).

The Chinese had a working calendar at least as early as the fourteenth or thirteenth century B.C., and probably much earlier. It is said that in 2159 B.C. two astronomers, Hi and Ho, were executed for failing to predict an eclipse, but modern authorities consider it more likely that their error was actually one of carelessness in the preparation of the official calendar. Some scholars have claimed that the Chinese had determined the length of the year at 365¼ days as early as the twelfth century B.C. About 350 B.C., the astronomer Shih Shen prepared what was probably the earliest star catalogue, containing about 800 entries. The Chinese also kept rather accurate records of comets, meteors, and fallen meteorites from 700 B.C. Records were made of sunspots visible to the naked eye and of what the Chinese called "guest stars," stars that are normally too faint to be seen but suddenly flare up to become visible for a few weeks or months (such a star is now called a *nova*). The

most significant of the Chinese observations of nova outbursts was of the great nova of 1054 A.D. in the constellation of Taurus. Today's remnant of that cosmic explosion is believed to be the Crab nebula, a chaotic, expanding mass of gas (see Chapters 25 and 31).

The Babylonians and Assyrians also knew the approximate length of the year several centuries B.C. The Babylonians used a lunar calendar of 12 months. The month of lunar phases is about 29½ days, so the lunar year has about 354 days, 11 days short of a tropical year; thus a thirteenth month had to be added from time to time as necessary to correct for the cumulative error of the short year.

In pre-Christian Egypt the astronomers were the priests. Although Egyptian astronomers had divided the sky into constellations and had named these star groups for deities in their mythology, their main astronomical function was in keeping the calendar. The Egyptians, like the Babylonians, used a lunar calendar at first, but by a few centuries B.C. they had adopted a year which began when the bright star Sirius could first be seen in the dawn sky, rising just before the sun. This *heliacal* rising of Sirius coincided fairly well with the average time of the annual flooding of the Nile, which gave the astronomer-priests the ability to predict very roughly when this economically important event could be expected to occur.

An important contribution of the Egyptians was the development of the rudiments of civil engineering. They made measurements with ropes and surveyed their land. Apparently they had developed some of the concepts of arithmetic, and primitive geometry has been found on papyri prepared between the nineteenth and sixteenth centuries B.C.

Both the Egyptians and Babylonians had learned to construct fairly accurate sundials for time keeping. The earliest sundial still preserved is an Egyptian one dating from the eighth century B.C.

Most of the peoples of antiquity believed in and practiced astrology. The origin of this religion was probably in Babylonia. The heavenly bodies were quite naturally regarded as different in substance from terrestrial matter, and were generally associated with deities. These gods, although not omnipotent, at least had influence on earthly and human affairs. Thus the particular configuration of the sun, moon, and planets in the sky at the time and place of one's birth was believed to be of significance to that person's character and destiny. Particularly important was the object in the ascendancy — that is, just rising — for that planet's influence must be strong in the air inhaled by the newborn child's first gasp. Astrology was greatly developed and expanded by the ancient Greeks; indeed, the earliest known systematic exposition of the subject is Ptolemy's *Tetrabiblos* in the second century A.D. Astrology is still widely accepted in some countries, and even in the Western nations of Judeo-Christian heritage it has a surprisingly large number of adherents.

The science of astronomy owes a great debt to astrology, because its practice required a detailed knowledge of the motions of the celestial bodies. This knowledge was needed not only because the destiny of a person was believed to be under the continual influence of the planets and their configurations, but in particular to construct his *horoscope* — a chart of the planets in the sky at the time and place of his birth, when their influence was at its greatest. Consequently, astronomers from the time of antiquity to the time of Galileo, even some who may not have believed in astrology per se, devoted most of their energy to observing carefully the positions and motions of the heavenly bodies, or to constructing models or schemes that would allow the accurate calculation of the planetary positions at arbitrary times in the future. Both Tycho Brahe and Kepler, great astronomers in the Shakespearian era, prepared horoscopes as part of their duties. Had they not been employed to do so, the development of the laws of planetary motion, and ultimately our modern technology, would certainly not have come when it did.

Thus the Babylonians observed the sun, moon, and planets, and learned, to some extent, to predict their motions, and even eclipses.

2.2 EARLY GREEK ASTRONOMY

The high point in ancient science was the Greek culture from 600 B.C. to 400 A.D. The earliest Greeks were not scientists in the modern sense. In general, their research took the form of solving academic problems, reasoning on a purely abstract level from axioms that were "obvious" to every Greek scholar. To make observations, perform experiments, or otherwise make contact with the physical world was not considered a "proper" form of investigation. Yet, as we shall see, in that great Greek reservoir of ideas and inspiration, some experiments were performed and some observations were carried out, with the result that science in general, and astronomy in particular, was raised to a level unsurpassed until the sixteenth century.

(a) Early Concepts of the Sky

THE CELESTIAL SPHERE

If we gaze upward at the sky on a clear night, we cannot avoid the impression that the sky is a great hollow spherical shell with the earth at the center. The early Greeks regarded the sky as just

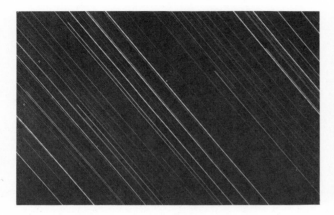

FIG. 2-1   Time exposure showing trails left by stars as a consequence of the apparent rotation of the celestial sphere. *(Lick Observatory.)*

such a *celestial sphere;* some apparently thought of it as an actual sphere of a crystalline material, with the stars embedded in it like tiny jewels. The sphere, they reasoned, must be of very great size, for if its surface were close to the earth, as one moved from place to place he would see an apparent angular displacement in the directions of the stars.

Of course, at any one time we see only a hemisphere overhead, but with the smallest effort of imagination we can envision the remaining hemisphere, that part of the sky that lies below the horizon. The fact that the sky appears as a full hemisphere and not merely part of one, convinced the ancients that the celestial sphere must be extremely large — infinite as far as the eye can tell — compared to the earth. If we watch the sky for several hours, we see that the celestial sphere is gradually and continually changing its orientation. The effect is caused simply by the rotation of the earth, which carries us under successively different portions of the sphere. Following along with us must be our *horizon,* that line in the distance at which the ground seems to dip out of sight, providing a demarcation between earth and sky.

(The horizon may, of course, be hidden from view by mountains, trees, buildings, or in large cities, smog.) As our horizon tips down in the direction that the earth's rotation carries us, stars hitherto hidden beyond it appear to rise above it. In the opposite direction the horizon tips up, and stars hitherto visible appear to set behind it. Analogously, as we round a curve in a mountain road, new scenery comes into view while old scenery disappears behind us.

The direction around the sky toward which the earth's rotation carries us is *east;* the opposite direction is *west.* The Greeks, unaware of the earth's rotation, imagined that the celestial sphere rotated about an axis that passed through the earth. As it turned, it carried the stars up in the east, across the sky, and down in the west.

CELESTIAL POLES

A careful observer will notice that some stars do not rise or set. As seen from the Northern Hemisphere, there is a point in the sky some distance above the northern horizon about which the whole celestial sphere appears to turn. As stars circle about that point, those close enough to it can pass beneath it without dipping below the northern horizon. A star exactly at the point would appear motionless in the sky. Today the star *Polaris* (the North Star) is within 1° of this pivot point of the heavens.

The Greeks regarded that pivot point as one end of the axis about which the celestial sphere rotates. We know today that it is the earth that spins about an axis through its North and South Poles. An extension of the axis would appear to intersect the sky at points in line with the North and South Poles of the earth but, because of the nearly infinite size of the celestial sphere, immensely far away. As the earth rotates about its polar axis, the sky appears to turn in the opposite direction about those *north* and *south celestial poles.*

An observer at the North Pole of the earth would see the north celestial pole directly overhead (at his *zenith*). The stars would all appear to circle about the sky parallel to the horizon, none rising or setting. An observer at the earth's equator, on the other hand, would see the celestial poles at the north and south points on his horizon. As the sky apparently turned about these points, all the stars would appear to rise straight up in the east and set straight down in the west. For an observer at an arbitrary place in the Northern Hemisphere (for example, in Greece) the north celestial pole would appear at a point between the zenith and the north point on his horizon, its location depending on his relative distances from the equator and North Pole of the earth (see Chapter 7). The stars that were not always above the horizon would rise at an oblique angle in the east, arc across the sky in a slanting path, and set obliquely in the west.

FIG. 2-2   Time exposure showing star trails in the region of the north celestial pole. The bright trail near the center was made by Polaris (the North Star). *(Yerkes Observatory.)*

RISING AND SETTING OF THE SUN

The sun is always present at some position on the celestial sphere. When the apparent rotation of the sphere carries the sun above the horizon, the brilliant sunlight scattered about by the molecules of the earth's atmosphere produces the blue sky that hides the stars that are also above the horizon. The early Greeks were aware that the stars were there during the day as well as at night.

ANNUAL MOTION OF THE SUN

The Greeks were also aware, as were the Chinese, Babylonians, and Egyptians before them, that the sun gradually changes its position on the celestial sphere, moving each day about 1° to the east among the stars. Of course, the daily westward rotation of the celestial sphere (or eastward rotation of the earth) carries the sun, like everything else in the heavens, to the west across the sky. Each day, however, the sun rises, on the average, about 4 minutes later with respect to the stars; the celestial sphere (or earth) must make just a bit more than one complete rotation to bring the sun up again. The sun, in other words, has an independent motion of its own in the sky, quite apart from the daily apparent rotation of the celestial sphere.

In the course of 1 year the sun completes a circuit of the celestial sphere. The early peoples mapped the sun's eastward journey among the stars. This apparent path of the sun is called the *ecliptic* (because eclipses can occur only when the moon is on or near it — see Chapter 9). The sun's motion on the ecliptic is in fact merely an illusion produced by another motion of the earth — its annual revolution about the sun. As we look at the sun from different places in our orbit, we see it projected against different stars in the background, or we would, at least, if we could see the stars in the daytime; in practice, we must deduce what stars lie behind and beyond the sun by observing the stars visible in the opposite direction at night. After a year, when we have completed one trip around the sun, it has apparently completed one circuit of the sky along the ecliptic. We have an analogous experience if we walk around a campfire at night; we see the flames appear successively in front of each of the people seated about the fire.

It was also noted by the ancients that the ecliptic does not lie in a plane perpendicular to the line between the celestial poles, but is inclined (today) at an angle of about 23½° to that plane. This angle is called the *obliquity* of the ecliptic, and was measured surprisingly accurately by several ancient observers. The obliquity of the ecliptic, as we shall see, is responsible for the seasons (Chapter 7) and also for the invariable tilt in the axes of terrestrial globe maps.

FIXED AND WANDERING STARS

The sun is not the only moving object among the stars. The moon and each of the five planets visible to the unaided eye — Mercury, Venus, Mars, Jupiter, and Saturn — change their positions in the sky from day to day. The moon, being the earth's nearest celestial neighbor, has the fastest apparent motion; it completes a trip around the sky in about 1 month. We are referring now to the independent motions of these objects among the stars; these motions are superimposed upon the daily rotation of the celestial sphere. The ancient Greeks distinguished between what they called the *fixed stars,* the real stars that appeared to maintain fixed patterns among themselves throughout many generations, and the *wandering stars* or *planets.* The word *planet* means "wanderer." Today, we do not regard the sun and moon as planets, but the Greeks applied the term to all seven of the moving objects in the sky. Because the seven "planets" were of great astrological significance, much of ancient astronomy was devoted to observing and predicting their motions. In fact,

they give us the names for the seven days of our week; Sunday is the sun's day, Monday the moon's day, and Saturday is Saturn's day. We have only to look at the names of the other days of the week in the Romance Languages to see that they are named for the remaining planets.

THE ZODIAC

The individual paths of the moon and planets in the sky all lie close to the ecliptic, although not exactly on it. The reason is that the paths of the planets about the sun, and of the moon about the earth, are all in nearly the same plane, as if they were marbles rolling about on the top of a table. The planets and moon are always found in the sky within a narrow belt centered on the ecliptic, called the *zodiac* (the "zone of the animals"). The zodiac was divided into twelve parts or *signs:* Aries, Taurus, Gemini, Cancer, Leo, Virgo, Libra, Scorpio, Sagittarius, Capricornus, Aquarius, and Pisces. The apparent motions of the planets in the sky result from a combination of their actual motions and the motion of the earth about the sun, and consequently they are somewhat complex.

CONSTELLATIONS

The backdrop for the motions of the "wanderers" in the sky is the canopy of stars themselves. Like the Chinese and the Egyptians, the Greeks had divided the sky into *constellations,* apparent configurations of stars. Modern astronomers still make use of these constellations to denote approximate locations in the sky, much as geographers use political areas to denote the locations of places on the earth. The boundaries between the modern constellations are imaginary lines in the sky running north-south and east-west, so that every point in the sky falls in one constellation or another. The *signs* of the zodiac, incidentally, are evenly spaced regions which do *not* correspond to the constellations bearing the same names. There was a rough correspondence some 2000 years ago, but that has long since changed due to precession (Chapter 6). Many of the 88 recognized constellations are of Greek origin and bear names which are Latin translations of those given them by the Greeks. Today, the lay person is often puzzled because the constellations seldom resemble the people or animals for which they were named. In all likelihood the Greeks themselves did not name groupings of stars because they resembled actual people or objects, but rather named sections of the sky in *honor* of the characters in their mythology, and then fitted the configurations of stars to the animals and people as best they could.

(b) The Earliest Astronomers

THE IONIAN SCHOOL
The earliest Greek scientists were the Ionians, who lived in what is now Asia Minor. They had the advantage of direct contact with the Mesopotamians and Egyptians. Foremost among the Ionians was Thales (about 640–545 B.C.). His most important contribution was the introduction of the concepts of geometry and surveying from Egypt. He also knew of the ecliptic, the lengths of the year and the seasons, and the position of the sun on the ecliptic during the different seasons.

Also of the Ionian school were Anaximander (611–547 B.C.), who may have been the first to speculate on the relative distances of the sun, moon, and planets; Anaximenes; Heraclitus; and others. In general, the Ionians held rather fanciful views of the universe and of the nature of the sun and earth.

THE PYTHAGOREANS
Pythagoras (died about 497 B.C.) was originally an Ionian, but he later founded a school of his own in southern Italy. He pictured a series of concentric spheres, in which each of the seven moving objects — the planets, the sun, and the moon — was carried by a separate sphere from the one that carried the stars, so that the motions of the planets resulted from independent rotations of the different spheres about the earth. The friction between them gave rise to harmonious sounds, the *music of the spheres,* which only the most gifted ear could hear.

Pythagoras also believed that the earth, moon, and other heavenly bodies are spherical. It is doubtful that he had a sound reason for this belief, but it may have stemmed from the realization that the moon shines only by reflected sunlight, and that the moon's sphericity is indicated by the curved shape of the *terminator,* the demarcation line between its illuminated and dark portions. If he had so reasoned that the moon is round, the sphericity of the earth might have seemed to follow by analogy. The belief that the earth is round never disappeared from Greek thought.

Another member of the Pythagorean school was Philolaus, who lived in the following century. He was the first, as far as we know, to introduce the concept that the earth is in motion. Apparently he held that it is too base to occupy the center of the universe, and assigned a *central fire* to that position. About the fire revolved the earth and other planets. The earth and seven planets or "wandering stars" made eight moving objects, and the sphere of "fixed" stars made a ninth object. Philolaus, however, as a confirmed Pythagorean, believed 10 to be the most perfect number because it is the sum of 1, 2, 3, and 4 (and perhaps also because he had 10 fingers); therefore, he believed there must be another body. This, he proposed, was a *counter earth,* or *antichthon,* which revolved around the central fire exactly between it and the earth — thus hiding the fire from view from any place on earth. (The central fire itself seems not to have been counted in the numbering of celestial objects.) The period of revolution of earth and antichthon was 1 day, and the earth rotated as it revolved about the fire so as to keep Greece always turned away from it and the antichthon. Philolaus regarded the celestial sphere as motionless and its apparent rotation as the result of the revolution and rotation of the earth. He proposed that the sun, moon, and planets moved in their respective spheres outside the orbit of the earth. It was an imaginative concept based entirely on fancy and superstition and cannot be regarded as a forerunner of the heliocentric theory.

Nevertheless, the concept of a moving earth had been introduced, although in a completely erroneous manner. It was a bold idea that may have had some influence on later Greek thought. Other Greek philosophers of the sixth to fourth centuries B.C. who are said to have believed in a moving earth are Hicetas, Heracleides, and Ecphantus. Centuries later Copernicus, in his *De Revolutionibus* (Chapter 3), quoted the Pythagoreans as authorities for his own doctrines.

THE SPHERES OF EUDOXUS

Eudoxus of Cnidus (408–355? B.C.) proposed a system of cosmology worthy of note. He represented the motions of the celestial bodies by a combination of rotating spheres. The stars, according to Eudoxus, were carried on the celestial sphere according to accepted theory. In addition, a separate series of concentric spheres was required for each of the planets, the sun, and the moon. Each sphere in a series was pivoted at two points on opposite sides of the next outer sphere; the innermost carried the planet (or sun or moon) itself. The pivot points of the various spheres in a series were not in line with each other, and the spheres all rotated at different rates. By giving each sphere an appropriate rate of rotation, and just the proper inclination of axis, Eudoxus was able to reproduce approximately the rather complicated apparent motions of the heavenly bodies. In all, his system required 27 spheres, 1 for the stars, 3 each for the

sun and moon, and 4 each for the five planets. Later, Callippus further refined the scheme by adding 7 more spheres, bringing the total to 34.

(c) "To Save the Phenomena"

In their invention of cosmological schemes, the Greeks did not always necessarily attempt to describe what they regarded as reality. A knowledge of reality, in the eyes of many of them, was available only to the gods; at any rate, mortal man could hardly expect to understand the real nature of the universe. Rather, Greek philosophers were often trying to find a scheme — a model — that would *describe the phenomena* and would predict events (eclipses, configurations of the planets, and so on). It is difficult to say to what extent this philosophy applies to all the Greek astronomical theories. Quite likely, however, the spheres of Eudoxus and Callippus were intended as a mere mathematical representation of the motions of the planets. It was a scheme that "saved the phenomena" better than ones before it, and in this respect it was successful. The epicycles of Ptolemy, developed later, may similarly be regarded as mathematical representations not intended to describe reality.

Modern science does no more. The laws of nature "discovered" by science are merely mathematical or mechanical models that describe *how* nature behaves, not *why*, nor what nature "actually" is. Science strives to find representations that accurately describe nature, not absolute truths. This fact distinguishes science from religion.

(d) Aristotle

Aristotle (384–322 B.C.), most famous of the Greek philosophers, wrote encyclopedic treatises on nearly every field of human endeavor. Unfortunately, his expositions in the physical sciences in general, and astronomy in particular, were less sound than some of his other work. Although his writings well summarized the total of Greek knowledge to that date, the greatest of the Greek accomplishments in the fields of physics and astronomy were yet to come. Aristotle, however, was accepted as the ultimate authority during the medieval period, and his views were upheld by authority even much later, when research had begun to make them untenable.

THE MOON'S PHASES

In spite of their errors, the writings of Aristotle contain several discussions that represent a substantial advance. First, we find among them clear and correct explanations of the phases of the moon and of eclipses, although these phenomena were almost certainly understood at an earlier date (the explanation is often attributed to Anaxagoras, sometime before 430 B.C.). The basic concepts of eclipses and lunar phases are so important to the development of astronomy that we shall consider them here rather than in the later chapters that deal more directly with these subjects.

Aristotle recognized that the progression of the moon's phases — that is, its changing shape during the month — results from the fact that the moon is not itself luminous but is illuminated by sunlight. Because of its sphericity, only half of the moon is illuminated, that is, having daylight, at one time — the half turned toward the sun. The apparent shape of the moon in the sky depends simply on how much of its daylight hemisphere is turned to our view.

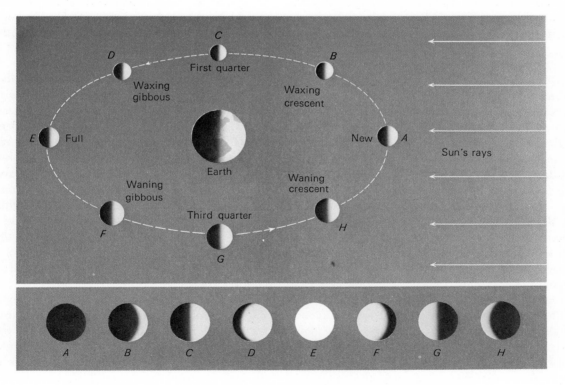

FIG. 2-3   Phases of the moon. The moon's orbit is viewed obliquely. Below, the appearance of the moon as seen from the earth is shown for several positions.

It was generally accepted by Aristotle and earlier Greek astronomers that the sun is more distant than the moon. This was surmised from the sun's slower apparent motion among the stars on the celestial sphere and also from the fact that the moon occasionally passes exactly between the earth and sun and temporarily hides the sun from view (*solar eclipse*). Thus when the moon is in the same general direction from earth as the sun (position *A* in Figure 2-3), its daylight side is turned away from the earth. Because its night side — the side turned toward us — is dark and invisible, we do not see the moon in that position. The phase of the moon is then *new*. (Perhaps it would seem more reasonable to call it "no moon" instead of "new moon," for we do not see any moon at all.) To appear silhouetted in front of the sun, producing a solar eclipse, the moon must be at the new phase and must also lie on the line joining the earth and sun (see Chapter 9). A solar eclipse does not occur at every new moon because the plane of the moon's orbit is inclined slightly to the plane of the ecliptic, so that the new moon usually lies above or below the earth-sun line.

A few days after new moon, the moon reaches position *B,* and from the earth we see a small part of its daylight hemisphere. The illuminated crescent increases in size on successive days as the moon moves farther and farther around the sky away from the direction of the sun. During these days the moon is in the *waxing crescent* phase. About a week after new moon, the moon is one quarter of the way around the sky from the sun (position *C*) and is at the *first quarter* phase. Here the line from the earth to the moon is at right angles to the line from the earth to the sun and half of the moon's daylight side is visible — it appears as a half moon. (The moon is seen as half full at this point because the sun is very much farther away than the moon, and the sun's rays that illuminate the moon and earth are essentially parallel.)

During the week after the first quarter phase we see more and more of the moon's illuminated hemisphere, and the moon is in the *waxing gibbous* phase (position *D*). Finally, about 2 weeks after new moon, the moon (at *E*) and the sun are opposite each other in the sky; the side of the moon turned toward the sun is also turned toward the earth; we have *full moon*. During the next 2 weeks the moon goes through the same phases again in reverse order — through *waning gibbous, third* (or *last*) *quarter,* and *waning crescent.* Occasionally the full moon passes through the earth's shadow, which of course extends outward in space in the direction opposite the sun. This is a *lunar eclipse.*

(If the student finds difficulty in picturing the phases of the moon from this verbal account, he should try a simple experiment: He should stand about 6 feet in front of a bright electric light outdoors at night and hold in his hand a small round object such as a tennis ball or an orange. If the object is then viewed from various sides, the portions of its illuminated hemisphere that are visible will represent the analogous phases of the moon.)

The lunar phases are thus easily understood. Nevertheless, this understanding was a notable accomplishment for observers at the time of Aristotle and before, who had no a priori knowledge that the moon was spherical, that it was not self-luminous, much less that it derived its light from the sun, and who had no original information whatsoever regarding the relative distances of the sun and moon.

THE SPHERICAL SHAPE OF THE EARTH

A second important topic discussed by Aristotle is the shape of the earth. He cited two convincing arguments for the earth's sphericity. First is the fact that during a lunar eclipse as the moon enters or emerges from the earth's shadow, the shape of the shadow seen on the moon is always round (see Figure 2-4). Only a spherical object always pro-

Jupiter. *(Lick Observatory.)*

Saturn. *(Lick Observatory.)*

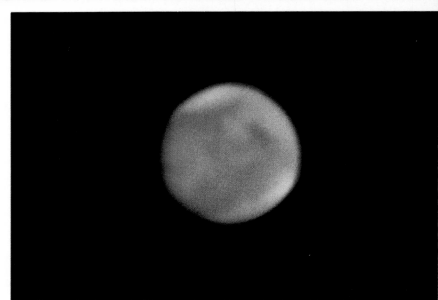

Mars. *(Lick Observatory.)*

Orion Nebula. *(Mount Wilson and Palomar Observatories.)*

FIG. 2-4   Partially eclipsed moon moving out of the earth's shadow.
Note the curved shape of the shadow. *(Yerkes Observatory.)*

duces a round shadow. If the earth were a disk, for example, there would be some occasions when the sunlight would be striking the disk edge on, and the shadow on the moon would be a line.

As a second argument, Aristotle explained that northbound travelers observed hitherto invisible stars to appear above the northern horizon and other stars to disappear behind the southern horizon. Southbound travelers observed the opposite effect. The only possible explanation is that the travelers' horizons had tipped to the north or south, respectively, which indicates that they must have moved over a curved surface of the earth. As a third piece of evidence that the earth is round, Aristotle mentioned that elephants had been observed to the east in India and also to the west in Morocco; evidently, those two places must not be far apart! He also listed a number of other proofs of an even less objective nature.

THE MOTION OF THE EARTH

Finally, Aristotle considered the possibility of the motion of the earth. The apparent daily motion of the sky, he pointed out, can be explained by a hypothesis of the rotation of either the celestial sphere or the earth. He rejected the latter explanation.

He also considered the possibility that the earth revolves about the sun rather than the sun about the earth. He discarded this *heliocentric* hypothesis in the light of an argument that has been used many times since. Aristotle explained that if the earth moved about the sun we would be observing the stars from successively different places along our orbit and their apparent directions in the sky would then change continually during the year.

Any apparent shift in the direction of an object as a result of motion of the observer is called *parallax*. An annual shifting in the apparent directions of the stars that results from the earth's orbital motion is called *stellar parallax*. For the nearer stars it is observable with modern telescopes (see Chapter 18), but it is impossible to measure with the naked eye because of the great distances of even the nearest stars. Aristotle, thus, was unaware of stellar parallax and incorrectly concluded that the earth is stationary.

2.3 LATER GREEK ASTRONOMY

The early Greeks, as we have seen, were aware of, and to some extent understood, the phenomena of the sky. Real progress in Greek astronomy, however, waited until after Aristotle, when the first quantitative astronomical measurements were made. We turn now to the school of astronomers centered in Alexandria, where Greek science attained its greatest heights.

(a) Aristarchus of Samos

Aristarchus of Samos (310–230? B.C.), the first famous astronomer of the Alexandrian school, devised a most ingenious method to find the relative distances from the earth to the sun and moon. His procedure rests on three assumptions: (1) the moon goes about the earth in a perfectly circular orbit, (2) the moon's orbital velocity is perfectly uniform, and (3) the sun, although more distant from the earth than the moon, is near enough so that its rays travel along diverging paths to different parts of the moon's orbit. As it happens, all three of these assumptions are incorrect. The moon's orbit is not a circle but an ellipse — a slightly "flattened" circle (Chapter 3) — and both the moon's geocentric distance and its orbital velocity vary. Furthermore, the sun is about 400 times more distant than the moon, and thus light from the sun strikes all parts of the moon's orbit along virtually parallel lines. At the time, however, Aristarchus' assumptions seemed perfectly reasonable; in any case, no one had made observations critical enough to demonstrate their lack of validity.

Aristarchus then reasoned as follows: The moon appears exactly *half full* (first and last quarters) when the terminator — the line dividing the light and dark halves — is a perfectly straight line as viewed from the earth. But the moon is spherical, and the terminator, being a line upon its surface, must be curved. Thus, the only way it can appear straight is for us to view it exactly edge on. That is, the plane of the terminator must contain the line of sight from the earth to the center of the moon. If that is true, the line from the moon to the earth must be at right angles to the line from the moon to the sun. In Figure 2-5 these right angles are EMS and $EM'S$. Now we see that because the sun is not infinitely far away, by Aristarchus' assumption, the points M', E, and M do not lie along a straight line. Hence the moon, moving at a uniform rate, should require a shorter time to go from M' to M than from M to M'. The difference between these intervals from third quarter to first quarter moon and from first quarter to third quarter determines the angle $M'EM$. For example, if the periods were equal, $M'EM$ would be a straight line, and the sun would have to be infinitely far away; if the period from M to M' were, say, twice that from M' to M, the angle $M'EM$ would be a third of a circle, or 120°.

Aristarchus then attempted to observe the exact instants of first and third quarter moons and determine the difference between the two intervals. Even with our modern equipment of the mid-twentieth century we could not observe the exact instants of quarter moon with sufficient accuracy to measure the angle $M'EM$, because of the sun's great distance. At the time Aristarchus made his observations, however, the moon's elliptical orbit may have been so oriented with respect to the sun that the distance the moon had to move between the first and third quarter phases, combined with the effects of the moon's variable speed, resulted in an observable difference between the periods. In any case, Aristarchus determined that the interval from first to third quarter moon was about 1 day longer than from third to first.

The angle $M'EM$ could then be constructed inside a circle representing the moon's orbit; the lines MS and $M'S$, drawn tangent to the circle at M and M', intersect as S, thus determining the position of the sun and hence its distance in terms of the size of the moon's orbit. Aristarchus determined that the sun was from 18 to 20 times more distant than the moon. This figure was accepted and used by other investigators for centuries after. Although it is about 20 times too small, it did show the sun

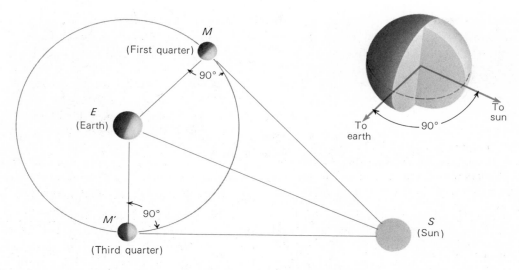

FIG. 2-5   Aristarchus's method of measuring the relative distances of the sun and moon.

to be much more distant than the moon, and its method of determination was a worthy scientific approach.

THE RELATIVE SIZES OF THE EARTH, MOON, AND SUN

Having determined the relative distances to the sun and moon, Aristarchus devised an equally ingenious technique, one later refined and used by Hipparchus, to determine the relative sizes of the sun, moon, and earth. As Aristotle had noted, the shape of the earth's shadow on the moon during a lunar eclipse is curved. During such an eclipse, from the time it took for the moon to cross the earth's shadow, Aristarchus estimated that the shadow, out there at the distance of the moon, is eight thirds the size of the moon itself.

It is well known that the sun and moon appear about the same *angular size* in the sky. By angular size we mean the angle subtended by the diameter of an object, that is, the angle of intersection between two lines drawn from a point on the earth (for example, the observer's eye) to opposite ends

of a diameter of the object. The sun and moon each has an angular size of about ½°. If, as Aristarchus had determined, the sun is 20 times as distant as the moon, it must also be 20 times as big to appear the same size. Aristarchus grossly overestimated the angular sizes of the sun and moon to be about 2° each. However, with these data at hand (even though sadly inaccurate), he could find the relative sizes of the earth, moon, and sun by geometrical construction.

We illustrate the geometrical principles of the construction in Figure 2-6. First, at E, which represents the center of the earth, we draw two lines that intersect at an angle of ½°. During a lunar eclipse the sun and moon are opposite in the sky; thus in direction s the ½° angle can be considered as representing the angular diameter of the sun, and in direction m the angular diameter of the moon. The sun S, and moon M, can now be drawn in, and at arbitrary distances from E as long as the distance ES is 20 times the distance EM. Now at M, the diameter of the earth's shadow, AA', can be constructed ⅛ times the size of the moon.

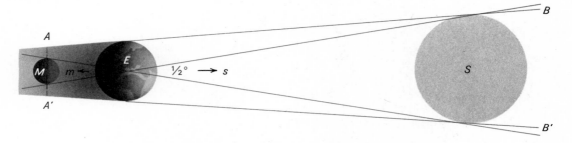

FIG. 2-6   Aristarchus's method of measuring the relative sizes of
the sun, moon, and earth.

Because the rays of sunlight, in which the earth casts its shadow, travel in straight lines, the lines *AB* and *A'B'*, drawn tangent to the sun at *B* and *B'*, must also be tangent to the earth. Thus, finally, the sphere of the earth can be drawn in to proper scale at *E*.

We have now constructed a scale drawing of the earth, moon, and sun. We need only measure with a ruler to obtain their relative sizes and distances. The only data required are the distance from earth to the sun compared to that of the moon, the angular size of each, and the compara-

TABLE 2.1 Sizes and Distances in Terms of the Earth's Diameter

	ARISTARCHUS	MODERN
Moon's distance	10	30
Moon's diameter	⅓	0.272
Sun's distance	200	11,700
Sun's diameter	7	109

tive sizes of the moon itself and the earth's shadow at the moon's distance, the latter being obtained in an eclipse observation. Although the data obtained and used by Aristarchus were inaccurate and his results somewhat far from the truth, at least they were better than any available before; indeed, they were the first objective measures of astronomical dimensions, and they were based on logical reasoning. The approximate values found by Aristarchus, and the actual values, all in units of the earth's diameter, are given in Table 2.1.

ARISTARCHUS' HELIOCENTRIC BELIEF

Perhaps it was his finding that the sun was seven times the earth's diameter that led Aristarchus to the conclusion that the sun, not the earth, was at the center of the universe. At any rate, he is the first person of whom we have knowledge who professed a belief in the heliocentric hypothesis — that the earth goes about the sun. He also postulated that the stars must be extremely distant to account for the fact that their parallaxes could not be observed.

(b) Measurement of the Earth by Eratosthenes

Aristarchus had measured some astronomical dimensions but only in terms of the size of the earth. The latter was not measured by him. The first fairly accurate determination of the earth's diameter was made by Eratosthenes (276–195 or 198 B.C.), who was another astronomer of the Alexandrian school.

To appreciate Erastosthenes' technique for measuring the earth, which is in principle the same as many modern methods, we must understand that the sun is so distant from the earth compared to its size, even by Aristarchus' estimate, that the sun's rays intercepted by all parts of the earth approach it along sensibly parallel lines. Imagine a light source near the earth, say at position *A* in Figure 2-7. Its rays strike different parts of the earth along diverging paths. From a light source at *B*, or at *C*, still farther away, the angle between rays that strike extreme parts of the earth is smaller.

The more distant the source, the smaller the angle between the rays. For a source *infinitely* distant, the rays travel along parallel paths. The sun is not, of course, infinitely far away, but light rays striking the earth from a point on the sun diverge from each other by at most an angle of less than ⅓ of a minute of arc ('), far too small to be observed with the unaided eye. As a consequence, if people all over the earth who could see the sun were to point at it, their fingers would all be pointing in the same direction — they would all be parallel to each other. The concept that rays of light from the sun, planets, and stars approach the earth along parallel lines is vital to the art of celestial navigation — the determination of position at sea.

Eratosthenes noticed that at Syene, Egypt (vicinity of the modern Aswân), on the first day of summer, sunlight struck the bottom of a vertical well at noon, which indicated that Syene was on a direct line from the center of the earth to the sun. At the corresponding time and date in Alexandria, 5000 stadia north of Syene (the *stadium* was a Greek unit of length), he observed that the sun was not directly overhead but slightly south of the zenith, so that its rays made an angle with the vertical equal to 1/50 of a circle (about 7°). Yet the sun's rays striking the two cities are parallel to each other. Therefore (see Figure 2-8), Alexandria must be 1/50 of the earth's circumference north of Syene, and the earth's circumference must be 50 × 5000, or 250,000, stadia. The figure was later revised to 252,000, so that each degree on the earth's surface would have exactly 700 stadia.

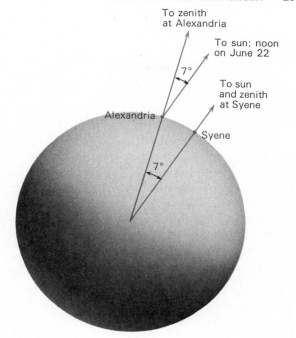

FIG. 2-8 Eratosthenes' method of measuring the size of the earth.

It is not possible to evaluate precisely the accuracy of Eratosthenes' solution because there is doubt as to which of the various kinds of Greek stadia he used. If it was the common Olympic stadium, his result was about 20 percent too large. According to another interpretation, he used a stadium equal to about 1/10 mi, in which case his figure was within 1 percent of the correct value of 24,900 mi. The diameter of the earth is found from the circumference, of course, by dividing the latter by π.

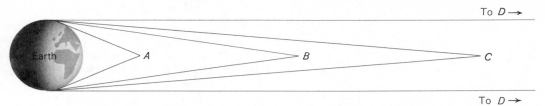

FIG. 2-7 The more distant an object, the more nearly parallel are the rays of light coming from it.

(c) Hipparchus

The greatest astronomer of pre-Christian antiquity was Hipparchus, who was born in Nicaea in Bithynia. The dates of his life are not accurately known, but he carried out his work at Rhodes, and possibly also at Alexandria, in the period 160–127 B.C. Until the student has read the later chapters in this text, he probably will not fully appreciate the subtlety of the phenomena Hipparchus detected, or the difficulty of the measurements he made — all without optical aid.

Arthur Berry† has classified his accomplishments, besides his individual astronomical discoveries, into four categories:

(1) He either invented, or at least highly developed, trigonometry, which makes it possible to solve for the elements of a triangle without geometrical construction — a process that is not feasible with the long, "skinny" triangles encountered in astronomical applications.

(2) He made extensive and systematic observations with the greatest accuracy that his instruments would permit.

(3) He made systematic and critical use of old observations, comparing them with later observations of his own to detect changes too slow to observe in a single lifetime. Many of his own observations, especially of the planets, were designed not for his own analysis but for use by astronomers in future generations.

(4) He devised a geometrical representation, involving *eccentrics* and *epicycles,* that described the motions of the sun and moon more precisely than had any previous scheme.

---

†*A Short History of Astronomy,* Charles Scribner's Sons, New York, 1898, and Dover Publications, Inc., New York, 1961.

HIPPARCHUS' STAR CATALOGUE

Hipparchus erected an observatory on the island of Rhodes and built instruments with which he measured as accurately as possible the directions of objects in the sky. He compiled a star catalogue of about 850 entries. He designated for each star its celestial coordinates, that is, quantities analogous to latitude and longitude that specify its position (direction) in the sky. He also divided the stars according to their apparent brightness into six categories, or *magnitudes,* and specified the magnitude of each star. In the course of his observations of the stars, and in comparing his data with older observations, he made one of his most remarkable discoveries — that the position in the sky of the north celestial pole had altered over the previous century and a half. Hipparchus correctly deduced that the direction of the axis about which the celestial sphere appears to rotate continually changes. The real explanation for the phenomenon is that the direction of the earth's rotational axis changes slowly because of the gravitational influence of the moon and sun, much as a top's axis describes a conical path as the earth's gravitation tries to tumble the top over. This variation of the earth's axis, called *precession,* requires about 26,000 years for one cycle (Chapter 6).

OTHER MEASUREMENTS

Hipparchus, refining the technique first applied by Aristarchus, also obtained a good estimate of the moon's size and distance. He used the correct value of ½° for the angular diameters of the sun and moon and (like Aristarchus) the value $\frac{8}{3}$ for the ratio of the diameter of the earth's shadow to the diameter of the moon. He tried several values for the relative distances of the sun and moon, including the value 20 found by Aristarchus, but found that the exact distance assumed for the sun, as long as it was large, did not have much effect on the figures he derived for the moon. He found the moon's distance to be 29½ times the earth's diameter; the correct number is 30.

He determined the length of the year to within 6 minutes, and even analyzed his possible errors, estimating that he could not be farther off than about 15 minutes. He also carefully observed the motions of the sun, moon, and planets, and found a method by which he could predict the position of the sun on any date of the year with an accuracy equal to the best observations and the position of the moon with somewhat less accuracy. His work made possible the reliable prediction of eclipses, and with the information he left, any astronomer thereafter could predict a lunar eclipse to within an hour or so. Hipparchus was also apparently the first to deal with a special problem relating to solar eclipses — that involving parallax. During a lunar eclipse, the moon enters the earth's shadow and darkens, and everyone on the earth who can see the moon can see the eclipse. On the other hand, a solar eclipse occurs when the moon passes directly in front of the sun. The sun and moon, however, are almost exactly the same angular size, and the moon can just barely cover the sun; consequently, an observer, to see a solar eclipse, must be at a place on the earth that is exactly in line with the sun and moon. Figure 2-9 (not to scale) illustrates the situation. An observer at A sees an eclipse, but to one at B the moon does not appear in front of the sun. The moon is close enough so that from different places on the earth it appears in substantially different directions; the maximum

difference is about 2°. The effect is another example of parallax. Hipparchus showed how to take account of parallax to determine at what places on earth a solar eclipse would be visible.

THE MOTIONS OF THE SUN AND MOON

Hipparchus' study of the motion of the sun deserves special mention. The earth's true orbit around the sun is not a circle but an ellipse; the earth's distance from the sun and its orbital speed both vary slightly. Now we can account for the apparent motion of the sun by imagining it to move around the earth in an elliptical path of exactly the same shape as the earth's orbit. This apparent path of the sun, as we have seen, is the ecliptic. Because we see the sun's apparent orbit edge on (from the inside), the ecliptic is a circle around the sky. Moreover, the sun's eastward rate of motion on the ecliptic will vary, exactly as the earth's orbital speed varies. The variation in speed is slight but is observable.

Eudoxus accounted for the sun's motion approximately by representing it with a series of rotating spheres. Later the mathematician Apollonius of Perga (latter half of the third century B.C.) suggested that the motions of all the heavenly bodies could be represented equally well by a combination of uniform circular motions. By uniform circular motion is meant a motion at a uniform rate of speed about the circumference of a circle. Because the circle is the simplest geometrical figure, and because uniform motion seemed

FIG. 2-9   Effect of parallax on the visibility of eclipses.

the most natural kind, Hipparchus, following the suggestion of Apollonius, attempted to find a combination of uniform circular motions that would account for the sun's apparently irregular behavior.

The plan he adopted was to represent the sun's orbit by an *eccentric,* a circle, but with the earth slightly off center (Figure 2-10). The scheme was highly successful because the true orbit of the earth is very close to a circle with the sun just off center. Now, one effect of the sun's variable speed on the ecliptic is to produce an inequality in the lengths of the seasons. Although the inequality had been known before, Hipparchus remeasured the small differences between the seasons' durations and from them deduced that the earth's distance from the center of the sun's orbit must be $\frac{1}{24}$ of the sun's distance. He found further that the earth and sun were nearest each other in early December, which was correct at that time. (The date has changed over the thousands of years because of precession and, to a lesser extent, because of a slow motion of the long axis of the earth's elliptical orbit; the closest approach now occurs in early January.)

Hipparchus pointed out that he could also have represented the sun's apparent motion by presuming it to move on the circumference of a portable circle called an *epicycle,* whose center, in turn, revolves about the earth in a circle called a *deferent* (Figure 2-11). He considered the eccentric a simpler and thus preferable system.

The moon's motion is more complicated, and Hipparchus was not quite so successful in finding a geometrical scheme to describe it. According to the model he adopted, the moon went in a circle about a point near the earth (an eccentric), but the center of the eccentric also revolved slowly about the earth. Hipparchus measured the 9-year period of this revolution, as well as a 19-year period during which the intersections of the moon's orbit with the ecliptic slide completely around the ecliptic, and the 5° inclination between the moon's orbit and the ecliptic. The apparent motions of the planets are even more complicated than that of the moon. Hipparchus thus declined to fit the planets into a cosmological scheme but rather made careful observations of their positions for use by later investigators.

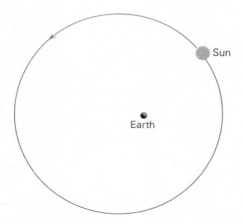

FIG. 2-10   The eccentric.

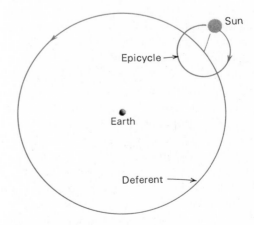

FIG. 2-11   The deferent and epicycle.

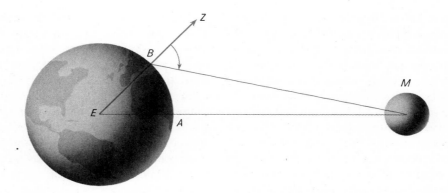

FIG. 2-12   Ptolemy's method of measuring the distance to the moon.

(d) Ptolemy

The last great Greek astronomer of antiquity was Claudius Ptolemy (or Ptolemeus), who flourished about 140 A.D. He compiled a series of 13 volumes on astronomy known as the *Almagest*. All of the *Almagest* does not deal with Ptolemy's own work, for it includes a compilation of the astronomical achievements of the past, principally of Hipparchus. In fact, it is our main source of information about Greek astronomy. The *Almagest* also contains accounts of the contributions of Ptolemy himself.

One of Ptolemy's accomplishments was the measurement of the distance to the moon by a technique essentially identical to the one used today. The method he used, the principle of which is illustrated in Figure 2-12, makes use of the moon's parallax, discussed by Hipparchus in connection with solar eclipses. Suppose we could observe the moon directly overhead. We would have to be, then, at position *A* on the earth, on a line between the center of the earth *E*, and the center of the moon *M*. Suppose that at the same time someone else at position *B* were to observe the angle *ZBM* between the moon's direction and the point directly over his head, *Z*. The angle *MBE* would then be determined in the triangle *MBE* (it is 180° − angle *ZBM*). The distance from *A*

to *B* determines the angle *BEM*. For example, if *A* is one twelfth of the way around the earth from *B*, the angle *BEM* is 30°. The side *BE* is of course the radius of the earth. We therefore know two angles and an included side in the triangle *MBE*. It is now possible to determine, either by trigonometry or geometrical construction, the distance *EM* between the centers of the earth and moon. This is an example of the principle of *surveying*. We shall discuss it further later.

In practice, we do not need another observer at *B*, for the rotation of the earth will carry us over there in a few hours anyway, and we can observe the angle *ZBM* then. We shall have to correct, however, for the motion of the moon in its orbit during the interval between our two observations; the moon's motion being known, the correction is a detail easily accomplished. Using the principle described, Ptolemy determined the moon's distance to be 59 times the radius of the earth or $29\frac{1}{2}$ times the earth's diameter — very nearly the correct value.

PTOLEMY'S SCHEME OF COSMOLOGY

Ptolemy's most important original contribution was a geometrical representation of the solar system that predicted the motions of the planets with considerable accuracy. Hipparchus, having determined by observation that earlier theories of the motions of the planets did not fit their actual behavior, and not having enough data on hand to solve the problem himself, instead massed observational material for posterity to use. Ptolemy supplemented the material with a few observations of his own and with it produced a cosmological hypothesis that endured until the time of Copernicus.

The complicating factor in the analysis of the planetary motions is that their apparent wanderings in the sky result from the combination of their own motions and the earth's orbital revolution. Notice, in Figure 2-13, the orbit of the earth and the orbit of a hypothetical planet farther from the sun than the earth. The earth travels around the sun in the same direction as the planet and in nearly the same plane, but has a higher orbital

speed. Consequently, it periodically overtakes the planet, like a faster race car on the inside track. The apparent directions of the planet, seen from the earth, are shown at successive intervals of time along lines $AA'A''$, $BB'B''$, and so on. In the right side of the figure we see the resultant apparent path of the planet among the stars. From positions B to D, as the earth passes the planet, it appears to drift backward, to the *west* in the sky, even though it is actually moving to the *east*. Similarly, a slowly moving car appears to drift backward with respect to the distant scenery when we pass it in a faster-moving car. As the earth rounds its orbit toward position E, the planet again takes up its usual eastward motion in the sky. The temporary westward motion of a planet as the earth swings between it and the sun is called *retrograde* motion. (During and after its retrograde motion

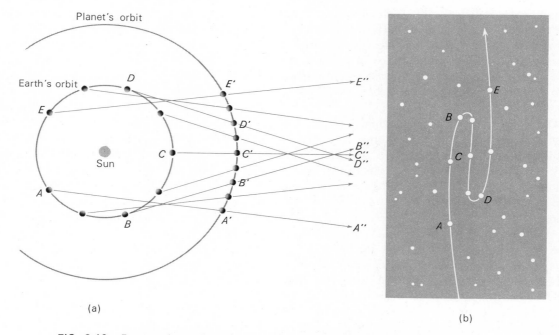

(a)

(b)

FIG. 2-13 Retrograde motion of a superior planet. (a) Actual positions of the planet and earth; (b) the apparent path of the planet as seen from the earth, against the background of distant stars.

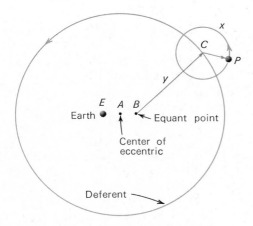

FIG. 2-14 Ptolemy's system of deferent, epicycle, eccentric, and equant.

the planet's apparent path in the sky does not trace exactly over itself because of the slight inclinations between the orbits of the earth and other planets. Thus, the retrograde path is shown as an open loop in Figure 2-13.) Obviously, it is difficult to find an explanation for retrograde motion on the hypothesis that the planet is revolving about the earth.

Ptolemy solved the problem by having a planet P (Figure 2-14) revolve in an epicyclic orbit about C. The center of the epicycle C in turn revolved in the deferent about the earth. When the planet is at position x, it is moving in its epicyclic orbit in the same direction as the point C moves about the earth, and the planet appears to be moving eastward. When the planet is at y, however, its epicyclic motion is in the opposite direction to the motion of C. By choosing the right combination of speeds and distances, Ptolemy succeeded in having the planet moving westward at the right speed at y and for the correct interval of time. However, because the planets, like the earth, travel about the sun in elliptical orbits, their actual behavior cannot be represented accurately by so simple a scheme of uniform circular motions. Consequently, Ptolemy made the deferent an eccentric, centered

not on the earth, but slightly away from the earth at A. Furthermore, he had the center of the epicycle, C, move at a uniform angular rate, not around A, or E, but a point B, called the *equant,* on the opposite side of A from the earth.

It is a tribute to the genius of Ptolemy as a mathematician that he was able to conceive such a complex system to account successfully for the observations. His hypothesis, with some modifications, was accepted as absolute authority throughout the Middle Ages, until it finally gave way to the heliocentric theory in the seventeenth century. In the *Almagest,* however, Ptolemy made no claim that his cosmological model described reality. He intended his scheme rather as a mathematical representation to predict the positions of the planets at any time. Modern astronomers do the same thing with algebraic formulas. Our modern mathematical methods were not available to Ptolemy; he had to use geometry.

2.4 OTHER GREEK SCIENCE

It would be wrong to leave ancient Greece without mention of Archimedes of Syracuse (287–212 B.C.). He was not an astronomer, but we are probably justified in calling him the founder of experimental physics — a forerunner of Galileo. He was also a great mathematician and geometrician. One of his accomplishments was to establish the value of π to within one part in several hundred.

In the physical sciences, he experimented with fluids, mirrors, and compound pulleys, and made machines of war. He is famous for his measurements of the densities of bodies, and for "Archimedes' principle," that the weight of a floating body equals the weight of the water it displaces. His most important contribution to astronomy may be the law of the lever. He found that a small force applied a long distance from the fulcrum of a lever can balance a large force a short distance from the fulcrum. The discovery is of great mechanical importance but also applies to the mutual revolution of two bodies, such as double stars, a planet and the sun, or a planet and a satellite. The distance of each of the two bodies from the point between them about which they revolve is found from exactly the same law that applies to the lever (Chapter 5).

2.5 INDIAN AND ARABIAN ASTRONOMY

In the next 13 centuries after Ptolemy the only significant astronomical investigations were carried out by the Hindus and Arabs. References have been made to Indian observations as far back as 1200 B.C., but there was apparently no substantial development as a science in India until about 300 A.D., after the influence of Greek and Babylonian science had reached there. In the period following, the Hindus made some creditable observations, among them the determination of the moon's distance as $64\frac{1}{2}$ earth radii (the actual figure is nearer 60). The most important contribution of the Hindus to the science of our culture is very likely our system of numbers and place counting.

The Arabs brought the Hindu system of numbers to Europe and developed algebra. They also had access to some of the records of the Greek astronomers. Their greatest contribution was probably to provide some continuity between ancient astronomy, on the one hand, and the development of modern astronomy in the Renaissance.

The best astronomer of the Arab world was Muhammad al Battani, also known as Albategnius, who worked in the 40-year period ending 918. He made new and accurate observations of his own, compiled tables of the positions of the sun and planets, predicted eclipses, and recalculated the rate of precession. Rather complete records of eclipse observations were kept by the astronomer Ibn Junis or Yũnus (near the year 1000).

Other ancient peoples, especially the Polynesians and the Maya, are noted for their astronomical observations. We shall not discuss these here, because the astronomy of these people had no direct influence on the development of science in the western world. We shall, however, describe the Maya calendar in Chapter 8.

We have dealt at some length with ancient astronomy, not simply to record the history of the science, but to introduce concepts regarding celestial phenomena as they were first conceived and investigated, which we shall reinvestigate in the light of modern knowledge. We have hoped to show something of the spirit of scientific inquiry, the detective nature of the work, and the growth of the scientific method.

Exercises

1. Where on earth would all the stars be visible at one time or another?
2. Where on earth would only half the sky ever be seen?
3. Why was Philolaus' hypothesis not a scientific one?
4. Show by a diagram how a solar eclipse *can* occur at new moon but does not usually occur.
5. About what time of day or night does the moon rise when it is full? When it is new?
6. Why can an eclipse of the moon never occur on the day following a solar eclipse?
7. Give some everyday examples of parallax.

8. Suppose Aristarchus had found that the interval from third quarter to first quarter moon was 1 week, and that the interval from first quarter to third quarter moon was 3 weeks. Then what distance would he have derived for the sun (in terms of the moon's distance)?

Answer: 1.414 times

9. Suppose Eratosthenes had found that at Alexandria at noon on the first day of summer the line to the sun makes an angle of 30° with the vertical. What then would he have found for the earth's circumference?

Answer: 60,000 stadia

10. Why would Eratosthenes' method not have worked if the earth were flat, like a pancake?

11. You are on a strange planet. You note that the stars do not rise or set but circle around parallel to the horizon. Then you travel over the surface of the planet in a straight line for 10,000 mi and find that at this new place the stars rise straight up from the horizon in the east and set straight down in the west. What is the circumference of the planet?

Answer: 40,000 mi

12. Suppose that the planets actually revolved around the sun at a uniform rate in exactly circular orbits centered on the sun. Show that their motion can be represented by placing the earth at the center and having the sun and planets revolve around it, each planet moving in an epicyclic orbit and taking exactly 1 year to move around its epicycle.

Apollo 11 astronaut Edwin E. Aldrin, Jr., prepares to deploy laser beam reflector and seismometer on the surface of the moon. Note the depth of the footprints in the ejecta as Aldrin walks next to a small crater off to the left of the picture. *(NASA)*.

CHAPTER 3

The Heliocentric
Hypothesis

3.1 MEDIEVAL ASTRONOMY

In medieval Europe no new astronomical investigations of importance were made. Rather than turning to scientific inquiry, the medieval mind rested in acceptance of authority and absolute dogma. In the realm of cosmology, this dogma involved belief in the crystalline spheres of Pythagoras (as perpetuated by Aristotle). To reconcile this view with the epicycles necessary to Ptolemaic technical astronomy, Purbach (1423–1461) revived a scheme, going back to Ptolemy himself, whereby the crystalline spheres were hollowed out to make room for the epicycles! Jerome Fracastor, writing in 1538, went even further. Desiring to eliminate such irregularities as eccentrics and equants altogether, he invented a system that required no less than 79 separate spheres.

Such was the state of affairs when Copernicus' *De Revolutionibus* was published in 1543.

3.2 COPERNICUS

Nicolas Copernicus (also Coppernicus, or Koppernig, 1473–1543) was born in Thorn on the Vistula in Poland, although since then the region often has been in the domain of Prussia. His training was in law and medicine, but Copernicus' main interest was in astronomy and mathematics. By the time he had reached middle age, he was well known as an authority on astronomy.

Copernicus was not principally an observer; his forte was mathematics. His most important contribution to science was a critical reappraisal of the existing theories of cosmology and the development of a new model of the solar system. His unorthodox and heretical idea that the sun, not the earth, is the center of the solar system had become known by 1530, chiefly through a manuscript circulated by him and his friends.

His ideas were set forth in full detail in his great book, *De Revolutionibus*, published in the year of his death, 1543. Supervision over the publi-

FIG. 3-1 Nicolas Copernicus. *(Yerkes Observatory.)*

cation of the book fell into the hands of one Osiander, a Lutheran preacher, who was probably responsible for the augmented title of the work — *De Revolutionibus Orbium Celestium* (On the Revolutions of the Celestial Spheres). Osiander wrote a preface, which he neglected to sign, expressing the (modern) view that science presented only an abstract mathematical hypothesis, and implying that the theory set forth in the book was only a convenient calculating scheme. The preface is almost certainly in contradiction to Copernicus' own feelings.

In *De Revolutionibus*, Copernicus sets forth certain postulates from which he derives his system of planetary motions. His postulates include the assumptions that the universe is spherical and that the motions of the heavenly bodies must be made up of combinations of uniform circular motions. We see, thus, that the work of Copernicus was not free from all preconceived notions.

(a) Relative Motion, and the Motion of the Earth

One concept that Copernicus dealt with at some length is that of relative motion. The point is that a person moving uniformly is not necessarily aware of his motion. This principle of relative motion is illustrated in Figure 3-2. Suppose that an observer at O sees an object at A. Now let the object move from A to A'. The direction of the object from O has changed from OA to OA', that is, through an angle α. On the other hand, exactly the same effect will result if the object remains stationary and the *observer* moves from O to O'. Now the direction of the object has changed from OA to $O'A$, but since $O'A$ is parallel to OA', the observer will still see the object apparently changing direction through an angle α. Thus, there is no a priori way of knowing whether the observer or the object has moved.

Copernicus argued from this principle that the apparent annual motion of the sun about the earth could be equally well represented by a motion of

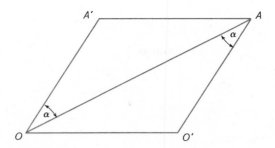

FIG. 3-2   Relative motion.

the earth about the sun. By a simple extension of the principle, Copernicus showed that the apparent rotation of the celestial sphere could be accounted for by assuming that the earth rotates about an axis and that the celestial sphere is fixed.

Contrary to popular belief, Copernicus did not *prove* that the earth is in motion, but he did offer substantial evidence to show that the hypothesis of a moving earth involved fewer *ad hoc* assumptions than that of a stationary one. To the objection that if the earth rotated about an axis it would fly into pieces, Copernicus answered that if such motion would tear the earth apart, the even faster motion (because of its greater size) of the celestial sphere required by the alternative hypothesis would be even more devastating to it.

(b) The Planetary Motions According to Copernicus

The most important concept set forth by Copernicus in *De Revolutionibus* is that the earth is but one of six (then known) planets that revolve about the sun. Through this principle, he was able to work out the correct general picture of the solar system. He placed the planets, starting nearest the sun, in the order Mercury, Venus, Earth, Mars, Jupiter, and Saturn. Further, he deduced that the nearer a planet is to the sun, the greater is its orbital speed. Thus the retrograde motions of the planets (Section 2.3d) were easily understood without the necessity for epicycles. Also, Copernicus worked out the correct approximate scale of

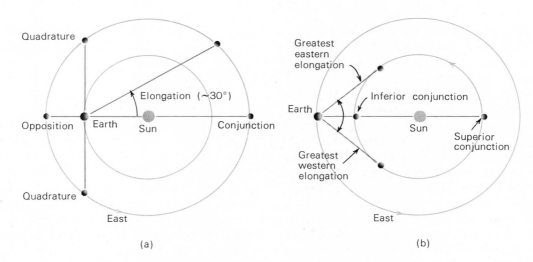

FIG. 3-3 (a) Configurations of a superior planet;
(b) configurations of an inferior planet.

the solar system. Before we describe his procedure, however, it will be helpful to define a few terms that describe the positions of planets in their orbits. These are illustrated in Figure 3-3.

A *superior planet* is any planet whose orbit is larger than that of the earth, that is, a planet that is farther from the sun than the earth (Mars, Jupiter, and Saturn). An *inferior planet* is a planet closer to the sun than the earth (Venus and Mercury).

Every now and then, the earth passes between a superior planet and the sun. Then that planet appears in exactly the opposite direction in the sky from the sun — or at least as nearly opposite as is allowed by the slight differences of inclination among the planes of the orbits of the other planets and the earth. At such time, the planet rises at sunset, is above the horizon all night long, and sets at sunrise. We look one way to see the sun, and in the opposite direction to see the planet. The planet is then said to be in *opposition*.

On other occasions, a superior planet is on the other side of the sun from the earth. It is then in the same direction from the earth as the sun is, and of course is not visible. At such time,

the planet is said to be in *conjunction*.

In between these extremes (but not halfway between), a superior planet may appear 90° away from the sun in the sky, so that a line from the earth to the sun makes a right angle with the line from the earth to the planet. Then the planet is said to be at *quadrature*. At quadrature, a planet rises or sets at either noon or midnight.

The angle formed at the earth between the earth-planet direction and the earth-sun direction is called the planet's *elongation*. In other words, the elongation of a planet is its angular distance from the sun as seen from the earth. At conjunction, a planet has an elongation of 0°, at opposition 180°, and at quadrature 90°.

An inferior planet can never be at opposition, for its orbit lies entirely within that of the earth. The greatest angular distance from the sun, on either the east or west side, that the inferior planet can attain is called its *greatest eastern elongation* or *greatest western elongation*.

When an inferior planet passes between the earth and sun, it is in the same direction from earth as the sun and is said to be in *inferior conjunction*. When it passes on the far side of the sun from the earth, and is again in the same direction as the sun, it is said to be at *superior conjunction.*

SIDEREAL AND SYNODIC PERIODS OF A PLANET

Copernicus recognized the distinction between the *sidereal period* of a planet — that is its actual period of revolution about the sun with respect to the fixed stars — and its *synodic period,* its apparent period of revolution about the sky with respect to the sun. The synodic period is also the time required for it to return to the same configuration, such as the time from opposition to opposition or from conjunction to conjunction.

Consider two planets, *A* and *B*, *A* moving faster in the smaller orbit (Figure 3-4). At position (1), planet *A* passes between *B* and the sun S. Planet *B* is at opposition as seen from *A*, and *A* is at an inferior conjunction as seen from *B*. When *A* has made one revolution about the sun and has returned to position (1), *B* has, in the meantime, moved on to position (2). In fact, *A* does not catch up with *B* until both planets reach position (3). Now planet *A* has gained one full lap on *B*.

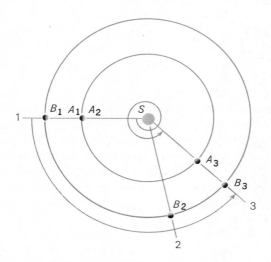

FIG. 3-4   Relation between the sidereal and synodic periods of a planet.

Planet *A* has revolved in its orbit through 360° *plus* the angle that *B* has described in traveling from position (1) to position (3) in its orbit. The time required for the faster-moving planet to gain a lap on the slower-moving one is the synodic period of one with respect to the other. If *B* is the earth and *A* an inferior planet, the synodic period of *A* is the time required for the inferior planet to gain a lap on the earth; if *A* is the earth and *B* a superior planet, the synodic period of *B* is the time for the earth to gain a lap on the superior planet.

What is observed from the earth is the synodic, not the sidereal period of a planet. By reasoning along the lines outlined in the last paragraph, however, Copernicus was able to deduce the sidereal periods of the planets from their synodic periods. Let a planet's sidereal period be *P* years and its synodic period *S* years. In *S* years, the earth, completing one revolution per year, must make *S* trips around the sun. (The quantity *S*, of course, can be less than 1, in which case the earth would complete less than one circuit.) The other planet, completing one revolution in *P* years, would make, in *S* years, *S/P* trips around the sun. Consider first an inferior planet. It has made one more trip around the sun during its synodic period than has the earth, so $S + 1 = S/P$, which, by rearrangement of terms, can be written

$$\frac{1}{P} = 1 + \frac{1}{S} \qquad \text{(for an } inferior \text{ } planet\text{)}.$$

In the case of the superior planet, it is the earth that gains the extra lap, and $S = S/P + 1$, which can be written

$$\frac{1}{P} = 1 - \frac{1}{S} \qquad \text{(for a } superior \text{ } planet\text{)}.$$

As an example, consider Jupiter, whose synodic period is 1.09211 years. Since Jupiter is a superior planet, $1/P = 1 - 1/1.09211 = 1 - 0.91566$, or $1/P = 0.08434$. Thus, $P = 1/0.08434 = 11.86$ years.

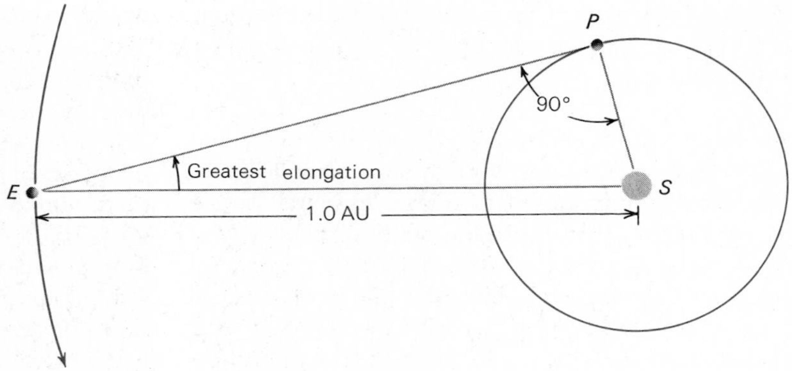

Fig. 3-5 Determination of the relative distance of an inferior planet from the sun.

RELATIVE DISTANCES OF THE PLANETS

Having determined the sidereal periods of revolution of the planets, Copernicus was able to find their relative distances from the sun. The problem is particularly simple for the inferior planets. When an inferior planet is at greatest elongation (Figure 3-5), the line of sight from the earth to the planet, EP, must be tangent to the orbit of the planet, and hence perpendicular to the line from the planet

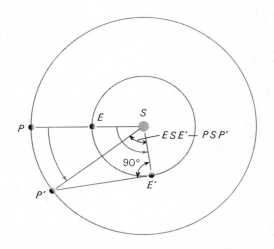

FIG. 3-6 Determination of the relative distance of a superior planet from the sun.

to the sun, PS. We have, therefore, a right triangle, EPS. The angle PES is observed (it is the greatest elongation) and the side ES is the earth's distance from the sun. The planet's distance from the sun can then be found, in terms of the earth's distance, by geometrical construction, or by trigonometric calculation.

As an illustration of the procedure by which the distance of a superior planet can be found, suppose (Figure 3-6) the planet P is at opposition. We can now time the interval until the planet is next at quadrature; the planet is then at P' and the earth at E'. With a knowledge of the sidereal periods of the planet and the earth, we can calculate the fractions of their respective orbits that have been traversed by the two bodies. Thus the angles PSP' and ESE' can be determined, and subtraction gives the angle $P'SE'$ in the right triangle $P'SE'$. The side SE' is the earth's distance from the sun, so enough data are available to solve the triangle and find the planet's distance from the sun, $P'S$ (again in terms of the earth's distance), by construction or calculation.

The values obtained by Copernicus for the distances of the various planets from the sun, in units of the earth's distance, are summarized in Table 3.1. Also given are the values determined by modern measurement.

TABLE 3.1 Distances of Planets from the Sun

PLANET	COPERNICUS	MODERN
Mercury	0.38	0.387
Venus	0.72	0.723
Earth	1.00	1.00
Mars	1.52	1.52
Jupiter	5.22	5.20
Saturn	9.17	9.54

So far, we have discussed the Copernican theory as though Copernicus regarded the planets as having circular orbits centered on the sun. However, we recall that Hipparchus and Ptolemy had introduced, centuries ago, epicycles, eccentrics, and equants to account for those minor irregularities that arise because of deviations from uniform circular motion (actually because the true orbits of planets are ellipses). Copernicus rejected the equants of Ptolemy as unworthy of the perfection of heavenly bodies and instead introduced a system of eccentrics and small epicycles to take care of the irregularities. In all, his system required 34 circles: 4 for the moon, 3 for the earth, 7 for Mercury, and 5 each for the other planets.

His excellence as a mathematician enabled Copernicus to work out the details of his planetary system so that it accounted for the apparent motions of the sun and planets with fair accuracy. The Copernican system is certainly closer to our modern view of the solar system than those that preceded it. On the other hand, Copernicus' use of epicycles, apparently intended as a literal description of nature, and his acceptance of received principles lead us to wonder whether his system of logic was any closer to that of modern science than were those of Hipparchus and Ptolemy.

3.3 TYCHO BRAHE

Three years after the publication of *De Revolutionibus,* Tycho Brahe (1546–1601) was born of a family of Danish nobility. Tycho (as he is generally known) developed an early interest in astronomy and as a young man made significant astronomical observations.

FIG. 3-7   Tycho Brahe. *(Yerkes Observatory.)*

In 1572 he observed a nova or "new star" (now believed to be a supernova — see Chapter 25) that rivaled the planet Venus in brilliance. Tycho observed the star for 16 months until it disappeared from naked-eye visibility. Now, we have seen (Section 2.3c) that the moon exhibits a diurnal parallax, or apparent displacement in direction, because of the rotation of the earth, which constantly shifts our position of observation. The effect is the same whether we regard it as being caused by the earth's rotation or a rotation of the celestial sphere carrying the moon about us. Tycho, despite the most careful observations, was unable to detect any parallax of his nova and accordingly concluded that it must be more distant than the moon. This conclusion was of the utmost impor-

FIG. 3-8 Tycho Brahe at Uraniborg, observing with his great mural quadrant. *(Yerkes Observatory.)*

king, Christian IV, lost patience with the astronomer and eventually discontinued his support. Thus, in 1597 Tycho was forced to leave Denmark. He took up residence near Prague, taking with him some of his instruments and most of his records. There Tycho Brahe spent the remaining years of his life analyzing the data accumulated over 20 years of observation. In 1600, the year before his death, he secured the assistance of a most able young mathematician, Johannes Kepler.

(a) Tycho's Observations

Tycho, like others of his time and before him, believed that comets were luminous vapors in the earth's atmosphere. In 1577, however, a bright comet appeared for which he could observe no parallax. Tycho concluded that the comet was at least three times more distant than the moon, and guessed that it probably revolved about the sun, in contradiction to his earlier beliefs. Other comets were observed by him or his students in 1580, 1582, 1585, 1590, 1593, and 1596.

Tycho is most famous for his very accurate observations of the positions of the stars and plan-

tance, for it showed that changes can occur in the celestial sphere, generally regarded as perfect and unchanging, apart from the regular motions of the planets.

The reputation of the young Tycho Brahe as an astronomer gained him the patronage of Frederick II, and in 1576 Tycho was able to establish a fine astronomical observatory on the Danish island of Hveen. The chief building of the observatory was named *Uraniborg*. The facilities at Hveen included a library, laboratory, living quarters, workshops, a printing press, and even a jail. There, for 20 years, Tycho and his assistants carried out the most complete and accurate astronomical observations yet made.

Unfortunately, Tycho was both arrogant and extravagant, and after Frederick II died, the new

FIG. 3-9 Tycho Brahe's observatory on the island of Hveen. *(Yerkes Observatory.)*

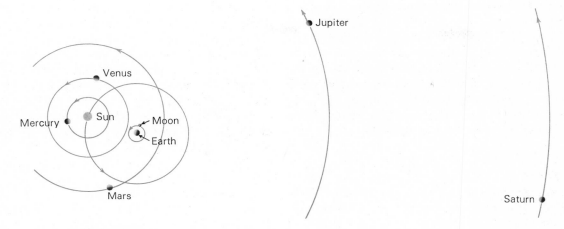

FIG. 3-10   Tycho's model for the solar system.

ets. With instruments of his own design he was able to make observations accurate to the very limit of vision with the naked eye. The positions of the nine fundamental stars in his excellent star catalogue were accurate in most cases to within 1′. Only in one case was he off by as much as 2′, and this was a star whose position was distorted by atmospheric refraction (see Chapter 11).

Tycho's observations included a continuous record of the positions of the sun, moon, and planets. His daily observations of the sun, extending over years and comprising thousands of individual sightings, led to solar tables that were good to within 1′. He reevaluated nearly every astronomical constant and determined the length of the year to within 1 second. His extensive and precise observations of planetary positions enabled him to note that there were variations in the positions of the planets from those given in published tables, and he even noted regularities in the variations.

(b) Tycho's Cosmology

Tycho rejected the Copernican heliocentric hypothesis on what seemed at the time to be very sound grounds. First, he found it difficult to reconcile a moving earth with certain Biblical statements; nor could he even imagine an object as heavy and "sluggish" as the earth to be in motion. The fact that he could not detect a parallax for even a single star, moreover, meant that the stars would have to be enormously distant if the earth revolved around the sun. The great void that would be required between the orbit of Saturn and the stars would alone have been enough to make him doubt the motion of the earth; even more convincing to Tycho was the fact that he believed that he could measure the angular sizes of stars. The brightest of them he thought to be 2′ across. Now, the farther away an object is, the larger must be its true size in order that it have a given angular diameter. Tycho could not detect as much as 1′ of parallax for any star, so it followed that the stars were so distant that, to have angular diameters of 2′, their actual sizes would have to be twice the size of the entire orbit of the earth. If they were still farther away, their diameters would have to be proportionally greater. (Later telescopic observations showed that the stars, unlike the planets, appear as luminous points; their disklike appearance to the naked eye is illusory.)

Tycho did, however, suggest an original system of cosmology, although it was not worked out in full detail. He envisioned the earth in the center, with the sun revolving about the earth each year, and with the other planets revolving about the sun in the order Mercury, Venus, Mars, Jupiter, and Saturn (see Figure 3-10).

3.4 KEPLER

Johannes Kepler (1571–1630) was born in Weilder-Stadt, Württemberg (southwestern Germany). He studied for a theological career, and attended college at Tübingen, where he learned the principles of the Copernican system. He became an early convert to the heliocentric hypothesis and defended it in arguments with his fellow scholars.

FIG. 3-11 Johannes Kepler. *(Yerkes Observatory.)*

In 1594, because of his facility as a mathematician, he was offered a position teaching mathematics and astronomy at the high school at Gratz. As part of his duties at Gratz, he prepared almanacs that gave astronomical and astrological data. Eventually, however, the power of the Catholic church in Gratz grew to the point where Kepler, a Protestant, was forced to quit his post. Accordingly, he went to Prague to serve as an assistant to Tycho Brahe.

Tycho set Kepler to work trying to find a satisfactory theory of planetary motion—one that was compatible with the long series of observations made at Hveen. After Tycho's death, Kepler succeeded him as mathematician to the Emperor Rudolph and obtained possession of the majority of Tycho's records. Their study occupied most of Kepler's time for the next 25 years.

(a) The Investigation of Mars

Kepler's most detailed study was of Mars, for which the observational data were the most extensive. He published the first results of his work in 1609 in *The New Astronomy*, or *Commentaries on the Motions of Mars*. He had spent nearly a decade trying to fit various combinations of circular motion, including eccentrics and equants, to the observed motion of Mars, but without success. At one point he found a hypothesis that agreed with observations to within 8′ (about one quarter the diameter of the full moon), but he believed that Tycho's observations could not have been in error by even this small amount, and so, with characteristic integrity, discarded the hypothesis. Finally, Kepler tried to represent the orbit of Mars with an oval, and soon discovered that the orbit could be fitted extremely well by a curve known as an *ellipse*.

PROPERTIES OF THE ELLIPSE

Next to the circle, the ellipse is the simplest kind of closed curve. It belongs to a family of curves known as *conic sections* (Figure 3-12). A conic

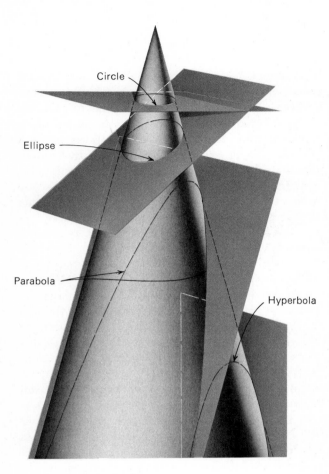

FIG. 3-12 Conic sections.

section is simply the curve of intersection between a hollow cone and a plane that cuts through it. If the plane is perpendicular to the axis of the cone (or parallel to its base), the intersection is a circle. If the plane is inclined at an arbitrary angle, but still cuts completely through the surface of the cone, the resulting curve is an ellipse. If the plane is parallel to a line in the surface of the cone, it never quite cuts all the way through the cone, and the curve of intersection is open at one end. Such a curve is called a *parabola*. If the plane is inclined at an even smaller angle to the axis of the cone, an open curve results that is called a *hyperbola*. The ellipse, then, ranges from a circle

at one extreme to a parabola at the other. The parabola separates the family of ellipses from the family of hyperbolas.

An interesting and important property of an ellipse is that from *any point* on the curve, the sum of the distances to two points inside the ellipse, called the *foci* of the ellipse, is the same. This property suggests a simple way to draw an ellipse. The ends of a length of string are tied to two tacks pushed through a sheet of paper into a drawing board, so that the string is slack. If a pencil is then pushed against the string, so that the string is held taut, and then slid against the string around the tacks (Figure 3-13), the curve that results is an ellipse; at any point where the pencil may be, the sum of the distances from the pencil to the two tacks is a constant length — the length of the string. The tacks, of course, are at the two foci of the ellipse.

The maximum diameter of the ellipse is called its *major axis*. Half this distance, that is, the distance from the center of the ellipse to one end, is the *semimajor axis*. The *size* of an ellipse depends on the length of the major axis. The *shape* of an ellipse depends on how close together the two foci are compared to the major axis. The ratio of the distance between the foci to the major axis is called the *eccentricity* of the ellipse. If an ellipse

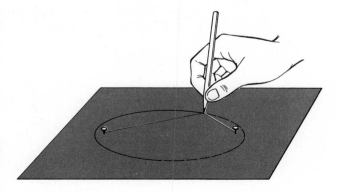

FIG. 3-13 Drawing an ellipse.

is drawn as described above, the length of the major axis is the length of the string, and the eccentricity is the distance between the tacks divided by the length of the string (Figure 3-14). If the foci (or tacks) coincide, the ellipse is a circle; a circle is, then, an ellipse of eccentricity zero. Ellipses of various shapes are obtained by varying the spacing of the tacks (as long as they are not farther apart than the length of the string). If one tack is removed to an infinite distance, and if enough string is available, "our end" of the resulting, infinitely long ellipse is a parabola. A parabola has an eccentricity of 1. An ellipse is completely specified by its major axis and its eccentricity.

Kepler found that Mars has an orbit that is an ellipse and that the sun is at one focus (the other focus is empty). The eccentricity of the orbit of Mars is only about $\frac{1}{10}$; the orbit, drawn to scale, would be practically indistinguishable from a circle. It is a tribute to Tycho's observations and to Kepler's perseverance that he was able to determine that the orbit was an ellipse at all.*

THE VARYING SPEED OF MARS
Before he saw that the orbit of Mars could be

_____
* Tycho did not die in Hveen.

represented accurately by an ellipse, Kepler had already investigated the manner in which the planet's orbital speed varied. After some calculation, he found that Mars speeds up as it comes closer to the sun and slows down as it pulls away from the sun. Kepler expressed this relation by imagining that the sun and Mars are connected by a straight, elastic line. As Mars travels in its elliptical orbit around the sun, in equal intervals of time the areas swept out in space by this imaginary line are always equal (Figure 13-15). This relation is commonly called the *law of areas*.

HOW KEPLER DETERMINED THE ORBIT OF MARS
At this point it is instructive to see how Kepler determined the distance between Mars and the sun at various positions of the planet in its orbit. In Figure 3-16, S represents the sun and M represents Mars at some point in its path around the sun. Suppose we observe Mars when the earth is at E_1. The angle SE_1M at the earth between Mars and the sun is observable. Since the sidereal period of Mars is 687 days, after 687 days Mars will return to point M. The earth, meanwhile, will have completed nearly two full revolutions around the sun and will be at E_2. Angle SE_2M can now be observed. In exactly 2 years, or $730\frac{1}{2}$ days, the earth will have returned to E_1. The earth is short by $730\frac{1}{2}$ —

FIG. 3-14   (a) Ellipses of the same major axis but various eccentricities;
(b) ellipses of the same eccentricity but various major axes.

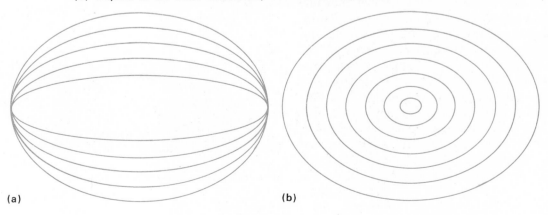

(a)   (b)

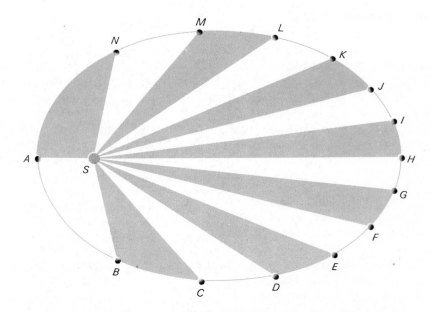

FIG. 3-15   Law of equal areas. A planet moves most rapidly on its elliptical orbit when it is at position *A*, nearest the focus of the ellipse, *S*, where the sun is. The planet's orbital speed varies in such a way that in equal intervals of time it moves distances *AB, BC, CD*, and so on, so that regions swept out by the line connecting it and the sun (shaded and clear zones) are always the same in area.

687 = 43½ days of completing two revolutions about the sun. Thus the angle E_1SE_2 is known — it is the angle through which the earth moves in 43½ days. Lines SE_1 and SE_2 are each the earth's distance from the sun. Thus two sides and an included angle of the triangle E_1SE_2 are known, and the triangle can be solved for the side E_1E_2, in terms of the distance from the earth to the sun, and for the angles SE_1E_2 and SE_2E_1.

Subtraction of angles SE_1E_2 and SE_2E_1 from SE_1M and SE_2M, respectively, gives the angles E_2E_1M and E_1E_2M, both in the triangle $E_1M_2E_2$. In that latter triangle, since two angles and an included side are now known, sides E_1M and E_2M and the third angle can be found. Finally, the distance of Mars from the sun (but in terms of the earth's distance) can be found from either triangle SE_1M or SE_2M.

Kepler found the distance of Mars from the sun at five points along its orbit by choosing from

Tycho's records the elongations of Mars on each of five pairs of dates separated from each other by intervals of 687 days.

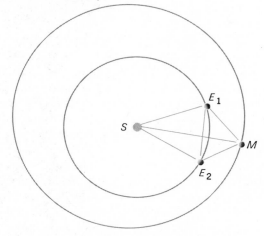

FIG. 3-16   Kepler's method of triangulating the distance to Mars.

KEPLER'S FIRST TWO LAWS OF PLANETARY MOTION SUMMARIZED

We may summarize the most important contributions in *The New Astronomy*, or *Commentaries on the Motions of Mars*, by stating what are now known as Kepler's first two *laws of planetary motion:*

KEPLER'S FIRST LAW: *Each planet moves about the sun in an orbit that is an ellipse, with the sun at one focus of the ellipse.*

KEPLER'S SECOND LAW (THE LAW OF AREAS): *The straight line joining a planet and the sun sweeps out equal areas in space in equal intervals of time.*

At the time of publication of the *New Astronomy* (1609), Kepler appears to have demonstrated the validity of these two laws only for the case of Mars. However, he expressed the opinion that they held also for the other planets.

(b) The Harmony of the Worlds

Although an excellent mathematician, Kepler was also a mystic, and he indulged freely in wild speculation and the occult. In his endeavor to find an underlying harmony in nature, he constantly searched for numerological relations in the celestial realm. It was a great personal triumph, therefore, that he found a simple algebraic relation between the lengths of the semimajor axes of the planets' orbits and their sidereal periods. Because planetary orbits are elliptical, the distance between a given planet and the sun varies. Now, the major axis of a planet's orbit is the sum of its maximum and minimum distances from the sun. Therefore, half of this sum, the semimajor axis, can be thought of as the *average* distance of a planet from the sun. In a circular orbit, the semimajor axis is simply the radius of the circle.

Kepler published his discovery in 1619 in *The Harmony of the Worlds*. The relation is now known as his *third,* or *harmonic, law.*

KEPLER'S THIRD LAW: *The squares of the sidereal periods of the planets are in direct proportion to the cubes of the semimajor axes of their orbits.*

It is simplest to express Kepler's third law with the algebraic equation.

$$P^2 = Ka^3,$$

where P represents the sidereal period of the planet, a is the semimajor axis of its orbit, and K is a numerical constant whose value depends on the kinds of units chosen to measure time and distance. It is convenient to choose for the unit of time the earth's period — the year — and for the unit of distance the semimajor axis of the earth's orbit. This latter quantity is known as the *astronomical unit* (AU). Note that the values found for the distances of the planets from the sun, both as

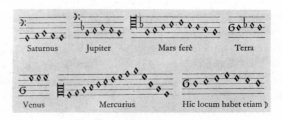

FIG. 3-17   Notes written by Kepler, representing the music "sung" by the planets (from his *Harmony of the Worlds*).

determined by Copernicus and by Kepler, are in astronomical units. With this choice of units, $K = 1$, and Kepler's third law can be written

$$P^2 = a^3.$$

We see that to arrive at his third law it was not necessary for Kepler to know the *actual* distances of the planets from the sun (say, in miles), only the distance in units of the earth's distance, the astronomical unit. The length of the astronomical unit in miles was not determined accurately until later (see Chapter 18).

As an example of Kepler's third law, consider Mars. The semimajor axis of Mars' orbit, *a*, is 1.524 AU. The cube of 1.524 is 3.54. According to the above formula, the period of Mars, in years, should be the square root of 3.54, or 1.88 years, a result that is in agreement with observations. Table 3.2 gives for each of the six planets known to Kepler the values of *a*, *P*, a^3, and P^2. To the limit of accuracy of the data given, we see that Kepler's law holds exactly, except for Jupiter and Saturn, for which there are very slight discrepancies. Decades later, Newton gave an explanation for the discrepancies, but within the limit of accuracy of the observational data available in 1619, Kepler was justified in considering his formula to be exact.

TABLE 3.2 Observational Test of Kepler's Third Law

PLANET	SEMIMAJOR AXIS OF ORBIT, *a* (AU)	SIDEREAL PERIOD (YEARS)	a^3	P^2
Mercury	0.387	0.241	0.058	0.058
Venus	0.723	0.615	0.378	0.378
Earth	1.000	1.000	1.000	1.000
Mars	1.524	1.881	3.537	3.537
Jupiter	5.203	11.862	140.8	140.7
Saturn	9.534	29.456	867.9	867.7

The harmonic law is practically the only thing of value in *Harmony of the Worlds*. Most of the book deals with mysticism. Kepler even went so far as to derive notes of music played by the planets as they move harmoniously in their orbits. The earth, for example, plays the notes *mi, fa, mi*, which he took to symbolize the "*miseria* (misery), *fames* (famine), *miseria*" of our planet.

(c) The Epitome

In 1618, 1620, and 1621, Kepler published (in installments) his text on astronomy, the *Epitome*

of the Copernican Astronomy. The book includes accounts of discoveries, both by himself and by Galileo (see Section 3.5), and firmly supports the Copernican view. Here, for the first time, Kepler implies that his first two laws had been tested and found valid for the other planets besides Mars (including the earth) and for the moon. Also, he states that the harmonic (third) law applies to the motions of the four newly discovered satellites of Jupiter as well as to the motions of the planets about the sun.

KEPLER'S ESTIMATE OF THE DISTANCE TO THE SUN We recall (Section 2.3) that both Hipparchus and Ptolemy had rather accurately measured the moon's distance to be about 60 times the radius of the earth. Earlier, Aristarchus (Section 2.3a) had found that the sun was about 20 times farther away than the moon, which placed it at a distance of about 1200 earth radii. This figure survived until the seventeenth century, when Kepler revised it upward. Kepler was not able to detect any diurnal parallax of Mars — that is, any apparent shift in direction caused by the rotation of the earth carrying the observer from one side of the earth to the other. Now the distance to Mars was known only in terms of the earth's distance from the sun — the astronomical unit; the actual assumed distance to Mars would be proportional to whatever value was assumed for the astronomical unit. If the latter were only 1200 earth radii, Mars should be near enough to allow observation of such a daily parallax. Kepler concluded that the astronomical unit must be at least three times the accepted figure — still a value seven times too small but nevertheless an improvement.

3.5 GALILEO

Galileo Galilei (1564–1642), the great Italian contemporary of Kepler, was born in Pisa. Galileo, like Copernicus, began training for a medical ca-

FIG. 3-18   Galileo Galilei. *(Yerkes Observatory.)*

reer, but he had little interest in the subject and later switched to mathematics. In school he incurred the wrath of his professors by refusing to accept on faith dogmatic statements based solely on the authority of great writers of the past. From his classmates he gained the nickname "Wrangler."

For financial reasons, Galileo was never able to complete his formal university training. Nevertheless, his exceptional ability as a mathematician gained him the post, in 1589, of professor of mathematics and astronomy at the university at Pisa. In 1592 he obtained a far better position at the university at Padua, where he remained until 1610, when he left to become mathematician to the Grand Duke of Tuscany. While at Padua he became famous throughout Europe as a brilliant lecturer and as a foremost scientific investigator.

(a) Galileo's Experiments in Mechanics

Galileo's greatest contributions were in the field of mechanics. In his time the principles of mechanics outlined by Aristotle were still regarded as absolutely authoritative. Most of Aristotle's notions, however, were based upon intellectual assumptions rather than actual experiment. Although the seeds of experimental science had been sown by certain of the later Greek scholars, notably Archimedes, the concept of performing experiments to learn physical laws was virtually unheard of in Galieo's time; scholars believed that all that was to be known could be found within the pages of Aristotle.

Galileo experimented with pendulums, with balls rolling down inclined planes, with light and mirrors, with falling bodies, and many other objects. Aristotle had said that heavy objects fall faster than lighter ones. Galileo showed that if a heavy and light object were dropped together, even from a great height, both would hit the ground at practically the same time. What little difference there is could easily be accounted for by the resistance of the air.

LAWS OF MOTION

In the course of his experiments, Galileo discovered laws that invariably described the behavior of physical objects. The most far-reaching of these is the *law of inertia* (now known universally as Newton's first law). The inertia of a body is that property of the body that resists any change of motion. It was familiar to all persons then as it is to us now that if a body is at rest it tends to remain at rest, and requires some outside influence to start it in motion. Rest was thus generally regarded as the *natural state of matter*. Galileo showed, however, that rest was no more natural than motion. If an object is slid along a rough horizontal floor, it soon comes to rest, because friction between it and the floor acts as a retarding force. However, if the floor and object are both highly polished, the body, given the same initial speed, will slide farther before coming to rest. On a smooth layer of ice, it will slide farther still. Galileo noted that the less the retarding force, the less the body's tendency to slow down, and he reasoned that if all resisting effects could be removed (for example, the friction of the floor or ground, and of the air) the body would continue in a steady state of motion indefinitely. In fact, he argued, not only is a force required to start an object moving from rest, but a force is also required to slow down, stop, speed up, or change the direction of a moving object.

Galileo also studied the way bodies accelerated, that is, changed their speed, as they fell freely, or rolled down inclined planes. He found that such bodies accelerate uniformly, that is, in equal intervals of time they gain equal increments in speed. Galileo formulated these newly found laws in precise mathematical terms that enabled one to predict, in future experiments, how far and how fast bodies would move in various lengths of time. It remained for Newton to incorporate and generalize Galileo's principles into a few simple laws so fundamental that they have become the basis of a great part of our modern technology (Chapter 4).

The great contribution of Galileo, then, was to lay the foundations of experimental science and to bring about a renaissance of the scientific method — a philosophy that was all but lost since its infancy in ancient Greece.

(b) Galileo's Astronomical Contributions

Sometime in the 1590s Galileo adopted the Copernican hypothesis of the solar system. In Roman Catholic Italy, this was not a popular philosophy, for the Church authorities still upheld as absolute truth the ideas of Aristotle and Ptolemy. It was primarly because of Galileo that in 1616 the Church issued a prohibition decree which stated that the Copernican doctrine was "false and absurd" and was not to be held or defended.

The prevailing notion of the time was that the celestial bodies belonged to the realm of the heavens where all is perfect, unchanging, and incorruptible. Perpetual circular motion, being the "perfect" kind of motion, was regarded as the natural state of affairs for those heavenly bodies. Once Galileo had established the principle of inertia — that on the earth bodies in undisturbed motion remain in motion — it was no longer necessary to ascribe any special status to the fact that the planets remain perpetually in orbit. By the same token, even the earth could continue to move, once started. What *does* need to be explained is why the planets move in curved paths around the sun rather than in straight lines. Evidently, Galileo was sufficiently imbued with Aristotelian concepts that he accepted uniform circular celestial motion without subjecting the planets to the same objective scrutiny that he applied in his terrestrial experiments.

In answer to the common objection that objects could not remain on the earth if it were in motion, Galileo noted that if a stone is dropped from the masthead of a moving ship it does not fall behind the ship and land in the water beyond its stern, but rather lands at the foot of the mast, for the stone already has a forward inertia gained from its common motion with the ship before it is dropped. In an analogous way, objects on the earth would not be swept off and left behind if the earth were moving, for they share the earth's forward motion.

GALILEO'S TELESCOPES

It is not certain when the principle was first conceived of combining two or more pieces of glass to produce an instrument that enlarged distant objects, making them appear nearer. Claims for the discovery exist as early as the time of Roger Bacon (thirteenth century). At any rate, the first telescopes that attracted much notice were made by the Dutch spectacle-maker Hans Lippershey in 1608. Galileo heard of the discovery in

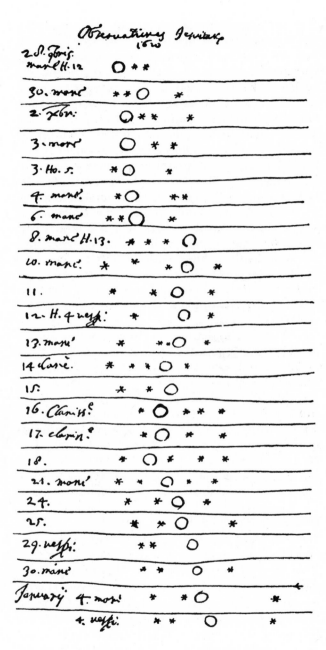

FIG. 3-19 Galileo's drawings of Jupiter and its satellites. *(Yerkes Observatory.)*

1609, and without ever having seen an assembled telescope he constructed one of his own with a three-power magnification — that is, that made distant objects appear three times nearer and larger. He quickly built other instruments, his best with a magnification of about 30.

SIDEREAL MESSENGER

It was a fairly obvious step to apply the newly invented telescope to celestial observations. The idea may have occurred to others about the same time as it did to Galileo, or possibly sooner. Galileo, however, rightly deserves the honor of having been the first to make significant astronomical telescopic observations, for he realized the importance of careful and persistent study of the objects he viewed. In 1610 he startled the world by publishing a list of his remarkable discoveries in a small book, *Sidereal Messenger* (*Sidereus Nuncius*).

Galileo found that many stars too faint to be seen with the naked eye became visible with his telescope. In particular, he found that some nebulous blurs resolved into many stars (for example, the Praesepe in Cancer) and that the Milky Way was made up of multitudes of individual stars. He found that Jupiter had four satellites or moons revolving about it with periods ranging from just under 2 days to about 17 days (eight other satellites of Jupiter have been found since). This discovery was particularly important because it showed that there could be centers of motion that in turn are in motion. It had been argued that if the earth were in motion the moon would be left behind, because it could hardly keep up with a rapidly moving planet. Yet here were Jupiter's satellites doing exactly that!

PHASES OF VENUS

Another important telescopic discovery that strongly supported the Copernican view was the fact that Venus goes through phases like the moon. In the Ptolemaic system, Venus is always closer to the earth than the sun, and thus, because Venus never has more than about 40° elongation, it would never be able to turn its fully illuminated surface to our view — it would always appear as a crescent. Galileo, however, saw that Venus went through both crescent and gibbous phases, and concluded that it must travel around the sun, passing at times behind and beyond it, rather than revolving directly around the earth (Figure 3-20). Mercury also goes through all phases.

IRREGULARITIES IN THE HEAVENS

Galileo's observations revealed much about our nearest neighbor, the moon. He saw craters, mountain ranges, valleys, and flat dark areas that he guessed might be water (the dark *maria*, or "seas," on the moon were thought to be water until long after Galileo's time). Not only did these discoveries show that the heavenly bodies, regarded as perfect, smooth, and incorruptible, do indeed have irregularities, as does the earth, but they showed the moon to be not so very dissimilar to the earth, which suggested that the earth, too, could belong to the realm of celestial bodies.

Of course, Galileo's conclusions did not go uncriticized. Ludovico delle Colombe, one of his critics, argued that the valleys and mountains on the moon were actually submerged beneath an invisible sea of a crystalline material whose outer surface was perfectly smooth. Galileo answered that this was a splendid suggestion with which he completely agreed, and that furthermore there were mountains of this same invisible substance rising 10 times higher than any mountains that could be seen on the moon!

Galileo also found that Saturn did not appear round, although his telescope was not good enough to show the true nature of the planet. It was not until 1655 that Huygens described the magnificent system of rings about Saturn (Chapter 14).

One of Galileo's most disturbing observations, to his contemporaries, was of spots on the sun, showing that this body also had "blemishes." Sun-

spots are now known to be large, comparatively cool areas on the sun that appear dark because of their contrast with the brighter and hotter solar surface (Chapter 24). Sunspots are temporary, lasting usually only a few weeks to a few months. Large sunspots actually had been observed before, with the unaided eye, but were generally regarded either as something in the earth's atmosphere or as planets between the earth and sun silhouetting themselves against the sun's disk in the sky. In fact, some of Galileo's critics attempted to explain the spots as satellites revolving about the sun.

Galileo observed the spots to move, day by day, across the disk of the sun. He also noted that they moved most rapidly when near the center of the sun's disk and increasingly slowly as they approached the limb (the sun's limb is its apparent "edge" as we see it in the sky). Often, after about 2 weeks, the same spots would reappear on the

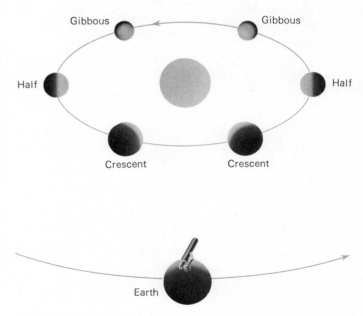

FIG. 3-20   Phases of an inferior planet.

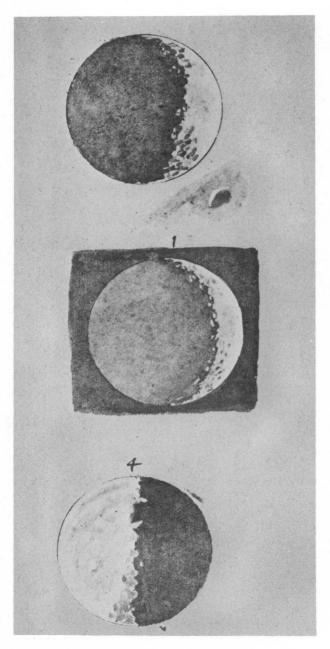

FIG. 3-21 Galileo's drawings of the moon. *(Yerkes Observatory.)*

opposite limb, and move slowly at first, then more rapidly toward the center of the disk. Galileo explained that the spots must be either on the surface of the sun or very close to it and that they were carried around the sun by its own rotation. Their variable speed, he showed, is an effect of foreshortening; when near the center of the sun's disk they are being carried directly across our line of sight, but near the limb most of their motion is either toward or away from us. He determined the sun's period of rotation to be a little under a month.

(c) Dialogue on the Great World Systems

As we have seen, Galileo had accumulated a great deal of evidence to support the Copernican system. By the decree of 1616 he was forbidden to "hold or defend" the odious hypothesis, but he still hoped to convert his countrymen to the heliocentric view — to help them to "see the light." He finally prevailed upon Pope Urban VIII, an old friend, to

FIG. 3-22 Galileo's drawings of sunspots. *(Yerkes Observatory.)*

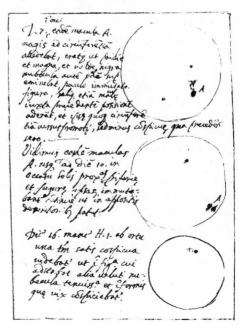

allow him to publish a book that explained fully all arguments for and against the Copernican system, not for the purpose of extolling it, but merely to examine it, and to show those of other nationalities that Italians were not ignorant of new theories.

The book appeared in 1632 under the title, *Dialogue on the Great World Systems (Dialogo dei Massimi Sistemi)*. The *Dialogue* is written in Italian (not Latin) to reach a larger audience, and is a magnificent and unanswerable argument for Copernican astronomy. It is in the form of a conversation, lasting 4 days, among three philosophers: Salviati, the most brilliant and the one through whom Galileo generally expresses his own views; Sagredo, who is usually quick to see the truth of Salviati's arguments; and Simplicio, an Aristotelian philosopher who brings up all the usual objections to the Copernican system, which Salviati promptly shows to be absurd.

It is pointed out in the preface to the *Dialogue* that the arguments to follow are merely a mathematical fantasy, and that divine knowledge assures us of the immobility of the earth. This was thinly cloaked irony, however, and Galileo's enemies acted quickly to build a case against him. He was called before the Roman Inquisition on the charge of believing and holding doctrines that are false and contrary to the Divine Scriptures. Urban VIII, apparently won over by Galileo's enemies, refused to come to his aid. Galileo was forced to plead guilty and deny his own doctrines. His sentence was very light, for that time, consisting of confinement to his own home, under close watch, for the last 10 years of his life. He was then nearly seventy.

The *Dialogue*, meanwhile, took its place along with Copernicus' *De Revolutionibus* and Kepler's *Epitome* on the *Index of Prohibited Books,* where it remained until 1835.

Exercises

1. Which (if any) of the following can never appear at opposition? At conjunction? At quadrature?

(a) Jupiter

(b) Earth

(c) Sun

(d) Venus

(e) Saturn

(f) Mars

(g) Mercury

(h) Moon

2. Does full moon occur at intervals of a sidereal or synodic revolution of the moon about the earth? Why?

3. What is the phase of the moon when it is at quadrature?

4. The synodic period of Mars is approximately 2 years. Assume that it is *exactly* 2 years and find Mars' sidereal period.

5. What would be the sidereal period of an inferior planet that appeared at greatest western elongation exactly once a year?

6. The synodic period of Saturn is 1.03513 sidereal years. What is its sidereal period?

> *Answer:* 29.46 years

7. What would be the distance from the sun, in astronomical units, of an inferior planet that has a greatest elongation of 30°? Assume circular orbits for the planets.

8. Suppose a superior planet has a synodic period of 2 years. It is at opposition on January 1 and is next at quadrature on May 1. What is its distance from the sun in astronomical units? Assume circular orbits. Could such a planet actually exist?

9. What is the major axis of a circle?

10. What is the eccentricity of the orbit of a planet whose distance from the sun varies from 180,000,000 to 220,000,000 mi?

11. The earth's distance from the sun varies from 91,500,000 to 94,500,000 mi. What is the eccentricity of its orbit?

> *Answer:* 0.016

12. Why does $K = 1$ when a is measured in astronomical units and P in years, in the equation for Kepler's third law given in Section 3.4?

13. What would be the period of a planet whose orbit has a semimajor axis of 4 AU?

14. What would be the distance from the sun of a planet whose period is 45.66 days?

> *Answer:* 0.25 AU

15. Draw a diagram showing Galileo's argument in favor of sunspots being on or very near the solar surface.

CHAPTER 4

Newton's Laws and Gravitation

Isaac Newton (1643–1727) was born at Woolsthrope, in Lincolnshire, England, almost exactly 1 year after the death of Galileo. (Newton was born on Christmas Day, 1642, according to the calendar in use at his time, but by the modern Gregorian calendar his birth date was January 4, 1643.) Many would agree with his scientific successor, Lagrange, that "Newton was the greatest genius that ever existed." It was Newton who formulated the basic laws of modern mechanics and showed them to be universal throughout the solar system, applying to the motions of the celestial objects as well as to objects on the earth. Furthermore, Newton's greatest contributions in the fields of astronomy, mechanics, optics, and mathematics were conceived by the time he had reached the age of 24.

Newton entered Trinity College at Cambridge in 1661 and 8 years later was appointed Lucasion Professor of Mathematics, a post that he held during most of his productive career. As a young man in college, he became interested in natural philosophy. He had originally worked out many of his ideas on mechanics and optics during the plague years of 1665 and 1666. Unfortunately, however, he did not publish the results of his investigations at that time, partially because his interests were easily diverted from one subject to another and partially because of outstanding problems and inconsistencies that he had not yet solved. As a result, others from time to time worked out and published the solutions to some of the problems that Newton had actually solved first.

Eventually Newton's friend, Edmund Halley, prevailed on him to collect and publish the results of his investigations in mechanics and gravitation. The result was Newton's *Philosophiae Naturalis Principia Mathematica,* one of the most important documents ever written. The *Principia,* as the book is generally known, was published at Halley's expense in 1687.

As in most great contributions, some of the ideas in the *Principia* were developed independently by others at about the same time, or even earlier. In the *Principia,* however, for the first time a very few natural laws are shown to be general, and to apply to a wide range of phenomena and experiences. Much of the immense variety of the universe became unified under these extraordinarily simple principles.

FIG. 4-1 Isaac Newton. *(Yerkes Observatory.)*

4.1 NEWTON'S LAWS OF MOTION

Newton's entire system is based on three propositions, now known as his three laws of motion. The laws of motion were not entirely new; Galileo had stated the first law and, in a less general form, the second. In the *Principia,* however, the laws are stated as a fundamental basis for all mechanics, and we find there, too, the first clear discussion of the concepts of *mass* and *force.*

(a) The First Law — Inertia

Newton's first law of motion states that in the absence of any outside influence an object at rest tends to remain at rest, and an object in motion tends to remain in motion, in a straight line, and at a constant speed. This property of a body that resists a change in motion is called its *inertia.* The first law of motion is therefore often called the law of inertia. Newton postulated that the law is general and holds for all objects, whether on earth or in space.

MOMENTUM

A measure of the inertia of an object is its *momentum.* Newton defined the momentum of a body as being proportional to its velocity. The constant of proportionality is that property of the body that gives it its inertial characteristics — that enables it to resist a change in its state of motion. This property is called *mass.*† The momentum of a body, then, is the product of its mass and velocity; that is,

$$momentum = mass \times velocity,$$

or, in the more simple algebraic notation,

$$p = mv,$$

where the symbols p, m, and v stand for momentum, mass, and velocity, respectively. Momentum involves two important concepts: *velocity* and *mass.*

----

†Strictly, this is called *inertial mass,* as opposed to *gravitational mass* (Section 4.3c).

VELOCITY

Velocity is a description of the speed a body moves and also its *direction* of motion. The velocity of a body changes if either its speed or direction changes; thus a racing car traveling at a uniform rate around a circular track constantly changes its velocity.

Velocity is often represented graphically by a *vector.* A vector that describes velocity is a line whose length is proportional to the speed and whose direction represents the direction of motion (see Figure 4-2).

Any change in the velocity of an object, either to start it, stop it, speed it up, slow it down, or change its direction, is *acceleration.*

Frequently the direction of motion of a body may not be relevant to the problem at hand. In such a case the velocity often is specified simply by the speed of the object. For example, we might say that the orbital velocity of the earth is 18 mi/sec and not trouble to indicate its direction of motion. Strictly speaking, therefore, we have not really specified completely the earth's velocity; properly we should have said that its speed is 18 mi/sec. This use of the word "velocity" as a synonym for "speed" is a careless habit, but is so widespread that we shall not attempt to overcome it here. It is important to remember, however, that *any* change in velocity, either of speed *or* direction, involves acceleration.

MASS

Mass is a concept first clearly defined by Newton. It is important that we thoroughly understand at the outset what is meant by "mass."

We all have an intuitive notion of mass. Whenever we speak of the *weight* of an object, we almost always mean its *mass.* There is an important distinction between weight and mass. The weight of an object is the *force of attraction* between that object and the earth (or some other body if the object is not on the earth). When we consider

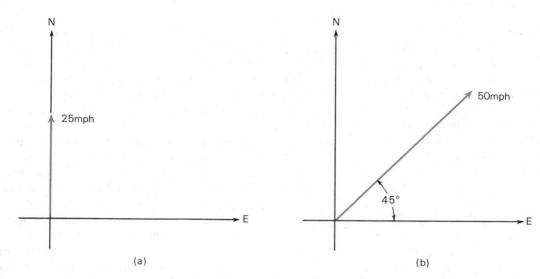

(a)   (b)

FIG. 4-2   Two vectors, representing velocities of 25 and 50 mi/hr, respectively, the first to the north, the second to the northeast.

weight later in the chapter, we shall find that its value depends upon the location of the object in question. On the surface of the moon, a person would weigh only about one sixth of what he does on earth; far off in interstellar space he would be virtually weightless. When we speak of weight in everyday conversation, however, we generally regard it as a property of an object itself; we should say "mass."

Mass is a property that characterizes the actual amount of material of a body. That property, its total content of matter, depends in no way upon where it is. If a 300-lb man goes to the moon, he may only *weigh* 50 lb there, but he has not changed his mass — he is as obese as ever. We might think of the mass of something as a measure of the total number of subatomic particles that compose it. Changing its shape, vaporizing it, freezing it, or sending it to the moon will not change its mass. By convention, we take the mass of an object to be numerically equal to its weight at sea level on the earth. Thus, 10 lb of lead weighs 10 lb here

and nothing in remote space, but it has a mass of 10 lb everywhere.

Actually, to say that the mass of an object is a measure of its total amount of material, while giving us an intuitive feeling for the concept, is not a satisfactory operational definition. In reality we have no means of counting directly the total number of subatomic particles in an object. To speak of mass as a measure of something we must define it in such a way that it can be experimentally determined. In the context of Newton's laws of motion, we can define mass as *that property of a body that gives it inertia,* that is, that *resists acceleration.* This is the *inertial* definition of mass. It will be clear how mass can be measured when we discuss Newton's second and third laws. The most common units of mass are the *gram* (the mass of 1 cubic centimeter of water) and the pound.

MASS, VOLUME, AND DENSITY

It is important not to confuse mass, volume, and density. *Volume* is simply a measure of the physical space occupied by a body, say in cubic inches, cubic feet, liters, and so on. In short, the volume is the "size" of an object — it has nothing to do with its mass. A lady's wristwatch and an inflated balloon may both have the same mass, but they have very different volumes.

The watch and balloon are also very different in *density,* which is a measure of how much mass is contained within a given volume. Specifically, it is the ratio of mass to volume:

$$\text{density} = \frac{\text{mass}}{\text{volume}}$$

or

$$D = \frac{m}{V},$$

where D and V are density and volume. The units of density could be pounds per cubic inch ($lb/in.^3$), grams per cubic centimeter (gm/cm^3), ounces per cubic mile, and so on. Sometimes density is given in terms of the density of water ($62\ lb/ft^3$ or 1 gm/cm^3), in which case it is called *specific gravity.* Iron has a specific gravity of 7.9 or a density of 7.9 gm/cm^3; gold has a specific gravity of 19.3. To sum up, then, *mass* is "how much," *volume* is "how big," and *density* is "how tightly packed."

THE FIRST LAW — SUMMARY

With the concepts of mass and momentum firmly established, we can now state Newton's first law of motion as follows: In the absence of any outside influence, the momentum of any body remains constant — that is, is *conserved.*

(b) The Second Law — Force

The second law of motion deals with changes in momentum. It states that if a force acts on a body it produces a *change* in the momentum of the body that is in the direction of the applied force. The second law, then, defines the concept of *force.*

Specifically, the magnitude or strength of a force is defined as the *rate at which it produces a change* in the momentum of the body on which it acts.

Some familiar examples of forces are the pull of the earth, the friction of air slowing down objects moving through it, the friction of the ground or a floor similarly slowing bodies, the impact of a bat on a baseball, the pressure exerted by air, and the thrust of a rocket engine.

Note that Newton's first law of motion is consistent with his second; when there is no force, the change in momentum is zero.

There are three ways in which the momentum of a body can change. Its velocity can change, or its mass, or both. Most often the mass of a body does not change when a force acts upon it; a change in momentum usually results from a change in velocity. Thus, in the vast majority of examples, the second law can be written as the simple formula

$$\text{force} = \text{mass} \times \text{acceleration},$$

because acceleration is the rate at which velocity changes. Let us consider acceleration in more detail.

We recall that acceleration occurs whenever the velocity of a body changes, and that the body's velocity changes when either the speed or the direction of its motion is altered. If the acceleration occurs in the same direction as the velocity, the body simply speeds up; if the acceleration occurs in the opposite direction to the velocity, the body slows down. If acceleration occurs exactly at *right angles* to the velocity, only the direction of motion of the body, and not its speed, changes.

Galileo had investigated falling bodies indirectly by experimenting with bodies rolling down inclined planes. The acceleration of falling bodies is downward (in the direction toward which the gravitational pull of the earth is acting). It accelerates a body in the direction it is already moving, and so simply speeds it up. Galileo determined that the speed of a falling body increases *uniformly* with time. One second after a body is dropped it moves 32 ft/sec. At the end of the second second it moves 64 ft/sec; at the end of the third second, 96 ft/sec; and so on. In other words, every second a falling body increases its speed by 32 ft/sec. The rate of change of its velocity, its *acceleration,* is then 32 ft/sec/sec.

If a body is slid along a rough horizontal surface, it slows down uniformly in time. It is therefore accelerated in a direction opposite to its velocity. The acceleration is produced by the force of friction between the moving body and the rough surface.

In general, both the speed *and* direction of a body may change. Suppose, for example, its velocity is the vector **v** lying in direction *AB* in Figure 4-3. Now suppose that for a short time Δ*t* a force with direction *BC* acts on the body. The force produces a change in velocity Δ**v**, in direction *BC*, and the new velocity **v′** is in direction *AC*.

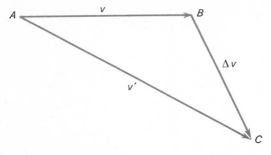

FIG. 4-3   Change in velocity.

(The Greek letter Δ — capital delta — is commonly used to denote a small change in a quantity.) Acceleration is the *rate* at which the velocity changes, so we reason that if the velocity of a body changes by an amount Δ**v** in time Δ*t*, the rate at which it changes, that is, its acceleration, is Δ**v**/Δ*t*. If the force (and hence acceleration) were not constant in magnitude and direction, Δ**v**/Δ*t* is only the average acceleration over the time Δ*t*. However, if

the interval Δ*t*, and hence Δ**v**, is very small, Δ**v**/Δ*t* becomes a good approximation to the *instantaneous* acceleration. (Instantaneous rates of change are the subject of *differential calculus,* invented by Newton to handle such problems in mechanics.)

Of course, a body accelerates, that is, its momentum changes, only while a force acts upon it. The force of gravity is constantly pulling falling bodies downward; thus, they continually accelerate. On the other hand, once a baseball ceases to be in actual contact with the bat, it continues forward in a straight line at a constant speed and would do so forever if air friction did not slow it, the earth pull it down, or an eager outfielder intercept it.

Whenever the mass of an object is subject to change, we cannot assume that force = mass × acceleration. The variation of mass must be taken into account in calculating the acceleration produced by a force. An important and familiar example is a rocket. Exhaust gases are ejected from the back to accelerate the rocket forward, with its remaining fuel. Thus, as the rocket accelerates, its mass continually diminishes.

Two equal accelerations may correspond to entirely different forces. Consider the forces required to accelerate an automobile and a bicycle each to a speed of 20 mi/hr in 20 seconds. Clearly, because of the car's greater mass, a proportionately greater force will be required to produce the necessary acceleration. Similarly, once the bodies are both moving at that speed, a far greater force is needed to *stop* the automobile as quickly as the bicycle.

COMBINATIONS OF FORCES

Newton showed how to compute the effect of two or more forces acting at once on an object. We shall illustrate the procedure when there are two forces, f_1 and f_2, acting together on a body at *O*. We can represent the forces by vectors just as we can velocities (Figure 4-4). The directions of the

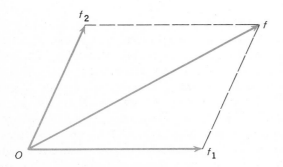

FIG. 4-4   Parallelogram of forces.

vectors show the directions of the forces and the lengths show the magnitudes of the forces (say, so many dynes to the inch). The arrows representing the vectors both start from the object on which the two forces are acting. We note now that the two arrows define two sides of a parallelogram, Of_1ff_2. The diagonal of the parallelogram, Of, is the resultant force acting on the body; the body accelerates as if acted upon by just that one force. The figure shown is called a *parallelogram of forces*.

STATIC EQUILIBRIUM

Of special interest is the situation in which two or more forces act upon a body in such a way as to cancel each other. Then the body is said to be in a state of *static equilibrium*, and does not accelerate. An example is a ball hanging from a rubber band — the rubber stretches until its tension exactly balances the force of gravity pulling the ball downward. Another example is the column of mercury in a barometer; the mercury stands in static equilibrium between its own weight and the pressure of the air.

(c) The Third Law — Reaction

Newton's third law of motion was a new idea. It states that all forces occur as *pairs* of forces that are mutually equal to and opposite each other. If a force is exerted upon an object, it must be exerted by something else, and the object will exert an equal and opposite force back upon that something. All forces, in other words, must be mutual forces acting *between* two objects or things. (Both need not be material objects; for example, a force exists between a magnetic field and an electrically charged particle.)

If a man pushes against his car, the car pushes back against him with an equal and opposite force, but if the man has his feet firmly implanted on the ground, the reaction force is transmitted through him to the earth. Because of its enormously greater mass, the earth accelerates far less than the car. Suppose a boy jumps off a table down to the ground. The force pulling him down is a mutual gravitational force between him and the earth. Both he and the earth suffer the same total change of momentum because of the influence of this mutual force. Of course, the boy does most of the moving; because of the greater mass of the earth, it can experience the same change of momentum by accelerating only a negligible amount.

A more obvious manifestation of the mutual nature of forces between objects is familiar to all who have played baseball. The recoil of the bat shows that the ball exerts a force on the bat during the impact, just as the bat does on the ball. The momentum imparted to the bat by the ball is transmitted through the batter to the earth, so the acceleration produced is far less than that suffered by the ball. Similarly, when a rifle is discharged, the force pushing the bullet out the muzzle is equal to that pushing backward upon the gun and marksman.

If a swimmer dives from the side of a rowboat, the force he exerts upon himself to spring out away from the boat is equal to a force that accelerates the craft backward away from him. Here, in fact, is the principle of rockets — the force that discharges the exhaust gases from the rear of the rocket is accompanied by a force that shoves the rocket forward. The exhaust gases need not push against air or the earth; a rocket operates best of all in a vacuum. Incidentally, passengers on a rocket ship would have to take care how they disposed of waste material. For example, if garbage were discharged through a port in the missile, the reaction force would accelerate the rocket slightly off course.

In all the cases considered above, a mutual force acts upon the two objects concerned; each object always experiences the same total change of momentum, but in opposite directions. Because momentum is the product of velocity and mass, the object of lesser mass will end up with proportionately greater velocity. Today we are developing atomic rocket engines. Atoms are shot from the rear of such engines. The masses of the atoms are extremely small, but they are propelled outward with speeds near that of light, so they carry momentum that is not completely negligible, and if enough atoms are shot out, the engine can measurably accelerate a rocket.

GENERALITY OF THE THIRD LAW

Newton phrased his third law by stating that for every action there is an equal and opposite reaction. In fact, a system may be far more complex than just a pair of bodies exerting mutual forces on each other. Suppose, for example, a rocket in space were to explode into thousands of pieces. It is then pointless to speak of two bodies exerting mutual forces on each other; we generalize by saying that each of the remnants of the explosion has exerted a force on each other in such a way that all the forces balance each other. Each particle has suffered a change in momentum, but the sum total of all changes in momentum of particles accelerated one way will be equal and opposite to the total change in momentum associated with all the particles going the opposite direction. All these changes in momentum balance each other, so that the momentum of the entire system, as long as it has not been acted on by *external* forces, is the same as before the explosion.

In other words, Newton's third law can be thought of as a generalization of his first. The first law states that in the absence of a force, a body's momentum is conserved. The third law means that if we isolate an entire system from outside forces, the total momentum of the system is conserved. *Internal* forces in the system may result in changes of momentum within it, but these are always accompanied by equal and opposite changes. The total momentum of a rocket, for example, does not change, as long as we always include the momentum of its exhaust gases and any other object (such as the atmosphere of the earth) that may have come in contact with these gases.

MEASURING MASS

We are now in a position to see how we can measure mass. Initially, some object must be adopted as a standard and said to have *unit mass* — for example, 1 cubic centimeter of water, or perhaps the mass of some king's favorite gemstone. Then the mass of any other object can be compared to it by measuring the relative accelerations produced when the same force acts on each of the two. We assure that each is subjected to the same force by isolating them from other objects and producing equal and opposite changes in their momenta with an *internal* force — say, by separating them with a compressed spring placed between them and released by an internal triggering device, or by detonating a small charge between them. Thus, the third law permits an *operational definition* of mass.

Having found a way to measure mass, we can now express the value of a force numerically. The most common units of force are *dynes* and *poundals*. A dyne is the force needed to give a mass of 1 gm an acceleration of 1 cm/sec/sec; a poundal is the force needed to give a mass of 1 lb an acceleration of 1 ft/sec/sec. At the surface of the earth the force due to gravity acting on a 1-lb mass is 32 poundals.

(d) The Laws of Motion Summarized

For convenience, we now summarize Newton's three laws of motion:

NEWTON'S FIRST LAW: *In the absence of outside forces, the momentum of a system remains constant.*

NEWTON'S SECOND LAW: *If a force acts upon a body, the body accelerates in the direction of the force, its momentum changing at a rate numerically equal to that force.*

NEWTON'S THIRD LAW: *Forces are always mutual; thus if a force is exerted upon a body, that body reacts with an equal and opposite force upon whatever exerts the force upon it.*

Newton's three laws, as we have seen, contain, respectively, definitions of the three concepts, momentum, force, and mass.

4.2 ACCELERATION IN A CIRCULAR ORBIT

It might be assumed that some force or power is required to keep the planets in motion. However, Galileo argued from the principle of inertia (Newton's first law of motion) that once started, the planets would remain in motion — that the state of motion for planets was as natural as for terrestrial objects. What does require explanation, however, is why the planets move in nearly circular orbits rather than in straight lines (the latter motion would eventually carry them away from the vicinity of the solar system). Galileo had not considered this problem.

By Newton's time, a number of investigators had considered the problem of circular motion. The correct solution to the problem was first published (in 1673) by the Dutch physicist Christian Huygens (1629–1695). However, Newton had found the solution independently in 1666.

Clearly, for a body to move in a circular path rather than in a straight line, it must continually suffer an acceleration toward the center of the circle. Such an acceleration is called *centripetal acceleration*. The central force that produces the centripetal acceleration (*centripetal force*) is, for a planet, an attraction between the planet and the sun. For a stone whirled about at the end of a string, the centripetal force is the tension in the string. With the help of Newton's laws of motion and some elementary mathematics we can calculate how great that central force has to be. We find that if a particle of mass m moves with a speed v on the circumference of a circle of radius r, the centripetal force is given by the formula

$$F = \frac{mv^2}{r}.$$

(a) *Derivation of the Centripetal Force*

Suppose the particle moves on the circumference of a circle centered at O (Figure 4-5). At some instant it is at position D and has a velocity of magnitude v and direction DE. By Newton's first law, if there were no forces acting upon the particle, it would continue to move in direction DE.

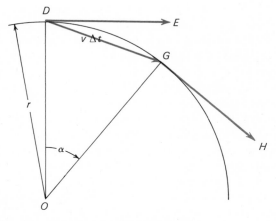

FIG. 4-5   Motion on a circular orbit.

After a brief interval of time Δt, it is at G, a distance along the circumference of the circle from D equal to its speed times Δt (for distance = rate × time); that is, G is a distance $v \times \Delta t$ along the arc from D. The particle's velocity still has magnitude v but is now in the direction GH. If the interval Δt is considered to be very short, the distance $v \Delta t$ along the arc is essentially the straight-line distance DG along the chord of the arc. (The angle α in Figure 4-5 is greatly exaggerated.)

The particle has accelerated, for it did not continue to move along direction DE. Because the particle is closer to the center of the circle than it would have been at E, the force that accelerated it must have acted in the direction toward the center. Since acceleration is the change of velocity per second, the acceleration the particle suffered during the time Δt is just the change of its velocity from direction DE to direction GH, divided by Δt. The angle through which the velocity changed must be the same as the angle α at the center of the circle subtended by the arc DG along which the particle moved, because its velocity at any point is always in a direction tangent to the circle, and the tangent to a circle at any point is always perpendicular to a line from that point to the center of the circle.

We can represent a change in velocity when that change is only in direction by a diagram such as in Figure 4-6. The original velocity has magnitude v and direction OA. After time Δt, the velocity has magnitude v and direction OB. The line representing the velocity has turned,

in time Δt, through an angle α at O. The magnitude of the acceleration of the particle, its change in velocity during 1 second, is thus the change in velocity Δv, that is, the vector AB, divided by Δt.

Since the triangle OAB in Figure 4-6 is exactly similar to the triangle DOG in Figure 4-5, we have the simple proportion

$$\frac{\Delta v}{v} = \frac{v \, \Delta t}{r}.$$

The acceleration of the particle is then

$$a = \frac{\Delta v}{\Delta t} = \frac{v^2}{r}.$$

If the particle has mass m, the central force required to produce this acceleration is $m \times a$, or

$$F = \frac{mv^2}{r},$$

as stated above.

Clearly, the above analysis is precise only if the angle α and the quantities Δv and Δt are extremely small. Here is another example in which Newton employed differential calculus (which he called *fluctions*); he obtained the result we derived, but rigorously, by allowing Δv and Δt to become infinitesimal; that is, he found the instantaneous rate of change of v.

4.3 UNIVERSAL GRAVITATION

It is obvious that the earth exerts a force of attraction upon all objects at its surface. This is a mutual force; a falling apple and the earth are pulling on each other. Newton wondered whether this force of attraction between the earth and objects on or near its surface might extend as far as to the moon and produce the centripetal acceleration required to keep the moon in its orbit. Newton wondered further whether there might be a general force of attraction between *all* material bodies. If so, the attractive force between the sun and each of the planets could provide the centripetal acceleration necessary to keep each in its respective orbit.

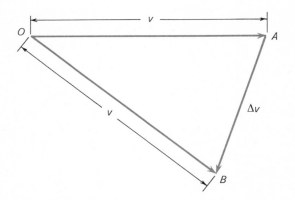

FIG. 4-6   Centripetal acceleration.

Thus Newton hypothesized that there is a universal attraction between all bodies everywhere in space. Here is an excellent example of the workings of the scientific method. First, certain *phenomena* were observed, or discovered — the motions of the planets and the moon, Kepler's laws of planetary motion, and Galileo's laws of falling bodies. Second, a *hypothesis* was suggested to describe these phenomena, in this case the hypothesis of universal attraction. Finally, Newton had to determine the mathematical nature of the attraction and test the hypothesis by using it to predict *new* phenomena. We shall now see how Newton formulated his law of universal gravitation.

(a) *The Mathematical Description of Gravitation**

For mathematical simplification we make the assumption that planets revolve around the sun in perfectly circular orbits. A more complicated analysis can be made to apply to the actual elliptical orbits. Using the results of Section 4-2 we find that the centripetal force that the sun must exert upon a planet of mass m_p, moving with speed v in a circular orbit of radius r, is

$$\text{force} = \frac{m_p v^2}{r}.$$

Now the period P of the planet, that is, the time required for the planet to go completely around the sun, is the circumference of its orbit ($2\pi r$) divided by its speed, or

$$P = \frac{2\pi r}{v}.$$

Solving the above equation for v, we find

$$v = \frac{2\pi r}{P}.$$

On the other hand, from Kepler's third law we know that the square of the period of a planet is in proportion to the cube of its distance from the sun. Because the sun is observed to be almost at the center of the planet's orbit, that distance is very nearly the radius of the orbit, r, and we have

$$P^2 = Ar^3,$$

where A is a constant of proportionality whose value depends on the units used to measure time and distance. Combining the last two equations, we find

$$v^2 = \frac{4\pi^2 r^2}{P^2} = \frac{4\pi^2 r^2}{Ar^3} = \frac{4\pi^2}{Ar},$$

that is,

$$v^2 \propto \frac{1}{r},$$

where the symbol $\propto$ means "proportional to."

If we substitute the above formula for v^2 into the one expressing the sun's centripetal force on the planet, we obtain

$$\text{force} \propto \frac{m_p}{r^2}.$$

The centripetal force exerted on the planet by the sun must therefore be in proportion to the planet's mass and in inverse proportion to the square of the planet's distance from the sun. According to Newton's third law, however, the planet must exert an equal and opposite attractive force on the sun. If the gravitational attraction of the planet on the sun is to be given by the same mathematical formula as that for the attraction of the sun on the planet, the planet's force on the sun must be

$$\text{force} \propto \frac{m_s}{r^2},$$

where m_s is the sun's mass. Since this is a mutual force of attraction between the sun and planet, it must be proportional to both the mass of the sun and the mass of the planet; therefore, the attractive force between the two has the mathematical form

$$\text{force} \propto \frac{m_s m_p}{r^2}.$$

Both the sun and a planet revolving around it experience the same change of momentum as a result of this mutual force between them. The fact that the sun is observed to remain more or less at the center of the solar system while the planets revolve around it is evidence that the sun's mass must be enormously greater than that of the planets. Therefore, its acceleration is relatively small, but it is actually observable.

(b) The Law of Universal Gravitation

For Newton's hypothesis of universal attraction to be correct, there must be an attractive force between all pairs of objects everywhere whose value is given by the same mathematical formula as that above for the force between the sun and a planet. Thus the force F between two bodies of masses m_1 and m_2, and separated by a distance d, is

$$F = G \frac{m_1 m_2}{d^2}.$$

Here G, the constant of proportionality in the equation, is a number called the *constant of gravitation*, whose value depends on the units of mass, distance, and force used. The actual value of G has to be determined by laboratory measurement of the attractive force between two material bodies (see Section 4.3e). If metric units are used (grams for mass, centimeters for distance, and dynes for force), G has the numerical value 6.67×10^{-8}.*

The above equation expresses Newton's law of universal gravitation, which is stated as follows:

Between any two objects anywhere in space there exists a force of attraction that is in proportion to the product of the masses of the objects and in inverse proportion to the square of the distance between them.

Not only is there a force between the sun and each planet, but also between any two planets. Because of the sun's far greater mass, the dominant force felt by any planet is that between it and the sun. The attractive forces between the planets have relatively little influence. Similarly, there is a gravitational attraction between any two objects on earth (for example, between two flying airplanes, or between the kitchen sink and a tree outside the house), but this force is insignificant compared to the force between each of them and the very massive earth.

Before we see how Newton tested his law of gravitation, let us investigate some of its other consequences.

(c) Weight

Newton hypothesized a force of attraction between all pairs of bodies. The earth is a large spherical mass, however, that can be thought of as being composed of a large number of component parts. An object, say a man, on the surface of the earth feels the simultaneous attractions of the many parts of the earth pulling on him from many different directions. Exactly what is the resultant gravitational effect of the many parts of a sphere, each pulling independently upon a mass outside the surface of the sphere? Here was a difficult problem to which Newton had to find a solution before he could test his law of gravitation.

To solve the problem, Newton had to calculate the force between an object on the surface of the earth and each infinitesimal piece of the earth, and then calculate how all of these forces combined. It was necessary for him to invent and use a new method of mathematics which he called *inverse fluxions* (today we call it *integral calculus*). Fortunately, the solution to the problem gives a beautifully simple result. A spherical body acts gravitationally as though all its mass were concentrated at a point at its center (Figure 4-7).† This meant that Newton could consider the earth, moon, sun, and planets as geometrical points as far as their gravitational influences are concerned.

The gravitational force, then, between an object of mass m on the earth and the earth itself, of mass M, is equal to the constant of gravitation times the product of the masses m and M, divided

---

*The notation 6.67×10^{-8} means 0.0000000667; see Appendix 3.

†Strictly, the statement is correct only if the density distribution within the body is spherically symmetrical.

 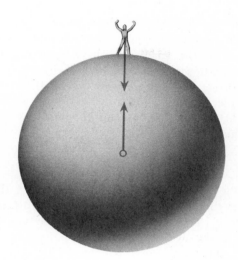

FIG. 4-7   Attraction of a sphere is as though all its mass were concentrated at its center.

by the square of the distance from the object to the center of the earth. The latter distance is just the radius of the earth, R. This gravitational force between the earth and a body on its surface is the body's *weight*. Algebraically, the weight W of a body is given by

$$W = G \frac{mM}{R^2}.$$

We see that the weight of a body is proportional to its mass. This circumstance gives us another method of measuring mass, namely, by measuring the weight of a body. Whenever the mass of an object is determined by its gravitational influence, as by measuring its weight, the mass so determined is defined as a *gravitational mass.* So far, the results of all experiments indicate that gravitational mass is exactly equivalent to *inertial mass,* measured as described in Section 4.1c.

Because the gravitational attraction between two bodies decreases as the square of their separation, an object weighs less if it is lifted above the surface of the earth. A person actually weighs less

at the top of a step ladder than on the ground. Careful laboratory experiments are able to detect such subtle changes in gravitational force. At 4000 mi above the earth's surface, a body is twice as far from the earth's center as it would be at the surface, so an object there would weigh 2^2, or 4, times less than at the surface of the earth. Far out in space, a person's weight drops virtually to zero. In the vicinity of another gravitating body, his weight is determined by the attraction between him and that body.

If an object is dropped from a height, the downward acceleration is equal to the force acting on it, that is, its weight, divided by its mass:

$$\text{acceleration} = \frac{W}{m} = G \frac{M}{R^2}.$$

Thus, although the gravitational force between the earth and a massive object is greater than between the earth and a less-massive one, all objects experience the same acceleration and fall at the same rate, as Galileo had found.

(d) Test of Gravitation: The Apple and the Moon

Suppose an apple is dropped from a height above the surface of the earth. We have seen that it accelerates 32 ft/sec/sec. If the hypothesis of gravitation is correct, the accelerations toward the earth of the apple (at the earth's surface) and of the moon (239,000 mi away) should both be given by the equation

$$\text{acceleration} = G\frac{M}{D^2},$$

where M is the mass of the earth and D is the distance of the object in question from the center of the earth. The acceleration, in other words, should be inversely proportional to the square of the distance from the earth's center. The apple's distance is about 4000 mi, and the moon's distance is 239,000 mi, about 60 times farther. Thus the acceleration of the moon should be 60^2, or 3600, times less than the apple's.

If we assume that the moon's orbit about the earth is a perfect circle (its orbit is, indeed, very nearly circular), we can use the formula for centripetal acceleration found in Section 4.2 to calculate the moon's acceleration:

$$\text{acceleration} = \frac{v^2}{D}.$$

The moon's distance (the radius of its orbit) is 239,000 mi, or about 1.26×10^9 ft. Its orbital speed is about 3350 ft/sec. If we substitute these numbers in the above formula we find

$$\begin{aligned}
\text{moon's acceleration} &= \frac{(3350)^2}{1.26 \times 10^9} \\
&= 0.0089 \text{ ft/sec}^2 \\
&= 0.107 \text{ in./sec}^2.
\end{aligned}$$

The acceleration predicted by the law of gravitation is

$$\begin{aligned}
\text{moon's acceleration} &= \frac{32}{(60)^2} \\
&= 0.0089 \text{ ft/sec}^2 \\
&= 0.107 \text{ in./sec}^2.
\end{aligned}$$

The law of gravitation predicts that because the moon is 60 times farther from the center of the earth than an apple, its acceleration should be 60^2 times less; this is exactly the acceleration that we observe for the moon. The test gives results that are consistent with Newton's law of universal gravitation.

(e) The Mass of the Earth and the Determination of G

Note that in the formula for the weight of an object at the surface of the earth, or for its acceleration toward the earth, the constant of gravitation, G, always occurs multiplied by the mass of the earth. If the latter were known, G could be evaluated at once from the known acceleration of gravity:

$$G = \frac{R^2 g}{M_E},$$

where R and M_E are the radius and mass of the earth and g is the known acceleration of gravity at the earth's surface — 32 ft/sec/sec (980 cm/sec/sec in metric units). Conversely, if G were known, the mass of the earth could be found:

$$M_E = \frac{R^2 g}{G}.$$

Hence the determination of G is equivalent to the determination of the mass of the earth.

There are various methods of finding the mass of the earth. All depend on a comparison of the gravitational attraction between some object of known mass and the earth, with the attraction between two objects of known mass.

The earth's mass can be determined simply and accurately with the use of a delicate balance. This method was applied by P. von Jolly in Munich in 1881. The principle of the balance method is shown in Figure 4-8.

Two equal weights, A and A', are placed in the two pans of the balance. Both weights have the same mass and are the same distance from the

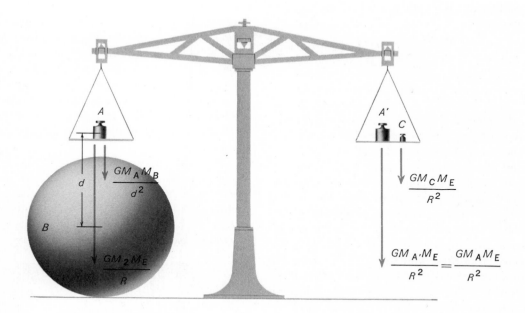

FIG. 4-8 "Weighing" the earth with a balance.

center of the earth, so the earth has the same attraction for both, and the system is in perfect balance. A very large mass B is then placed under the pan with the mass A, with its center only a small distance d from A. Now, in addition to the force between A and the earth, there is a small gravitational force between A and B. However, B is far enough from the weight A' in the other pan so that the attraction between B and A' is negligible. Thus A feels a greater downward force than A', and the balance is upset.

To restore the balance, a very small weight, C, is placed in the pan with A'. Now, the attraction between the earth and C is just enough to counteract the attraction between weight A and the large ball, B, and the pans hang evenly again. Since the system is again in perfect balance, the downward force on both pans must be the same. We let M_E and R be the mass and radius of the earth and

$M_A = M_{A'}$, M_B, and M_C be the masses of the weights A, B, and C. The total downward force on the left pan, the force between A and the earth *plus* the force between A and B, is

$$G\frac{M_E M_A}{R^2} + G\frac{M_A M_B}{d^2},$$

and the total downward force on the right pan, the force between A' and the earth *plus* the force between C and the earth, is

$$G\frac{M_E M_{A'}}{R^2} + G\frac{M_E M_C}{R^2}.$$

If we now equate these two forces (and remember that $M_A = M_{A'}$), we obtain

$$G\frac{M_E M_A}{R^2} + G\frac{M_A M_B}{d^2} = G\frac{M_E M_A}{R^2} + G\frac{M_E M_C}{R^2}.$$

After canceling out the common terms, and solving for M_E, we find

$$M_E = \frac{R^2}{d^2} \frac{M_A M_B}{M_C}.$$

Since all the masses and distances on the right side of the last equation can be measured directly, the mass of the earth can be calculated.

A more accurate modern method of "weighing" the earth is with a torsion balance. In a torsion balance two spheres are connected by a horizontal light rod that is suspended in the middle by a thin quartz fiber. Then another large mass is placed near one of the spheres. The gravitational force between the large mass and the sphere causes the sphere to move toward it, twisting the quartz fiber. The amount of twist, or torsion, in the fiber is an accurate measure of the force between the two gravitating masses. This force can then be compared to that between an object and the earth to obtain the earth's mass.

The results of the best determinations give for the mass of the earth 5.98×10^{27} gm, or 6.6×10^{21} English tons.

4.4 LIMITATIONS TO NEWTON'S LAWS

There are many examples to demonstrate the generality of Newton's laws. During the eighteenth and nineteenth centuries, Newton's laws seemed so powerful that they were generally believed to apply to *all* natural phenomena. However, by the twentieth century, certain experiments had been performed which yielded results that did not conform to Newtonian mechanics. From these experiments grew two new branches of physics, based on new postulates. These new disciplines provide descriptions of physical phenomena within realms in which Newton's laws are found not to apply. We describe them only briefly.

(a) Relativity

Particularly perplexing were the results of an experiment carried out by Michelson and Morley in 1887, which was intended to reveal the absolute speed of the earth in space by measuring the velocity of light in different directions. Much to the surprise of all, the experiment gave null results — it completely failed to reveal any motion of the earth at all! Other similar experiments gave similar results.

In 1905, Albert Einstein (1879–1955) published a paper in which he outlined his *special theory of relativity.* The theory was based on two postulates: (1) there is no absolute reference system with respect to which we can measure "absolute" motions in space; the best we can hope

FIG. 4-9   Albert Einstein. *(Yerkes Observatory.)*

to measure is the *relative* motion of one object with respect to another; and (2) the speed of light, with respect to all observers, is always the same. There is much experimental evidence to support the latter postulate, particularly evidence obtained from the orbits of double stars. The first postulate seems a perfectly reasonable one.

From these two postulates, Einstein showed that two observers, moving with respect to each other, will disagree on their measurements of length, time, velocity, and mass. Einstein derived equations that showed by exactly how much the two observers would disagree. Contrary to popular belief, the special theory of relativity is not extremely difficult or obscure. It can be mastered by any competent student in college physics. Not only did the theory seem sound and reasonable, but it completely explained the results of perplexing experiments such as those of Michelson and Morley. Today the results of a vast number of observations and experiments are in complete accord with special relativity. Among these is the equivalence of mass and energy predicted by the special theory through the famous equation $E = mc^2$.

In 1916 Einstein generalized the theory to include the effects of accelerations between various observers. This more far-reaching theory is the *general theory of relativity*. It is much more difficult than the special theory, mathematically. The general theory of relativity predicts new laws of motion and a new law of gravitation. Einstein proposed three astronomical tests for this new law of gravitation. They are: (1) the orbit of Mercury should undergo an observable perturbation — the orientation of its major axis should slowly change in space at a rate that is different from that predicted by the perturbation theory of Newtonian mechanics (Chapter 14). (2) Light passing near the surface of the sun should be deflected slightly in direction. The effect should be observable during a total solar eclipse, when the sun's brillance is blotted out and stars appearing in nearly the same direc-

tion as the sun can be observed (Chapter 9). (3) The wavelengths of light leaving certain extremely dense stars (white dwarfs) should be slightly lengthened (Chapter 23). All three of the predicted effects have been observed, although there is still some controversy over some of the observations. Furthermore, although the results of these three tests are compatible with relativity, they are also compatible with other theories. However, most modern physicists accept the general theory of relativity, as well as the special theory, as an accurate description of nature.

The predictions of special relativity differ from those of Newtonian mechanics only when speeds are encountered that are very near the speed of light. Einstein's more general laws predict results identical to Newton's when the relative velocities encountered are within our ordinary range of experience.

(b) The Quantum Theory

All matter is composed of atoms. Atoms, in turn, are composed of still smaller subatomic particles — protons, electrons, neutrons, and so on. Experiments concerned with the behavior of these subatomic particles and the nature of radiation show that Newtonian mechanics is not adequate to describe their motions. There gradually developed in the present century a theory based on new physical principles that describes the behavior of these tiny particles to the degree of accuracy of present-day observations. This is the *quantum theory*. Only when it is applied to these tiny particles does the quantum theory predict results that are measurably different from those expected from Newton's laws. When applied to problems encountered in everyday experience, the quantum theory, like relativity, predicts the same results that we obtain from Newtonian theory.

Later, especially in Chapter 10, we shall have occasion to look more closely at some of the consequences of the quantum theory.

(c) Newton's Laws Wrong?

In view of these limitations to Newton's laws, we might be inclined to say that Newtonian theory is in error. If we do, we are forgetting the significance of natural laws. Laws are the means by which science attempts to *describe* nature, not *explain* it. The mistake is only to apply them to situations that are outside their range of validity. The very simple principles of Newtonian mechanics, well within the grasp of any high school student, give an amazingly accurate representation of a great many natural phenomena. Among these are the motions of the planets and other bodies in the solar system, and even of stars and systems of stars. The application of Newtonian mechanics to the motions of celestial bodies is *celestial mechanics,* the subject of Chapters 5 and 6.

Exercises

1. What is the momentum of a body if its velocity is zero? Does the first law of motion, stated as: "In the absence of a force the momentum of a body is constant," include the case of a body at rest?

2. Can a quantity of lead be more massive than a pile of feathers? Can it be less massive? Can it have a greater volume? A smaller volume? A greater density? A smaller density? Give examples.

3. How many accelerators are there in a standard passenger car? Explain.

4. Why is it nonsense to speak of "force of forward motion"?

5. Suppose two billiard balls impact each other. Ignoring frictional forces, describe their velocities and accelerations.

6. Sometimes a cue ball hits another billiard ball and after the impact the other ball moves quickly forward while the cue ball stands motionless, or moves only slightly. How can this be explained in terms of Newton's third law?

7. What does a pan balance measure? A spring balance?

8. Did Newton need to know the size of the earth in miles to test his theory of gravitation on the apple and moon? Why or why not?

9. Invent and describe an experiment by which you might measure the inertial mass of a small solid object in terms of the mass of another object chosen as a standard.

*10. Calculate the gravitational attraction between a 200-lb man and a 100-lb woman 10 feet apart, and compare it to the attraction between the man and the earth.

11. How much would a 200-lb man weigh 4000 mi above the surface of the earth? How much would he weigh 16,000 mi above the surface of the earth?

Answer to second part: About 8 lb.

CHAPTER 5

Celestial Mechanics

Newton's laws of motion and gravitation enable us to predict the motions of bodies under the influence of their mutual gravitation. In this chapter we shall consider the interactions of two bodies, both of which are either point masses or spherically symmetrical, so that they act (gravitationally) as point masses. The subject is called the *two-body problem*.

5.1 CENTER OF MASS*

According to Newton's third law, the total momentum of an isolated system is conserved; that is, all changes of momentum within it are balanced. We can, therefore, define a point within the system that remains fixed (or moves uniformly) as if the entire mass of the system were concentrated at that point. It is called the *center of mass*. It can be shown that the center of mass of a complex body (which is a collection of point masses joined rigidly together) is that point at which the body balances when placed near a gravitating body; thus it is also often called the *center of gravity*.

The center of mass (or gravity) for two bodies is given a special name: the *barycenter*. It must lie on a line connecting the centers of the bodies. We now derive the location of the barycenter relative to the bodies. For simplicity we shall assume that each revolves about it in a circular orbit. With somewhat more advanced mathematics, we would find that the result we derive is correct for any kind of motion of the two bodies — as long as they are acted on only by mutual forces between them.

Let the two bodies, of masses m_1 and m_2, revolve about and on opposite sides of the point O on the line between their centers (Figure 5-1). The distances of the two bodies from O are r_1 and r_2. To accelerate body 1 into a circular orbit, the force upon it must be $m_1 v_1^2 / r_1$. The force on body 2 is $m_2 v_2^2 / r_2$. As in Section 4.3, we can write the orbital speed of each body in terms of its period of revolution and its distance from the center of its orbit, that is,

$$v_1 = \frac{2\pi r_1}{P} \quad \text{and} \quad v_2 = \frac{2\pi r_2}{P}.$$

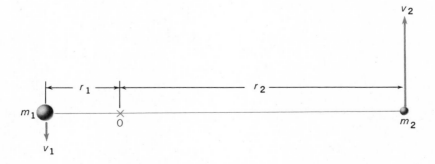

FIG. 5-1 Center of mass; the two bodies, 1 and 2, mutually revolve about point *O*.

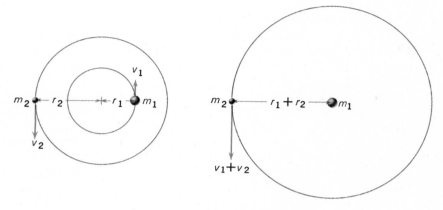

FIG. 5-2   Relative orbit.

Of course, both bodies have the same period — the period of their mutual revolution.

The same force — the gravitational attraction between the bodies — produces the centripetal acceleration of each. Therefore, the centripetal force acting on each of them must be the same. Thus we can write

$$\frac{m_1 v_1^2}{r_1} = \frac{m_2 v_2^2}{r_2},$$

or

$$\frac{m_1 4\pi^2 r_1^2}{P^2 r_1} = \frac{m_2 4\pi^2 r_2^2}{P^2 r_2}.$$

After cancellation of common factors, the above equation becomes

$$\frac{r_1}{r_2} = \frac{m_2}{m_1}.$$

We see that the point of mutual revolution of two bodies on a line between them is located so that the distance of each body from that point is in inverse proportion to its mass.

The concept of center of mass reminds us of the law of the lever investigated by Archimedes (Chapter 2). The problem may be illustrated by two boys on a seesaw. For the seesaw to balance properly, the fulcrum must be located proportionately closer to the boy of greater mass. Sim-ilarly, the barycenter of two revolving bodies lies proportionately closer to the body of greater mass.

The center of mass of a system of revolving bodies consisting of the sun and a planet, because of the sun's far greater mass, lies very close to the center of the sun — in most cases, within its surface.

(a) Relative Orbits

As two bodies revolve mutually about their barycenter they must always maintain the same relative distances from it; if one body doubles its distance so must the other, and if one comes closer, the other must lessen its distance by a proportional amount. We see, then, that the orbits of the two bodies are exactly similar to each other (that is, are the same shape), the sizes of the orbits being in inverse proportion to the masses of the bodies.

The similarity of the two orbits makes it easy to refer to the orbit of one object relative to the other. It is conventional to choose the center of the more massive body as reference. Of course neither body is actually fixed — both move about the barycenter — but it is often convenient to *regard* one (the more massive) as origin, and consider the motion of the other with respect to it. This is called the *relative orbit*. The earth and moon, for example, revolve about their barycenter on a line between their centers. The earth is *not* at the center of the moon's orbit, but we can and often do speak of the relative orbit of the moon about the earth's center.

The size of the relative orbit is the sum of the sizes of the individual orbits about the barycenter. If the bodies move in circular paths (Figure 5-2), the relative orbit is a circle with a radius equal to the sum of the distances of the two bodies from the barycenter (it is also equal to the distance between the bodies). If the orbits are ellipses, the relative orbit is an ellipse of major axis equal to the sum of the major axes of the elliptical orbits of the two bodies; all three ellipses have the same eccentricity.

We now turn to the motions of planets, satellites, and other celestial objects.

5.2 ORBITAL MOTION EXPLAINED

If an object is dropped toward the ground, at the end of 1 second it has accelerated to a speed of 32 ft/sec. Its average speed during that second is 16 ft/sec. Thus it drops, in 1 second, through a

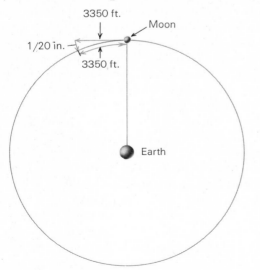

FIG. 5-3   Velocity and acceleration of the moon.

distance of 16 ft. The moon, which accelerates only 1/3600 as much, drops toward the earth only about 16/3600 ft in 1 second, or about $\frac{1}{20}$ in. In other words, as the moon moves forward in its orbit for 1 second and travels 3350 ft (about $\frac{2}{3}$ mi) it falls $\frac{1}{20}$ in. toward the earth. However, because of the earth's curvature, the ground has fallen away under the moon by that same distance of $\frac{1}{20}$ in., so the moon is still the same distance from the earth. In this way the moon literally "falls around the earth." In the period of about 1 month it has "fallen" through one complete circuit of the earth and is back to its starting point (Figure 5-3).

Orbital motion is thus easily understood in terms of Newton's laws of motion and gravitation. At any given instant, the orbital speed of the moon would tend to carry it off in space in a straight line tangent to its orbit. The gravitational attraction between the earth and moon provides the proper centripetal force to accelerate the moon into its nearly circular path. Consequently, the moon continually falls toward the earth without getting any closer to it. The orbital motions of the planets are similarly explained.

(a) The Fable of "Centrifugal Force"

Orbital motion is often incorrectly described as being a balance between two forces, the force of gravitation pulling the moon toward the earth, and an outward "centrifugal force" that keeps the moon from falling into the earth. This explanation, however, is more confusing than explanatory.

If one whirls a rock around over his head at the end of a string, he feels a "tug" in the string. This tug is the tension in the string. It is the result of the mutual pull between his hand and the stone. He provides, through the string, the centripetal force needed to accelerate the stone into a circular path. The stone tugs with a mutual force on the person's hand. However, there is no outside force pulling the stone away from him. If there were, and the string were to break, removing the central or centripetal force, the stone would move radially away from the person. Actually, if the string broke, the stone would fly off in a straight line in a direction *tangent* to its former circular path, continuing in the direction it was moving at the instant the string broke, in accord with Newton's first law of motion.

As a further example, if one rounds a corner in a rapidly moving car he feels his body being shoved against the side of the car on the outside of the curve. The car has pushed against him to accelerate him into a curved path. The force he feels is the mutual force between him and the car that provides his centripetal acceleration. If the side of the car were to dissolve suddenly, he would continue moving *forward* in the direction he was going at that instant, and would not fly out at right angles to the side of the car.

In short, there is no such thing as centrifugal (or outward) force in orbital motion. No force is pulling outward on the moon to balance the attraction between the moon and earth. If there were such a balance of forces, the moon would either not move at all or would move in a straight line, as predicted by Newton's first law. It is, in fact, just the *unbalanced* gravitational force that causes the moon to move in its nearly circular path around the earth. Of course, this discussion of the moon's motion applies equally to the planets' motions about the sun. (A "fictitious" centrifugal force, however, which arises from the introduction of a rotating coordinate system, is sometimes useful in the solution of difficult problems.)

(b) The Mutual Forces and the Moon's Mass

Many authors explain that "centrifugal force" is a reaction force to gravity, predicted by Newton's third law. This too is incorrect. The reaction to the earth's gravitational force on the moon is the moon's equal but opposite gravitational attraction for the earth. In other words, both pull mutually and centrally *toward* each other; there is no *outward* "reaction"

force. The earth and moon suffer the same change of momentum as they revolve around each other, but the earth, being much more massive than the moon, accelerates far less. The center of mass about which they mutually revolve can be located by making careful observations of the other planets — especially of Mars — and, more recently, by tracking space probes. Small monthly periodic variations in the apparent motion of Mars result from the earth's monthly revolution about the barycenter. From the size of those variations we find that the barycenter is about 3000 mi from the center of the earth, or about 1000 mi below its surface. The moon is about 81 times as far from the barycenter and so is correspondingly less massive than the earth (Chapter 9).

5.3 NEWTON'S DERIVATION OF KEPLER'S LAWS

Kepler's laws of planetary motion are *empirical* laws; that is, they describe the way the planets are *observed* to behave. Kepler himself did not succeed in finding more fundamental laws or relationships from which his three laws of planetary motion would follow. On the other hand, Newton's laws of motion and gravitation were proposed by him as the basis of all mechanics. Thus it should be possible to derive Kepler's laws from them. Newton did, in fact, show that the motions of the planets, as described by Kepler, followed from his fundamental postulates.

(a) *Kepler's First Law* *

Consider a planet of mass m_p at a distance r from the sun moving with a speed v in a direction at right angles to the line from the planet to the sun. The centripetal force needed to keep the planet in a *circular* orbit, that is, at constant distance from the sun, is

$$\text{force} = \frac{m_p v^2}{r}.$$

Now suppose the gravitational force between the planet and the sun happens to be greater than the force given by

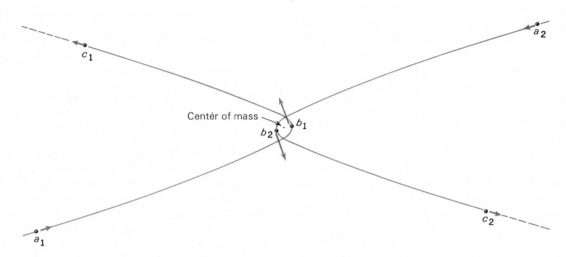

FIG. 5-4 Relative hyperbolic orbits.

the above equation. Then the planet will receive more acceleration than is necessary to keep it in a circular orbit, and it will move in somewhat closer to the sun. As it does so, its speed will increase, just as the speed of a falling stone increases as it approaches the ground.

Because of the planet's increased speed and decreased distance from the sun, a greater centripetal force is required to keep it at a constant distance from the sun. Eventually, as the planet continues to sweep in closer to the sun at higher and higher speed, a point will be reached at which the gravitational force between the two is no longer sufficient to produce enough centripetal acceleration to keep the planet from moving out away from the sun. Thus the planet will move outward as it rounds the sun until it has reached a position where the gravitational acceleration is again greater than the circular centripetal acceleration, and the process is repeated.

If the situation were reversed, and the planet were moving fast enough for the centripetal force required for circular motion to be greater than the gravitational attraction, the planet would move outward and consequently slow down until the gravitational force could pull it back again.

Thus we see, qualitatively, how a planet may follow an elliptical orbit. If, however, a planet had a high enough speed, the gravitational force between it and the sun might never be enough to provide sufficient centripetal force to hold the planet in the solar system, and the planet would move off into space. Its orbit would then be a *hyperbola* rather than a closed, elliptical path (Figure 5-4). There is a certain critical speed, which depends on the planet's distance from the sun, at which the planet can just barely escape the solar system along a *parabolic* orbit. This critical speed is called the *parabolic velocity*, or the *velocity of escape*.

To prove rigorously that the gravitational force between the sun and a planet can result in an orbit for the planet that is either a circle, ellipse, parabola, or hyperbola, is beyond the power of elementary algebra. Newton, in solving the problem, made use of his new *fluxions*, which we now know as differential calculus. He showed, in fact, that the gravitational interaction between *any* two bodies would result in an orbital motion of each body about the other that is some form of a *conic section* (see Section 3.4).

Circular and parabolic orbits require theoretically precise speeds that would not be expected to occur in nature; thus we would not expect to find a planet (or other object) with *exactly* a circular or parabolic orbit. The latter divides the family of elliptical (closed) from the family of hyperbolic (open) orbits that actually do occur. We shall see what conditions determine the type of orbit in Section

5.4. The planets, of course, do not have hyperbolic orbits or they would long since have receded into interstellar space; their orbits, then, must be elliptical, as found by Kepler.

(b) Kepler's Second Law*

In the preceding paragraphs we saw how a planet speeds up as it approaches the sun and slows down as it pulls away, in qualitative agreement with Kepler's second law. Newton derived the second law rigorously with a simple geometrical proof of this law of equal areas, which we repeat here.

Consider a planet at A revolving about the sun at S (see Figure 5-5). In a short interval of time, the planet's forward velocity would ordinarily carry it to B. However, the gravitational pull between it and the sun accelerates it to C. Since we are considering a very brief interval of time, we can regard the acceleration of the planet as being along a direction BC, parallel to AS, the direction from the planet to the sun at the beginning of the instant. The planet now has a velocity along the direction AC. In the next brief interval of time, equal in length to the first interval, the planet would ordinarily continue moving in a straight line at a constant speed, and would end up at D, along the extension of AC, so that the distance CD was equal to the distance AC. However, again the sun accelerates the planet toward it (now in direction CS), so the planet actually moves along CE.

Consider AC and CD to be the bases of the triangles ASC and CSD, respectively. Since AC = CD, the two triangles have equal bases. They also have the same altitude — the perpendicular distance of S from AD or its extension. Thus triangles ASC and CSD, having equal bases and altitudes, have equal areas.

Now SC is a common base of triangles SEC and SDC. Those triangles also have equal altitudes, the distance between the parallel lines SC and ED. Thus triangles SEC and SDC are equal in area.

Because triangles ASC and CSE are both equal in area to triangle SCD, they are equal in area to each other. These are the areas swept out by a line from the planet to the sun in two successive equal intervals of time. Many such brief intervals of time can be combined to show that the areas swept out in any two equal intervals of time are equal. Thus Kepler's second law is verified.

The argument given is rigorous only if the brief time intervals are infinitesimal. But we can imagine triangles ASC and CSE to be as small in area as we like, and the proof is still valid. Here is another example in which Newton employed differential calculus in deriving the result of letting time intervals approach zero.

In fact, we did not need to restrict ourselves to a planet moving around the sun. If any two objects revolve about each other under the influence of a central force, the law of equal areas will apply. The law of areas happens to be the geometrical manifestation of the principle of mechanics known as the *conservation of angular momentum.*

(c) Kepler's Third Law*

We may now use the ideas we have developed to demonstrate a simple derivation of Kepler's third law, but again only for circular orbits. Newton derived the same form of the law for elliptical orbits.

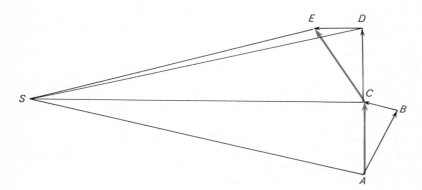

FIG. 5-5 Geometrical proof of the law of areas.

For each of two mutually revolving bodies, the gravitational attraction between the two provides the centripetal acceleration to keep them in circular orbits. If the bodies have masses m_1 and m_2 and distances r_1 and r_2 from their common center of mass, they are separated by a distance $r_1 + r_2$ and we can equate the gravitational force to the centripetal force for each body. Equating the formulas derived in Sections 4.3 and 5.1 for centripetal force, we obtain, for body 1:

$$\frac{Gm_1m_2}{(r_1 + r_2)^2} = \frac{m_1 4\pi^2 r_1}{P^2},$$

and, for body 2:

$$\frac{Gm_1m_2}{(r_1 + r_2)^2} = \frac{m_2 4\pi^2 r_2}{P^2}.$$

If we cancel out the masses common to each side of each equation and add the two, we obtain

$$\frac{G(m_1 + m_2)}{(r_1 + r_2)^2} = \frac{4\pi^2}{P_2}(r_1 + r_2),$$

or

$$(m_1 + m_2)P^2 = \frac{4\pi^2}{G}(r_1 + r_2)^3.$$

Since $r_1 + r_2$ is the distance between the two bodies, we recognize it, in the case of a planet going around the sun in a circular orbit, as the semimajor axis, *a*, of the orbit. Then the above equation looks the same as the formula we gave in Chapter 3 for Kepler's third law, except for the factor $m_1 + m_2$ and the factor $4\pi^2/G$. The latter is simply a constant of proportionality. If the proper units are chosen for distance and time, *G* will take on such a value that $4\pi^2/G$ will equal unity. We shall see in a moment why Kepler was unaware of the factor $(m_1 + m_2)$.

First, let us illustrate two systems of units such that the constant of proportionality $4\pi^2/G = 1$. One such system is to measure the sum of the masses of the revolving bodies, $m_1 + m_2$, in units of the combined mass of the sun and earth; the period in years; and the separation of the bodies in astronomical units. Then if the equation is applied to the mutual revolution of the earth and sun, everything in the equation other than the factor $4\pi^2/G$ is equal to unity, so it must be unity also. It is easy to see that another suitable choice of units is the combined mass of the earth and moon for the sum of the masses, the sidereal month (the moon's period of revolution about the earth) for the period, and the moon's mean distance for the separation of the bodies.

Newton derived his equation, not only for the planets moving about the sun, but also for any pair of mutually revolving bodies — two stars, a planet and a satellite, or even a plate and a spoon revolving about each other in space.

Newton's version of Kepler's third law differs from the original in that it contains as a factor the sum of the masses of the two revolving bodies. It will become clear why Kepler was not aware of that term if we note that we can consider the sun and earth to be a pair of mutually revolving bodies. We shall see in Chapter 22 that the sun has a mass of about 330,000 times that of the earth. Thus the combined mass of the sun and the earth is, to all intents and purposes, the mass of the sun itself, the earth's mass being negligible in comparison. Suppose we choose the mass of the sun for our unit of mass. Then, in the earth-sun system, $m_1 + m_2 = 1$. Furthermore, the sum of the masses of the sun and any other planet is also very nearly unity. Even Jupiter, the most massive planet, has only 1/1000 of the mass of the sun; for the sun and Jupiter, $m_1 + m_2 = 1.001$, a number so nearly equal to 1.000 that Kepler was unable to detect the difference from Tycho's observations. (However, the fact that the masses of Jupiter and Saturn are not completely negligible compared to the sun accounts, in part, for the slight discrepancies in Kepler's version of his third law as applied to Jupiter and Saturn; see Table 3.2.) Thus, if we apply the equation Newton derived to the mutual revolution of the sun and a planet, and choose years and astronomical units for the units of time and distance, and the solar mass for the unit of mass, Newton's equation reduces to

$$(m_1 + m_2)P^2 = (1)P^2 = P^2 = a^3,$$

in agreement with Kepler's formulation of the law.

(d) Kepler's Laws Restated — Summary

We now restate Kepler's three laws of planetary motion in their more general form, as they were derived by Newton:

KEPLER'S FIRST LAW: *If two bodies interact gravitationally, each will describe an orbit that is a conic section about the common center of mass of the pair. In particular, if the bodies are permanently associated, their orbits will be ellipses. If they are not permanently associated, their orbits will be hyperbolas.*

KEPLER'S SECOND LAW: *If two bodies revolve about each other under the influence of a central force (whether or not in a closed elliptical orbit), a line joining them sweeps out equal areas in the orbital plane in equal intervals of time.*

KEPLER'S THIRD LAW: *If two bodies revolve mutually about each other, the sum of their masses times the square of their period of mutual revolution is in proportion to the cube of the semimajor axis of the relative orbit of one about the other.*

In metric units, the algebraic formulation of Newton's version of Kepler's third law is

$$(m_1 + m_2)P^2 = \frac{4\pi^2}{G} a^3,$$

where $G = 6.67 \times 10^{-8}$ (cgs units). If either of the sets of units shown in Table 5.1 is used, the law becomes

$$(m_1 + m_2)P^2 = a^3.$$

TABLE 5.1 Systems of Units for Which $4\pi^2/G = 1$

	I	II
Units of $(m_1 + m_2)$	Sun's mass	Earth's mass + moon's mass
Units of P	Sidereal year	Sidereal month
Units of a	Astronomical unit	Mean distance of moon from earth

5.4 ENERGY OF A TWO-BODY SYSTEM — THE *VIS VIVA* EQUATION*

It can be shown that for any two mutually revolving bodies, of masses m_1 and m_2, and with a relative orbit of semimajor axis a, the magnitude of the velocity, v, of one body with respect to the other at an instant when the bodies are at a distance r apart, is given by the equation

$$v^2 = G(m_1 + m_2)\left(\frac{2}{r} - \frac{1}{a}\right).$$

This equation is called the *vis viva equation*, or, because it expresses the conservation of energy of the system, the *energy equation*. If the relative orbit of one body around the other is a circle, $r = a$, and the equation gives for the *circular velocity*

$$v^2 = G(m_1 + m_2)\frac{1}{r}.$$

If the relative orbit is a parabola, the bodies escape from each other, and the *vis viva* equation gives for the *parabolic* or *escape velocity*,

$$v^2 = G(m_1 + m_2)\frac{2}{r},$$

because a parabolic orbit can be considered an ellipse of eccentricity 1, with $a = \infty$ (infinity). Note that the velocity of escape is equal to the circular velocity times the square root of 2.

If two bodies have the most minute sideways motion with respect to each other, they cannot fall into each other but will move in elliptical orbits about each other. If that sideways velocity is just great enough so that the centripetal force required for a circular orbit is exactly equal to the bodies' mutual gravitational attractive force, they will move about each other in circular orbits. This critical speed is the *circular velocity* given by the second to last equation above. A still higher sideways motion will produce elliptical orbits of larger major axes than the diameters of the circular orbits. A sideways velocity equal to the velocity of escape of one body with respect to the other will result in parabolic orbits, and still higher velocities give hyperbolic orbits.

The *vis viva* equation is quite general; the motion of one body with respect to the other need not be sideways. If v is the speed of one body with respect to the other, whatever the direction, the equation gives the corresponding value of a, the semimajor axis of the relative orbit. For any closed orbit (circle or ellipse), a must be positive and finite. A value of v greater than the parabolic velocity results in open or hyperbolic orbits. The semimajor axis of a hyperbolic orbit is taken as negative.

It is important to note that if two objects approach each other from a great distance in space, they can never "capture" each other into elliptical orbits. Their mutual attraction will speed them up so that they pass each other with a relative speed greater than their mutual velocity of escape, and they will swing out away from each other again, moving in orbits that are hyperbolas (Figure 5-4). As an example, it is impossible to send a rocket to the moon, and to cause it to move on an elliptical orbit about the moon, without slowing it down when it is in the lunar vicinity. That is why a rocket intended for lunar orbit carries a *retrorocket* designed to reduce its speed at an appropriate time so that it can enter an elliptical orbit about the moon. Otherwise, it would bypass the moon on a hyperbolic orbit.

5.5 MASSES OF PLANETS AND STARS

Our only means of measuring the masses of astronomical bodies is to study the way in which they react gravitationally with other bodies. Newton's derivation of Kepler's third law, which includes a term involving the sum of the masses of the revolving bodies, is most useful for this purpose.

Consider a planet, such as Jupiter, that has one or more satellites revolving about it. We can select one of those satellites, and regard it and its parent planet as a pair of mutually revolving bodies. We measure the period of revolution of the satellite (say, in sidereal months), and the distance of the satellite from the planet (in terms of the distance of the moon from the earth), and insert those values into the equation

$$m_1 + m_2 = \frac{a^3}{P^2}.$$

Since both a and P are observed, we can immediately calculate the combined mass of the planet and its satellite. Obviously most of this mass belongs to the planet, its satellites all being very small compared to it. Thus $m_1 + m_2$ is, essentially, the mass of the planet in terms of the mass of the earth.

As a numerical example, Deimos, the outermost satellite of Mars, has a sidereal period of 1.262 days and a mean distance from the center of Mars of 14,600 mi. In sidereal months, the period of the satellite is 1.262/27.3 = 0.0463. In units of the distance of the moon from the earth, Deimos has a distance from the center of Mars of 14,600/239,000 = 0.0611. Thus the mass of Mars plus the mass of Deimos is given by

$$m_{\text{Mars}} + m_{\text{Deimos}} = \frac{(0.0611)^3}{(0.0463)^2} = \frac{2.28 \times 10^{-4}}{2.14 \times 10^{-3}}$$

$$= 0.11 \text{ earth mass.}$$

Since Deimos is a very tiny satellite (only about 5 mi across), its mass can be neglected compared to that of Mars, and we find that Mars has a mass of just over one tenth that of the earth.

In Chapter 22 we shall see that we use the same mathematical technique to determine the masses of stars that are members of binary-star systems (a binary star is a pair of stars that revolve around each other). In fact, we can use Newton's version of Kepler's third law to estimate the mass of our entire Galaxy (Chapter 27), or even of other galaxies (Chapter 32).

5.6 ORBITS OF PLANETS

A classical problem in celestial mechanics is to compute the orbit of a planet (or minor planet) from observations of its directions at various times as seen from the earth. We have seen (Section 3.4a) how Kepler determined the orbit of Mars geometrically. With the additional knowledge of Newton's laws of motion and gravitation it is possible to find the orbit of a planet with far fewer observational data. In 1801 Karl Friedrich Gauss (1777–1855) invented a method of determining the orbit of an object moving around the sun from observations of it extending over only a few weeks (Chapter 15). Since Gauss's time various mathe-

The launching of Surveyor I
by an Atlas-Centaur rocket. *(NASA.)*

Astronaut Edward H. White II performing a space walk during
the third orbit of Gemini IV in May, 1965. *(NASA.)*

The Moon during a lunar eclipse.
(William Close.)

The Milky Way in the direction of the galactic cent[e]
(U.S. Naval Observatory; Flagstaff Statio[n]

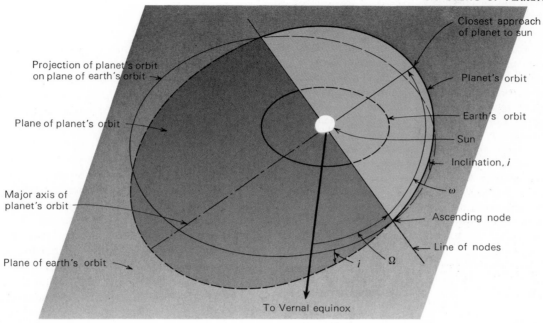

FIG. 5-6   Elements of an orbit.

matical techniques have been developed to handle the same problem. The complication is that we observe the other planets from a moving earth. Accurate positions in the sky of the object on each of at least three different dates, preferably separated by more than a week, must be known to calculate its detailed path around the sun. The orbit, when finally determined, is usually designated by a set of quantities known as its elements.

(a) Elements of Orbits*

The orbit of a planet or some other body moving about the sun can be specified uniquely by six items of information, or *elements.*

Two elements are needed to describe the size and shape of an orbit. We have already seen, for example (Section 3.4a), that the size and shape of an elliptical orbit can be specified by the semimajor axis and eccentricity of the ellipse. In fact, the same two quantities serve to specify the size and shape of *any* conic section, and hence of any orbit. Three other data are required to specify the orientation of the orbit with respect to some reference system, say, one defined by the earth's orbit. A final element is needed to specify where the object is in its orbit at some particular time, so that its location at other times can be computed. A total of six such orbital elements are sufficient if the object is in orbit around the sun and if it has a mass that can be neglected in comparison with the sun's mass. If, however, the sum of the masses of two mutually revolving bodies is not known, a seventh datum is needed to specify their orbit completely. If the relative orbit is an ellipse, the period of mutual revolution of the bodies suffices for this seventh element. If it is a hyperbola, the *areal velocity* replaces the period. The areal velocity is the rate at which an imaginary line between the two bodies sweeps out an area in space with respect to one of the bodies.

The six (or seven) orbital elements can be specified in a multitude of ways. In Table 5.2 is summarized the set of elements that is most conventional for describing the orbit of an object revolving about the sun. It must be emphasized, however, that other sets of data can be used for the elements of an orbit and, indeed, often are used in modern practice. The elements described in Table 5.2 are illustrated in Figure 5-6.

If the set of elements given in Table 5.2 is used, the inclination and longitude of the ascending node, i and Ω, describe the orientation of the orbital plane. The argument of perihelion, ω, gives the orientation of the orbit in its plane. The semimajor axis a and eccentricity e give the size and form of the orbit. The time of perihelion passage T and period P are the data required to calculate the position of the object in its orbit. If the object is one of small mass circling the sun, the period is superfluous, for it can be obtained from the semimajor axis with the use of Kepler's third law. If the orbit is not an ellipse, the areal velocity rather than the period can be used.

Some of the elements of the orbits of the planets in the solar system are given in Appendix 9.

TABLE 5.2 Elements of an Orbit

NAME	SYMBOL	DEFINITION
Semimajor axis	a	Half of the distance between the apsides of the conic that represents the orbit (usually measured in astronomical units).
Eccentricity	e	Distance between the foci of the conic divided by the major axis.
Inclination	i	Angle of intersection between the orbital planes of the object and the earth.
Longitude of the ascending mode	Ω	Angle from the vernal equinox (where the ecliptic and celestial equator intersect with the sun crossing the equator from south to north), measured to the east along the ecliptic plane, to the point where the object crosses the ecliptic traveling from south to north (the ascending node).
Argument of perihelion	ω	Angle from the ascending node, measured in the plane of the object's orbit and in the direction of its motion, to the perihelion point (its closest approach to the sun).
Time of perihelion passage	T	One of the precise times that the object passed the perihelion point.
Period	P	The sidereal period of revolution of the object about the sun.

5.7 ARTIFICIAL SATELLITES

An artificial satellite is a manmade object that is in orbit around the earth. It is an astronomical body in its own right. If some of the artificial satellites that have been launched are temporary astronomical objects, it is because they dip into the atmosphere of the earth during some portions of their revolutions. The friction of the air causes a satellite to lose energy so that eventually it spirals into the denser part of the atmosphere where friction heats it until it burns up completely. If an artificial satellite is launched so that its entire orbit is outside the earth's atmosphere, it will remain in orbit indefinitely as an astronomical body.

To illustrate how a satellite can be launched, imagine a man on top of a high mountain, firing a rifle in a direction exactly parallel to the surface of the earth (Figure 5-7 — adapted from a similar one in Newton's *Principia*). Imagine, further, that the friction of the air could be removed, and that all hindering objects, such as other mountains, buildings, and so on, were absent. Then the only force that acts on the bullet after it leaves the muzzle of the rifle is the gravitational force between the bullet and earth.

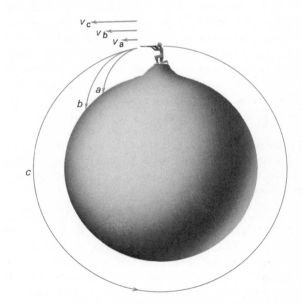

FIG. 5-7 Firing a bullet into a satellite orbit.

If the bullet is fired with muzzle velocity v_a, it will continue to have that forward speed, but meanwhile the gravitational force acting upon it will accelerate it downward so that it strikes the ground at a. However, if it is given a higher muzzle velocity v_b, its higher forward speed will carry it farther before it hits the ground, for, regardless of its forward speed, its downward gravitational acceleration is the same. Thus this faster-moving bullet will strike the ground at b. If the bullet is given a high enough muzzle velocity, v_c, as it accelerates toward the ground, the curved surface of the earth will cause the ground to tip out from under it so that it remains the same distance above the ground, and "falls around" the earth in a complete circle. This is another way of saying that at a critical speed v_c the gravitational force between the bullet and earth is just sufficient to produce the centripetal acceleration needed for a circular orbit about the earth. The speed v_c, the *circular satellite velocity* at the surface of the earth, is about 5 mi/sec, or 18,000 mi/hr.

Novelist Jules Verne anticipated earth satellites long ago. In one of his stories an enemy force was planning to bomb a city with a gigantic cannon ball. However, the cannon ball was propelled with too great a speed — in fact, the circular satellite velocity — so it passed harmlessly over the city and on into a circular orbit around the earth.

(a) Possible Satellite Orbits*

Suppose that a missile is shot up to an altitude of a few hundred miles, then turned so that it is moving horizontally, and finally given a forward horizontal thrust. It will proceed in an orbit the size and shape of which depend critically on the exact direction and speed of the missile at the instant of its "burnout," that is, the instant when the thrust supplied by its fuel is shut off. First, suppose that it is moving exactly horizontally, or parallel to the ground, at burnout. The possible kinds of orbits it can enter are shown in Figure 5-8.

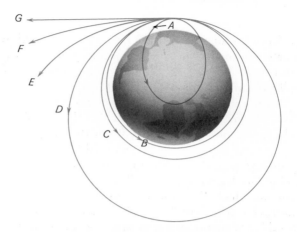

FIG. 5-8 Various satellite orbits that result from different burnout velocities but that are all parallel to the earth's surface.

If the missile's burnout speed is less than the circular satellite velocity, its orbit will be an ellipse, with the center of the earth at one focus of the ellipse. The *apogee* point of the orbit, that point that is *farthest* from the center of the earth, will be the point of burnout; the *perigee* point (closest approach to the center of the earth) will be halfway around the orbit from burnout.

If the burnout speed is substantially below the circular satellite velocity, most of its elliptical orbit will lie beneath the surface of the earth (orbit A), where, of course, the satellite cannot travel; consequently, it will traverse only a small section of its orbit before colliding with the surface of the earth (or more likely, burning up in the dense lower atmosphere of the earth). If the burnout speed is just slightly below the circular satellite velocity, the missile may clear the surface of the earth (orbit B), although its orbit will probably lie too low in the atmosphere for the satellite to be long-lived.

If the burnout speed were exactly the circular satellite velocity, a circular orbit centered on the center of the earth would result (orbit C). It is extraordinarily unlikely that a missile could be given so accurate a direction and speed that a perfectly circular orbit could be achieved. A slightly greater burnout speed will produce an elliptical orbit with *perigee* at burnout point and apogee halfway around the orbit (orbit *D*).

A burnout speed equal to the velocity of escape from the earth's surface, that is, the parabolic velocity (about 7 mi/sec), will put the missile into a parabolic orbit that will just enable the vehicle to escape from the earth into space (orbit *E*). A still higher burnout speed will produce a hyperbolic orbit in which the missile escapes the earth with energy to spare (orbit *F*). The higher the burnout speed, the nearer will the orbit be to a straight line (orbit *G*).

We can apply the *vis viva* equation to the orbit of a satellite moving about the earth. Let us measure speed in terms of the circular-satellite velocity at the earth's surface, the masses in terms of the earth's mass, and r and a in units of the radius of the earth. In these units, the constant G is equal to unity, and the equation simplifies to

$$v^2 = \left(\frac{2}{r} - \frac{1}{a}\right).$$

Suppose a satellite is launched from a point near the earth's surface (say within about 200 mi); r is 1.05, and v at that point is the burnout speed. Then the semimajor axis of the orbit, a (a measure of the size of the orbit), can easily be calculated if the burnout speed is known:

$$\frac{1}{a} = \frac{2}{r} - v^2.$$

Negative values of a correspond to hyperbolic orbits.

As an example, suppose the burnout speed is 6 mi/sec, or about 1.222 in units of the circular-satellite velocity. Then, we find for a

$$\frac{1}{a} = \frac{2}{1.05} - (1.222)^2 = 1.904 - 1.496,$$

or

$$a = 2.45 \text{ earth radii.}$$

Such a satellite would have an apogee distance of about 9700 mi from the center of the earth, or about 3750 mi above the surface.

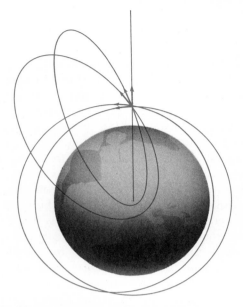

FIG. 5-9 Various satellite orbits that result from the same burnout speed but in different directions. All these orbits have the same major axis.

The *vis viva* equation holds regardless of the direction the two bodies are moving with respect to each other. Note that there is no term in the equation for an earth satellite that involves the direction in which a missile is moving at burnout. Thus, even if the missile were not moving parallel to the ground at burnout, the major axis of its orbit would depend only on its burnout speed (see Figure 5-9). However, the *eccentricity*, or shape, of the orbit does depend on the direction of motion of the missile. We see in Figure 5-9 that for a missile launched into a satellite orbit near the surface of the earth, unless the burnout direction is nearly parallel to the ground, the resulting orbit will be too eccentric to clear the surface of the earth.

(b) Ballistic Missiles

A ballistic missile is a rocket or missile that is given an initial thrust and then allowed to coast in an orbit to its target. Such missiles can be considered earth satellites and temporary astronomical bodies.

A ballistic missile travels in an elliptical orbit with the center of the earth at one focus.

Most of the orbit, however, lies beneath the surface of the earth, as in orbit *A* in Figure 5-8. One of the two intersections of the orbit with the earth's surface is at the launching point. The point is to give the missile the correct burnout speed and direction so that the other intersection of the orbit with the surface of the earth will occur at the target. The missile, then, travels along that part of its orbit that lies outside the earth's surface until it collides with the earth at the calculated point (Figure 5-10).

The calculation of precise trajectories for ballistic missiles, as with regular earth satellites, is a complicated task. Account must be taken of various perturbations introduced by the slight asphericity of the earth (Chapter 6), the drag of the earth's atmosphere, and complicating effects due to the earth's rotation. However, the basic principles are quite simple, as we have seen, and were completely described by Newton.

(c) First Artificial Satellites

For the International Geophysical Year 1957–1958, it was proposed to launch small earth satellites that would carry instruments to investigate the conditions just outside the earth's atmosphere.

FIG. 5-10   Orbit of a ballistic missile.

The first successful launching of an artificial earth satellite, an important milestone in the history of human technology, was accomplished by the Soviet Union on October 4, 1957. This first Soviet satellite, called *Sputnik I* (the Russian word for "satellite"), had an overall weight of about 4 tons, and a scientific instrumentation package weighing 184 lb. The instrument package and launching rocket traveled about the earth in a period of 96 minutes in an elliptical orbit that ranged from 142 to 588 mi above the earth's surface. Since Sputnik I was never completely outside the earth's atmosphere, it gradually lost energy, and finally burned up in the denser lower atmosphere on January 4, 1958. On November 3, almost a month after Sputnik I was put in orbit, the Russians launched a second satellite that survived in orbit until April 14, 1958. This second satellite was very similar to the first except that it had a heavier instrumentation payload that included a live dog.

The first successful American satellite was launched by an Army Jupiter C missile on January 31, 1958. By this time, the "space age" was well under way. Satellites carry instruments that radio back to earth data about our immediate space environment. Ballistic rockets (which are satellites in a sense, as we have explained) carry instruments 100 or more miles above the earth's surface. One advantage of rockets that return to the ground is that instrument payloads can sometimes be recovered intact, so that photographs can be obtained of the earth, the sun, the solar spectrum, and so on.

Among the kinds of information obtained by various rockets and space probes are data on the density and temperature of the earth's outer atmosphere, the density of micrometeorites in space, cosmic radiation, the density of charged particles near the earth, the nature of the earth's outer magnetic field, the nature of the solar radiation in the ultraviolet, the direct appearance of the sun in ultraviolet light, the appearance of the back side of the moon, the gravitational field of the earth, and ultraviolet and x-ray radiation from beyond the solar system. We shall discuss these data where appropriate in the following chapters. Information is so rapidly accumulating from satellite and rocket research, however, that we cannot be completely up to date.

5.8 INTERPLANETARY PROBES*

We have now learned the principles of space travel. Rockets, once they have left the earth, are astronomical bodies. They obey the same laws of celestial mechanics as the planets and natural and artificial satellites. In other words, rockets or space probes travel in orbits. If the space vehicles carry auxiliary rocket engines and extra fuel, it may be possible to alter their orbits at will, but the principles remain the same.

We shall illustrate one particular kind of space trajectory by showing one of the many possible ways to reach each of the planets Mars and Venus. The orbits to Mars and Venus we show are those that require the expenditure of the least energy as the rocket leaves the earth and are thus the most economical of fuel. The orbits of the successful United States *Mariner* Venus and Mars probes, and of the similar Soviet probes, were nearly of this type.

Suppose, for simplicity, that the orbits of Venus, Earth, and Mars are circles centered on the sun (when the slight ellipticity of planetary orbits is taken into account, the problem is similar but slightly more complicated). The least-energy orbit that will take us to Mars is an ellipse tangent to the earth's orbit at the space vehicle's *perihelion* (closest approach to the sun) and tangent to the orbit of Mars at the vehicle's *aphelion* (farthest from the sun) (see Figure 5-11).

The earth is traveling around the sun at the right speed for a circular orbit. For us on the earth to enter the elliptical orbit to Mars, we must achieve a speed, in the same direction as the earth is moving, that is slightly greater than the earth's circular velocity (which is about 18.5 mi/sec). To calculate this speed, we employ the *vis viva* equation. The major axis of the elliptical orbit we want to achieve is the sum of the radii of the orbits of the earth and Mars. Half of this major axis is the value *a*. The appropriate value of *a* is 1.26 AU. The value *r* is,

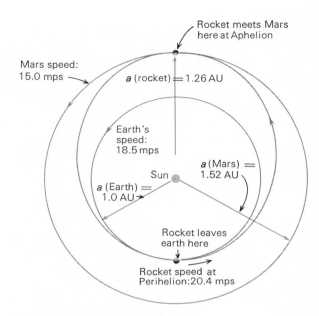

FIG. 5-11   Least-energy orbit to Mars.

of course, the earth's distance from the sun, and $m_1 + m_2$ is the combined mass of the sun and the spaceship (the latter is negligible). The required speed turns out to be slightly over 20 mi/sec. Since the earth is already moving 18.5 mi/sec, we need to leave the earth with the proper speed and direction so that when we are far enough from it that its gravitational influence on us is negligible compared to the sun's, we are still moving in the same direction as the earth with a speed relative to it of just under 2 mi/sec.

We have now entered an orbit that will carry us out to the orbit of Mars. The time required for the trip can be found from Kepler's third law, because our spaceship is a planet. The period required to traverse the entire orbit is $a^{3/2}$ years if a is measured in astronomical units. The entire period of the orbit is thus $(1.26)^{3/2} = 1.41$ years. The time required to reach the aphelion point (Mars' orbit) is half of this, or about 8½ months. The trip will have to be planned very carefully so that when we reach the aphelion point of the least-energy orbit, Mars will be there at the same time. Space probes generally carry rocket engines that can be activated by radio command from the earth, so that minor corrections in their trajectories can be made, as necessary, for them to achieve their missions.

The return trip from Mars to the earth is of the same type as the trip from the earth to Venus, to which we now turn our attention. The orbit to Venus is very similar to the orbit to Mars, except now it is at the *aphelion* point that the trajectory ellipse is tangent to the earth's orbit, and at the *perihelion* point that it is tangent to the orbit of Venus (Figure 5-12). The semimajor axis of this orbit is half the sum of the radii of the orbits of the earth and Venus, which is 0.86 AU. From the *vis viva* equation we find that the speed at the aphelion point in the orbit is 17 mi/sec, about 1½ mi/sec *less* (rather than more) than the earth's speed. The space vehicle would have to leave the earth, as before, with enough speed so that when it has left the earth's vicinity it has a speed with respect to the earth of 1.5 mi/sec but in a direction *opposite* that of the earth's motion. Then, relative to the sun, the vehicle is moving at the required 17 mi/sec and will reach the orbit of Venus along the desired elliptical orbit. The travel time to Venus, found as before from Kepler's third law, is about 5 months. Returning from Venus to earth is similar to traveling from earth to Mars.

FIG. 5-13   The Jet Propulsion Laboratory's Mariner II spacecraft which made a successful flight to the vicinity of Venus in 1962. (*Jet Propulsion Laboratory.*)

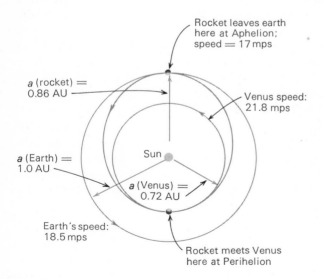

FIG. 5-12   Least energy orbit to Venus.

From the foregoing discussion it is obvious that the earth and the planet to be visited must be at a critical configuration at the time of launch, in order that the space vehicle meet the planet at the other end of the vehicle's heliocentric orbit. These critical configurations occur at intervals equal to the synodic period of the planet. In practice it is not necessary, and is seldom feasible, to launch the rocket at exactly the proper instant to achieve the least-energy orbit. However, there is a short range of time (typically a few weeks) during which a *nearly* least-energy orbit can be achieved. The length of this time period, called a "window" in space jargon, depends on the thrust capabilities of the available rockets (that is, on how much energy, above the least possible needed, can be supplied by the rocket). "Windows" for Mars journeys occur at intervals of about 780 days; those for Venus trips at intervals of about 584 days.

Exercises

1. How far is the barycenter from a star of three times the mass of the sun in a double-star system in which the other star has a mass equal to the sun's and a distance of 4 AU from the first star?

2. What is the period of mutual revolution of the two stars described in Exercise 1?

3. Why does Newton's version of Kepler's third law have the form

$$(m_1 + m_2)P^2 = a^3,$$

with the constant of proportionality equal to unity, if $m_1 + m_2$ is in units of the combined mass of the earth and moon, P in sidereal months, and a in units of the moon's distance? Find another such set of units, other than those given in the text, for which the constant of proportionality is unity.

4. A cow attempted to jump over the moon but ended in an orbit around the moon instead. Describe how the cow could be used to determine the mass of the moon.

5. What would be the period of an artificial satellite in a circular orbit around the earth with a radius equal to 60,000 mi? (Assume that the moon's distance and period are 240,000 mi and 27⅓ days, respectively.)

Answer: About ½ week

6. Why is it easier to get a space probe to escape the earth than to put a satellite into a nearly perfectly circular orbit?

7. As air friction causes a satellite to spiral inward closer to the earth, its orbital speed *increases*. Why?

8. If a lunar probe is to be launched from the earth's surface into an elliptical orbit whose apogee point is at the moon, why must the eccentricity of the orbit be nearly 1?

9. How could you calculate the period of an artificial satellite if its perigee and apogee altitudes above the earth's surface were known?

*10. Verify the periods given in the text for the times required to reach the planets Venus and Mars along least-energy orbits.

*11. Show why the times at which a space vehicle can be sent to a planet on a least-energy orbit occur at intervals of the synodic period of the planet.

*12. Describe how a space vehicle must be launched if it is to fall into the sun.

*13. If a satellite has a nearly circular orbit at a critical distance from the earth's center, it will have a period or revolution equal to 1 day and thus can appear stationary in the sky above a particular place on earth. Calculate the radius of the orbit of such a *synchronous* satellite.

Answer: About 26,000 mi

CHAPTER 6

The Problem of More Than Two Bodies

Until now we have considered the sun and a planet, or a planet and one of its satellites, as a pair of mutually revolving bodies. Actually, the planets (and different satellites of a planet) exert gravitational forces upon each other as well. These interplanetary attractions cause slight variations from the orbits that would be expected if the gravitational forces between planets were neglected. Unfortunately, the problem of treating the motion of a body that is under the gravitational influence of two or more other bodies is very complicated, and is called the *multibody* or *n-body problem*.

6.1 THE *n*-BODY PROBLEM — PERTURBATION THEORY

The *n*-body problem is, in general, that of describing the motion of any body in a collection (or cluster) of many objects all interacting under the influence of their mutual gravitation.

If the exact position of each other body is specified at any given instant, we can calculate the combined gravitational effect of the entire ensemble on any one member of the group — it is merely an extension of the application of the parallelogram of forces (Section 4.1b). Knowing the force on the body in question, we can find how it will accelerate; a knowledge of its initial velocity, therefore, is enough to calculate how it will move in the next instant of time, and thus to follow its motion. However, the problem is complicated by the fact that the gravitational acceleration of one body depends on the positions of all the other bodies in the system. Since they, in turn, are accelerated by all the members of the cluster, we must simultaneously calculate the acceleration of

each particle produced by the combination of the gravitational attractions of all the others to follow the motions of all of them, and hence of any one. Such extremely complex calculations have been carried out, with electronic computers, to follow the evolution of hypothetical clusters of up to at least 100 members.

Although computations of the type just described can be carried out, in principle, to study the motion of any one member of a group or cluster of bodies, it is not possible to write an equation that will describe the trajectory (or orbit) of that body for all time, as it is in the two-body problem (in which the orbits are always conic sections). Consequently, the *n*-body problem is often said to have no solution. Actually, by numerical calculation many problems can be solved to the desired precision, although for some problems of importance, such as the evolution of the solar system, even the biggest electronic computer is not adequate. In principle, however, the *n*-body problem is not solvable only in the sense that a single equation does not describe the motion.

Calculations have been performed to follow the evolution of hypothetical clusters, but they are not feasible for the study of all problems. It is, however, possible to derive certain properties of a cluster of particles interacting gravitationally in order to study it statistically — that is, to specify the average behavior of its members. We shall discuss some of these applications in Chapter 28, which deals with star clusters.

(a) Perturbation Theory*

Fortunately, the many-body problem can be solved rather accurately when a given body feels predominantly the gravitational force of one other mass. The motion of a planet around the sun, for example, is determined mainly by the gravitational force between it and the sun, the force between it and any of the other planets being very small in comparison. Thus the influences of the other planets can be regarded as small corrections to be applied to the two-body solution; these corrections are called *perturbations.*

There are two approaches to perturbation theory. One is to calculate directly the actual gravitational force on a planet (or minor planet or satellite) due to the combined attraction of all perturbing bodies, and knowing how each other planet is moving, to calculate in detail how the object in question will move. This is called the method of *special perturbations.* The procedure is especially tractable with modern computers and is the one most often used to calculate the orbit of a lunar or planetary probe moving under the combined influences of the various members of the solar system that it passes near.

The other approach is the method of *general perturbations.* Here the position at any moment of a planet (or minor planet or satellite) with respect to the sun (or earth), in combination with its velocity at that instant, is used to calculate the elliptical orbit it would follow if there were no perturbations. In other words, its orbit is computed on the basis of the two-body theory. This orbit is only a temporary one for the object because it will gradually be perturbed by other bodies than the sun (although it may well represent the body's actual motion for a considerable time); it is called an *osculating orbit.* Algebraic formulas can then be derived which express, with tolerable accuracy, how the elements of the osculating orbit will change with time, owing to the perturbations of the other planets. This procedure was usually employed before high-speed computers were developed, which facilitated the calculation of special perturbations. General perturbations are still widely used in many applications.

We shall see in Chapter 14 how perturbation theory led to the discovery of the planet Neptune.

(b) Special Solutions — The Restricted Three-Body Problem*

As stated above, the *n*-body problem can be solved in general only by laborious numerical calculation. There are, however, some special circumstances in which there exist solutions, or partial solutions, in the form of algebraic equations. Usually, these solutions apply only when the collection of particles has a very particular (and unlikely) configuration. One case, however, has partial solutions with very interesting applications. This is the *restricted three-body problem,* first considered by the French mathematician and astronomer, Louis Lagrange (1736–1813).

Lagrange investigated the behavior of a small particle moving in the gravitational field of two objects revolving about each other in circular orbits. The restriction is that the particle must have too small a mass to have any gravitational influence on the other two bodies. Lagrange found that there ar five positions relative to the two objects in mutual circular revolution (Figure 6-1) where the small mass, once placed, will move on a circular orbit always maintaining a fixed orientation with respect to the two greater masses. The three points marked *A* are unstable, in the sense that if the small body is displaced slightly from one of them it will leave its circular orbit. Because small perturbations are always likely to occur, we would not expect to find many examples in nature in which three bodies revolve exactly in those configurations.

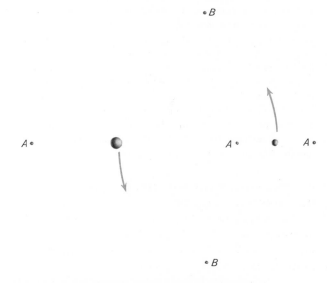

FIG. 6-1   Lagrangian points, at each of which a body of infinitesimal mass moves in a circular orbit, maintaining a fixed orientation with respect to two bodies mutually revolving in circular orbits.

The points marked *B* in Figure 6-1, however, are stable; the small object at one of those positions would not be forced away by slight perturbations. Note that the two bodies of larger mass and either of the *B* points are at the corners of an equilateral triangle. We do, in fact, find natural examples of this kind of motion. The best known is the equilateral configuration defined by the sun, the planet Jupiter, and the two groups of *Trojan minor planets* (Chapter 15). (The sun and Jupiter move in nearly circular paths around each other, and the minor planets have negligible mass in comparison, so the conditions of the problem are approximately met.) It has been suggested long ago that small particles could similarly revolve about the earth 60° ahead of and/or behind the moon. Some observers have even reported sighting faint patches of light that could be clouds of such particles reflecting sunlight to us, but these sightings have never been confirmed.

Lagrange's solution to the restricted three-body problem also specifies the regions of space within which the small particle can move relative to the two larger ones. We shall see (Chapter 30) that in at least one stage in the evolution of a typical star, it becomes a giant, greatly distending its outer layers. Now there are many double-star systems in which the two stars revolve about each other in nearly circular orbits. If the two stars are relatively close together, and if one evolves to a large enough giant, the atoms of its outer distended layers, having negligible mass, move about in the system in the manner predicted by Lagrange. We find, thus, that during the evolution of stars in binary systems, matter can flow from one star to another, or can flow in an orbit around one or both stars, or can even flow into space, escaping the two stars altogether. Examples will be encountered in Chapter 25.

6.2 THE GRAVITATIONAL EFFECTS OF NONSPHERICAL BODIES

Bodies with spherical symmetry act, gravitationally, as point masses, for which the gravitational influences are easily calculated. In nature, however, most bodies are not exactly spherical, and the simple two-body theory does not give precise results. If the shape of a body deviates only slightly from a sphere, we usually approximate its gravitational influence by that produced by a point mass and treat the small effects of its asphericity as perturbations. A common cause of the deformation of a star or planet from a perfect sphere is its rotation. Rapidly rotating planets, such as Jupiter, are noticeably flattened. The rotational flattening of the earth is slight but is important.

(a) The Shape of the Earth

Because of the earth's rotation, the inertia of its constituent parts tends to make them fly off tangentially into space. Therefore, each particle on and in the earth must be undergoing a constant *centripetal* acceleration to keep it in place. Of course, it is the earth's gravitation that provides this acceleration. Only a small part of the gravitational force on an object at the earth's surface is required to provide the centripetal acceleration that keeps it on the ground; the remainder is the object's actual *weight*. In other words, the rotation of the earth slightly reduces the weights of objects on the earth's surface (and the interior parts of the earth itself as well), because some part of the gravitational force is used up in providing the centripetal acceleration.

On the other hand, a body exactly on the axis of the earth has zero speed and suffers no centripetal acceleration. Therefore, at the poles of the earth (they are the ends of its rotational axis) the full force of gravity goes into the weight of objects. As a person travels away from a pole, his weight gradually diminishes and is least at the equator, where the centripetal acceleration is greatest. The effect is slight; a 300-lb man would lose only about 1 lb this way.

Newton showed that the result of this effect should be a distortion of the earth from a purely spherical shape. Suppose the earth were spherical (Figure 6-2). The direction of the gravitational pull upon a constituent rock in the earth near its surface is toward the center of the earth. This force must produce the combined effects of the centripetal acceleration on the rock and its weight. Now

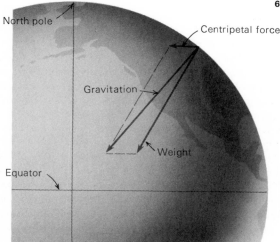

FIG. 6-2   Direction of the weight of a body on a spherical earth.

the centripetal force is not directed toward the center of the earth (except for objects at its equator) but rather is toward the center of the path on which the rock is moving. Thus the centripetal force is directed perpendicularly toward the earth's axis of rotation. If we once again apply the principle of the parallelogram of forces, we see that the centripetal force must represent one side, the gravitational force the diagonal, or resultant, and the weight of a body the other side of the parallelogram. In other words, the weight of the rock in question is *not directed toward the earth's center* and is not in a direction perpendicular to the surface of the earth. Thus the surface rocks will slide toward the equator, adjusting their distribution until the weight of each *is* perpendicular to the surface of the earth at its location.

Analysis shows that the resulting shape of the earth should be nearly that of an *oblate spheroid*, a slightly flattened sphere that has an elliptical cross section. That is, the earth should be flattened at the poles, whereas the equatorial regions should be slightly bulged out. Consequently, a person at one of the poles would be nearer the earth's center than he would be at the equator. Here is an additional effect that causes the weight of an object to

be less at the equator. Both effects together, the rotation of the earth and the earth's oblate shape, reduce the weight of an object at the equator compared to its weight at a pole by 1 part in 190.

The gravitational acceleration, in other words, varies slightly over the surface of the earth. The variation was first measured by John Richer, who was sent on a scientific expedition from Paris to Cayenne (in French Guiana) in the years 1671–1673. Richer found that at Cayenne (5° north latitude) a pendulum beat slower than it did at Paris. In 1673 Huygens showed that the period, or swinging rate, of a pendulum is proportional to the square root of the ratio of its length to the gravitational acceleration. Thus, Richer's measurements showed the variation of the acceleration of gravity between Paris and Cayenne. Later these results were used in a theoretical analysis to compute the shape of the earth.

Another consequence of the earth's oblateness is to make the distance along the surface of the earth corresponding to one degree of latitude longer in the vicinity of the poles than near the equator (Figure 6-3). The effect is easily ob-

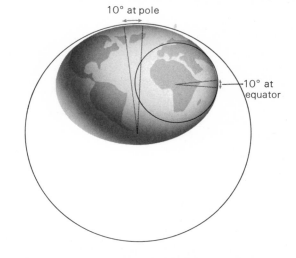

FIG. 6-3   Length of 10° of latitude at the equator and at the North Pole.

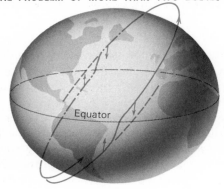

FIG. 6-4   Effect of the earth's equatorial bulge
on an earth satellite.

servable, and the length of a degree at various
latitudes has been measured. It varies from 68.7
statute miles at the equator to 69.4 mi at the poles.

The actual oblateness of the earth is small.
Its diameter from pole to pole is only 27 mi less
than through the equator, about 1 part in 298.

(b) Perturbations on Earth Satellites*

The equatorial bulge of the earth is responsible for a
deformation in the earth's gravitational field from that
which would be produced by a point mass or spherically
symmetrical earth. These deformations are especially important near the surface of the earth and produce conspicuous perturbations in the orbits of low-altitude earth
satellites.

As an example, we illustrate the perturbation on the
orientation of the *line of nodes* — the line of intersection
between the plane of the earth's equator and that of a

satellite orbit. A rigorous discussion of this perturbation
will not be attempted here; however, a physical feeling
for how the asphericity of the earth can produce such an
effect can be gained through the following oversimplified
description.

Consider a satellite moving eastward and crossing
the equator plane from south to north (Figure 6-4). As it
approaches the equator plane, the equatorial bulge pulls
it slightly northward. As it passes north of the equator
plane, the bulge pulls it slightly southward. The orbit of
the satellite has thus been displaced a little westward.
On the other side of the earth, as the satellite passes from
north to south it is pulled first to the south and then to
the north; again its orbit is displaced slightly westward.

Actually, the changes are not abrupt, as suggested here
and in Figure 6-4. In the nonspherical gravitational field of
the oblate earth the satellite is continually being accelerated from the two-body orbit. Gradually, then, as it
revolves about the earth, its orbit slides in a direction
opposite to that of its motion, and the line of nodes
slowly rotates to the west, or *regresses* (if the motion
of the satellite is to the east). For a satellite with a period
of 2 hours, an orbital eccentricity of 0.2, and with an
orbit inclined at 45° to the equatorial plane, the line of
nodes rotates 3.4° per day.

Many other perturbations of satellite orbits exist. The
exact nature of these effects, many of them subtle, depends critically on the precise shape of the earth — or,
more precisely, on the distribution of mass within the
earth.

(c) Satellite Investigations of the Earth's Shape*

Since the motions of earth satellites depend on the precise

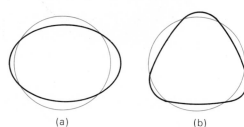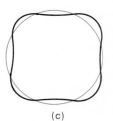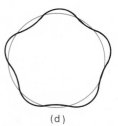

FIG. 6-5   From (a) to (d), respectively: The form of the second, third, fourth,
and fifth spherical harmonics — part of a series of surfaces superimposed
to represent the mean shape of the earth.

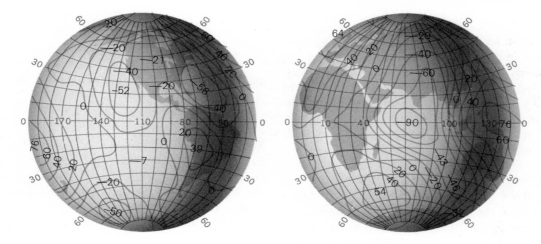

FIG. 6-6   Height, in meters, of the mean surface of the earth above
(positive numbers) and below (negative numbers) the surface of an oblate
spheroid with a flattening of 1 part in 298.25. *(Courtesy of William Kaula, UCLA.)*

shape of the earth, careful studies of perturbations on satellite orbits should enable us to derive the earth's shape rather accurately.

The deformations of the earth's shape from that of a perfect sphere are usually represented by a series of what are called *spherical harmonics.* The most important deformation is the equatorial bulge, which is represented by the *second* zonal harmonic (Figure 6-5), a slightly oblate figure superposed on the spherical earth. The kind of deformations represnted by the third, fourth, and fifth zonal harmonics are also shown schematically in Figure 6-5. All these shapes are really three-dimensional and are symmetrical about the axis of the earth's rotation; only cross sections are shown in the figure. In addition, there are harmonics that are not symmetric about the rotation axis. The general forms of the deformations represented by the various harmonics are correctly indicated, but the actual deformations of the earth are exceedingly small. The relative importance of the various harmonics is revealed by satellite observations. When a large number of harmonics, properly scaled, are superimposed on one another the shape of the earth is obtained.

For example, observations of the earliest Navy Vanguard satellite first revealed a deformation represented by the third harmonic. This showed that there is a very slight amount of "pear shape" superimposed on the oblate earth, as shown in Figure 6.5(b). A complete representation of the gravity field requires an infinite set of harmonics. Satellites have measured a few dozen harmonics.

Analyses of satellite perturbations, combined with sensitive measures of surface gravity over the earth, lead to an accurate knowledge of the form of the earth and permit the preparation of maps like the one in Figure 6-6, by William Kaula at UCLA. The earth is first represented by an oblate spheroid (a flattened sphere with elliptical cross section) in which the equatorial diameter exceeds the polar diameter by only 1 part in 298.25. The figures on the map are the heights, in meters, of the geoid above (positive numbers) or below (negative numbers) the surface of that oblate spheroid. The geoid is the same as mean sea level over the oceans; it is a surface everywhere perpendicular to the direction of gravity. The *topography* — the ups and downs of continents, mountains, and other crustal features — affects the geoid only indirectly through slight gravitational attraction.

6.3 DIFFERENTIAL GRAVITATIONAL FORCES

Two perfectly spherical bodies, as has been explained, attract each other as if they were point masses, located at their own centers. If however, a body deviates from perfect sphericity, even slightly, it no longer acts as a point mass, and we have seen in Section 6.2 how satellite orbits are consequently perturbed. The shape of a planet can be affected by rotation. Another contribution to asphericity is supplied by the *differential gravitational forces* that two neighboring bodies exert on each other. These forces, in turn, result in such phenomena as *tides* and *precession*.

(a) One Body's Attraction on Two Others

A *differential gravitational force* is the *difference* between the gravitational forces exerted on two neighboring particles by a third more distant body. As an example, consider Figure 6-7, in which three bodies are shown in a line. These are either point masses or perfectly spherical objects whose gravitational effect on external objects is the same as that produced by point masses. To the left is a large body of mass M. To the right are two bodies, each of whose masses we shall assume, for ease of calculation, to be unity — say, each has a mass of 1 gm. The first of the small bodies, body 1, is at a distance R from the large one; the other, body 2, is at a distance $R + d$.

The force of attraction between the large mass and body 1 is

$$F_1 = \frac{GM}{R^2},$$

and that between the large mass and body 2 is

$$F_2 = \frac{GM}{(R + d)^2}.$$

Note that F_2 is slightly *smaller* than F_1 because of the greater distance between the large mass and body 2. The difference $F_1 - F_2$ is the differential gravitational force of the large mass on the two smaller masses.

In Figure 6-8(a) the forces F_1 and F_2 are shown as vectors pointing toward the large mass to the left. Because the force on body 1 is greater than on body 2, the *differential* force tends to separate the two bodies.

The center of mass of two small bodies is halfway between them. We can write the force between the large mass and the center of mass of the small ones as

$$F_{CM} = \frac{2GM}{(R + \tfrac{1}{2}d)^2}.$$

The gravitational attraction between the large mass and the center mass of the two small ones is less

FIG. 6-7   Attraction of a large mass and two smaller ones.

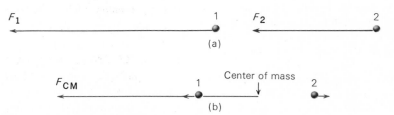

FIG. 6-8   Forces on the smaller masses shown as vectors.

than the force between the large mass and body 1 but is greater than that between the large mass and body 2. With respect to the center of mass, therefore, both body 1 and body 2 feel themselves pulled *outward*. Figure 6-8(b) shows the force acting on each of the small bodies, *with respect to the center of mass of the two*. If the bodies are free to move, they will separate unless their mutual gravitational attraction (not shown in Figure 6-8) is greater than the differential force.

(b) *Calculation of Differential Gravitational Force**
In the example described in the preceding paragraphs, the differential gravitational force ΔF was found to be

$$\Delta F = F_1 - F_2 = \frac{GM}{R^2} - \frac{GM}{(R+d)^2}.$$

Combining the two terms of ΔF, we find, with simple algebra,

$$\Delta F = GM\frac{(R+d)^2 - R^2}{R^2(R+d)^2} = GM\frac{2Rd + d^2}{R^2(R+d)^2}$$

$$= GM\frac{d(2R+d)}{R^2(R+d)^2}.$$

Now let us suppose that the distance R is very much greater than the distance d. In this case, $R + d$ is so nearly equal to R that we can write

$$R + d \approx R.$$

Similarly,

$$2R + d \approx 2R.$$

With this approximation, our equation for ΔF becomes

$$\Delta F = 2GM\frac{dR}{R^4} = 2GM\frac{d}{R^3}.$$

Now let us denote by δF the differential force corresponding to a unit separation of the two small bodies, that is, for the case where $d = 1$. Then[†]

$$\delta F = \frac{2GM}{R^3},$$

and the total differential force is

$$\Delta F = d \times \delta F.$$

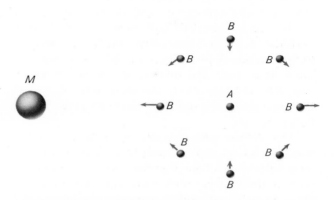

FIG. 6-9 Vector differences between forces of attraction of mass M on each of masses A and B.

In the foregoing calculations it was assumed that the three bodies are in a line. In general, the bodies are not lined up, and the differential gravitational force between two of them is not simply the arithmetic difference between the forces exerted on each by the third body. Since a force is a *vector* (for it has both magnitude and direction), the difference must be calculated according to the rules of vector subtraction (see Section 4.1b). In Figure 6-9 is shown a mass M (to the left) attracting each of two masses, A and B. Mass B is shown in various orientations with respect to the line between A and M. In each case is shown the *vector difference* between the force of attraction of M on B and on A. Usually, this differential gravitational force acts in such a way as to tend to separate A and B. However, when B is nearly at right angles to the line joining M and A, M's force on the two is in slightly different directions and tends to pull them closer together; then the differential force on B is directed more or less toward A.

———————

[†]This result could have been obtained immediately with differential calculus by differentiating the gravitational force; thus

$$\frac{dF}{dR} = \frac{d}{dR}\left(\frac{-GM}{R^2}\right) = \frac{2GM}{R^3}.$$

The minus sign in the gravitational force denotes that the force acts in such a direction as to decrease R.

(c) Differential Forces and Perturbations

Many pertubations can be looked upon as an effect of differential gravitational forces. A third body perturbs the orbital motion of two bodies because its gravitational force on both bodies is not the same. If it were, the two bodies would be accelerated by the third body the same amount, and in the same direction, and their relative motion would be unchanged.

As an example, consider the influence of the sun on the mutual revolution of the earth and moon. Both are about the same distance from the sun, so both are accelerated approximately the same amount and follow a nearly common orbit about the sun. The relative motion of the earth and moon, then, depends mainly on their mutual attraction and not on the sun. For the sun to play no role at all in the orbital revolution of the earth and moon about each other, however, its differential force upon them would have to be zero. Actually, at new moon the moon is accelerated more strongly toward the sun, and at full moon the earth feels the stronger acceleration. Therefore, the gravitational force between earth and moon is slightly altered by the differential force of the sun on the earth-moon system. This is the most important cause of those irregularities in the motion of the moon described in Chapter 9.

6.4 TIDES

Early in history it was realized that tides were related to the moon, because the daily delay in high tide ("high water") is the same as the daily delay in successive transits of the moon across the local meridian. A satisfactory explanation of the tides, however, awaited the theory of gravitation, supplied by Newton.

(a) Earth Tides

First, we shall consider the effects of the moon's attraction on the solid earth. For the moment, we ignore the flattening of the earth due to its rotation. Our planet can be regarded as being composed of a large number of particles, each of unit mass, all bound together by their mutual gravitational attraction and cohesive forces. The gravitational forces exerted by the moon at several arbitrarily selected places in the earth are illustrated in Figure 6-10. These forces differ slightly from each other because of the earth's finite size; all parts are not equally distant from the moon, nor are they all in exactly the same direction from the moon. If the earth retained a perfectly spherical shape, the resultant of all these forces would be that of the force on a point mass, equal to the mass of the earth, and located at the earth's center. Such is approximately true, because the earth is nearly spherical, and it is this resultant force on the earth that causes it to accelerate each month in an elliptical orbit about the barycenter of the earth-moon system.

The earth, however, is not *perfectly* rigid. Consequently, the differential force of the moon's attraction on different parts of the earth causes the earth to distort slightly. The side of the earth nearest the moon is attracted toward the moon more strongly than is the center of the earth, which, in turn, is attracted more strongly than is the side of the earth opposite the moon. Thus, the differential force tends to "stretch" the earth slightly into a *prolate spheroid* with its major axis pointed toward the moon. That is, the earth takes on a shape such that a cross section whose plane contains the line between the centers of the earth and moon is an ellipse with its major axis in the earth-moon direction.

In Figure 6-11 are shown the forces (as vectors) that are acting at several points on the surface of the earth. In each case, the forces are shown with respect to the earth's center. The dashed vectors represent the forces due to the earth's gravity — that is, the weights of various parts of the earth. The solid vectors (much exaggerated in

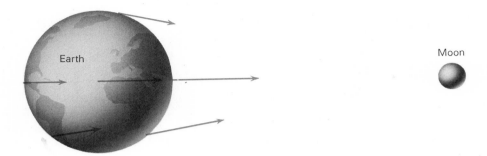

FIG. 6-10   Moon's attraction on different parts of the earth.

length) represent the differential gravitational forces due to the varying attraction of the moon on different parts of the earth. They are called the *tidal forces*. Those parts of the earth closer to the moon than the earth's center are attracted more strongly toward the moon than parts of the earth near its center. Thus the tidal forces are directed *toward* the moon. Those parts on the opposite side of the earth are attracted less strongly than are parts at the earth's center. The tidal forces there are directed *away* from the moon.

In each case, the vector representing the force can be broken into two components, one in the *vertical* direction, that is, away from the direction of the earth's gravity, and one in the *horizontal* direction, along the surface of the earth. The effect of the vertical component of the tidal force is to change slightly the weight of the surface rocks of the earth. The effect of the horizontal component is to attempt to cause the surface regions of the earth to flow horizontally.

If the earth were perfectly spherical, its gravitational attraction for objects on its surface would be in a *vertical* direction, toward the center of the earth. The actual earth, however, distorts under the influence of the tidal forces and is not quite spherical (we are still ignoring distortion due to rotation). Consequently, the earth's gravitational pull upon objects on its surface is not exactly in a direction perpendicular to the surface; there is a slight *horizontal* component in the gravitational pull of the earth upon its surface regions (see Figure 6-12).

If the earth were fluid, like water, it would distort until all the horizontal components of the tidal forces were exactly balanced by the horizontal pull of the earth at all points throughout it. Then the net force upon an object at the earth's surface, its *weight*, would be in a vertical direction. It would depend on two factors, those components of the earth's gravitational attraction and of the tidal force that are normal to the surface of the earth at that point.

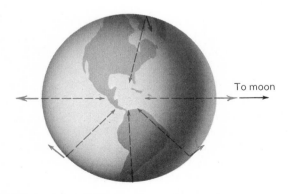

FIG. 6-11   Gravitational and tidal forces at various places on the earth's surface.

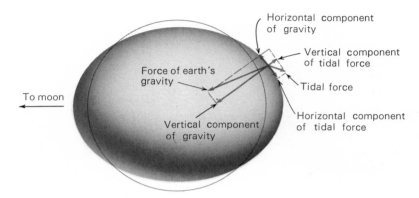

FIG. 6-12   Deformation of the solid earth under the influence of tidal forces (much exaggerated).

Measures have been made to investigate the actual deformation of the earth. It is found that the solid earth does distort, as would a liquid, but only about one third as much, because of the high rigidity of the earth's interior. In fact, the rigidity of the earth must exceed that of steel to account for the small degree of its tidal distortion, a result in agreement with the results obtained from seismic studies (Chapter 13). The maximum tidal distortion of the solid earth amounts at its greatest to only about 9 in.

As the earth rotates, different parts of it are continually being carried under the moon, so the direction and magnitude of the tidal force acting at any given place on the earth's surface are constantly changing. If the earth were viscous, its distortional adjustment would lag somewhat as the tidal forces on it change. However, direct observation shows that the earth readjusts its shape under the influence of the changing tidal force almost instantaneously. This circumstance implies that the earth is not only almost perfectly rigid, but also *highly elastic.*

In summary, tidal forces on the earth, that is, the differential gravitational forces of the moon's attraction on different parts of the earth with respect to its center, cause the solid earth to distort continually from a spherical shape, rising up and down and tilting as a fluid surface would do, but by only about one third the amount. Furthermore, these deformations are nearly instantaneous, changing just as quickly as the tidal forces change due to the earth's rotation. These facts show the earth to be more rigid than steel and to be highly elastic.

Rotation, of course, also distorts the earth's shape. The equatorial bulge of the earth that results from the tidal distortions described above is superimposed on its rotation. This polar flattening of the earth is, of course, a very much greater distortion than the distortion due to the tides.

(b) Ideal Ocean Tides

In Figure 6-13 are shown vectors representing tidal forces (relative to the earth's center) at various points on the earth's surface. These forces are directed generally toward the moon on the side of the earth under the moon and away from the moon on the opposite side. Within a zone around the earth that is roughly the same distance from the moon as the earth's center, the tidal forces are directed more or less toward the center of the earth. At these points, the attraction toward the moon is the same in magnitude as it is at the center of the earth, but because of the relatively small distance of the moon, it is in a direction that tends to pull those points closer to the earth's center. Each of the vectors (solid arrows) representing a tidal force in Figure 6-13 is resolved into components perpendicular to and parallel to the earth's surface (dashed arrows). We have seen that if the earth were fluid it would take on a shape such that points on its surface would feel no horizontal component of force. However, the earth is sufficiently rigid to be distorted from a sphere by only about one third of the amount required to remove these horizontal forces. Consequently, objects at the surface of the earth that are not restrained from horizontal motion, for example, the waters in the oceans, are free to flow in the direction of the horizontal components of the tidal forces. We shall assume, first, that the earth is covered uniformly by a deep ocean, and investigate the nature of the tides produced in it.

The actual accelerations of the ocean waters caused by the horizontal components of the tidal force are very small. These forces, acting as they

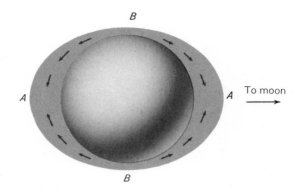

FIG. 6-14 Tidal bulges in the "ideal" oceans.

do over a number of hours, however, produce motions of the water that result in measurable tidal bulges in the oceans. Water on the lunar side of the earth is drawn toward the sublunar point (the point on the earth where the moon appears in the zenith), piling up water to greater depths on that side of the earth, with the greatest depths at the sublunar point. On the opposite side of the earth, water is drawn in the *opposite* direction, producing a tidal bulge on the side of the earth opposite the moon (see Figure 6-14).

It is important to understand that it is the horizontal components of the tidal forces that produce the tidal bulges in the oceans. At the two opposite points on the earth where the moon is at the zenith and at the nadir, the tidal forces are exactly radial — that is, directed away from the earth's center. At those points the horizontal components are zero, and there is no acceleration causing the water to flow along the surface of the earth. The tidal forces serve only to reduce very slightly the weight of the water, but because of the low compressibility of water, its physical expansion (because of its reduced weight) is completely negligible. Thus, at these points, the tidal forces play virtually no role at all, even though that is where the water is piled up the most.

In a beltlike zone around the earth from which the moon appears on the horizon, the direction of

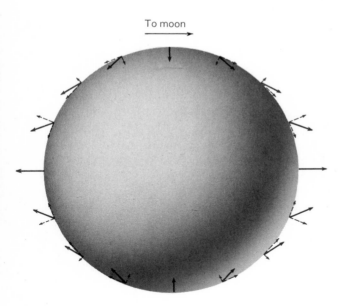

FIG. 6-13 Components of the tidal forces.

the tidal force is *inward,* toward the center of the earth, and again there are no horizontal components. The weight of the water is increased very slightly, but it is not appreciably compressed. Here, also, the tidal forces have no effect, although it is in this belt that the ocean level is lowest.

The tidal bulges in the oceans, then, do not result from the moon compressing or expanding the water, nor from the moon lifting the water "away from the earth." Rather, the tidal bulges result from an actual flow of water over the earth's surface, toward the regions below and opposite the moon, causing the water to pile up to greater depths at those places. It is the horizontal components of the tidal forces that produce this flow; those components, or "tide-raising forces," are greatest in regions of the earth intermediate between those from which the moon appears at the zenith or the nadir (points *A* in Figure 6-14) and on the horizon (points *B*).

The tidal bulge on the side of the earth *opposite* the moon often seems mysterious to students who picture the tides as being formed by the moon "lifting the water away from the earth." What actually happens, of course, is that the differential gravitational force of the moon on the earth tends to stretch the earth, elongating it slightly toward the moon. The solid earth distorts slightly, but because of its high rigidity, not enough to reach complete equilibrium with the tidal forces. Consequently, the ocean, moving freely over the earth's surface, flows in such a way as to increase the elongation and piles up at points under and opposite the moon.

In this section we have regarded the earth as though its ocean waters were distributed uniformly over its surface. In this idealized picture, only approximately realized even in the largest oceans, the tides would cause the depths of the ocean to range through only a few feet. The rotation of the earth would carry an observer at any given place alternately into regions of deeper and shallower water. As he was being carried toward the regions under or opposite the moon where the water was deepest, he would say, "the tide is coming in"; when carried away from those regions, he would say, "the tide is going out." During a day, he would be carried through two tidal bulges (one on each side of the earth) and so would experience two "high tides" and two "low tides" each day.

The two high tides during a day need not be equally "high," however. For example, in northern or southern temperate latitudes during summer or winter, the axis of the tidal bulges is inclined to the equator. The observer in the Northern Hemisphere (shown in Figure 6-15) would find the high tide on the side of the earth under the moon much higher than the high tide half a day later. An observer in the Southern Hemisphere would find the opposite effect. In extreme cases there may appear to be only one "high tide" a day.

(c) Tides Produced by the Sun

The sun also produces tides on the earth, although the sun is less than half as effective a tide-raising agent as the moon. Actually, the gravitational at-

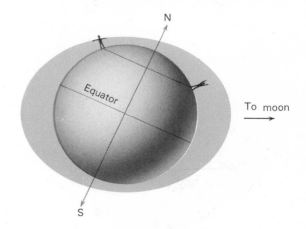

FIG. 6-15   Inequality of the two "high tides" during a day.

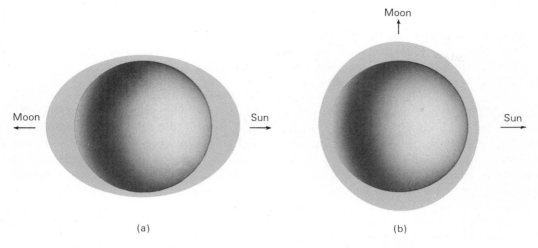

FIG. 6-16   (a) Spring tides; (b) neap tides.

traction between the sun and the earth is about 150 times as great as that between the earth and the moon. We recall, however, that the tidal force is the differential gravitational force of a body on the earth. The sun's attraction for the earth is much greater than the moon's, but the sun is so distant that it attracts all parts of the earth with almost equal strength. The moon, on the other hand, is close enough for its attraction on the near side of the earth to be substantially greater than its attraction on the far side. In other words, its *differential gravitational pull* on the earth is greater than the sun's, even though its total gravitational attraction is less.

If there were no moon, the tides produced by the sun would be all we would experience, and the tides would be less than half as great as those we now have. The moon's tides, therefore, dominate. On the other hand, when the sun and moon are lined up, that is, at new moon or full moon the tides produced by the sun and moon reinforce each other and are greater than normal. These are called *spring tides*. Spring tides (which have nothing to do with Spring) are approximately the same, whether at new moon or full moon, because tidal bulges occur on both sides of the earth — the side *toward* the moon (or sun) and the side away from the moon (or sun).

In contrast, when the moon is at first quarter or last quarter (at quadrature), the tides produced by the sun partially cancel out the tides of the moon, and the tides are lower than usual. These are *neap tides*. Spring and neap tides are illustrated in Figure 6-16.

Although spring tides are the highest type of tides, they are not all equally high, because the distances between the earth and sun and the earth and moon (and hence the tide-raising effectiveness of these bodies) both vary. The moon's distance varies by about 10 percent, and its tide-raising effectiveness varies by about 30 percent. The highest spring tides occur at those times when the moon is also at perigee.

(d) The Complicated Nature of Actual Tides
The "simple" theory of tides, described in the preceding paragraphs, would be sufficient if the earth were completely surrounded by very deep oceans. However, the presence of land masses stopping the flow of water, the friction in the oceans and between oceans and the ocean floors, the rotation of the earth, the variable depth of the ocean,

winds, and so on, all complicate the picture.

Both the times and the heights of high tide vary considerably from place to place on the earth. The tidal flow of waters over shallow parts of the ocean, and into irregular coastal regions, causes the tides there to range generally far more than the 2 or 3 ft that they range in the deep oceans. In particular, where the tidal flow of water is funneled shoreward by a V-shaped inlet that narrows back away from the sea, the difference between high and low water may be especially large. Such a place is the Bay of Fundy between New Brunswick and Nova Scotia in Southeast Canada, where the tidal range sometimes exceeds 50 ft.

The prediction of the actual time of high tide at any given place is even more complicated. The earth's rapid rotation causes the tide-raising forces within a given mass of water to vary too rapidly for the water to adjust completely to them. These forces, however, recurring periodically, set up forced oscillations in the ocean surfaces, so that the water over a large area rises and lowers in step. Consequently, the highest water does not necessarily occur when the moon is on the meridian (or in the lower half of the meridian, extended below the horizon through the nadir), but rather when the oscillations of the ocean, produced by the tidal forces acting upon it, pile up the water to its greatest depth at that location. The latter depends critically upon the shape and depth of the adjacent ocean basin. The time, in hours and minutes, by which high tide lags behind the moon's passage of the meridian, either above or below the horizon, is called the *establishment of the port*. The establishment of the port is different for different places but is very nearly constant for a given place. The United States Coast and Geodetic Survey prepares and publishes each year the *Tide Tables*, which give the times and heights of tides at principal ports throughout the world.

Tides also occur in the atmosphere. These atmospheric tides are complicated by weather phenomena, but in principle they are the same as earth and ocean tides. They are, obviously, of importance to meteorologists.

(e) Measurement of Ideal Tides

In view of the complexity of tides, the problem of measuring the tidal forces produced by the sun and moon might seem very difficult indeed. However, they were measured simply and directly by Michelson in 1913. He laid out a horizontal pipe, 500 ft long, half filled with water. Tides were produced naturally in the water in the pipe, causing periodic fluctuations in the water level that could be measured through sealed windows at both ends of the pipe. The range in depth of the water was less than $\frac{1}{1000}$ in., but it was measured with a microscope to within 1 percent. The experiment was performed with pipes running both east and west and north and south.

The tides produced in the pipes were only 69 percent as great as would be expected from calculations based on the assumption that the earth is completely rigid. It is from this experiment that the extent of the tides produced in the solid earth could be inferred.

(f) Other Tides

The tides produced by the moon and sun upon the earth are not the only tides in nature. The earth exerts a tidal force upon the moon that is stronger than the one the moon exerts upon the earth. The fact that the moon keeps one face directed toward the earth can hardly be accidental, but must have resulted from the earth's tidal forces upon the moon.

In fact, all bodies in the universe exert a tidal force on all other bodies, just as they exert a general gravitational attraction. In most cases these tidal forces are too small to produce observable effects. For example, the tides produced by planets on each other, and on the sun, are entirely negligible. On the other hand, we find instances of binary stars, in which the two stars are so close together as to produce very substantial tidal distortion, many times greater than that produced by the earth and moon upon each other (see Chapter 22).

(g) Criterion for a Satellite*

We can now delineate the criterion for the maximum and minimum distances that a satellite can have from a planet. The former depends on the differential gravitational force of the sun and the latter on the tidal force of the planet itself. For a satellite to remain always in a closed orbit about a planet, its orbital velocity, with respect to that planet, must always be less than its velocity of escape, or the parabolic velocity. The formula for velocity of escape, given in Section 5.4, assumes that the only force between two bodies is their mutual gravitational attraction. However, this attraction must be corrected for the differential gravitational force between the two if a third body is present.

A related problem is to find the minimum distance a satellite can be from its planet. At smaller distances the satellite could not withstand the differential, or tidal, forces exerted on it by the planet and would be torn apart. E. Roche investigated the problem in 1850 and found that if the constituent parts of a satellite are held together only by their mutual gravitation, as, for example, in a liquid body, and if the satellite has the same density as its planet, the critical distance is 2.44 times the planet's radius. At a greater distance, the satellite suffers only tidal distortion, but holds together. At a smaller distance it is torn apart by the tidal forces, for they are greater than the gravitational forces holding the satellite together. If the satellite has high rigidity, so that cohesive forces aid to gravitational ones in binding it together, it could survive at a somewhat smaller distance from the planet. The critical distance at which a satellite can survive tidal destruction is called *Roche's limit*. The rings of Saturn are particles that are closer to the planet than the distance at which a large solid body can survive — that is, they are within Roche's limit.

6.5 PRECESSION

The earth, because of its rapid rotation, is not perfectly spherical but has taken on the approximate shape of an oblate spheroid, its equatorial diameter being 27 mi greater than its polar diameter. As we have seen, the plane of the earth's

FIG. 6-17   Bay of Fundy at high and low tides. *(National Film Board of Canada.)*

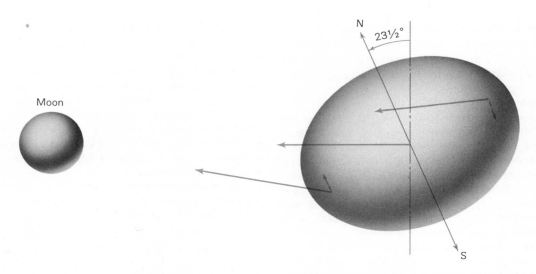

FIG. 6-18   Differential force of the moon on the oblate earth tends to "erect" its axis.

equator, and thus of its equatorial bulge, is inclined at about 23½° to the plane of the ecliptic, which, in turn, is inclined at 5° to the plane of the moon's orbit. The differential gravitational forces of the sun and moon upon the earth not only cause the tides but also attempt to pull the equatorial bulge of the earth into coincidence with the ecliptic.

The latter should be clear from inspection of Figure 6-18. The solid arrows are vectors that represent the attraction of the moon on representative parts of the earth. The part of the earth's equatorial bulge nearest the moon is pulled more strongly than the part farthest from the moon, and the earth's center is pulled with an intermediate force. The dashed arrows show the differential forces with respect to the earth's center. Note how they tend not only to "stretch" the earth toward the moon, but also to pull the equatorial bulge into the plane of the ecliptic. The differential force of the sun, although less than half as effective, does the same thing. Thus, the gravitational attractions of the sun and moon upon the earth act in such a way as to attempt to *change the direction of the earth's axis of rotation,* so that it would stand perpendicu-

lar to the orbital plane of the earth. To understand what actually takes place, we must digress for a moment to consider what happens when a similar force acts upon a top or gyroscope.

(a) Precession of a Gyroscope
Consider the top (a simple form of gyroscope) pictured in Figure 6-19. If the top's axis is not perfectly vertical, its weight (the force of gravity between it and the earth) tends to topple it over. The actual force that acts to change the orientation of the axis of rotation of the top is that component of the top's weight that is perpendicular to its axis. We know from watching a top spin that the axis of the top does not fall toward the horizontal, but rather moves off in a direction *perpendicular to the plane defined by the axis and the force tending to change its orientation.* Until the spin of the top is slowed down by friction the axis does not change its angle of inclination to the vertical (or to the floor), but rather describes a conical motion (a cone about the vertical line passing through the pivot point of the top). This conical motion of the top's axis is called *precession.*

(b) Qualitative Explanation of Precession*

The surprising phenomenon of precession can be understood in terms of Newton's laws of motion. Consider, for simplicity, the jack-shaped gyroscope in Figure 6-20(a), consisting of four masses supported at the ends of rigid light rods perpendicular to each other and to the axis of rotation. As the gyroscope spins, the masses move in the plane indicated. Suppose now that a force F is applied to the axis in a direction perpendicular to the plane defined by the axis of the jack and the line between masses 2 and 4. The force is transmitted through the rods to each of the four masses. Mass 1 feels a force tending to raise it (in the orientation of the diagram), and mass 3 feels a force tending to lower it; only masses 2 and 4 do not feel forces in the vertical direction. Masses 2 and 4 tend to continue moving in the same plane as before the force was applied. Mass 1 accelerates upward, but because of its forward motion it moves along the path ab. Similarly, mass 3 accelerates downward, but because of its forward motion follows path cd. Thus, after a part of a revolution, the masses are in the positions shown in Figure 6-20(b). The axis of rotation has changed, not in the direction of the applied force, but at right angles to it.

The above discussion is not a very rigorous description of precession; it is intended only to give the reader some feeling for the fact that the axis of a spinning top does not yield in the direction of a force acting on it. When we consider how each of the constituent parts of the top should behave under the influence of the applied force we can understand the apparently strange motion of the axis of the whole spinning body in terms of Newton's laws. It can be shown, however, by a rigorous mathematical treatment, that if a force is applied to the axis of any spinning body, the axis itself will move in a plane perpendicular to that defined by the force and the instantaneous axis of rotation.

(c) Precession of the Earth

The differential gravitational force of the sun on the earth tends to pull the earth's equatorial bulge into the plane of the ecliptic, and that of the moon tends to pull the bulge into the plane of the moon's orbit, which is nearly in the ecliptic. These forces, in other words, tend to pull the earth's

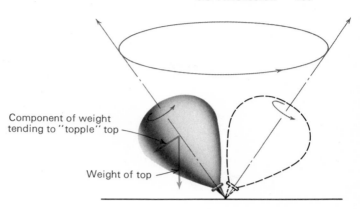

Component of weight tending to "topple" top

Weight of top

FIG. 6-19   Precession of a top.

axis into a direction approximately perpendicular to the ecliptic plane. Like a top, however, the earth's axis does not yield in the direction of these forces, but precesses. The obliquity of the ecliptic remains approximately $23\frac{1}{2}°$. The earth's axis slides along the surface of an imaginary cone, perpendicular to the ecliptic, and with a half-angle at its apex of $23\frac{1}{2}°$ (see Figure 6-21). The precessional motion is exceedingly slow; one complete cycle of the axis about the cone requires about 26,000 years.

Precession is this motion of the axis of the earth. It must not be confused with *variation in latitude* (Chapter 7), which is caused by a slight wandering of the terrestrial poles with respect to the earth's surface. Precession does not affect the cardinal directions on the earth or the positions of geographical places that are measured with respect to the earth's rotational axis, but only the orientation of the axis with respect to the celestial sphere.

Precession does, however, affect the positions among the stars of the celestial poles, those points where extensions of the earth's axis intersect the celestial sphere. In the twentieth century, for example, the north celestial pole is very near Polaris. This was not always so. In the course of 26,000 years, the north celestial pole will move on the celestial sphere along an approximate circle of about $23\frac{1}{2}°$ radius, centered on the pole of the

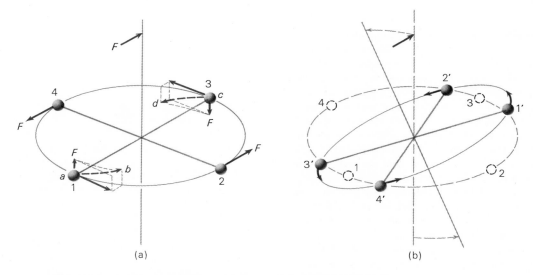

(a)   (b)

FIG. 6-20   (a) Force applied to the axis of a simple gyroscope;
(b) the new orientation taken by the gyroscope.

ecliptic (where the perpendicular to the earth's orbit intersects the celestial sphere). This motion of the pole is shown in Figure 6-22. In about 12,000 years, the celestial pole will be fairly close to the bright star Vega.

As the positions of the poles change on the celestial sphere, so do the regions of the sky that are circumpolar; that is, that are perpetually above (or below) the horizon for an observer at any particular place on earth. The Little Dipper, for example, will not always be circumpolar as seen from north temperate latitudes. Moreover, 2000 years ago, the Southern Cross was sometimes visible from parts of the United States. It was by noting the very gradual changes in the positions of stars with respect to the celestial poles that Hipparchus discovered precession in the second century B.C. (Section 2.3c).

(d) Nutation

If the differential gravitational attractions of the sun and moon upon the earth's equatorial bulge were always exactly the same, precession of the earth's axis would be the smooth conical motion we have described in the pre-

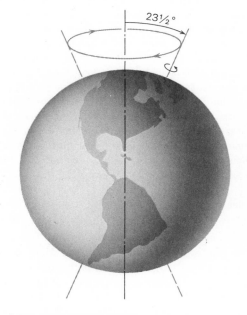

FIG. 6-21   Precession of the earth.

ceding sections. However, the effect of the differential forces on the orientation of the earth's axis depends on the directions of the sun and moon with respect to the direction of its 23½° tilt. These directions change as the earth and moon move in their respective orbits. Moreover, the moon's orbit is inclined at about 5° to the ecliptic. Not only is that 5° inclination slightly variable itself, but the intersections of the moon's orbit with the ecliptic slide around the ecliptic in 18.6-year intervals (the regression of the nodes).

The *average* effect of the sun and moon on the earth's equatorial bulge is to produce the relatively smooth precession we have described. We define the mean pole of rotation of the celestial sphere as a fictitious one that describes this smooth precessional motion. The motion of the actual celestial pole varies slightly around the motion of the mean pole. These variations, which are quite small,

can be fairly well represented by an elliptical orbit of the actual pole about the mean pole with a semimajor axis of 9".2, and a period of about 19 years. In other words, the motion of the celestial pole about the ecliptic pole is not quite a perfect circle, but a slightly wavy circle, with the "waves" having amplitudes of about 9 seconds of arc (") — small compared to the 23½° radius of the precessional orbit of the pole in the sky. This slight "nodding" of the pole about a smooth circle is called *nutation.*

(e) *Planetary Precession**

Up to now we have implied that the plane of the earth's orbit is fixed in space. The earth's orbit, however, is constantly being perturbed by the gravitational attractions of the other planets upon the earth. These perturbations are very slight, but they do measurably alter the plane of the earth's orbit and hence the position of the pole of the

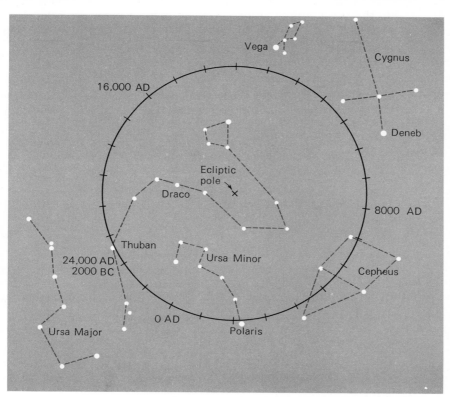

FIG. 6-22   Precessional path of the north celestial pole among the northern stars.

ecliptic on the celestial sphere. This motion of the pole of the ecliptic, only a fraction of a second of arc per year, adds to the complications of precession.

The motion of the mean celestial pole with respect to the ecliptic pole is called *lunisolar precession*. The motion of the ecliptic pole, because of planetary perturbations of the earth's orbital motion, is called *planetary precession;* the ecliptic pole moves only about $\frac{1}{40}$ as fast as the celestial pole. The two kinds of motion combined give *general precession.*

Exercises

1. Find the separation d between two small bodies, each of unit mass, lined up with a large body of mass M, at a distance R from the nearest of the small bodies, such that the gravitational attraction between the small bodies is just equal to the differential gravitational force between them caused by their attraction to the large body. The answer should be in terms of G, M, and R.

2. If the three bodies described in the last exercise are free to move and no other bodies or forces are present, how may their motion be described? How do the various forces change as the bodies move?

3. Strictly speaking, should it be a 24-hour period during which there are two "high tides"? If not, what should the interval be?

4. Compute the relative tide-raising effectiveness of the sun and moon. For this approximate calculation, assume that the earth is 80 times as massive as the moon, and 300,000 times less massive than the sun, and assume that the sun is 400 times as distant as the moon.

Answer: Moon is $\frac{8}{3}$ times as effective

5. Explain why the north celestial pole moves in the sky along a circle centered on the pole of the ecliptic, rather than some other point.

6. What will be the principal north circumpolar constellations as seen from Los Angeles (latitude 34° north) in the year 18,000?

7. In the year 13,000, will Orion be circumpolar as seen from the North Pole? Explain.

8. What would be the annual motion of the equinoxes along the ecliptic if the entire precessional cycle required only 360 years?

9. Describe how perturbations on the earth's motion by Mars can be considered as due to a differential gravitational force.

10. Does a bicycle offer another example of precession? Explain. (*Hint*: Consider how a rider can steer by leaning to one side.)

CHAPTER 7

Earth and Sky

In the preceding chapters we were concerned with the mechanics that dictate the motions of celestial bodies. Now we turn our attention to the motion of the earth, and the relation between earth and sky. We shall consider the physical properties of the earth as a planet in Chapter 13.

The most apparent of the earth's motions are its *rotation* and *revolution*. In astronomy "rotation" is used to describe the turning of a body about an axis running through it and "revolution" to describe the motion of a body around an exterior point. Thus the earth rotates on its axis and revolves around the sun.

7.1 ROTATION OF THE EARTH

We have seen that the apparent rotation of the celestial sphere could be accounted for either by a daily rotation of the sky around the earth, or by the rotation of the earth itself. Copernicus, Kepler, Galileo, and Newton had piled up convincing circumstantial evidence in favor of a rotating earth, but it was not until the nineteenth century that simple direct proofs of the rotation of the earth were devised.

(a) The Foucault Pendulum

In 1851 the French physicist Jean Foucault suspended a 200-ft pendulum weighing 60 lb from the domed ceiling of the Pantheon in Paris. He started the pendulum swinging evenly by drawing it to one side with a cord and then burning the cord. The direction of swing of the pendulum was recorded on a ring of sand placed on a table beneath its point of suspension. At the end of each swing a pointed stylus attached to the bottom of the bob cut a notch in the sand. Foucault had taken great care to avoid air currents and other influences that would disturb the direction of swing of the pendulum. Yet, after a few moments it became apparent that the plane of oscillation of the pendulum was slowly changing with respect to the ring of sand, and hence with respect to the earth.

The only force acting upon the pendulum was that of gravity between it and the earth, and, of course, this force was in a downward direction. If the earth were stationary, there would be no force that could cause the plane of oscillation of the pendulum to alter, and, in accord with Newton's first law, the pendulum should continue to swing in the same direction. The fact that the pendulum slowly changed its direction of swing with respect to the earth is proof that the earth rotates.

It is comparatively easy to visualize a Foucault pendulum experiment at the North Pole. Here we can imagine the plane of swing of the pendulum maintaining a fixed direction in space with respect to the stars, while the earth turns under it every day. Thus, at the North (or South) Pole, a pendulum would *appear* to rotate its plane of oscillation once completely in 24 hours (actually, 23 hours 56 minutes — see Chapter 8). At other places than the poles, the problem is complicated because the pendulum must always swing in a vertical plane that passes through the center of the earth. That plane of oscillation obviously must change with respect to the stars.

We must think of the pendulum as measuring the rate at which the earth turns around directly

Titicaca in the Andes. *(Gemini IX photograph; NASA.)*

The Southern Cross. *(John B. Irwin.)*

Moon and clouds over the western Pacif[ic]
(Gemini VII photograph; NASA.)

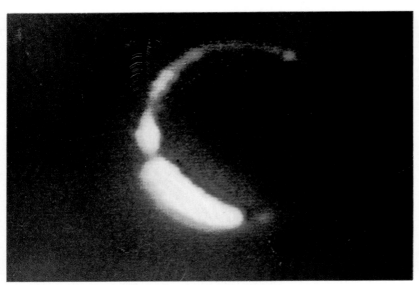

Start of solar eclipse produced by the earth, photographed from the moon by Surveyor III, April 24, 1967. The color is obtained by reconstruction from three black and white television photographs made through different filters and transmitted to earth. *(NASA.)*

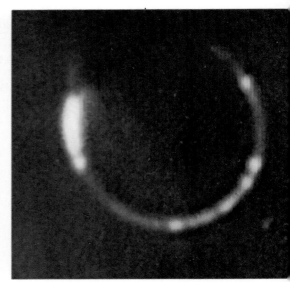

Same eclipse, 38 minutes later. The bright ring is sunlight refracted into the earth's shadow and onto the moon by the earth's atmosphere. *(NASA.)*

beneath it — that is, the rate of rotation of the earth about an imaginary line from the center of the earth out through the point of the pendulum's suspension.† It is about this line that the plane of swing of the pendulum does not rotate. If we imagine ourselves looking down upon the earth's North Pole, we can "see" the earth spinning beneath us like a phonograph record. On the other hand, if we imagine ourselves looking down on the earth's equator, we do not see a rotation of the earth beneath us, only a west-east translational

---

†The following is a brief derivation of the period of a Foucault pendulum: Let the angular velocity of the earth about its polar axis be ω (see the figure). Then the component of this angular velocity about a radius vector through a point on the earth's surface at latitude ϕ is $\omega \sin \phi$. The time required for one full rotation about this radius vector is $360°/\omega \sin \phi$. Since $\omega = 360°/(23 \text{ hours } 56 \text{ minutes})$, the period of a pendulum at latitude ϕ is

$$P = \frac{23^{\text{h}}56^{\text{m}}}{\sin \phi}.$$

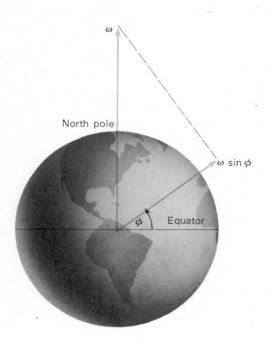

FIG. 7-1   A Foucault pendulum. *(Griffith Observatory.)*

motion. At intermediate latitudes we see beneath us a combination of west-east motion and a certain degree of rotation. At the equator, therefore, a pendulum would not appear to change direction of swing, while at latitudes intermediate between the equator and the poles its period — the time required for it to change its apparent plane of oscillation through 360° — would have a value somewhere between 24 hours and infinity, depending on the exact latitude. For example, at a latitude of 34° (the latitude of Los Angeles), the Foucault pendulum has a period of just under 43 hours. This is the time required for the earth to turn around a line from its center through Los Angeles, or the time required for a spectator to be carried completely around the pendulum by the turning earth.

It should be noted that the turning earth also turns the support system for the pendulum, and consequently the wire and bob of the pendulum itself. However, the rotation of the wire and bob of the pendulum does not alter the direction of swing. Try the following simple experiment. Improvise a small pendulum, say a watch and watch chain. Swing the watch to and fro, holding the end of the chain in the fingers. Now twist the chain in the fingers; the watch will twist with the chain, but will *not* change the direction of swing.

(b) The Coriolis Effect

The apparent rotation of the plane of oscillation of the Foucault pendulum is an excellent demonstration of the rotation of the earth underneath a freely moving body. Any such apparent deflection in the motion of a body, resulting from the earth's rotation, is called the *coriolis effect*. The moving body need not be the bob of a pendulum. Any object moving freely over the surface of the earth appears to be deflected to the right in the Northern Hemisphere (to the left in the Southern Hemisphere) because of the rotation of the earth beneath it. As an example of the effect, consider a projectile fired to the north from the equator.

The projectile starts its northward trip with an *eastward* velocity that it shares with the turning earth just before it is fired (Figure 7-2); at the equator this eastward velocity is about 1000 mi/hr.

There is no westward force on the projectile to slow it down, so it continues to move eastward after being fired. Proceeding northward over the curved surface of the earth, however, it comes closer to the axis of the earth's rotation. To conserve its angular momentum, the projectile's linear speed to the east must increase if its distance from the axis of rotation decreases. (Figure skaters make use of this principle by drawing in their outstretched arms to spin faster.) Meanwhile the ground beneath the northbound projectile moves

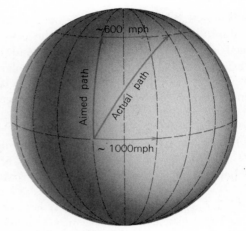

FIG. 7-2   Coriolis effect.

eastward progressively slower, because that ground, closer to the earth's axis, has less far to move in its daily rotation. We see, then, that the eastward speed of the projectile increases and that of the ground beneath it decreases. Thus, relative to the ground, the missile veers off to the east, that is, to the right for one looking in the direction of its motion.

A similar analysis would show that no matter in what direction a projectile moves, in the Northern Hemisphere it veers off to the *right,* and in the Southern Hemisphere to the *left* of its target. This effect must be corrected for in the firing of long-range artillery and of course, in the launching of missiles.

Winds, blowing toward a low-pressure area similarly veer off to the right of this area (left in the Southern Hemisphere). However, the force continually trying to equalize the pressure of the air accelerates the wind toward the low-pressure area. The wind, rather than "falling" directly into the low center, is caused to circle *around* the low center by the inertia of the forward moving air (Figure 7-3). If it were not for the earth's rotation, winds would blow directly into low-pressure regions, but because the winds veer off and miss the lows, they end up with a *cyclonic* motion. In the Northern Hemisphere, the winds always blow around storm centers in a *counterclockwise direction,* whether they be hurricanes, tornados, or those great general cyclonic storms that blow into our west coast from the Pacific. In the Southern Hemisphere, the winds are *reversed* — that is, they move around storm centers in a *clockwise* direction.

Falling bodies display a similar phenomenon caused by the rotation of the earth. Suppose a body is dropped down a deep vertical well located, say, at the equator. The eastward velocity of the body increases as it falls toward the axis of rotation. Progressively deeper down the well, on the other hand, the rocks, being closer to the axis of rotation, are moving to the east more and more slowly. Thus, the body is deflected toward the east wall of the well. The principle dictates that if the well were deep enough, the body would strike the east wall.

(c) Qualification of the Above "Proofs"

We have discussed "proofs" that the earth rotates. However, the proofs described are valid only if Newton's laws of motion are assumed to hold.

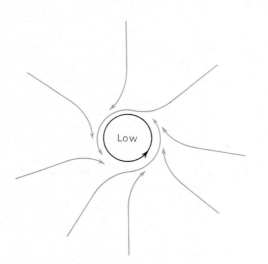

FIG. 7-3 Circulation of winds about a low-pressure area.

These laws are the postulates that form the basis of mechanics — they must not be thought of as absolute truths. They are valid only insofar as they provide an adequate model or representation of nature. We have not really proved that the earth rotates, but have shown that the assumption of its rotation is the only hypothesis consistent with the mechanical laws we use to represent nature.

7.2 RELATION OF EARTH AND SKY

(a) Positions on the Earth

To denote positions of places on the earth, we must set up a system of coordinates on the earth's surface. The earth's axis of rotation (that is, the locations of its North and South Poles) is the basis for such a system.

A *great circle* is any circle on the surface of a sphere whose center is at the center of the sphere. The earth's *equator* is a great circle on the earth's surface halfway between the North and South Poles. We can also imagine a series of great circles that pass *through* the North and South Poles. These circles are called *meridians;* they intersect the equator at right angles.

A meridian can be imagined passing through an arbitrary point on the surface of the earth (see Figure 7-4). This meridian specifies the east-west location of that place. The longitude of the place is the number of degrees, minutes, and seconds of arc along the equator between the meridian passing through the place and the one passing through Greenwich, England, the site of the old Royal Observatory. Longitudes are measured either to the east or west of the Greenwich meridian from 0 to 180°. The convention of referring longitudes to the Greenwich meridian is of course completely arbitrary. As an example, the longitude of the bench mark in the clock house of the Naval Observatory in Washington, D.C., is 77°03′56″.7 W. Note in Figure 7-4 that the number of degrees along the equator between the meridians of Greenwich and

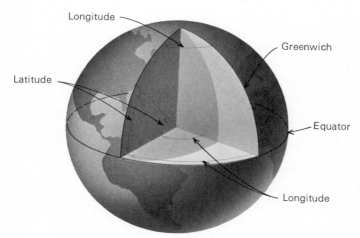

FIG. 7-4   Latitude and longitude of Washington, D.C.

Washington is also the angle at which the planes of those two meridians intersect at the earth's axis.

The latitude of a place is the number of degrees, minutes, and seconds of arc measured along its meridian to the place from the equator. Latitudes are measured either to the north or south of the equator from 0 to 90°. As an example, the latitude of the above-mentioned bench mark is 38°55'14".0 N. Note that the latitude of Washington is also the angular distance between it and the equator as seen from the center of the earth.†

(b) Variation of Latitude

With astronomical observations it is possible to measure the latitude of a place to within about 0".01, or about 1 ft. It is found that the latitude of

---

†Strictly, this is *geocentric* latitude. Because of the earth's oblate shape, there are several ways to define latitude. The *geodetic* (or *geographical*) latitude commonly used is defined by the angle between the equatorial plane and the perpendicular to the mean "sea-level" surface of the earth at the place in question. It may differ by several minutes from geocentric latitude. *Astronomical* latitude and longitude, obtained directly from astronomical observations, may differ from geodetic latitude and longitude by a few seconds of arc, owing to the deflection of the plumb bob by mountains or other crustal irregularities (Section 4.3e).

any given place on earth shows a periodic variation of several hundredths of a second of arc or several feet. Apparently, this variation is caused by a slight shifting of the solid earth with respect to its axis.

Systematic observations over the earth show that the exact positions of the poles — that is, the ends of the axis of the earth's rotation — wander about with respect to the ground. These wanderings seem to be composed of two independent motions. The first is a motion of each pole along the circumference of an approximate circle about 20 ft in diameter, in a period of 1 year. This motion is believed to be caused by seasonal changes in the distribution of air masses over the earth. The second motion has a period of about 14 months and is also a nearly circular wandering of the poles, but the diameter of this circle has varied from 10 to 50 ft. Its cause may be a natural oscillation of the earth in response to irregular variations in atmospheric density and motions.

The variation of latitude can be thought of as a motion of the terrestrial poles over the ground. However, it is actually the earth itself that does the shifting, while the direction of its axis of rotation remains fixed relative to the stars; that is, the *celestial* poles are unaffected.

It should be emphasized that the variation in latitude is very slight — only a few feet. The annual and 14-month wobbles do not seem to be connected in any way with such phenomena as the ice ages. However, there is superimposed a small drift of the poles in one direction (a *secular* drift) of about 4 in./year, which may be related to such long-term changes in the earth.

[Variation of latitude should not be confused with precession (Section 6.5c), which is a slow shifting of the earth's rotation axis with respect to the stars.]

(c) Positions in the Sky

In denoting positions of objects in the sky, it is often convenient to make use of the fictitious *celestial sphere,* a concept, we recall, that many early peoples accepted literally. We can think of the celestial sphere as being a hollow shell of extremely large radius, centered on the observer. The celestial objects appear to be set in the inner surface of this sphere, so we can speak of their positions *on* the celestial sphere. We can devise coordinate systems, analogous to latitude and longitude, to designate these positions. Of course we are really only denoting their *directions* in the sky.

The point on the celestial sphere directly overhead an observer (defined as opposite to the direction of a plumb bob) is his *zenith.* Straight down, 180° from his zenith, is the observer's *nadir.* Halfway between, and 90° from each, is his *horizon.* (This is the *celestial* horizon and will not necessarily coincide with the apparent horizon, which may be interrupted with such things as mountains, buildings, and trees.) Note that observers at different places have different zeniths, nadirs, and horizons.

CELESTIAL EQUATOR AND POLES

The apparent rotation of the sky takes place about an extension of the earth's axis of rotation. That is, the sky appears to rotate about points directly over the North and South Poles of the earth — the *north celestial pole* and the *south celestial pole.* Halfway between the celestial poles, and thus 90° from each, is the *celestial equator,* a great circle on the celestial sphere that is in the same plane as the earth's equator; it would appear to pass directly through the zenith of a person on the equator of the earth. Great circles passing through the celestial poles and intersecting the celestial equator at right angles (analogous to meridians on the earth) are called *hour circles.*

THE CELESTIAL MERIDIAN

The great circle passing through the celestial poles and the zenith (and also through the nadir) is called the observer's *celestial meridian.* It coincides with the projection of his terrestrial meridian, as seen from the earth's center, onto the celestial sphere. The celestial meridian intercepts the horizon at the *north* and *south* points on the horizon. Halfway between these north and south points are the *east* and *west* points on the horizon.

As the earth turns, the observer's terrestrial meridian moves under the celestial sphere; therefore, his celestial meridian sweeps continuously around the celestial sphere. Another way of putting it is to say that as the sky turns around the earth, the stars pass by the observer's stationary celestial meridian.

It helps to visualize these circles in the sky if we imagine that the earth is a hollow transparent spherical shell with the terrestrial coordinates (latitude and longitude) painted on it. Then we imagine ourselves at the center of the earth, looking out through its transparent surface to the sky. The terrestrial poles, equator, and meridians will be superimposed upon the celestial ones.

ALTITUDE AND AZIMUTH

The most obvious coordinate system is based on the horizon and zenith of the observer. Great circles passing through the zenith (*vertical circles*) intersect the horizon at right angles. Imagine a vertical circle through a particular star (Figure 7-5). The

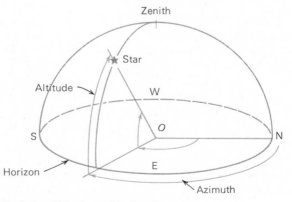

FIG. 7-5   Altitude and azimuth.

altitude of that star is the number of degrees along this circle from the horizon up to the star. It is also the angular "height" of the star as seen by the observer.

The *azimuth* is the number of degrees along the horizon to the vertical circle of the star from some reference point on the horizon. In astronomical tradition, azimuth formerly was measured from the south point on the observer's horizon, but in modern practice azimuth is measured from the north point, in conformity with the convention of navigators and engineers. In any case, azimuth is measured to the east (clockwise to one looking down from the sky) along the horizon from 0 to 360°.

RIGHT ASCENSION AND DECLINATION

The principal disadvantage of the altitude and azimuth system (the *horizon system*) is that as the earth turns, the coordinates of the celestial objects are constantly changing. It is desirable, therefore, to devise a coordinate system that is attached to the celestial sphere itself, just as the system of latitude and longitude is permanently attached to the earth. Then the positions of the stars remain fixed rather than changing rapidly as the earth's rotation causes the sky to rotate. A system which comes close to meeting these requirements is *right ascension* and *declination*, or the *equator system*.

Right ascension and *declination* bear the same relation to the celestial equator and poles that longitude and latitude do to the terrestrial equator and poles. *Declination* gives the arc distance of a star (or other point on the celestial sphere) along an hour circle north or south of the celestial equator. *Right ascension* gives the arc distance measured eastward along the celestial equator to the hour circle of the star from a reference point on the celestial equator. That reference point is the *vernal equinox*, one of the points on the celestial sphere where the celestial equator and the ecliptic inter-

sect (Section 7.4a). Because of precession, both the celestial equator and the vernal equinox slowly move with respect to the stars (Section 7.4c); thus the right ascension and declination of a star continually change, but the changes are so gradual as not to be important, for most purposes, over a period of 1 year or so. The lack of constancy of right ascension and declination makes the system less than ideal, but it is still the most convenient one available, for it is based on the celestial equator and is thus symmetrical with respect to the earth's axis of rotation. Right ascension and declination are therefore very useful for pointing telescopes and moving them to follow the daily motions of the stars (Chapter 11).

There exist several celestial coordinate systems in common use. Each has its advantages for special purposes. It is important for astronomers and navigators to understand these systems, but their value to students taking a survey course in astronomy may not be great unless practical exercises or sessions at the telescope are included. We have, therefore, not attempted to explain astronomical coordinate systems with particular thoroughness in the text. They are, instead, carefully defined in Appendix 7.

(d) The Orientation of the Celestial Sphere

The next step is to determine the orientation of the celestial sphere with respect to the zenith and horizon of a particular observer on the earth. At the North (or South) Pole, the problem is very simple indeed. The north celestial pole, directly over the earth's North Pole, appears at the zenith. The celestial equator, 90° from the celestial poles, lies along the horizon. An observer at one of the terrestrial poles would never see more than half the sky.

At the equator the problem is almost as simple. The celestial equator, in the same plane as the earth's equator, passes through the zenith, and since it runs east and west, it intersects the horizon at the east and west points. The celestial poles, being 90° from the celestial equator, must be at the north and south poles on the horizon. Evidently at points on the earth between the equator and poles, one of the celestial poles must be a certain distance above the horizon.

Now consider an observer at an arbitrary latitude on the earth. In Figure 7-6 the latitude is shown as north of the equator, but for a southern latitude the case would be exactly analogous. Since the terrestrial North Pole is on the observer's terrestrial meridian, the north celestial pole will have to be on his celestial meridian, at some altitude above the north point on the horizon. Suppose the angle from the observer's zenith down to the north celestial pole is z. Being on the celestial sphere, the north celestial pole is so distant that an observer at the center of the earth would see it in the same direction as our observer. Thus the angle at the

center of the earth between the observer and the north celestial pole is also z. (See Figure 7-5; the angles of intersection between each of two parallel lines and a third line are equal.)

We recognize that z is just 90° *minus* the observer's latitude. But also z is 90° (the altitude of the zenith) *minus* the altitude of the north celestial pole. Thus we see that *the altitude of the north (or south) celestial pole is equal to the observer's north (or south) latitude.*

Finally, since the celestial equator is 90° from the celestial poles, it must cut through the east and west points on the horizon, tilt southward as it extends up above the horizon, and cross the celestial meridian a distance south of the zenith that is also equal to the observer's latitude.

(e) The Motion of the Sky As Seen From Different Places On Earth

Imagine an observer at the earth's North Pole. The celestial north pole is at his zenith and the celestial equator along his horizon. As the earth rotates, the sky turns about a point directly overhead. The stars neither rise nor set; they circle parallel to the horizon. Only that half of the sky that is north of the celestial equator is ever visible to this observer. Similarly, an observer at the South Pole would only see the southern half of the sky (see Figure 7-7).

The situation is very different for an observer at the equator (Figure 7-8). There the celestial poles, the points about which the sky turns, lie at the north and south points on the horizon. All stars rise and set; they move straight up from the east side of the horizon and set straight down on the west side. During a 24-hour period, all stars are above the horizon exactly half the time.

For an observer between the equator and North Pole, say at 34° north latitude, the situation is as depicted in Figure 7-9. Here the north celestial pole is 34° above the observer's northern horizon.

FIG. 7-6   The altitude of the celestial pole equals the observer's latitude.

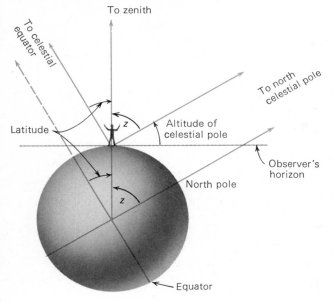

To zenith

To celestial equator

To north celestial pole

Latitude

Altitude of celestial pole

Observer's horizon

North pole

Equator

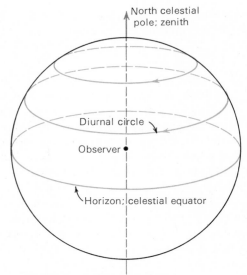

FIG. 7-7   Sky from the North Pole.

The south celestial pole is 34° *below* the southern horizon. As the earth turns, the stars appear to circle around and around, parallel to the celestial equator. The whole sky seems to pivot about the north celestial pole. For this observer, stars within 34° of the North Pole can never set. They are always above the horizon, day and night. This part of the sky is called the *north circumpolar zone* (*perpetual apparition*) at the latitude of 34° N. To observers in the United States, the Big and Little Dippers and Cassiopeia are examples of star groups that are in the north circumpolar zone. On the other hand, stars within 34° of the south celestial pole never rise. That part of the sky is the *south circumpolar zone* (*perpetual occultation*). To most U.S. observers, the Southern Cross is in that zone. At the North Pole, half the sky is in perpetual apparition and half in perpetual occultation.

Stars north of the celestial equator, but outside the north circumpolar zone, in the greater parts of their daily paths, or *diurnal circles,* lie above the horizon; hence they are up more than half the time. Stars *on* the celestial equator are up exactly half

the time, for their diurnal circle is the celestial equator. Because it is a great circle, exactly half of it must be above the horizon. Stars south of the celestial equator, but outside the south circumpolar zone, are up less than half the time.

(f) *Celestial Navigation*

The concepts that we have just dealt with have been applied by navigators for centuries to determine position at sea (and more recently, in air as well). The celestial coordinates of the sun, moon, planets, and brighter stars are computed for years in advance (in this country at the U.S. Naval Observatory), and are tabulated day by day in such publications as the *Nautical Almanac.* To find his position, a navigator must measure the altitudes of at least two celestial objects with a sextant (or one object at at least two different times), have available the celestial coordinates of those objects, say, from the *Almanac,* and for each of his observations know the exact time at some known place on earth — usually Greenwich, England. He obtains the time from a *chronometer,* an accurate clock set to keep Greenwich time. In modern practice, the chronometer is checked regularly with radio broadcasts of the accurate Greenwich or Washington time.

The basic principle of *celestial navigation* is that of Eratosthenes' method of determining the size of the earth (Chapter 2). The celestial bodies are so distant that the direction to any of them from all places on earth must

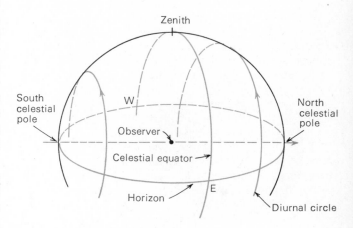

FIG. 7-8   Sky from the equator.

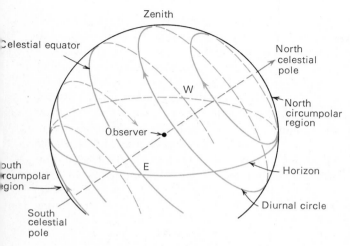

FIG. 7-9   Sky from latitude 34° N.

be the same. One can easily calculate where on earth a particular star or planet will appear at the local zenith. Let that place be denoted S (Figure 7-10). Now the angle in the sky between the navigator's zenith and that star must be the same as the angle at the center of the earth between him and the *substellar point, S.* Thus, the navigator knows that he is somewhere along a circle on the surface of the earth whose radius, in degrees, is equal to the zenith angle of the star. The circle is called a circle of *position.* By observing a *second* star, which must be at the zenith of a place denoted S', the navigator finds that in addition he must be somewhere along a second circle of position, centered at S'. Since there are only two points of intersection of the two circles of position, there are only two places on earth where the ship can be. Usually, the navigator has a clear enough idea of where he is to eliminate one of those places (it might, for example, lie in a different ocean, or on land). If not, observation of a third star settles the matter.

In present-day practice, the procedure of celestial navigation is roughly as follows: First, the navigator assumes a position (latitude and longitude) for his ship (or airplane). He can usually make an intelligent guess of his position by having kept track of the ship's speed and direction since his last *fix* (determination of his location). Second, with the aid of the data in the *Almanac,* and from

a knowledge of Greenwich time, he calculates exactly what the altitude should be for each of three or more stars at an instant a few minutes in the future, assuming that his guess of his position is accurate. Third, he measures the altitudes of those stars at the instant of time he used in his calculations.

If he has actually guessed his position correctly, the observed and computed altitudes for the stars should agree. Generally, there are small discrepancies that reflect the error in his assumed position. The amount of the discrepancies tells him just how far off his guess was and how to correct it to obtain his actual position.

(g) Nomenclature of Stars

While we are discussing the celestial sphere, it is in order to describe briefly the system for naming the stars and constellations. As has been stated (Section 2.2a) the ancients designated certain apparent groupings of stars in honor of characters or animals in their mythology. We retain most of these *constellations* today, although their number has been augmented to 88 (they are listed in Appendix 16). By action of the International Astronomical Union in 1928, the boundaries between

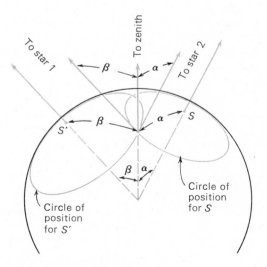

FIG. 7-10   Circles of position.

constellations were established as east-west lines of constant declination and north-south segments of hour circles. Because of precession (Section 6.5c), over the years the constellation boundaries have gradually tilted slightly from precisely north-south and east-west. Although all constellation boundaries ran north-south and east-west, they nevertheless jogged about considerably, so that the modern constellations still contain most or all of the brighter stars assigned to them by the ancients. Consequently, the boundaries often delineate highly irregular regions of the sky, reminding one of the boundaries between Congressional districts that result from the gerrymandering practices of many state legislatures. At any rate, because every position on the celestial sphere (or every direction in the sky) lies in one or another constellation, we commonly use constellations today to designate the places of stars or other celestial objects.

Many of the brighter stars have proper names. Often these are Arabic names that describe the positions of the stars in the imagined figures that the Greek constellations represented. For example, *Deneb,* is Arabic for "tail" and is the star that marks the tail of *Cygnus* (the Swan), and *Denebola* is the star at the tail of *Leo* (the Lion). Some star names, however, are of modern origin; for example, *Polaris,* the "pole star."

A superior designation of stars was introduced by a Bavarian, J. Bayer, in his *Atlas* of the constellations, published in 1603. He assigned successive letters of the Greek alphabet to the more conspicuous stars in each constellation, in approximate order of decreasing brightness. The full star designation is the Greek letter, followed by the genitive form of the constellation name. Thus Deneb, the brightest star in Cygnus, is α *Cygni*, and Denebola, the second brightest in Leo, is β *Leonis.* Bayer's ordering of stars by brightness was not always correct, and on occasion he deviated from the scheme altogether and assigned letters to stars of comparable brightness according to their geometrical arrangement in the constellation figure, for example, in the Big Dipper. Bayer's nomenclature is still in wide use today.

Fainter stars were subsequently given number designations, with the numbers increasing in order of the stars' right ascensions. The majority of stars, however, too faint to see without a telescope, are designated, if at all, only by their numbers in various catalogues. The most famous and extensive catalogue, which contains one third of a million stars, was compiled in the years following 1837 by F. W. Argelander at the Bonn Observatory. Stars in this Bonn Catalogue, or *Bonner Durchmusterung,* are known by their *BD numbers.* The Bonner Durchmusterung was later extended to part of the sky too far south to observe at Bonn, and was eventually supplemented with a catalogue of the southernmost stars made at Cordoba in Argentina.

Many other catalogues, with ever-increasing accuracy of the star positions they record, have been and are being compiled. Many star catalogues are prepared for special purposes, or list only certain types of stars, or stars in certain regions of the sky. A commonly used catalogue produced by the Harvard College Observatory gives the spectral types of the stars (Chapter 21); stars in this *Henry Draper Catalogue* are denoted by their *HD numbers.* Data for the nearest stars, listed in Appendix 12, are seen, from the variety of nomenclature, to have been selected from several different catalogues. Many stars, of course, are listed in more than one catalogue and thus bear various names — a circumstance that has sometimes confused even astronomers.† The vast majority of stars, however,

---

†A theoretical astronomer in France published, some years ago, a paper in which he reported calculations he had performed to derive some of the internal properties of a number of stars. His calculations were based on observational data selected by him from various catalogues. One of the stars he gave results for is Sirius; another is α Canis Majoris. He did not know that these were the same star!

too faint and numerous to measure and catalogue, remain nameless.

7.3 THE REVOLUTION OF THE EARTH

We have seen (Chapter 2) that the earth's revolution about the sun produces an apparent annual motion of the sun around the sky. The sun's apparent eastward journey among the stars on the celestial sphere is along the *ecliptic*. However, we have also seen that the sun's apparent revolution about the earth can be explained either by a motion of the sun or of the earth. What evidence is there that it is the earth, not the sun, that moves?

(a) Proofs of the Earth's Revolution

As in discussing the earth's rotation, we can only *prove* that the earth revolves if we are willing to accept certain postulates. If we adopt Newton's laws of motion, it follows simply and directly that the earth must revolve about the sun and not vice versa.

It is obvious that *either* the earth goes around the sun or the sun around the earth. Thus we have a system of *two mutually revolving bodies*. The problem is simply to determine where the common center of revolution is. In Chapter 22 we shall see how the mass of the sun is determined. It is calculated that the sun is about 330,000 times as massive as the earth. Thus, if we use the equation for finding the center of mass (Section 5.1), we see that the common center of mass of the earth-sun system must be less than 1/300,000 of the distance from the center of the sun to the center of the earth. This puts it well inside the surface of the sun. Essentially, then, the earth revolves around the sun.

There are also some geometrical consequences of the earth's revolution that would be very difficult to explain if the earth were assumed to be stationary. The simplest of these is stellar parallax, an apparent shift in the directions of the nearer stars, resulting from the fact that we look at them from successively changing positions as the earth moves in its orbit (Chapter 2). For centuries, the *absence* of observable stellar parallaxes was considered proof that the earth does *not* move. However, in the nineteenth century, the effect was observed for the first time. The first observation of stellar parallax is usually accorded to Bessel, who measured the parallax of the star 61 Cygni in 1838 (Chapter 18). The stars are so distant that the parallax of even the nearest star is very small — less than 1″. With modern telescopes, however, the effect is observable.

Another effect of the earth's revolution is the *aberration of starlight*. The familiar analogy given is that of a man walking in the rain with a straight stovepipe (Figure 7-11). If the stovepipe is held vertically, and if the raindrops are assumed to fall vertically, they will fall through the length of the pipe only if the man is standing still. If he walks forward, he must tilt the pipe slightly forward, so that drops entering the top will fall out the bottom without being swept up by the approaching inside wall of the pipe. If the raindrops fall with a speed

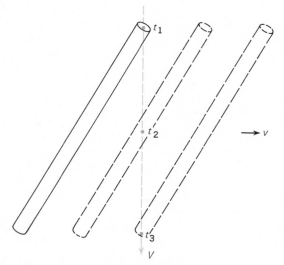

FIG. 7-11 Raindrops falling through a drain pipe.

V, and if the man walks with a speed v, the distance by which the top of the pipe precedes the bottom, divided by the vertical distance between the top and bottom of the pipe, must be in the ratio v/V.

Similarly, because of the earth's orbital motion, if starlight is to pass through the length of a telescope, the telescope must be tilted slightly forward in the direction of the earth's motion. In other words, the apparent direction of a star is displaced slightly from its geometrical direction, and the displacement is in the direction of the earth's orbital motion. This aberration of starlight was first explained by Bradley in 1729. The effect is greatest when the earth is moving at right angles to the direction to the star, and disappears when the earth moves directly toward or away from the star. A star that is on the ecliptic appears to shift back and forth by a small amount in a straight line during the year, for during part of the year the earth is moving in one direction compared to the star's, and during the rest of the year the earth is

moving in the opposite direction. A star in a direction perpendicular to the earth's orbit appears to describe a small circle in the sky, for its apparent direction is constantly displaced in the direction of the earth's orbital motion from the direction it would have as seen from the sun. Stars in between these extremes appear to shift their apparent directions along tiny elliptical paths (Figure 7-12).

The actual amount by which a star's direction seems displaced is the angle between the star's geometrical direction and the direction that the telescope has to be pointed to observe the star. Analogous to the tilt of the stovepipe, this forward tilt of the telescope is in the ratio of the speed of the earth to the speed of light. The speed of light is about 10,000 times that of the earth's orbital motion, so the angle through which a telescope must be tilted forward varies during the year by about 1 part in 10,000, or about 20½″. Thus, 20.″5 is the semimajor axis of the small orbit through which the direction of a star seems to shift due to aberration.

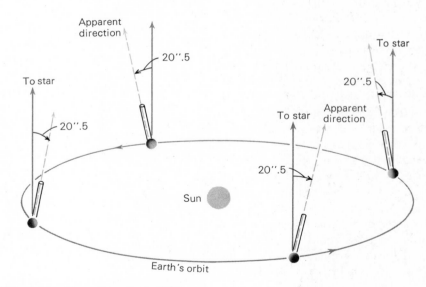

FIG. 7-12 Aberration of starlight.

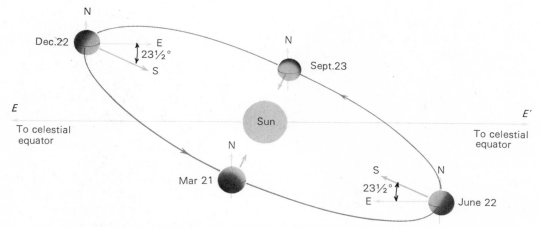

FIG. 7-13   The seasons are caused by the inclination of the plane of the earth's orbit to the plane of the equator.

7.4 THE SEASONS

The earth's orbit around the sun is an ellipse, its distance from the sun varying by about 3 percent. However, the changing distance of the earth from the sun is *not* the cause of the seasons. The seasons result because the plane in which the earth revolves is not coincident with the plane of the earth's equator. The planes of the equator and ecliptic are inclined to each other by about 23½°. This angle of 23½° is called the *obliquity of the ecliptic.*

Globes of the earth are usually mounted with the earth's axis tilted from the vertical. This tilt is the same angle of 23½°, for that is the angle the earth's axis must make with the perpendicular to the plane of its orbit around the sun. The result of the obliquity of the ecliptic is that the Northern Hemisphere is inclined *toward* the sun in June and away from it in December.

(a) The Seasons and Sunshine

Figure 7-13 shows the earth's path around the sun. The line *EE'* is in the plane of the celestial equator. In the figure the earth appears to pass alternately above and below this plane, but the celestial sphere is so large, and the celestial equator so far away,

that a line from the center of the earth through the earth's equator always points to the celestial equator.

We see in the figure that on about June 22 (the date of the *summer solstice*), the sun shines down most directly upon the northern hemisphere of the earth. It appears 23½° *north* of the equator and thus on that date passes through the zenith of places on the earth that are at 23½° north latitude. The situation is shown in detail in Figure 7-14, which shows the earth on the date of the summer solstice. To an observer on the equator, the sun appears 23½° north of the zenith at noon. To a person at a latitude 23½° N, the sun is overhead at noon. This latitude on the earth, at which the sun can appear at the zenith at noon on the first day of summer is called the *Tropic of Cancer*. We see also in Figure 7-14 that the sun's rays shine down past the North Pole; in fact, all places within 23½° of the pole, that is, at a latitude greater than 66½° N, have sunshine for 24 hours on the first day of summer. The sun is as far north on this date as it can get; thus, 66½° is the southernmost latitude where the sun can ever be seen for a full 24-hour period (the *midnight sun*); that circle of latitude is called the *Arctic Circle*.

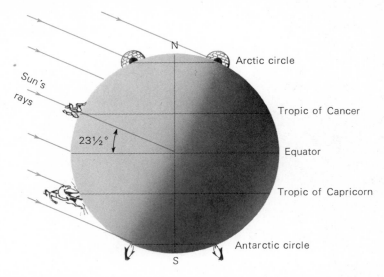

FIG. 7-14
The earth on June 22.

On the other hand, the sun's rays shine very obliquely on the Southern Hemisphere. In fact, all places within 23½° of the South Pole — that is, south of latitude 66½° S (the *Antarctic Circle*) — have no sight of the sun for the entire 24-hour period.

The situation is reversed 6 months later, about December 22 (the date of the *winter solstice*), as is shown in Figure 7-15. Now it is the Arctic Circle that has a 24-hour night and the Antarctic Circle that has the midnight sun. At latitude 23½° S, the *Tropic of Capricorn*, the sun passes through the zenith at noon. It is winter in the Northern Hemisphere, summer in the Southern.

Finally, we see in Figure 7-13 that on about March 21 and September 23 the sun appears to be in the direction of the celestial equator, and, on these dates, the equator itself is the diurnal circle for the sun. Every place on the earth then receives exactly 12 hours of sunshine and 12 hours of night. These points, where the sun crosses the celestial equator, are called the *vernal* (spring) *equinox* and *autumnal* (fall) *equinox. Equinox* means "equal night."

FIG. 7-15
The earth on December 22.

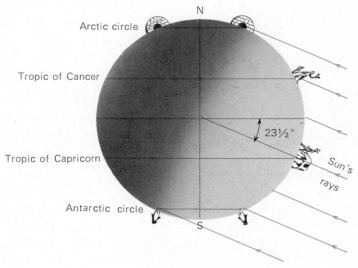

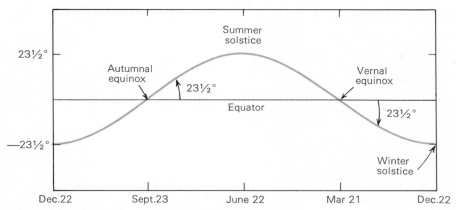

FIG. 7-16   Plot of the ecliptic around the celestial equator.

Figure 7-16 is a map in which the sky is shown flattened out, as in a Mercator projection of the earth. The equator runs along the middle of the map, and the ecliptic is shown as a wavy line crossing the equator at the two equinoxes. Both equator and ecliptic are, of course, great circles, but they cannot both be shown as straight lines on a flat surface. Notice that the ecliptic intersects the equator at an angle of 23½°, and that its northernmost extent is 23½° north of the equator (the summer solstice) and its southernmost extent is 23½° south of the equator (the winter solstice).

Figure 7-17 shows the aspect of the sky at a typical latitude in the United States. During the spring and summer, the sun is north of the equator and is thus up more than half the time. A typical spot in the United States, on the first day of summer (about June 22), receives about 14 or 15 hours of sunshine. Also, notice that the sun appears *high* in the sky, and so in these seasons the sunlight is more direct, and thus more effective in heating than in the fall and winter when the sun appears at a lower altitude in the sky.

In the fall and winter the sun is south of the equator, where most of its diurnal circle is below the horizon, and so it is up less than half the time. On about December 22, a typical city at, say 30 to 40° north latitude receives only 9 or 10 hours of sunshine. Also, the sun is low in the sky; a bundle of its rays is spread out over a larger area on the ground (Figure 7-18) than in summer; because the energy is spread out over a larger area, there is less for each square foot, and so the sun at low altitudes is less effective in heating the ground.

(b) The Seasons at Different Latitudes

At the equator all seasons are much the same. Every day of the year, the sun is up half the time, so there are always 12 hours of sunshine at the

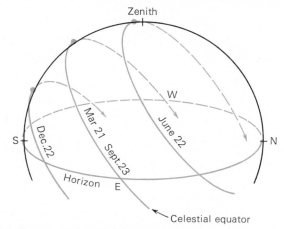

FIG. 7-17   Diurnal paths of the sun for various dates at a typical place in the United States.

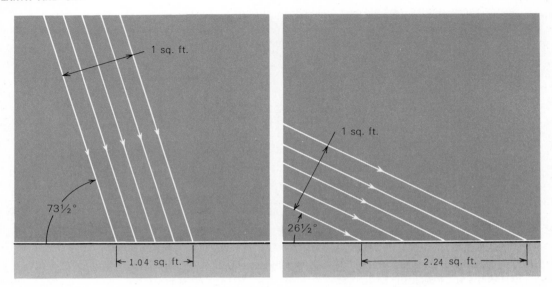

FIG. 7-18   Effect of the sun's altitude. When the sun is low in the sky, its rays are more oblique to the ground and are spread over a larger area than when the sun is high in the sky.

equator. About June 22, the sun crosses the meridian 23½° north of the zenith, and about December 22, 23½° south of the zenith.

The seasons become more severe as one travels north or south of the equator. At the Tropic of Cancer, on the date of the summer solstice, the sun is at the zenith at noon. On the date of the winter solstice, the sun crosses the meridian 47° south of the zenith. At the Arctic Circle, on the first day of summer the sun never sets, but at midnight can be seen just skimming the north point on the horizon. About December 22, the sun does not quite rise at the Arctic Circle, but just gets up to the south point on the horizon at noon. Between the Tropic of Cancer and the Arctic Circle, the number of hours of sunshine and the noon altitude of the sun range between these two extremes.

We recall that at the North Pole, all celestial objects that are north of the celestial equator are always above the horizon and, as the earth turns, circle around parallel to it. The sun is north of the celestial equator from about March 21 to September 23, and so at the North Pole the sun rises when

it reaches the vernal equinox and sets when it reaches the autumnal equinox. There are 6 months of sunshine at the Pole. The sun has its maximum altitude of 23½° about June 22; before that date it climbs gradually higher each day, and after that it drops gradually lower. A navigator can easily tell when he is at the North Pole, for there the sun circles around the sky parallel to the horizon, getting no higher or lower (except gradually as the days go by).†

In the Southern Hemisphere, the seasons are similar, except that they are reversed from those in the north. While we are having summer in the United States, in Australia it is winter. Furthermore, in the Southern Hemisphere, the sun crosses the meridian generally to the *north* of the zenith.

---

†It is said that one botanist considered the North Pole to be an excellent place to raise sunflowers, because there were so many hours of sunshine during the summer months. He accordingly planted some there, and they did quite well for a while. However, sunflowers like to *face* the sun, and as they followed the sun around and around the sky, they ended by wringing their own necks!

In Buenos Aires, you would want a house with a good *northern* exposure.

The earth, in its elliptical orbit, reaches its closest approach to the sun about January 4. It is then said to be at *perihelion*. It is farthest from the sun, at *aphelion,* about July 5. We see, then, that the earth is closest to the sun when it is winter in the north. However, it is summer in the Southern Hemisphere when the earth is at perihelion, and the earth is farthest from the sun during the Southern Hemisphere's winter. Therefore, we might expect the seasons to be somewhat more severe in the Southern Hemisphere than in the Northern. However, there is more ocean area in the Southern Hemisphere; this and other topographical factors are more important in their influence on the seasons on the earth than is the earth's changing distance from the sun. We shall see that for Mars, whose orbit is considerably more eccentric than the earth's, the same kind of situation does have a pronounced effect upon the seasons.

(c) Precession of the Equinoxes

As the earth's axis precesses in its conical motion (Section 6.5c), the equatorial plane retains (approximately) its 23½° inclination to the ecliptic plane; that is, the obliquity of the ecliptic remains constant. However, the intersections of the celestial equator and the ecliptic (the equinoxes) must always be 90° from the celestial poles (because all points on the celestial equator are 90° from the celestial poles). Thus, as the poles move because of precession, the equinoxes slide around the sky, moving westward along the ecliptic. This motion is called the *precession of the equinoxes.* The angle through which the equinoxes move each year, the *annual precession,* is 1/26,000 of 360°, or about 50″. Each year as the sun completes its *eastward* revolution about the sky with respect to, say, the vernal equinox, that equinox has moved *westward,* to meet the sun, about 50″. Since it takes the sun about 20 minutes to move 50″ along the ecliptic (or, more

accurately, because it takes the earth that long to move through an angle of 50″ in its orbit about the sun), a *tropical year,* measured with respect to the equinoxes, is 20 minutes shorter than a *sidereal* year, measured with respect to the stars.

Precession has no important effect on the seasons. The earth's axis retains its inclination to the ecliptic, so the Northern Hemisphere is still tipped toward the sun during one part of the year and away from it during the other. Our calendar year is based on the beginnings of the seasons (the times when the sun reaches the equinoxes and solstices), so spring still begins in March (in the Northern Hemisphere), summer in June, and so on. The only effect is that as the precessional cycle goes on, a given season will occur when the earth is in gradually different places in its orbit with respect to the stars. In the twentieth century, for example, Orion is a *winter constellation;* we look out at night, away from the sun, and see Orion in the sky during the winter months. In 13,000 years, half a precessional cycle later, it will be summertime when we look out in the same direction, away from the sun, and see Orion. Similarly, Scorpio is a summer constellation now, whereas in the year 15,000 it will be a winter constellation (see Figure 7-19).

The vernal equinox is sometimes called the *first point of Aries,* because about 2000 years ago, when it received that name, it lay in the constellation of Aries. Now, because of precession, the vernal equinox has slid westward into the constellation of Pisces.

(d) Twilight

We all know that the sky does not immediately darken when the sun sets. Even after the sun is no longer visible from the ground, the upper atmosphere of the earth can catch some of the rays of the setting sun and scatter them helter-skelter, illuminating the sky. Gases in the earth's atmosphere are dense enough to scatter appreci-

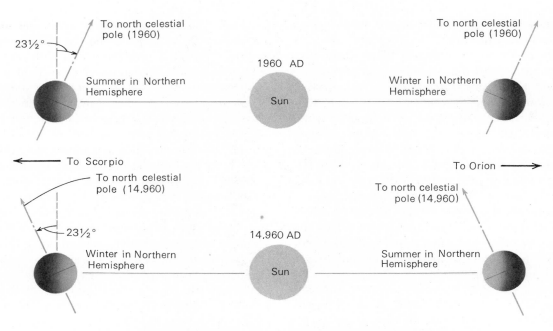

FIG. 7-19   Summer and winter constellations change due to precession.

able sunlight up to altitudes of about 200 mi. The sun must be at least 18° below the horizon for all traces of this postsunset or presunrise sky light *(twilight)* to be absent. At latitudes near the equator, where the sun rises and sets nearly vertically to the horizon, twilight lasts only a little over 1 hour. However, at far northern and southern latitudes, the sun rises and sets in a much more oblique direction and takes correspondingly longer to reach a point 18° below the horizon, so twilight may last for 2 hours or more. In the far northern countries twilight lasts all night in the summertime. At the North Pole there are 6 weeks of twilight in the late winter before sunrise and again in the early fall after sunset.

(e) Lag of the Seasons

Places in the Northern Hemisphere receive the most sunlight about June 22, and the least about December 22. Yet these are not the dates of the hottest and coldest weather of the year, respectively. The hottest and coldest weather generally lags the opening of summer and winter by several weeks.

The actual amount of heat received from the sun by a given area on the earth is called its *insolation.* In the Northern Hemisphere, the insolation is greatest on the first day of summer and decreases thereafter. During the preceding winter, however, the hemisphere has cooled considerably and large deposits of ice and snow have formed. During the spring, as the insolation increases, these snow deposits slowly melt and the hemisphere gradually warms up. This warming-up process generally continues somewhat past the date of the summer solstice. In other words, during the early weeks of summer, a good portion of the insolation is going into the melting of ice and the heating of the land and oceans. The temperatures do not reach their maxima until the portions of the earth that contribute to the climate of a particular place have thawed out as much as they are going to. This usually occurs in August (in the Northern Hemisphere), although the date varies from place to place with the local topography.

Similarly, the coldest time of year is not at the winter solstice, even though that is the time of the least insolation, because the land and ocean, having been warmed up in the previous summer, are still cooling down. They reach their maximum chilling sometime in midwinter.

7.5 THE MANY MOTIONS OF THE EARTH

In this chapter we have discussed in detail two of the earth's motions: rotation and revolution. However, there are many other motions of the earth, dealt with in other sections. Here, for completeness, these motions are summarized:

(1) The earth *rotates* daily on its axis.
(2) The earth periodically shifts slightly with respect to its axis of rotation *(variation in latitude)* (Section 7.2b).
(3) The earth *revolves* about the sun.
(4) The gravitational pull of the sun and moon on the earth's equatorial bulge causes a very slow change in orientation of the axis of the earth called *precession* (Section 6.5).
(5) Because the moon's orbit is not quite in the plane of the ecliptic, and because of a slow change in orientation of the moon's orbit, there is a small periodic motion of the earth's axis superimposed upon precession, called *nutation* (Section 6.5d).
(6) Actually, it is the center of mass of the earth-moon system, or *barycenter,* that revolves about the sun in an elliptical orbit. Each month the center of the earth revolves about the barycenter (Section 5.1).
(7) The earth shares the motion of the sun and the entire solar system among its neighboring stars. This *solar motion* is about 12 mi/sec (Chapter 19).
(8) The sun, with its neighboring stars, shares in the general *rotation of the Galaxy.* Our motion about the center of the Galaxy is about 150 mi/sec (Chapter 27).
(9) All other galaxies are observed to be in motion. Therefore, our *Galaxy is in motion* with respect to other galaxies in the universe (Chapter 33).

In view of the many motions of the earth, one might wonder what the absolute speed of the earth is in the universe. To determine the absolute velocity of the earth was the object of the Michelson-Morley experiment, which we have seen (Chapter 4) gave null results. A fundamental postulate of Einstein's special theory of relativity is that it is not possible to define an absolute coordinate system with respect to which the absolute speed of an object in space can be determined.

Exercises

1. Show that the apparent deflection in the direction of swing of a Foucault pendulum in the Southern Hemisphere would be to the *left* rather than to the *right,* as in the Northern Hemisphere.

2. The radius of curvature of the earth at the poles or at the equator is the radius of a sphere whose surface matches the curvature of the earth at that point (as shown in Figure 6-3). How much greater is the radius of curvature of the earth at the poles than at the equator?

3. What is the latitude of
(a) The North Pole?
(b) The South Pole?
(c) A point halfway between the equator and the North Pole?

4. Why has longitude no meaning at the North or South Pole?

5. Draw a diagram to show that for an observer south of the equator the altitude of the south celestial pole is equal to his latitude south.

6. Why is exactly half of any great circle in the sky above the horizon at once?

7. Prove that the celestial equator must pass through the east and west points on the horizon.

8. Prove that if vertically falling raindrops dropping with a speed of V are to fall through a drain pipe, the pipe must be tilted forward so that its top precedes its bottom by a distance which, when divided by the vertical extent of the pipe, is in the ratio v/V, where v is the speed of the drain pipe.

9. How can we tell that stars are displaced in the direction of the earth's motion (aberration), since all stars in a given part of the sky appear shifted by the same amount?

10. Where on earth is it possible for the ecliptic to lie along the horizon?

11. Explain why New York has more hours of daylight on the first day of summer than does Los Angeles?

12. In far northern countries such as Canada and Scotland, the winter months are so cloudy that astronomical observations are nearly impossible. Why is it that good observations cannot be made in those places during summer nights?

CHAPTER 8

Time and the Calendar

One of the most ancient uses of astronomy was the keeping of time and the calendar. From the earliest history in virtually every center of civilization — China, India, Mesopotamia, Egypt, Greece, even the Mayan and Aztec civilizations in the Western Hemisphere — man kept track of the motions in the heavens to regulate his time and date.

8.1 TIME OF DAY

The measurement of time is based on the rotation of the earth. As the earth turns, objects in the sky appear to move around us, crossing the meridian each day. Time is determined by the position in the sky, with respect to the local meridian, of some reference object on the celestial sphere. The interval between successive meridian crossings or *transits* of that object is defined as a *day*. The actual length of a day depends on the reference object chosen; several different kinds of days, corresponding to different reference objects, are defined. Each kind of day is divided into 24 equal parts, called *hours*.

(a) The Passage of Time; Hour Angle

Time is reckoned by the angular distance around the sky that the reference object has moved since it last crossed the meridian. The motion of that point around the sky is analogous to the motion of the hour hand on a 24-hour clock. The angle measured to the west along the celestial equator from the local meridian to the hour circle passing through any object (for example, a star) is that object's *hour angle*. (An hour circle is a great circle on the celestial sphere running north and south

through the celestial poles — see Chapter 7.) *Time can be defined as the hour angle of the reference object.*

As an example, suppose that the star *Rigel* is chosen as the reference for time. Then when Rigel is on the meridian it is $0^h0^m0^s$, "Rigel time." Twelve *Rigel hours* later, Rigel is halfway around the sky, at an hour angle of 180°, and the Rigel time is $12^h0^m0^s$. When Rigel is only 1° east of the meridian, and one *Rigel day* is nearly gone, the star is at an hour angle of 359°, and the Rigel time is $23^h56^m0^s$.

Time can be represented graphically by means of a *time diagram*, as in Figure 8-1. Here we imagine ourselves looking straight down on the north celestial pole from *outside* the celestial sphere.

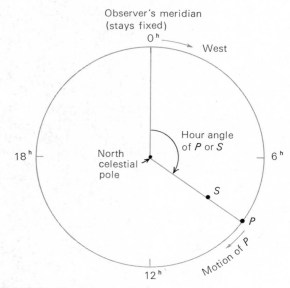

FIG. 8-1   Time diagram.

The pole appears as a point in the middle of the diagram, and the celestial equator appears as a circle centered on the pole. As the earth turns to the east, the local meridian of an observer sweeps around the sky, so that its intersection with the celestial equator would move *counterclockwise* around the circle in the time diagram. However, it is customary to represent the observer's meridian as fixed, intersecting the equator, say, at the top of the diagram. Then the celestial sphere must be regarded as rotating *clockwise* with respect to the meridian. Let the reference object be denoted S. Its hour circle intersects the equator at P, and as the celestial sphere rotates, the point P moves clockwise around the circle, like the hour hand of a clock. The hour angle of S (or P) increases uniformly with the rotation of the celestial sphere. In the diagram it is shown as about $120°$ (the time would then be about 8 hours). Since the celestial equator intersects the horizon at the east and west points, P is below the observer's horizon in the time interval from 6 to 18 hours; the same is not true of S unless it happens to lie on the equator.

Because of the relation between hour angle and time, it is often convenient to measure angles in time units. In this notation, 24 hours corresponds to a full circle of $360°$, 12 hours to $180°$, 6 hours to $90°$, and so on. One hour equals $15°$, and $1°$ is 4 minutes of time.

Here we must distinguish between minutes and seconds of *time* (subdivisions of an hour), denoted ^{m} and ^{s}, respectively, and minutes and seconds of *arc* (subdivisions of a degree), denoted $'$ and $''$, respectively. The conversion between units of time and arc is given in Table 8.1.

TABLE 8.1 Conversion Between Units of Time and Arc

TIME UNITS	ARC UNITS
24^h	$360°$
1^h	$15°$
4^m	$1°$
1^m	$15'$
4^s	$1'$
1^s	$15''$

Because we base time on the rotation of the earth, it might seem that the earth's rotation with respect to the stars would be the nearest we could come to defining the "true rotation rate" of the earth. We could then measure the passage of time by the hour angle of some fixed point on the celestial sphere. The interval between two successive meridian transits of this point would define such a "stellar day." This is the kind of rotation period of the earth that would be measured, say, by the Foucault pendulum. (Actually, of course, the stars are in motion and do not define a perfectly fixed reference system; however, they are so distant that over moderate periods of time they remain sensibly fixed on the celestial sphere.)

We could measure time by such a scheme, but it turns out that other kinds of time are more convenient. The most common kinds of time in use are *sidereal time* and *solar time*, which are based on the sidereal day and the solar day.

(b) The Sidereal and Solar Day

The *sidereal* day is defined as the interval between successive transits of the *vernal equinox*, the point on the celestial sphere where the sun, in its apparent annual path around the sky, crosses the celestial equator to the north on the first day of spring. Because of *precession* (Section 6.5), the vernal equinox is not quite a fixed point on the celestial sphere, but slowly shifts its position. Technically, therefore, the term "sidereal day" is a misnomer, for its length is not based strictly on the earth's rotation with respect to the stars. However, because the motion of the vernal equinox in the sky is very slow, a sidereal day is very close to the true period of rotation of the earth with respect to the stars — within one hundredth of a second.

The *solar day* is the period of the earth's rotation with respect to the sun. The solar and sidereal days are *not* the same, as a study of Figure 8-2 will show. Suppose we start counting

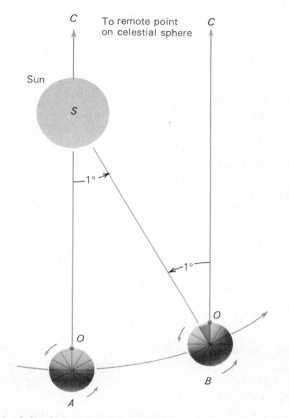

FIG. 8-2 Sidereal and solar day.

complete a *solar* day it must turn a little more to bring the sun back to the meridian.

In other words, a solar day is slightly *longer* than a sidereal day, or one complete rotation of the earth. There are about 365 days in a year and 360° in a circle; thus the daily motion of the earth in its orbit is about 1°. This 1° angle, *ASB*, is nearly the same as the additional angle over and above 360° through which the earth must turn to complete a solar day. It takes the earth about 4 minutes to turn through 1°. A solar day, therefore, is about 4 minutes longer than a sidereal day.

Each kind of day is subdivided into hours, minutes, and seconds. A unit of solar time (hour, minute, or second) is longer than the corresponding unit of sidereal time by about 1 part in 365. In units of solar time, one sidereal day is $23^h56^m4^s.091$. The period of the earth's rotation with respect to the stars is $23^h56^m4^s.099$ in solar time units.

(c) Sidereal Time

Sidereal time is based on the sidereal day with its subdivisions of sidereal hours, minutes, and seconds. It is defined as the *hour angle of the vernal equinox.* The sidereal day begins at sidereal *noon*, $0^h0^m0^s$ sidereal time, with the vernal equinox on the meridian.

Sidereal time is very useful to astronomy and navigation. The common coordinate system used to denote positions of stars and planets on the celestial sphere (right ascension and declination) is referred to the celestial equator and the vernal equinox, much as latitude and longitude on the earth are referred to the earth's equator and the meridian of Greenwich, England (see Appendix 7). Therefore, the position of a star in the sky with respect to the observer's meridian is directly related to the sidereal time. Every observatory maintains clocks that are rated to keep accurate sidereal time.

We regulate our everyday lives, however, by the sun, not the vernal equinox; for example, the working day is usually determined according to

a day when the earth is at *A*, with the sun on the meridian of an observer at point *O* on the earth. The direction from the earth to the sun, *AS*, if extended, points in the direction *C* among stars on the celestial sphere. After the earth has made one rotation with respect to the stars, the same stars in direction *C* will again be on the local meridian to the observer at *O*. However, because the earth has moved from *A* to *B* in its orbit about the sun during its rotation, the sun has not yet returned to the meridian of the observer but is still slightly to the east. The vernal equinox is so nearly fixed among the stars that the earth has completed, essentially, one *sidereal* day, but to the daylight hours. It is far more desirable to use solar time for ordinary purposes.

(d) Apparent Solar Time

Just as sidereal time is reckoned by the hour angle of the vernal equinox, so *apparent solar time* is determined by the hour angle of the sun. At midday, apparent solar time, the sun is on the meridian. The hour angle of the sun is the time *past midday* (*post meridiem,* or P.M.). It is convenient to start the day not at noon, but at midnight. Therefore the elapsed apparent solar time since the beginning of a day is the hour angle of the sun *plus* 12 hours. During the first half of the day, the sun has not yet reached the meridian. We designate those hours *before midday* (*ante meridiem,* or A.M.). We customarily start numbering the hours after noon over again, and designate them P.M., to distinguish them from the morning hours (A.M.). On the other hand, it is often useful to number the hours from 0 to 24, starting from the beginning of the day at midnight. For example, in various conventions, 7:46 P.M. may be written as 19^h46^m, 19:46, or simply 1946.

On about September 23, the sun passes through the autumnal equinox, halfway around the sky from the vernal equinox. On that date, at midnight, when the day begins, the vernal equinox is on the meridian, and so solar time and sidereal time are in agreement. With each succeeding day, however, sidereal time gains 3^m56^s on solar time, and the two kinds of time do not agree again until the daily difference between them accumulates to a full 24 hours — 1 year later.

Apparent solar time, defined as the hour angle of the sun plus 12 hours, is the most obvious and direct kind of solar time. It is the time that is kept by a sundial. In a sundial, a raised marker, or *gnomon,* casts a shadow whose direction indicates the hour angle of the sun. Apparent solar time was the time kept by man through many centuries.

The exact length of an apparent solar day, however, varies slightly during the year. Recall that

the difference between an apparent solar day and a sidereal day, if time is counted from noon on one day, is the extra time required, after one rotation of the earth with respect to the vernal equinox, to bring the sun back to the meridian. The length of this extra time depends on how far *east* of the meridian the sun is after the completion of 1 sidereal day. The earth rotates to the east at a constant rate of 1° every 4 sidereal minutes. Thus, if the sun were exactly 1° east of the meridian, about 4 sidereal minutes would be needed to bring it to the meridian. (Actually, it is just over 4 sidereal minutes, because the earth is still advancing in its orbit during that period, which moves the sun another 10″ to the east along the ecliptic.) If the sun were

FIG. 8-3 A sundial constructed so that a gnomon parallel to the earth's axis casts a shadow on the inner side of a ring that lies parallel to the earth's equatorial plane. *(Griffith Observatory.)*

more or less than 1° east of the meridian, the extra time required would be a little more or less than 4 sidereal minutes.

The length of the apparent solar day would be constant if the eastward progress of the sun, in its apparent annual journey around the sky, were precisely constant. However, there are two reasons why the amount by which the sun shifts to the east is not the same every day of the year.

The first reason is that the earth's orbital speed varies. In accord with Kepler's second law — the law of areas — the earth moves fastest when it is nearest the sun (perihelion) in early January and slowest when it is farthest from the sun (aphelion) in July. However, its rate of rotation is nearly constant. Consequently, it moves *farther* in its orbit during a sidereal day in January than in July (Figure 8-4). The sun's apparent motion along the ecliptic is just the reflection of the earth's revolution, so the sun's daily progress to the east reflects the inequalities of the earth's daily progress in its orbit. We see, then, that the extra amount by which the earth must turn after a sidereal day to complete a rotation with respect to the sun is not always exactly the same.

The second reason for the variation in the rate of the sun's eastward progress, and the consequent nonuniformity in the length of the apparent solar day, is that the sun's path — the ecliptic — does not run exactly east and west in the sky, along the celestial equator, but is inclined to the equator by 23½°. Even if the earth's orbit were circular, so that the sun moved uniformly along the *ecliptic,* the amount by which it moved to the east would vary slightly throughout the year. The situation is illustrated in Figure 8-5, which shows the celestial sphere, the celestial equator, and the ecliptic. To make the effect more obvious, the obliquity of the ecliptic is grossly exaggerated. Now suppose the sun moved equal distances along the ecliptic near March 21 and June 22; such equal distances are marked off on the ecliptic in the figure. Near the equinox, part of the sun's motion is northward, and it progresses less far to the *east* than it does along the ecliptic. At the solstice, on the other hand, not only is the sun moving due east, but it is also north of the equator where the hour circles converge, so that a 1° advance on the ecliptic is *more* than 1° advance to the east. A similar analysis shows the sun would also make more eastward progress near

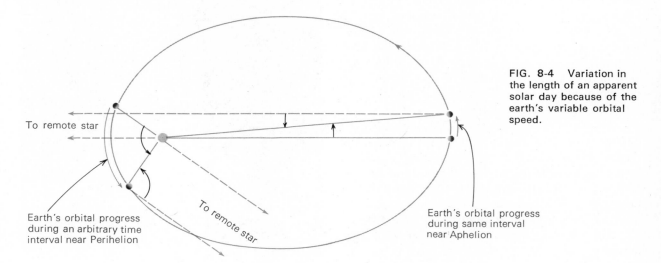

FIG. 8-4 Variation in the length of an apparent solar day because of the earth's variable orbital speed.

To remote star

Earth's orbital progress during an arbitrary time interval near Perihelion

To remote star

Earth's orbital progress during same interval near Aphelion

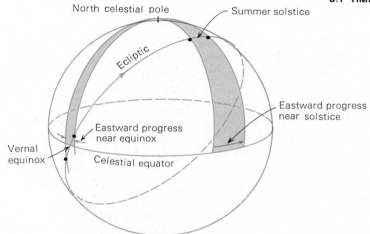

FIG. 8-5 The sun's apparent eastward daily progress varies because of the obliquity of the ecliptic.

North celestial pole

Summer solstice

Ecliptic

Eastward progress near solstice

Eastward progress near equinox

Vernal equinox

Celestial equator

the winter solstice than near the autumnal equinox. With the actual 23½° obliquity, it turns out that a 1° advance on the ecliptic corresponds to 0°.92 advance to the *east* at the equinoxes and 1°.08 advance to the east at the solstices. Thus, even if the sun did move uniformly on the ecliptic, its eastward progress would be variable.

The apparent solar day is always *about* 4 minutes longer than a sidereal day, but because of the sun's variable progress to the east, the precise interval varies by up to ½ minute one way or the other. The variation can accumulate after a number of days to several minutes. After the invention of clocks that could run at a uniform rate, it became necessary to abandon the apparent solar day as the fundamental unit of time. Otherwise, all clocks would have to be adjusted to run at a different rate each day.

(e) Mean Solar Time

Mean solar time is based on the *mean solar day,* which has a duration equal to the *average* length of an apparent solar day. Mean solar time is defined as the hour angle of the mean sun plus 12 hours, where the *mean sun* is a fictitious point in the sky that moves uniformly to the east along the *celestial equator,* with the same average eastern rate as the true sun. In other words, mean solar time is just apparent solar time averaged uniformly.

The irregular rate of apparent solar time causes it to run alternately ahead of and behind mean solar time. The difference between the two kinds of time can accumulate to about 17 minutes. The difference between apparent solar time and mean solar time is called the *equation of time,* shown graphically in Figure 8-6. One can read from the plot, for any date of the year, the correction to apply to mean solar time to obtain apparent solar time; that is, when the equation of time is positive, apparent time is *ahead* of mean time. Often the equation of time is plotted on globes of the earth as a nomogram, shaped like the figure eight and placed in the region of the South Pacific Ocean.

Although mean solar time has the advantage of progressing at a uniform rate, it is still inconvenient for practical use. Recall that it is defined as the hour angle of the mean sun. But hour angle is referred to the local celestial meridian, which is different for every longitude on earth. Thus, observers on different north-south lines on the earth have a different hour angle of the mean sun and hence a different mean solar time. If mean solar time were strictly observed, a person traveling east or west would have to reset his watch continually as his longitude changed, if it were always to read the local mean time correctly. For instance, a commuter traveling from Oyster Bay to New York City would have to adjust his watch as he rode through the East River tunnels, because Oyster Bay time is actually a few seconds more advanced than that of Manhattan.

(f) Standard and Zone Time

Until near the end of the last century, every city and town in the United States kept its own local mean time. With the development of railroads and telegraph, however, the need for some kind of standardization became evident. In 1883 the nation was divided into four time zones. Within each zone, all places keep the same time, the local mean solar time of a standard meridian running more or less through the middle of each zone. Now a traveler resets his watch only when the time change has amounted to a full hour. For local convenience, the boundaries between the four time zones are chosen to correspond, as much as possible, to divisions between states. Mean solar time, so standardized, is called *standard time*. The standard time zones in the United States (not including Alaska and Hawaii) are Eastern Standard Time (EST), Central Standard Time (CST), Mountain Standard Time (MST), and Pacific Standard Time (PST), which respectively keep the mean times of the meridians of 75, 90, 105, and 120° west longitude. Hawaii and Alaska both keep the time of the meridian 150° west longitude, 2 hours less advanced than Pacific Standard Time.

In 1884, largely under the impetus of two papers by Sanford Fleming, an international conference was held in Washington, D.C., in which 26 nations were represented. At that conference it was agreed to establish a system of 24 international time zones around the world. Each time zone, on the average, is 15° wide in longitude, although the zone divisions are usually irregular over land areas to follow international boundaries. At sea, the zone time of any place is the mean time of the standard meridian running through the center of the zone of that place. The zones are numbered consecutively from the Greenwich meridian; those west of Greenwich are denoted (+) and those east are denoted (−). The Eastern Standard time zone is zone number +5.

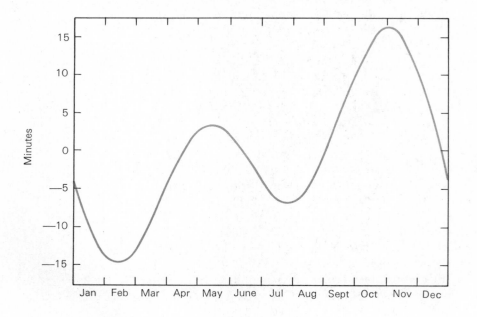

FIG. 8-6 Equation of time (apparent *minus* mean time).

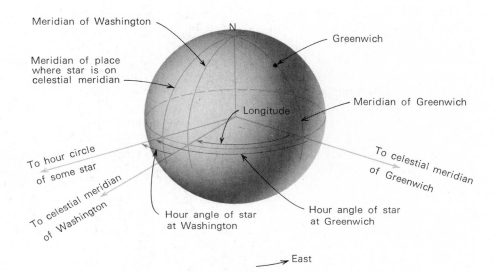

Meridian of Washington

Greenwich

Meridian of place where star is on celestial meridian

N

Meridian of Greenwich

Longitude

To hour circle of some star

To celestial meridian of Greenwich

To celestial meridian of Washington

Hour angle of star at Washington

Hour angle of star at Greenwich

East

FIG. 8-7   A difference in hour angle equals the difference in longitude.

(g) Daylight Saving Time

To take advantage of the maximum amount of sunlight during waking hours, most states in this country, as well as many foreign nations, keep what is called *daylight saving time* during the spring and summer. Daylight saving time is simply the local standard or zone time of the place *plus* 1 hour. Thus, on a summer evening, when it would ordinarily grow dark about 8:00 P.M. standard time, it is light until 9:00 P.M. daylight saving time.

It might seem that daylight saving time is needed more in the winter than the summer. It is not practical in the winter, however, for if clocks were set an hour ahead in December, in many parts of the country it would still be dark at 7:30 A.M. while people are on their way to work.†

(h) Time Around the World; The International Date Line

The direction of the earth's rotation is to the east. Therefore, places to the east of us must always have

---

†Daylight saving time has been compared to the action of a woman who cut 1 ft off the top of her blanket and sewed it on to the bottom to make it longer.

a time *more advanced* than ours. The celestial meridian of New York sweeps under the sun on the celestial sphere about 3 hours earlier than does the celestial meridian of San Francisco. Thus in New York the hour angle of the sun, the local time there, is 3 hours ahead of San Francisco time. The times of places halfway around the world from each other differ by 12 hours.

In general, the difference in the local time of any two places on earth is equal to their difference in longitude, if longitude is measured in time units rather than degrees. The more easterly place always has the more advanced time. The statement is true of any kind of time, sidereal, apparent solar, mean solar, and so on, because any kind of local time is defined as the hour angle of some reference object. The difference in the hour angles of the same object seen at two different places is just the difference in the directions of the local meridians of the two places, that is, the difference in their longitudes. Thus, the determination of longitude is equivalent to finding the difference between local time and the time at some known place on earth — say the time at Greenwich (Figure 8-7).

The local mean solar time of the meridian running through Greenwich, England, is called *universal time.* Data are usually given in navigational tables, such as the *Nautical Almanac,* for various intervals of universal time.

The fact that time is always more advanced to the east presents a problem. Suppose a man moving eastward travels around the world. He passes into a new time zone, on the average, for about every 15° of longitude he travels and, each time, he dutifully sets his watch ahead an hour. By the time he has completed his trip, he has set his watch ahead through a full 24 hours, and has thus *gained a day* over those who stayed home. Let us look at the problem another way. It is 3 hours later in New York than in Berkeley, California. In London, it is 8 hours later than in Berkeley; in Tokyo, still farther to the east, it is 17 hours later. Because San Francisco is about 100° east of Tokyo, it follows that in San Francisco, a few miles to the west across the San Francisco Bay from Berkeley, it is 7 hours later than at Tokyo, or 24 hours − 1 day − later than at Berkeley! One might suppose that by going around the world to the east or west enough times, we could go into the future or past as far as we desired, creating the equivalent of a time machine.

The solution to the dilemma is the *international date line,* set by international agreement along the 180° meridian of longitude. The date line runs about down the middle of the Pacific Ocean, although it jogs a bit in a few places to avoid cutting through groups of islands and through Alaska. By convention, at the date line, the date of the calendar is changed by 1 day. If a person crosses the date line from west to east, so that he is advancing his time, he compensates by decreasing his date; if he crosses from east to west, he increases his date by 1 day.

If an ocean liner crosses the international date line from west to east just after its passengers have been served a Christmas dinner, the crew can begin the preparation of a second feast, for the next day will be Christmas again. On the other hand, one could skip Christmas altogether by crossing the date line from east to west.

(i) Summary of Time

We may briefly summarize the story of time as follows:

(1) The ordinary day is based on a rotation of the earth with respect to the sun (not the stars).

(2) The length of the apparent solar day is slightly variable because of the earth's variable orbital speed and the obliquity of the ecliptic.

(3) Therefore, the average length of an apparent solar day is defined as the mean solar day.

(4) Time based on the mean solar day is mean solar time. It is defined as the local hour angle of the mean sun plus 12 hours. The mean sun is an imaginary sun that revolves annually on the celestial equator at a perfectly uniform eastward rate.

(5) The mean solar time at any one place, called *local mean time,* varies continuously with longitude, so that two places a few miles east and west of each other have slightly different times.

(6) Therefore time is standardized, so that each place in a certain region or zone keeps the same time — the local mean time of the standard meridian in that zone.

(7) In many localities, standard time is advanced by 1 hour in spring and summer to take advantage of the maximum number of hours of sunshine during the waking hours. This is daylight saving time.

The procedure for determining standard time from apparent solar time, as read, say, from a sundial, is illustrated in the following example: At Los Angeles (118° west longitude) the apparent solar (sundial) time on March 16 was 11:30 A.M. From the equation of time we note that on March

16 apparent solar time is 9 minutes behind mean solar time. Thus the local mean time is 11:39 A.M. Now, Los Angeles is in the Pacific Standard Time zone, which keeps the time of the meridian at 120° west longitude. Los Angeles is 2° east of that meridian, so its local time is 8 minutes *more advanced* than that of the 120° meridian. Pacific Standard Time is thus 11:31 A.M.

(j) The Measurement of Time

At the United States Naval Observatory in Washington, D.C., and in certain other observatories throughout the world, time is measured as a routine procedure. Specially designed telescopes measure the exact instants when stars cross the **meridian**. Since the position of the mean sun on the celestial sphere is known with respect to the stars, the time of the meridian transit of a star gives indirectly the mean solar time. The time, determined at the Naval Observatory, is broadcast both by the Navy and by the National Bureau of Standards. Time can be determined by astronomical observations to about $\frac{1}{100}$ second.

The first accurate mechanical device for keeping time was the pendulum clock. The idea that a pendulum, because of its very regular rate of swing, could be used to regulate the rate of a clock dates at least from the sixteenth century. Later, in 1656, Huygens published a comprehensive treatise, *The Pendulum Clock*, that outlined many of the properties of a swinging pendulum. The period of oscillation of a pendulum is very regular and depends only on the length of the pendulum and the acceleration of gravity. These circumstances make it possible to make a pendulum clock of high accuracy by constructing the pendulum itself of a special alloy that expands only very slightly with changes in temperature and by keeping the clock in a temperature-controlled chamber.

Today, observatories often use quartz-crystal clocks. In such a clock, the natural frequency of oscillation of a vibrating quartz crystal is used to regulate the frequency of a current that runs an electric clock. The vibration of a quartz crystal is far more constant than that of a pendulum, or of the usual 60-cycle alternating current produced by commercial power-generating stations. However, even the rate of quartz-crystal clock changes very gradually with age.

More accurate still are modern atomic clocks, in which the frequency standard is the frequency of electromagnetic radiation absorbed (or emitted) by atoms or molecules. This frequency standard is

FIG. 8-8 A photographic zenith tube, which automatically takes pictures of objects as they cross the meridian. *(U.S. Navy.)*

used to regulate the frequency of oscillation of a less accurate standard, such as a quartz crystal. In the ammonia clock, for example, a source of radio waves is set to a frequency of 2.4×10^{10} vibrations per second by a vibrating quartz crystal. This radio radiation is then passed into a tube containing molecules of ammonia gas, which absorb exactly that frequency, so no radiation gets through. However, if the quartz crystal gets off frequency slightly, the beam of radiation is not absorbed by the ammonia molecules, but passes through the tube. This transmitted radiation then activates a servomechanism which corrects the frequency of oscillation of the quartz crystal. Atoms of cesium gas in an electric field also absorb at certain radio frequencies and can be used similarly to control the frequency of a vibrating crystal.

Even more accurate timing is potentially available with the future development of "nuclear clocks." The German physicist R. L. Mössbauer has found a technique that makes it possible to fix the exact amount of energy in x-rays emitted from an isotope of iron (^{57}Fe) obtained from the radioactive decay of cobalt. The x-rays will be absorbed by a target of the same isotope of iron only if they possess the correct energy to within 1 part in 1 trillion (10^{12}); thus extremely slight variations in the x-ray energy can be detected. If the detection of such minute variations can be adapted to the regulation of timekeeping devices, such clocks can be expected to keep time to within 1 second in 30,000 years.

It should be noted that these time standards (pendulums, vibrating crystals, atoms, and so on) are devices for measuring accurately the *passage* of time. They do not in themselves define what time it is. For this we must still use astronomical observations of the rotation of the earth.

(k) Variations in the Earth's Rotation

Since the innovation of quartz crystals, time standards have been accurate enough to measure irregularities in the rotation of the earth. Slight expansions and contractions in the earth can change its rotation rate enough to produce accumulated clock errors of ½ minute or more. Furthermore, slight seasonal variations have been detected in the length of the day; these probably result from the shifting of ice and snow deposits and air masses over the earth. Superposed upon these slight fluctuations of the earth's rotation rate is a gradual slowing down of the rotation in the amount of about $\frac{1}{1000}$ second per century. This gradual slowing is generally attributed to tidal friction (Chapter 9).

(l) Ephemeris Time*

The mechanics of the solar system is now understood well enough to allow accurate prediction of the positions of the sun, moon, and planets for years in advance. To state exactly where some object will be on the celestial sphere at some future instant of time, however, we need a precise knowledge of the time interval between the last accurate observation of that object and the future instant in question. Because of the irregularities in the earth's rotation rate, the mean solar day varies slightly in length over the years and does not provide an accurate enough measure of time to use in the prediction of future positions of celestial objects to the precision that is possible with gravitational theory.

Consequently, tables which give positions of the sun, moon, and planets are now based on a kind of time called *ephemeris time* (abbreviated E. T.). Ephemeris time progresses at a precisely uniform rate. An *ephemeris second* is equal to the length of 1 mean solar second at the beginning of the year 1900 (strictly, an ephemeris second is 1/31,556,925.97474 of the length of the *tropical* year at that date). Ephemeris and universal times were in agreement near the beginning of the century.†

---

†For convenience, ephemeris time was originally adjusted to agree with the times for which the positions of the sun as given in Simon Newcomb's tables (prepared near the start of the twentieth century) were correct. For example, suppose the sun at some instant was observed at point P on the celestial sphere, and according to Newcomb's tables the sun should have reached P at time t; t, then, was chosen as the ephemeris time for that instant.

Changes in the earth's rotation rate, however, have now caused the two kinds of time to differ (in 1970) by more than ½ minute.

Ephemeris time is determined from observations of the sun, moon, or planets, by calculating when, according to the rate of passage of ephemeris time, one of these bodies should reach its observed position among the stars on the celestial sphere. (Usually observations of the moon are used for determining ephemeris time.) Local mean (or universal) time, which depends on the earth's current rate of rotation, is determined from observations by timing when certain stars cross the meridian. The difference between ephemeris and universal times is obtained frequently and is tabulated in standard astronomical publications. The difference cannot be predicted for future times, except by uncertain extrapolation, because the future variations in the earth's rotation are not known.

8.2 THE DATE OF THE YEAR

The natural units of the calendar are the day, based on the period of rotation of the earth; the month, based on the period of revolution of the moon about the earth, and the year, based on the period of revolution of the earth about the sun. The difficulties in the calendar have resulted from the fact that these three periods are not commensurable — that is, one does not divide evenly into any of the others.

The period of revolution of the moon with respect to the stars, the *sidereal month,* is about 27⅓ days. However, the interval between corresponding phases of the moon, the more obvious kind of month, is the moon's period of revolution with respect to the sun, the *synodic month,* which has about 29½ days (Chapter 9).

There are at least three kinds of year. The period of revolution of the earth about the sun with respect to the stars is called the *sidereal year.* Its length is 365.2564 mean solar days, or $365^d6^h9^m10^s$.

The period of revolution of the earth with respect to the vernal equinox, that is, with respect to the beginnings of the various seasons, is the *tropical* year. Its length is 365.242199 mean solar days, or $365^d5^h48^m46^s$. Our calendar, to keep in step with the seasons, is based on the tropical year. Because of precession (Section 6.5), the tropical year is slightly shorter than the sidereal year.

The third kind of year is the *anomalistic year,* the interval between two successive perihelion passages of the earth. Its length is 365.2596 mean solar days, or $365^d6^h13^m53^s$. It differs from the sidereal year, because the major axis of the earth's orbit slowly shifts in the plane of the earth's orbital revolution. Perturbations of the other planets cause this shift.

(a) *The Week**

The week is an independent unit arbitrarily invented by man, although its length may have been based on the interval between the quarter phases of the moon. The seven days of the week are named for the seven planets (including the sun and moon) recognized by the ancients. In order of supposed decreasing distance from the earth, these seven objects are Saturn, Jupiter, Mars, the sun, Venus, Mercury, and the moon. (It was believed that the faster-moving objects were the nearest. The assumption is not necessarily correct — Mercury, the fastest-moving planet, does not come as close to the earth as Venus.)

Each hour of the day was believed to be ruled by one of the planets in the order named. A particular day of the week was named for the planet that ruled it during the first hour of that day. The first day, Saturday, was ruled by Saturn during the first hour, Jupiter during the second, Mars during the third, and so on. Saturn thus ruled, in addition to the first hour, the eighth, fifteenth, and twenty-second. Jupiter and Mars were allotted the twenty-third and twenty-fourth hours of Saturday, leaving the first hour of the second day, Sunday, for the sun. If the scheme is continued, it is found that the moon rules the third day (Monday), and Mars, Mercury, Jupiter, and Venus the fourth, fifth, sixth, and seventh days. Our Anglo-Saxon names for the fourth through the seventh days come from the Teutonic equivalents of the Roman gods for which the planets Mars, Mercury, Jupiter, and Venus were named. The connection between those planets and the days of the week is obvious, however, if we look, for example, at the Italian names for Tuesday, Wednesday, Thursday, and Friday: *Martedi, Mercoledi, Giovedi,* and *Venerdi.*

(b) The Roman Republican Calendar

The roots of our modern calendar go back to the Roman republican calendar, which derives from earlier Roman and Greek calendars dating from at least the eighth century B.C. The earliest Roman calendar probably had 10 months, the last 4 of which have given us the names of our months, September, October, November, and December. But by the first century B.C., 2 additional months, January and February, had been added.

The original Roman calendar was a lunar one in that the months were based on the moon's synodic period; each month began with a new moon. To give the months an average length of 29½ days (the lunar synodic period) the months had 29 and 30 days alternately. The difficulty with the lunar calendar is that 12 lunar months add up to only 354 days, whereas the tropical year has about 365¼ days. After about 3 years, the difference accumulates to a whole month. To keep their year in step with the year of the seasons, the ancients adopted the policy of intercalation — that is, they simply inserted a thirteenth month every third year or so. The normal 12-month years were "empty years" and the 13-month years were "full years."

The Roman republican calendar, in use by about 70 B.C., had 12 months. These months, and their duration in days, were Martius (31); Aprilis (29), Maius (31), Iunius (29), Quintilis (31), Sextilis (29), September (29), October (31), November (29), December (29), Ianuarius (29), and Februarius (28). The year thus had 355 days. From the middle of the second century B.C. January (Ianuarius) 1 officially marked the beginning of the year, although in the popular view the year ended with February 23. When an extra month had to be intercalated every 2 to 4 years to bring the average length of the year to 365¼ days, it was added immediately after February 23. Then followed the last 5 days of February, and March.

Unfortunately, the management of the calendar was left to the discretion of the priests, who greatly abused their authority by declaring as full years those in which their friends were in public office. The intercalation process became such a political football that by the time of the reign of Julius Caesar, a Roman traveler going from town to town could find himself going from year to year! Thus, in 46 B.C., Julius Caesar instigated a calendar reform.

FIG. 8-9   Julius Caesar.
(New York Public Library.)

(c) The Julian Calendar

At the advice of the Alexandrian astronomer Sosigenes, Caesar adopted a new calendar, which had 12 more or less equal months averaging about 30½ days in length rather than 29½. The features of the Julian calendar reform of 46 B.C. were as follows:

(1) The lunar synodic month was abandoned as a basic unit in the calendar. Instead, each year contained 12 months, which contained a total of 365 days. Caesar distributed the 10 extra days among the 12 months, but there appears to be a difference of opinion among historians as to which months originally had how many days.

(2) The calendar was to be based on the tropical year, whose length had at that time been determined to be 365¼ days. Of course, one fourth of a day could not be "tacked on" to the end of the calendar year. Therefore, *common* years were to contain only 365 days. However, after 4 years, this ¼ day per year adds up to 1 full day. Thus every fourth year was to have 366 days (a *leap year*), the extra day being added to February. The *average* length of the year would then be 365¼ days. Note that the process of leap year is analogous to the intercalation of extra months in "empty years" to make them "full years" of 13 months.

(3) To bring the date of the vernal equinox, which had fallen badly out of place in the Roman republican calendar, back to its traditional date of March 25, Caesar intercalated 3 extra months in the year 46 B.C., bringing its length to 445 days. Forty-six B.C. was known as the "year of confusion." The Julian Calendar was introduced, then, on January 1, 45 B.C.

After Caesar's death in 44 B.C., the month Quintilis (the fifth month in the original Roman calendar) was renamed in his honor (thus our name, July). Later, the Roman senate did some further juggling with the Julian calendar. Sextilis (originally the sixth month) was renamed in honor of Augustus Caesar, successor to Julius, and the present number of days for each month resulted.

(d) The Council of Nicaea

The dates of observance of Easter and certain other religious holidays were fixed by order of the Council of Nicaea (Nice) in 325 A.D. Easter, according to the rule adopted, falls on the first Sunday after the fourteenth day of the moon (almost full moon) that occurs on or after March 21. At that time March 21 was the date of the vernal equinox. (The Sunday *after* full moon was specified intentionally to avoid the possibility of an occasional coincidence with the Jewish Passover.)

Note that between 45 B.C. and 325 A.D., the date of the vernal equinox had slipped back from March 25 to March 21. This was because the Julian year, with an average length of 365¼ days, is 11^m14^s longer than the tropical year of $365^d5^h 48^m46^s$. The slight discrepancy had accumulated to just over 3 days in those four centuries.

(e) The Gregorian Calendar

By 1582, that 11 minutes and 14 seconds per year had added up to another 10 days, so that the first day of spring was occurring on March 11. If the trend were allowed to continue, eventually Easter and the related days of observance would be occurring in early winter. Therefore, Pope Gregory XIII instituted a further calendar reform.

The Gregorian calendar reform consisted of two steps. First, 10 days had to be dropped out of the calendar to bring the vernal equinox back to March 21, where it was at the time of the Council of Nicaea. This step was expediently accomplished. By proclamation the day following October 4, 1582, became October 15.

The second feature of the new Gregorian calendar was that the rule for leap year was changed so that the average length of the year would more closely approximate the tropical year. In the Julian calendar, every year divisible by four was a leap year, so that the average year was 365.250000 mean solar days in length. The error between this and the tropical year of 365.242199 mean solar days accumulates to a full day every 128 years. Ideally, therefore, one leap year should be made a common year, thus dropping 1 day, every 128 years. Such a rule, however, is cumbersome.

Instead, Gregory decreed that three out of every four century years, all leap years under the Julian calendar, would be common years henceforth. The rule was that only century years divisible by 400 should be leap years. Thus, 1700, 1800, and 1900, all divisible by 4, and thus leap years in the old Julian calendar, were *not* leap years in the Gregorian calendar. On the other hand, the years 1600, and 2000, both divisible by 400, are leap years under both systems. The average length of this Gregorian year was 365.2425 mean solar days, and was correct to about 1 day in 3300 years.

The Catholic countries immediately put the Gregorian reform into effect, but countries under control of the Eastern Church and most Protestant countries did not adopt it until much later. It was 1752 when England and the American colonies finally made the change. The year 1700 had been a leap year in the Julian calendar but not the Gregorian; thus the discrepancy between the two systems had become 11 days. By parliamentary decree, September 2, 1752, was followed by September 14. Although special laws were passed to prevent such breaches of justice as landlords collecting a full month's rent for September, there were still riots, and people demanded their 11 days back. To make matters worse, in England it had been customary to follow the ancient practice of starting the year on March 25, originally the date of the vernal equinox. In 1752, however, the start of the year was moved back to January 1, so in England and the Colonies 1751 had no months of January and February and had lost 24 days of

March! We celebrate George Washington's Birthday on February 22, 1732, but at the time of his birth, a calendar would have read February 11, 1731. Russia did not abandon the Julian calendar until the time of the Bolshevik revolution. The Russians then had to omit 13 days to come into step with the rest of the world.

The Gregorian calendar has now been modified slightly to come into better conformity with the tropical year: the years 4000, 8000, 12,000, and so on, all leap years in the original Gregorian calendar, are now common years. The calendar is thus accurate to 1 day in about 20,000 years.

At a meeting of the Congress of the Orthodox Oriental Churches at Constantinople in 1923, a slightly improved version of the Gregorian calendar was adopted for the Eastern churches. This Eastern calendar is shorter than the Julian by 7 days every 900 years, rather than 3 days every 400 years. The rule for leap year is that century years, when divided by 900, will be leap years only if the remainder is either 200 or 600. The years 2000 and 2400 will be leap years in both the Gregorian and Eastern Orthodox calendars. The years 2100, 2200, 2300, 2500, 2600, and 2700 will not be leap years in either. The two calendars will not diverge until 2800, which will be a leap year in the Gregorian calendar but not in the Eastern Orthodox. The Eastern Orthodox calendar year has an average length of 365.2422 mean solar days, very nearly that of the tropical year. Its error is only 1 day in 44,000 years.

(f) The Julian Day

The complexities of the calendar can be avoided by the use of the so-called Julian day, abbreviated J.D. The Julian day corresponding to a particular date is simply the number of days that have elapsed since January 1, 4713 B.C. This date was chosen for astrological reasons but is close to the one that in medieval times was believed to have marked the creation of the earth. The Julian Day 2,438,396 began at noon, universal time, January 1, 1964. The Julian date is used rather commonly to refer to astronomical events.

(g) The Mayan Calendar*

Of the various calendar systems of other ancient civilizations, one of the most interesting was that of the Maya, whose civilization, flourishing in the Yucatan area in Central America, was contemporary with the early European civilizations. The Mayan calendar was later adopted, at least in part, by the conquering Aztecs.

The Mayan calendar was more sophisticated and complicated than either the Roman or Julian calendar. Apparently, the Maya did not attempt to correlate their calendar accurately with the length of the year or lunar month. Rather, their calendar was a system for keeping track of the passage of days and for counting time far into the past or future. Among other purposes, their calendar was useful for predicting astronomical events — for example, the positions of Venus in the sky.

The Mayan calendar consisted of three simultaneous systems for counting days. The first was the sacred almanac, called the Tzolkin by modern archaeologists, which was somewhat analogous to our week of 7 named days that recur in specified order perpetually. However, the Tzolkin had 20 named days rather than 7; moreover, each day's name was accompanied by a number. The numbers ran from 1 to 13 and were then repeated. The first day of the sacred almanac was 1 *Imix*, the second 2 *Ik*, and so forth, up to the thirteenth day, which was 13 *Ben*. The next day, *Ix*, was accompanied by the number 1 again, that is, 1 *Ix*. The twentieth day was 7 *Ahau*. Then the day names started over with *Imix* again, but this time *Imix* appeared with the number 8: 8 *Imix*. In other words, the numbers were always out of phase with the day names. After 13×20, or 260, days, 1 *Imix* appeared again, after which the whole series was repeated, and so on, indefinitely. Thus, the Tzolkin was a counting system containing 260 combinations of numbers and names.

The second counting system of the Mayan calendar was a 365-day period that is approximately equal to the year. However, there was no intercalation of extra days, or "leap-year" scheme, so the 365-day period did not remain fixed with respect to the seasons. This 365-day period was divided into 18 *uinals* (analogous to months, but not equal to the moon's period) of 20 days each, with 5 "unlucky" days tacked on as a nineteenth uinal. Each day was numbered according to its position in its uinal; thus the Maya would speak of 17 *Yaxkin*, much as we would say July 23. (Of course, there is no simple correspondence between the dates in our calendar and theirs.) To give both the Tzolkin day name and uinal date, the Maya might say, for example, 7 *Ik* 15 *Yaxkin*; analogously, we might say Thursday, January 11. In our calendar, January 11 can fall on a Thursday every several years (it would be every 7 years if it were not for leap year), but a date in the Mayan calendar, specified like 7 *Ik* 15 *Yaxkin*, occurred exactly once every 18,980 days, or about every 52 years.

Finally, to specify completely a particular date, the Maya made use of what is called the *long count*, a perpetual tally of the days that have elapsed since a particular date about 3000 years in the past. This system is analogous to that of the Julian day, described in Section 8.2f. The starting date, however, was not meant to be that of "the beginning" — it was merely an arbitrary starting point, from which days could be counted. The significant feature of the long count is that it employed

FIG. 8-10   Aztec calendar. *(Mexican National Tourist Council.)*

a vigesimal number system, that is, one based on 20 (rather than 10, as is our decimal system). The useful property of our decimal system is not, particularly, that it is based on the number 10 but that it employs the *zero*, without which arithmetic would be extremely tedious. (If the reader questions this statement, he should try multiplying 53,498 by 627 in Roman numerals.) The Maya made use of the zero in counting and arithmetic and employed a method of place value, analogous to our own method of writing numbers, many centuries before the Arabs introduced the concept to Europe.

(h) Further Calendar Reform

Further reforms proposed for our present calendar are not intended to improve its approximation to the tropical year, but to make it more symmetrical and to arrange that any particular date of the year will always fall on the same day of the week. The two reforms suggested most frequently are the 13-month calendar and the *world calendar.*

The 13-month calendar contains 13 months of 28 days each. Each month begins on a Sunday and ends on a Saturday; therefore, all months are identical. However, $13 \times 28 = 364$, so that it is necessary to add an extra day at the end of the thirteenth month — a year-end holiday or extra Saturday. On leap year, another Saturday or leap-year day would have to be inserted. One feature of this calendar is that it contains 13 days that are Friday the thirteenth.

The world calendar is a less radical departure from our present one. It includes the same 12 months we have now. The four quarters of the year are identical. The first month in each quarter has 31 days and begins on a Sunday. The second and third months of the quarter have 30 days each, and the third month ends on a Saturday. Year-end Day, December Y, follows December 30 every year. Leap Year Day, June L, follows June 30 each leap year. Each month of the world calendar has the same number of week days (counting Saturdays), and, of course, as in the 13-month calendar, any date of the year always falls on the same day of the week. Five of the months are not changed at all from their present form; in no case are the dates of months shifted by more than 2 days from their present arrangement. The world calendar would probably be an advantage for fiscal and budgetary purposes. On the other hand, it has the monotony of Christmas always falling on a Monday.

The world calendar is sponsored by the World Calendar Association, New York. Because it is supported by many organizations throughout the world, this calendar reform probably has a better chance than most of eventually being adopted.

Exercises

1. Suppose the moon were adopted as a reference for time. About how much longer would a "lunar day" be than a sidereal day (for an earthbound observer)?

2. How many more sidereal days per year are there than solar days? Why?

3. (a) On what date, approximately, do sidereal and solar time agree?

(b) On what date, approximately, is sidereal time 4 hours ahead of solar time? (*Hint:* When is the sun on the vernal equinox? When is it halfway around the sky from the vernal equinox? Where must the sun be with respect to the vernal equinox for the sidereal and solar days to begin simultaneously?)

4. If a star rises at 8:30 P.M. tonight, approximately what time will it rise 2 months from now?

5. At New Orleans, Louisiana (longitude 90° W), on February 1, a sundial reads 10:25 A.M. What is the approximate Central Standard Time? (Use Figure 8-6.)

   Answer: 10:38 A.M.

6. At Boston, Massachusetts (longitude 71° W), on November 1, a sundial reads 3:20 P.M. What is the approximate Eastern Standard Time?

   Answer: 2:47 P.M.

7. If it is 1:00 A.M., July 17, at longitude 165° W, what are the time and date at longtitude 165° E?

8. What is the greatest number of Sundays possible in February for the crew of a vessel making weekly sailings from Siberia to Alaska?

   Answer: 10

9. If the local mean time is 2:30 P.M. and the Universal Time is 10:30 A.M., what is the longitude?

10. If the sun is on the meridian on October 10, and the Universal Time is 15:30, what is the longitude? (Ignore, here, the equation of time.)

   Answer: 52½° W

11. Show that the Gregorian calendar will be in error by 1 day in about 3300 years.

12. If the earth were to speed up in its orbit slightly, so that a tropical year were completed in exactly 365^d3^h, how often would we need to have a leap year? What rule, if any, would there be about century years?

Following page: Earthrise from the moon, photographed by Apollo 8 astronauts, December 1968. *(NASA.)*

CHAPTER 9

Aspects and Motions of the Moon; Eclipses

The moon, the earth's nearest celestial neighbor and its only known natural satellite, accompanies the earth in its annual revolution about the sun. Although the moon shines only by reflected sunlight, it is nevertheless the second most brilliant object in the sky, and is one of the three celestial objects (the sun, the moon, and Venus) that can be seen in broad daylight. Many of the ancients regarded the moon as "ruler of the night" just as the sun is "ruler of the day." It was one of the first objects to be studied with the telescope and is the celestial body about which we know the most.

Having examined, in the last two chapters, the motions of the earth and how those motions affect the appearance of the sky, we now turn our attention to the apparent and real motions of the moon and certain other of its aspects. The physical nature of the moon itself, however, is discussed in Chapter 13.

9.1 ASPECTS OF THE MOON

The moon, because of its proximity, appears to move more rapidly in the sky than any other natural astronomical object, except meteors, which are within the earth's atmosphere. The moon appears the same size as the sun in the sky, subtending ½° of arc. Its larger surface features are easily visible to the unaided eye and form the facial markings of the "man in the moon." As it travels about the earth each month, it displays different parts of its daylight hemisphere to our view and progresses through its cycle of phases.

(a) Moonlight

The most conspicuous property of the moon is its light. The amount of light we receive from the moon varies immensely with its phase. When the moon is full, its light is nearly bright enough to read by; we receive only about 10 percent as much

FIG. 9-1   The earth-lit, the full, and the totally eclipsed moon. *(Yerkes Observatory.)*

light from the moon at first and last quarter, and only one thousandth of the light of the full moon when the moon appears as a thin crescent 20° from the sun in the sky.

Despite the brilliance of the full moon, it shines with less than 1/400,000 the light of the sun. Even if the entire visible hemisphere of the sky were packed with full moons, the illumination would be only about one fifth or less of that in bright sunlight.

Because the moon shines by reflected sunlight, we can calculate the moon's reflecting power from its apparent brightness. The moon and earth are at about the same distance from the sun; consequently, the moon receives as much sunlight per square inch of its surface as does the earth. Calculation shows that if all this light were reflected back into space, the full moon would appear about 14 times as bright as it actually is. The fraction of incident light that is reflected by a body is called its *albedo*. The average albedo of the moon is thus about 0.07. The moon absorbs most of the sunlight that falls upon it; its surface is quite dull. The absorbed energy heats up the surface of the moon until the energy is radiated away again as radiant heat.

When the bright part of the moon appears as only a thin crescent, the "night" side of the moon often appears faintly illuminated. Leonardo da Vinci (1452–1519) first explained this illumination as *earthshine*, light reflected by the earth back to the night side of the moon, just as moonlight often illuminates the night side of the earth.

(b) Sidereal and Synodic Months

The moon's sidereal period of revolution about the earth, that is, the period of its revolution with respect to the stars, is $27^d7^h43^m11^s5$ (27.32166 days). However, during this period of the moon's sidereal revolution, the earth and moon together revolve about $\frac{1}{13}$ the way around the sun, or about 27°. The sun, therefore, appears to move 27° to

the east on the celestial sphere during the period of the moon's sidereal revolution. In other words, the moon would not, in its sidereal period, have completed a revolution about the sky with respect to the sun, and consequently would not have completed a cycle of phases. To complete a revolution with respect to the sun, the moon requires, on the average, $29^d12^h44^m2^s8$ (29.53059 days). We have, then, two kinds of month: the *sidereal month,* the period of revolution of the moon with respect to the stars, and the *synodic month,* the period with respect to the sun (Figure 9-2). (Compare with the concept of sidereal and solar days, Chapter 8.)

(c) The Moon's Apparent Path in the Sky

If the moon's position among the stars on the celestial sphere is carefully noted night after night, it is seen that the moon changes its position rather rapidly, moving, on the average, about 13° to the east per day. In fact, during a single evening the moon can be seen visibly creeping eastward among the stars. The moon's apparent path around the celestial sphere is a great circle (or very nearly so) that intersects the sun's path, the ecliptic, at an angle of about 5°.

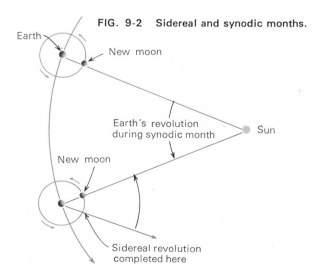

FIG. 9-2 Sidereal and synodic months.

The moon's path intersects the ecliptic at two points on opposite sides of the celestial sphere. These points are called the *nodes* of the moon's orbit. The node at which the moon crosses the ecliptic while moving northward is called the *ascending node,* and the node at which the moon crosses the ecliptic moving southward is the *descending node.*

The moon's orbit is constantly and gradually changing because of perturbations, just as the orbits of artificial satellites change. The most important of the perturbations are caused by the gravitational attraction of the sun. One of the effects of the perturbations on the moon's orbit is that the nodes slide westward along the ecliptic, completing one trip around the celestial sphere in about 18.6 years. This motion is called the moon's *regression of the nodes.* Perturbations also cause the inclination of the moon's orbit to the ecliptic to vary from 4°57′ to 5°20′; the average inclination is 5°9′.

If it were not for the regression of the nodes, the moon's orbit would maintain a nearly fixed angle to the celestial equator. It maintains a nearly fixed angle of about 5° to the ecliptic, but because the nodes constantly shift, its angle of inclination to the equator varies from 23½ + 5°, or about 28½°, to 23½ − 5°, or about 18½° (23½° is the inclination of the ecliptic to the celestial equator).

(d) Delay in Moonrise from Day to Day

We have seen that the moon's average eastward motion with respect to the stars is about 13° per day. The sun, on the other hand, apparently moves to the east about 1° per day. With respect to the sun, therefore, the moon moves eastward about 12° per day. As the earth turns on its axis, the moon, like other celestial objects, appears to rise in the east, move across the sky, and set in the west. But because of its daily eastward motion on the celestial sphere, it crosses the local meridian each day about 50 minutes later, on the average, than on the previous day. We could define this interval of 24^h50^m

(see also Section 8.1) as the average length of an apparent lunar day.

Conditions similar to those that cause the length of an apparent solar day to vary also cause apparent lunar days to vary in length. The moon's true orbit is an ellipse; the moon's orbital speed consequently varies, in accordance with Kepler's law of areas. The moon's eastward progress in its orbit is therefore not constant. Moreover, since the moon's orbit is inclined to the celestial equator, the moon's *eastward* progress in the celestial sphere (that is, the projection of the lunar motion on the celestial equator) would not always be uniform even if the moon did move uniformly in its orbit. The daily retardation in successive transits of the moon across the local meridian ranges from 38 to 66 minutes.

Moonrise (and moonset) is similarly retarded from day to day. If the moon did not move with respect to the sun, it would rise at nearly the same time from one day to the next. At moonrise, the moon occupies some particular place on the celestial sphere. Approximately 24 hours later, the same place on the celestial sphere rises again, but the moon in the meantime has moved off to the east, so moonrise does not occur until a little later. At the equator, the daily delay is the same as the moon's delay in crossing the meridian. However, at other latitudes the moon and stars rise obliquely to the horizon, rather than in a direction perpendicular to it. Consequently, the time required for the earth to turn the sky westward through the angle representing the moon's eastward motion is not necessarily the same as the daily delay in moonrise. The phenomenon of the harvest moon, discussed below, provides an excellent example. In the northern parts of the United States, the daily delay in moonrise can vary from a few minutes to well over 1 hour.

(e) The "Harvest Moon"

The *harvest moon* is the full moon that occurs nearest the autumnal equinox. Because the moon, when full, is opposite the sun in the sky, it must rise as the sun sets. When the sun is near the autumnal equinox, the full moon is near the vernal equinox, so at the time of the harvest moon the vernal equinox is rising with the full moon. When the vernal equinox is rising, the ecliptic makes its minimum angle with the horizon for an observer in intermediate northern latitudes.

Since the moon's orbit lies within 5° of the ecliptic, it is evident that at the same hour on successive nights the moon's apparent motion, being nearly parallel to the horizon, will not change the moon's relation to the eastern horizon appreciably. In Figure 9-3, the full moon (position 1) is shown rising at sunset. At the same time on the next night it has moved about 12° along its orbit; but it is not very far below the horizon and will rise by moving along the line AB, parallel to the celestial equator. The earth will not have to turn far to bring up the moon on this second night. Thus, for several nights near full moon in late September or early October there will be bright moonlight in the early evening — a traditional aid to harvesters. The phenomenon of the harvest moon is most striking in far northern latitudes.

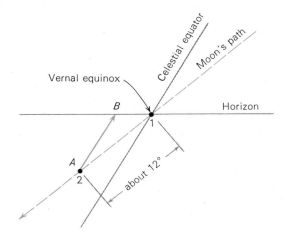

FIG. 9-3 The harvest moon.

(f) The Progression of the Moon's Phases

The phases of the moon (explained in Section 2.2d) were thoroughly understood by the ancients. The relation between any phase of the moon and the moon's corresponding position in the sky at any time of day should now be clear. Except at extreme northern or southern latitudes, one can easily tell where to look for the moon in the sky, if he knows its phase, from a consideration of Figure 9-4.

In Figure 9-4 we imagine ourselves looking down upon the earth and the moon's orbit from the north. The moon is shown in several positions in its monthly circuit of the earth: 1, 2, 3, and so on. The sun is off to the right of the figure at a distance so great that its rays approach the earth and all parts of the moon's orbit along essentially parallel paths. The daylight sides of the earth and moon — the sides of those bodies turned toward the sun — are indicated. For each position of the moon, its phase, that is, its appearance *as viewed from the earth,* is shown just outside its orbit. Several observers are at various places on the earth, *A, B, C,* and so on. The time of day, indicated for each observer, depends on the position of the sun in the sky with respect to his local meridian or, equivalently, on his position on the earth with respect to the meridian, where it is noon.

For person *A* it is 3:00 P.M. If he sees the moon on his meridian, it must be in the waxing crescent phase (the moon can be seen easily at noon when in this phase). If the moon is in the waning crescent phase it is setting, for it lies on his western horizon. West is the direction away from which the turning earth carries the observer, and his horizon lies in a plane tangent to the surface of the earth at the point where he is standing. If the moon is new it is in about the same direction as the sun in the western sky, and if it is at first quarter it is in the eastern sky. If the moon is rising it must be in the waxing gibbous phase.

For person *B* it is 6:00 P.M. (Person *B* could be person *A* 3 hours later. During a period of even

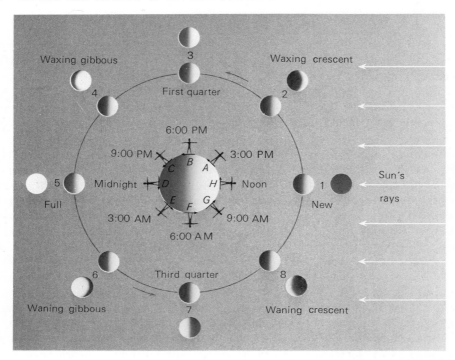

FIG. 9-4 Phases of the moon and the time of day. (The outer series of figures shows the moon at various phases as seen in the sky from the earth's surface.)

a full day the moon does not move enough for its phase to change appreciably.) For *B* the moon is setting if new, and rising if full. If it is a waxing crescent or gibbous it is in the western or eastern sky, respectively. If it is in the first quarter phase, it appears on the meridian.

For person *D* it is midnight. If the moon is full, it rose at sunset, and is now on the meridian. The first or third quarter moon is just setting or rising, respectively. The waxing or waning gibbous moons must appear in the western or eastern sky.

By studying Figure 9-4 we can see that as the moon goes about the earth during the month and gradually changes its phase, its approximate position in the sky is completely determined for any time of day or night. For example, the full moon rises at sunset and sets at sunrise. The first quarter moon rises at noon and sets at midnight, and so on.

Let us take an example. Suppose the waning crescent moon, 30° west of the sun, is observed to rise. We desire to find the time of day. First, we

draw a picture of the earth and indicate the direction of the sun and the earth's daylight side (Figure 9-5). Next we draw in the moon 30° west of the sun. To see the moon rising, it must be in the direction of the eastern horizon. (We have exaggerated the size of the earth in Figures 9-4 and 9-5. Actually the direction along the horizon to the moon is very nearly parallel to the line from the center of the earth to the moon.) The only place where that is possible is for the observer shown, where the time is about 4:00 A.M.†

In the foregoing discussion we have ignored the dependence of the moon's position on the latitude of the observer. However, for the most places

---

†Suppose your best beau told you how beautiful the moon was as it rose the other "night," and that you know the moon was a waning crescent that couldn't have risen until about 4:00 A.M. If you learn something of the habits of the moon, you may learn something of the habits of your best beau!

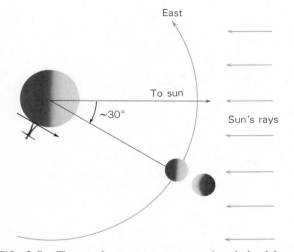

FIG. 9-5   The waning crescent moon when it is rising.

on earth the scheme outlined illustrates well enough where one should look for the moon in the sky.

(g) Configurations of the Moon

The configurations of the moon are named like those of the planets. The first or third quarter moon is at *quadrature*. The full moon is at *opposition,* and the new moon is at *conjunction.*†

(h) The Rotation of the Moon

Even naked-eye observation is sufficient to determine that as the moon goes about the earth it keeps the same side toward the earth. The same facial characteristics of the "man in the moon" are always turned to our view. Because the moon always presents the same side to us, it is sometimes said not to rotate on its axis. However, this statement is incorrect. In Figure 9-6 the arrow on the moon represents some lunar feature. If the moon did not rotate, as in (a), we would see that feature part of the time, and part of the time we would see the other side of the moon. Actually the moon rotates on its axis with respect to the stars in the same period as it revolves about the earth, and so always turns the same side toward us (b).

_____

†Either full or new moon is also called *syzygy.* It is a term seldom used by astronomers, but it is often encountered in crossword puzzles.

FIG. 9-6   If the moon did not rotate, (a) it would turn all its sides to our view; actually it does rotate, (b) in the same period as it revolves, so we always see the same side.

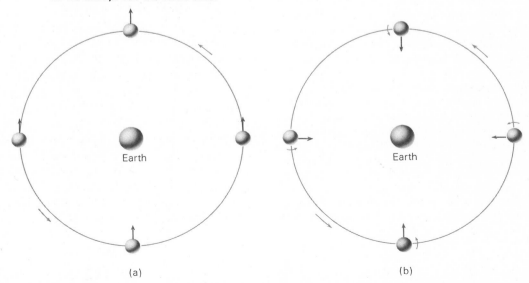

(a)   (b)

The coincidence of the periods of the moon's rotation and revolution can hardly be accidental. It is believed to have come about as a result of the earth's tidal forces on the moon (Section 6.4).

We often hear the back side of the moon (the side we do not see) called the "dark side." Of course, the back side is dark no more frequently than the front side. Since the moon rotates, the sun rises and sets on all sides of the moon. The back side of the moon is receiving full daylight at new moon; the dark side is then turned toward the earth.

(i) Librations*

Actually, for several reasons, the moon does not always keep exactly the same face turned toward a particular terrestrial observer, but presents a little more than half its surface to our view over a period of time. The motions of the moon and earth that allow us to see slightly different parts of its surface at different times are called *librations*. Librations were first observed and explained by Galileo.

There are three librations that are geometrical in character. These *geometrical librations* are libration in longitude, libration in latitude, and diurnal libration.

Libration in longitude is a consequence of the fact that while the moon *rotates* at a uniform rate, its angular velocity in its elliptical orbit is variable. Therefore, although the moon rotates and revolves at the same *average* rate, the two motions get slightly out of step with each other. As a consequence, we on earth are able to see, as the moon moves, first a little around its west side and then a little around its east side. The extra amount of the moon that we can observe ranges up to nearly 8° either way.

Libration in latitude occurs because the moon's axis of rotation, like the earth's, is not exactly perpendicular to the plane of its orbit, but is tilted at an angle of about 6½°. The orientation of the moon's axis remains constant during its revolution about the earth, just as the direction of the earth's axis remains constant during the earth's revolution about the sun. This results in something quite like the way in which the sun's rays alternately fall upon the earth's north and south poles during the course of the year. In the case of the moon, we on earth

see as much as 6½° beyond the moon's north pole when it is on one side of its orbit, and the same amount beyond the moon's south pole 2 weeks later.

Diurnal libration is caused by the rotation of the earth carrying the observer back and forth over a base line of several thousand miles. When we see the moon rising we are looking at it from a point some 4000 mi to one side of the earth's center, and can see farther around the moon's western "edge" than when the moon is on the meridian. At moonset, our vantage point has shifted 4000 mi in the opposite direction, and we see a little around the moon's eastern "edge." This libration amounts to about 1° in either direction.

Because of the geometrical librations, only 41 percent of the moon's surface is never presented to the earth; 41 per cent is always presented, and 18 percent is alternately visible and invisible.

The moon also undergoes very slight *physical librations,* which are actual irregularities in its rotation. These irregularities arise because the moon is not perfectly spherical but is technically a *triaxial ellipsoid,* that is, it has three different diameters in three mutually perpendicular directions. The longest diameter of the moon is directed toward the earth, and the earth's tidal attraction on this "bulge" perturbs the moon's rotation, causing it to oscillate slightly to and fro. Despite the moon's triaxial shape, it is more nearly a perfect sphere than the earth; its maximum and minimum diameters differ by less than 2 miles. The physical librations, therefore, are very small, and displace features on the moon's surface from their mean positions by only a mile or so.

9.2 THE MOON'S DISTANCE AND SIZE

The moon's distance is only about 30 times the diameter of the earth; consequently, the direction of the moon differs slightly as seen from various places on the earth. We have seen (Section 2.3d) how Ptolemy made use of this principle to determine geometrically the distance of the moon. Modern methods of geometrically determining the moon's distance are essentially the same as Ptolemy's. Today, however, the moon's distance can be determined more accurately by other methods.

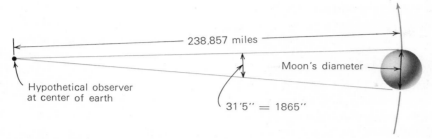

FIG. 9-7   Measuring the moon's diameter.

(a) Radar Measures of the Moon's Distance

The technique of radar is that of transmitting radio waves to a distant object and then detecting those waves that are reflected by that body back to the original source. The direction from which the reflected waves come indicates the direction of the body that reflects them. The time the waves spend making the round trip (the speed of transmission of the radiation being known) indicates the distance of that object.

We know the moon's direction from optical observations, and so are interested only in the time required for the radar signals to travel to the moon and back. The radar signals, being radio waves (a form of electromagnetic radiation), travel with the speed of light — an accurately known quantity. The interval of time between the instant that a wave is broadcast to the moon and the instant it returns can be measured electronically to within a few millionths of a second. The distance to the moon is, of course, the speed of the waves multiplied by half the time required for the round trip.

The first radar contact with the moon was achieved by the United States Army Signal Corps in 1946. A very accurate determination of the moon's distance by this technique was made by the Naval Research Laboratory in 1957. The latter measure gives for the mean distance from the center of the earth to the center of the moon the value 384,403 km, or 238,857 mi. The NRL scientists consider their measure correct to within about 1 mi.

(b) The Moon's Diameter

The mean angular diameter of the moon is 31' 5''. From its angular size and its distance, the linear size of the moon can easily be found. Because the method is the same as that applied to measure the diameters of planets and other astronomical objects that subtend a measurable angle, we shall explain the procedure in detail.

Notice in Figure 9-7 that because the moon's angular size is relatively small, its linear diameter is essentially a small arc of a circle, with the observer as center and with a radius equal to the moon's distance. Obviously, the moon's diameter is the same fraction of a complete circle as the angle subtended by the moon is of 360°. A complete circle contains 1,296,000'' (there are 60'' per minute, 60' per degree, and 360° in a circle). As seen from the center of the earth, the moon's mean angular diameter of 31'5'', or 1865'', is thus $\frac{1}{695}$ of a circle. The moon's diameter, therefore, is $\frac{1}{695}$ of the circumference of a circle of radius 238,857 mi. Since the circumference of a circle is 2π times its radius, we have

$$\text{diameter of moon} = \frac{2\pi(238,857)}{695} = 2160 \text{ mi.}$$

This type of calculation can be generalized if we note that the distance along the arc of a circle of radius R subtended by 1'' is $2\pi R/1,296,000 = R/206,265$. Thus, if an object, say a planet, subtends an angle of α seconds, and has a distance D, its linear diameter d is given by

$$d = \frac{\alpha D}{206,265}.$$

Accurate calculations of the moon's diameter give as a result 3475.9 km, or 2159.86 mi, with an uncertainty of a few hundredths of a mile. This is the moon's equatorial diameter in a direction perpendicular to its direction from the earth. We have already seen (Section 9.1i) that the diameter differs slightly in different directions. The moon's diameter is a little over a quarter that of the earth's, and its volume is about $\frac{1}{49}$ of the earth's.

In practice, of course, we do not observe the moon from the center of the earth but from its surface. Both the angular size and direction of the moon differ for different observers on the earth. However, because the moon's distance and the size of the earth are known, the measured angular size of the moon can easily be corrected to the value it would have if the moon were viewed from the earth's center. The standard reference tables, such as the *Astronomical Ephemeris,* tabulate the moon's position in the sky, its distance, and its angular diameter as seen from the earth's center. To use these tables the astronomer or navigator must correct the tabulated values to those appropriate to his nongeocentric position. To aid in these corrections, the tables also give the moon's *horizontal parallax,* the angle at the moon subtended by the equatorial radius of the earth.

9.3 THE TRUE ORBIT OF THE MOON

When the moon's distance is measured during different times of the month, it is found to vary by more than 10 percent. In obedience to Kepler's first law, the moon's orbit, basically, is an ellipse with the earth at one focus. The sun, however, produces strong perturbations on the moon, and the elements of its orbit can be stated only in an average sense. Indeed, the moon's orbit changes so rapidly that if the moon's positions over a month, relative to the center of the earth, are plotted carefully, even on a sheet of standard typing paper, the orbit is seen to not close on itself. The celestial mechanics of the moon's motion (lunar theory) is very complex; here we can only mention briefly a few of these complexities.

It is easy to visualize one of the effects of the sun's perturbations. The sun's tidal force on the earth-moon system tends, on the average, to separate the two. At new moon, the sun pulls on the moon more strongly than on the earth, tending to pull them apart, and at full moon the sun's greater pull on the earth has the same result. At first and third quarters the sun tends to converge the earth and moon together slightly (because the sun is in slightly different directions as seen from the earth and moon), but this effect is weaker than the tendency to separate the two at full and new moon. Consequently the month is about 53 minutes longer than it would be in the sun's absence.

The tidal force of the sun also acts to speed up and slow down the moon in different parts of its orbit. The net result of these accelerations is to cause the *line of apsides* (the major axis of the moon's orbit) to rotate toward the east in the orbital plane in a period of 8.85 years. Thus the position in the sky of the moon's *perigee* (closest approach of the moon to the earth) advances to the east about the sky in this period.

The plane of the moon's orbit is inclined at about 5°8' to the ecliptic plane, and one component of the sun's force on the moon is toward the ecliptic plane, thus trying to pull the moon into it. Analogous to precession, however, the moon's orbit does not tilt back into the plane of the ecliptic, but rather the line of nodes regresses (that is, the nodes slide westward), moving around the ecliptic in 18.6 years.

The solar perturbations on the moon's motion depend critically on the average distance of the sun. Because of planetary perturbations on the earth the eccentricity of the earth's orbit slowly oscillates in value. In the present era the eccentricity is decreasing, resulting in a slight and slow increase in the mean distance of the sun. One effect is to shorten slightly the length of the month. The effect was discovered by Edmund Halley in the course of his investigation of ancient eclipse records. The explanation was later supplied by Laplace.

Because of the above (and other) perturbations on the moon, the eccentricity of its orbit changes even during 1 month. The mean value, over many years, is 0.0549. The moon, however, may come as close as 221,463 mi from the earth's center and pass as far from it as 252,710 mi. At its farthest, however, the moon is still only about $\frac{1}{370}$ as far as the sun, and its orbital speed of about $\frac{3}{5}$ mi/sec is only about $\frac{1}{30}$ of the earth's orbital speed about the sun. Thus, if the moon's actual path with respect to the sun is plotted, it is seen to vary only minutely from the earth's orbit; in fact, the moon's orbit is always concave to the sun.

The rapid changes in the moon's orbit make it useful to define two additional kinds of months, which are especially useful in the prediction of eclipses. They are the *nodical* or *draconic month* of 27.2122 days, the time required for the moon to make two successive passages of the same node, and the *anomalistic month* of 27.555 days, the time between two successive perigee passages.

(a) Mutual Revolution of the Earth-Moon System

We saw in Sections 4.4c and 4.4e that one body does not strictly revolve about another, but that the two bodies mutually revolve about their center of mass, or the *barycenter*. It is the barycenter of the earth-moon system that revolves annually in an elliptical orbit about the sun, while the earth and moon simultaneously revolve about the barycenter in a shorter period — the sidereal month.

The elliptical orbit of the center of the earth about the barycenter constitutes an independent

motion of the earth. The motion can be detected by careful observations of the nearer planets, or better yet, of near-approaching minor planets (Chapter 15). The motion of Mars, for example, shows monthly oscillations, carrying it a little ahead and then a little behind its regular orbital motion. This oscillation is only apparent and results from the motion of the earth, carrying us first to one side and then to the other side of the barycenter. When Mars is at its closest, the apparent displacements caused by the orbital motion of the earth around the barycenter amount to about 17″. The corresponding mean distance of the center of the earth from the barycenter is 2903 mi. Thus, the earth and moon jointly revolve about a point approximately 1000 mi below the surface of the earth.

(b) The Mass of the Moon

In Section 5.1 we saw that the distances of two bodies from their barycenter are inversely proportional to their masses. The 2903-mi distance of the earth from the barycenter is 1/82.3 of the distance from the earth to the moon; hence the moon is 81.3 times as far from the barycenter as the earth and so is only 1/81.3 as massive. Because the earth's mass is 6.6×10^{21} tons (Section 4.3e), the moon's mass is $(6.6 \times 10^{21})/81.3 = 8.1 \times 10^{19}$ English tons (7.35×10^{19} metric tons).

The best modern determinations of the moon's mass are obtained by measuring the accelerations the moon produces on space probes — either those sent to the moon itself, or that pass it by on interplanetary missions. Analysis of the motion of the Mariner V probe, which passed about 6000 mi from Venus on October 19, 1967, gives for the earth-moon mass ratio the value 81.3004 ± 0.0007; this is the best determination at the time of writing.

(c) *Friction in the Tides and Tidal Evolution*

One force that is believed to affect, over long times, the mutual revolution of the earth-moon system is friction in the tides. As the earth rotates, the tidal waters are continually flowing back and forth over each other, across ocean floors, and in and out over coastal shallows. The resulting friction within the water and between the water and the solid earth draws a considerable amount of energy from the kinetic energy of rotation of the earth and expends this energy in the form of heat. Even friction in the tidal distortions of the solid earth and in atmospheric tides may play a role. The earth, consequently, is slowing in its rate of rotation, and the day is gradually lengthening at the rate of about 0.0016 second per century. The continuous dissipation of the rotational energy of the earth is calculated to be approximately 2 billion horsepower.

The slow increase in the length of the day is very small compared to the short-period fluctuations in the earth's rotation that were described in Section 8.1k; however, those rapid changes in the earth's rotation are *periodic,* which means that at the end of each cycle of changes the earth returns to its original rotation rate. The slow increase in the length of the day due to tidal friction, on the other hand, is a *secular* change, which means that the day continues to lengthen as time goes on, and the changes accumulate until they become noticeable. Suppose a clock keeping accurate mean solar time had been started 2000 years ago. Today that clock would be out of step with the position of the sun in the sky by about 3 hours. This "clock error" was discovered by comparing the locations on earth where ancient eclipses actually occurred and where they would be predicted to have occurred if the earth were not slowing. Those locations depend on the exact times the eclipses occurred. Modern eclipses, predicted from the times of ancient ones, consistently take place too early. A slowing of the rotation of the earth, and a consequent increase in our unit of time, is the only explanation consistent with gravitational theory.

It can be shown that whereas the earth slows down in its axial spin as it loses kinetic energy, the angular momentum of the earth-moon system must be conserved. According to a theory of tidal evolution worked out by Sir George Darwin (son of the naturalist), the moon must slowly spiral outward away from the earth, thus maintaining the constancy of the total angular momentum of the earth and moon. (But this is superimposed on a temporary *decrease* of the moon's distance caused by a lowering of the eccentricity of the earth's orbit.) As the

moon's distance from the earth increases, its period of revolution must also increase and its orbital speed must decrease, in accord with Kepler's third law. The day and the month, in other words, will both lengthen. Eventually the day will catch up with the month in length, when both require about 47 of our present days.

With the day and month of equal duration, the earth and moon will present the same faces to each other. The earth, therefore, will no longer rotate under the tidal bulge produced by the moon, and friction due to lunar tides will have ceased. The solar tides, however, will still be at work. The small friction due to these remaining solar tides on the earth will cause the day to lengthen still more, so that it will then be *longer* than the month. The friction of the lunar tides will consequently return and this time will tend to accelerate the earth's rotation rather than retard it. The friction of solar and lunar tides, in other words, will work against each other. According to the theoretical predictions, the angular momentum associated with the mutual revolution and rotation within the earth-moon system can then be gradually transferred to that associated with the orbital motion of the system as a whole about the sun. As a result, the moon should spiral back toward the earth again.

It has been suggested that eventually, if this course of evolution should proceed to its ultimate end, the strong tidal forces of the earth on the moon would tear the moon apart, leaving a ring of particles revolving about the earth, like the rings of Saturn. It must be emphasized that these changes are exceedingly slow, and even if the predictions and assumptions on which they are based are correct, the entire cycle involves billions of years. It is doubtful, to say the least, that the earth and moon will remain in their present forms long enough for such a course of evolution to take place.

9.4 SHADOWS AND ECLIPSES

Important phenomena associated with the motions of the earth-moon system are eclipses. Eclipses occur whenever any part of either the earth or the moon enters the shadow of the other. When the moon's shadow strikes the earth, people on earth within that shadow see the sun covered at least partially by the moon; that is, they witness a *solar*

FIG. 9-8 Shadows in light from a point source.

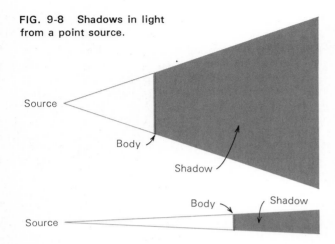

eclipse. When the moon passes into the shadow of the earth, people on the night side of the earth see the moon darken — a *lunar eclipse*.

A shadow is a region of space within which rays from a source of light are obstructed by an opaque body. Ordinarily, a shadow is not visible. Only when some opaque material, which will show the contrast between lighted and unlighted areas, intersects the shadow and the surrounding area does the shadow become visible.

(a) Shadow from a Point Source

If a source of illumination is a point source, the shadow cast by an opaque body has sharp boundaries. From a point inside the shadow, the light source is not visible; outside the shadow it is. The boundaries of the shadow diverge radially from the source. The more distant the source, the smaller is the angle of divergence. If the source is infinitely distant, the boundaries of the shadow are parallel (see Figure 9-8).

(b) Shadow from an Extended Source

If a light source is not a point source but presents a finite angular size as seen from the opaque body, as is almost always the case, the shadow cast by the body is not limited to the inner part of the shadow, the *umbra*, where complete darkness prevails. From within the umbra, of course, no part of the light source is visible. At any point completely outside the shadow, there is no obscuration of light, and from such places the entire light source is visible. Between the umbra and the region of full light lies a space of partial illumination, within which the illumination ranges from complete darkness at the boundary of the umbra to full illumination at the outer boundary of the entire shadow. This transition region of the shadow is called the *penumbra*. From any point within it, a part, but not all, of the light source is visible.

As an illustration, consider the shadow cast by a spherical body, such as the earth, moon, or a planet, in sunlight (Figure 9-9). It is obvious from

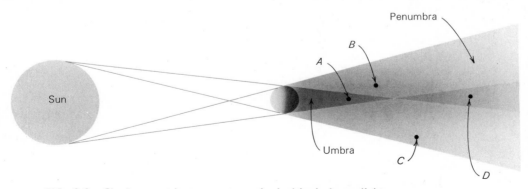

FIG. 9-9 Shadow cast by an opaque spherical body in sunlight.

the figure that because the umbra includes that region of space from which no part of the sun is visible, it must have the shape of a cone, pointing away from the sun. The umbra of the moon or a planet is sometimes called the *shadow cone* of that body. Everybody on the night side of the earth is within the umbra, or shadow cone, of the earth.

The penumbra, on the other hand, is that region from any point within which only part of the sun is covered by the eclipsing body. It is clear from the figure that the penumbra has the shape

of a truncated cone pointed *toward* the sun, and that it includes the umbra, as a reversed cone symmetrical about the same axis. The appearance of the sun, as seen from points *A, B, C,* and *D* in the shadow, is shown in Figure 9-10.

The size of the umbra cast by an opaque sphere in sunlight depends on the size of the sphere and its distance from the sun. The length of the umbra is directly proportional to the distance of the sphere from the sun. Figure 9-11 illustrates this relation.

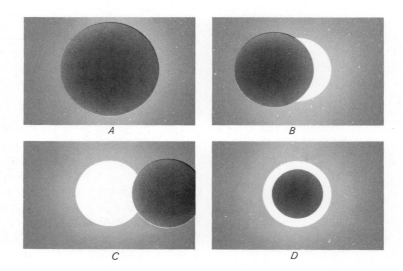

FIG. 9-10   Appearance of the sun from various parts of a shadow.

FIG. 9-11   Shadow lengths at various distances from the sun.

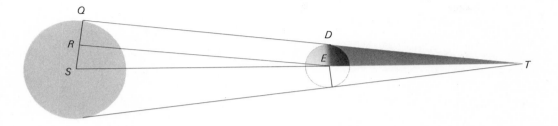

FIG. 9-12   Calculation of a shadow length.

(c) The Lengths of Shadows*

The length of the shadow cast by a sphere in sunlight is easily computed; the principle is illustrated in Figure 9-12. The edge of the shadow *DT* is an extension of line *QD,* which is tangent to both the sun and the sphere. We construct line *RE,* parallel to *QD,* and note that the two triangles *SRE* and *EDT* are similar. From this fact, the equation

$$\frac{ET}{SE} = \frac{ED}{SR}$$

follows directly. Line *SE* is the distance between the centers of the sphere and sun. Line *ED* is the radius of the sphere, and line *SR* is the difference between the radii of the sun and the sphere. Line *ET,* of course, is the desired shadow length. Thus, we have

$$\text{shadow length} = \frac{\text{distance from sun} \times \text{radius of sphere}}{\text{radius of sun} - \text{radius of sphere}}$$

As an example, let us calculate the average length of the earth's shadow. The radii of the sun and earth are 432,000 and 3963 mi, respectively, and the mean distance of the sun is about 93 million mi. We have, therefore,

$$\text{length of earth's shadow} = \frac{93,000,000 \times 3963}{432,000 - 3963}$$
$$= 860,000 \text{ mi.}$$

(d) Eclipse Seasons

For the moon to appear to cover the sun and thus to produce a solar eclipse, it must be in the same direction as the sun in the sky, that is, it must be at the *new* phase. For the moon to enter the earth's shadow and produce a lunar eclipse, it must be opposite the sun; that is, it must be at the *full* phase. Eclipses occur, therefore, only at new moon and at full moon. If the orbit of the moon about the earth lay exactly in the plane of the earth's orbit about the sun — in the ecliptic — an eclipse of the sun would occur at every new moon and a lunar eclipse at every full moon. However, because the moon's orbit is inclined at about 5° to the ecliptic, the new moon, in most cases, is not *exactly* in line with the sun, but is a little to the north or to the south of the sun in the sky. Similarly, the full moon usually passes a little south or north of the earth's shadow.

However, if full or new moon occurs when the moon is at or near one of the *nodes* of its orbit (where its orbit intercepts the ecliptic), an eclipse can occur. The line through the center of the earth that connects the nodes of the moon's orbit is called the *line of nodes.* If the direction of the sun lies along, or nearly along, the line of nodes, new or full moon occurs when the moon is near a node, and an eclipse results. The situation is illustrated in Figure 9-13. The orientation of the moon's orbit, and the line of nodes, *nn',* remains relatively fixed during a revolution of the earth about the sun. There are, therefore, just two places in the earth's orbit, points *A* and *B,* where the sun's direction lies along the line of nodes. It is only during the

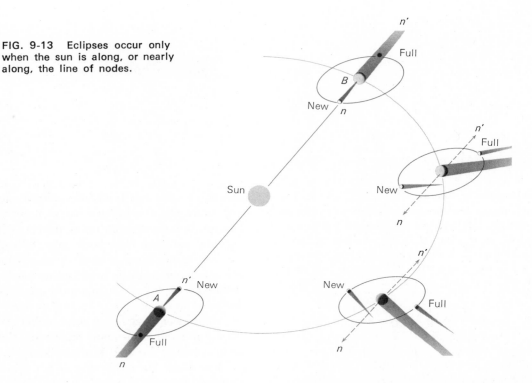

FIG. 9-13 Eclipses occur only when the sun is along, or nearly along, the line of nodes.

times in the year, roughly 6 months apart, when the earth-sun line is approximately along the line of nodes that eclipses can occur. These times are called *eclipse seasons.*

We have seen (Section 9.3) that because of perturbations of the moon's orbit, the line of nodes is gradually moving westward in the ecliptic, making one complete circuit in 18.6 years. Therefore, the eclipse seasons occur earlier each year by about 20 days. In 1970 the eclipse seasons are near February and August; in 1975 they are near November and May.

9.5 ECLIPSES OF THE SUN
One of the most surprising coincidences of nature is that the two most prominent astronomical objects, the sun and the moon, have so nearly the same apparent size in the sky. Although the sun is about 400 times as large in diameter as the moon, it is also

about 400 times farther away, so both the sun and moon subtend about the same angle — about ½°.

The apparent or angular sizes of both sun and moon vary slightly from time to time, as their respective distances from the earth vary. The average angular diameter of the sun (as seen from the center of the earth) is 31′ 59″, and the average angular diameter of the moon is slightly less, 31′ 5″. However, the sun's apparent size can vary from the mean by about 1.7 percent and the moon's by 7 percent. The maximum apparent size of the moon is 33′ 16″, larger than the sun's apparent size, even at its largest. Therefore, if an eclipse of the sun occurs when the moon is somewhat nearer than its average distance, the moon can completely hide the sun, producing a *total solar eclipse.* An equivalent way of stating it is to say that a total eclipse of the sun occurs whenever the umbra of the moon's shadow reaches the surface of the earth.

(a) Geometry of a Total Solar Eclipse

The geometry of a total solar eclipse is illustrated in Figure 9-14. The earth must be at a position in its orbit such that the direction of the sun is nearly along the line of nodes of the moon's orbit. Furthermore, the moon must be at a distance from the surface of the earth that is less than the length of the umbra of the moon's shadow. Then, at new moon, the moon's umbra intersects the ground at a point X on the earth's surface. Anyone on the earth at X (within the small area covered by the moon's umbra) will not see the sun and will witness a total eclipse. The moon's penumbra, on the other hand, covers a larger area of the earth's surface. Any person within the penumbra will see part but not all of the sun eclipsed by the moon — a partial solar eclipse. The regions of total and partial eclipse correspond to points A and B in Figures 9-9 and 9-10.

As the moon moves eastward in its orbit at about 2400 mi/hr, its shadow sweeps eastward across the earth at the same speed. The earth, however, is rotating eastward at the same time, so the speed of the shadow with respect to a particular place on earth is less than 2400 mi/hr. At the equator, where the rotation of the earth carries places eastward at about 1040 mi/hr, the shadow moves relative to the earth with a speed of about 1060 mi/hr. In higher latitudes the speed is greater. In any case, the tip of the truncated cone of the umbra of the moon's shadow sweeps along a thin band across the surface of the earth, and the total solar eclipse is observed successively along this band (refer to Figure 9-14). This path across the earth within which a total solar eclipse is visible (weather permitting) is called the *eclipse path*. Within a zone about 2000 mi on either side of the eclipse path, a partial solar eclipse is visible — the observer, inside this limit, being located in the penumbra of the shadow.

Because the moon's umbra just barely reaches the earth, the width of the eclipse path, within which a total eclipse can be seen, is very small. Under the most favorable conditions, the path is only 167 mi wide in regions near the earth's equator. At far northern or southern latitudes, because the moon's shadow falls obliquely on the ground, it can cover a path somewhat more than 167 mi wide.

It does not take long for the moon's umbra to sweep past a given point on earth. The duration of totality may be only a brief instant. It can never exceed about 7½ minutes.

(b) Appearance of a Total Solar Eclipse

A total solar eclipse is one of the most spectacular of natural phenomena. If a person is anywhere near the path of totality of a solar eclipse, it is well worth his while to move into the eclipse path so that he may witness this rare and impressive event.

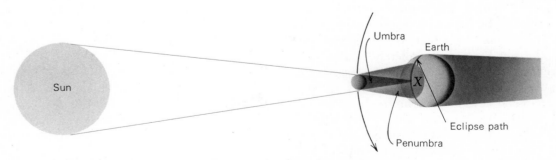

FIG. 9-14   Geometry of a total solar eclipse (not to scale).

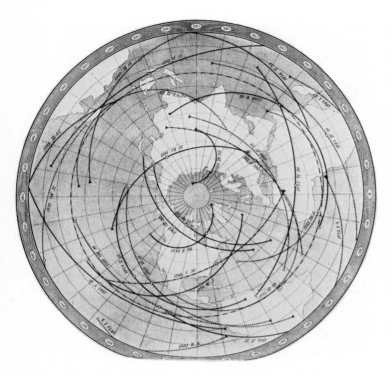

FIG. 9-15   Eclipse path of total solar eclipses occurring between 1963 and 1984. *(Yerkes Observatory.)*

The very beginning of a solar eclipse is the *first contact,* when the moon just begins to silhouette itself against the edge of the sun's disk. The *partial phase* of the eclipse is the period following first contact, during which more and more of the sun is covered by the moon. *Second contact* occurs from 1 to 2 hours after first contact, at the instant when the sun becomes completely hidden behind the moon. In the few minutes before second contact (the beginning of totality) the sky noticeably darkens; some flowers close up, and chickens may go to roost. Because the diminished light that reaches the earth must come solely from the edge of the sun's disk, and consequently from the higher layers in its atmosphere (see Chapter 24), the sky and landscape take on strange colors. In the last instant before totality, the only parts of the sun that are visible are those that shine through the lower valleys in the moon's irregular profile and line up along the periphery of the advancing edge of the moon — a phenomenon called *Baily's beads.* During totality, the sky is quite dark and the brighter stars are visible.

As Baily's beads disappear and the bright disk of the sun becomes entirely hidden behind the moon, the *corona* flashes into view. The corona is the sun's outer tenuous atmosphere, consisting of sparse gases that extend for millions of miles in all directions from the apparent surface of the sun. It is ordinarily not visible because the light of the corona is very feeble compared to that from the underlying layers of the sun that radiate most of the solar energy into space. Only when the brilliant glare from the sun's visible disk is blotted out by the moon during a total eclipse is the pearly white corona, the sun's outer extension, visible. Recently, however, it has become possible to photograph the inner, brighter, part of the corona with an instrument called a coronagraph, a telescope in which a black disk in the telescope's focal plane produces an artificial eclipse, enabling at least the brighter part of the corona to be studied at any time.

Also, during a total solar eclipse, the chromosphere can be observed — the layer of gases just above the sun's visible surface. Prominences, great jets of gas extending above the sun's surface, are sometimes viewed. These outer parts of the sun's atmosphere are discussed more completely in Chapter 24.

The total phase of the eclipse ends, as abruptly as it began, with *third contact,* when the moon begins to uncover the sun. Gradually the partial phases of the eclipse repeat themselves, in reverse order. At *last contact* the moon has completely uncovered the sun.

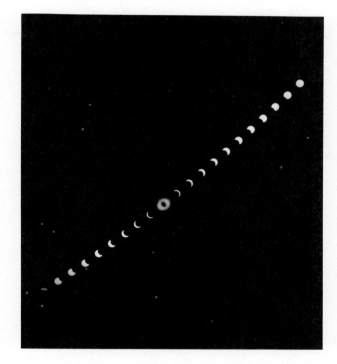

FIG. 9-16   Time-lapse photograph showing the moon passing in front of the sun during a total solar eclipse. *(American Museum of Natural History.)*

FIG. 9-17   Baily's beads, photographed by Van Biesbröck, May 20, 1947. *(Yerkes Observatory.)*

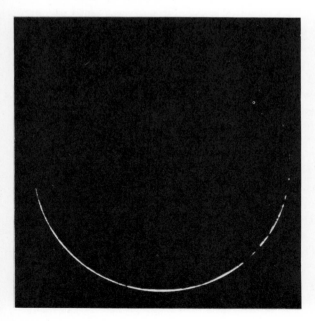

FIG. 9-18   The corona of the sun, photographed during the eclipse of June 8, 1918. *(Mount Wilson and Palomar Observatories.)*

According to ancient Chinese legend, an eclipse was an occasion when a giant dragon started to swallow the sun. People would therefore beat sticks and make noise to frighten the dragon away. Their noisemaking always did the job; sooner or later, the dragon would disgorge the sun, leaving it as whole as before.

(c) Value of Total Solar Eclipses

In addition to being beautiful to watch, total eclipses of the sun have considerable astronomical value. Many data are obtained during eclipses that are otherwise not accessible. For example, during an eclipse we can determine the exact relative positions of the sun and moon by timing the instants of the four contacts. We can take direct photographs and make spectrographic observations of the sun's outer atmosphere and prominences. We can measure the light and heat emitted by the corona. We can determine how meteorological conditions are affected by solar eclipses and can learn something about the light-scattering properties of the earth's atmosphere.

Historically, one of the most important of eclipse observations concerns the apparent positions of stars in the sky near the sun during totality. According to the general theory of relativity, light rays should be slightly deflected when passing near a massive body such as the sun. This means that those stars whose light rays pass very close to the disk of the sun in the sky should appear in a slightly displaced direction (Figure 9-19). It is possible to observe stars whose directions are close to that of the sun only during total solar eclipses. At other times, the bright glare of sunlight hides the stars, even from telescopic observation. The procedure for applying this test of general relativity is to photograph the star field near the sun during total eclipse and then to compare the measured positions of the star images with those observed when the same stars are photographed, by the same instruments, at other times of the year, when the sun is elsewhere in the celestial sphere. Unfortunately, the expected deflections of starlight passing near the sun are small and are consequently very difficult to measure. Positive results of this test of general relativity have, however, been reported. Most experts feel that it has been established that this gravitational deflection of starlight exists, although there is still some argument over the amount. At least the observations have been shown to be not inconsistent with general relativity.

(d) Annular Eclipses of the Sun

More than half the time the moon does not appear large enough in the sky to cover the sun completely, which means that the umbra of its shadow does not reach all the way to the surface of the earth. The geometry of the situation is illustrated in Figure 9-20. When the moon's shadow cone does not reach the earth, a total eclipse is not possible. However, if the boundaries of the umbra of the

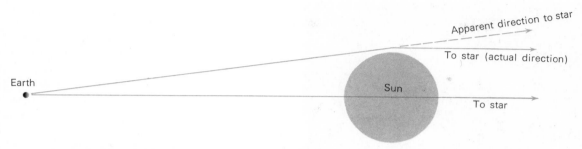

FIG. 9-19 Gravitational deflection of starlight passing near the sun — a test for general relativity.

shadow are extended until they intersect the earth's surface, they define a region on the ground within which the moon can be seen completely silhouetted against the sun's disk, with a ring of sunlight showing around the moon. This kind of eclipse is called an *annular eclipse,* from the Latin word *annulus,* meaning "ring." The extension of the moon's umbra, within which an annular eclipse is visible, sweeps across the ground in a path much like the path of totality of a total eclipse. As is true for total eclipses, a partial eclipse is visible within a region of 2000 mi or more on either side of the annular eclipse path. In this case, the regions of annular and partial eclipse correspond to positions *D* and *C,* respectively, in Figures 9-9 and 9-10.

An annular eclipse begins and ends like a total eclipse. However, because the sun is never completely covered by the moon, the corona is not visible, and although the sky may darken somewhat, it does not get dark enough for stars to be seen. An annular eclipse is not so spectacular, nor has it the scientific value of a total eclipse.

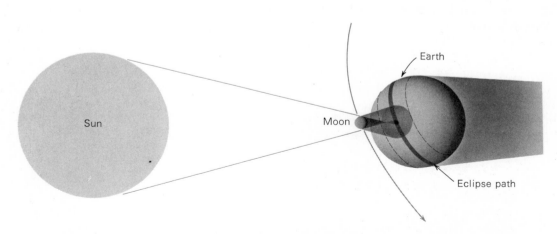

FIG. 9-20 Geometry of an annular eclipse (not to scale).

FIG. 9-21   An annular eclipse of the sun.
Bright spots are sunlight streaming through
lunar valleys. *(Lick Observatory.)*

Sometimes, because of the earth's finite size and sphericity, the umbra of the moon's shadow may be long enough to reach the surface of the earth only near the time when the moon is most nearly in a line between the centers of the earth and sun. Then it falls short of the earth's surface over the beginning and end of the eclipse path (see Figure 9-22). These mixed annular and total eclipses are relatively rare.

(e) Partial Eclipses of the Sun

A *partial eclipse* of the sun is one in which only the penumbra of the moon's shadow strikes the earth. During such an eclipse, the moon's umbra passes north or south of the earth, and from nowhere can the sun appear to be covered completely by the moon. Also, a total or annular eclipse appears partial from regions outside the eclipse path but within the zone of the earth that is intercepted by the moon's penumbra. Few people, therefore, have seen total or annular solar eclipses, whereas most have had the opportunity to see the sun partially eclipsed. The moon seems to "skim" across the northern or southern part of the sun. How much of the sun can appear covered depends, of course, on how close the observer is to the path of totality or annularity. Partial eclipses are interesting but not spectacular. The progress of the eclipse can be observed conveniently through heavily smoked glass or densely exposed photographic film. Only if the observer is within a few hundred miles of the eclipse path will he see the sky darken appreciably.

9.6 ECLIPSES OF THE MOON

A lunar eclipse occurs when the moon, at the full phase, enters the shadow of the earth. There are

FIG. 9-22   Mixed annular and total eclipses.

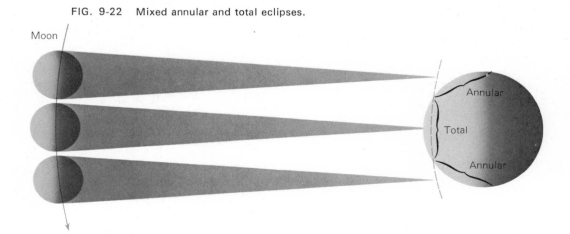

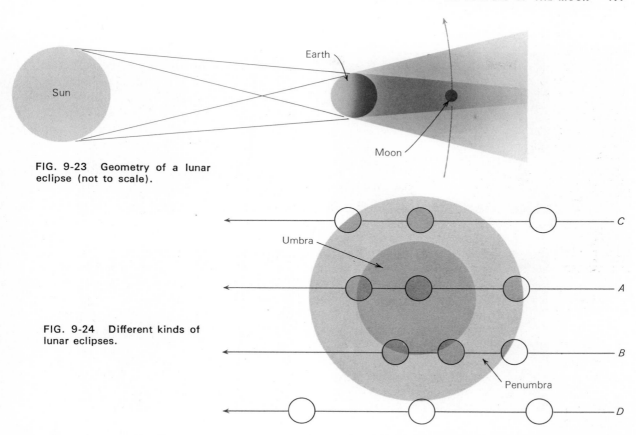

FIG. 9-23   Geometry of a lunar eclipse (not to scale).

FIG. 9-24   Different kinds of lunar eclipses.

three kinds of lunar eclipses: *total, partial,* and *penumbral.*

(a) Geometry of a Lunar Eclipse

The geometry of lunar eclipses is shown in Figures 9-23 and 9-24. Unlike a solar eclipse, which is visible only in certain local areas on the earth, a lunar eclipse is visible to everyone who can see the moon. Weather permitting, a lunar eclipse can be seen from the entire night side of the earth, including those sections of the earth that are carried into the earth's umbra while the eclipse is in progress. A lunar eclipse, therefore, is observed far more frequently from a given place on earth than is a solar eclipse.

Figure 9-24 shows the cross section of the earth's shadow at the moon's distance. Since the cross-sectional diameter of a cone is proportional to the distance from the apex of the cone, the cross section of the earth's shadow cone at the moon's distance is in the same proportion to the size of the earth as the moon's distance from the end of the shadow is to the total length of the shadow. The umbra is thus found to be 5700 mi in diameter at the moon's distance. The value varies slightly from one eclipse to another because the earth-moon distance varies, and because the diameter of the umbra at the place where the moon enters the shadow depends on the moon's and the sun's distance at the time of the eclipse. The penumbra of the earth's shadow averages about 10,000 mi across at the moon's distance.

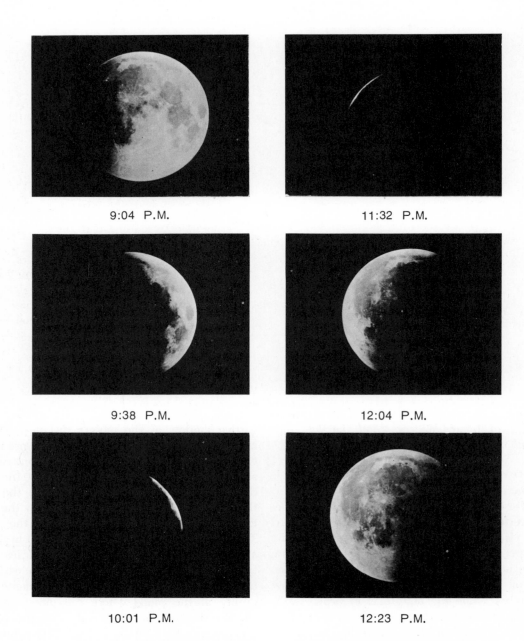

9:04 P.M.

11:32 P.M.

9:38 P.M.

12:04 P.M.

10:01 P.M.

12:23 P.M.

FIG. 9-25   Sequence of photographs of the total lunar eclipse of November 17/18, 1956. *(Photographed by Paul Roques, Griffith Observatory.)*

In Figure 9-24 are shown four of the many possible paths of the moon through the earth's shadow. A total lunar eclipse occurs when the moon passes completely into the umbra (path *A*). A *partial eclipse* occurs if only part of the moon skims through the umbra (path *B*), and a *penumbral eclipse* occurs if the moon passes through the penumbra, or partially through the penumbra, but does not come into contact with the umbra (paths *C* and *D*).

(b) Appearance of Lunar Eclipses

Penumbral eclipses usually go unnoticed even by astronomers. Only within about 700 mi of the umbra is the penumbra dark enough to produce a noticeable darkening on the moon. However, the diminished illumination on the moon's surface can be detected photometrically.

Every total or partial lunar eclipse must begin with a penumbral phase. About 20 minutes or so before the moon reaches the shadow cone of the earth, the side nearest the umbra begins to darken somewhat. At the moment called *first contact*, the limb of the moon (the "edge" of its apparent disk in the sky) begins to dip into the umbra of the earth. As the moon moves farther and farther into the umbra, the curved shape of the earth's shadow upon it is very apparent. In fact, Aristotle listed the round shape of the earth's shadow as one of the earliest proofs of the fact that the earth is spherical (Section 2.2d).

If the eclipse is a partial one, the moon never gets completely into the umbra of the earth's shadow but passes on by, part of it remaining in the penumbra, where it still receives some sunlight. At *last contact* the moon emerges from the umbra.

On the other hand, if the eclipse is a total one, at the instant of *second contact* the moon is completely inside the umbra, and the total phase of the eclipse begins. Even when totally eclipsed, the moon is still faintly visible, usually appearing a dull coppery red. Kepler explained this phenomenon in his treatise *Epitome*. The illumination on the eclipsed moon is sunlight that has passed through the earth's atmosphere and been refracted by the air into the earth's shadow (Figure 9-26).

FIG. 9-26   Illumination of the moon during a total eclipse by sunlight refracted by the earth's atmosphere into the earth's shadow.

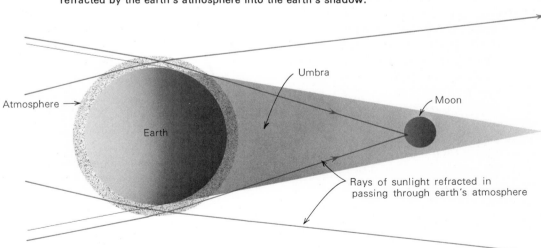

Atmosphere

Earth

Umbra

Moon

Rays of sunlight refracted in passing through earth's atmosphere

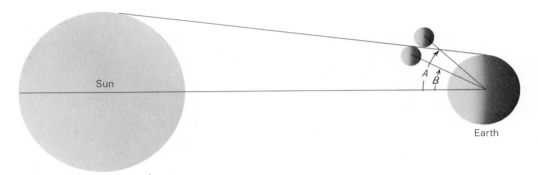

FIG. 9-27   Condition for a solar eclipse.

The eclipse will be darkest if the center of the lunar disk passes near the center of the umbra. The darkness of the lunar eclipse depends also on weather conditions around the terminator of the earth. It is only here, on the line between day and night on the earth, that sunlight passing through the atmosphere can be refracted into the shadow. Heavy cloudiness in that critical region will allow less light to pass through. The light striking the eclipsed moon inside the umbra is reddish because red light of longer wavelengths penetrates through the long path of the earth's atmosphere most easily.

Totality ends at *third contact,* when the moon begins to leave the umbra. It passes through its partial phases to *last contact,* and finally emerges completely from the penumbra. The total duration of the eclipse depends on how closely the moon's path approaches the axis of the shadow during the eclipse. The moon's velocity with respect to the shadow is about 2100 mi/hr; if it passes through the center of the shadow, therefore, about 6 hours will elapse from the time the moon starts to enter the penumbra until it finally leaves it. The penumbral phases at the beginning and end of the eclipse last about 1 hour each, and each partial phase consumes at least 1 hour. The total phase itself can last as long as 1 hour 40 minutes if the eclipse is central, or, of course, a shorter time if the moon does not pass through the center of the umbra.

9.7 ECLIPTIC LIMITS

We have seen that solar eclipses occur only when the moon is new and lunar eclipses when the moon is full. Furthermore, eclipses occur only when the new or full moon is near a node. If the sun, earth, and moon were geometrical points, the new or full moon would have to occur *exactly* at a node if there were to be an eclipse. Because of the finite sizes of the three bodies, however, part of the sun may appear covered by part of the moon for observers at certain places on earth even though at new moon the moon is not located *exactly* at the node. Similarly, in a lunar opposition, it is possible that, because of the large size of the earth's shadow, the full moon can pass partially through it or even pass completely into it, even if the full moon is not exactly at the node. Thus, the requirement that the moon be at a node can be relaxed slightly; there is some leeway within which eclipses can occur. The limits of this leeway are called *ecliptic limits.*

(a) *Solar Ecliptic Limits**

Note in Figure 9-27 that a partial eclipse of the sun is just barely visible somewhere on earth if the new moon encroaches on the conical surface enveloping the earth and sun. Under these conditions, it is a simple problem in geometry to find the angle A at the center of the earth between the centers of the sun and moon. Angle A turns out to be about 1½°. For an eclipse to be seen as *central,* that is, annular or total, the new moon must be completely inside the conical surface, in which case the angular separation, at the earth's center, between the centers of the moon and sun must be less than or equal to the angle *B*; *B* is about 1°.

In Figure 9-28 are shown the ecliptic and the moon's apparent path in the celestial sphere in the region of one of the nodes (the ascending node), which is indicated by *O*. The moon, *M*, is shown in four places in its path. The large circles, marked *C*, represent cross sections at the moon's distance of the cone tangent to the sun and earth. The centers of these circles, *Q*, *P*, *P'*, and *Q'*, are the positions of the sun's center at four different places along the ecliptic. Note that if new moon occurs at *R*, with the sun at *Q*, the centers of the sun and moon are separated by just the critical angle *A* for which some kind of eclipse is visible at some place on earth. Similarly, an eclipse can just barely occur if the sun and moon are at *Q'* and *R'*. If new moon occurs when the sun is outside the part of the ecliptic lying between *Q* and *Q'*, an eclipse cannot occur. If the new moon is at *S* or *S'*, and the sun at *P* or *P'*, respectively, the two bodies are separated by the critical angle *B* required for a central eclipse. Central eclipses cannot take place if the new moon occurs when the sun is not within the region on the ecliptic between *P* and *P'*. The angular distance *OQ* or *OQ'* is called the *ecliptic limit* for a solar eclipse. The sun must be within this distance of the node at new moon to be eclipsed.

The angular distance *OP* or *OP'* is called the *central ecliptic limit*; the sun must be within the central ecliptic limit of the node at new moon for an annular or total eclipse to occur. The angle of intersection between the moon's path and the ecliptic being known (about 5°), the actual values of the ecliptic limit and central ecliptic limit can be found easily by trigonometric calculation or geometrical construction. They are about 17 and 10 degrees, respectively.

The ecliptic limits can vary considerably, for they depend on the exact angular sizes of the sun and moon in the sky (and hence on their exact distances from the earth) and on the exact inclination angle of the moon's orbit to the ecliptic; all these are variable quantities. If we consider the extreme range of their variability, however, we can find the largest and smallest possible values for the ecliptic limits. These are called the *major* and *minor ecliptic limits*, respectively. The *major* and *minor solar ecliptic limits* are 18°31' and 15°21'. The *major* and *minor central ecliptic limits* are 11°50' and 9°55'. At the time of new moon, if the sun is farther from one of the moon's nodes than the major ecliptic limit, no eclipse can occur. If it is between the major and minor limits, an eclipse may occur; it is contingent upon the relative distances of the earth, sun, and moon at that instant, and upon the exact inclination of the moon's orbit. If the sun is within the minor ecliptic limit of a node at new moon, an eclipse must occur.

FIG. 9-28   Solar ecliptic limits.

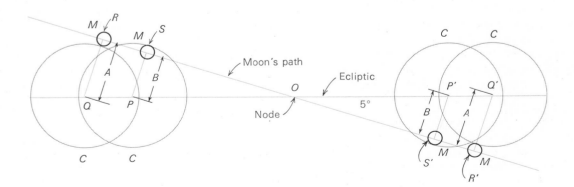

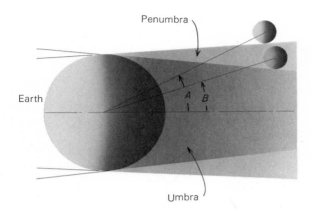

FIG. 9-29   Condition for a lunar eclipse.

(b) *Lunar Ecliptic Limits**

The *axis*, or center line, of the umbra of the earth's shadow is directed toward a point on the ecliptic exactly opposite the sun in the sky. If the moon is full while it is on or near the ecliptic (that is, near one of its nodes), it can enter the earth's shadow. In Figure 9-29 are shown the angles *A* and *B*, at the center of the earth, between the direction to the center of the moon and the axis of the umbra at which the moon can encroach, respectively, on the penumbra and umbra.

The penumbra shades gradually from the darkness of the umbra to full light outside; consequently, there is no observable phenomenon at the penumbra's outer boundary. The critical angle, *A*, for a penumbral eclipse, although well defined geometrically, has little real significance. Usually, we speak only of the limits of an umbral lunar eclipse (total or partial), for which the angle *B* applies. Angle *B* is a little less than 1°.

The *lunar ecliptic limit* is the maximum angle between the axis of the umbra and the direction of the node (or between the center of the sun and the opposite node) for which the moon can pass within an angular distance *B* of the umbral axis. The geometry of the situation, very similar to that of the solar ecliptic limit, is illustrated in Figure 9-30. If full moon occurs exactly at the node, *O*, the lunar eclipse is central. There will be no umbral eclipse if the moon passes the earth's shadow when it is farther from the node than at *Q* or *Q'*. The angle *OQ* or

OQ' is thus the ecliptic limit for an umbral lunar eclipse. The angle varies somewhat, as does the solar ecliptic limit, because of the variations in the distances of the sun and moon and of the inclination of the moon's orbit. The *major* and *minor lunar ecliptic limits* are 12°15' and 9°30', respectively.

(c) *Frequency of Eclipses**

The angle through which the sun moves along the ecliptic during a synodic month averages 29°6', and varies from this value only slightly. Between two successive new moons, therefore, the sun moves less than twice the minor solar ecliptic limit of 30°42' (2 × 15°21'). Consequently, it is impossible for the sun to pass through a node without at least one eclipse during that eclipse season. There must, then, be at least two solar eclipses during any one calendar year. If the sun is eclipsed within a few days after it reaches the western ecliptic limit, a second eclipse can follow at the next new moon, just before the end of the eclipse season. If the first season falls in January, five solar eclipses are possible in a single calendar year, for that same eclipse season will begin again in the following December. This happened in 1935. Twice the *central* solar ecliptic limit is narrower (at most 23°40'), and no more than one total or annular eclipse of the sun is possible during an eclipse season; the number during a calendar year varies from none to three.

Eclipses of the moon are about as frequent as those of the sun if penumbral eclipses are counted, the number per year ranging from two to five. Umbral eclipses, on the other hand, occur about as often as central solar eclipses; there are from none to three in any one year.

Penumbral eclipses are inconspicuous, so the total number of observable solar eclipses over a period of time outnumbers the observable lunar eclipses by nearly three to two. Lunar eclipses are more common at any one station, however, for they can be viewed from more than half of the globe, whereas solar eclipses, even partial ones, are visible only in limited areas. Total solar eclipses occur on an average about once every 1½ years, but they are visible only within narrow eclipse paths; their average frequency at any one place is about once every 360 years.

The maximum number of all kinds of eclipses (solar and lunar) in any one calendar year is seven.

9.8 RECURRENCE OF ECLIPSES*

Thousands of years ago the ancients noticed that similar eclipses occurred at regular intervals. This recurrence of eclipses made it possible for early astronomers to predict eclipses with fair accuracy.

(a) *Circumstances For Two Similar Eclipses**

In order that an eclipse may be followed, after a lapse of time, by another eclipse very similar to the first one, the following conditions must be met: (1) the moon must be at the same phase again (new for a solar, full for a lunar eclipse); (2) the moon, when at that phase, must be in the same place in its orbit with respect to the node; and (3) the sun and moon must have the same distances from the earth again. If, in addition, the two eclipses are to have similar eclipse paths on the earth, they must occur at about the same time of the year.

Let us consider requirements (1) and (2) first. For the moon to return to exactly the same phase again, an integral (or whole) number of synodic months must have elapsed. To return to the same place in its orbit again with respect to the node, the moon must have made an integral number of revolutions about its orbit with respect to the nodes; that is, there must be an integral number of *nodical* or *draconic months.* (The term "draconic" is derived from the ancient Chinese superstition that eclipses were caused by dragons swallowing the sun.)

Suppose the synodic and nodical months were exactly 30 and 27 days long, respectively. Then eclipses nearly identical to each other would occur every 270 days, because both 9 synodic months and 10 nodical months would add up to 270 days. However, the synodic month is actually 29.5306 days and the nodical month is 27.2122 days, and there exists no integral least common multiple of these two numbers.

There are, on the other hand, some periods of time that are very nearly integral multiples of both the synodic and the nodical months. For example, the interval of 47 synodic months is almost the same as the interval of 51 nodical months:

$$47 \text{ synodic months} = 1387.938 \text{ days};$$
$$51 \text{ nodical months} = 1387.822 \text{ days}.$$

As an illustration of the significance of this coincidence, suppose there is a solar eclipse. On the forty-seventh new moon following that eclipse, there will be another eclipse in which the moon is only about one tenth of a day's journey beyond the original place in its orbit relative to the node. Since the moon moves about 13° with respect to the node during 1 full day, this second eclipse will occur at a time when the sun and the moon are situated only a little over 1° from where they were, relative to the node, at the time of the eclipse 47 months earlier. Suppose the first eclipse had occurred just inside the western ecliptic limit. The second eclipse will take place a little farther inside the limit; the eclipse path on the earth will lie at latitudes slightly different from those of the first path. After another 47 synodic months, a third eclipse will occur, again about 1° further in from the western end of the ecliptic limit, and its path on the earth will again be displaced only slightly from that of the second eclipse. There will be a series of similar eclipses, occurring at intervals of 47 months, or about 1388 days. The

FIG. 9-30   Lunar ecliptic limits.

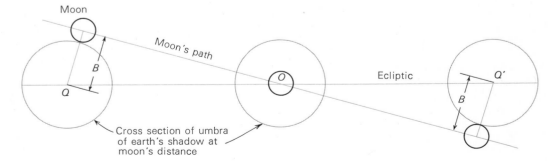

series of eclipses will continue until the position of the sun and moon during eclipse, shifting a little over a degree eastward from one eclipse to the next, has progressed along the ecliptic from the western end of the limit to the eastern end thereof, through a range of about 35°. There will be about 30 eclipses in this series.

Any two successive eclipses of the series described in the previous paragraph will have similar geometry but they will not be identical; some will be total and others annular. For identical eclipses, the relative distances of the earth, sun, and moon must be the same in both cases. Because the eccentricity of the earth's orbit about the sun is rather small, we may, as a first approximation, regard the moon's changing distance from the earth as the factor that determines the type of eclipse. The moon's closest approach to the earth (perigee) occurs at intervals of the anomalistic month of 27.55455 days. The anomalistic month differs from the sidereal and nodical months because the major axis of the moon's elliptical orbit (the line of *apsides*) gradually changes orientation in the plane of the moon's orbit (Section 9.3). The successive eclipses in an eclipse series are of about the same type if the integral number of synodic months is not only nearly equal to an integral number of nodical months but also is

TABLE 9.1   Intervals of Eclipses in
Four Representative Cycles

1	47 synodic months	= 1387.938 days
	51 nodical months	= 1387.822
	50 anomalistic months	= 1377.7275
2	223 synodic months	= 6585.321
	242 nodical months	= 6585.357
	239 anomalistic months	= 6585.538
3	3803 synodic months	= 112,304.83
	4127 nodical months	= 112,304.75
	4076 anomalistic months	= 112,312.35
4	4519 synodic months	= 133,448.73
	4904 nodical months	= 133,448.63
	4843 anomalistic months	= 133,448.69

nearly equal to an integral number of anomalistic months. In Table 9.1 the various intervals are given for four eclipse cycles.

(b) Eclipse Series; the Saros*

Any of the above cycles can be used to predict eclipses; it is simplest to use the synodic month as the unit of time. Note that the second and fourth cycles contain nearly integral numbers of nodical, anomalistic, and synodic months. In both of those cycles the successive eclipses are of about the same type. In the fourth cycle, however, the interval is so long that changes in the orbit of the moon become significant. Most useful for the prediction of eclipses, therefore, is the second cycle of 223 synodic months, just 10 or 11 days in excess of 18 years. This cycle is called the *saros*.

Nearly identical eclipses, both solar and lunar, recur at intervals of the saros. Successive eclipses, following each other at these 18-year intervals, are said to belong to the same series. Of course, many eclipses occur during any 18-year period; thus, many series of eclipses are in progress at once. Because the saros cycle contains so nearly an integral number of tropical years, at each successive eclipse in the cycle the earth's axis has almost the same orientation relative to the ecliptic plane (or to the earth-sun line); thus successive eclipses even occur at approximately the same latitudes and have similar paths over the earth's surface.

The first solar eclipse in any series belonging to the saros cycle is a partial eclipse that occurs when the sun and the moon are at an eastern ecliptic limit. If it is the descending node, the eclipse is just barely visible near the south pole of the earth. If the eclipse is at the eastern limit of the ascending node, it is just visible at the north pole. After 223 synodic months the next eclipse in the series occurs, with the moon just about 1/30 of a day's journey, or slightly under ½°, west of the same place relative to the node. Thus the eclipse occurs with the sun and moon about ½° farther inside the ecliptic limit. After about 70 successive eclipses separated by intervals of the saros (about 1200 years), the positions of the sun and moon at the times of the eclipses have shifted through the node to the western ecliptic limit, and the series ends. The first dozen eclipses in the series are only partial ones visible near one of the polar regions of the earth (north pole for the sun and moon at the ascending node, south pole for the sun and moon at the descending node). The last dozen eclipses are partial ones visible at the other polar region. The middle 45 eclipses of the series are total or annular, the eclipse paths gradually shifting in latitude from one pole to the opposite one during the series.

If the saros interval contained an exactly integral number of days, successive eclipses in a series would occur at the same time of day for each place on earth, and the eclipses would all be visible at nearly the same longitudes as well as at the same latitudes. As it is, however, the saros interval is 6585.32 days; the approximate ⅓-day remainder causes successive eclipses in the cycle to occur about one third of the way around the world from each other. However, every third eclipse in a series does follow a path lying in nearly the same terrestrial longitudes.

Series of lunar eclipses, each one similar to the preceding one, also occur at intervals of the saros. However, because the limits for umbral lunar eclipses are smaller than for solar eclipses, a series of lunar eclipses runs through only about 50 saroses, which requires about 870 years.

The saros interval contains a nearly integral number of anomalistic months, so any two successive eclipses in the same series will be of nearly the same type, that is, both total, both annular, and so on. Of course, the type of eclipse changes as the series progresses. Especially notable is the series of total solar eclipses to which belong the eclipses of 1937, 1955, and 1973, because the duration of totality of these eclipses is near the maximum possible.

Table 9.2 lists total solar eclipses of appreciable duration of totality that are visible from inhabited places on the earth during the interval 1950 to 2000 A.D. This interval contains nearly three saros cycles, and the similarity of successive eclipses in the same cycle is evident from inspection. All eclipses belonging to a given saros cycle are identified with the same letter in the table.

TABLE 9.2 Important Total Solar Eclipses in the Second Half of the Twentieth Century

SAROS CYCLE	DATE	DURATION OF TOTALITY (MIN)	WHERE VISIBLE
a	1952 Feb. 25	3.0	Africa, Asia
b	1954 June 30	2.5	North-Central U.S. (Great Lakes), Canada, Scandinavia, U.S.S.R., Central Asia
c	1955 June 20	7.2	Southeast Asia
e	1958 Oct. 12	5.2	Pacific, Chile, Argentina
f	1959 Oct. 2	3.0	Northern and Central Africa
g	1961 Feb. 15	2.6	Southern Europe
h	1962 Feb. 5	4.1	Indonesia
i	1963 July 20	1.7	Japan, Alaska, Canada, Maine
j	1965 May 30	5.3	Pacific Ocean, Peru
k	1966 Nov. 12	1.9	South America
a	1970 March 7	3.3	Mexico, Florida, parts of U.S. Atlantic coastline
b	1972 July 10	2.7	Alaska, Northern Canada
c	1973 June 30	7.2	Atlantic Ocean, Africa
d	1974 June 20	5.3	Australia
e	1976 Oct. 23	4.9	Africa, Indian Ocean, Australia
f	1977 Oct. 12	2.8	Northern South America
g	1979 Feb. 26	2.7	Northwest U.S., Canada
h	1980 Feb. 16	4.3	Central Africa, India
i	1981 July 31	2.2	Siberia
j	1983 June 11	5.4	Indonesia
k	1984 Nov. 22	2.1	Indonesia, South America
l	1987 March 29	0.3	Central Africa
a	1988 March 18	4.0	Philippines; Indonesia
b	1990 July 22	2.6	Finland, Arctic Regions
c	1991 July 11	7.1	Hawaii, Central America, Brazil
d	1992 June 30	5.4	South Atlantic
e	1994 Nov. 3	4.6	South America
f	1995 Oct. 24	2.4	South Asia
g	1997 March 9	2.8	Siberia, Arctic
h	1998 Feb. 26	4.4	Central America
i	1999 Aug. 11	2.6	Central Europe, Central Asia

9.9 PHENOMENA RELATED TO ECLIPSES

We have considered so far only eclipses that involve the sun, moon, and earth. However, there are phenomena with similar geometrical properties that involve other celestial bodies. Examples are *occultations* and *transits*.

(a) Occultations

The moon often passes between the earth and a star; the phenomenon is called an *occultation*. The stars are so remote that the shadow of the moon cast in the light of a star is extremely long and is, in fact, sensibly cylindrical. Because a star is virtually a point source, there is no penumbra. During an occultation, a star suddenly disappears as the eastern limb of the moon crosses the line between the star and observer. If the moon is at a phase between new and full, the eastern limb will not be illuminated and the star may appear to vanish mysteriously as the dark edge of the moon covers it. Because the moon moves through an angle about equal to its own diameter every hour, the longest time that an occultation can last is about 1 hour. It can have a much shorter duration if the occultation is not central. Geometrically, occultations are equivalent to total solar eclipses, except that they are total eclipses of stars other than the sun.

The sudden disappearance of a star behind the limb of the moon during an occultation is evidence that the moon has no appreciable atmosphere. If there were one, the star would fade gradually as the moon's limb approached it, because the starlight would traverse a long path of the lunar atmosphere. Occultations also demonstrate the extremely small angular sizes of the stars (owing to their great distances). If a star had an appreciable angular size it would require a perceptible time to disappear behind the moon, as is true during the partial phases of a total solar eclipse. Actually, the partial phases of occultations have been measured photoelectrically, but they are extremely brief, less than a few hundredths of a second. The angular sizes of stars cannot be observed directly in a telescope, but they can often be determined by various techniques (see Chapter 22). It has been possible to observe that stars of large computed angular size require longer to disappear behind the moon than those of small angular size. Observations of occultations of celestial radio sources have been useful in detecting the accurate positions and angular sizes of those objects.

In the past, occultations have been valuable for determining the exact position of the moon. Because the stars appear as points, their positions in the celestial sphere can be determined with high accuracy, whereas it is much more difficult to measure the exact position of the moon, which not only appears large, but reflects so much sunlight that the fainter stars around it are often invisible. On the other hand, if an occultation can be accurately timed, the exact direction of the moon as seen from the place on earth where the occultation is observed can be found. In modern times, occultations are less valuable for this purpose, because special cameras have been developed to photograph the moon against a background of comparison stars. Its position can thus be found with considerable precision, by measurement on the photographic plate.

Occultations of the brighter stars are listed in advance in various astronomical publications. The times and durations of the occultations and the places on earth from which they are visible are given. Also listed are the comparatively rare occultations of planets by the moon, and of stars by planets.

(b) Transits

A *transit* is a passage of an inferior planet (Mercury or Venus) across the front of the sun's disk at the time of inferior conjunction. Usually each of these planets appears to pass north or south of the sun, but it can pass in front of the sun if inferior conjunction occurs when the planet is near one of the nodes of its orbit — the points where its orbit crosses the ecliptic.

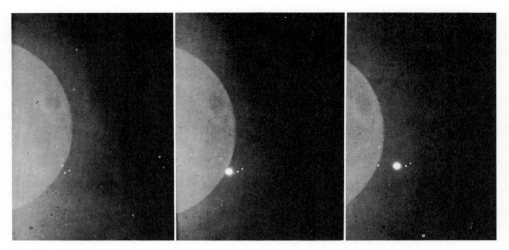

FIG. 9-31   Series of photographs showing *(left to right)* the emergence of Jupiter and three of its satellites after their occulation by the moon. *(Photographed by Paul Roques, Griffith Observatory.)*

The sun passes the nodes of Mercury's orbit in May and November; these are the seasons for transits of Mercury. However, transit "limits," analogous to ecliptic limits, are only a few degrees, and Mercury is seldom near enough to a node during the brief periods in May and November when transits of Mercury are possible. There are, on the average, about 13 transits of Mercury per century.

Transits of Venus are more rare than those of Mercury because the limits are narrower and because inferior conjunctions occur less frequently. Recurrences of transits occur just as recurrences of eclipses occur. Between two transits there must be an integral number of synodic periods of the planet, and there must be an integral number, or very nearly an integral number, of periods in which the planet has returned to the node. At present, transits of Venus occur in pairs separated by intervals of about 8 years. The last pair of Venusian transits were those of 1874 and 1882. The next pair will be on June 8, 2004, and June 6, 2012.

Transits are analogous to annular eclipses of the sun. The planets Venus and Mercury, seen from earth, are too small to cover the sun completely; their shadow cones fall far short of reaching the surface of the earth. The appearance of a transit is that of a black dot slowly crossing the disk of the sun from east to west. The silhouette of Mercury against the sun is too small to see without a telescope. That of Venus can be barely observed, without optical aid, if the sun is properly viewed by projection, or through dense filters to protect the eye.

(c) Eclipses on Other Planets

Eclipses would be visible from planets with satellites, other than the earth. For example, total solar eclipses would be common to observers living on the outside of the dense cloud layers that shroud the planet Jupiter. Several of Jupiter's 12 satellites regularly cast their shadows on the atmosphere of the planet. The dark spots on Jupiter where these shadows strike its atmosphere are easily visible through telescopes on earth. Only on the planets Mercury, Venus, and Pluto, which have no known satellites, are eclipses impossible.

(d) Stellar Eclipses

The most important kind of eclipses, to astronomers, are those of *eclipsing binary stars*. A binary system consists of two stars that revolve about each other. Such double stars are very common — many naked-eye stars are actually binary. When the plane of mutual revolution of such a double star happens to be so oriented that we see it edge on, each star periodically passes partially or entirely behind the other, thus being eclipsed. Then some or all of the light from the eclipsed star is prevented from reaching the earth, and the combined light received from the system is diminished. The most famous eclipsing binary star is the bright star *Algol* in the constellation of Perseus. Algol is really two stars revolving around each other, a bright one and a faint one. About every 2 days and 21 hours, the bright star passes partially behind the faint one and the apparent brightness of Algol drops down to less than half of normal — a change readily observable by the naked eye. A study of the light variation of such binary systems gives much information about the sizes and masses of their member stars. Eclipsing binary stars are described in Chapter 22.

Exercises

1. The moon shines by reflected sunlight, so how would the spectrum of moonlight compare with the solar spectrum?

2. When earthshine is brightest on the moon, what must be the phase of the earth, as seen from the moon?

3. When the moon is full, is there much "moonshine" on the earth?

4. How long would a synodic month be if the moon's sidereal month were 4 of our present calendar months? (*Hint:* See Chapter 3.)

5. How many more sidereal months than synodic months must there be in a year? Why?

6. If the moon revolved from east to west rather than from west to east, would a synodic month be longer or shorter than a sidereal month? Why?

7. Describe the daily delay in the rising of the full moon, both in the Northern and Southern Hemispheres, nearest the time when the sun is at the vernal equinox.

8. What is the phase of the moon if

(a) It rises at 3:00 P.M.?

(b) It is on the meridian at 7:00 A.M.?

(c) It sets at 10:00 A.M.?

9. What time does

(a) The first quarter moon cross the meridian?

(b) The third quarter moon set?

(c) The new moon rise?

10. Describe the phases of the earth as seen from the moon. At what phase of the moon (as seen from the earth) would the earth (as seen from the moon) be a waning gibbous?

11. Suppose you lived in the crater Copernicus on the moon.
(a) How often would the sun rise?
(b) How often would the earth set?
(c) Over what fraction of the time would you be able to see the stars?
12. The mean distance from the earth to the moon is 238,857 mi, and light travels at about 186,000 mi/sec. How long do radar waves take to make a round trip to the moon?

> *Answer:* 2.56 seconds

13. When Mars is at a distance of 35,000,000 mi, its angular diameter is 24.8″. What is its linear diameter?

> *Answer:* About 4200 mi

14. Suppose the moon's distance were exactly 240,000 mi from the earth and that the earth's distance from the barycenter turned out to be 60,000 mi. How much more massive than the moon would the earth be?
15. Sketch the shadow of a pencil cast by an electric lamp. Indicate the umbra and penumbra.
16. What is the angle at the apex of the shadow cone of the earth? (*Hint:* What is the angular size of the sun?)
17. Jupiter, having a diameter of about 86,000 mi, is more than twice the diameter of Neptune, with a diameter of only about 30,000 mi. Yet, the umbra of Neptune's shadow is about twice as long as the umbra of Jupiter's shadow. Explain.
18. What are the relative lengths of the shadow cones of the earth and moon? Assume that both objects are exactly the same distance from the sun. (Illustrate and show the method of your calculation, rather than looking up the figures in the chapter.)
19. What will be the eclipse seasons in 1981?
20. If a solar eclipse were annular over part of the eclipse path and total over part, would the total eclipse be of long or short duration? Why?
21. Does the longest duration of a solar eclipse occur when
(a) The sun is at its nearest and the moon at its nearest?
(b) The sun is at its farthest and the moon at its nearest?
(c) The sun is at its nearest and the moon at its farthest?
(d) The sun is at its farthest and the moon at its farthest?
22. Describe what an observer at the crater Copernicus on the moon would observe during what would be a total solar eclipse as viewed from the earth.
23. Verify that at the moon's distance the earth's shadow is about 5700 mi across. Draw a diagram.
24. Compare the actual ratio of the size of the earth's shadow to the moon's diameter to the value 8/3 estimated by Aristarchus (Chapter 2).
25. Does the moon enter the shadow of the earth from the east or west side? Explain why.

26. Describe the phenomenon observed by a spectator on the moon while the moon is being eclipsed.

27. If the penumbra of the earth's shadow is 10,000 mi across, and if the moon moves 2000 mi/hr with respect to the shadow, why does it take 6 hours instead of only 5 hours to get completely through the penumbra?

*28. Suppose there were five lunar and two solar eclipses during a calendar year. What would be the nature of the lunar eclipses? Explain.

*29. Can there be four solar and three lunar eclipses during a calendar year? If so, what are the natures of the lunar eclipses? Explain your answer.

30. Which planets can the moon never occult while at the full phase? Why?

31. If the earth and moon had their present distance from each other and their present period of mutual revolution, but if they were removed to the distance of Jupiter from the sun, would total solar eclipses be more common or less common at any one place? Explain.

CHAPTER 10

Electromagnetic Radiation and How It Reacts with Matter

Light is one form of *electromagnetic energy,* so called because the energy is carried by electric and magnetic waves through empty space. Actually, visible light constitutes only a small part of a wide range of different kinds of electromagnetic radiation, which differ only in *wavelength* but are called by different names: gamma rays, x-rays, ultraviolet, light, infrared, and radio waves.

10.1 NATURE AND PROPERTIES OF ELECTROMAGNETIC RADIATION

By its omnipresence, light is familiar to us all. Yet, to describe its nature, or to represent it with pictorial conceptions, is extremely difficult. The best we can do is to describe its properties with mathematical models. Most of the characteristics of electromagnetic radiation can be described adequately if it is represented as energy propagated in waves, although no medium is required to transmit the waves.

Any wave motion can be characterized by a *wavelength.* Ocean waves provide an analogy. The wavelength is simply the distance separating successive wave crests. Various forms of electromagnetic energy differ from each other only in their wavelengths. Those with the longest waves, ranging up to miles in length, are called *radio waves.* Forms of electromagnetic energy of successively shorter wavelengths are called, respectively, infrared radiation, light, ultraviolet radiation, x-rays, and finally the very short wave gamma rays. Short radio waves and long infrared waves are sometimes referred to as *radiant heat.* All these forms of radiation are the same basic kind of energy and could be thought of as different kinds of light. In this book, however, we shall reserve the term "light" to describe those wavelengths of electromagnetic radiation that, by their action upon the organs of vision, enable them to perform the functions of sight. Light is also called *luminous energy.*

Not all the properties of electromagnetic radiation can be described by this simple wave model. Experiments with electrons have shown that radiation of any given wavelength is always absorbed or emitted in quantities of energy that are whole multiples of some basic tiny quantity of radiant energy. It is as if electromagnetic energy were composed of many discrete "packets" of energy. These units of radiant energy are called *photons.* Photons must not be thought of as particles, for, as we have said, electromagnetic energy travels with a wave-like motion. Photons can be regarded as individual wave trains propagating through space, each spreading out in all directions from its source. This picture, however, describes only how electromagnetic radiation behaves, not what it actually "is."

Light, with other forms of electromagnetic radiation, is thus regarded both as a wave form of energy, and as composed of great numbers of photons, or packets of energy. The wave picture of light is adequate to describe many of the properties of light, including those principles of optics that are relevant to the design of telescopes. For some purposes, however, we shall have to call upon the photon representation of light. Different kinds of photons, corresponding to electromagnetic radiation of different wavelengths, differ from each other in their energies and are detected by different means.

Before we investigate the differences among photons, however, we should investigate those properties that they have in common.

(a) Propagation of Light — Inverse-Square Law

Photons propagate through empty space in a straight line. Their direction, however, may be altered in passing through or near matter, either by reflection, refraction, or diffraction. We shall discuss reflection and refraction of visible light later in this chapter. The phenomenon of *diffraction* is a deflection of light when it passes the edge of an opaque body. Light seems to spread itself out, slightly, into the shadow of an object it passes. (The study of diffraction belongs to the field of *physical optics* and will not be discussed further here; however, the bibliography includes references that deal with the subject.) Diffraction is important in many astronomical applications; among other things, it provides a theoretical limitation on the resolving power of a telescope.

An important property of the propagation of electromagnetic energy is the *inverse-square law*, a property that also belongs to the propagation of other kinds of energy, for example, sound. The amount of energy that would be picked up by a telescope, or an eye, or any other detecting device of fixed area, located at any given distance d from a light source O, is proportional to the amount of energy, or the number of photons, crossing each square inch of the surface of an imaginary sphere of radius d (Figure 10-1). A certain finite amount of energy, or number of photons, is emitted by the source in a given time interval. That energy spreads out over a larger and larger area as it moves away from the source. When it has moved out a distance d, it is spread over a sphere of area $4\pi d^2$ ($4\pi d^2$ is the surface area of a sphere of radius d). If E is the amount of energy in question, an amount $E/4\pi d^2$ passes through each square inch (if d is measured in inches).

We see, then, that the amount of energy pass-

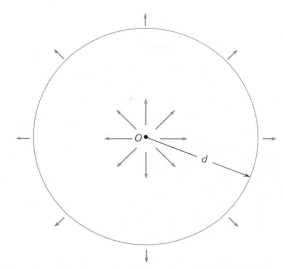

FIG. 10-1   Inverse-square law of light propagation.

ing through a unit area *decreases with the square of the distance from the source*. At distances d_1 and d_2 from a light source, the amounts of energy received by a telescope (or other detecting device), l_1 and l_2, are in the proportion:

$$\frac{l_1}{l_2} = \frac{4\pi d_2^2}{4\pi d_1^2} = \left(\frac{d_2}{d_1}\right)^2.$$

The above relation is known as the inverse-square law of light propagation. In this respect, the propagation of radiation is similar to the effectiveness of gravity, because the force of gravitation between two attracting masses is inversely proportional to the square of their separation. Figure 10-2 also illustrates the inverse-square law of propagation.

(b) Speed of Light

All electromagnetic energy propagates in a vacuum with a speed of 2.997929×10^{10} cm/sec, or about 186,000 mi/sec. Both theory and all available evidence show that this speed of light is the highest possible speed, and furthermore that an observer anywhere in space would observe light to move

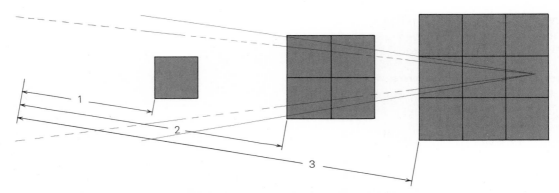

FIG. 10-2   As light energy radiates away from its source, it
spreads out, so that the energy passing through unit area decreases as
the square of the distance from the source.

with the same speed with respect to him. In fact, this universality of the observed speed of light is one of the postulates of the theory of relativity.

(c) *Measuring the Speed of Light**

Galileo attempted to measure the speed of light in the seventeenth century. By opening the dark slide on a lantern, he allowed a beam of light to flash to an assistant located some distance away. The assistant, upon observing the flash, then opened the slide on his own lantern. Galileo had hoped to determine the speed of light by timing the interval from the instant that he opened his lantern until he received the return light signal. He apparently tried the experiment with his assistant placed slightly less than a mile away and was forced to conclude that the speed of light was too great to permit its measurement by that particular experiment.

In 1675, the Danish astronomer, Ole Roemer, made observations that later provided a fairly accurate determination of the speed of light. Several of Jupiter's satellites are regularly eclipsed when they pass into the shadow of the planet. Roemer noticed periodic variations in the times of occurrences of these eclipses, which he attributed to the time it takes light to travel through space.

The principle of Roemer's observation is indicated in Figure 10-3. Suppose that at a certain time, t_0, a satellite of Jupiter enters Jupiter's shadow. We on earth are not aware of the event until somewhat later, at t_1, when news of the event, traveling at the speed of light, reaches the earth, say, at position E_1 in its orbit. The satellite completes a revolution around Jupiter and again enters eclipse at time t_2, and the news reaches the earth, at time t_3, when the earth is at position E_2. Because the earth has moved farther away from Jupiter during the revolution of the satellite, light must travel a greater distance to bring us the news of the second eclipse than it traveled to bring us news of the first one. Thus the *measured* time between eclipses, $t_3 - t_1$, is a longer interval than the actual time, $t_2 - t_0$. On the other hand, when the earth is on the other side of its orbit, and is moving toward Jupiter, the measured time between eclipses is *less* than the actual time.

Observations of this sort indicated that light requires about $16\frac{1}{2}$ minutes to traverse the diameter of the earth's orbit. The earth, traveling the circumference of its orbit in 1 year, travels a distance equal to the diameter of its orbit in $1/\pi$ years, about 10,000 times the time required for light to make the trip. Thus, light travels with a speed about 10,000 times as great as the earth's orbital speed. Later, when the earth's distance from the sun had been determined with reasonable accuracy, the earth's orbital speed was found to be $18\frac{1}{2}$ mi/sec. The speed of light is about 10,000 times that figure, or 186,000 mi/sec. Until 1849, observations such as Roemer's and the measurement of the aberration of starlight were the only known methods of determining the speed of light.

In 1849, the French physicist, Hippolyte Fizeau (1819–1896) invented a laboratory method of measuring the speed of light. His procedure was to send a beam of light to a distant mirror. On the way, the light had to pass through the gap between two teeth in a toothed wheel. If the wheel was stationary, or rotating only slowly, the mirror would reflect the light beam back through the same gap. If, however, the wheel was set in rapid rotation, by the time the reflected light beam reached the wheel a tooth would have moved into the location occupied by the gap when the light first passed the wheel. Fizeau's procedure was to find at what speed he had to rotate the wheel in order that the reflected light beam would be so eclipsed. Knowing the speed of rotation of the wheel, he could calculate the time required for a tooth to move into the position occupied by its adjacent gap. This was the same time that the light spent traversing the distance from the wheel to the mirror and back. In Fizeau's experiment, the mirror was a little over 5 miles from the toothed wheel, so that the light had to travel about 10.7 mi. His result for the speed of light was accurate to within about 4 percent. M. Cornu later applied the same method with an improved apparatus and measured the speed of light to an accuracy of 1 percent.

A superior method was invented by Foucault, famous for his pendulum experiment (Chapter 7). Foucault *reflected* light to a distant mirror from a mirror that was rapidly rotating. By the time the beam of light had been reflected back from the distant stationary mirror to the rotating one, the latter had turned slightly, and so did not reflect the light back to its original source but in a slightly different direction (Figure 10.4). By measuring the angle between the direction of the incident beam originally sent to the rotating mirror and the direction of the final beam reflected back from it, and knowing also the rate of rotation of the mirror, Foucault could determine the time light spent making the round trip between the mirrors. The speed he found for light was too small by about 1000 mi/sec. However, the same principle was later applied by the American physicist, Albert A. Michelson, who in 1878 measured the speed of light at 299,910 km/sec, very close to the accepted value today. More than 40 years later, in 1924–1926, Michelson made an even more accurate determination by using two stations near Los Angeles, separated by about 22 mi. He placed his light source and rotating mirror apparatus on Mount Wilson and the distant stationary mirror on Mount San Antonio. The distance between the two peaks was accurately surveyed to within a fraction of an inch.

The final result of the experiment made at Mount Wilson gave, for the speed of light in air, the value 299,729 km/sec. Light is slowed down (compared to its speed in a vacuum) in passing through a transparent substance. Even in air, its speed is retarded by nearly 70 km/sec.

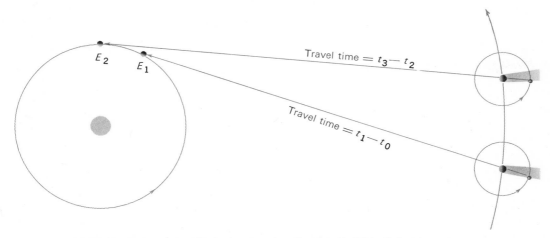

FIG. 10-3   Roemer's method of measuring the time it takes light to cross the earth's orbit; light from Jupiter's satellite takes longer to reach the earth at E_2 than at E_1.

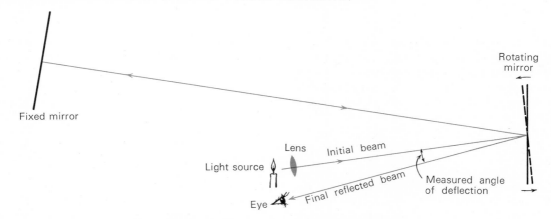

FIG. 10-4   Foucault's method of measuring the speed of light.

In the last few decades, techniques have been developed to measure the speed of light with very high accuracy, completely inside the laboratory, eliminating the necessity of sending light over long distances, which are difficult to measure accurately. In the precise laboratory procedures, the passage of light over very precisely measured distances is timed electronically. The value accepted today for the speed of light in a vacuum is 299,793 km/sec.

(d) Different Kinds of Electromagnetic Energy

We have seen that all forms of electromagnetic energy have certain characteristics in common: their wavelike method of propagation and their speed. Nevertheless, the various kinds of radiant energy, differing from each other in wavelength, are detected by very different means.

Radio waves have the longest wavelength, ranging up to miles in length. Those used in short-wave communication and in television have wavelengths ranging from inches to yards. When these waves pass a conductor, such as a radio antenna, they induce in it a feeble current of electricity, which can be amplified until it can drive a loudspeaker or recording apparatus.

The shortest wavelengths of radio radiation, less than about one millimeter ($\frac{1}{25}$ in.), merge into infrared radiation. Infrared photons can be detected with thermocouples, lead sulfide cells, and gas bulbs such as the Golay cell. These are devices that generate small currents when they absorb radiation of wavelengths to which they are sensitive. Infrared radiation of wavelengths less than about $\frac{1}{1000}$ in. can be photographed with special photographic emulsions. A simple device for detecting infrared radiation is a container of water. As the water absorbs the radiation its temperature rises.

Electromagnetic radiation with a wavelength in the range 0.000016 to 0.000028 in. comprises visible light. An appropriate unit to express the wavelength of visible light is the *angstrom*. One angstrom (abbreviated Å) is 1 hundred-millionth of a centimeter; 1 cm is about $\frac{2}{5}$ in. Visible light, then, has wavelengths that range from about 4000 to 7000 Å. The exact wavelength of visible light determines its *color*. Radiation with a wavelength in the range 4000 to 4500 Å or so gives the impression on the retina of the eye of the color violet. Radiations of successively longer wavelengths give the impression of the colors blue, green, yellow, orange, and red, respectively. The array of colors of visible light is called the *spectrum*. A mixture of light of all wavelengths, in about the same relative proportions as are found in the light emitted into space by the sun, gives the impression of white light.

Radiation of wavelengths too short to be visible to the eye is called *ultraviolet*. Radiations with wavelengths less than about a millionth of an inch comprise x-rays. Both ultraviolet radiation and x-rays, like visible light, can be detected photographically.

Electromagnetic radiation of the shortest wavelength, to less than a billionth of an inch, is called *gamma radiation*. Gamma rays are often emitted in the course of nuclear reactions and by radioactive elements. Gamma radiation is generated in the deep interiors of stars; it is gradually degraded into visible light by repeated absorption and reemission by the gases that comprise stars.

The array of radiation of all wavelengths, from radio waves to gamma rays, is called the *electromagnetic spectrum*.

(e) Relation Between Wavelength, Frequency, and Speed

Because of its wavelike character, the propagation of light can be compared to the propagation of ocean waves. While an ocean wave travels forward, the water itself is displaced only in a vertical direction. A stick of wood floating in the water merely bobs up and down as the waves move along the surface of the water. Waves that propagate with this kind of motion are called *transverse waves*.

Light also propagates with a transverse wave motion. At one time an invisible medium in space, called *ether*, was supposed to exist, suffering lateral displacements as light propagated through it. Today, the *ether hypothesis* is generally discarded. Apparently light travels with its highest possible speed through a perfect vacuum.

In this respect, light differs markedly from *sound*. Sound propagates as a physical vibration of matter. It does not travel at all through a vacuum. The displacements of the matter that carry a sound impulse are in a *longitudinal* direction, that is, in the direction of the propagation, rather than at right angles to it. Sound is actually a traveling wave of alternate compressions and rarefactions of the matter through which it moves. Of course, sound also travels far more slowly than electromagnetic radiation — only about $\frac{1}{5}$ mi/sec through air at sea level.

For any kind of wave motion, sound or light, we can derive a simple relation between wavelength and *frequency*. The frequency of light (or sound) is the rate at which wave crests pass a given point, that is, the number of wave crests that pass per second. Imagine a long train of waves moving to the right, past point O (Figure 10-5), at a speed c. If c is measured, say, in centimeters per second, we can measure back to a distance of c centimeters to the left of O and find the point P along the wave train that will just reach the point O after a period of 1 second. The frequency f of the wave train — that is, the number of waves that pass O during that second — is obviously the number of waves between P and O. That number of waves, times the length of each, λ, is equal to the distance c. Thus we see

FIG. 10-5   Relation between wavelength, frequency, and the speed of radiation.

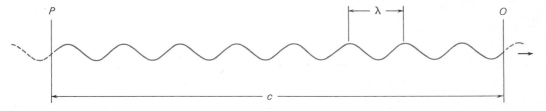

that for any wave motion, the speed of propagation equals the frequency times the wavelength, or, symbolically,

$$c = f \times \lambda.$$

(f) The Energy of Photons

According to modern quantum theory, each photon carries a certain discrete amount of energy that depends only on the frequency of the radiation it comprises. Specifically, the energy of a photon is proportional to the frequency; the constant of proportionality, denoted h, is called *Planck's constant,* named for Max Planck (1858–1947), the great German physicist who was one of the originators of the quantum theory. If metric units are used (that is, if energy is measured in ergs and frequency in cycles or waves per second), Planck's constant has the value $h = 6.62 \times 10^{-27}$ erg · sec. The energy of a photon, in algebraic notation, is

$$E = h \times f.$$

Since we have seen in Section 10.1e that the frequency times the wavelength is equal to the speed of light, we also note that the energy of a photon is inversely proportional to the wavelength; that is,

$$E = \frac{hc}{\lambda}.$$

Photons of violet and blue light are thus of higher energy than those of red light. The highest energy photons of all are gamma rays; those of lowest energy are radio waves.

10.2 LAWS OF GEOMETRICAL OPTICS

Having discussed the properties of electromagnetic radiation in general, we now turn our attention to those of visible light. The properties of light that are most important to the design and construction of telescopes and other astronomical instruments can be summarized simply in three laws of *geometrical optics* — the laws of *reflection, refraction,* and *dispersion.*

(a) The Law of Reflection

The law of reflection describes the manner in which light is reflected from a smooth, shiny surface. We shall speak frequently of the normal to a surface. The *normal* to a surface at some point is simply a line or direction perpendicular to that surface at that point. If light strikes a shiny surface, its direction must make a certain angle with the normal to the surface at the point where it strikes. That angle is the *angle of incidence.* The angle that the reflected beam of light makes with the normal is called the *angle of reflection.* The law of reflection states that the angle of reflection is equal to the angle of incidence and that the reflected beam lies in the plane formed by the normal and the incident beam (see Figure 10-6).

(b) The Law of Refraction

The law of refraction deals with the deflection of light when it passes from one kind of transparent medium into another. Every transparent substance can be characterized by its *index of refraction,* a measure of the degree to which the speed of light is diminished in passing through it. Specifically, the

FIG. 10-6 Law of reflection.

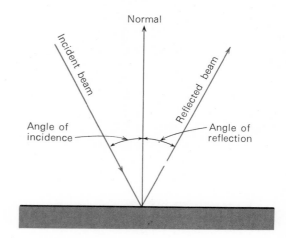

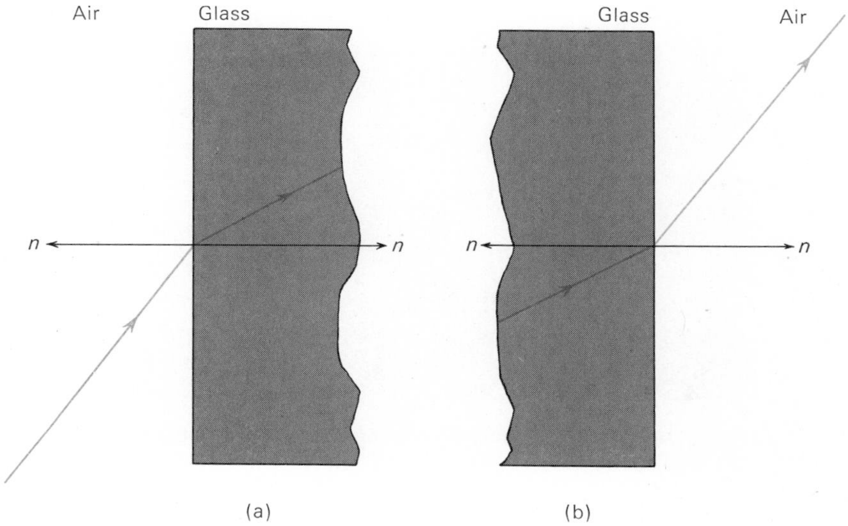

FIG. 10-7 Refraction of light between glass and air: (a) when light passes from air to glass; (b) when light passes from glass to air.

index of refraction is the ratio of the speed of light in a vacuum to that in the substance. Whenever light passes from a medium having one index of refraction into a medium having another index, such as from air into glass or from glass into air, it is always bent, or refracted, at the interface between the two media. Usually, media of higher densities have higher indices of refraction. Thus if light passes from air into glass, it passes from a medium of lesser into a medium of higher index of refraction.

The law of refraction states that if light passes from one medium into a second one of a different index of refraction, the angle the light beam makes with the normal to the interface between the two substances is always *less* in the medium of higher index. Thus, if light goes from air into glass or water, it is bent *toward* the normal to the interface, while if it goes from water or glass into air, it is bent *away* from the normal. The law of refraction is expressed mathematically by Snell's law.†

―――――――
†*Snell's law* is
$$n \sin \alpha = n' \sin \alpha',$$
where α and α' are the angles between a light beam and the normal to the interface between two media of indices of refraction n and n', respectively.

Water has an index of refraction of about $1\frac{1}{3}$; crown glass and flint glass have indices of about 1.5 and 1.6, respectively. Diamond has the very high index of refraction of 2.4.

It is the refraction of light when it passes from the water into the air that makes the handle of a spoon appear bent if a spoon is immersed in a glass of water. Similarly, light entering the earth's atmosphere from space is slightly bent. In a vacuum, the index of refraction is taken, by convention, to be exactly 1.0. The index of air at sea level is about 1.00029. The light from stars, planets, the sun, and the moon, is bent, upon entering the earth's atmosphere, in such a way as to make the object appear to be at a greater altitude above the horizon than it actually is. Atmospheric refraction is greatest for objects near the horizon. It raises the apparent altitude of objects on the horizon by about $\frac{1}{2}°$. Refraction of light, passing through both the glass and the earth's atmosphere, is illustrated in Figures 10-7 and 10-8.

The illusion that the moon (or sun) looks larger near the horizon than when it is high in the sky is *not* due to refraction. Actually, refraction raises the lower limb of the moon more than the upper, so that the moon really looks smaller and

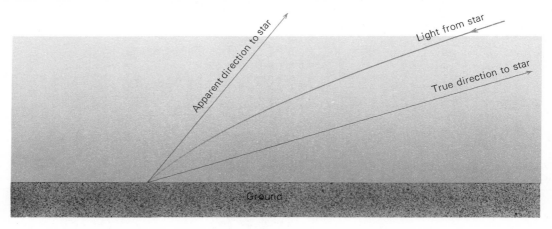

FIG. 10-8   Atmospheric refraction. Light is bent upon entering the
earth's atmosphere in such way as to make stars appear
at a higher altitude than they actually are. (The effect is grossly
exaggerated in this figure.)

oval near the horizon, not larger. The apparent enlargement of the moon or sun when seen near the horizon is a purely psychological effect that has been the subject of much discussion and investigation by psychologists.

FIG. 10-9   Dispersion at the interface of two media.

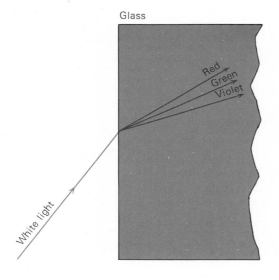

(c) The Law of Dispersion

Dispersion of light is the manner in which white light, a mixture of all wavelengths of visible light, can be decomposed into its constituent wavelengths or colors when it passes from one medium into another. The phenomenon of dispersion occurs because the index of refraction of a transparent medium is different for light of different wavelengths. In general, the index of refraction is greater for light of shorter wavelengths. Thus, whenever light is refracted in passing from one medium into another, the violet and blue light, of shorter wavelengths, is bent more than the orange and red light of longer wavelengths.

Figure 10-9 shows the way in which light of different wavelengths is bent different amounts in passing from, say, air into glass. Figure 10-10 shows how light can be separated into different colors with a prism, a triangular piece of glass. Upon entering one face of the prism, light is refracted once, the violet light more than the red, and upon leaving the opposite face, the light is bent again, and so is further dispersed. Even greater dispersion can be obtained by passing the light through a series of prisms. If the light leaving a prism is focused upon

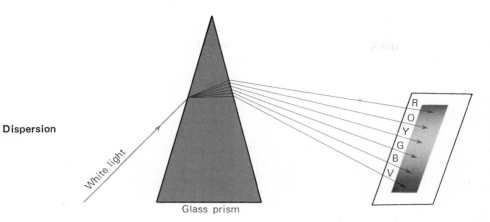

FIG. 10-10   Dispersion by a prism.

a screen, the different wavelengths or colors that compose white light are lined up side by side. The array of colors is a spectrum.

It has long been known that colors are produced when white light passes through glass, but before Newton's time it was believed that the glass itself produced the colors. Newton produced a spectrum with a glass prism, and then recombined the various colors by reflecting them with separate mirrors to a screen. When he had thus recombined the colors, as best he could, he observed that the resulting spot of light was white again, which indicated that the glass prism had not produced the colors but had merely separated them out from white light.

Atmospheric refraction occurs in varying amounts for different colors, also. This atmospheric dispersion combined with turbulence in the air often produces colorful effects when a bright star is seen low above the horizon, and its different colors are bent by the air in differing amounts. When astronomers measure the exact positions of stars in the sky, they must not only take account of the distortion of the position of the star by atmospheric refraction but also take care to note the color of the light with which the star is observed, because the correction for refraction is different for different colors.

(d) *Weather Optics* *

Nature provides an excellent example of the dispersion of light in the production of a rainbow. Raindrops, tiny spherical droplets of water in the air, act like prisms. Light from the sun entering a raindrop is bent, the blue and violet light being bent the most. This bent light strikes the inside rear surface of the drop and is reflected back toward the front surface. Thus the light leaves the raindrop by passing through the same side that it enters. But when it leaves the drop, it is again refracted, and again dispersed, just as when light leaves a glass prism. Thus

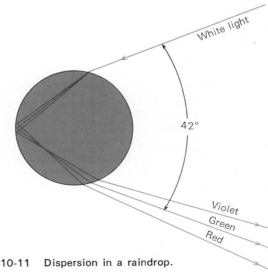

FIG. 10-11   Dispersion in a raindrop.

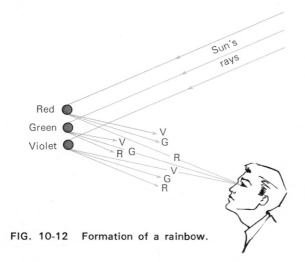

FIG. 10-12   Formation of a rainbow.

sunlight is spread into the rainbow of colors — the rainbow is nothing more than the spectrum of sunlight. Figure 10-11 shows how a raindrop produces a spectrum.

The light emerges from a raindrop at an angle of about 42° from the direction at which it enters. Thus (see Figure 10-12), to see a rainbow, the observer must have the sun behind him, at an altitude of less than 42° above the horizon. The rainbow then appears as an arc, with an angular radius of 42°, centered about a point exactly opposite the sun. Since the sun must be above the horizon to illuminate the raindrops, the center of the rainbow must be below the horizon; an observer on the ground can never see a rainbow as more than half a complete circle (observers in airplanes, or on mountains, may occasionally see a rainbow as more than a semi-circle).

Although the arc of a rainbow has a radius of approximately 42°, the different colors of sunlight are refracted and reflected by the droplets back to the observer in slightly different directions, so the band of color has a finite width. It may be seen in Figure 10-12 that from the upper drops it is the red light, bent the least, that enters the observer's eye, and from the lower drops it is the violet light, bent the most, that the observer sees. Thus, the top, or outside, of the arc of the rainbow appears red, and the bottom, or inside, of the arc appears violet. The other colors come from the drops in between.

Another natural phenomenon that involves the refraction and dispersion of light is that of a *halo* about the

sun or moon. A halo is a faint ring of light, of angular radius 22°, caused by the bending of light as it passes through the tiny ice crystals that form cirrus clouds at altitudes of more than 20,000 ft. As in the rainbow, the violet light is refracted most in passing through the crystals. Consequently, the outer edge of a solar or lunar halo appears violet, the inner edge red (Figure 10-13).

10.3 SPECTROSCOPY IN ASTRONOMY

The light from a star or other luminous astronomical body, as well as from the sun, can be decomposed into its constituent wavelengths, producing a spectrum. A device used to observe visually the spectrum of a light source is a *spectroscope;* one used to photograph a spectrum is a *spectrograph.* In Chapter 11 it will be explained how a spectroscope (or spectrograph) may be constructed and how it can be attached to a telescope so that we may study the spectra of astronomical objects.

(a) The Value of Stellar Spectra

If the spectrum of the white light from the sun and stars were simply a continuous rainbow of

FIG. 10-13   Formation of a halo around the sun or moon by ice crystals in cirrus clouds.

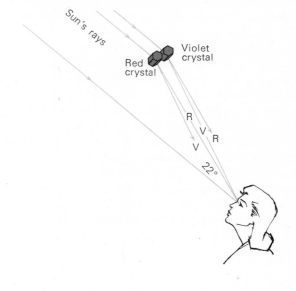

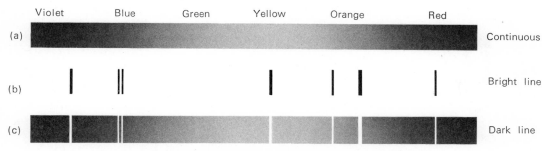

Violet   Blue   Green   Yellow   Orange   Red

(a)   Continuous

(b)   Bright line

(c)   Dark line

FIG. 10-14   Three kinds of spectra: (a) continuous; (b) bright line;
(c) dark line. (The spectra are shown as they appear on photographic negatives.)

colors, astronomers would not have such intensive interest in the study of stellar spectra. To Newton, the solar spectrum did appear as just a continuous band of colors. However, in 1802, William Wollaston (1766–1828) observed several dark lines running across the solar spectrum. He attributed these lines to natural boundaries between the colors. Later, in 1814–1815, the German physicist Joseph Fraunhofer 1787–1826), upon a more careful examination of the solar spectrum, found about 600 such dark lines. He noted the specific positions in the spectrum, or the wavelengths, of 324 of these lines. To the more conspicuous lines he assigned letters of the alphabet, with the letters increasing from the red to the violet end of the spectrum. Today we still refer to several of these lines in the solar spectra by the letters assigned to them by Fraunhofer.

Subsequently, it was found that such dark spectral lines could be produced in the spectra of artificial light sources by passing their light through various transparent substances. On the other hand, the spectra of the light emitted by certain glowing gases were observed to consist of several separate bright lines. A preliminary explanation of these phenomena (although grossly over-simplified in light of modern knowledge) was provided in 1859 by Gustav Kirchhoff (1824–1887) of Heidelberg. Kirchhoff's explanation is often given in the form of *three laws of spectral analysis:*

FIRST LAW: *A luminous solid or liquid emits light of all wavelengths, thus producing a continuous spectrum.*

It is often stated that a highly compressed gas also emits a continuous spectrum, but the statement is an oversimplification. It is usually true if the gas is opaque.

SECOND LAW: *A rarefied luminous gas emits light whose spectrum shows bright lines, and sometimes a faint superposed continuous spectrum.*

THIRD LAW: *If the white light from a luminous source is passed through a gas, the gas may abstract certain wavelengths from the continuous spectrum so that those wavelengths will be missing or diminished in its spectrum, thus producing dark lines.*

We distinguish, then, among three types of spectra. A *continuous* spectrum is an array of all wavelengths or colors of the rainbow. A *bright line* or *emission* spectrum appears as a pattern or series of bright lines; it is formed from light in which only certain discrete wavelengths are present. A *dark line* or *absorption spectrum* consists of a series or pattern of dark lines — missing wavelengths — superposed upon the continuous spectrum of a source of white light (see Figure 10-14).

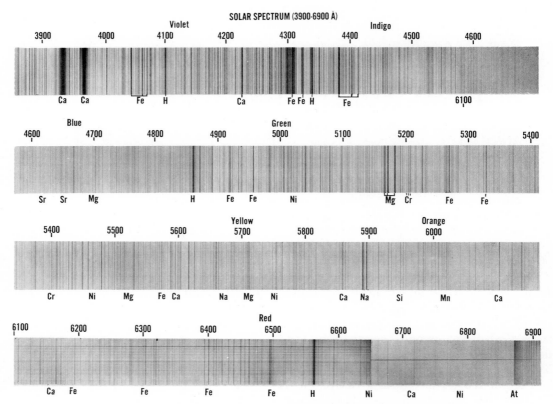

FIG. 10-15   The solar spectrum. Labels indicate the elements in the sun's photosphere that cause some of the dark lines. The wavelengths, in Angstroms, and the colors of the different parts of the spectrum are also labeled. *(Mount Wilson and Palomar Observatories.)*

The great significance of Kirchhoff's laws is that each particular chemical element or compound, when in the *gaseous form,* produces its own characteristic pattern of dark or bright lines. In other words, each particular gas can absorb or emit only certain wavelengths of light, peculiar to that gas. The presence of a particular pattern of dark (or bright) lines characteristic of a certain element is evidence of the presence of that element somewhere along the path of the light whose spectrum has been analyzed.

Thus the dark lines (Fraunhofer lines) in the solar spectrum give evidence of certain chemical elements between us and the sun, absorbing those wavelengths of light. It is easy to show that most of the lines must originate from gases in the outer part of the sun itself.

The wavelengths of the lines produced by various elements are determined by laboratory experiment. Most of the thousands of Fraunhofer lines in the sun's spectrum have now been identified with more than 60 of the known chemical elements.

Dark lines are also found in the spectra of stars and in stellar systems. Much can be learned from their spectra, in addition to evidence of the chemical elements present in a star. A detailed study of its spectral lines indicates the temperature, pressure, turbulence, and physical state of the gases in that star; whether or not magnetic and electric fields are present, and the strengths of those fields; how fast the star is approaching or receding from us; and many other data. Information about a star can also be obtained by studying its continuous spectrum.

The study of the spectra of celestial objects is the most powerful means at the astronomer's disposal for obtaining data about the universe.

(b) The Doppler Effect

In 1842 Christian Doppler (1803–1853) pointed out that if a light source is approaching or receding from the observer, or vice versa, the light waves will be, respectively, crowded closer together or spread out. The principle, known as the *Doppler principle* or *Doppler effect*, is illustrated in Figure 10-16. In (a) the light source is stationary with respect to the observer. As successive wave crests 1, 2, 3, and 4 are emitted, they spread out evenly in all directions, like the ripples from a splash in a pond. They approach the observer at a distance λ behind each other, where λ is the wavelength of

the light. On the other hand, if the source is moving with respect to the observer, as in (b), the successive wave crests are emitted with the source at different positions, S_1, S_2, S_3, and S_4, respectively. Thus, to observer A, the waves seem to follow each other by a distance *less* than λ, whereas to observer C they follow each other by a distance greater than λ. The wavelength of the radiation received by A is shortened; the wavelength of the radiation received by C is lengthened. Because the wavelengths seem to A to follow each other at an accelerated rate, A also observes the light at a higher frequency than if the source were stationary, while C receives the light at a diminished frequency. To observer B, in a direction at right angles to the motion of the source, no effect is observed. The effect is produced only by a motion *toward* or *away* from the observer, a motion called *radial velocity*. Observers between A and B or B and C would observe some shortening or lengthening of the light waves, respectively, for a component of the motion of the source is in their line of sight.

†For observer D in Figure 10-16(b), the component is Vcosθ.

FIG. 10-16 Doppler effect.

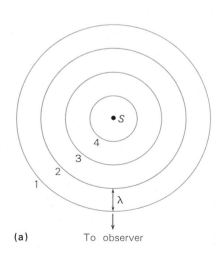

(a) To observer

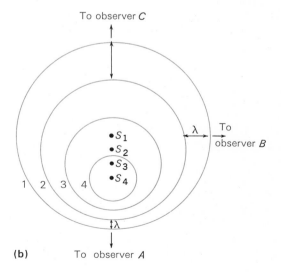

(b) To observer A

The Doppler effect is also observed in sound. We have all heard the apparent higher than normal pitch of the whistle of an approaching train and the lower than normal pitch of a receding one. In the case of sound, the precise formula for the amount of shortening or lengthening of the sound waves depends on whether the source or observer, or both, are in motion. For light, however, it follows from the special theory of relativity that there is no way of determining which is in motion, and the formula for the increase or decrease in the wavelength, the *Doppler shift,* is identical for each case. The wavelengths are shortened if the distance between the source and observer is decreasing, and they are lengthened if the distance is increasing.

The exact formula for the Doppler shift is:

$$\frac{\Delta\lambda}{\lambda} = \frac{\sqrt{1 + v/c}}{\sqrt{1 - v/c}} - 1,$$

where λ is the wavelength emitted by the source, $\Delta\lambda$ is the difference between λ and the wavelength measured by the observer, c is the speed of light, and v is the relative line of sight velocity of the observer and source, which is counted as positive if the velocity is one of recession and negative if it is one of approach. If the relative line of sight velocity of the source and observer is small compared to the speed of light, however, the formula reduces to the simple form, which is usually used:

$$\frac{\Delta\lambda}{\lambda} = \frac{v}{c}.$$

Solving this last equation for the velocity, we find†

$$v = c\frac{\Delta\lambda}{\lambda}.$$

---

†If v represents the space velocity at angle θ to the line of sight, as in Figure 10-16(b), then

$$\frac{\Delta\lambda}{\lambda} = \frac{v}{c} \cos\theta,$$

or

$$v \cos\theta = c\frac{\Delta\lambda}{\lambda}.$$

If a star approaches or recedes from us, the wavelengths of light in its continuous spectrum appear shortened or lengthened, as well as those of the dark lines. However, unless its speed is tens of thousands of miles per second, the star does not appear noticeably bluer or redder than normal. The Doppler shift is thus not easily detected in a continuous spectrum (except for very remote galaxies — see Chapter 33) and cannot be measured accurately in such a spectrum. On the other hand, the wavelengths of the absorption lines can be measured accurately, and their Doppler shift is relatively simple to detect. Generally, when the spectrum of a star or other object is photographed at the telescope, sometime during the exposure the light from an iron arc or some other emission-line source is allowed to pass into the same spectrograph, and the spectrum of the arc is then photographed just beside that of the star. The known wavelengths of the bright lines in the spectrum of the arc (or other laboratory source) serve as standards against which the wavelengths of the dark lines in the star's spectrum can be accurately measured. Further illustrations of the Doppler effect are given in later chapters.

10.4 RADIATION LAWS

If a body intercepts electromagnetic radiation, it generally reflects some of it, transmits some, and absorbs the rest. Except for transparent objects the transmitted radiation is negligible, if even measurable. The energy associated with the absorbed radiation heats the body, and would continue to heat it indefinitely if the body did not begin to radiate that energy away. A body (for example, Mars) that is exposed to an approximately constant flow of radiation (sunlight falling on Mars) will eventually reach an equilibrium temperature such that the average rate at which it reradiates energy just equals the rate at which it absorbs it. Therefore, an opaque body (Mars, or some other planet) may be observed both by the radiation it reflects and by the reradiated energy it previously absorbed.

The quality of the reflected radiation depends on the absorbing properties of the body. If, for example, it is exposed to white light and absorbs the same fraction of electromagnetic energy of all wavelengths, it appears, in reflected light, to be white. If it preferentially absorbs short wavelengths, the light it reflects is dominated by long wavelengths and it appears red. Thus the colors of objects depend on their absorbing properties.

Many complicating factors determine the quality of the energy reemitted by the body, but the most important is its temperature when the energy it reemits is in equilibrium with that which it absorbs. If a body with no internal source of energy were perfectly reflecting, it would absorb and reemit nothing and its temperature would be at absolute zero. Planets typically have temperatures of tens or hundreds of degrees absolute,† and reradiate the solar energy they absorb mostly at infrared and radio wavelengths. At room temperature on earth, almost all objects (including people) are radiating infrared radiation, with a maximum intensity near 0.01 mm wavelength.

(a) Perfect Radiators

Of particular interest is a hypothetical *perfect radiator*, also called a *black body*. This is an idealized body that absorbs all the electromagnetic energy incident on it. Its temperature then depends only on the total radiant energy striking it each second, and the quality of the energy it reradiates can be precisely predicted from the *radiation laws*.

A perfect radiator or black body is black in the sense that it is completely opaque to all wavelengths but of course need not look black. At room temperature the radiated energy is mostly

in the invisible infrared and a black body does look black, but a perfect radiator with a temperature of thousands of degrees appears very bright indeed.

Stars, for example, are usually fair approximations of black bodies, because they are composed of hot gases that are very opaque — that is, stellar material efficiently absorbs radiation. The high opacity of the gases of a star helps it hold in its internal heat. Deep in the interior of a star, where the temperatures are maintained at millions of degrees, thermonuclear reactions produce the star's energy. This energy very slowly filters out toward the surface of the star, being passed from atom to atom. Finally, the energy reaches a region in the star where the density and opacity of the gases have decreased to the point where photons have a chance of escaping from the star, through its tenuous outer atmosphere, into interstellar space. These escaping photons comprise the radiation that we observe from the star. They emanate, in a typical star, from a comparatively thin outer region or "shell" called the *photosphere;* in the sun, this zone from which light escapes and reaches the earth is only about 200 mi deep. The photosphere is the visible "surface" we see when we look at the sun, although it is not really a surface at all.

Because of this high opacity of the gases that constitute stars, the light from stars resembles, approximately, that from perfect radiators. The resemblance is not complete because the temperature increases inward through a stellar photosphere, and thus the stellar radiation comes from a variety of layers of slightly different temperature. Moreover, because the opacity of the photosphere to radiation depends on the wavelength of that radiation, the depth within the photosphere from which a given fraction of the emerging light comes depends on the color of that light. However, the spectral energy distribution of a star (the distribution of its light among various wavelengths) usually can be approximated by the energy emitted by a perfect radiator

---

†In astronomy, temperature is almost always expressed on the Kelvin or absolute scale — that is, in centigrade degrees above absolute zero ($-273°C$); see Appendix 5.

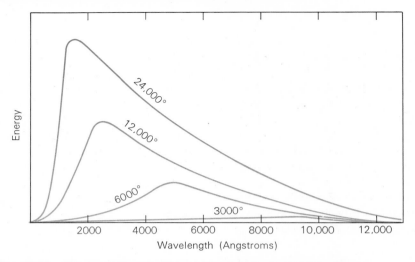

FIG. 10-17 Energy emitted at different wavelengths for black bodies at several temperatures.

at a temperature more or less representative of that of the star's photospheric layers. Thus, with fair accuracy, we can apply the laws of black-body radiation to a star.

A good laboratory approximation to a black body is an enclosed chamber with well-insulated walls that are painted black on the inside. The internal energy, nearly all absorbed and reemitted by the interior walls, can be observed through a tiny hole in the chamber. Such a device was used to derive experimentally the nature of black-body radiation and for testing the derived radiation laws.

(b) Planck's Radiation Law

The energy emitted from black bodies had been studied experimentally during the last century. It was only about the beginning of the 20th century, however, that the German physicist Max Planck (1858–1947) found a theoretical interpretation for black-body radiation. He succeeded in this by adopting the hypothesis that light energy is *quantized*, that is, it occurs only in discrete "packets," or *photons*, each of energy (in ergs) hc/λ (Section 10.1f). An equation derived by Planck gives the radiant energy emitted per second at any wave-

length from 1 cm² of the surface of a black body of any temperature.† In Figure 10-17, this energy distribution, computed from Planck's formula, is plotted against wavelength for several different temperatures. Three points, in particular, should be noted:

(1) A perfect radiator at any temperature emits *some* radiation at *all* wavelengths, but not in equal amounts.

(2) A hotter black body emits more radiation (per square centimeter) at *all* wavelengths than does a cooler black body.

(3) A hotter black body emits the largest proportion of its energy at shorter wavelengths than a cooler black body does. Hot stars appear blue because most of their energy is at short wavelengths; cool stars appear red because most of their energy is at long wavelengths.

---

†If T is the temperature in degrees Kelvin, λ the wavelength in centimeters, and k Boltzmann's constant (1.37×10^{-16}), $E(\lambda, T)$, the energy in ergs emitted per unit wavelength interval per second per square centimeter and into unit solid angle, is

$$E(\lambda, T) = \frac{2hc^2}{\lambda^5} \frac{1}{e^{hc/\lambda kT} - 1}.$$

(c) Wien's Law

We notice in Figure 10-17 that there is a particular wavelength, λ_{max}, at which a perfect radiator emits its maximum light. The higher the temperature of the body, the shorter λ_{max}; in fact, λ_{max} is inversely proportional to the temperature:

$$\lambda_{max} = \frac{constant}{T}.$$

If the wavelength is measured in centimeters and the temperature in degrees Kelvin, the constant has the value 0.2897. The foregoing relation is known as *Wien's law.*†

Since stars resemble black bodies, the wavelengths at which they emit their maximum light indicate their temperatures. For the sun, this wavelength is about 4.750×10^{-5} cm, which corresponds to a temperature of 6100°K. In practice, we do not have to observe the wavelength of maximum light; whenever we observe the color index of a star (Chapter 20), we are comparing the intensity of its radiation in two different wavelength regions. We can determine at what temperature a black body would have the same ratio of intensity in the same two wavelength regions, and thereby find the star's temperature. A more accurate comparison with a black body can be obtained if more than two spectral regions are observed. A stellar temperature determined by comparison of the spectral distribution of the star's radiation with that of a black body is called a *color temperature.*

(d) The Stefan–Boltzmann Law

If we sum up the contributions from all parts of the spectrum, we obtain the total energy emitted by a black body over all wavelengths. The total energy emitted per second per square centimeter by a black body at a temperature T is given by the equation known as the *Stefan–Boltzmann law:*

$$E(T) = \sigma T^4.$$

The constant σ, called the *Stefan–Boltzmann constant,* has the value 5.672×10^{-5}, if T is in °K and $E(T)$ is in ergs/cm²/sec. The Stefan–Boltzmann law was derived in the last century from thermodynamics, but it can also be derived from Planck's law.‡ Note that a blue, hot star emits more energy per unit area than does a cool, red star.

10.5 ABSORPTION AND EMISSION OF LIGHT BY ATOMS

The radiation from stars shows that even their outermost, coolest layers have temperatures of thousands of degrees. Stellar temperatures are above the boiling points of all chemical substances. We shall see in Chapter 29 that the material of a star must be completely gaseous throughout its entire structure. Moreover, except in the atmospheres of the cooler stars, the temperatures are too high for chemical compounds or even for simple molecules to exist. Consequently, the gases of stars are in the atomic state. These atoms are responsible for the absorption and emission of light that occurs.

We know that at least some of these atoms can emit a *continuous* spectrum, that is, light over a continuous range of wavelengths, because we see such continuous radiation coming from stars. We know also that at least some of the atoms can absorb radiation (the gas is opaque over large ranges of wavelengths), for there is no wavelength at which we can see into the interiors of stars. Further, we know that the atoms sometimes absorb radiation at certain discrete wavelengths much more strongly than at other wavelengths, for we observe the dark absorption lines in the spectra of the sun and stars,

---

†Wien's law can be derived from Planck's formula by finding the wavelength at which the derivative of Planck's formula equals zero.

‡By integrating Planck's formula.

at which wavelengths light escapes from the stars only with greatly reduced intensity. Finally, we know that atoms can also emit light of these same wavelengths, for we observe emission lines in the laboratory spectra of gases and in the spectra of gaseous nebulae (Chapter 26).

We now consider the structure of atoms and how they give rise to the absorption and emission processes described above.

(a) Structure of Atoms

Atoms are the "building blocks" of matter; they are the smallest particles into which a chemical element can be subdivided and still retain its chemical identity. From experiments in the physics laboratory, we have learned that an atom consists of two basic parts: a nucleus and a system of electrons. The nucleus of an atom contains practically all the atom's mass. The nucleus also carries a positive electric charge. The charge on a nucleus is always an integral (or whole) multiple of a certain small unit of charge.* The atoms of the different chemical elements differ from each other in the number of these charge units on their nuclei. The number of charge units on the nucleus of an atom is called its *atomic number.* An atom of hydrogen (the simplest kind of atom) has an atomic number of 1. The atomic numbers of helium and lithium are 2 and 3, respectively, of oxygen, 8, and of uranium, 92. Atoms of still higher atomic number have been produced in the nuclear physics laboratory, but most of them are unstable and in relatively short times they spontaneously decay into simpler atoms.

The mass of an atomic nucleus is also nearly equal to an integral multiple of a basic unit, called the *atomic mass unit* (amu). The atomic mass unit is about 1.66×10^{-24} g. The mass of a nucleus, in amu, is called its *atomic weight.* Most hydrogen nuclei have an atomic weight of about 1; a few, however, have atomic weights of 2 or even 3.

Atomic nuclei of the same atomic number but different masses are said to compose the different *isotopes* of that element. For example, an isotope of hydrogen in which the atomic nuclei have an atomic weight of 2 is called *deuterium,* or sometimes "heavy hydrogen." Atoms of the most common isotope of helium have a mass of approximately 4, and those of oxygen have a mass of 16.† Most of the heavier atoms have atomic masses 2 to 3 times their atomic numbers.

Nuclei can be decomposed into smaller particles. The most important of these are *protons,* which have a charge of one unit and a mass of approximately one unit (the nucleus of the most common isotope of hydrogen *is* a proton), and *neutrons,* which have a mass of about one unit, but are electrically neutral. The structure of the atomic nucleus will be discussed further in Chapter 29.

An electron has a mass of $\frac{1}{1835}$ amu and carries a *negative* charge that is numerically equal to the positive charge on a single proton. Under ordinary circumstances, an atom contains just as many electrons as its nucleus has positive units of charge, and so is electrically *neutral.* The electrons of an atom are clustered outside the nucleus and revolve about it. The orbital motion of an electron corresponds to a certain amount of energy possessed by the atom, and the larger the orbit, the more the energy. It is possible for an atom to absorb or emit energy (for example, in the form of electromagnetic energy), and in the process one (or more) of its electrons moves, respectively, into a larger or smaller orbit.

(b) Spectrum of Hydrogen

A clue to the structure of atoms came from the study of the spectrum of hydrogen, whose atoms

---

*The unit of charge is 4.80×10^{-10} electrostatic unit.

†In chemical notation the atomic mass unit is $\frac{1}{16}$ the mean mass of the oxygen atom. In the new physical notation, it is $\frac{1}{12}$ the mass of an atom of the most common isotope of carbon. The nearest integer to the atomic mass of a nucleus is called its *mass number.*

FIG. 10-18   Series of Balmer lines in the spectrum of hydrogen.

are the simplest in nature. The dark lines of hydrogen that can be observed in the spectra of many stars occur in an orderly spaced series of wavelengths. The bright lines of hydrogen that are observed in the laboratory spectrum of glowing hydrogen are observed in the same series of wavelengths. The Swiss physicist Balmer found that these wavelengths could be represented by the formula

$$\frac{1}{\lambda} = 0.001097\left(\frac{1}{2^2} - \frac{1}{n^2}\right),$$

where λ is the wavelength in Angstroms and n is an integer that can take any value from 3 on. If $n = 3$, the wavelength of the first line in the so-called *Balmer series* of hydrogen is obtained (at 6563 Å — in the red). For $n = 4$, 5, and so on, the wavelengths of the second, third, and higher Balmer lines are obtained. As n approaches larger and larger values, the wavelengths of the successive Balmer lines become more and more nearly equal. The lines of hydrogen in stellar spectra are observed to do just this; they approach a limit at about 3650 Å (Figure 10-18), corresponding to a value of $n = \infty$.

After Balmer's work, other series of hydrogen lines were found. The *Lyman series,* in the ultraviolet, approaches a limit at about 912 Å. The *Paschen series,* in the infrared, approaches a limit at about 8200 Å. Still farther in the infrared are found the *Brackett series,* the *Pfund series,* and so on. All these series (including the Balmer series) can be predicted by the more general formula, known as the *Rydberg formula:*

$$\frac{1}{\lambda} = R\left(\frac{1}{m^2} - \frac{1}{n^2}\right),$$

where m is an integer and n is any integer greater than m. The *Rydberg constant, R,* has the value

109,678 if λ is measured in centimeters. For the Lyman series, $m = 1$; $m = 2$, 3, 4, . . . for the Balmer, Paschen, Brackett, and so on, series.

(c) The Bohr Atom

The Danish physicist Niels Bohr (1885–1962) suggested that the hydrogen spectrum can be explained if the assumption is made that only orbits of certain sizes are possible for the electron in the hydrogen atom. By specifying those permissible sizes for the electron orbits, Bohr was able to compute the values of energy, corresponding to the orbital motion of the electron, that are possible for an individual atom. He assumed that an atom can change from one allowed state of energy to another state of higher energy if its electron moves from a smaller to a larger allowed orbit. Conversely, according to the hypothesis, if the electron moves from a larger to a smaller orbit, the atom changes from a higher to a lower state of energy. One way in which an atom can gain or lose energy is by absorbing or emitting light. Since light is composed of *photons* whose energies depend on their wavelengths (the energy of a photon is hc/λ), the only wavelengths of light that could be absorbed or emitted are those corresponding to photons possessing energies equal to differences between various allowed energy states of the hydrogen atom. Those energy states are given by a simple formula that Bohr derived from the orderly progression of wavelengths in each series of lines in the hydrogen spectrum. In other words, by assuming that the allowed sizes of the electron orbits and thus the possible energies of the hydrogen atom are *quantized,* Bohr was able to account for the spectrum of hydrogen.

For example, suppose a beam of white light (which consists of photons of all wavelengths) is passed through a gas of atomic hydrogen. A photon of wavelength 6563 Å has the right energy to raise an electron in a hydrogen atom from the second to the third orbit, and can be absorbed by those hydrogen atoms that are in their second to lowest energy states. Since the energy of a photon is inversely proportional to its wavelength, the shorter the wavelength of a photon, the higher its energy. Photons with higher energies corresponding to the other successively shorter wavelengths in the Balmer series, therefore, have the right energies to raise an electron from the second orbit to the fourth, fifth, sixth, and larger orbits, and can also be absorbed. Photons with intermediate wavelengths (or energies) cannot be absorbed. Thus, the hydrogen atoms absorb light only at certain wavelengths, and produce the spectral *lines*. Conversely, hydrogen atoms in which electrons move from larger to smaller orbits emit light — but again only light of those energies or wavelengths that correspond to the energy differences between permissible orbits.

The transfer of electrons giving rise to spectral lines is shown in Figure 10-20.

A similar picture can be drawn for kinds of atoms other than hydrogen. However, since they ordinarily have more than one electron each, the energies of the orbits of their electrons are much more complicated, and the problem of their spectra is much more difficult to handle theoretically.

(d) Energy Levels of Atoms and Excitation

Bohr's model of the hydrogen atom was one of the beginnings of the quantum theory, and was a great step forward in the development of modern physics and in our understanding of the atom. However, we know today that atoms cannot be represented by quite so simple a picture as the Bohr model. Even the concept of discrete orbits of electrons must be abandoned, because according to the modern quantum theory it is impossible at any

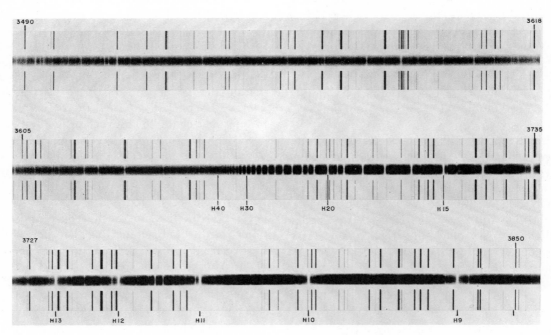

FIG. 10-19 Photograph of the spectrum of the star HD 193182, showing Balmer series down to and including the continuum. Several of the lines in the series are identified by number. *(Mount Wilson and Palomar Observatories.)*

instant to state the exact position and the exact velocity of an electron in an atom simultaneously. Nevertheless, we still retain the concepts that only certain discrete energies are allowable to an atom. These energies, called *energy levels,* can be thought of as representing certain mean or average distances of an electron from the atomic nucleus.

Ordinarily, the atom is in the state of lowest possible energy, which, in the Bohr model, would correspond to the electron being in the innermost orbit. However, an atom can absorb energy which raises it to a higher energy level (corresponding, in the Bohr picture, to the movement of an electron to a larger orbit). The atom is then said to be in an *excited state.* Generally, an atom remains excited only for a very brief time; after a short interval, typically 1 hundred-millionth of a second or so, it drops back down to its lowest energy state, with the simultaneous emission of light, unless it chances to absorb another photon first and go to a still higher state. (In the Bohr model, this corresponds to a jump by the electron back to the innermost orbit.) The atom may return to its lowest state in one jump, or it may make the transition in steps of two or more jumps, stopping at intermediate levels on the way down. With each jump, it emits a photon of the wavelength that corresponds to the energy difference between the levels at the beginning and end of that jump. An energy-level diagram for a hydrogen atom and several possible *atomic transitions* are shown in Figure 10-21; compare this figure with the Bohr model, shown in Figure 10-20.†

Because atoms that have absorbed light and

---

†Actually, according to the modern quantum theory, atomic energy levels are not *perfectly* discrete (that is, "sharp"), but are, rather, the *most probable* energies of an atom. The vast majority of atoms will have energies that lie in the immediate neighborhood of one of the allowed levels. At any time, however, a few atoms will possess energies that deviate somewhat from one of those values. Spectral lines caused when atoms jump from one energy level to another, therefore, are not perfectly sharp, but spread over a small, but finite, range of wavelengths. The effect is called *natural line broadening* (Chapter 21).

have thus become excited generally deexcite themselves and emit that light again, we might wonder why dark lines are ever produced in stellar spectra. In other words, why doesn't this reemitted light "fill in" the absorption lines? Some of the reemitted light actually *is* received by us and this light does partially fill in the absorption lines, but only to a slight extent. The reason is that the atoms reemit the light they absorb in random directions. Now the absorption of light we would otherwise observe is of that light "coming our way" from deeper, hotter levels in the star's atmosphere. The light those same atoms reemit, however, has approached them from all directions — from higher, cooler ones as well as deeper, hotter ones. Thus the reemitted light is less intense than that which was absorbed from the deeper levels, and the fraction of it that gets into the beam and enters our telescope is not sufficient to replace that from the hot gases that was absorbed. Therefore, the dark lines persist. We can observe the reemitted light as emission lines only if we can view the absorbing atoms from a direction from which no light with a continuous spectrum is coming — as we do, for example, when we look at gaseous nebulae (Chapter 26). Figure 10-22 illustrates the situation.

We can calculate from theory the allowable energies of the simplest kinds of atoms and thus the wavelengths of the absorption lines that can be produced by those atoms. For the more complicated atoms, however, the wavelengths of the spectral lines are determined empirically by laboratory experiment.

Atoms in a gas are moving at high speeds and are continually colliding with each other. They can be excited and deexcited, therefore, by these collisions as well as by absorbing and emitting light. The mean velocity of atoms in a gas depends upon its temperature (actually, *defines* the temperature), and if we know the temperature of the gas, we are able to calculate what fraction of those atoms, at any given time, will be excited to any given energy level. In the photosphere of the sun,

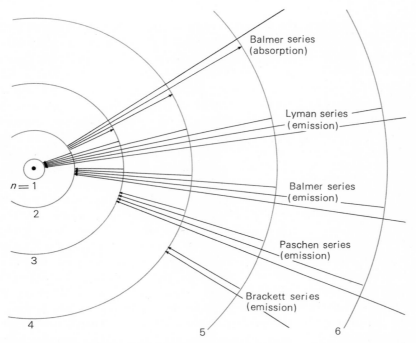

FIG. 10-20   Emission and absorption of light by the hydrogen atom according to the Bohr model.

for example, where the temperature is in the neighborhood of 6000°K, only about 1 atom of hydrogen in 100 million is excited to its second energy level by the process of collision. The Balmer lines of the hydrogen spectrum arise when atoms in this second energy level absorb light and rise to higher levels. At any given time, therefore, most atoms of hydrogen in the sun cannot take part in the production of the Balmer lines.

(e) Ionization

We have described how certain discrete amounts of energy can be absorbed by an atom, raising the atom to an excited state, and moving one of its electrons farther from its nucleus. If enough energy is absorbed, the electron can be removed completely from the atom. The atom is then said to be *ionized*. The minimum amount of energy required to ionize an atom from its lowest energy state is called its *ionization energy* or *ionization potential*. Still greater amounts of energy must then be ab-

sorbed by the ionized atom (called an *ion*) to remove a second electron. The minimum energy required to remove this second electron is called the *second ionization energy* or *potential*. The third, fourth, and fifth ionization potentials are the successively greater energies required to remove the third, fourth, and fifth electrons from the atom, and so on. If enough energy is available (in the form of very short wavelength photons or in the form of a collision with a very fast moving atom) an atom can become *completely* ionized, in which case it loses all of its electrons. A hydrogen atom, having only one electron to lose, can be ionized only once; a helium atom can be ionized twice, and an oxygen atom, eight times. Any energy over and above the energy required to ionize an atom can be absorbed also, and appears as energy of motion (kinetic energy) of the freed electron. The electrons released from atoms that have been ionized move about in the gas as free particles, just as the atoms and ions do.

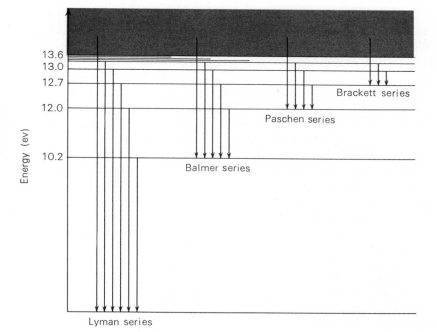

FIG. 10-21 Energy-level diagram for hydrogen. The shaded region represents energies at which the atom is ionized (See Section 10.5e for explanation of ionization.)

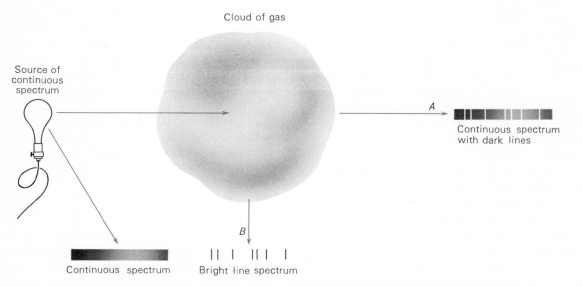

FIG. 10-22 The atoms in the gas cloud produce absorption lines in the continuous spectrum of the white light source when viewed from direction *A*, but produce emission lines (of the light they reemit) when viewed from direction *B*. The spectra are shown as they appear on photographic negatives.

An atom that has become ionized has lost a negative charge — that carried away by the electron — and thus is left with a net positive charge. It has, therefore, a strong affinity for a free electron, and eventually will capture one and become neutral (or ionized to one less degree) again. During the capture process, the atom emits one or more photons, depending on whether the electron is captured at once to the state corresponding to the lowest energy level of the atom, or whether it stops at one or more intermediate levels on the "way in." Any energy that the electron possessed as kinetic energy before capture can be emitted. Absorption or emission of light over a continuum of wavelengths therefore accompanies the process of ionization or recapture, at least over those wavelengths corresponding to energies higher than the ionization energy of the atom. Atoms of various kinds, being ionized and deionized, and absorbing and emitting light at various wavelengths, account for much of the continuous opacity and continuous spectrum of the sun and stars.

Just as the excitation of an atom can result from a collision between it and another atom, so also can ionization. The rate at which such collisional ionizations occur depends on the atomic velocities and hence on the temperature of the gas. The rate of recombinations of ions and electrons also depends on their relative velocities — that is, on the temperature — but in addition it depends on the density of the gas; the higher the density, the greater the chance for recapture, because the different kinds of particles are crowded closer together. From a knowledge of the temperature and density of a gas, it is actually possible to calculate the fraction of atoms that have been ionized once, ionized twice, and so on. In the photosphere of the sun, for example, we find that most of the hydrogen and helium atoms are neutral, whereas most of the atoms of calcium, as well as many other metals, are once ionized.

The energy levels of an ionized atom are entirely different from those of the same atom when it is neutral. In each degree of ionization, the energy levels of the ion, and thus the wavelengths of the spectral lines it can produce, have their own characteristic values. In the sun, therefore, we find lines of *neutral* hydrogen and helium, but of *ionized* calcium. Ionized hydrogen, of course, having no electron, can produce no absorption lines.

There is an additional mechanism by which ionized atoms can absorb and emit light in the continuous spectrum. When an electron passes near an ion, it is attracted by that ion's positive charge (because opposite electrical charges attract each other). If the electron is not captured, it will pass the ion in a hyperbolic orbit, much like two stars passing in interstellar space. While the electron is passing, however, the ion can absorb or emit a photon which accompanies a corresponding increase or decrease in the kinetic energy of the hyperbolic motion of the electron. The process is called *free-free* absorption or emission (because the electron is "free" of the ion both before and after the encounter), or *bremsstrahlung*. Photons of any wavelength can be absorbed or emitted in bremsstrahlung.

(f) Molecular Spectra*

Molecules, combinations of two or more atoms, can also absorb or emit light. In addition to undergoing electronic transitions, molecules can also rotate and vibrate, all of which involve energy. The quantum theory predicts that the energy of vibration and rotation of molecules is quantized, like the energy of atoms. The vibrational and rotational energies are generally low, but they add to or subtract from the energies corresponding to electronic transitions. Consequently, in place of each atomic energy level there is a series of closely spaced levels, each one corresponding to a different mode of vibration or rotation of the molecule. Many more different transitions between energy levels are possible, therefore, differing from each other only slightly in energy or wavelength. Molecules, in

FIG. 10-23   Series of closely spaced lines comprising *bands* in the spectrum of the molecule, titanium oxide. Note how the bands coalesce to form *band heads. (Mount Wilson and Palomar Observatories.)*

other words, produce series of closely spaced lines known as *molecular bands.* In spectra of those stars in which molecules exist, these many molecular lines within a band are often not resolved as separate, and only a single broad absorption feature is observed.

(g) Summary of Emission and Absorption Processes
Atoms are characterized by energy levels which correspond to various distances of their electrons from their nuclei. By absorbing or emitting radiant energy, an atom can move from one to another of these levels, thus raising or lowering its energy. Since only certain discrete energy levels exist for each kind of atom, the absorbed or emitted radiation occurs only at certain energies or wavelengths, producing dark or bright spectral lines.

An atom is said to be *excited* if it is in any but its lowest allowable energy level. It is said to be *ionized* if, by the absorption of energy, it has lost one or more of its electrons. It can be excited or deexcited by collisions as well as by the absorp-

tion and emission of radiation; it can be ionized by collision or by absorbing radiation. Any wavelength of light corresponding to an energy greater than the ionization energy of an atom can be absorbed or emitted in the process of ionization or of deionization when the ion captures an electron.

Ionized atoms (*ions*) can also absorb or emit light at continuous wavelengths, when free electrons pass near them. The relative energy of the ion and passing electron changes by the same amount as the energy absorbed or emitted.

The quantum theory is a complicated mathematical subject by which we have been able to account for the observed behavior of atoms and their interactions with radiation. It is based on assumptions that may seem arbitrary, but in this respect it serves as another example of how science operates. The justification of the quantum theory is that it "works" — that is, it successfully predicts the observed phenomena.

Exercises

1. How many times brighter or fainter would a star appear if it were moved to
(a) Twice its present distance?
(b) Ten times its present distance?
(c) One half its present distance?

2. "Tidal waves," or *tsunamis,* are waves of seismic origin that travel rapidly through the ocean. If tsunamis traveled at the speed of 400 mi/hr, and approached a shore at the rate of one wave crest every 15 minutes, what would be the distance between those wave crests at sea?

3. Because of refraction, the sun appears to rise before it is above the geometrical horizon and to set after it has dropped below the geometrical horizon. By how much does atmospheric refraction increase the hours of sunshine in a typical day?

Answer: About 4 minutes or more

4. Suppose that a spectral line of some element, normally at 5000 Å, is observed in the spectrum of a star to be at 5001 Å. How fast is the star moving toward or away from the earth?

Answer: About 37 mi/sec away from the earth

5. How could you measure the rotation rate of the sun by photographing the spectrum of light coming from various parts of the sun's disk?

6. How could you measure the earth's orbital speed by photographing the spectrum of a star at various times throughout the year?

7. Radiation with a spectral energy distribution similar to the spectral response of the human eye appears white. Discuss the color of sunlight.

8. What color would you expect light to appear that was equally intense at all visible wavelengths?

9. What is the temperature of a star with a wavelength of maximum light of 2.897×10^{-5} cm?

10. Only the Balmer series of hydrogen lines is ordinarily observed in the spectra of stars. Can you suggest an explanation for why the Lyman series, for example, is hard to observe? What about the other series of hydrogen lines?

11. Refer to the Rydberg formula and indicate to what series each of the following transitions in a hydrogen atom corresponds. State whether the spectral line produced is one of absorption or emission. List the lines in order of *increasing wavelength.*

(a) $m = 3$ to $n = 8$ (d) $m = 2$ to $n = 13$
(b) $m = 1$ to $n = 2$ (e) $n = 17$ to $m = 2$
(c) $m = 1$ to $n = 4$ (f) $n = 14$ to $m = 5$

12. Most hydrogen atoms in the sun are in their lowest state of energy, so what series of absorption lines would hydrogen produce most strongly in the sun? Do we observe these lines in the solar spectrum? If not, why not?

CHAPTER 11

Astronomical
Instruments

Astronomical instruments have been used since ancient times, but before Galileo they did not employ optical systems. Most of the instruments used by Hipparchus and Ptolemy, and the later, more refined ones of Tycho Brahe, were essentially calibrated "sighting" devices for measuring the directions in the sky of celestial objects. Brahe, for example, designed and built accurate sextants and quadrants, in which sights for observing a star were carried on a long pivoted arm. The end of the arm moved along an arc of a circle with a graduated scale on which the precise angle of the arm and hence of the sights could be read, indicating in turn the angle of the star being observed above the horizon (that is, the star's altitude). Sundials also date from early times.

It is to the astronomical telescope, however, beginning with Galileo (Section 3.5b), that we now turn our attention.

11.1 FORMATION OF AN IMAGE

The most important part of a telescope is the *objective,* the lens or mirror that produces an image. First, therefore, we consider how an image is formed.

(a) Formation of an Image by a Lens

Having examined the laws of geometrical optics, we are in a position to understand the formation of images by optical systems. Images were first produced by simple convex lenses. To illustrate the principle, let us imagine two triangular prisms, base to base, as in Figure 11-1. Now suppose we select two of the parallel rays of light from a dis-

tant object and allow one ray to enter each prism. The light rays are refracted by the prisms and meet at a point *F*.

This is the principle underlying the formation of an image by a lens. In Figure 11-2 a simple convex lens is shown. Parallel light, say, from a distant star or other light source, is incident upon the lens from the left. A convex lens is a lens thicker in the middle than at the edges. In cross section it is no more than a series of segments of prisms piled one on another, the different prisms being constructed with slightly different slopes of their sides. If the curvatures of the surfaces of the lens are correct, light passing through the lens will be refracted in such a way that it converges toward a point. Convex lenses whose surfaces are portions of spherical surfaces are easiest to manufacture. Such lenses will refract a parallel beam of light to a point as shown in Figure 11-2, if the curvature of the surfaces is slight.

The point where light rays come together is called the *focus* of the lens. At the focus, an *image* of the light source appears. The distance of the focus, or image, behind the lens is called the *focal*

FIG. 11-1 Principle of image formation.

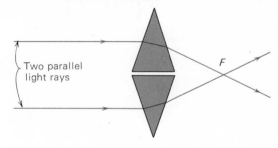

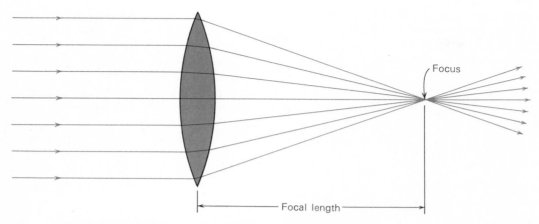

FIG. 11-2   Formation of an image by a simple convex lens.

length of the lens. A lens or other device that forms an image is called an *objective*.

We have seen how an image can be formed of a point source, say, a star. However, the image itself in that case is just a point of light. In Figure 11-3 we see how an image is formed of an extended source, for example, the moon. From each point on the moon, light rays approach the lens along parallel lines. However, from different parts of the moon, the parallel rays of light approach the lens from different directions. The light from each point on the moon strikes all parts of the lens (or *fills* the lens); these rays of light are focused at a point at a distance behind the lens equal to the focal length of the lens. If a screen, such as a white card, is placed at this distance behind the lens, a bright spot of light appears, representing that point on the moon. Light from other points on the moon similarly focuses at other points, producing bright spots on the card in different places. Thus, an entire image of the moon is built up at the focus of the lens. The plane in which the image is formed is called the *focal plane*.

Note that if part of the lens is covered up, or if the middle is cut out, or if ink or mud is splattered over it, as long as part of the lens is still

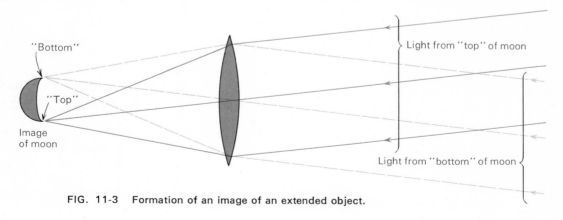

FIG. 11-3   Formation of an image of an extended object.

transparent to light, an entire image will be formed. All parts of the lens contribute to each part of the image. Covering up part of the lens cuts down the total amount of light that can strike each portion of the image and thus makes the image fainter, but nevertheless the whole image is formed. An ordinary camera lens produces an image at the focal plane (where the film is placed) just as is shown in Figure 11-3. Every photographer knows that he can cover up part of his camera lens by "stopping it down" with an iris diaphragm. In so doing he will cut down the *brightness* of the image (and hence the effective exposure on the film), but the outer parts of the image will not be removed. The part of the lens that remains uncovered still produces the entire image.

We have discussed the case only where the object whose image is formed is so distant that light from any point on it can be regarded as approaching the lens along parallel rays. This is always true when any astronomical body is observed. Nearby terrestrial objects may be so close that the assumption is not valid. Then the image is formed at a point farther from the lens than the focal length.

Finally, we note that the image formed is always inverted and reversed (upside down and left to right) with respect to the object.

FIG. 11-4   Formation of an image by a concave mirror.

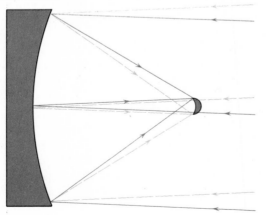

The human eye provides a good example of a simple convex lens. Light passing through the lens of the eye forms an image on the retina, a membrane at the rear of the eyeball that contains light-sensitive nerve endings. The eye lens forms inverted images on the retina, but the brain interprets the image so that it appears erect. The eye is equipped with an adjustable iris diaphragm that automatically enlarges or contracts to allow more or less light to enter the eye, so that an image of the optimum brightness can be produced on the retina. An ordinary camera is a nearly complete analogy to the eye.

(b) Formation of an Image by a Mirror

Rays of light can also be focused to form an image with a concave mirror — one hollowed out in the middle. Parallel rays of light, as from a star, fall upon the curved surface of the mirror (Figure 11-4), which is coated with silver or aluminum to make it highly reflecting. Each ray of light is reflected according to the law of reflection. If the mirror has the correct concave shape, all the rays are reflected back through the same point, the focus of the mirror. At the focus appears the image of the star. As in a lens, the distance from the mirror to the focus is its focal length.

Rays of light from an extended object are focused by a mirror, exactly as they are by a lens, into an inverted image of the object. The principal difference between image formation by a lens and by a mirror is that the mirror reflects the light back into the general direction from which it came, so that the image forms in front of the mirror. The image can be inspected, as with the lens, by allowing the light to illuminate a screen, such as a white card, held at the focus of the mirror. The card, of course, will block off part of the incoming light, but since, as in the lens, all parts of the mirror contribute to the formation of all parts of the image, the presence of the card will not produce a "hole" in the image but will merely reduce its brightness.

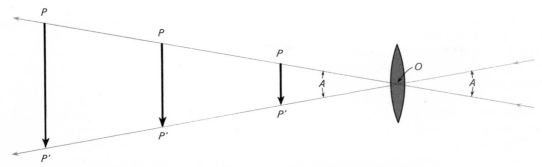

FIG. 11-5   The scale of an image is proportional to the focal length of the lens or mirror.

If the shape of the mirror is part of a concave spherical surface, it will produce a fair-quality image, provided that the entire mirror constitutes only a very small part of a sphere, so that its size is small compared to its focal length. We shall see that the curves of telescope mirrors are usually parabolic in cross section.

11.2 PROPERTIES OF AN IMAGE

The most important properties of an image are its *scale* or size, its *brightness,* and its *resolution.* We shall consider these in turn.

(a) Scale of an Image

The *scale* of an image is a measure of its size. In all astronomical applications we are dealing with objects whose sizes can be expressed in angular units, and it is generally convenient to express the scale of an image as the linear distance in the image that corresponds to a certain angular distance in the sky. For example, suppose that an image of the moon were produced that is exactly 1 in. across. The moon has an "apparent" or angular size of ½°; that is, it *subtends* ½° in the sky (see Section 2.3a). The scale of the image is thus ½° per inch or, equivalently, 2 in. per degree.

The scale of an image depends only on the focal length of the lens or mirror that produces it.

It can be shown that light rays passing through the center of a lens are not deflected. If two rays pass through the center of a lens at an angle A to each other (as from two stars separated by that angle in the sky, or as from two sides of an extended object, like the moon), they leave the back of the lens at the same angle A to each other. The distance behind the lens that the image forms is the focal length of the lens, which depends on the curvature of the lens surfaces. But wherever the image forms, the two rays will pass through the image at points P and P', corresponding to points in the sky separated by the angle A (Figure 11-5). Thus P, P', and O form a triangle with base PP' and altitude equal to the focal length of the lens. It is seen that the distance PP' in the image, corresponding to an angle A in the sky, is directly proportional to the focal length of the lens. The same is true of the image formed by a mirror.

Numerically, the distance *s*, in an image corresponding to 1° in the sky, is given by the equation

$$s = 0.01744f,$$

where *f* is the focal length of the lens or mirror. The scale of an image, then, can be computed from the above formula. The 200-in. mirror of the Hale telescope on Palomar Mountain has a focal length of 660 in. It produces images with a scale of 11.5

in. per degree, which corresponds to 12″3 per millimeter (for a definition of units of angular measure, see Appendix 3). Most astronomical telescopes have lenses or mirrors that give image scales in the range 10″ to 200″ per millimeter.

(b) Brightness of an Image

The *brightness* of an image is a measure of the amount of light energy that is concentrated into a unit area of the image, say, per square inch or square millimeter. The brightness of an image determines whether it is above the threshold of visibility or, alternatively, how long a time would be required to record the image photographically. When we consider the factors that determine the brightness of an image, we must distinguish between the image of an extended object and the image of a point source.

First we consider the brightness of an image of an extended object, such as the moon, a planet, a nebula, a galaxy, or the faint illumination of the night sky. The amount of light that is concentrated into a unit area of the image must be proportional to the area of the lens or mirror that gathers and focuses the light. Since the area of a circular aperture is equal to π times the square of its radius (or $\pi/4$ times its diameter squared), the image brightness must be proportional to the square of the diameter of the lens or mirror. But also, the amount of light per unit area of the image must be inversely proportional to the total area of the image over which the light must be spread. The area of the image is proportional to the square of its linear size, and the latter, as we have seen, is proportional to the focal length of the lens or mirror. Therefore, the brightness of the image of an extended object is proportional to the square of the diameter of the lens, which determines how much light gets into the entire image, and is inversely proportional to the square of the focal length, which determines over how large an area this total light must be spread.

To summarize, we can say that the brightness B of an extended image is given by

$$B = \text{constant} \times \left(\frac{a}{f}\right)^2,$$

where a is the diameter or aperture and f is the focal length of the lens or mirror. The constant of proportionality is a number that depends on the units chosen to measure the various quantities, and also on the amount of light actually leaving each unit area of the object itself. The quantity f/a is called the *focal ratio* or simply the *f ratio* of the lens. In common notation, if the focal length is, say, eight times the aperture of the lens, the focal ratio is written $f/8$, which should be interpreted as $a/f = \frac{1}{8}$. (Note that $f/8$ means $a = f/8$). Every photographer is familiar with the concept of focal ratio when applied to his camera lens. In all but the simplest cameras the clear aperture of the lens can be increased or decreased by adjusting an iris diaphragm, thereby changing the focal ratio of the lens. A typical 35-mm camera, for example, might have focal ratio adjustments varying from $f/3.5$ to $f/16$. Since the focal ratio determines the image brightness, it determines the effective exposure on the film for a given exposure time — that is, the time required to expose the image to a particular blackness. For this reason, the focal ratio is often called the "speed" of an optical system. In high-speed systems the focal length is short compared to the lens size. The focal ratio is thus smaller (that is, the ratio of aperture to focal length is a larger number) in high-speed lenses than in low-speed lenses.

Now consider the brightness of the image of a point source. Stars are effectively point sources, for even through the largest telescopes the stars still appear too small to show any apparent disks. Therefore a lens or mirror of good quality concentrates the starlight into a "point" image regardless of the focal length. For a point-source object such as a star, the amount of light in the image thus depends only on the amount of light gathered by the lens or mirror, and hence is proportional to the square of the aperture.

In summary, the brightness of the image of an extended object is proportional to the square of the ratio a/f of the lens or mirror, whereas the brightness of the image of a point source is proportional to the square of the aperture.

(c) Resolution

Resolution refers to the fineness of detail inherently present in the image. Even if the lens or mirror is of perfect optical quality, it cannot produce perfectly sharp and detailed images. Because of the phenomenon of *diffraction,* a point source does not form an image as a true point but as a minute spot of light surrounded by faint, concentric, evenly spaced rings. The angular size of that central spot of light, called the central *diffraction disk* of the image, is inversely proportional to the aperture of the lens or mirror, and directly proportional to the wavelength of the light observed. No detail can be resolved in the image if that detail is smaller than the diffraction disk. For example, the diameter predicted by geometrical optics for the image of a star produced by the lens of a telescope is much smaller than the size of the diffraction disk. Therefore, we do not see the geometrical image of a star itself but only a *diffraction pattern* that the telescope lens produces with the star's light. If a star is viewed telescopically under good conditions, the diffraction pattern, consisting of the bright central disk and faint surroundings, is clearly visible.

In the image of an extended source, the diffraction patterns of the various parts of the image, all overlapping each other, wash out the finest details. A feature on the surface of the moon or Mars that is smaller than the diffraction disk produced by the telescope lens may be visible, but its true size and shape will not be distinguishable. If two stars are so near each other that their images are closer together than the size of the diffraction disk of either, they will blend together to produce a single spot, or perhaps a slightly elongated diffraction pattern. The ability of an optical system (lens or mirror) to distinguish fine detail in an image it produces, or to produce separate images of two close stars, is called its *resolving power.* In astronomical practice, the resolving power of a lens or mirror is described in terms of the smallest angle between two stars for which separate recognizable images are produced. Two stars that lie less than 1′ from each other cannot be separated with the human eye. With a 6-in. lens of good quality, separate images are produced of two stars only 1″ apart. The 6-in. lens thus has higher resolving power than the eye.

We have said that the angular size of the central diffraction disk of a point source depends on the wavelength of light used and the diameter of the lens or mirror. If the wavelength and aperture are measured in the same units (for example, both in centimeters), the smallest angle (α) in seconds of arc that can be resolved by a lens or mirror of aperture d is given by the equation,

$$\alpha = 2.1 \times 10^5 \times \frac{\lambda}{d} \text{ seconds of arc,}$$

where λ is the wavelength. If λ is chosen as 5.5×10^{-5} cm (5500 Å) — near the middle of the visible spectrum — and if d is measured in inches, the formula becomes

$$\alpha = \frac{4.56}{d} \text{ seconds of arc.}$$

The above formulas do not hold for the eye because the coarse structure of the retina limits the resolution ideally obtainable with the eye lens. Also, as we shall see in Section 11.7, atmospheric turbulence usually degrades the actual resolving power of large telescopes below the theoretical value.

11.3 ABERRATIONS OF LENSES AND MIRRORS

An image produced by any optical system always has imperfections. These are called *aberrations*. A few of the more important kinds of aberrations will be discussed here. Aberrations are always most serious in telescopes of low focal ratios (high speed).

(a) Chromatic Aberration

Chromatic or *color aberration* is a consequence of dispersion. Consider the simple lens shown in Figure 11-6. Suppose parallel rays of white light from a remote star are incident upon the left side of the lens. Because the different wavelengths that comprise white light are refracted in different amounts upon entering and leaving the lens, they do not all focus at the same place. The shorter wavelengths (violet and blue light) are bent the most, and focus nearest the lens, while the longer wavelengths of orange and red light focus farther from it. The effect of this chromatic aberration is to produce color fringes in the image.

Chromatic aberration is less serious in lenses of large focal ratio. Some telescopes constructed in the late seventeenth century, therefore, utilized lenses of extremely long focal length to obtain as large a focal ratio as possible for a given aperture. Telescopes more than 100 ft long were in common use, and ones with lengths of up to 600 ft are said to

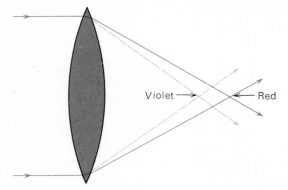

FIG. 11-6 Chromatic aberration.

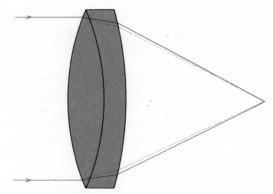

FIG. 11-7 Correction of chromatic aberration.

have been built. Such extremely long telescopes, however, were so difficult to use that they were, for the most part, impractical.

Chromatic aberration is now corrected by constructing a lens of two pieces of glass of different types (usually crown and flint glass) and thus of different indices of refraction. One of the pieces of glass, or *elements* of the lens, is a *concave* lens, that is, it is thinner in the middle than at the edges, so that it diverges the light rather than converging it. (The concave element of the lens diverges light only enough to correct for the chromatic aberration; the combined lenses still converge incident light to a focus.) The light of shorter wavelengths is diverged the most, just as it is converged the most in the usual convex lens. Dispersion, like refraction, is greater in the glass of higher index. It is possible to make the chromatic aberration completely cancel out any two given wavelengths, if exactly the right curvatures are chosen for the surfaces of the two lenses (Figure 11-7). Lenses can be designed so that this aberration is nearly canceled out, or is greatly reduced, over a considerable range of wavelengths.

When a lens is designed, a choice must be made as to the range of wavelengths for which the chromatic aberration is to be corrected. Lenses designed for photographic purposes are usually cor-

40-inch refracting telescope.
(...rkes Observatory.)

The 48-inch Schmidt telescope.
(Mount Wilson and Palomar Observatories.)

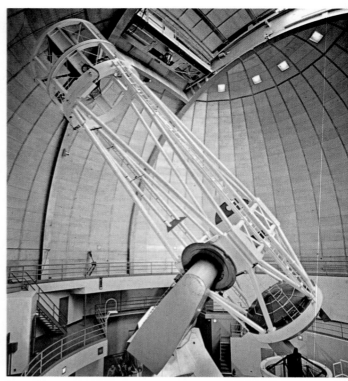

The 200-inch telescope dome photographed in moonlight. *(Mount Wilson and Palomar Observatories.)*

The 120-inch telescope. *(Lick Observatory.)*

Lick Observatory from the east. *(Lick Observatory.)*

The 200-inch telescope dome. *(Mount Wilson and Palomar Observatories.)*

Interior of the observer's cage at the prime focus of the 200-inch telescope. *(Mount Wilson and Palomar Observatories.)*

200-inch telescope. *(Mount Wilson and Palomar Observatories.)*

The earth from 22,300 miles, photographed with the Applicati
Technology Satellite in synchronous orbit on November 10, 1967. *(NAS*

rected for the blue spectral region, because the common photographic emulsions are most sensitive to blue light. On the other hand, lenses designed for visual observation are corrected for the green and yellow spectral region, to which the eye is most sensitive. If the image produced by a lens corrected for photographic purposes is viewed directly, disturbing color fringes are often observed.

When an image is produced by a mirror, the light never has to pass through any glass, and so is not dispersed. Thus an image produced by a mirror does not suffer chromatic aberration.

(b) Spherical Aberration

Spherical surfaces are the most convenient to grind and polish, whether they be the concave or convex surfaces of a lens or mirror. Unfortunately, however, a simple lens or mirror with spherical surfaces suffers from the imperfection called *spherical aberration*. Figure 11-8 illustrates spherical aberration. Light striking nearer the periphery focuses closer to the lens or mirror; light striking near the center focuses farther away.

In a lens, spherical aberration can be corrected or greatly reduced by constructing the lens of two pieces of glass or elements of different indices of refraction, just as in the correction of chromatic aberration. The elements are so designed that the spherical aberration introduced by one is canceled out by the other. Most lenses designed for astro-

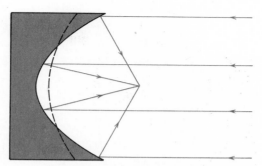

FIG. 11-9   Correction of spherical aberration with a concave parabolic mirror.

nomical purposes consist of two elements. The curvatures of the spherical surfaces of the elements are designed to reduce both the spherical and chromatic aberrations to a tolerable amount.

For a mirror, spherical aberration can be eliminated by grinding and polishing the surface not to a spherical shape, but to a *paraboloid of revolution*, that is, to a surface whose cross section is a parabola (Figure 11-9). A paraboloid has the property that parallel light rays striking all parts of the surface are reflected to the same focus. Similarly, light leaving the focus of a paraboloid is reflected at the surface in a parallel beam. An automobile headlight or searchlight, for example, has a parabolic reflector with the light source at the focus. Mirrors designed for astronomical telescopes generally have parabolic surfaces.

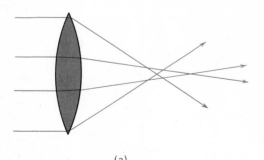

(a)

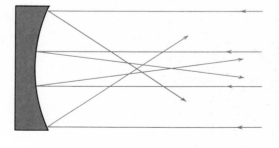

(b)

FIG. 11-8   Spherical aberration: (a) in lens; (b) in mirror.

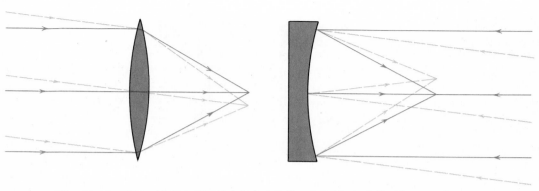

FIG. 11-10 "On axis" (solid lines) and "off axis" (dashed lines) rays of light striking a lens or mirror.

(c) Other Aberrations

The other most important aberrations of lenses and mirrors are the following.

COMA

Coma is an aberration that distorts images formed by light rays that do not strike the lens or mirror "square on" but at a glancing angle, or technically, those that approach the lens or mirror *off axis* (see Figure 11-10). This off-axis distortion is particularly serious in the images formed by the parabolic mirrors used in the largest existing telescopes. Photographs taken with such telescopes are reproduced in the latter part of this book. The reader should carefully inspect one of these (or should inspect Figure 11-11). He will find that the star images near the center of the picture appear as sharp round dots, whereas those near the corners, which are formed by light entering the telescope off axis, are distorted into tiny "tear drops" or "commas" pointing toward the center of the photograph. The cometlike shape of these images accounts for the name "coma" given to this particular aberration.

ASTIGMATISM

Astigmatism is an aberration produced in an optical system when rays of light approaching the lens in different planes do not focus at the same spot. In the example shown in Figure 11-12, the rays in a vertical plane focus farther from the lens than those in a horizontal plane. A geometrically perfect lens produces astigmatism only for off-axis rays.

FIG. 11-11 Photograph of the globular cluster M3 (at left) taken with a large reflecting telescope. Note how the distortion due to *coma* increases from center to edge. (The original photograph is centered on the cluster.) *(Mount Wilson and Palomar Observatories.)*

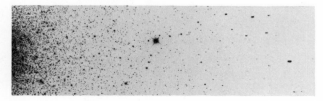

FIG. 11-12 Astigmatism.

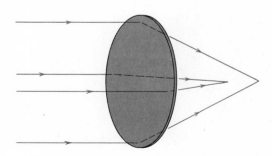

DISTORTION

Distortion is an aberration in which the image forms sharply but its shape is distorted. For example, straight lines in the object may image as curved lines.

CURVATURE OF FIELD

Curvature of field is an aberration in which the image is sharp but different parts of it are formed at different distances from the lens or mirror, so that the image cannot be formed on a flat screen or photographic plate.

11.4 THE SCHMIDT OPTICAL SYSTEM

An ingenious optical system that utilizes both a mirror and a lens was invented by the German optician Bernhard Schmidt of the Hamburg Observatory. We noted in the last section that images formed by a mirror do not suffer from chromatic abberation. To avoid spherical aberration, parabolic rather than spherical mirrors are used in practice. The principal disadvantage of the parabolic mirror is that it produces good images over only a relatively small field of view, that is, for light that approaches the mirror very nearly on axis.

On the other hand, for a spherical surface, any line reaching the surface through its center of curvature (that is, through the center of the sphere of which the surface is a part) is perpendicular, that is, "square on" to the surface. Thus, all rays of light that pass through an aperture (or opening) located at the center of curvature of a spherical surface must strike the surface "on axis." The Schmidt optical system, utilizing this principle, employs a spherical mirror that is allowed to receive light only through an opening located at its center of curvature (Figure 11-13). Thus there can be no off-axis aberration. The only trouble is that a spherical mirror, suffering as it does from spherical aberration, produces generally poor images for light coming from any direction. Schmidt solved the problem by introducing a thin correcting lens at the aperture at the center of curvature of the mirror. The lens is of the proper shape to correct the spherical aberration introduced by the spherical mirror but does not have to be thick enough to introduce appreciable aberrations of its own. Thus, the Schmidt optical system produces excellent images over a large angular field. A disadvantage of the Schmidt system is that the focal surface is not a plane but a sphere concentric with the spherical mirror.

Today, the Schmidt optical system, and modifications of it invented by Maksutov, Wright, and others, are widely used in science and industry.

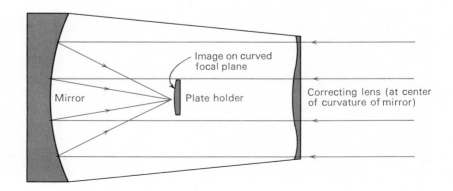

FIG. 11-13 Schmidt optical system. The mirror is generally larger than the lens, so images of unreduced intensity can be produced over an appreciable field of view.

Image on curved focal plane

Mirror

Plate holder

Correcting lens (at center of curvature of mirror)

11.5 THE COMPLETE TELESCOPE

Now that we have seen how an image can be produced, we are able to understand the operation of a telescope. Today there are two general kinds of astronomical telescopes in use. They are (1) *refracting* telescopes, which utilize lenses to produce images; and (2) *reflecting* telescopes, which utilize mirrors to produce images. As we have seen, Schmidt telescopes, and modifications thereof, are primarily reflecting telescopes but have refracting correcting lenses.

The refracting telescope is the most familiar. This is the kind of telescope that we can literally "look through." Ordinary binoculars are two refracting telescopes mounted side by side. A lens, generally consisting of two or more elements, is usually mounted at the front end of an enclosed tube. The tube is not really essential — its purpose is merely to block out scattered light — an open framework would suffice as well. In a refracting telescope, the objective, the optical part that produces the principal image, is the lens at the front of the tube. The image is formed at the rear of the tube, where various devices must be available to inspect it, photograph it, or otherwise utilize it.

The reflecting telescope was first conceived by James Gregory in 1663, and the first successful model was built by Newton in 1668. Here a concave mirror (usually a paraboloid) is used as an objective. The mirror is placed at the *bottom* of a tube or open framework. The mirror reflects the light back up the tube to form an image near the front end. Since, in a reflector, the image is formed directly in front of the mirror, it is more inconvenient to get at for observation than the image formed by a refracting telescope. However, this disadvantage is often offset by other advantages inherent in the reflecting telescope.

The principle of reflecting telescopes has been known since Newton's time. Because of the difficulty of precisely producing the reflecting surface, however, the reflecting telescope did not become an important astronomical tool until the time of William Herschel, a century later.

Schmidt telescopes are used for wide-angle photography in astronomy. The image is formed directly in front of the spherical mirror about halfway from the mirror surface to the correcting lens. Most often the entire system is enclosed in a light-tight tube, and the photographic emulsion, curved to fit the spherical focal surface, is inserted at the focus of the mirror in the center of the tube. The exposure can be started and stopped by uncovering and recovering the correcting lens.

(a) Getting at the Image — Various Focuses

The purpose of a telescope is to produce an image that must be accessible for inspection, study, or photography (Section 11.6). For some types of image analysis, bulky equipment must be placed at the telescope focus. There are various possible arrangements for getting at the focus. Which arrangement is chosen depends on the type of telescope and on the purpose for which the image is to be used.

REFRACTING TELESCOPE

In a refracting telescope, the image is formed behind the objective lens, at the rear of the telescope tube. Thus even with a large refractor there is no problem in gaining access to the image. All that is required is a suitable ladder or platform to lift the observer to the point from which he can conveniently reach the rear of the tube of the telescope.

REFLECTING TELESCOPE

With a reflecting telescope the problem of image accessibility arises because a concave mirror produces the image in front of the mirror, in the path of the incoming light. The place where the image is formed by the mirror is called the *prime focus*. If the image is to be photographed, a small plate or film holder can be suspended at the prime focus in the middle of the mouth of the telescope tube. The plate holder generally blocks out only a small fraction of the incoming light, so that the brightness of the image is only slightly dimmed. However, if the image is to be inspected visually or with a spectrograph (see Section 11.6), the observer or the necessary apparatus blocks out so much incoming light that the prime focus is often impractical. Thus the light is usually diverted before it comes to a focus so that the image is formed in a more convenient location.

NEWTONIAN FOCUS

In the Newtonian-type reflecting telescope, the problem is solved by a flat mirror mounted diagonally in the middle of the tube so that it intercepts the light just before it reaches the focus. The mirror reflects the light to the side so that it comes to a focus just outside the tube (see Figure 11-14).

CASSEGRAIN FOCUS

Another arrangement for a reflecting telescope is the Cassegrain system, in which a small convex mirror rather than a flat mirror is suspended in the

telescope tube. The convex mirror intercepts the light before it reaches the prime focus and reflects it back down the tube of the telescope. In many reflecting telescopes (for example, the 200-inch telescope) a small hole is provided in the center of the objective mirror so that the light reflected from the convex mirror can form an image behind the objective. If the objective has no hole (for example, the 100-inch telescope on Mount Wilson), the image must be reflected out to the side of the tube with a third, flat mirror, mounted diagonally just in front of it. These arrangements are illustrated in Figure 11-14.

The convex mirror diverges rays of light. It must be chosen with the proper curvature to diverge the light coming to a focus from the primary mirror only enough to diminish its rate of convergence so that it will come to a focus after traveling a longer distance. Thus, the effective focal length of a Cassegrain telescope is longer than that of a Newtonian reflector. In effect, the focal length is the same as it would be if the image were produced by a lens or mirror of the same diameter as the objective of the telescope but converging

FIG. 11-14 Various focus arrangements: (a) prime focus; (b) Newtonian focus; (c) and (d) two types of Cassegrain focus; (e) Coudé focus.

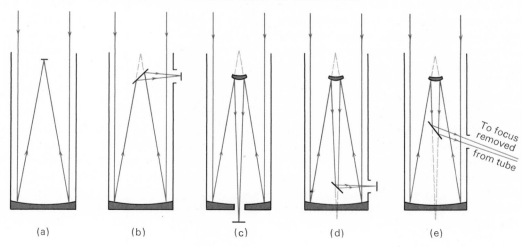

(a)   (b)   (c)   (d)   (e)

To focus removed from tube

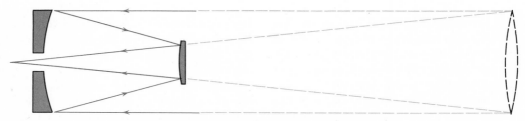

FIG. 11-15   Effective focal length of a Cassegrain telescope.

the light at the same rate as it actually converges after leaving the convex secondary mirror (see Figure 11-15). It can be shown that the exact shape of the convex mirror must be a hyperboloid: its cross section is a hyperbola.

COUDÉ FOCUS

In the Coudé arrangement, a convex mirror intercepts the light just before the prime focus is reached, and the light is reflected back down the tube until it reaches one of the pivot points about which the telescope tube can be rotated to point to various parts of the sky. There it is intercepted by a flat mirror that reflects the light outside the tube to a stationary observing station. Because the station is not attached to the moving part of the telescope, heavy equipment can be used there. The details of the Coudé focus are illustrated in the diagram of the 200-inch telescope (Figure 11-16).

Most large reflectors in use in astronomical observatories are equipped with auxiliary flat and convex mirrors so that several of the possible focus arrangements can be used.

(b) Telescope Mountings

The telescope tube must be mounted in such a way that it can turn to any direction in the sky. Nearly all astronomical optical telescopes have *equatorial mounts*. An equatorial mount (see Figure 11-17) allows the telescope to turn to the north and south about one axis and to the east and west about another. The axis for the east-west motion of the telescope is parallel to the axis of the earth's rota-

tion, and the other axis, about which the telescope can rotate to north or south, is perpendicular to this axis.

The two axes of motion of a telescope with an equatorial mount allow the telescope to be turned directly in right ascension and declination, the celestial coordinates in which astronomical positions are generally tabulated. Most telescopes are

FIG. 11-16   The 200-inch Hale telescope, sectional view through dome. *(Drawing by Russell W. Porter, Mount Wilson and Palomar Observatories.)*

equipped with setting circles that indicate the direction or coordinates in the sky toward which the telescope is pointing. An important advantage of the equatorial mount is that a simple slow motion of the telescope about its axis parallel to the earth's axis — the *polar axis* of the telescope — is sufficient to compensate for the apparent motions of the stars across the sky that result from the earth's rotation. A mechanical or electric mechanism called a *clock drive* is usually employed to drive the telescope automatically about its polar axis, so that it will follow the stars accurately as they move across the sky.

(c) Housings for Astronomical Telescopes

The conventional housing for an astronomical telescope is a hemispherical dome. The dome usually has an oblong window or slot on one side, extending from the spring line of the dome (its "base") to the top of the dome. Some kind of shutter can close over the window to protect the telescope during daylight and bad weather, when it is not in use. The dome is generally mounted on rails so that the window can be turned to any direction. In a modern observatory, the dome usually turns automatically to keep the slot always oriented in the direction in which the telescope points. The domes of the largest telescopes are usually well insulated against heat, so that the interior can be maintained at nighttime temperatures, thus preventing rapid temperature changes from distorting the critical shape of the telescope mirrors or lenses.

(d) Advantages of Various Kinds of Telescopes

Refracting, reflecting, and Schmidt telescopes all have their special advantages. The choice of the kind of telescope to build or use depends on the type of project to be worked on.

Refracting telescopes are usually constructed with long focal lengths and relatively slow speeds ($f/12$ to $f/16$). In such instruments comatic aberration and distortion can be kept small, resulting in a larger field of view for these telescopes than for typical reflectors. Refractors are best suited for

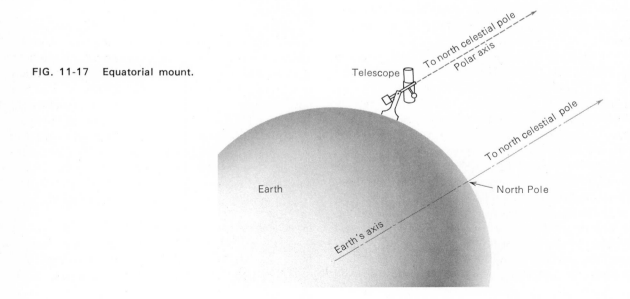

FIG. 11-17   Equatorial mount.

astrometry, the measurement of accurate positions and small angles in the sky. Also, because the image is formed at the back of the tube of a refractor, it is convenient for visual observations. On the other hand, refractors suffer from chromatic aberration, and can be constructed to perform well only for a limited spectral region. Because a refracting lens must contain at least two elements to produce good images, there are at least four surfaces of glass to be ground and polished. Furthermore, a lens can be supported only along its periphery; a very large lens sags of its own weight and distorts its shape.

A reflecting telescope, on the other hand, utilizes a mirror that has only one optical surface to be perfected. Because the light does not go through the mirror, this need not be made of optically perfect glass; furthermore, a mirror is easier to support, for it can be braced at all points along its back. Thus it is feasible to make much larger mirrors than lenses; the largest objective lens in use is 40 in. in diameter, whereas the largest mirror of optical quality is 200 in. across. (At the time of writing, the 236-inch telescope in the Soviet Union is not yet complete.) Because it is easier to construct reflecting telescopes, most homemade telescopes built by amateur astronomers are of this type.

Reflecting telescopes with more than twice the aperture of the 200-inch telescope have been built (at Mount Wilson and elsewhere), but these are special-purpose instruments designed merely to detect light from certain stars or to search for cool stars that emit most of their energy in the invisible infrared. For detection purposes it is not essential that there be good image quality, but it must be possible to gather radiation from very faint sources. Consequently these special-purpose telescopes have large mirrors of very low quality. One ingenious technique used to construct such a mirror is to spin a large circular tray of liquid epoxy about a vertical axis. The surface of the liquid takes the shape of a paraboloid, and when the epoxy hardens it maintains that parabolic surface, which can then be coated with aluminum or silver to make a reflecting telescope mirror. Another scheme is to mount many small concave mirrors on a frame that holds them to a large parabolic surface. Still another is to support a thin film of aluminized mylar on an accurately circular ring with a plate sealed to the back side. When the pressure is reduced inside this box, the mylar film forms a paraboloid. Such mirrors do not, of course, possess an optical-quality surface, and the images produced are not adequate for ordinary astronomical research. They are adequate for special purpose, however, for even a very large fuzzy image of a star can be identified with infrared detecting devices at the focus of the mirror.

Because the ordinary reflecting telescope is free of chromatic aberration, it is better suited to study of the spectra of stars. Reflecting telescopes of good optical quality can be made with relatively short focal lengths — that is, with high optical speeds — so they can photograph faint surfaces in shorter times than refracting telescopes can. However, high-speed reflectors have serious coma; only a small region near the center of the field is in good focus.

The Schmidt telescope is a compromise between reflectors and refractors. Schmidts can be made to produce excellent images over a large field of view. The 48-inch Schmidt on Palomar Mountain can photograph an area of the sky the size of the bowl of the Big Dipper on a single photograph, whereas with the 200-inch reflector, also at Palomar, about 400 photographs are required to cover the same area. Schmidt telescopes are consequently especially useful for survey purposes. However, the correcting lens of a standard Schmidt system has nonspherical surfaces and is difficult to make.

(e)   Some Famous Telescopes

The United States, until recently, has had a monopoly on the world's largest telescopes. The "size" of a telescope refers to the diameter of its lens or mirror, or of the correcting lens in the case of a Schmidt. Thus, a 6-inch telescope has a lens or mirror 6 inches in diameter.

The largest existing refractor is the 40-inch refracting telescope at the Yerkes Observatory of the University of Chicago. The Yerkes Observatory is located at Williams Bay, Wisconsin. The second largest refractor is the 36-inch telescope at the Lick Observatory of the University of California, on Mount Hamilton, California. These big refractors were built long ago, before the technique of building large reflectors was perfected.

Some of the largest reflectors are in California. The 200-inch Hale telescope on Palomar Mountain

has a focal length of 660 in. It can be used at the prime focus, Cassegrain focus, or Coudé focus. At the prime focus there is a cage 6 ft across in which the observer can ride while he makes his observations. The Hale telescope is one of the few existing telescopes in which the observer is carried "inside" the telescope. The 17-ft Pyrex mirror weighs 14½ tons; the entire instrument has a weight of about 500 tons.

Until the completion of the 200-inch telescope in 1949, the largest reflector was the 100-inch telescope at the Mount Wilson Observatory. (The Mount Wilson and Palomar Observatories are jointly operated by the California Institute of

FIG. 11-18   The 200-inch mirror of the Hale telescope, shown on the polishing machine at the California Institute of Technology. *(Mount Wilson and Palomar Observatories.)*

Technology and the Carnegie Institution of Washington; they are under a single director and astronomical staff.)

In 1960, the 120-inch telescope of the Lick Observatory went into operation. It has about the same focal length as the 200-inch telescope, and many of the design features are similar to those of the large reflector at Palomar. The United States National Observatory at Kitt Peak, near Tucson, Arizona, has an 84-inch reflector, and a telescope of about 158-in. aperture is under construction. A similar 158-in. reflector is planned for the Southern Station of the National Observatory at Cerro Tololo in the Western Andes in Chile. An organization of European nations also plans a large reflector (about 140-in. aperture) in Chile. An additional 150-inch telescope to be installed in Australia is now in the planning stages. The 98-inch Newton telescope is being installed at the Royal Greenwich Observatory at Herstmonceux, England. East Germany has a 79-in. universal reflector at the Tautenburg Observatory, and a new 79-in. reflector has been installed at Ondrejev, near Prague, Czechoslovakia. The Soviet Union has a 104-in. reflector in use at the Crimea Observatory and a 236-in. reflector under construction. The latter, when complete, will be the world's largest telescope of optical quality. There are more than a dozen other reflectors of aperture between 70 and 90 in. scattered over the world and several larger reflectors of nonoptical quality for special purposes.

One of the largest Schmidt telescopes is the 4-foot Schmidt of the Palomar Observatory.† Its correcting lens actually has an aperture of 49 in. rather than the 48 in. usually quoted. The focal length of the telescope is 120 in., and the over-all tube length is about 25 ft. It can photograph an area of the sky

6°6 square on a single 14-in. square plate of thin glass (which must be curved to fit a sphere of 10-ft. radius during exposure). Because of the high speed of the Palomar Schmidt ($f/2.5$) it can reach its photographic limit (the exposure that reveals the faintest stars that can be photographed) with short time exposures — from about 10 minutes to 1 hour, depending on the type of photographic emulsion used — and it is ideal for surveying the sky. The first main project assigned to the Schmidt after its completion in 1949 was to produce a photographic atlas of the sky. This sky survey, financed by the National Geographic Society, and called the National Geographic Society–Palomar Observatory Sky Survey, took 7 years to complete. It provides the most comprehensive coverage of the sky available. Copies of the 1870 photographs that comprise the survey have been distributed to more than 100 astronomical institutions throughout the world. Large reflectors, like the 200-inch, can probe deeper in space, but only in small regions at a time. It would take the 200-inch telescope at least 10,000 years to complete a survey of the sky comparable to the one completed by the Palomar Schmidt telescope.

11.6 ASTRONOMICAL OBSERVATIONS

The popular view of the astronomer is a man in a cold observatory peering through his telescope all night. Most astronomers do not live at observatories but near the universities or laboratories where they work. A typical astronomer might spend a total of only a few weeks a year observing at the telescope, and the rest of his time measuring or analyzing his data. Some astronomers work at purely theoretical problems and never observe at a telescope. Actually, professional astronomers seldom inspect telescopic images visually except to center the telescope on a desired region of the sky or to make adjustments. On the contrary, the image is generally utilized in one of several other ways.

---

†The universal telescope of the Tautenburg Observatory can be used as a Schmidt by inserting a removable 53-in. correcting lens; it then becomes the world's largest Schmidt.

(a) Projection on a Screen

The most direct way to use the image is to place a screen, for example, a white card, at the focus of the telescope and allow the image to fall on it. This is the most convenient way to view the sun, for example. However, a visual image does not provide a permanent record that can be measured and studied later in the laboratory. Therefore, it is far more common to photograph the image instead of viewing it directly.

(b) Telescopic Photography

To photograph the image, the screen is replaced by a photographic plate or film. The image then forms on a light-sensitive coating which, when developed, provides a permanent record of the image — one that can be measured, studied, enlarged, published, and inspected by many individuals. When used for photography, a telescope becomes nothing more than a large camera; the lens or mirror of the telescope serves as the camera lens. The 200-inch telescope is just a large camera.

The most important advantage in using a telescope as a camera is that photographic emulsions can accumulate luminous energy and build up an image during a long exposure. Most astronomical objects of interest are very remote, and hence the light we receive from them is very feeble. However, long time exposures can be made. Astronomical exposures commonly run into hours in length, and occasionally over several successive nights. The longer the exposure, the more faint light gradually accumulates to help build up the photographic image. Objects can be photographed that are about a hundred times too faint to see by just "looking through" a telescope. The layman is often disappointed when he takes his first look through an astronomical telescope, for what he can see visually is nothing compared to the spectacular photographs, such as those reproduced in this book, that are the result of long time exposures.

A photographic emulsion consists of a thin layer of gelatin, usually mounted on a base of celluloid or glass. Within the emulsion are suspended silver compounds that undergo a chemical reaction when activated by light, resulting in the formation of metallic silver. If the emulsion is then immersed in an appropriate chemical solution (that is, if it is developed), more silver is formed where light started the process and those molecules of silver compounds that were not activated by light are chemically removed, leaving grains of deposited silver in the parts of the emulsion that were exposed.

The astronomer, like any photographer, has his problems. He must determine the proper exposure to use and the best way to develop his photographic plates. (For scientific purposes, photographic emulsions on glass plates are usually employed rather than on celluloid film, for the glass will not curl or stretch and is better for accurate measurements.) During a time exposure, he must carefully "guide" his telescope to keep it centered on the object he is photographing. As mentioned earlier, telescopes are generally driven automatically to follow the gradual motion of the stars that results from the earth's rotation. However, minor irregularities in the driving mechanism of the telescope and motions of the image caused by atmospheric disturbances must be continually corrected. The astronomer sights a star through an auxiliary telescope mounted securely alongside the main telescope tube, or sometimes he uses a star image produced by the main telescope objective, but one that is outside the field of view that he is photographing. He keeps this star centered on illuminated cross hairs by moving the telescope slightly from time to time, thus assuring that the main telescope is tracking properly.

(c) Visual Inspection with an Eyepiece

Occasionally it is desirable to inspect a telescopic image visually, doing what is commonly called "looking through" a telescope. To best inspect detail

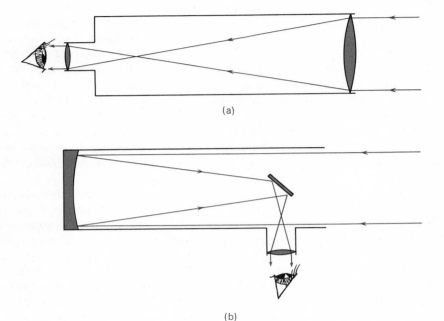

(a)

(b)

FIG. 11-19 Use of an eyepiece with (a) a refracting telescope; (b) a reflecting telescope.

in the image, it is customary to view it with a magnifying lens. A common hand magnifier could be used for this purpose. A telescopic *eyepiece* or *ocular* is simply a high-quality magnifying lens that is used to view the image. Figure 11-19 shows how an eyepiece is used in a refracting telescope and in a Newtonian reflector. In practice, eyepieces usually have two or more elements to reduce aberrations.

When an extended celestial object is viewed through a telescope equipped with an eyepiece, it appears enlarged, that is, closer, than when viewed naturally. The factor by which an object appears larger (or nearer) is called the *magnifying power* of the telescope. For example, the moon appears to subtend an angle of ½° when viewed with the naked eye. If, when viewed through a particular telescope, the moon appears to subtend 10°, the *magnifying power* of the telescope is 20.

Everyone knows that if he has to read fine print he must use a magnifying glass of higher power than when he reads the print in a newspaper. A higher power magnifying glass is one of shorter focal length. Similarly, eyepieces of different focal lengths, used in conjunction with the same telescope objective, produce different image magnifications. It is the purpose of the objective of a telescope to produce an image; it is the purpose of the eyepiece to magnify the image to the point where details in it can be viewed. In principle, any desired magnification can be obtained if an eyepiece of sufficiently short focal length is used. Therefore, it does not make sense to ask an astronomer what "power" his telescope is. The power can be changed at will by using different eyepieces. At every observatory there is a collection of eyepieces of different focal lengths that can be interchanged for various magnifications.

The exact value of the magnifying power can

easily be calculated by dividing the focal length of the objective of the telescope (f_o) by the focal length of the eyepiece (f_e); that is,[†]

$$\text{magnifying power} = \frac{f_o}{f_e}.$$

The highest useful magnification is that with which the finest details resolved in the image can be viewed. Additional magnification reveals no further detail but merely enlarges the blur of the diffraction pattern. Atmospheric disturbances further limit useful magnification, especially in large telescopes (Section 11.7).

(d) Measurement of an Image

One way of measuring the image produced by a telescope is with a *filar micrometer*. In this device,

two illuminated parallel cross hairs are placed at the focus of the telescope where the image is formed. The image with the cross hairs is viewed through a high-power magnifying eyepiece. The separation of the cross hairs can be varied by turning a precision screw, and the exact distance between them is indicated by the position of the screw. Thus the diameter of the image of a planet, or the distance between the images of two stars, can be measured with the instrument. The linear size of an image depends on its angular size in the sky and on the focal length of the telescope (see Section 11.2a). Therefore, the filar micrometer provides a means of measuring small angular sizes and separations in the sky. The instrument can also be used to measure the direction of one celestial object from another nearby one. The filar micrometer is particularly useful for following the revolution of visual binary stars about each other (Chapter 22).

(e) Spectroscopy

Spectroscopy occupies at least half of the available observing time of most large telescopes. The technique of spectroscopy is made possible by the phenomenon of dispersion of light into its constituent wavelengths. We saw in Chapter 10 that white light is a mixture of all wavelengths and that these can be separated by passing the white light through a prism, producing a spectrum.

A spectrum can also be formed by passing light through, or reflecting it from, a *diffraction grating*.

[†]The above formula for magnifying power can be verified from an inspection of the accompanying figure. As stated in Section 11.2a and seen in Figure 11-5, light rays passing through the center of a lens are not deflected. Thus the angle θ subtended by the image at the objective is the same as the angle subtended by the object (presumed to be very distant) as seen from the objective (or to the naked eye). As viewed through the eyepiece, however, the image appears to have an angular diameter ϕ. From inspection,

$$\sin \phi = \frac{AB}{2f_e} \quad \text{and} \quad \sin \theta = \frac{AB}{2f_o}.$$

For small angles, $\sin \theta = \theta$, and $\sin \phi = \phi$; therefore,

$$\frac{\phi}{\theta} = \frac{f_o}{f_e}.$$

By definition, ϕ/θ is the magnifying power.

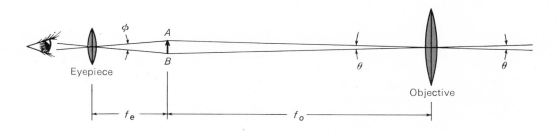

These are of two kinds: A *transmission grating* is a transparent sheet or plate (such as glass) that is ruled with many fine, parallel, closely spaced scratches (typically, thousands per inch), so that light can pass through the material only between the scratches. A *reflection grating* consists of a shiny reflecting surface, say a mirror, upon which are ruled similar fine scratches or grooves, so that light can be reflected only from the places on the surface between the scratches. In either kind of grating, light emerges as if from a series of long, parallel, finely spaced slits. Because of diffraction, the light fans out in all directions as it passes through these "slits." The waves of light from different slits interfere with each other in such a way that at any given point beyond the grating, the waves cancel each other out except at one specific wavelength. With an appropriate lens system, successive wavelengths of light can be focused at successive points along a line parallel to the surface of the grating, and in a direction perpendicular to the scratches

in it. A spectrum can thus be produced by a grating as well as by a prism. (Space here does not permit a description of the details of how the light waves from a grating interfere with each other to produce the spectrum.)

A *spectrograph* is a device with which the spectrum of a light source can be photographed. Attached to a telescope, the spectrograph can be used to photograph the spectrum of the light from a particular star. The construction of a simple spectrograph is illustrated in Figure 11-20. Light from the source enters the spectrograph through a narrow slit and is then collimated (made into a beam of parallel rays) by a lens. In Figure 11-20 the collimated light is shown entering a prism, but it can just as well pass through or be reflected from a grating. Different wavelengths of light leave the prism (or grating) in different directions, because of dispersion (or diffraction). A second lens placed behind the prism forms an image of the spectrum on the photographic plate. If the light from the

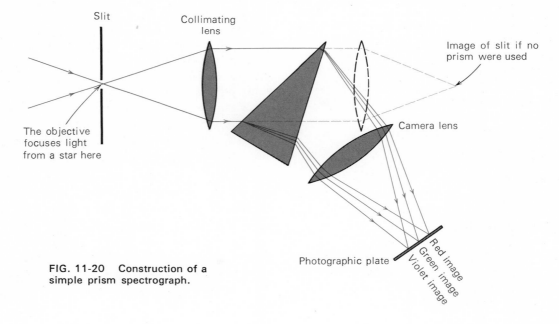

FIG. 11-20   Construction of a simple prism spectrograph.

collimating lens had been passed directly into the second lens, rather than through the prism first, the second lens would simply produce an image of the slit through which the light entered the spectrograph (dashed lines in Figure 11-20). However, because of the dispersion introduced by the prism, light of different wavelengths enters the second lens from slightly different directions, and consequently the lens produces a different image of the slit for each different wavelength. The multiplicity of different slit images of different colors lined up at the photographic emulsion is the desired spectrum. If the photographic plate is removed and an eyepiece is used to inspect the spectrum, the whole instrument becomes a *spectroscope.*

All of the parts of a spectrograph — slit, 2 lenses, prism (or grating), and photographic plate — must be attached to a rigid framework so that they do not shift during long exposures. The whole spectrograph may weigh hundreds of pounds and require a lot of space behind the focus.

"SINGLING OUT" THE STAR

The images of many stars appear in the focal plane of a telescope. However, it is no problem to single out the star whose spectrum is desired. The entire spectrograph is attached to the telescope; a metal plate containing the small slit of the spectrograph is centered on axis at the focus. The desired star image must now lie on that slit so that its light will pass into the spectrograph. In front of the slit a flat mirror can usually be flipped into the telescope beam at 45°, so that the astronomer can examine the pattern of stars and pick out the one he wants. Then the whole telescope can be moved slightly so that that star's image will be centered on the slit.

(f) The Objective Prism

An objective prism is a special plate of glass that is occasionally placed in front of the objective lens of a telescope. This plate is relatively thin; both of its faces are flat but at a slight angle to each other, so that, in effect, it is a thin prism. Starlight entering the telescope must first pass through the objective prism and is slightly dispersed. With the objective prism, if a photograph is taken of a field of stars, the image of each star is not a point but rather a small spectrum. In this way, the spectra of many stars can be photographed at once. The spectra are too small and of too low a dispersion (the colors are not widely spread out) to reveal much detail, and a comparison spectrum (a spectrum of a laboratory light source) cannot be photographed along side each stellar spectrum to provide a wavelength standard. Nevertheless, enough of the major features of the spectra show to enable the stars to be classified roughly according to their spectra.

(g) Photoelectric Photometry

An important datum about a star is the amount of light we receive from it, or its *magnitude.* The matter will be discussed in detail in Chapter 20. At present, the most accurate device for measuring stellar radiation received at the telescope is the *photomultiplier,* a light-sensitive electron tube placed at the focus of the telescope. Light from a star is gathered by the telescope and is focused on a light-sensitive surface just inside the glass envelope of the tube. Photons striking this *photocathode* dislodge electrons, which are attracted to a positively charged element in the tube, thus creating a feeble current. The current is amplified at a number of successive stages within the tube, and finally in an external amplifier. A measure of the amount of current produced in the tube is a measure of the amount of light striking the photocathode — that is, a measure of the star's brightness.

Other radiation-sensitive devices in which incident photons induce electric currents are the thermocouple, thermopile, and lead sulfide cell. These detectors are used to measure infrared radiation from celestial objects.

(h) Recent Photoelectric Techniques

Much research is being carried out today to find ways to improve the efficiency of image recording by our present telescopes by using photoemissive surfaces, that is, surfaces that emit electrons when exposed to light. One of the devices now in use is the *image tube.* The image tube utilizes a photoemissive surface that is placed at the focus of the telescope. Starlight, striking the surface, generates a flow of electrons that are focused electronically onto a phosphor screen (such as a television screen). The illuminated screen can be photographed directly, or the electrons can be focused instead, onto a photographic plate placed some distance behind the photoemissive surface. It takes an average of about 100 photons to render a grain in a photographic emulsion developable, so that it deposits a silver grain and produces a photographic image. On the other hand, in the image tube, a single electron can result in making a grain in the emulsion developable, and on the average only about two photons are required to dislodge an electron from the photoemissive surface. The electrons are focused on the phosphor screen or photographic plate in the positions corresponding exactly to those in which the photons strike the photoemissive surface. Thus an astronomical photograph is obtained up to 50 times faster than by the usual procedure.

Such devices, many still in the experimental stage, hold promise that we may soon be able to accomplish in a single night at the telescope what hitherto has required weeks. We are, in effect, increasing the sizes of our telescopes by learning to use them more efficiently.

(i) What a Telescope Does — Summary

We summarize this section by reviewing the three basic things that a telescope does: it *resolves,* it *magnifies,* and it *gathers light.* The greater resolution of a telescope as compared to the human eye enables us to detect details in celestial bodies that would otherwise remain indistinguishable. The magnifying power of a telescope, when used with an eyepiece, enables us to inspect these details visually. The most important property of a telescope is its ability to gather light so that we can detect objects too remote or faint to see with the unaided eye and to collect enough energy to measure and analyze the light from celestial objects. The main reason that we build bigger and bigger telescopes is to obtain more and more light-gathering power. The 200-inch Hale telescope on Palomar Mountain can photograph objects 10 million times fainter than those we can see without optical aid.

11.7 ATMOSPHERIC LIMITATIONS

The earth's atmosphere, so vital to us all, is the biggest headache to the observational astronomer. In at least five ways the air imposes limitations upon the usefulness of telescopes.

(1) The most obvious limitation is that of weather — clouds, wind, rain, and the like. Even on a clear, windless night, however, the atmospheric conditions may be such as to render a telescope virtually useless for many types of work.

(2) Even in visible wavelengths the atmosphere filters out a certain amount of the starlight that strikes it and dims the stars slightly. To wavelengths longer than the near infrared out to the radio wavelength of about 1 cm, the air is almost opaque, and it is opaque also to all wavelengths shorter than 2900 Å. Thus, most of the electromagnetic spectrum remains invisible to ground-based observatories.

(3) In the daytime, the air molecules scatter about so much sunlight that the resulting blue sky hides all celestial objects except the sun, moon, and Venus. At night the sky is darker, but never completely dark. Near cities the air scatters about the glare from city lights, producing an illumination that hides the faintest stars and limits the depths that can be probed by telescopes as well as by the naked eye.

Even starlight scattered by the air contributes to the brightness of the night sky.

(4) To make matters worse, the air emits light of its own. Charged particles and x-rays from the sun and outer space, bombarding the upper atmosphere, set the air aglow. When large numbers of these particles are funneled into the atmosphere in the polar regions by the earth's magnetic field, auroras result. In the absence of recognizable auroras, the night air glow still remains. Its brightness does vary from time to time. During times of maximum solar activity, the night sky brightness may be two or three times normal (see Chapter 24). The faint illumination of the night sky limits the time that a telescopic photograph can be exposed without fogging it completely and thus also limits the faintness of the stars that can be recorded.

(5) When the air is unsteady, star images are blurred. In astronomical jargon, the measure of the atmospheric stability is the *seeing*. When the seeing is good, the stars appear as sharp points and fine detail can be seen on the moon and planets. When the seeing is bad, images of celestial objects are blurred and distorted by the constant twisting and turning of light rays by turbulent air. It is seeing, in fact, that causes stars to "twinkle."

Because stars present too small an angular size for existing telescopes to resolve their disks, their images should be geometrical points or rather tiny diffraction patterns. But in bad seeing, a star may appear as a "mothball" or as a dancing firefly. The star image usually looks "bigger" in bad seeing, but not because the telescope has magnified it, only because the air has distorted it. Use of higher magnification only magnifies these atmospheric disturbances. Under typical observing conditions a star image appears to be from 1″ to 4″ in diameter. This is called the "seeing disk" of the star, and of course has nothing to do with the star's actual size. Under the very best conditions obtainable, star images seldom have diameters less than about ¼″. This is about the angle that can be resolved by a 20-in. telescope. The theoretical resolving power is therefore almost never realized in telescopes of larger aperture than about 30 in.; the seeing always limits their resolution. The purpose of building larger telescopes is not to resolve more, or make things appear nearer or bigger, but to gather more light and see fainter and more remote objects.

In view of the limitations set by the earth's atmosphere, the advantages of a lunar observatory are obvious. More immediately feasible, however, are telescopes on rockets and satellites. From space vehicles and rockets, important astronomical observations have been made which could not have been obtained from the ground. Also, balloons have carried men and instruments, or instruments alone, to altitudes of 100,000 ft for the purpose of obtaining astronomical observations. At those altitudes, most of the atmosphere is below the observer, and the seeing is much improved. From balloons have been obtained some of the best photographs of the sun (Chapter 24) and evidence for water vapor in the atmosphere of Venus (Chapter 14).

11.8 RADIO TELESCOPES

In 1931 K. G. Jansky of the Bell Telephone Laboratories was experimenting with antennas for long-range radio communication when he encountered interference in the form of radio radiation coming from an unknown source. He discovered that this radiation came in strongest about 4 minutes earlier on each successive day, and correctly concluded that since the earth's sidereal rotation period is 4 minutes shorter than a solar day, the radiation must be originating from some region of the celestial sphere. Subsequent investigation showed that the source of the radiation was the Milky Way.

In 1936 Grote Reber built the first antenna specifically designed to receive these cosmic radio waves. In 1942 radio radiation was first received from the sun at radar stations in England. After World War II, the technique of making astronomical observations at radio wavelengths developed rapidly, especially in Australia, the Netherlands, England, and more recently the United States.

Radio waves have now been received from many astronomical objects — the sun, moon, some planets, gas clouds in our galaxy, other galaxies, and various other objects. The technique of radio astronomy has become an extremely important tool in observational astronomy.

(a) Detection of Radio Energy from Space

First, it is important to understand that radio waves are not "heard"; they have nothing whatever to do with sound. Although in commercial radio broadcasting, radio waves are modulated or "coded" to carry sound information, the sound itself is not transmitted. The radio waves merely carry the information that a radio receiver must "decode" and convert into sound by means of a loudspeaker or earphones. Sound is a physical vibration of matter; radio waves, like light, are a form of electromagnetic radiation. We can also code visible light to carry sound information, as is done, for example, by the sound track on a movie film.

Many astronomical objects emit all forms of electromagnetic radiation — radio waves as well as light, infrared and ultraviolet radiation, and so on. The radio waves we can receive from space are those that can penetrate through the ionized layers of the earth's atmosphere — those with wavelengths in the range from about 8 mm to about 17 m. The human eye and photographic emulsions are not sensitive to radio waves; we must detect this form of radiation by different means. Radio waves have the property that they induce a current in conductors of electricity. An antenna is such a conductor; it intercepts radio waves which induce a feeble current in it. The current is then amplified

in a radio receiver until it is strong enough to measure or record.

If we lay a photographic plate out on the ground in daylight, it will be exposed by sunlight, indicating the presence of a light source in the sky. We can place various color filters in front of the plates and detect the presence of various colors in the light that exposes them. Such an experiment, however, does not indicate the direction in the sky of the light source.

Similarly, a radio antenna can be strung up outside, and currents induced in it indicate the presence of a source of radio radiation. Electronic filters in the radio receiver can be "tuned" to amplify only a certain frequency at a time, and thus can determine what frequencies or wavelengths are present in the radio radiation. The earliest astronomical observations of radio energy were detected in this way. As in the case of a photographic plate laid out in sunlight, however, a single antenna does not indicate the direction of the source.

(b) Radio Reflecting Telescopes

Radio waves are reflected by conducting surfaces just as light is reflected from an optically shiny surface and according to the same law of reflection. A radio reflecting telescope consists of a parabolic reflector, analogous to a telescope mirror. The reflecting surface can be solid metal, or a fine mesh such as chicken wire. In the professional jargon, the reflecting paraboloid is called a "dish." Radio dishes are usually mounted so that they can be steered to point to any direction in the sky, and gather up radio waves just as an optical reflecting telescope can be directed in any direction to gather up light. The radio waves collected by the dish are reflected to the focus of the paraboloid, where they form a radio image. In an optical telescope, a photographic plate, photomultiplier, spectrograph, or some other device is placed at the focus to utilize the image. In a radio telescope, an antenna or wave guide is placed at the focus of the dish. Radio waves focused on the antenna induce in it a current. This current is conducted to a receiver, not unlike ordinary home radio receivers in principle, where the current is amplified. As the optical astronomer chooses the type of photographic emulsion that is sensitive to the colors of light he wants to detect, the radio astronomer tunes his receiver to amplify the wavelengths he wants to receive from space.

One advantage of astronomical observations at radio wavelengths is that some of the important atmospheric effects discussed in Section 11.7 are not bothersome. In particular, radio observations are not affected by atmospheric seeing (although there is a similar *scintillation* effect due to clouds of ions in interplanetary space). They are less affected by weather and sky brightness, and at some wavelengths can even be made throughout the entire 24-hour day. Manmade radio interference, however, is a serious problem.

(c) What a Radio Telescope Does
Like an optical telescope (when used without an eyepiece), a radio telescope both gathers radiation and resolves.

"LIGHT"-GATHERING POWER
The ability of a radio telescope to gather radiation depends on its size. The radio energy received from most astronomical bodies is very small compared to the energy in the optical part of the electromagnetic spectrum. Hence radio dishes are usually built in large sizes; few are under 20 ft across. At first thought, the problem of constructing a large parabolic reflecting surface to a point of sufficient accuracy might seem prohibitive. The 200-in. mirror of the Hale telescope has a surface accurate to about 2 millionths of an inch, about one eighth of the wavelength of visible light. Ideally, radio reflectors should be built to similar accuracy — $\frac{1}{8}$ wavelength. However, a radio dish designed to receive radio waves of a length of 8 in., for example, need be accurate to only about 1 in.

RESOLVING POWER
With a radio reflecting telescope, radio radiation can be detected from a particular direction, and the direction of the source in the sky can be determined. We have seen that optical telescopes can resolve images of very small angular size. A 6-in. telescope can determine the location of a star to within about 1″. The formula given in Section 11.2 for the smallest resolvable angle,

$$\alpha = 2.1 \times 10^5 \frac{\lambda}{a},$$

where a is the aperture of the telescope, holds for radio telescopes as well as optical telescopes. Thus we can easily compute the resolving power of a radio telescope.

The main difficulty with radio telescopes is that the wavelength of radio radiation, λ in the above formula, is far greater than for visible light, so the resolving power for a telescope of a given size is correspondingly less. Radio waves of 8 in., for example, are some 400,000 times longer than waves of visible light, so to resolve the same angle, a radio telescope would have to be 400,000 times larger than an optical telescope. To resolve 1″ at 8-in. wavelength, a radio telescope would have to be nearly 40 mi across. The largest steerable radio telescopes in use today are only 200 or 300 ft in diameter. At a wavelength of 8 in. it is capable of resolving two points about 800 mi apart on the moon. If it were not for atmospheric "seeing," the 200-inch telescope could resolve, in visible light, two points 200 ft apart on the moon. The human eye can resolve points on the moon separated by about 72 mi. Thus the largest radio telescopes have far poorer resolving power than even the human eye. Consequently, it is very difficult to determine accurately the positions of radio sources in the sky.

THE INTERFEROMETER
The situation is improved by the use of the radio interferometer. Here two radio dishes (or bare antennas without focusing dishes) are placed some

distance apart. Unless the source of radio radiation happens to lie along a perpendicular bisector of the line between the antennas, the radio waves will strike one antenna a brief instant before the other, so that the two antennas will receive the same waves at slightly different times and thus become "out of phase" with each other; that is, the antennas receive different parts of a given wave. The difference in phase between the waves detected at the two antennas can be measured electronically. Because this phase difference depends on the angle the direction to the source makes with the line between the antennas, that angle can be determined. If the two antennas are due east and west of each other, an observation of this sort gives a much more accurate measure of the east-west posi-

tion of the source in the sky than could be obtained with a single radio telescope. If the antennas are placed due north and south of each other, the other coordinate of the source can be found. There are several types of radio interferometers, some, consisting of large arrays of antennas, more complex than the type described here.

Recently, very wide interferometers have been made by simultaneous recording in two radio telescopes thousands of miles apart. Both stations must also record a standard frequency (like "ticks" of an accurate clock) to measure the phase difference in the records of the radio signals from the one object they are observing. In this way resolution of a small fraction of a second of arc has been achieved.

The principle of the radio interferometer is like that of the optical interferometer of the Michelson type, which has been used to measure small angles optically. The method of operation of the interferometer will not be described here, but some results of measures made with a Michelson interferometer are given in Chapter 22.

(d) Radar Astronomy

Radar is the technique of transmitting radio waves to an object and then detecting the radio radiation that the object reflects back to the transmitter. The time required for the radio waves to make the round trip can be measured electronically, and because they travel with the known speed of light the distance of the object is determined. The value of radar in navigation, whereby surrounding objects can be detected and their presence displayed on a screen, is well known.

In recent years the radar technique has been applied to the investigation of the solar system. Radar observations of the moon and some of the planets have yielded our best knowledge of the distances of those worlds. In addition, as will be discussed in later chapters, radar observations have determined the rotation periods of Venus and Mer-

FIG. 11-21   The twin 90-foot radio telescopes of the Radio Observatory of the California Institute of Technology. *(California Institute of Technology.)*

cury and some information about the nature of the surfaces of those objects.

(e) Famous Radio Telescopes

Radio astronomy is still a new and rapidly developing technique and large radio telescopes are being built at a fast pace. Among the famous large steerable telescopes (that is, those that can be pointed to various positions in the sky) are the 250-foot dish at Jodrell Bank, England, the 210-foot Parkes dish in Australia, and the 300-foot dish at the National Radio Astronomy Observatory in Greenbank, West Virginia. Several institutions have pairs of large dishes that can be separated to various distances and are used as interferometers. Among these installations is Caltech's Owen's Valley (California) radio observatory, with two movable 90-ft dishes and one 130-ft dish.

Several important radio telescopes are large "bowls" carved out of the ground and lined with reflecting metal meshes. These bowls observe radio sources as they pass overhead, although some directional flexibility is provided by moving the antenna around in the vicinity of the focus of the bowl. One of the best-known installations of this type is the 3300-ft bowl at Arecibo, Puerto Rico.

Still larger installations exist which consist of arrays of antennas spread out over the ground, some more than a mile across. These act as interferometers as well as receivers. There are large arrays in Australia, France, Canada, Italy, the Netherlands, the Soviet Union, and the United States.

11.9 ROCKET AND SPACE OBSERVATIONS

Since World War II, rockets, satellites, and space probes have provided means of carrying instruments above the earth's atmosphere, and even to the vicinity of the moon, Venus, and Mars, making possible astronomical observations that could never have been made from the earth's surface. Most of these observations are described in the appropriate chapters to follow. An account of astronomical instruments would be grossly incomplete, however, without a brief summary of some of these recent techniques.

In 1946 the United States Naval Research Laboratory launched a rocket to observe the far ultraviolet radiation from the sun. Since then, many rockets have been launched to make ultraviolet and x-ray photographs of the sun. The photographs were recovered from the rockets after they parachuted back to earth. Spectra have also been obtained of the sun and some other stars at wavelengths in the ultraviolet and in the region of the lower-energy x-rays. High-energy x-rays and gamma rays can be detected (outside the earth's atmosphere) by the ionization they produce in gas chambers. Several sources of x-rays have been found; some are gaseous nebulae, some appear to be associated with particular stars, and some are not yet identified with optically visible objects. In addition, a general background of x-ray radiation of unknown origin has been found. At the time of writing, no sources have been found that emit gamma rays in sufficient intensity to be detected.

Many observations have been obtained from earth satellites — both of the earth and of its space environment. In particular, NASA's orbiting solar observatories (OSOs) and orbiting astronomical observatories (OAOs) are designed exclusively for astronomical investigations. The first OSO was launched on March 4, 1962; subsequent ones were launched in 1965 and in 1967.

Probes to the moon, Venus, and Mars have carried instruments to investigate those worlds and the interplanetary space on the way. Television cameras have given us excellent views of the moon and Mars. Other instruments carried on these probes have measured magnetic fields, the intensity of microwave (shortwave radio) radiation, corpuscular radiation (atomic nuclei from the sun), micrometeorite densities in space, and other phenomena.

Exercises

1. What would be the size of the image of the moon produced by the 200-inch telescope (focal length = 660 in.)?

    Answer: About 5.75 in.

What size would be the image of the moon formed by a telescope of the same focal length but with an aperture of only 100 in.?

2. If the moon can be photographed in 1 second with a telescope of 20-in. aperture and 40-in. focal length, how long a time would be required to photograph the moon with a telescope of 40-in. aperture and 160-in. focal length?

3. Suppose that all stars emitted exactly the same total amount of light and that stars were distributed uniformly throughout space. Show that the depth in space to which stars could be observed would be proportional to the aperture of the telescope used.

4. What is the smallest angle that could theoretically be resolved by the 200-inch telescope at a wavelength of 5000 Å?

5. What kind of telescope would you use to take a color photograph entirely free of chromatic aberration? Why?

6. Suppose one wants the moon to appear 45° across when viewed through a telescope of 50-in. focal length. What focal-length eyepiece must be used?

7. Ordinary 7 × 50 binoculars magnify seven times (magnifying power = 7) and have objective lenses of 50-mm aperture (about 2 in.). For light of 5000 Å, what is the smallest angle that can be resolved by the lenses of binoculars? Could two stars separated by this angle actually be seen as separate stars when viewed through 7 × 50 binoculars? Why?

8. Because of the night airglow, the sky is a faint luminous surface. What determines how long it would take for this sky illumination to expose a telescopic photograph to a given blackness? What determines how intense or black the image of a star on the same photograph will be? After a photographic plate has been exposed to a certain blackness, further exposure merely fogs the plate, rendering faint star images invisible on the photograph. Show that the limiting faintness of stars that can be photographed with a given telescope depends only on the focal length of that telescope.

9. When Mars is at its closest, it subtends about 24″ in the sky. The diameter of Mars is about 4200 mi. In principle, how close together could two features on Mars be and still be distinguished with the 200-inch telescope? In practice, because of the limitations set by the earth's atmosphere, how close together can they be? With a 24-in. telescope, could a person be sure that "canals" on Mars were only 2 mi wide? Why?

10. If the 200-inch telescope were used as a radio telescope to observe radio waves of 8-in. wavelength, how accurately could it "pinpoint" the direction of a radio source?

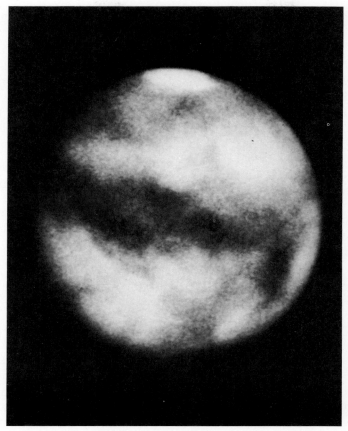

CHAPTER 12

The Solar System
in General

The ancient observer, who considered the earth to be central and dominant in the universe, regarded the sun, moon, and planets as luminous orbs that moved about on the celestial sphere through the zodiac.

Our solar system is indeed dominated by one body, but it is the sun, not the earth. Our sun, so important to us, is merely an ordinary, "garden-variety" star. Only careful scrutiny at close range would reveal the tiny planets to an imaginary interstellar visitor. First Jupiter, the largest, would be seen; then Saturn; and perhaps only with the greatest difficulty, the earth and other planets. Almost 99.9 percent of the matter in the system *is* the sun itself; the planets comprise most of what is left — the earth scarcely counts among them. The countless billions of other objects that inhabit the solar system, mostly unknown to the ancients, would probably remain unnoticed by a casual traveler passing through the solar neighborhood.

We turn now to those worlds of the solar system. They will be considered individually in detail in later chapters. Here we take only a brief look at some of the general characteristics of the solar system, and remark on a few of the properties that its constituent worlds have in common.

12.1 INVENTORY OF THE SOLAR SYSTEM

The solar system consists of the sun and a large number of smaller objects gravitationally associated with it. These other objects are the planets, their satellites, the comets, the minor planets or asteroids, the meteoroids, and an interplanetary medium of very sparse gas and microscopic solid particles. The relative prominence of the various kinds of members of the solar system can be seen from Table 12.1, which lists the approximate distribution of mass among the bodies of the solar system. The last four entries in the table are order of magnitude guesses only.

TABLE 12.1 Distribution of Mass in the Solar System

OBJECT	PERCENTAGE OF MASS
Sun	99.86
Planets	0.135
Satellites	0.00004
Comets	0.00003 (?)
Minor planets	0.0000003 (?)
Meteoroids	0.0000003 (?)
Interplanetary medium	<0.0000001 (?)

(a) The Sun

The sun, practically, *is* the solar system. It is a typical star — a great sphere of luminous gas. It is composed of the same chemical elements that compose the earth and other objects of the universe, but in the sun (and other stars) these elements are heated to the gaseous state. Tremendous pressure is produced by the great weight of the sun's constituent layers. The high temperature of its interior and the consequent thermonuclear reactions keep the entire sun gaseous. There is no distinct "surface" to the sun; the apparent surface we observe is optical only — the layer in the sun at which its gases become opaque, preventing us from seeing deeper into its interior. The temperature of that region is about $6000°K$ ($11,000°F$). Relatively sparse outer gases of the sun extend for millions of miles into space in all directions. The visible part of the sun is 864,000 mi across, which is 109 times the diameter of the earth. Its volume is 1⅓ million times that of the earth. Its mass of 2×10^{33} gm exceeds that of the earth by 333,000 times. The sun's energy output of 4×10^{33} ergs/sec, or about 500 sextillion horsepower, provides all the light and heat for the rest of the solar system. The sun derives this energy from thermonuclear reactions deep in its interior, where temperatures exceed 12 million °K. We shall describe the sun, a typical star, in Chapter 24.

(b) The Planets

Most of the material of the solar system that is not part of the sun itself is concentrated in the planets. The nine known planets include the earth, the five other planets known to the ancients: Mercury, Venus, Mars, Jupiter, and Saturn; and the three discovered since the invention of the telescope: Uranus, Neptune, and Pluto. In contrast to the sun, the planets are small, relatively cool, and (at least mostly) solid. They give off no light of their own but shine only by reflected sunlight.

We should say something of the nomenclature given certain groups of planets. We have seen (Chapter 3) that the two planets nearer the sun than the earth (Mercury and Venus) are called inferior planets, and the ones with orbits outside the earth's are called superior planets. The four innermost planets (Mercury through Mars) are also called *inner planets;* Jupiter, Saturn, Uranus, Neptune, and Pluto are then the *outer planets.* Finally, the four largest planets (Jupiter, Saturn, Uranus, and Neptune) are often called the *Jovian planets* (after Jupiter) or, occasionally, *major planets,* and the others, including Pluto, are the *terrestrial* (earthlike) *planets.*

The masses of the planets, in terms of the mass of the earth, range from 0.05 (Mercury) to 318 (Jupiter). The mass of Jupiter is greater than that of all the other planets combined. In diameter, the planets range from 3000 mi (Mercury) to 86,000 mi (Jupiter). Most, but not all, of the planets are surrounded by gaseous atmospheres. All but three of the planets are known to have natural satellites; Jupiter leads with 12 known moons.

All the planets revolve about the sun in the same direction — from west to east (counterclockwise as viewed from the north). Their mean distances from the sun range from 0.39 AU (36 million mi) for Mercury to 39.46 AU (3670 million mi) for Pluto. Their periods of orbital revolution range from 88 days for Mercury to 248 years for Pluto; the corresponding mean orbital velocities range from about 30 to about 3 mi/sec. The orbits of the planets are all in nearly the same plane. Pluto is the planet whose orbit is inclined at the greatest angle (17°) to that of the earth; next is Mercury, whose orbit is inclined to the earth's by 7°. The orbits of the planets are also nearly circular; all have eccentricities of less than 0.1 except for Pluto ($e = 0.25$) and Mercury ($e = 0.21$).

The planets all rotate as they revolve about the sun. By *rotation* is meant a turning of an object on an axis running through it, as distinguished from *revolution,* which refers to a motion of the object as a whole about another object or point. The Jovian planets are all rapid rotaters; Jupiter rotates most rapidly, in a period of 9^h50^m. Mercury rotates 1½ times during its 88-day revolution about the sun. Venus rotates still more slowly, in 243 days, but from east to west, reverse to the rotation of most of the planets (that is, *retrograde*). Some of the planets, especially Jupiter and Saturn, show marked oblateness or flattening because of their rapid rotation.

(c) Satellites

The satellites of the planets are the next most prominent members of the solar system. Only Mercury, Venus, and Pluto do not have known satellites. Jupiter has 12, Saturn has 10, Uranus has 5, Neptune and Mars 2 each, and earth 1. Six of the 32 known natural satellites of the solar system are about as large, or larger, than our moon, although none is so large as the moon in comparison with its primary planet. (The earth-moon system is sometimes referred to as a "double planet.") Only

Titan, Saturn's largest satellite, is known to have an atmosphere. Most of the satellites revolve about their planets from west to east, and most have orbits that are approximately in the equatorial planes of their primary planets. (The moon is an exception — its orbit is nearly in the ecliptic plane — Section 9.3.)

(d) Comets

Comets are loose swarms of small particles that revolve about the sun in orbits of very high eccentricity — that is, very elongated elliptical orbits. Comets spend most of their time in those parts of their orbits that are very far from the sun, where they receive negligible radiant energy. However, as a comet moves in closer to the sun, it warms up and some of the particles vaporize to form a cloud of gas, or *coma,* around the swarm of particles. The particles and surrounding coma make up the *head* of a comet. When a comet is only a few astronomical units from the sun, the pressure of the sun's radiation and solar corpuscular radiation sometime force particles and gases away from the head to form a *tail.* The heads of comets average from 10,000 to 100,000 mi across, and tails sometimes grow to lengths of many millions of miles. The entire mass of a typical comet, however, is probably less than a billionth — perhaps as little as a trillionth — that of the earth. A comet is, therefore, an extremely flimsy entity.

More than a thousand comets have been observed, and an average of 5 to 10 new ones are discovered telescopically each year. Most newly discovered comets have periods of thousands or even millions of years, and have never before been observed in recorded history. It has been hypothesized that there may be a vast cloud of a hundred billion or so comets surrounding the sun at a distance of trillions of miles, and that occasional perturbations from passing stars start some of those comets moving in toward the sun. Unlike the planets, which move in orbits that are all nearly coplanar, comets approach the sun along orbits that are inclined at all angles to the plane of the earth's orbit.

(e) Minor Planets

The *minor planets,* or *asteroids* as they are often called are small planets, differing from the nine principal planets primarily only in size. There are probably tens of thousands of them large enough to observe with existing telescopes. Ceres, the largest minor planet, has a diameter of about 500 mi. Only a few hundred minor planets are over 25 mi across; most are only a few miles or less in diameter. None of these objects is large enough to retain an atmosphere.

Like the principal planets, the asteroids move from west to east in elliptical orbits of small eccentricity and lie close to the plane of the earth's orbit. Typical minor planets have mean distances of about 2.5 to 3 AU from the sun and have periods in the range 4 to 6 years; thus, they occupy a region of the solar system that lies between the orbits of Mars and Jupiter. A few of the minor planets, however, have orbits that bring them close to the earth; several minor planets pass closer to the earth than any other known objects except the moon and meteoroids.

(f) Meteoroids

The numbers of small solid objects revolving about the sun that are too small to observe with telescopes are very great indeed, and the number seems to be greater and greater for objects of smaller and smaller size. These tiny astronomical bodies, too small to observe individually as each travels unhindered in its orbit, are called *meteoroids.* Their presence becomes known only when they collide with the earth, and plunging through the earth's atmosphere, heat with friction until they vaporize. The luminous vapors that are produced look like stars moving quickly across the sky and are popularly known as "shooting stars." Such phenomena are correctly called *meteors.* On a typical clear dark night, about half a dozen meteors can be seen per hour from any given place on earth. The total number of meteoroids that collide with the earth's atmosphere during a 24-hour period is estimated at 200 million. Recently, meteoroids have also been recorded as they strike space vehicles.

Meteoroids fall into two categories. Those of the first type revolve about the sun, like the major and minor planets, in orbits of relatively small eccentricity that lie near the principal plane of the planets. These meteoroids resemble the minor planets except that they are smaller. The other and far more common type of meteoroids seems to be associated with comets. Meteoroids of this latter category travel in orbits that are very enlongated (of high eccentricity). They approach the earth from any direction, at any angle to the plane of the earth's orbit. Frequently, meteoroids of the latter type collide with the earth in groups or *showers,* and some such showers have been found to have orbits in common with known comets.

Rarely, a meteoroid of the first type survives its flight through the earth's atmosphere and lands on the ground. It is then called a *meteorite.* A number of such fallen meteorites can be inspected in various museums. The largest known meteorites have masses of about 50 tons. Most are the size of pebbles. Chemical analysis reveals them to be formed of the same chemical elements that exist on earth and elsewhere in the cosmos.

The smallest meteoroids are slowed down by the atmosphere before they have a chance to heat up and vaporize. There are microscopic in size and are known as *micrometeorites.* They eventually settle to the ground as meteoritic dust. Micrometeorites are the most common kind of meteoroids and are the kind most frequently encountered by space rockets.

(g) The Interplanetary Medium

The interplanetary medium is composed of two components: *interplanetary dust* and *interplanetary gas.*

Interplanetary dust can be thought of as a sparse distribution of micrometeorites throughout the solar system, or at least throughout the main disk that contains the orbits of the planets. Individual particles have been detected as they strike rockets. Collectively, the particles can be observed by the sunlight they reflect. On a dark clear night a faint band of light can be seen circling the sky along the ecliptic. This band of light is generally brightest near the sun and is best seen in the west within a few hours after sunset or in the east within a few hours before sunrise. Sometimes, however, it can be seen as a complete band across the sky. Because this light is confined to the region of the ecliptic or zodiac, it is called the *zodiacal light.* Spectrographic analysis of the zodiacal light shows it to be reflected sunlight. It is presumed to be due to reflection from microscopic solid particles.

There is also a very tenuous distribution of gas spread through the solar system, as well as solid particles. It is thought that the earth's atmosphere, rather than ending abruptly, must gradually thin out into the interplanetary gas. Evidence of interplanetary gas also comes from space probes, whose instruments have recorded rapidly moving atoms and charged atomic particles. High-altitude rockets, carrying cameras that photograph in the far ultraviolet, have recorded a faint illumination that is apparently light emitted by hydrogen gas, present either high in the earth's atmosphere or in interplanetary space. Practically all of the interplanetary gas consists of ions and electrons ejected into space from the sun. This flow of corpuscular radiation from the sun is called the *solar wind.* It is, in fact, a sort of continuation of the solar corona (see Chapter 24).

We conclude that the region of interplanetary space contains minute, widely spread particles, and very sparse gas. In the neighborhood of the earth, there are only a few ions per cubic centimeter. This is a far better vacuum than can be produced in any terrestrial laboratory.

12.2 THE PLANETS

The other planets of the solar system are of special interest, for they, like the earth, are worlds that revolve about the sun and derive their light and warmth from it. The planets are considered individually in Chapter 13. Here we shall summarize how certain information about them can be obtained.

(a) Some Basic Characteristics

The basic characteristics of a planet are its mass, its size, and its distance from the sun. From these data alone we can predict to some extent many of its other characteristics, for example, whether or not it can be expected to have an atmosphere, what kind of atmosphere, what its temperature is likely to be, and many other physical data.

We determine the distances of planets from the earth, and from the sun, by essentially the same method used by Kepler and described in Section 3.4a, although today we employ more sophisticated mathematical techniques, such as that of Gauss (Section 5.6). The method involves geometrically surveying the distance to the planet by sighting it from different places in the earth's orbit. This procedure, of course, gives the distance to the planet only in terms of the size of the earth's orbit — that is, in astronomical units. The evaluation of the astronomical unit in, say, miles — that is, the determination of the scale of miles of the solar system — is a far different problem; it will be discussed in Chapter 18.

The mass of a planet must be determined by measuring the gravitational acceleration that it produces on other objects. Four different methods have been used: (1) observing the acceleration a planet produces upon one of its satellites, (2) observing the perturbations a planet produces upon the motions of other planets, (3) observing the perturbations a planet produces upon the motion of a close-approaching minor planet, and (4) observing the effects a planet produces on manmade space

probes. The latter technique has been especially important in determining good values for the masses of Venus and Mars.

If a planet had a single satellite, the problem would be particularly simple, for the planet and satellite would comprise a pair of mutually revolving bodies. Direct observations would give the period of mutual revolution and the mean separation of the two bodies in angular measure (that is, degrees, minutes); when the distance of the planet had been determined, the angular radius of the orbit could be converted to linear measure (say, miles). Then the combined mass of the two bodies, which would exceed only slightly the mass of the planet itself, could be calculated with the use of Newton's formulation of Kepler's third law (see Section 5.3). As it happens, no other planet in the solar system (besides the earth) is known to possess only a single satellite. On the other hand, the other satellites of the solar system (besides the moon) are so small in comparison to the planets about which they revolve that their masses are certainly negligible compared to those of their primaries. Consequently, the gravitational interactions between the satellites of a particular planet can be neglected; the dominant gravitational force acting upon each satellite is that between it and its primary planet. To a very good approximation, therefore, a planet and any one of its satellites can be regarded as a pair of mutually revolving bodies, just as the sun and any planet can be considered a pair of mutually revolving bodies (Section 5.3c). The mass of the planet is thus obtained from Newton's formulation of Kepler's third law, just as if the planet had only a single satellite of negligible mass.

If a planet does not have a satellite, its mass cannot be calculated quite so conveniently, but nevertheless can be found, sometimes very accurately, by noting the acceleration its gravitational field produces upon other objects in the solar system — that is, by studying the perturbations it produces upon their orbits. Observations of planetary motions must be extended over a long period of time to detect the small changes brought about by these perturbations. More than a century ago, however, gravitational theory had been refined to the point that just such observations were used to predict the existence of the planet Neptune.

Finally, if a minor planet happens to pass close to a planet, or a space probe is sent close to it, the path of the smaller object may be substantially altered. Although that body is close to the planet, the two can be considered, approximately, a two-body system; they will describe hyperbolic orbits about each other. Because of the planet's relatively greater mass, it will be perturbed in its own orbit only negligibly, whereas the minor planet might suffer a large change in its motion. A comparison of the orbital elements of the minor planet or space probe before and after the encounter reveals how much it was accelerated by the planet, and hence the planet's mass. In principle, measures of the perturbations produced by a planet upon the motion of a comet can also be used to calculate the planet's mass.

The diameter of a planet is found from its angular diameter and its distance, exactly as the diameter of the moon is calculated. The procedure was described in greater detail in Section 9.2b.

(b) Rotation of the Planets

All the planets are observed to rotate. At least four techniques have been employed to determine planetary rotation rates.

The most direct method to observe a planet's rotation is to watch permanent surface features move across its disk. If a particular surface feature is observed for a large number of rotations, a very accurate value for the time required for one rotation is obtained. In this way, for example, the rotation period of Mars has been well determined.

If surface features cannot be seen, or are indistinct, the rotation period of a rapidly rotating planet can be found from the Doppler effect in the lines in the spectrum of sunlight reflected from the planet. Since one side of the planet is turning toward us and the other side away, light reflected from the approaching limb is Doppler-shifted to slightly shorter wavelengths (relative to light reflected from the center of the planet's disk) and light from the opposite limb is shifted to longer wavelengths. Suppose the planet's image is focused across the slit of a spectrograph at the telescope; if the apparatus is oriented so that the slit runs along the equator of the planet's image, all the lines in the spectrum will appear tilted (the top of the spectrum being light from one limb and the bottom being light from the other). A measure of the amount of tilt reveals the rotation rate (see Figure 12-1). The rotation periods of Uranus and Neptune are found in this way.

If the planet is too small or distant to show an appreciable disk through the telescope, its rotation may be detected from variations in its brightness. Pluto, for example, seems to have surface irregularities so that different sides reflect light with different intensity. Thus as the planet rotates, it varies in brightness (as seen from the earth) with a period equal to the rotation period.

The most recent technique is that of *radar*. Radio waves beamed to a planet from earth are reflected back but are Doppler-shifted according to the relative line of sight velocity of the earth and planet. The waves reflected from the limb of the planet turning toward us are shifted to shorter wavelengths than those that are reflected from the center of the planet's disk, and radio waves reflected from the receding limb are shifted to longer wavelengths. Thus, although the transmitted beam may consist of waves of a very narrow wavelength range, those reflected back will comprise a somewhat larger range of wavelengths. Measurement of this range of wavelengths (or *bandwidth*) gives the planet's rate of rotation. In this way, the rotation rates of Mercury and Venus have been found.

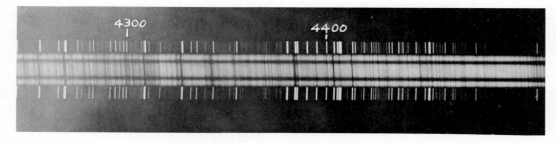

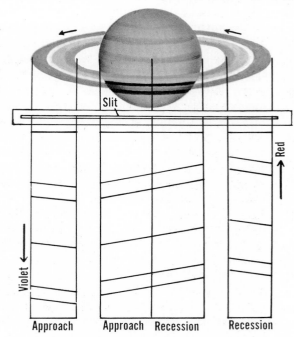

FIG. 12-1 Spectra of Saturn and its rings, showing rotation. *(Lowell Observatory.)*

Radar has been used to determine the rotation of Venus in still another way. Venus apparently contains several surface features that reflect radio waves better than surrounding regions on the surface of the planet. These features are observed, by radar, to move across the planet much as we see optically visible features move across the disks of planets as they rotate. Accurate timing of the radar waves reflected by these "radar features" not only confirms the rotation obtained with bandwidth measurements, but actually gives a very accurate value for the rotation period.

(c) Atmospheres

All the planets except Mercury and probably Pluto are surrounded by appreciable gaseous atmospheres. A planetary atmosphere manifests itself in several ways. In most cases, opaque clouds in the atmosphere reflect light brilliantly. Among the planets possessing atmospheres, only Mars has an atmosphere thin enough so that we can see through it to examine features on the surface of the planet. Even the Martian atmosphere, however, blurs the markings on the planet, especially when viewed in blue or violet light of the shorter wavelengths, because gas molecules scatter short wavelengths more effectively than long ones. The atmosphere of Venus scatters considerable sunlight around the terminator into its night side, producing an easily observable twilight zone.

Spectrographic analysis of sunlight reflected from a planet can reveal what some of the gases are that compose its atmosphere. We have seen (Section 10.5) how atoms of gas abstract certain wavelengths from the light that passes through the solar atmosphere, leaving dark lines in the solar spectrum that are characteristic of those atoms. Because the sunlight must pass through the earth's atmosphere before reaching the telescope, some light is also absorbed by the gases in our own atmosphere, producing additional, *telluric,* absorption lines. Sunlight reflected from the surface or cloud layers of a planet must pass through some of the outer gases in its atmosphere as well. Thus, if such light from a planet is observed spectroscopically, additional dark lines appear, produced by the gases on that planet.

By carefully comparing the spectrum of direct sunlight and the spectrum of sunlight reflected from a planet, the astronomer can ascertain which of the multitude of dark lines in the latter spectrum must actually originate in that planet's atmosphere. The identification is further aided by the fact that the planet, generally, has a radial velocity, that is, a component of velocity along the line of sight. The Doppler effect thus shifts the spectral lines arising in the planet's atmosphere with respect to the lines produced in the atmospheres of the sun and earth. Proper identification of the lines in the spectrum of a planet reveals the identity of the gases in its atmosphere that produced them.

Not all gases in the atmosphere of a planet can be detected spectrographically. Any oxygen, water vapor, carbon dioxide, methane, or ammonia that is abundant in the planet's atmosphere produces lines that are easy to detect. On the other hand, at the prevailing temperatures of planets, the gases hydrogen, helium, and nitrogen do not produce conspicuous lines in the observable spectrum — that is, at wavelengths at which radiation can penetrate the earth's atmosphere.

Which gases are present in a planet's atmosphere must depend, at least in part, on how that planet was formed and on its subsequent history. Most of the gases in the earth's atmosphere, for example, may have escaped (that is, *outgassed*) from the earth's crustal rocks, and have changed subsequently due to photosynthesis and other chemical activity, whereas the constituents of Jupiter's atmosphere may reflect the original composition of the material from which Jupiter was formed.

FIG. 12-2   The Pioneer site at the Goldstone Tracking Station where radar signals were received in the 1961 Venus radar experiments. The radar signal was transmitted from the Echo site at Goldstone. *(Jet Propulsion Laboratory.)*

On the other hand, the kind of atmosphere a planet has also depends on its ability to hold the various gases. The molecules of a gas are always in rapid motion; if their speeds exceed the velocity of escape of a planet (Section 5.4), that kind of gas can gradually "evaporate" from the planet into space. To investigate the conditions under which certain kinds of gases can be retained in a planetary atmosphere, we turn now to the kinetic theory of gases.

(d) Kinetic Theory

The *kinetic theory of gases* is the study of the behavior of the particles that compose a gas. Gases, like all matter, are composed of units of matter called *molecules*. Molecules are composed, of course, of the still smaller *atoms*. Molecules of pure elements consist of one or more atoms of that element. Molecules of a chemical compound consist of two or more atoms of two or more different kinds bound together. For example, a molecule of the compound *water* consists of one oxygen atom and two hydrogen atoms. The atomic structure of the molecules determines the identity of the substance they compose.

Molecules of a gas are in rapid motion, darting this way and that, frequently colliding with each other. At sea level on the earth there are some 10^{19} such molecules bouncing about in each cubic centimeter of air. The *kinetic energy* (or energy of motion) of a moving object is defined as $\frac{1}{2}mv^2$, where m is the mass of the object and v is its speed. Ordinary temperature, *kinetic temperature,* is a measure of the mean (or average) kinetic energy of molecules. Specifically, the absolute temperature of a gas, T, is related to the mean energy of the molecules comprising it by the formula

$$\overline{\tfrac{1}{2}mv^2} = \tfrac{3}{2}kT,$$

where the bar over the term on the left indicates that it is the average kinetic energy of the gas molecules and k is Boltzmann's constant (in metric units, $k = 1.38 \times 10^{-16}$ erg/deg.). If the above formula is solved for the mean speed of molecules in a given kind of gas, there results

$$\bar{v} = \sqrt{\frac{3kT}{m}}.$$

Thus we see that the mean speed of molecules in any particular gas is proportional to the square root of the temperature and inversely proportional to the square root of the mass of a single molecule of that gas. The higher the temperature, the faster the molecules move; at a given temperature, molecules of greater mass move slower, on the average, than those of smaller mass.

The temperature is measured from absolute zero; at $T = 0$, the mean energy of the molecules is zero. Here is the meaning of absolute zero: As gases are cooled, their molecules move more and more slowly, and at absolute zero all molecular motion ceases. This occurs at $-273°C$ $(-459°F)$. Absolute, or Kelvin, temperature is thus the centigrade temperature $+273°$. (For a discussion of conversion from one kind of temperature to another, see Appendix 5.)

At any given time, some molecules move at less than the average speed, and others move faster. A few are moving at several times the average speed. An individual molecule, suffering frequent collisions, constantly exchanges energy with other molecules. Sometimes it moves relatively slowly; at other times it may get a good "jolt" in a collision and move far faster than the average. From kinetic theory we can calculate the relative numbers of molecules moving at various speeds if we know the kind of gas (and hence the mass of each molecule) and the temperature. We consider now what determines the temperature in a planetary atmosphere.

(e) Effective Temperatures of Planets

The effective temperature of a planet is the temperature of a black body, or perfect radiator, of the same size as the planet that receives the same amount of radiation from the sun that the planet absorbs. The effective temperature corresponds roughly to the kinetic temperature at the surface of the planet or at the level in its atmosphere from which it reradiates most of the energy it absorbs back into space. For Mars, whose atmosphere is nearly transparent, the effective temperature is approximately that of its solid surface, on the average. Venus and Jovian planets, on the other hand, have opaque atmospheres, and their effective temperatures correspond to kinetic temperatures high in their atmospheres, from which is emitted the radiation we observe from them.

Just outside the earth's atmosphere, 1.37×10^6 ergs/sec of solar radiation are incident on each square centimeter. This is called the *solar constant* (Chapter 24). The earth's distance from the sun is 1 AU. Thus, from the inverse-square law of propagation of electromagnetic radiation (Section 10.1a), we see that at a planet of distance r AU from the sun, the incident energy per square centimeter is $1.37 \times 10^6/r^2$ ergs/sec. The total solar energy intercepted by a planet is $1.37 \times 10^6/r^2$ times the cross-sectional area of the planet, which is just the area of a circle of radius R equal to the planet's, that is, πR^2. Thus the planet intercepts $\pi R^2 \times 1.37 \times 10^6/r^2$ ergs/sec. Now it absorbs some of this energy and reflects the rest. The fraction of incident energy reflected is called the planet's *albedo*, denoted A. If it reflects a fraction A, it must absorb a fraction $1 - A$. Thus, finally, the total energy absorbed each second by a planet is

$$\text{energy absorbed} = \frac{(1 - A)\pi R^2 \times 1.37 \times 10^6}{r^2}.$$

On the average, a planet must reradiate energy at exactly the rate it absorbs it; otherwise, it would heat up or cool off until this equilibrium is achieved. Now the energy emitted by a black body at temperature T is σT^4 ergs/sec from each square centimeter (Section 10.4). Thus, for a spherical black body of radius R cm, the total energy radiated per

second is $4\pi R^2 \sigma T^4$ ($4\pi R^2$ is the area of the surface of a sphere of radius R). Because the effective temperature of a planet is that of a black body that absorbs (and hence reradiates) as much energy as the planet, that effective temperature is found by equating the energy absorbed by the planet to that emitted by a similar black body; that is,

$$\frac{(1 - A)\pi R^2 \times 1.37 \times 10^6}{r^2} = 4\pi R^2 \sigma T^4,$$

from which we find

$$T = \left[\frac{(1 - A) \times 1.37 \times 10^6}{4\sigma r^2} \right]^{1/4}.$$

The albedo of a planet is found by comparing its observed brightness to that calculated for a perfectly reflecting body of the same size and distance from the sun.

As an example, consider Saturn. Saturn is 9.54 AU from the sun, so r^2 in the above formula is $(9.54)^2$, or 91. The Stefan–Boltzmann constant, σ, is 5.67×10^{-5}. Saturn absorbs about 43 percent of the incident energy on it and reflects the rest; thus A is 0.57. The above formula gives, for the effective temperature of Saturn, $72°K$, or about $-330°F$.

Most of the energy emitted by planets is in the infrared. We can observe this energy directly at the telescope by using infrared detecting devices. This gives a separate check and is an independent way of finding the albedo of a planet.

Planetary temperatures derived in this way, of course, hold only in an average sense. A planet receives all its solar energy on its daylit hemisphere. If it does not rotate rapidly it may not reemit the energy it absorbs very uniformly in all directions. Mercury and the moon, for example, are hot on their sunlit sides and cold on their night sides. Moreover, it must be remembered that the effective temperatures do not correspond to temperatures at the solid surfaces of planets with opaque atmospheres. Thus, Venus is much hotter at its surface than it is high in its atmosphere, from which most of its radiated energy escapes into space (Chapter 14).

(f) Retention of Atmospheres

A molecule moving with a speed greater than the velocity of escape from a planet will not, in general, actually escape, because collisions with molecules above it will deflect it downward. Molecules at high enough levels in the atmosphere, on the other hand, which are moving in the right direction with enough speed will escape. Because some molecules are moving with much higher than average speeds, it is not necessary for the mean molecular speed in a gas to be equal to the escape velocity of a planet for that gas to be lost to space. Calculations show that if the mean molecular speed is as much as one third the velocity of escape, the planet will lose half of that gas from its atmosphere in a few weeks. If the molecular speed is even one fifth the escape velocity, that gas will disperse into space in a few hundred million years.

To be conservative, we shall suppose that for a planet to hold a particular gas in its atmosphere for several billion years, its velocity of escape must be at least six times the mean molecular speed for that gas.

In summary, to see whether a planet can hold a particular kind of gas in its atmospheres, we must (1) estimate a representative temperature appropriate to the planet's atmosphere (by the method described in the last subsection), (2) calculate the mean speed of the molecules in the gas at that temperature (Section 12.2d), and (3) compare six times that speed to the velocity of escape from the planet. The latter is $\sqrt{2GM/R}$, where G is the gravitational constant (6.67×10^{-8}) and M is the planet's mass. (Care must be taken to use the same units of speed in all calculations.) The condition that a gas, whose molecules have mass m each, be retained is thus

$$6\sqrt{\frac{3kT}{m}} < \sqrt{\frac{2GM}{R}}.$$

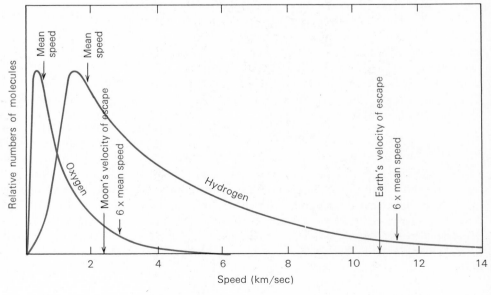

FIG. 12-3 Distribution of speeds of oxygen and hydrogen molecules at a temperature appropriate for the earth and moon.

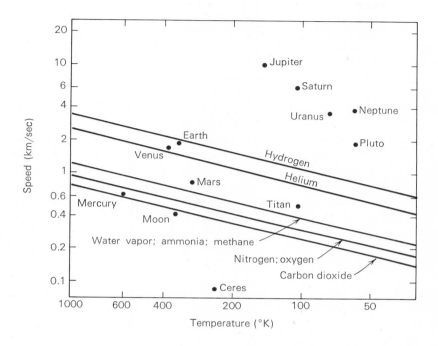

FIG. 12-4 Molecular speeds of common gases at various temperatures.

As an illustration, Figure 12-3 shows, roughly, the relative distributions of molecular speeds of oxygen and hydrogen that would be expected at a temperature of about 300°K — which would be representative of the earth and moon if they both had appreciable atmospheres. The solid curves show the relative numbers of molecules of each gas that would be moving at various speeds. In each case, the mean speed and six times the mean speed are indicated. It is seen that neither earth nor moon should be able to hold hydrogen, for six times its mean molecular speed would exceed the escape velocity of both. Oxygen, however, would be expected to be retained by the earth but not by the moon.

In Figure 12-4 are shown plots of the mean molecular speeds of various common gases as a function of the absolute temperature in degrees Kelvin. On the same plot are shown points representing the planets, the moon, the satellite Titan, and the minor planet Ceres. The position of the point for each body indicates one sixth of its velocity of escape and a temperature that is probably a minimum value of that appropriate to the levels where an atmospheric constituent can escape (generally higher than the effective temperature of the planet). A gas is not to be expected to be retained by a planet if the line representing its molecular speeds passes above the point for that planet. Moreover, it must be recognized that the escape of gases occurs at the uppermost levels of a planet's atmosphere, where the temperature (as a result of ionization by ultraviolet solar radiation) may be many times higher than the typical value — as is true for the earth. Thus the points representing the planets should probably be shifted to the left in the plot, to the high temperatures appropriate to the "escape layers" of their atmospheres. A particular atmospheric gas may well escape from a planet even if the line representing its molecular speed passes a little below the point for that planet.

It is apparent from the figure that Mercury, Ceres, and the moon would not be expected to retain any of the common gases, even if they ever did possess atmospheres. Uranus, Neptune, and Pluto would be able to retain any gas, except that all but hydrogen and helium would probably be in the liquid or solid state because of the low temperatures there. Pluto's low albedo (Chapter 14) suggests that it has no atmosphere at all, whereas the spectra of Uranus and Neptune show strong absorption bands of methane; apparently parts of their atmospheres are warm enough to allow some methane to sublimate to the gaseous state.

Mars could not hold hydrogen and helium but would be expected to retain the heavier gases. The giant planets, Jupiter and Saturn, could retain any gas. Hydrogen (although it is difficult to detect spectroscopically) and probably also helium are abundant in the atmospheres of all the Jovian planets. One would not expect to find atmospheric oxygen on these planets, for the free hydrogen present would combine chemically with oxygen to form water, which in turn would freeze out of the atmospheres. Hydrogen is observed on those planets, both in molecular form and in combination with carbon in the form of methane (CH_4). Ammonia (NH_3) is also present in the atmosphere of Jupiter.

Hydrogen could not be retained for billions of years in the atmospheres of Venus or Earth. Nor could appreciable helium be retained at the high temperatures expected for the escape layers. Moreover, substantial amounts of hydrogen and helium may never have been present in the gaseous state on the earth and Venus. If, as is widely hypothesized, the planets coalesced from a preplanetary material, that material may have been swept clear of gas before the planets formed. Moreover, since the velocity of escape from a given mass diminishes as the square root of the distance from the center of that mass, a young, still evolving planet (depending on its temperature) might have a harder time acquiring an atmosphere of light gases than a finished planet would have holding on to that atmosphere. In other words, the fact that a planet may now be able to retain a particular gas in its atmosphere does not necessarily mean that that gas is present.

(g) Interiors of Planets

The study of the internal structure of planets is a very difficult subject. Our knowledge of the interior of the sun is far more advanced than our knowledge of the cold worlds that revolve around it. Stars, like the sun, are completely gaseous; the physical laws that pertain to the behavior of gases are relatively simple and hence are better understood than those that pertain to the nature of solid or liquid matter under high pressure.

Yet, some clues enable us to draw a few conclusions about the interiors of the planets. A knowledge of the mass and radius of a planet are sufficient for calculation of its mean density. Knowing the density, we can say something about the material that composes the planet. Clues to the distribution of mass in a planet can be gathered from the degree to which the planet flattens under the influence of its rotation, and from the perturbative effects that its flattening has on its satellites and on the planet's rate of precession. With reference to the earth, additional valuable information is obtained from the transmission and reflection of seismic waves at various levels in the interior (Chapter 13).

Some of the conditions in the interior of a planet can be inferred from the fact that the planet is in *hydrostatic equilibrium;* that is, it is neither contracting nor expanding rapidly. The various layers of matter are all attracted toward the planet's center by the gravitational force exerted upon them by the deeper regions. These weights of the overlying layers must be supported by the pressures of the layers beneath. If we know how the mass of a planet is distributed throughout its interior, we can calculate how great these pressures must be at

Mercury   Earth

Ceres   Venus   Mars

Sun

Moon   Jupiter   Saturn   Titan

Uranus   Neptune   Pluto

4 largest
(Galilean) satellites

FIG. 12-5   Relative sizes of the sun and planets.

every point from the surface to the center. Even without information on the mass distribution we can calculate lower limits, below which the central pressures of planets cannot lie.

As far as their internal structures are concerned, the planets fall naturally into two groups: the terrestrial planets, Mercury, Venus, Earth, Mars, and Pluto (sometimes the moon is also included); and the Jovian planets, Jupiter, Saturn, Uranus, and Neptune. The terrestrial planets are relatively small and dense and are believed to be composed mostly of rocky and metallic material. The Jovian planets are relatively large and of low mean density (indeed, Saturn has a mean density less than that of water). The chemical composition of the Jovian planets is probably more nearly typical of the general cosmic abundances of the various chemical elements — for example, the relative abundances found in the sun. Hydrogen and helium, the most abundant elements in the universe, probably dominate in the Jovian planets. Jupiter and Saturn, at least, are almost certainly

composed mostly of hydrogen compressed to a solid state (see Chapter 14).

(h) Summary

The various planets of the solar system differ greatly, and in many ways. These differences can be largely understood, however, in terms of the respective distances of the planets from the sun and their respective sizes and masses. These three basic parameters are important to an understanding of Chapters 13 and 14, which deal with the planets individually.

In Appendices 9 and 10 are tabulated various data pertaining to the planets. However, many precise figures are often more confusing than helpful to the nonscientific reader. Therefore, in Table 12.2 we have summarized, very roughly, a few of the most important data concerning the planets. It is easier to remember that Jupiter is about 5 AU from the sun than to remember the figure 778,730,000 km. The relative sizes of the planets and sun are shown in Figure 12-5.

TABLE 12.2 Some Approximate Planetary Data

	MERCURY	VENUS	EARTH	MARS	JUPITER	SATURN	URANUS	NEPTUNE	PLUTO
Distance from sun (AU)	$\frac{2}{5}$	$\frac{3}{4}$	1	$1\frac{1}{2}$	5	10	20	30	40
Mass in terms of earth's mass	$\frac{1}{20}$	$\frac{4}{5}$	1	$\frac{1}{9}$	318	95	15	17	$\frac{1}{10}(?)$
Radius in terms of earth's radius	$\frac{2}{5}$	1	1	$\frac{1}{2}$	11	9	4	4	$\frac{1}{2}(?)$
Rotation rate	59 days	243 days retrograde	24 hours	$24\frac{1}{2}$ hours	10 hours	10 hours	11 hours	16 hours	6 days
Velocity of escape (mi/sec)	$2\frac{1}{2}$	$6\frac{1}{2}$	7	3	37	22	13	14	(?)
Approx effective temperature† (°K)	450	235	240	220	100	75	50	40	40
Observed gases in atmosphere	—	CO_2, H_2O, HCl, HF, $O_2(?)$	N_2, O_2, A, CO_2, H_2O, etc.	CO_2, H_2O	H_2, CH_4, NH_3	H_2, CH_4	H_2, CH_4	H_2, CH_4	—
Number of known satellites	0	0	1	2	12	10	5	2	0

†Temperatures on the sunlit sides of planets, or at their surfaces, may be much higher. The sunlit side of Mercury, for example, has a temperature of over 600°K, and the surface temperature of Venus is also near 600°K (Chapter 14).

Exercises

1. From the data given in the chapter, show that Kepler's third law holds for Pluto.

2. What would be the period of a comet whose orbit has a semimajor axis of 10,000 AU?

3. It was once suggested that the minor planets and meteoroids may have originated from a planet that once broke up. In what way would that planet have had to differ from the other major planets?

4. A double planet has a period of mutual revolution of about $27\frac{1}{3}$ days, and the two bodies are separated by about 239,000 mi. What is the combined mass of the two? If they are in the solar system, what are their names?

5. The velocity of escape of Mars is only a little greater than that of Mercury. Why then does Mars have an appreciable atmosphere while Mercury does not?

6. On the earth we have free nitrogen. Why then do we not find ammonia (NH_3) in our atmosphere as we do in the atmosphere of Jupiter?

7. Calculate the sidereal periods corresponding to each of the distances of the planets from the sun given in Table 12.2. Compare your results with the accurately observed values tabulated in Appendix 9.

8. Suppose there were a planet 100 AU from the sun, of mass and radius like the earth's. Would it be expected to have helium in its atmosphere? How about ammonia? Explain in each case.

9. Explain why observations of the infrared radiation emitted by a planet can help us learn its albedo.

CHAPTER 13

The Earth-Moon System

Having described the general properties of the solar system, we turn our attention to its individual members. Except for the sun, which we shall discuss in Chapter 24, the planets are the most important solar system objects. Most of the planets have satellites, and several of the satellites are larger than the moon. Of all satellites, however, the moon is the largest in comparison to its primary; the earth-moon system has, in fact, often been called the "double planet." Moreover, the earth and moon are the objects we have explored most directly, so it is fitting that we should begin our investigation of the solar system with them.

13.1 APPEARANCE FROM SPACE

As we saw in Section 9.1a, the moon's albedo at full moon is 0.07. It is, in other words, a rather black object, absorbing 93 percent of the sunlight incident on it. Knowing the moon's reflecting power and its size, we can easily calculate how bright it would appear from the other planets.

If we knew the earth's reflecting power, or albedo, we could predict how bright it would appear as seen from the other planets as well. One way to determine the earth's albedo is by measuring the brightness of earthshine on the moon (Section 9.1a). When the earth is near the full phase, as seen from the moon, the night side of the moon is illuminated by light reflected to it from the earth. Measures of this illumination show that the full earth provides an average of about 80 times as much light on the moon as the full moon provides for us. The exact ratio varies between 60 and 100 times, because it depends on the cloud covering over the earth. On the other hand, the

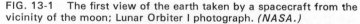

FIG. 13-1   The first view of the earth taken by a spacecraft from the vicinity of the moon; Lunar Orbiter I photograph. *(NASA.)*

earth presents 13½ times as much area to reflect light. It turns out, therefore, that the earth reflects light from 4½ to 7½ times as efficiently as the moon and has a mean albedo near 0.4.

When the earth is at opposition as seen from Venus, it is at the full phase, its entire illuminated hemisphere being presented to that planet. At this point the apparent brightness of the earth, from Venus, would be from 10 to 20 times as great as that of Venus as we see it. Venus reflects more sunlight than the earth does, but we see Venus at its brightest only as a crescent. The moon would appear about as bright from Venus as Jupiter does to us, and, as the months go by, the moon would seem to swing back and forth around the earth, reaching a maximum separation of just over ½°. From Mars and the more distant planets, the earth would appear brightest when in the crescent phase; because of its greater distance from the sun, and a reflecting power lower than that of Venus, the earth would appear only about one fourth as bright as Venus does from earth.

If astronomers on Venus could observe the earth through telescopes like the best we have on earth, they would certainly note that the earth abounds in surface detail. The fineness of detail that they could observe would depend on their own atmospheric conditions. Even if there were no atmosphere, the smallest features they could theoretically resolve would be about 2 to 3 miles across; it is doubtful if they could actually resolve finer markings than 30 or so miles in diameter. It seems unlikely that they could be aware of human activity on the earth.

13.2 GROSS PROPERTIES OF THE EARTH AND MOON

In previous chapters it has been explained how the mass of the earth (Section 4.3e) and of the moon (Section 9.3b) are found. The mass of the earth is 5.98×10^{24} kg (about 6.6×10^{21} English

FIG. 13-2   The nearly full earth photographed by Lunar Orbiter V. *(NASA.)*

tons); the moon's mass, smaller by a factor of 81.3, is 7.35×10^{22} kg.

(a) Sizes

It was shown in Section 2.3a how Eratosthenes measured the circumference of the earth about 250 B.C. Some modern methods of determining the dimensions of the earth are based essentially on the same principles as those employed by Eratosthenes.

The problem of finding the circumference of the earth is equivalent to determining the angle at the center of the earth between the radii of two places of known linear separation along its surface. Eratosthenes used observations of the sun's direction at noon on the first day of summer to find the angle at the earth's center between Alexandria and Syene, known to be 5000 stadia apart.

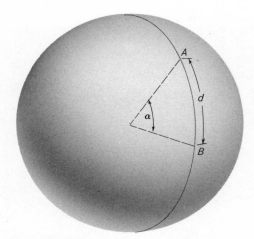

FIG. 13-3   Relation between angle and arc on a sphere.

For the calculation, two kinds of observations are needed. First, the distance between the two stations must be measured by surveying, or triangulation, the principle of which was explained in Chapter 2. Second, by means of celestial observations similar to those Eratosthenes made of the sun, the angular separation (at the center of the earth) of the two places must be measured.

Suppose it has been determined that two places, A and B, on the earth (Figure 13-3) are separated by d miles along its surface and by α degrees at its center. Then, the distance d is to the earth's circumference C as the angle α is to 360°; that is, $C = d \times 360°/\alpha$. Also, the distance along the earth's surface corresponding to 1° (the "length" of 1°) is d/α.

In 1671, in Paris, the French astronomer Jean Picard (1620–1682) determined the length of 1° to within a few yards. We have seen (Section 6.2d) that the earth is not perfectly spherical. Consequently, the exact length of 1° varies slightly over the earth. Along the earth's equator 1° is about 60 nautical miles (a nautical mile is defined as $\frac{1}{60}$° of latitude on the earth, that is, $1'$). A nautical mile is 1.1516 statute miles, so the equatorial circumfer-

ence of the earth is $360 \times 60 \times 1.1516$, or 24,900 statute miles. The diameter of a sphere is its circumference divided by π. The earth's equatorial diameter is 7927 mi, or 12,756 km.

It was described in Section 9.2b how the moon's size is measured. Its diameter is 3476 km (2160 mi), or 0.273 times that of the earth.

(b) Densities

The volume of a sphere is $4\pi R^3/3$, where R is its radius. The radii of the earth and moon are known, so their volumes can be computed. The volume of the earth is 1.08×10^{27} cm³. The mean, or average, density of the earth is the ratio of its mass (5.98×10^{27} gm) to its volume (see Section 4.1); that is,

$$\text{density of earth} = \frac{5.98 \times 10^{27}}{1.08 \times 10^{27}} = 5.5 \text{ gm/cm}^3.$$

The earth thus outweighs its volume in water by 5½ times. The mean density of the moon, less than the earth's, is only 3.34 g/cm³.

(c) Surface Gravities and Escape Velocities

In Section 4.3 we found that the acceleration of gravity at the surface of a spherical body of mass M and radius R (or the weight of a unit mass on the surface of that body) is given by the formula

$$a = G\frac{M}{R^2},$$

where G is the gravitational constant. At the earth's surface the acceleration of gravity is 980 cm/sec². Since the moon's mass is 0.0123 of the earth's and its radius is 0.273 of the earth's radius, we find that the acceleration of gravity at the surface of the moon is in proportion to that at the surface of the earth in the amount

$$\frac{a(\text{moon})}{a(\text{earth})} = \frac{0.0123}{(0.273)^2} = 0.165,$$

or about one sixth. On the moon, therefore, objects weigh about one sixth what they do on the surface of the earth, and if they are dropped at the surface of the moon, they accelerate to the ground only one sixth as fast as on earth.

The velocity of escape from a planet or satellite is $\sqrt{2GM/R}$ (Chapter 12). The escape velocity from the earth is 11.19 km/sec (about 7 mi/sec), but from the moon is only 2.38 km/sec (about 1.5 mi/sec). For comparison, the sun's escape velocity is 618 km/sec.

13.3 ATMOSPHERES OF THE EARTH AND MOON

From the discussion in Section 12.2f and with a knowledge of the escape velocities from the earth and moon, we would expect the earth to be able to retain all but the lightest gases in its atmosphere and the moon to have no atmosphere at all. Such is, in fact, the case.

We, on earth, live at the bottom of the ocean of air that envelops our planet. The atmosphere, weighing down upon the surface of the earth under the force of gravitation, exerts a pressure which at sea level amounts to nearly 15 lb/in.^{2}. If the weight of the air over 1 in.^{2} is 15 lb, the total weight of the atmosphere may be found by multiplying this figure by the surface area of the earth in square inches. We find, thus, that the total weight of the atmosphere is about 6×10^{15} tons, so the atmosphere has a mass of about a millionth that of the earth.

(a) Upper Extent of the Earth's Atmosphere

There probably is no real upper limit to the earth's atmosphere. At higher and higher altitudes the air thins out more and more until it disappears into the extremely sparse gases of interplanetary space. We have observational evidence that thin vestiges of the atmosphere extend to heights of at least 1000 mi. *Auroras,* discharges in the upper atmosphere caused when charged particles from the sun bombard the upper air, have been observed to 600 mi, and analysis of drag on earth satellites reveals

evidence for some atmosphere at heights of 1000 to 1300 mi.

We can measure the height of an aurora by the same principle that we use to survey the distance to the moon or any other object. Two observers, at A and B (Figure 13-4), simultaneously measure the directions and angular altitudes of the same feature on an auroral display (usually by means of photography). From station A the direction to the aurora (Figure 13.4) would determine the angle BAC, and the angular altitude is angle DAC. From station B, the direction and altitude give angles ABC and DBC. Thus, in triangle ABC, two angles, ABC and BAC, and an included side, the known distance AB between the observers, are determined, and the triangle can be solved for, say, the distance AC from the observer A to the point C directly under the aurora. Now, in triangle ACD the side AC, the angle DAC, and the right angle ACD determine CD, the height of the auroral display.

The most conspicuous auroras occur at heights of 50 to 100 mi, but occasional ones have been observed to 600 mi.

(b) Composition of the Atmosphere

The chemical composition of the earth's atmosphere is determined by quantitative chemical analysis. At the earth's surface, the constituent gases are found to be 78 percent nitrogen, 21 percent oxygen, and 1 percent argon, with traces of water (in the gaseous form), carbon dioxide, and other gases. At lower elevations, variable amounts of dust particles and water droplets are also found suspended in the air.

The gases nitrogen and argon are relatively inert chemically. It is the oxygen that sustains animal life on earth, by allowing animals to oxidize their food to produce energy. Of course, oxygen is also required for all forms of combustion (rapid oxidation) and thus is necessary for most of our heat and power production. Plants absorb carbon

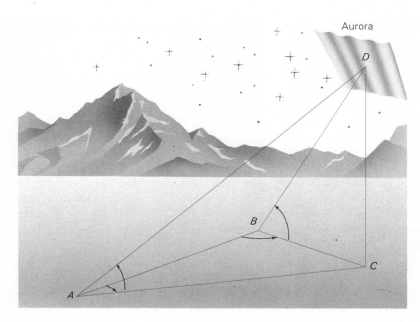

Aurora

FIG. 13-4   Triangulation of aurora altitude.

(c) Levels of the Atmosphere

dioxide in the process of photosynthesis, and release oxygen, which helps to replenish the oxygen consumed by man and the other animals.

The density of the air drops rapidly at higher and higher elevations. In fact, half of the total atmosphere is packed down within 3½ mi of the earth's surface. The gravitational attraction between the atmosphere and the earth crowds the air close to the ground. The pressure exerted by the rapid motions of the molecules that comprise air prevents it from collapsing completely and supports a small fraction of the atmosphere even to great heights.

In the lower part of the atmosphere, the temperature also drops rapidly with increasing elevation. It is this region, the *troposphere,* where all weather occurs. The upper extent of the troposphere varies from about 5 mi at the poles to about 10 mi at the equator.

Above the troposphere, and extending to a height of about 50 mi, is the *stratosphere.* The tem-

perature through most of the stratosphere is fairly constant — a little lower than 60° below zero Fahrenheit. Between 25 and 40 mi up, however, there is a warm zone; in 1947, V-2 rockets launched at White Sands, New Mexico, first recorded above freezing temperatures in this region.

The hot layer is due to the presence of ozone in that level of the atmosphere. Ozone is a heavy form of oxygen, having three atoms per molecule instead of the usual two. It has the property of being a good absorber of ultraviolet light. In absorbing the sun's short-wavelength ultraviolet light, the ozone is heated up and warms the parts of the atmosphere where it is present. Incidentally, the protective ozone layer helps to prevent some of the sun's dangerous ultraviolet radiation from penetrating the earth's surface. The region of the stratosphere extending upward from the ozone layer is sometimes called the *mesosphere.*

From 40 to 50 mi, the temperature drops to below −60° F again. Above 50 mi the temperature rises rapidly and at 250 to 300 mi reaches values in the thousands of degrees Fahrenheit. In this higher region of the atmosphere, molecules of oxygen and nitrogen break up into individual atoms of those elements. Radiation and high-speed charged particles from the sun and space bombard the atoms and ionize many of them. Therefore, the upper part of the atmosphere (above 50 mi) is called the *ionosphere.*

It is in the ionosphere that auroral discharges occur, and meteoroids burn up and produce meteors. The density of the air in the ionosphere is extremely low. Studies of the effects of atmospheric drag on satellites indicate that at 300 mi there are only about 200 million atoms per cubic inch, as compared to 10^{19} molecules per cubic centimeter at sea level. An air density of a billion atoms per cubic inch is considered a very good vacuum by laboratory standards.

Sharp increases in the densities of ionized or charged particles in the ionosphere occur at several levels known as the D, E, F_1, and F_2 layers. These layers reflect radio waves of wavelengths greater than about 17 m and make long-range radio communication possible. Radio waves can travel around the world by reflecting alternately between the ground or ocean and the ionized layers in the ionosphere. Wavelengths shorter than 17 m, however, (sometimes up to 30 meters) can penetrate the ionosphere and those from space are detected with radio telescopes.

(d) Evidence for the Lack of a Lunar Atmosphere

Telescopic observations of the moon, as well as observations from lunar probes, confirm its expected lack of an appreciable atmosphere. On the earth the air scatters the sunlight around into a certain portion of its night side, producing a twilight zone. On the moon there is no evidence of such a twilight zone. Furthermore, whenever the moon occults (passes in front of) a star, the star's light is observed to blink out suddenly, rather than to dim gradually, as it would if it had to shine through an atmosphere around the moon.

FIG. 13-5   The full moon. *(Mount Wilson and Palomar Observatories.)*

It has been pointed out by some investigators that the moon would be able to retain some heavy gases. Careful observations have been made in an attempt to detect evidence of a hazy appearance of certain lunar features that might result from a very thin atmosphere of such heavy gases. A few observers have reported slight evidence that thin "mists" might exist temporarily in the floors of some of the lunar craters, but these observations have not been confirmed.

The space between the planets, on the other hand, is not a perfect vacuum but contains a sparse distribution of atoms of gas — mostly ionized hydrogen. The moon must be continually colliding with this interplanetary gas and attracting occasional atoms of it to its surface. To be sure, these captured atoms must soon escape the moon's gravitational attraction, but nevertheless there might be an equilibrium concentration of sparse gases around the moon — a balance between the moon's rate of capture and rate of loss of its atmosphere. However, observations of radio waves passing near the moon's surface from the *Crab Nebula* (a strong source of radio waves — see Chapter 25) have been made to see if the waves were slightly deflected, as if by very thin ionized gases. These observations show that the density of this lunar gas, if it is present at all, is less than 10^{-13} times the density of the atmosphere at the earth.

Even if the moon should have thin gases around it, their density at the moon's surface must represent an extremely high vacuum by terrestrial laboratory standards. To all intents and purposes, we can regard the moon as being completely devoid of an atmosphere.

(e) The Moon's Lack of Water

In the absence of air, the moon can have no water on its surface. It is well known that water boils more easily (at a lower temperature) at high altitudes in the mountains than at sea level. If all air pressure could be removed above water, it would boil away. On the moon, therefore, if there ever were any liquid water, it would have evaporated and then, as a gas, dispersed into space with the rest of the moon's atmosphere.

In the absence of air and water, there can be no weather on the moon. There are no clouds, winds, rain, snow, or even smog, which accounts for many of the differences between the moon and earth. Weather on the earth and the running of water over its surface have been major sources of erosion that have washed and worn away entire mountain ranges, and even the faces of continents many times over during geologic time. On the moon, where air and water cannot contribute to erosion, features are more nearly permanent. There are formations of all ages on the moon, standing side by side; features that were formed probably billions of years ago are often still intact, standing with those formed in the recent past. We find the history of the moon written upon its face.

Although there can be no surface water on the moon, we cannot rule out the possibility of underground water. Some astronomers suggest that the moon could have a considerable supply of water trapped beneath its surface rocks. Occasional catastrophic events, such as meteoritic bombardment, they suggest, could release some of this trapped water, which would subsequently evaporate into space. It is interesting to speculate what erosive effects such temporary water at a place on the lunar surface might have. There is not yet any definite evidence that the moon does, in fact, have any underground water "tables."

13.4 TEMPERATURES ON THE EARTH AND MOON

Because of the moon's lower albedo, it absorbs a greater fraction of incident sunlight than the earth does and would be expected to have a higher effective temperature. Actually because of the moon's lack of atmosphere its temperature range between day and night is rather extreme. The earth's atmosphere serves to reflect and disperse much of the sun's radiation during the day and to blanket heat in near the surface at night.

The moon rotates with respect to the sun in about 29½ days. Therefore, for any one place on the moon the sun is shining for about 2 weeks at a time. Without any atmosphere to absorb or reflect some of the sun's radiation, the surface temperature rises to a high value, as more than 90 percent of the incident energy from the sun is absorbed. Eventually this energy is reradiated into space, mostly as infrared radiation. The rate at which the moon radiates electromagnetic energy at various wavelengths depends on the temperature of its surface (Section 12.2e). Thus, measures of the infrared energy from the moon give its surface temperature.

Infrared radiation can be measured with a heat

or infrared-sensitive device, such as a lead sulfide cell or thermopile, placed at the focus of a telescope. These measures indicate that the moon's temperature ranges from just above the boiling point of water, where the sun is shining, down to about 100°K (−173°C) on the dark side of the moon. The *Surveyor* lunar probes that landed on the moon in 1966 and 1967 carried instruments to record the moon's temperature and to relay the information back to earth by radio. The values obtained with earth-based telescopes were confirmed.

The rate at which the moon's temperature drops when sunlight is blocked off has been measured during lunar eclipses. As the moon enters the earth's shadow, its surface temperature has been observed to drop more than 150°C in 1 hour. Certain craters, however, cool more slowly than their surroundings, giving eclipse "hot spots." This rapid temperature drop can be explained partially by the moon's lack of an atmosphere to act as a blanket for the heat. However, if the moon's surface were covered with solid rocks like those over much of the earth's surface, the temperature drop would occur more slowly. Such rocks, when exposed to sunlight, conduct heat well into their interiors and take a long time to cool. This evidence suggested that most of the moon's surface must be covered with fine dust particles or granules, which conduct heat very poorly and therefore heat up only at their surfaces and so can cool rapidly. The prediction was later confirmed with the Surveyor experiments.

Observations of the moon at radio wavelength have also been used to measure its temperature, since the emission of radio energy at various wavelengths by a body depends, like infrared radiation, on its temperature. The radio waves, however, come from deeper beneath the surface of the moon and indicate temperatures from a lower depth in the moon's surface material. The radio observations suggest that the temperatures a few feet beneath

the surface of the moon are nearly constant, and average between −10 and −40°C.

13.5 THE MAGNETIC FIELDS OF THE EARTH AND MOON

The earth has a magnetic field similar to that produced by a bar magnet. Nearly everyone is familiar with the way iron filings align themselves along the lines of force that extend between the north and south poles of a magnet. The magnetic poles of the earth are located at about latitude 78° north and south, or about 825 miles from its geographical poles; the north magnetic pole is in Northeast Canada. Between the magnetic poles of the earth stretch lines of force along which compass needles align. The positions of the magnetic poles themselves, and the orientations of the lines of magnetic force, gradually shift around. During periods of solar activity (Chapter 24) rapid short-period fluctuations in the earth's magnetic field also occur. The over-all strength of the field is fairly weak. Its origin is believed to be in the earth's core. Fluid motions in the electrically-conducting core are thought to cause it to act like a dynamo. The rotation of the earth causes the magnetic field to be aligned approximately with the rotation axis. The energy source of this dynamo is still uncertain.

(a) Satellite Measures

Both U.S. and Soviet satellites and space probes have carried magnetometers into space to measure the strength of the earth's magnetic field far from its surface. The field strength is observed to decrease rapidly with distance from the surface and is roughly inversely proportional to the cube of the distance from the center of the earth, as predicted from theory. The exact strength and extent of the field is also observed to vary greatly from time to time. The fluctuations are probably associated with the ejection of charged particles from the sun during times of solar activity (Chapter 24).

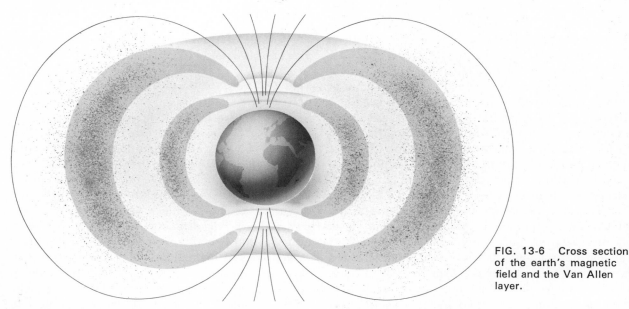

FIG. 13-6 Cross section of the earth's magnetic field and the Van Allen layer.

(b) The Van Allen Layer

An important discovery made by the early artificial satellites was the presence of a zone of highly energetic charged particles high above the earth. The inner part of this radiation zone was detected by instruments carried aboard the Army satellite Explorer I, launched on January 31, 1958. More information about the nature and extent of the belt of radiation was obtained with instruments on Explorer III, Explorer IV, and Pioneer I. In December 1958, the Army space probe Pioneer III revealed an additional region of radiation high above the one first discovered. The radiation-detecting experiments on Explorer I were under the direction of James A. Van Allen, physicist at the State University of Iowa; the radiation zone is called, in his honor, the Van Allen layer.

Figure 13-6, showing a cross section of the magnetic field surrounding the earth, displays the approximate location and extent of the most intense part of the Van Allen layer. It surrounds the earth, rather like a broad doughnut. The inner intense region is centered about 2000 mi above the surface of the earth and has a thickness of 3000 mi or more. The outer region of the Van Allen layer is about 10,000 to 12,000 mi from the earth's surface and has a thickness of from 4000 to 6000 mi.

The Van Allen radiation zone consists of rapidly moving charged particles trapped in the earth's magnetic field. Most of the fastest moving particles are electrons, although the inner part of the Van Allen layer contains high energy protons as well. The particles in the inner region are moving with speeds far greater than those in the outer one. Some are moving with energies as great as those possessed by electrons that have been accelerated through an electric field of billions of volts. The radiation strengths, heights, and thicknesses of the regions in the Van Allen layer, especially of the outer one, vary considerably from time to time. These regions are not really sharply defined; trapped charged particles must exist throughout the earth's magnetic field. The inner and outer parts of the Van Allen belt may be thought of as regions where the density of the moving particles is much greater than elsewhere.

The approach to the Apollo 11 landing site, photographed from an altitude of 63 miles. The landing zone is in the center of the picture, near the top. *(NASA.)*

Lunar terrain near lunar module. Apollo 11 mission, July 20, 1969. *(NASA.)*

India and Ceylon, looking north with the Bay of Bengal on the right.
Gemini XI photograph taken from an altitude of 410 nautical miles. *(NASA.)*

The origin of the trapped particles is not known with certainty. It seems likely, however, that the outer region consists mostly of particles captured from the "wind" of corpuscular radiation from the sun (see Chapter 24). It has been suggested that the protons and electrons in the inner region may be produced in interactions between molecules of the earth's atmosphere and cosmic rays (Chapter 31).

The Van Allen layer may be a hazard to space travelers, for upon striking the walls of a space vehicle, the fast-moving particles would produce x-rays that are very dangerous to man. Space travelers will either have to pass through this region very rapidly, so as to be exposed to the radiation for as short a time as possible, or else leave the earth along a polar route, for we observe (Figure 13-3) that the Van Allen belt does not extend over the polar regions of the earth.

(c) The Lunar Magnetic Field

Various lunar probes, especially the lunar orbiters launched in 1966 and 1967, have carried magnetometers to measure any magnetic field around the moon. So far no such field has been detected. If present, it must be very weak. Nor have trapped charged particles been detected; that is, the moon has no "Van Allen" layer.

The lack of a lunar magnetic field is not surprising. The moon rotates slowly and, because of its relatively low mass, is not believed to have a fluid core — factors thought to be associated with the origin of the earth's field.

13.6 INTERNAL STRUCTURES OF THE EARTH AND MOON

(a) Clues Obtained from the Earth's Mean Density

Our knowledge of the mean density of the earth tells us something about its interior. By sampling surface rocks, we have found that the average density of the outer crust of the earth is about 2.7 gm/cm^3. On the other hand, we have seen that the mean density of the earth as a whole is 5.5 gm/cm^3. Because of the low surface density but high average density, we must conclude that the interior of the earth is very dense indeed.

The weights of the various layers of the earth bearing down upon its interior causes the pressure to increase inward. Calculations indicate that the pressure at the earth's center must be close to 50 million $lb/in.^2$. When subjected to this great pressure, matter is highly compressed and heated. We can expect the central regions of the earth, then, to be hot and dense.

(b) Seismic Studies

Stresses that build up gradually in the crust of the earth are often released by slippages along fissures or *faults*. The energy released in these sudden movements in the earth's crust results in earthquakes. Vibrations are sent out and travel to all parts of the earth. Some types of the vibrations travel along the surface; others pass directly through the interior.

The study of these waves that originate in earthquakes is *seismology*. The seismic vibrations are picked up and recorded by delicate instruments stationed around the earth's surface. An analysis of the time required for seismic waves to reach these various stations provides information about the earth's interior.

The transmission of the waves through most of the earth's interior indicates that the major part of the earth is extremely rigid — more so than steel. This major part of the earth is called its *mantle*. The density in the mantle increases downward from about 3½ to 5½ gm/cm^3. It is believed to be composed mostly of basic silicate rocks.

At the upper boundary of the mantle is the *crust,* a shell extending on the average about 20 mi inward from the earth's surface under the continents. The crust consists of surface rocks such as granite and basalt, with overlying oceans and sedimentary rocks. Evidence from seismic studies suggests that the crust extends somewhat deeper under the continents than under the ocean floors, where it may be only a few miles thick.

At the inner boundary of the mantle, about 1800 mi below the surface, is the *core* of the earth. The outer part of the core acts like a liquid, for it does not transmit certain kinds of seismic waves. The innermost part of the core is extremely dense and hot (in the thousands of degrees).

(c) Temperature

The temperature increases downward in the crust of the earth about 1°C for every 100 ft. As the core slowly cools, conduction gradually transmits heat from the earth's interior out though the mantle and crust. However, the conductivity of the interior is relatively low, and cooling of the hot core must occur at a very slow rate. Calculations have shown that most of the heat in the crust, perhaps over 80 percent, may not come from the deep interior but from the radioactive decay of uranium, thorium, and potassium — unstable radioactive elements in the crust itself.

(d) Age of the Earth

The age of the earth can be estimated from the degree of radioactive decay of certain elements in the crust. Atoms of uranium and thorium disintegrate spontaneously but very gradually through various elements, including radium, and end as atoms of lead. The exact rate of this radioactive decay from one element to another has been measured in the laboratory. The decay is found to occur at a very constant and regular rate.

Studies of the relative proportions of lead, uranium, and radium in mineral deposits containing these elements indicate how long the disintegration process has been at work and give the ages of these

surface rocks, from which we derive the age of the earth itself. The results of different investigations differ somewhat, but all confirm that the age of the earth is in the billions of years. The oldest rocks on the surface have ages of 3.5 billion years. The age of the earth itself is estimated at about 4.5 billion years, on the basis of more subtle studies of the ratios of those isotopes of lead that do not originate from radioactive decay.

(e) Interior of the Moon

The moon's mass and mean density are both lower than the earth's. Because of the lower weight of lunar material, it should not compress as much as terrestrial matter, and internal pressures in the moon are expected to be far less than in earth. It is doubtful that the moon's central pressures have been great enough to have heated the matter appreciably and have produced a liquid core.

Analysis of the perturbations on orbits of the lunar orbiting satellites gives some indication of the mass distribution within the moon. It is found that the moon is, indeed, much less centrally concentrated than the earth is. The structure of the moon has not yet been studied with seismic experiments.

13.7 THE SURFACE OF THE MOON

The most conspicuous of the moon's surface features are easily visible to the unaided eye. In medieval Europe these markings were generally regarded as colorations rather than actual irregularities in the surface of the moon. Galileo's telescope observations, however, revealed mountains, craters, valleys, and what appeared to be seas. The idea developed that the moon might be a world, perhaps not unlike our own. Many other lunar observers followed in Galileo's lead. In 1647 John Hevel of Danzig (1611–1687) published his comprehensive treatise on the moon, *Selenographia*. In a series of carefully prepared plates, Hevel identified many of the moon's features and gave them names, in many cases in honor of similar features on the earth. The names he gave the mountain ranges on the moon survive today.

The early lunar observers regarded the moon as having continents and oceans and as being a possible abode of life. We know today, however, that the resemblance of lunar features to terrestrial ones is superficial. The moon's lack of air and water makes most of its features unlike anything we know on earth.

(a) Modern Observations

Many excellent photographs of the moon have been obtained with earth-based telescopes, but today most of the lunar photographs obtained from earth have been rendered obsolete by high-quality photographs obtained with the spectacular Ranger and Surveyor lunar probes and with the lunar Orbiters. These pictures are transmitted back to earth by means of a television scanning technique. Soviet lunar probes have also produced high-quality photographs, and in fact a Soviet space vehicle was the first to send to earth photographs of the moon's far side. Moreover, the Soviet Luna 9 made the first soft landing on the moon on January 31, 1966, and transmitted close-up photographs to earth for 3 days. The American Ranger, Surveyor, and Orbiter missions, however, have been far more fruitful.

In the Ranger series in 1964 and 1965, Rangers 7, 8, and 9 made successful "hard" landings on the moon; that is, they crashed directly onto the moon as planned. On the way, however, a great many views of the moon were televised directly back to earth, with resolution on the lunar surface as high as 1 ft. The best observations from earth telescopes gave a resolution of about ½ mi on the moon.

The first successful unmanned American soft landing on the moon was by Surveyor I, launched on May 30, 1966. Surveyor I alone sent over 11,000 photographs back to earth. In the next year and a half, Surveyors III, V, VI, and VII made successful landings. Each had television cameras which showed both the lunar landscape and close-up views of experiments done on the lunar "soil" by equipment carried on the probe. In addition to obtaining photographs, the Surveyors analyzed the chemical nature of lunar surface material and studied its physical and magnetic properties. Observations were even made of part of the solar corona far from the sun's surface (Chapter 24).

The five lunar Orbiters were all highly successful. The first was launched on August 10, 1966, and the last on August 1, 1967. Each was sent near the moon and then put into a lunar orbit by firing a retrorocket; a final maneuver changed its orbit to make its closest approach to the moon (perilune) conveniently near the surface (usually about 28 mi). Each Orbiter carried a 260-ft roll of 70-mm film, on which were made simultaneous exposures with a wide angle camera and with a second camera with a telephoto lens. The film was processed automatically and then scanned with a television camera, and the picture information was transmitted to earth. The entire "front" side of the moon (the side turned toward us) was photographed with a resolution averaging 10 times better than that of the best photographs obtained from earth. About 99.5 percent of the far side of the moon was photographed with better resolution than that of the best earth photographs of the near side. In addition, the Orbiters delineated several possible landing sites for astronauts in the Apollo project, photographed the earth, made measures of micrometeorites and the moon's magnetic field (not detected), and measured charged particles in the vicinity of the moon.

(b) The Lunar Seas

The largest of the lunar features are the so-called seas, still called *maria* (Latin for "seas"). It is they that form the features of the "man in the moon." The maria are of course dry land. They are great plains, with relatively smooth, flat floors that appear darker than the surrounding regions.

FIG. 13-7   The western hemisphere of the moon. *(Lick Observatory.)*

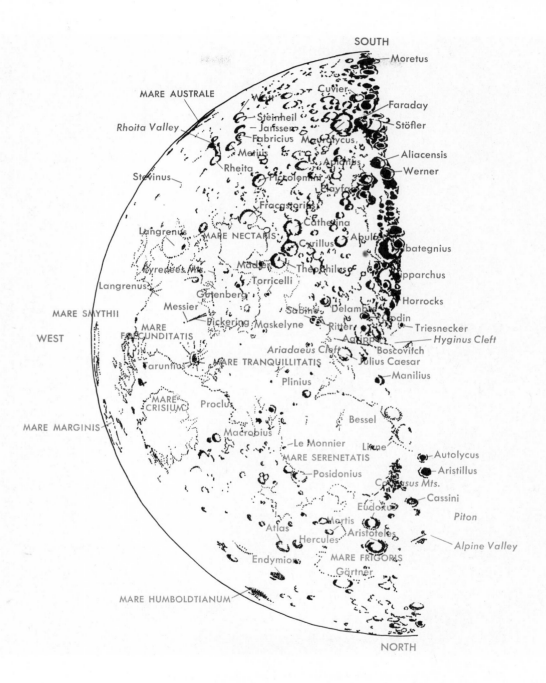

SOUTH

Moretus

MARE AUSTRALE

Cuvier
Weif
Faraday
Rhoita Valley
Steinheil
Stöfler
Janssen
Fabricius
Maurolycus
Metius
Aliacensis
Rheita
Apianus
Werner
Stevinus
Piccolomini
Playfair
Fracastoria
Catherina
Langrenus
MARE NECTARIS
Cyrillus
Albategnius
Abulf
Pyrenees Mts.
Madler
Theophilus
Hipparchus
Langrenus
Torricelli
Messier
Gutenberg
Horrocks
MARE SMYTHII
Sabine
Delamb
Godin
Triesnecker
Bickering
Maskelyne
Ritter
WEST
MARE
FECUNDITATIS
Agippa
Hyginus Cleft
Ariadaeus Cleft
Boscovitch
Tarunfius
Julius Caesar
MARE TRANQUILLITATIS
Manilius
Plinius
MARE
CRISIUM
Proclus
Bessel
MARE MARGINIS
Macrobius
Le Monnier
Licne
Autolycus
MARE SERENETATIS
Aristillus
Posidonius
Caucasus Mts.
Cassini
Eudoxu
Piton
Atlas
Mortis
Aristoteles
Hercules
Alpine Valley
Endymion
MARE FRIGORIS
Gärtner
MARE HUMBOLDTIANUM

NORTH

FIG. 13-8 The eastern hemisphere of the moon. (Lick Observatory.)

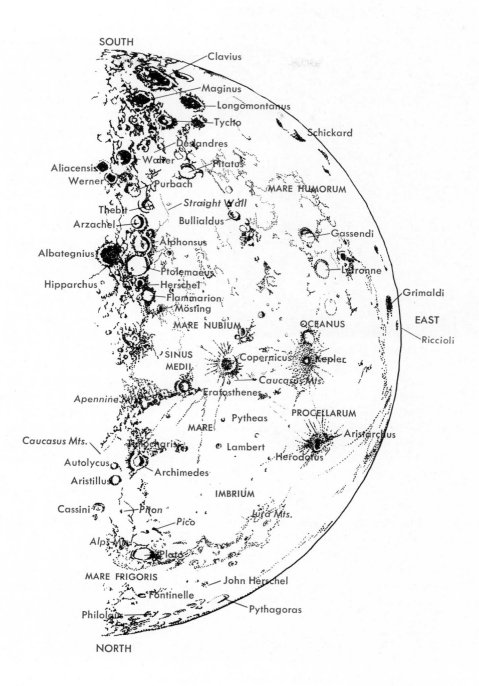

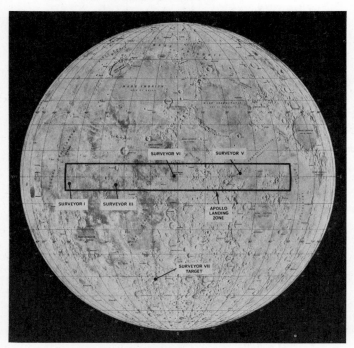

FIG. 13-9   Lunar spacecraft landing areas. *(NASA.)*

The largest of the 14 lunar "seas" on the moon's visible hemisphere is *Mare Imbrium* (the "Sea of Showers"), about 700 mi across. The other maria have equally fanciful names, such as Mare Nubium ("Sea of Clouds"), Mare Nectaris ("Sea of Nectar"), Mare Tranquilitatis ("Tranquil Sea"), Mare Serenitatis ("Serene Sea"), and so on. Most of the maria are roughly circular in shape, although many of them are interconnected or overlap slightly, and all have irregularities and baylike inlets, such as Sinus Iridum ("Bay of Rainbows") on the north "shore" of Mare Imbrium.

Careful inspection of good photographs shows that the maria do not have perfectly smooth floors, as they appear to have at first glance. They are speckled with thousands of tiny craters resembling potholes ranging in size down to a few inches or less across, and inside some maria there are large craters. Some of the mare floors show wavelike

ripples when the sunlight strikes them at a glancing angle. In a few places cliffs are found, suggesting slippages of the moon's crust along faults — fissures or lines of weakness in the crust. The best example is the "Straight Wall" in Mare Nubium, which is a cliff 600 ft high and 80 mi long.

The Surveyors landed in several different maria and found them to be remarkably similar to each other. The mare floors are covered with a layer of material, possibly debris that has fallen down from the highlands, that appears to range from 5 to 10 m thick. The material consists mostly of fine grains with a variety of sizes, but which are typically about $\frac{1}{1000}$ in. in diameter. However, mixed in are numerous aggregates or "clots" of fine grains and some hard rocks. The material is, on the average, very porous and is compressible, but Surveyor measurements show it to be able to support easily the weight of an astronaut. The mean density of the material is between 0.7 and 1.2 gm/cm³.

(c) The Lunar Craters

Even from earth there can be observed on the moon some 30,000 craters, circular depressions ranging in size from less than 1 mile to over 100 miles across. The largest of the craters are *Clavius* and *Grimaldi*, both nearly 150 miles in diameter. Following the custom started by John Riccioli in 1651, craters are generally named after famous scientists and philosophers. Other famous craters are Tycho, Copernicus, Kepler, Aristarchus, and Plato. Craters are sometimes still named after outstanding selenologists (astronomers who study the moon) today.

Although craters can be seen on the moon at almost any time, even with binoculars, the best time to view them is when the moon is near the first or last quarter phase. Then the terminator, dividing day and night on the moon, runs about down the middle of the apparent disk of the moon in the sky. The sun strikes craters near the terminator at a glancing angle and they cast long shadows, which makes them stand out in bold relief.

In general, craters are found over those regions of the moon that are not covered by the seas. There are comparatively few craters in the maria themselves, which suggests that the maria may be more recent features on the moon than most of the craters. The existence of crater tops barely visible in some of the mare floors supports this view. Most of the craters are found in the rugged regions of the moon. There they occur in all sizes and probably all ages as well. Frequently they overlap, one piled on top of another. An especially rugged region of overlapping craters is found in the vicinity of Tycho, near the moon's south pole.

FIG. 13-10   Part of the crater Tycho, photographed by Lunar Orbiter V. *(NASA.)*

FIG. 13-11   Part of the moon's far side, photographed by Lunar Orbiter V. *(NASA.)*

The largest craters are often called *walled plains*. Examples are Clavius, Plato, and Ptolemaeus. A walled plain often appears to be a sunken region with little or no outside wall, sloping up to the crater rim. Some of the walled plains do not have circular walls, but are irregular in shape.

Most of the craters of moderate size are quite circular and have outside walls. In a few cases the crater floors are higher than the surrounding landscape, but in the majority of cases they are lower. Their inside walls are almost always steeper than the outside walls and rise to heights of as much as 10,000 ft above the crater floors. Many of these craters have mountain peaks in their centers. They somewhat resemble craters left by bomb explosions and may have had a violent origin. An excellent example is the crater Copernicus.

The most numerous craters are the smallest ones, those ranging in size down to craters only a few feet across photographed by the lunar probes. Often craters, especially the smaller ones, occur in clumps or clusters or in long lines or rows of craters. In some cases lines of craters seem to follow shallow cliffs or what appear to be faults.

(d) The Lunar Mountains

There are several mountain ranges on the moon. Most of them bear the names of terrestrial ranges — the *Alps, Apennines, Carpathians,* and so on. The similarity of mountains on the moon and the earth ends, however, with their names. Because of the absence of water, the lunar mountain ranges are devoid of the drainage features so characteristic of our own mountain ranges, and the lack of weather erosion on the moon results in a different local appearance of mountains there.

The heights of lunar mountains above the surrounding plains can be determined by measuring the lengths of the shadows they cast. In Figure 13-13 is illustrated one of the possible procedures of determining the elevation of a lunar feature. A mountain peak, M, near the terminator and near

FIG. 13-12   Part of the southern half of the moon's far side, photographed by Lunar Orbiter II. *(NASA.)*

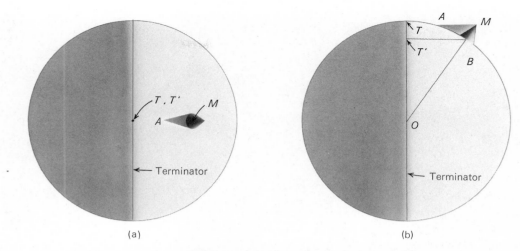

FIG. 13-13. Measuring the height of a lunar mountain. (a) The apparent disk of the moon as it appears in the sky; (b) the geometry of the situation seen at right angles to the line of sight from the earth.

the center of the moon's disk, casts its shadow over the distance *AB* from its base toward the terminator. The length of the shadow *AM* and the projected distance of the mountain from the terminator, *T'B*, can be measured in seconds of arc, and these measures can be converted to miles or kilometers.

The geometry of the problem is shown in Figure 13-13(b), in which the moon is seen from a direction at right angles to the line from the moon to the earth. The line *AM* is perpendicular to *OT*, and *AB* is perpendicular to *OM*. Therefore, the two right triangles *OT'B* and *ABM* are similar, and we have the proportion

$$\frac{BM}{AM} = \frac{T'B}{OB}.$$

Since *OB* is the known radius of the moon, and *T'B* and *AM* are measured quantities, the height of the mountain, *BM*, can be calculated.

Pains were taken that the photographs made with lunar Orbiters would be at such angles that

the features on the moon would have shadows — enabling the determination of the heights of those features.

The highest lunar peaks range up to elevations of 25,000 ft or more, comparable to the highest peaks in the Himalayas on the earth. It is not certain, however, that any mountain on the moon is quite as high as Mount Everest. (Of course, on earth mountain heights are measured from sea level rather than from the elevations of the surrounding plains.)

(e) Other Lunar Features

There are many other lunar features besides the seas, craters, and mountains. The *Alpine Valley*, for example, is a deep straight gorge cutting through the Alps Mountains. Also on the moon are many crevasses or clefts, half a mile or so across, called *rilles*.

Especially interesting are the *rays* — bright streaks that seem to radiate out from certain of the craters that appear to be of explosive origin, notably Tycho and Copernicus. The rays have

FIG. 13-14   The crater Copernicus, viewed from the south; photographed by Lunar Orbiter II. *(NASA.)*

FIG. 13-15.   A region of craters and domes on the moon, photographed by Lunar Orbiter II. *(NASA.)*

widths of 5 or 10 mi or more and in many cases extend for distances of hundreds of miles. Some of the rays from Tycho seem to extend completely around the visible hemisphere of the moon. They follow nearly great circles on the moon's surface, and sometimes are *tangent* to the craters from which they originate rather than coming from the exact centers of the craters. They cast no shadows and are seen best near full moon, when the sun shines down on them most directly. At full moon the ray system around Tycho gives the moon almost the appearance of a peeled orange. Ranger photographs show some of the rays to be associated with lines of craterlets.

(f) Changes on the Moon

From time to time one reads of reported changes on the moon. Often these changes are reported by nonprofessional astronomers. Many of the lunar features do change strikingly in appearance as the phase of the moon changes and the angle of illumination by the sun varies. There is no authenticated record, however, of any permanent recent change on the moon.

On the other hand, evidence of temporary activity on the moon has been reported by the Soviet astronomer N. A. Kozyrev, of Pulkovo Observatory. On the night of November 3, 1958, Kozyrev obtained a spectrogram of sunlight reflected from the mountain peak in the middle of the crater Alphonsus near the center of the moon's disk. Superimposed upon the continuous spectrum of sunlight, his photograph revealed a series of bright broad bands of the sort emitted by gases composed of carbon compounds. The obvious interpretation of Kozyrev's spectrogram of Alphonsus is that a volcanic eruption occurred in the mountain peak and that hot gases released in the process added their own light to that of reflected sunlight. So far, no other similar eruptions have been observed on Alphonsus or other mountains, nor has there been confirmation by any other reputable astronomer of the eruption reported on November 3. The possibility of experimental error has not been ruled out, and many lunar experts do not regard the observations as definitive.

Other temporary phenomena reported on the moon include bright, apparently luminous red spots, observed telescopically by several astronomers in Arizona who were engaged in mapping lunar features. From time to time other temporary luminescences have been reported on the moon, but the reality of these phenomena has not been universally accepted. One possible explanation that has been proposed is that luminous areas on the moon are regions of its surface caused to glow by a bombardment of solar corpuscular radiation particles.

Some real changes on the moon must certainly occur. One piece of definite evidence comes from a lunar Orbiter photograph that shows a 60-ft boulder that apparently rolled into and out of a depression, leaving its obvious track in the lunar surface.

(g) Composition of the Lunar Surface

Three of the Surveyor soft landers, Surveyors V, VI, and VII, each carried an experiment to analyze the chemical composition of the material on the lunar surface. The device shot a collimated beam of alpha particles (nuclei of helium atoms) into the material to be analyzed. The alpha particles interacted with the atomic nuclei of lunar material and produced many protons and other alpha particles that were sprayed out. The particular distribution of energies of these secondary protons and alpha particles depends on the kinds of atoms in the bombarded material. The energy distributions were measured by the instrument, and the data were transmitted back to earth. From these measures,

FIG. 13-16   The Alpine Valley, photographed by Lunar Orbiter V. *(NASA.)*

FIG. 13-17 A region of many craters, including Tycho; photographed by Lunar Orbiter V. *(NASA.)*

the composition of the lunar surface rocks and grains was derived.

The results from all three Surveyors, in widely separated places on the moon, were strikingly similar. The most abundant element is oxygen, accounting for about 58 percent of the atoms present. Most of the oxygen atoms are chemically united with silicon, next most abundant, and accounting for about 20 percent of the atoms. Aluminum is third and provides about 6 to 8 percent of the atoms. Calcium, iron, and magnesium are present in small amounts; carbon and sodium are not abundant enough to have been detected by the experiments and must contribute at most 2 or 3 percent each. This composition, within experimental accuracy, is close to the averages for basaltic rocks on earth. Evidently, some, at least, of the terrestrial and lunar surface materials have had a similar chemical history.

(h) Origin of the Lunar Features

Much has been written concerning the origin of the lunar features, but most of the hypotheses suggested are still controversial. We shall mention here only a few.

The maria give the impression that they may have originated from great flows of molten rock streaming into lower regions on the moon. There are many places where neighboring features appear to have been inundated by the maria, and in some of the mare floors partially submerged craters can be seen. Some selenologists are of the opinion that the maria are, indeed, seas of solidified lava covered with the layer of fine particles found by the Surveyors. But, if so, the origin of the buried molten material is unknown. It has been pointed out that the moon's inner layers are not heavy enough to have produced sufficient heating to form a molten core. If the source of the lava were internal, it

FIG. 13-18   Lunar Orbiter photograph of the track of a boulder that rolled in and out of a depression on the moon. (NASA.)

would probably have been produced by radioactivity in the moon's crust. Another possibility is that bombardment of the moon by very large meteroids or small minor planets may have generated enough heat at the points of impact to produce the lava.

The present surfaces of the maria, as we have seen, are of a granular nature and not solid rock exposed directly to space. Whether or not there is frozen lava beneath the surface material, most lunar experts regard the present maria as "dust bowls" filled by fragments of rock broken down from the surrounding highlands by gradual erosion caused by temperature changes, cosmic rays, corpuscular radiation from the sun, and meteoritic bombardment.

The origin of the large lunar craters is now believed to be due to meteoritic impacts. Craters 50 miles in diameter can be produced by the infall of a body with a mass of about 8×10^{12} tons (or about 10 mi in diameter). The very largest craters,

the walled plains, give the appearance of regions that have sunk in the moon's surface. The walled plains, however, are among the older of the lunar craters, as is evidenced by the many smaller craters formed in them and on their rims. It may be that the walled plains once appeared like other craters, but that over the hundreds of millions of years erosive forces of the type mentioned above have broken down the wallls along their rims until the craters appear only as sunken depressions. Although most experts lean to the meteoritic hypothesis for the origin of the large craters, some or many of the smaller ones, those less than 10 mi. in diameter, may have had a volcanic origin.

If meteoritic collisions with the moon have formed so many craters there, one might wonder why we do not find such features on the earth. There may indeed have been many thousands of large craters of meteoritic origin on the earth in the past, but they would have been washed away long since by the erosion of wind and rain; only those formed in the very recent past would still be intact. A few meteorite craters are known on earth; one of the largest of suspected meteoritic origin is the Reiss Kessel crater, which is 15 miles across. The best known is the Barringer (or Great Meteor) Crater, near Winslow, Arizona, which is nearly 1 mile across (Chapter 17).

The origin of the rays is probably associated with the formation of the craters from which they radiate. It is possible that they may have been formed from secondary particles blown out over the moon's surface at the time of formation of the craters.

An interesting hypothesis for the origin of the rilles is that they (or some of them) were once lunar rivers. It has been suggested that a meteoritic impact forming a crater might release some of the water (if it exists) trapped beneath the surface. Water exposed to a vacuum immediately boils, and the heat of vaporization lost to the boiling water cools it while it boils. Soon the boiling water cools

FIG. 13-19   View from the right-hand window of the Apollo 11 Lunar Module. *(NASA).*

so much that it actually freezes — at least the top part exposed to the vacuum. Thus, it is suggested, water from beneath the moon's surface, spilling out of a newly formed crater, might soon become a river of water covered with a thick layer of ice. This ice-capped river can flow slowly downhill, possibly cutting out the gorges we now recognize as the rilles. The ice and water eventually evaporate and disperse into space, but the "river" could possibly last 100 or so years — long enough to dig a rille. Although calculations show this model to be feasible, it is still only one of several possible hypotheses and some aspects need to be verified. Nevertheless, it serves as an example of modern thinking in this field of research and suggests interesting possibilities.

(i) The Moon's Far Side

Because the moon rotates as it revolves about the earth, turning always the same face toward us (Section 9.1h), only one hemisphere (actually slightly more, as a result of librations — see Section 9.1g) is accessible to telescopic observations from earth. Our knowledge of the far side had to await space technology.

The first photograph of the moon's far side was obtained from a Soviet space rocket launched on October 4, 1959. The photograph was televised to

FIG. 13-20   Moonscape near Surveyor VII, which landed in the southern highlands of the moon in the vicinity of the crater Tycho. Note the lunar rocks littering the area. *(NASA.)*

FIG. 13-21   View to the northeast in a mosaic of photographs made with
the television camera on Surveyor VII, from a position about
18 miles north of the crater Tycho. (NASA.)

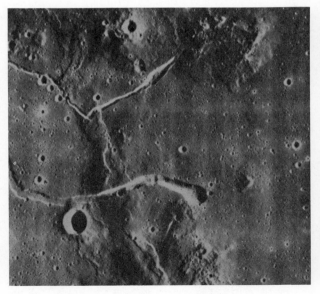

FIG. 13-22   Lunar Orbiter V photograph of
a lunar rille. (NASA.)

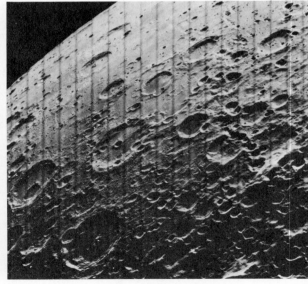

FIG. 13-23   Lunar Orbiter V photograph of
part of the moon's far side. (NASA.)

earth by radio signals from a distance of about 275,000 miles. Although this first epoch-making photograph is not of the highest resolution and understandably does not reveal the sort of fine detail later obtained with the lunar Orbiters, it was sufficient to show that the moon's far side is similar, in most characteristics, to the hemisphere we observe from earth. Some of the more prominent features have been named by the Russians.

The far side of the moon is now mapped with quality matching the best available of the near side before the space program. Although both hemispheres are similar to each other in small details, it is rather striking that the far side has a relatively small part of its surface covered by maria.

13.8 A BASEBALL GAME ON THE MOON

A fitting way to summarize a description of conditions on the moon is to imagine a baseball game on that world.

Because of the moon's lower surface gravity as compared to the earth, both players and ball would accelerate to the ground more slowly. The runners could take larger steps, and the ball, when thrown or batted, would fly longer before striking the ground. The size of the diamond would have to be increased two or three times.

With a good hit, a batter could easily bat a ball half a mile. The outfielder, however, using binoculars, could follow the ball and even with his clumsy pressurized suit, could be under it with a few great strides, if he didn't sink too deeply in the lunar surface material.

We would never have to issue rain checks on the moon, for there is no weather. There should be no need to call a game because of darkness, either. If a game were started in early morning, there would be two weeks of sunshine in which to play. One satisfaction of the spectators would be lost, however, for they could not bawl out the umpire; there is no air to carry the sound of their voices to his ears.

Exercises

1. What would a 300-lb man weigh on the moon?

2. If the moon had its present radius but four times its present mass, what would be the velocity of escape from its surface?

3. At what phase of the moon could we most conveniently measure the temperature of the night side? Why?

4. Suppose a mountain is observed on the moon 100 mi from the terminator. Its shadow is 40 mi long. How high is the mountain?

Answer: About 3.7 mi, or 19,500 ft

*5. Suppose a mountain peak on the night side of the moon rises just high enough to catch some of the rays of the rising sun and shines like a bright spot of light. If the mountain is just 100 mi from the terminator, what is its height?

Answer: About 4.6 mi

6. On which planets might you expect to find strong magnetic fields, and possible Van Allen belts, and why? On which planets might you *not* expect to find magnetic fields and why?

7. Compare the acceleration of gravity 1000 mi above the surface of the earth with that 1000 mi above the surface of the moon.

*8. Consult an appropriate reference to learn the molecular weights of various gases and suggest one or two gases that the moon could hold for 1 billion years or more.

*9. Suppose that the moon has 30,000 craters more than 1 mile across, and the earth has only 3. Assume the earth's radius to be four times the moon's. If craters have been formed on the earth and moon at a constant rate for the past 5 billion years, and if the moon's craters are never destroyed by erosion, how old, roughly, are those craters on earth?

Answer: About 3×10^4 years or less

CHAPTER 14

The Other Planets

14.1 MERCURY

Mercury is one of the brightest objects in the sky. At its brightest it is inferior in brilliance only to the sun, the moon, the planets Venus, Mars, and Jupiter, and the star Sirius. Yet, most people — including even Copernicus, it is said — have never seen Mercury. The planet's elusiveness is due to its proximity to the sun. Its orbit is only about one third the size of the earth's; it can never appear further from the sun than about 28°. It is visible to the unaided eye, for a period of only about 1 week, at times when it is near eastern elongation and can appear above the western horizon just after sunset and also when it is near western elongation, and rises in the east shortly before sunrise. Because the synodic period of Mercury is 116 days, the intervals of its visibility as an "evening star" after sunset, and as a "morning star" before sunrise, occur about three times a year. However, Mercury sets so soon after the sun (or rises so shortly before the sun), that only rarely can it appear above the horizon when the sky is completely dark; generally, one must look for it in twilight.

Mercury was well known to the ancients of many lands. The earliest observers, however, did not recognize it as the same object when it appeared as an evening star and as a morning star. The early Greeks, for example, called it Mercury when it was visible in the evening and Apollo when it was seen in the morning twilight. The corresponding names given it by the Egyptians were Horus and Set, and the Hindus called it Raulineya and Buddha.

(a) Mercury's Orbit

Mercury is the nearest to the sun of the nine major planets and, in accordance with Kepler's third law, has the shortest period of revolution about the sun — 88 of our days; it is appropriately named for the fleet-footed winged messenger god. Its mean orbital speed is nearly 30 mi/sec.

The semimajor axis of the orbit of Mercury, that is, the planet's median distance from the sun, is 36 million miles, or 0.39 AU. However, because Mercury's orbit has the high eccentricity of 0.206, its actual distance from the sun varies from 28.6 million miles at perihelion to 43.4 million miles at aphelion. Pluto is the only major planet with a more eccentric orbit. Furthermore, the 7° inclination of the orbit of Mercury to the plane of the ecliptic is also greater than that of the orbit of any other planet except Pluto.

(b) General Properties and Structure

The mass of Mercury, which has no known satellite, is determined from the perturbations its gravitational influence produces on other bodies, such as occasional comets or minor planets that pass near it. The minor planet Icarus, for example, passed within about 10 million miles of Mercury in April 1968. Mercury's mass can also be found from its effects on the Venus space probes, for example, Mariners II and V. The best estimate today of the planet's mass is about $\frac{1}{18}$ that of the earth. Mercury is the least massive planet in the solar system.

Mercury is also the smallest of the planets, having a diameter of only about 3030 mi, or less than half that of the earth. Its mean density is about 5½ times that of water — about the same as the mean density of the earth. Such a high density for Mercury is surprising; if its chemical makeup is the same as the earth's, its density should be somewhat less, for its weaker gravitational force would compress it less strongly. Possibly Mercury has a greater admixture of the heavier elements (for example, iron) than does the earth.

(c) Telescopic Appearance

Mercury, like Venus, presents different portions of its illuminated hemisphere to us as it revolves about the sun and hence goes through phases similar to those of the moon (see Section 3.5b). The alternate crescent and gibbous shape of Mercury is its only conspicuous telescopic characteristic.

Observations of Mercury are very difficult because of its nearness to the sun. At night it can be seen only close to the horizon, and then for but a short while. Its light must traverse a long path through the earth's atmosphere, which not only dims the planet, but subjects the rays to disturbances in the air ("seeing") which are especially troublesome for all observations of objects near the horizon. Consequently, nearly all telescopic observations of the planet are carried out in daylight, when it is high in the sky. Unfortunately, daytime observational conditions are seldom good. Nevertheless, experienced observers have been able to detect permanent surface markings on Mercury. They appear simply as darker areas, something like the maria on the moon, but less conspicuous.

(d) Rotation

Visual studies of Mercury's indistinct surface markings seemed to indicate that the planet kept one face to the sun, and for many years it was widely believed that Mercury's rotation period equaled its period of revolution about the sun of 88 days.

Radar observations of Mercury in the mid 1960s, however, showed conclusively that Mercury does rotate with respect to the sun (Section 12.2b). Its sidereal period of rotation (that is, with respect to the distant stars) is about 59 days. G. Colombo first pointed out that this is very nearly two thirds

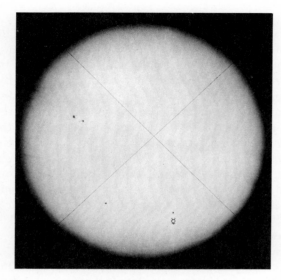

FIG. 14-1   Photograph of Mercury in transit across the disk of the sun. *(Royal Observatory, Greenwich, England.)*

of the planet's period of revolution, and subsequently Goldreich and Peale showed that there are theoretical reasons for expecting that Mercury can rotate stably with a period *exactly* two thirds of its revolution — 58.56 days. They argue that if Mercury were not perfectly spherical, but were deformed such that one dimension through the equator were longer than the others, the sun's force on that bulge should force the long axis of the planet to point to the sun when it is at perihelion where the sun's differential force on Mercury is strongest. This condition would be met if the planet rotated with its revolution period, but also with certain other rotation periods, the most likely being two thirds the revolution period, so that at successive perihelions alternate ends of the long axis of the planet are pointed toward the sun. [The actual deformation of the planet, of course, is at most very small; Mercury is spherical to a high approximation. Recall (Section 9.1) that the moon is also a triaxial ellipsoid.]

(e) Temperature

The temperature of the middle of the daylight hemisphere of Mercury has been measured at the Mount Wilson Observatory with a thermopile. Infrared radiation focused onto the thermopile by the telescope generates a feeble electric current whose strength is proportional to the amount of infrared energy radiated by the planet (radiation from the sun that is merely reflected to the earth by the planet is very weak at those wavelengths). From the thermopile current reading, the temperature can be calculated (Section 12.2e). The lighted side of Mercury has a temperature of about 610°K (640°F), hot enough to melt tin and lead. The dark side of the planet is too cold to radiate a measurable amount of radiation in the infrared. A recent observation places an upper limit of 150°K on the night-side temperature — that is, it is colder than −123°C. Since all sides of Mercury receive sunlight during its revolution (the period from noon to noon is 176 days), we conclude that its surface material cannot hold heat well; heat is not conducted well into depths beneath its surface. Like the moon, the surface material of Mercury has low conductivity.

(f) Atmosphere of Mercury

Because of the high temperature of the illuminated side of Mercury and the low velocity of escape, we would not expect it to retain an appreciable atmosphere. The lack of a Mercurian atmosphere seems to be confirmed by its low albedo; it reflects only 6 percent of the light incident upon it — less than any other planet. This low reflecting power is characteristic of a solid surface, like the moon's; an atmosphere would reflect light far more efficiently. Moreover, there is no visible twilight zone on Mercury. An atmosphere would scatter light a little distance around the planet into the night side. No trace of an atmosphere is revealed in the spectrum of sunlight reflected from the planet.

A few visual observers, notably Antoniadi and Schiaparelli, have reported what appeared to be a transient haze on Mercury, and consequently the possibility of a thin atmosphere on the planet has never been entirely ruled out. More recently (1952) the French astronomer A. Dollfus found weak evidence for a thin Mercurian atmosphere from measures of the polarization of sunlight reflected from its surface. The measures are difficult, however, and at best marginal.

Nevertheless, it is worth mentioning why polarization measures can be relevant. We have seen (Chapter 10) that light can be regarded as a motion of waves, each vibrating in a direction perpendicular to the direction of propagation. Ordinarily, in a beam of light, waves are present that are vibrating in all possible orientations. Under some circumstances, however, the orientations of all the wave vibrations are parallel or aligned. Then the light is said to be *polarized*. If there is a preponderance of vibrations in certain directions over others, the light is said to be partially polarized. One way to polarize light is to pass it through a filter made of one of various polarizing crystalline substances — a Nicol prism, for example — or through *Polaroid,* a trade name for a manufactured polarizing material. Such a filter transmits only those light waves that are vibrating in a particular plane. Now, suppose that light which is already polarized is passed through such a filter. If the filter is oriented in a direction so that the plane of the vibrations it passes is coincident with the plane of polarization of the light beam, the light will pass through. If the filter is slowly rotated, however, less and less of the light will be transmitted, until the plane of vibration that the filter will pass is at right angles to the plane of vibration of the polarized light beam; then no light will be transmitted. Thus the degree to which a light beam is polarized can be detected.

Light scattered by molecules, as in the earth's atmosphere, is strongly polarized, whereas that reflected from an opaque surface like the moon's is only weakly polarized. Dollfus seemed to find that the light reflected from Mercury was polarized to an extent that would not be expected if Mercury were not surrounded by gases, which led him to suspect that Mercury could have a very thin atmosphere. As stated above, however, the observations are far from definitive, and are not generally accepted.

Mercury cannot be expected to hold an atmosphere permanently; if there are gases there, they must represent an equilibrium between the capture of gases by Mercury as it sweeps through interplanetary space (and the possible release of gases from its surface rocks or from its interior), and the loss of those gases by their subsequent escape. The haze observed on Mercury by some observers, if real, may be fluorescence in its atmosphere — that is, gases set aglow by ultraviolet radiation from the sun. We emphasize, however, that there is no firm evidence for any gas around Mercury.

14.2 VENUS

Venus, named for the goddess of love and beauty, is sometimes called the "earth's sister," for it is most like the earth in mass and size of all the planets. It approaches the earth more closely than any other planet — at its nearest it is only 25 million miles away.

Venus is a beautiful object in the night sky; its brillance is exceeded only by that of the sun and moon, save for very rare comets and bright meteors. At night, it can cast a shadow; at its brightest it can be seen easily in broad daylight if one knows exactly where to look for it.

Like Mercury, Venus is an inferior planet (nearer the sun than the earth); consequently, it appears to swing back and forth in the sky, during its synodic period, from one side of the sun to the other. Like Mercury, therefore, it appears sometimes as an "evening star" and sometimes as a "morning star." The early Greeks, thinking it to be

two objects, called it *Phosphorus* and *Hesperus* when it was seen in the morning and in the evening, respectively. Pythagoras, in the sixth century B.C., is credited with being the first to recognize that Phosphorus and Hesperus were one and the same planet.

Because Venus is farther than Mercury from the sun, it reaches much greater eastern and western elongations, and can appear as far as 47° from the sun. It can only be seen in the west in the early evening or in the east before sunrise; however, it is visible for a longer time than Mercury, after sunset or before sunrise. Indeed, it is often so conspicuous in the evening sky that people surprised at its brilliance have called observatories to inquire whether it is the star of Bethlehem.

(a) Venus's Orbit

The median distance of Venus from the sun is 67,270,000 mi. Its orbit is the most nearly circular of any of the planetary orbits, having an eccentricity of only 0.007. Its distance from the sun, consequently, varies by only about 1 million miles.

To complete its revolution about the sun in 225 days, Venus moves with a mean orbital speed of about 22 mi/sec. Its synodic period — the interval between successive conjunctions or oppositions — is 584 days. The inclination of its orbit to the ecliptic is 3°24'.

(b) Physical Characteristics

The mass of Venus, which, like Mercury, has no satellite, must be calculated from the perturbations it produces upon the motions of the earth and other bodies. The best value, found from the Mariner V perturbations, is 0.82 times the mass of the earth. Venus has a radius of 6056 km, only about 200 mi less than the earth's. Its mean density is 5.2 times that of water.

It is possible to calculate what the approximate internal pressures and densities of Venus are if we assume that the compressibility of its material is

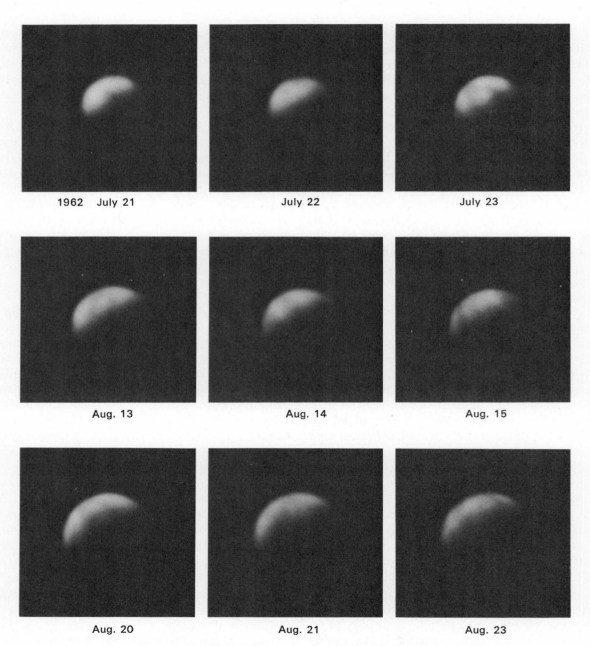

FIG. 14-2   Ultraviolet photographs of Venus in 1962. *(Lick Observatory.)*

the same as that of the material in the earth, the latter being known from seismic studies. At the center of Venus, the pressure exerted by the weight of the overlying rocks must be near 2.6×10^{12} dynes/cm^2, or about 2½ million times the pressure of the earth's atmosphere. The central density must be about 11 times that of water. Most investigators believe that the cores of Venus and other terrestrial planets are largely metallic, probably mostly iron and nickel.

(c) Telescopic Appearance

Venus, like Mercury, goes through phases. Galileo was the first to report the phases of Venus, a discovery that had great importance in disproving the validity of the Ptolemaic cosmology (Section 3.5b).

Venus is not observable at the full phase, because then it is at superior conjunction — on the other side of the sun from the earth — and is too nearly in line with the sun. Even if it could be seen, however, Venus would not show its greatest brilliance because at that time it is also at its greatest distance from the earth and subtends an angle of only 10″. Venus has its largest angular size of 64″ when it is nearest the earth, but then it is at the new phase (at inferior conjunction) and, again, is unobservable. Venus is at its greatest brilliance when it is a crescent, with an elongation of about 39°; this occurs about 36 days before and after inferior conjunction.

The surface of Venus is not visible because it is shrouded by dense clouds (or at least an opaque atmosphere). These clouds reflect sunlight very efficiently; measures of the albedo of the planet indicate that somewhere between one half and three quarters of the incident sunlight is reflected, a circumstance that contributes greatly to its brightness. Some rather indistinct dark markings can be observed on Venus. They show up best when the planet is photographed in violet light. They appear as broad streaks across the disk of the planet and may exhibit slow changes. They doubtless originate in the clouds or upper atmosphere.

(d) Rotation

The rotation period of Venus, like that of Mercury, has been determined by radar observations. The cloud markings are too indistinct and not permanent enough for us to observe the planet rotating visually, and it has long been established that Venus' rotation is too slow to be detected by means of optical Doppler shifting of the light from one limb relative to the other. The latter fact alone showed that the rotation period could not be shorter than about 1 month.

The first radar observations of the planet's rotation were made in the early 1960s. Surprisingly, they showed Venus to rotate from east to west — in the reverse direction from the rotation of most other planets — in a period of about 250 days. R. Carpenter, at the Jet Propulsion Laboratory's Goldstone Radar Tracking Station in Southern California, has identified several features on the planet that show up in the radar (reflected radio) signals. These are apparently localized topographical features on Venus that have high reflectivity at radio wavelengths. In any case, these features seem to be permanent and have been observed for several rotation periods of Venus. The rate at which they cross the planet's disk has given a rather accurate determination of its rotation period — retrograde (east to west) at 242.9 days, with an uncertainty of about one tenth of a day.

(e) Temperature

Venus, being closer than the earth to the sun, receives about twice as much solar energy per unit area and should, therefore, have a warmer climate than ours. If its albedo is 0.76 (a recent determination by Kuiper), we should expect an equilibrium temperature of 234°K (see Section 12.2e). Actual measures of the radiation from the planet, made at the

Mount Wilson and Palomar Observatories, indicate temperatures, for both sunlit and dark hemispheres, in the range 230 to 240°K (-27 to $-45°$F). These temperatures are based on measures of infrared radiation and refer to the upper atmospheric layers of Venus, from which that radiation originates. But a heated body emits all wavelengths of electromagnetic radiation, radio waves as well as infrared rays. The strength of the radio radiation from Venus at wavelengths greater than 3.2 cm has also been measured and is found to correspond to that expected from a body at a temperature of about 700°K (nearly 900°F). Since radio waves would penetrate the Venusian cloud layer, they presumably originate from a region at or near the planet's surface. This high surface temperature for Venus was indicated also by measures made from the Mariner II Venus probe in 1962. The Soviet Venus probe, Venera 4, which landed on the planet on October 18, 1967, carried instruments that recorded a temperature of about 550°K. There is a question, however, whether this temperature was recorded at the planet's surface.

(f) Atmosphere

That Venus has an appreciable atmosphere was established long ago by the following observations:

(1) The high albedo of Venus is characteristic of clouds or gases; an exposed rocky or dusty surface, such as that of Mercury or the moon, has a very low albedo.

(2) When Venus is at the crescent phase, the horns of the crescent extend a bit around the limb of the planet to the dark side, indicating that slightly more than a hemisphere is illuminated. This is evidence of the existence of an atmosphere that scatters sunlight past the terminator, producing a twilight zone.

(3) The spectrum of light reflected from the Venusian clouds contains absorption lines due to gases lying above those clouds.

The first gas to be identified spectrographically in the atmosphere of Venus was carbon dioxide, in 1932. More recently, traces of hydrogen flouride and hydrogen chloride have been detected.

It is difficult to detect water vapor on a planet by means of spectrographic observations from the surface of the earth, because water in the earth's atmosphere absorbs light and contaminates the planetary spectrum. In 1959 the spectrum of Venus was photographed from a balloon at an altitude high above most of the terrestrial water vapor; an analysis of that spectrum shows evidence for the presence of water in the atmosphere of the planet. Subsequent observations by telescopes on earth confirmed a small concentration of water vapor.

A partial chemical analysis was performed on the Venusian atmosphere by Venera 4, the Soviet probe launched on June 11, 1967, which passed Venus on October 18 of that year and dropped, by parachute, an instrument package to the surface of the planet. The Venera 4 data are reported to show that the atmosphere of Venus is 90 to 95 percent carbon dioxide, roughly 1 percent water vapor, and from 0.4 to 0.8 percent oxygen. Nitrogen may be present but is difficult to detect; at most it comprises 7 percent of the atmosphere, and it may be absent altogether. The relatively high concentrations reported for water, and especially for oxygen, are at variance with spectroscopic data and are open to serious question. The Venera 4 data also indicate that the atmosphere is very dense; its surface pressure is probably at least 20 times that of the earth's atmosphere at sea level and may be even much higher.

The Soviet data are in general agreement with the results of theoretical models of the Venusian atmosphere calculated by astronomers years earlier. A dense atmosphere is required, in fact, to account for the planet's high surface temperature. The American Mariner V space probe, launched to Venus about the same time as the Soviet Venera 4 and passing about 2500 miles from the planet's surface on October 19, 1967, also carried instruments to measure the conditions in the atmosphere of Venus and in its immediate space environment. Although Mariner V made no chemical analysis of the Venusian atmosphere, it did transmit radio waves through it to the earth, and the attenuation of those signals provided information on the density and some other properties of the atmosphere of Venus.

Other instruments carried by Mariner V attempted to measure any magnetic field around Venus, the density of hydrogen in its outermost atmosphere, and the flow of solar corpuscular radiation near the planet. No magnetic field was detected, indicating that if there is any field at all it is at most a few thousandths as intense as the field around the earth. On the other hand, evidence for a dense ionosphere in the upper levels of Venus' atmosphere was found. Ionized hydrogen was found in the outermost part of the atmosphere of Venus; there is also ionized hydrogen surrounding the earth's outermost atmosphere. The ionosphere in the Venusian atmosphere is denser than that in the earth's atmosphere, and even though Venus has no measurable magnetic field, solar corpuscular radiation is deflected around the planet by its ionosphere, rather than striking its surface directly, as on the moon.

The nature of the Venusian "clouds" is still a mystery. The polarization of sunlight reflected from them only approximately resembles that of light reflected from water droplets. Furthermore, the clouds show a slight yellowish tinge, whereas water droplets should be white. Yet, it is still possible that the clouds are water clouds, perhaps contaminated by impurities that give them their color; some experts consider it *probable* that they are water clouds. On the other hand, there are also other kinds of droplets or crystals that could display the characteristics of the Venusian cloud layer. Perhaps there are no clouds at all; if Venus' atmosphere is dense enough, the scattering of sunlight by molecules themselves or dust particles might produce the cloudlike effect.

What the surface of Venus is like is an even bigger guess. Because of the high surface temperature and because of the low abundance of water vapor in the atmosphere of Venus, most investigators regard the surface as being hot and dry. Some even envision vast dust storms blowing across the planet.

14.3 MARS

Mars is sometimes called the "newspaper planet"; it is the most favorably situated of the planets for observation from the earth and has excited more interest and comment than any other. Only on Mars and Mercury can we see the solid surface, and Mercury is too near the sun to observe easily. Venus comes a little closer to the earth, but at its closest it is at the new phase when it cannot be studied. Mars, on the other hand, is sometimes favorably situated for observation all night long.

(a) Orbit of Mars

The median distance of Mars from the sun is 141,690,000 mi, but its orbit is somewhat eccentric (eccentricity of 0.093) and its heliocentric distance varies by 26 million mi. The sidereal period of revolution is 687 days. The orbit is inclined to the ecliptic plane by 1°51'.

At intervals of the synodic period of Mars, 780 days, the earth passes between it and the sun. Then Mars, at opposition, is above the horizon all night. It is also at its nearest to the earth and is most favorably disposed for observation. All oppositions of Mars are not equally favorable, however; at such times its distance from earth can be anywhere from 35 to 63 million mi, depending on where the planet is in its orbit when the earth passes it. This circumstance is due to the eccentricity of Mars' orbit and, to a lesser extent, that of the earth's orbit (see Figure 14-3). The most favorable oppositions are those that occur when Mars is near perihelion —

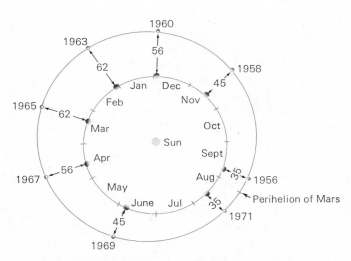

FIG. 14-3 Orbits of the earth and Mars, with the positions of the planets shown for several oppositions.

its closest approach to the sun and hence to the earth's orbit. These are the oppositions that take place during our late summer or early fall. Such optimum opportunities to investigate Mars arise about once or twice every 15 to 17 years; one occurs in 1971.

(b) Satellites

Centuries ago, Kepler, hearing of Galileo's discovery of four satellites of Jupiter, speculated that Mars should have two moons. There was, of course, no scientific justification for this speculation. Again, in 1726, Jonathan Swift in his satire *Gulliver's Travels* described Gulliver's visit to the land of Laputa, where he found scientists engaged in many interesting investigations. In one of them, Gulliver reported, Laputian astronomers had discovered

". . . [two] satellites, which revolve about Mars, whereof the innermost is distant from the centre of the primary planet exactly three of the diameters, and the outermost five; the former revolves in the space of ten hours, and the latter in twenty one

and an half; so that the squares of their periodical times are very near in the same proportion with the cubes of their distance from the centre of Mars, which evidently shows them to be governed by the same law of gravitation, that influences the other heavenly bodies."

It is an interesting coincidence that in 1877, 150 years later, Asaph Hall of the Naval Observatory actually discovered two small satellites of Mars that closely resemble those described by Swift. Those moons are named *Phobos* and *Deimos,* meaning "fear" and "panic" — appropriate companions of the god of war, Mars. Phobos is 5800 mi from the center of Mars, and revolves about it in 7h39m; Deimos has a distance of 14,600 mi and a period of 30h18m. The "month" of Phobos is less than the rotation period of Mars; consequently, Phobos would appear to an observer on Mars to rise in the west.

Both satellites are too small to present measurable disks; their sizes must be estimated from their brightnesses. If they have albedos similar to our moon or Mercury, Phobos must have a diameter of between 10 and 20 mi and Deimos must be about half that size.

(c) Physical Properties

The mass of Mars can be found conveniently and accurately from the distances and periods of its satellites applied to Newton's formulation of Kepler's third law (Section 4.5a) and also from Mars' perturbations on interplanetary probes; it is only 0.107 times as massive as the earth. The diameter of the planet is about 4200 mi, which gives it a mean density of approximately four times that of water and a surface gravity of 38 percent of the earth's. A 200-lb man would weigh about 75 lb on Mars.

The interior structure can be determined if, as in the case of Venus, we assume the same compressibility at a given pressure as is found for the material of the earth's interior. The central pressure

FIG. 14-4   Photographs of Mars and the city of San Jose, both taken from Mount Hamilton. One pair of photographs *(left)* was taken in violet, and the other *(right)* in infrared light, to illustrate the difference in atmospheric penetration at the two wavelength regions. *(Lick Observatory.)*

of Mars is calculated to be about 400,000 times the earth's atmospheric pressure; the central density is about 8.6 times that of water.

(d) Telescopic Appearance

Until Mariner IV, launched November 28, 1964, flew by Mars 228 days later and televised to earth photographs obtained at an altitude of 6000 mi, our only views of the Martian surface were through ground-based telescopes. The layman, having heard so much about Mars, is usually disappointed when he first sees it through a telescope. At its nearest, Mars is 150 times the moon's distance; a feature 1 mi wide on the moon can be resolved about as well as a feature 150 mi wide on Mars. At its nearest, the planet subtends only about 25″.

Since Mars is farther than the earth from the sun, it does not go through crescent phases. When it is near quadrature, however, it does appear distinctly gibbous. Except for its occasional gibbous shape, Mars usually resembles a shimmering orange ball — the shimmering is caused, of course, by the earth's atmosphere. Most of the Martian surface appears yellowish orange or red. Often white areas, the *polar caps,* can be seen at one or both of the planets' poles. Under good conditions large dark areas can be seen as well. They have a grayish color and somewhat resemble oceans; they were once believed to be bodies of water and so were named *maria,* like the "seas" on the moon.

No evidence has ever been found for steep mountain ranges, which might be recognized by the shadows they would cast. Many observers, however, have seen haze and what appear to be occasional clouds, some of which move over the planet's surface at about 20 mi/hr. Moreover, all expert ob-

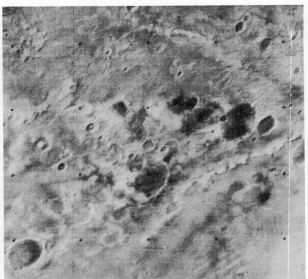

FIG. 14-5   Mars, photographed from a distance of 293,200 mi. with the television system carried by the Jet Propulsion Laboratory's Mariner 7 space vehicle that flew by the planet in August 1969. The television observations of Mars from Mariners 4, 6, and 7 were all under the direction of R. B. Leighton of the California Institute of Technology. *(NASA)*.

FIG. 14-6   Part of the Martian south polar cap photographed by Mariner 7 from an altitude of about 2000 mi. Except near the terminator the exposure on the cap has been suppressed by the automatic gain control of the television system in order that the "snow"-covered craters would stand out with better contrast. *(NASA)*

servers are agreed that under the best observing conditions an enormous amount of detail is visible. The difficulty is that the fine details can be seen only in brief glimpses during fleeting instants of steadiness of the earth's atmosphere. The best Martian photographs obtained with terrestrial telescopes reveal some of the finer markings, but experienced observers, taking advantage of rare moments of good seeing, have recorded far more.

Because of the difficulty in observing fine details on Mars, it is small wonder that no two observers agree exactly on what they see. Particularly controversial are the "canals" — what appear to be straight streaks; they were first observed in 1877 by the Italian astronomer Schiaparelli, who called them *canali* ("channels"). Many observers since have seen straight-appearing markings; the more

prominent of them show on photographs. One of the most famous Martian observers was the American astronomer Percival Lowell, who charted more than 400 "canals," about 50 of which he thought were double, and many of which intersected at what he called "oases."

On the other hand, other excellent observers have never been able to see what they would definitely class as *fine straight lines*. In the first place, the largest existing telescopes could not even theoretically resolve features on Mars that are less than 5 mi across. In practice, atmospheric "seeing" prevents us from resolving markings on Mars that are closer together than about 30 to 50 mi. Smaller features might be visible, but their true sizes and forms could not be discerned. Indeed, some observers report that what appear to be straight fea-

o 11 astronaut Edwin Aldrin deploying the seismometer. The laser reflector is to his
and further in the background. In the center background is the lunar module. *(NASA.)*

Mars, photographed with the 200-inch telescope. *(Mount Wilson and Palomar Observatories.)*

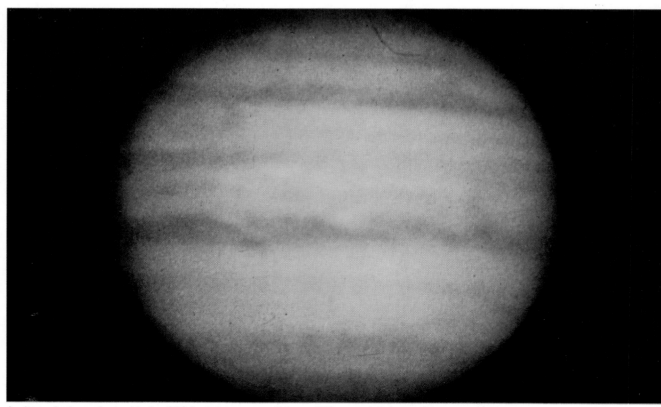

Jupiter, photographed with the 200-inch telescope. *(Mount Wilson and Palomar Observatories.)*

tures break up, under the very best conditions, into many smaller, disconnected spots.

Finally, there is a psychological tendency of the eye to connect features whose separations are near the limit of resolution. This propensity of the eye to simplify and smooth out very fine, chaotic detail may help "produce" the narrower "canals." There is no question about the appearance of the larger Martian surface features or of the presence of finer markings, but the existence of fine straight lines has never been confirmed.

The Mariner 4, 6, and 7 Mars probes broadcast to earth many photographs of the planet's surface. Although the detail of those photographs far surpasses that of telescopic observations obtained from earth, they did not resolve all of the mysteries. A great amount of detail was revealed, but no "canals" were found. On the other hand, the surface of the planet was seen to be covered with thousands of craters superficially resembling those of the moon. A magnetometer carried by Mariner 4 failed to detect a Martian magnetic field.

(e) Rotation and Seasons

The permanence of the surface features of Mars enables us to determine its rotation period with great accuracy; its sidereal day is $24^h37^m23^s$, very near the rotation period of the earth. This high precision is not obtained by watching Mars for a single rotation, but by noting how many turns it makes in a long period of time. Good observations of Mars date back for more than 200 years, during which period tens of thousands of Martian days have passed. The value accepted today is accurate to within a few hundredths of a second.

The equator of Mars is inclined to its orbital plane by about 25° — very close to the 23½° angle between the earth's orbit and its equator. Thus

FIG. 14-7 Surface features on Mars photographed from Mariner 6 on July 30, 1969, at an altitude of 2150 mi. At left is a wide-angle view and at right a narrow-angle view, covering N–S distances over the planet's surface of about 430 mi. and 45 mi., respectively. What appears to be a mountain range extending NW from the large crater on the right is actually part of the rim of a much larger crater seen near the eastern edge of the wide-angle photograph on the left. The regularly-spaced black dots are reference points within the TV system. *(NASA)*

each of Mars' poles is alternately tipped toward and away from the sun, and the planet goes through seasons, much like those on earth. Because of the longer Martian year, however, seasons there each last about six of our months. We recall that the earth is at perihelion in January (during summer in the Southern Hemisphere) and that, therefore, differences between the seasons south of the equator would be more extreme than those in the Northern Hemisphere, were it not for the unequal distribution of land and water over the earth. The same situation exists with Mars; however, because its orbit is more eccentric than earth's, seasonal climatic variations in its southern hemisphere are considerably more pronounced than those in the northern hemisphere.

Certain of the Martian seasonal changes can be observed telescopically. Most conspicuous are the changes in the polar caps. The southern polar cap, in winter, reaches a maximum diameter of about 3700 mi, and extends halfway to the planet's equator. In the spring, the cap shrinks, and often disappears entirely in summer. The north polar cap, in a region where the range of seasonal temperature is less extreme, does not reach so large a size, and never quite disappears. It is estimated that if the white caps were frozen water they would have to be very thin — probably less than a few feet — to melt or evaporate so quickly under the influence of the relatively small amount of solar heat. There are certain places where isolated patches of white from a receding polar cap linger longest as summer approaches; these are generally thought to be areas of higher elevation, where the weather is cooler.

The maria also are reported by many to change with the seasons, not much in size, but in intensity of color. During Martian spring and summer, most observers describe them as a conspicuous gray, and some call them green or even blue. In the fall and winter, they fade to less intense colors, some becoming scarcely distinguishable from the surrounding reddish regions.

(f) Martian Temperatures

The temperatures determined for Mars are based on measures of infrared radiation from its surface, obtained not only from earth-based observatories, but also with detectors carried by Mariners 6 and 7. The maximum equatorial temperatures are slightly over 300°K (about 80°F). The maria tend to be from 10 to 30°F warmer than the surrounding regions. The sunrise and sunset temperatures at the equator, although more uncertain, may be around −4° and 40°F, respectively. At night, the temperature drops to nearly −100°F.

The temperature at the south polar cap was measured by Mariner 7 and was found to range down to about −250°F, which is near the frost point for carbon dioxide. The temperature of the north polar cap is believed to be less extreme, but it probably does not get above −100°F.

(g) The Martian Atmosphere

The existence of a Martian atmosphere is easily demonstrated by the way it scatters sunlight. Measures of the sunlight reflected from Mars reveal the presence of a Martian atmosphere; measures by all three Mariners showed that the atmospheric pressure at the surface of the planet is everywhere less than 10 millibars; this is somewhat under 1 percent of the sea-level pressure of the earth's atmosphere. The relatively low albedo of Mars (0.15) confirms the fact that its atmosphere is not dense. It is dense enough, however, to produce a distinct twilight zone. Also, as in the earth's atmosphere, light of shorter wavelengths is scattered more than that of longer wavelengths. The surface markings on Mars, therefore, show up much better on photographs made in red light than in blue.

The prevalence of haze and scattered clouds is further evidence of an atmosphere. In fact, an atmospheric haze usually completely obscures the Martian surface on photographs made in violet light. On rare occasions this so-called violet layer dissipates, allowing the features below to be seen in light of short wavelengths. This phenomenon, called the *blue clearing*, usually is observed near the time when Mars is at opposition.

The first gas to be identified spectrographically in the Martian atmosphere was carbon dioxide, which is estimated to comprise about 80 percent of the gases present. Spectrograms obtained at Mt. Wilson in 1963 revealed evidence for a very low concentration of water vapor on Mars. Oxygen is absent except for minute traces in the atomic form. Mariners 6 and 7 found no trace of nitrogen.

(h) Interpretation of the Martian Features

POLAR CAPS

Kuiper has found that the infrared spectrum of light reflected from the white polar caps is similar to that from frost or snow. Dollfus finds the polarization also agrees with the hypothesis that the caps are frozen water. However, water molecules that evaporate from the caps would eventually be expected to escape Mars; the concentration of water in the atmosphere is observed to be low. Water may possibly be released from the rocks, on the other hand, and it had been assumed until recently that the polar caps were some kind of frozen water. Modern studies, however, indicate that carbon dioxide would be exected to solidify in the polar regions of the planet in winter. Consequently, it is now believed that part, if not all, of the polar caps are composed of solid carbon dioxide — that is, Dry Ice.

CLOUDS

The polarization of the light reflected from the white clouds is like that reflected from ice crystals; the transient scattered white clouds, then, might be like our cirrus clouds. Some of the Martian clouds have a bluish tint. They are usually observed near Martian sunrise or sunset. They are probably not composed of ice crystals, but could be made up of small water droplets. It has also been suggested that the clouds may be particles of frozen carbon dioxide. Occasionally, large yellow clouds are seen moving across the surface of the planet. These are thought to be dust clouds.

BLUE HAZE

The violet layer or haze cannot be produced by water drops, ice crystals, or dust, and is one of the most perplexing of the Martian mysteries. One hypothesis is that it is a faint illumination emitted by atmospheric atoms about 60 mi above the planet's surface that have been ionized by charged particles from the sun.

MARIA

The seasonal color changes of the maria suggests the growth and decay of vegetation. However, the spectrum of the light they reflect shows conclusively that their color cannot be due to chlorophyll. On the other hand, some simple plant organisms, such as lichen, often have coloring pigments other than chlorophyll, so the color of the maria has no particular significance with regard to the hypothesis that they are vegetation. Important evidence which would indicate that the maria are areas of plant life would be if the infrared spectrum of their reflected light displays features similar to those produced by materials containing the carbon-hydrogen (CH) radical, among which are those materials that compose living organisms. A search for infrared spectral features due to the CH radical was made by William Sinton at the 200-inch telescope during the 1956 opposition of Mars. He did find a feature in the radiation reflected from the maria (but not from the red deserts) that resembled that produced by the CH radical, and he tentatively identified CH in the maria. Subsequent investigations, however, have led Sinton to suspect that the effect could have been produced in the earth's atmosphere, and the original interpretation is now open to serious question.

One of the most ardent supporters of the hypothesis of "life on Mars" was Percival Lowell, who built the Lowell Observatory in Arizona and devoted much of his life to observing Mars. He believed that the many straight markings he observed were too nearly geometrically perfect to be natural; he envisioned them as artificial waterways — actual canals — built by intelligent beings to carry the melting waters from the polar caps across the desert to irrigate Martian crops. Few, if any, astronomers today take Lowell's views seriously. Even if the polar caps were water, there would be very little of it there — not enough to fill a good-sized lake on earth. Moreover, at the prevailing temperatures on Mars, the canals would have to be artificial iceways, not waterways, and most astronomers now discount the existence of the canals, anyway. We know of no form of animal life that can survive without oxygen or under the other hostile conditions that prevail on Mars. We cannot rule out the possibility of unknown forms of animal metabolism, but certainly there is no evidence in favor on animal life there. Even evidence for plant life is at best very circumstantial, and although some astronomers acknowledge the possibility of vegetation on Mars, its presence has never been established. Indeed, many alternative interpretations to those mentioned here have been advanced to account for the Martian features. Hopefully, future space probes may provide answers to some of the Martian riddles, but for the present Mars remains a mysterious and provocative world.

14.4 JUPITER

Jupiter is well named for the leader of the gods. Next to the sun, it is the largest and most massive object in the solar system.

(a) Orbit of Jupiter

The median distance of Jupiter from the sun is 484 million mi, 5.2 times that of the earth. Its orbit has an eccentricity of 0.048, so its distance from the sun

varies by 47 million mi. Jupiter's mean orbital speed of 8.1 mi/sec carries it once around its orbit in 11.96 years. Its orbital plane is inclined 1°18′ to the ecliptic.

(b) Satellites

Jupiter has twelve known satellites, the largest number of any planet in the solar system. The four largest moons were discovered by Galileo; they are so bright that they would be visible to the unaided eye on a clear dark night were it not for their proximity to the bright planet. Since the four *Galilean* satellites present observable disks, their angular sizes can be measured directly with a filar micrometer, and their linear sizes calculated from the known distance of Jupiter. Their diameters can also be determined from a knowledge of their orbital velocities and from the time it takes them to enter or leave the umbra of Jupiter's shadow. The two largest are about 3000 mi across, 1½ times the size of our moon; the other two are about the same size as the moon. Their distances from Jupiter's center range from about ¼ million to more than 1 million mi; their corresponding periods run from $1^d 18\frac{1}{2}^h$ to $16^d 16\frac{1}{2}^h$. Their discovery provided strong support for the Copernician theory (Section 3.5b) The timing of the intervals between their eclipses led to the first measurement of the velocity of light (Section 10.1b)

Subsequently, eight additional satellites have been discovered, the latest in 1951. They are all much smaller than the Galilean moons, and their sizes must be estimated from their brightnesses and assumed albedos. Their probable diameters range from about 15 to 150 mi. They are known by the numbers 5 through 12, assigned to them in order of their discovery. The most distant is about 15 million mi from Jupiter, and has a period of more than 2 years.

The five inner satellites of Jupiter (including the four Galilean) have nearly circular orbits that lie near the planet's equatorial plane. The outer seven, however, have rather eccentric orbits, some of which have large inclinations to Jupiter's equator. The four most distant satellites revolve from east to west, contrary to the motions of most of the objects in the solar system. None of the moons of Jupiter has an observable atmosphere.

(c) Physical Properties

The mass of Jupiter has been calculated both from the periods of its satellites and the sizes of their orbits and from the perturbations it produces on the orbits of passing minor planets. It is found to be 318 times as massive as the earth, a value which is very close to $\frac{1}{1000}$ the mass of the sun. A recent observation of the perturbations Jupiter produced on the orbit of the minor planet ⑥⑤ Cybele led to a new determination of Jupiter's mass; the ratio of the mass of the sun to that of Jupiter was found to be 1047.39. Jupiter is more massive than all the other objects in the solar system combined, except the sun itself.

Jupiter is also the largest planet, having a diameter of 87,000 mi, or about 11 times the earth's diameter. If it were as near to the sun as Mars, it would be the brightest planet in the sky and would sometimes be visible in daylight.

Jupiter's surface gravity is higher than the earth's; a body at its surface would weigh 2.64 times what it would weigh on earth. Jupiter has the highest velocity of escape of any planet, 37 mi/sec; it can easily retain all kinds of gases in its atmosphere. Its mean density, however, is much less than the earth's — only 1.33 times the density of water.

Because of the tremendous gravitational attraction of Jupiter for its constituent parts, it would most certainly be compressed to a far greater mean density unless it were composed almost entirely of hydrogen and helium — the lightest and most common elements in the universe. Even these elements are probably compressed to the solid state — like ice — throughout practically all of Jupiter's interior. The largest uncertainty in the calculation of a model for Jupiter's internal structure is due to the incompleteness of our knowledge regarding the compressibility of solid hydrogen and helium at the various temperatures and pressures that exist inside Jupiter. A model, computed by Wendell DeMarcus, predicts a central pressure of over 100 million times the earth's sea-level atmospheric pressure, and a density of about 31 times that of water. DeMarcus finds, further, that Jupiter must be at least 78 percent hydrogen, by mass.

It is interesting to note that Jupiter has very nearly the maximum possible size for a body of "cold" hydrogen — that is, one that is not radiating its own light and heat. Less massive bodies than Jupiter would occupy a smaller volume. More massive bodies, by virtue of their greater gravitation, would also be compressed to a smaller volume than Jupiter's. In this latter case, the extreme internal pressures would force the electrons to abandon the atoms that compose those bodies. Then the atoms would obey laws of physics different from those that are appropriate to matter with which we are ordinarily familiar.

(d) Telescopic Appearance

Because Jupiter is so far from the sun, it can appear only very slightly gibbous, and usually presents a nearly full disk to the telescopic viewer. Its relatively high albedo (0.51) makes it quite bright, despite the small amount of sunlight that reaches it. Its four brightest satellites are easy to see, even with binoculars; the shadows of some of the satellites occasionally can be observed crossing the planet's disk.

The disk of the planet itself is crossed with alternate light and dark cloud bands parallel to its equator. Details in their cloud bands are numerous, and exhibit gradual changes, showing that we are looking at atmospheric phenomena, not the solid surface of the planet. The light bands are not pure white, but show various colored regions: yellow, red, and sometimes blue.

One of the most striking surface features of Jupiter is the Great Red Spot that was seen telescopically at least as far back as 1831. It may have been the "spot" on Jupiter described by Cassini in 1660 and used by him to determine the planet's rotation period. It changes in size and shape slightly and also in the intensity of its color. It has been as large as 30,000 mi across. It has been described as a floating "island" of liquid or solid particles; however, there is no suggestion as to what the particles may be, and there are strong hydrodynamic arguments against the hypothesis. A modern theory for the Red Spot is that it is the top of a *Taylor column*, a vertical flow of gases over a large topographical feature on the solid surface of the planet.

(e) Rotation and Flattening

Distinct details in the cloud patterns on Jupiter allow us to determine its rotation rate; it is the most rapidly spinning of all the planets. However, it does not rotate quite as a solid body, or at least the clouds that we observe do not. For most of the planet, the rotation rate is fairly constant, averaging one rotation every 9^h55^m. A broad band about 10,000 to 15,000 mi wide along the equator, however, moves about 200 mi/hr faster than the surrounding regions, and completes a rotation in 9^h50^m. The rotation of the planet can be measured spectroscopically as well as visually. Jupiter's equator is inclined at only 3° to its orbit plane, so the planet has no appreciable seasons.

Jupiter's rapid rotation has caused it to become noticeably oblate, just as the earth is slightly flattened because of its rotation. When observed visually through a telescope, Jupiter's flattening is readily apparent; its polar diameter is less than its equatorial diameter by 1 part in 15. The polar and equatorial diameters of the earth differ by only 1 part in 298.

(f) Atmosphere

The temperature of Jupiter, calculated from the intensity of radiation that comes, probably, from its visible clouds, is about 130°K, or −220°F, just about what we would expect for a planet of Jupiter's albedo and distance from the sun. Because of the high velocity of escape from Jupiter, we would expect it to retain even the lightest gases; hydrogen and helium, therefore, should be abundant constituents of its atmosphere. Nitrogen, neon, and other inert gases also may be fairly abundant. The gases on Jupiter that are observed most easily by spectrographic means are, however, methane and ammonia; these gases are abundant but still comprise only a fraction of 1 percent of the total mass of Jupiter's atmosphere. Hydrogen, when cold, is difficult to observe spectrographically because its principal spectral lines lie in the far ultraviolet, to which the earth's atmosphere is opaque. However, *molecular* hydrogen (the H_2 molecule) does produce a series of absorption lines in the infrared.

A recent spectrographic analysis of the gases in Jupiter's atmosphere shows that the amount of hydrogen above each square inch in the opaque layer of Jupiter's clouds is from 10 to 20 times as great as the entire amount of atmosphere over each square inch of the earth's surface at sea level. Jupiter's clouds are already high in its atmosphere, so the total amount of atmospheric hydrogen there is immense. Methane has an abundance of about $\frac{1}{1000}$ that of the hydrogen, and ammonia is only $\frac{1}{20}$ as plentiful as methane. The ratio of hydrogen to carbon (the latter in combination with hydrogen in the methane) in Jupiter's atmosphere is found to lie between 1200 and 1800. The higher figure is close to the corresponding ratio in the sun.

The white cloud bands on Jupiter are usually interpreted as being associated with gases surging upward through the atmosphere. One guess is that ammonia crystals in them reflect most of the observed light; at temperatures prevalent on Jupiter, most of the ammonia solidifies. The dark bands are believed to be downcurrents in the atmosphere, through which we see to deeper, darker levels.

Theoretical studies indicate that the total atmosphere cannot extend downward more than about 1 percent of the planet's radius. Within 100 or 200 mi below the visible clouds, the pressures are high enough to liquefy hydrogen. The atmosphere is pictured as gradually becoming thicker and "slushier" with depth, until it finally is condensed to an "ocean" of liquid hydrogen and other elements. The ocean probably gradually hardens to the solid planet in about another 150 or so miles down.

(g) "Van Allen" Layers

In the late 1950s radio energy was observed from Jupiter that is more intense at *longer* than at shorter wavelengths — just the reverse of what is expected from a body radiating away its heat. It is typical, however, of the radiation emitted by electrons accelerated by a magnetic field. Observations made at the Radio Observatory of the California Institute of Technology, near Bishop, California, showed that the radio energy originated from a region surrounding the planet whose diameter is several times that of Jupiter itself. Furthermore, the radio energy was found to be polarized, another characteristic of radiation from accelerated electrons (radio waves can be polarized, like visible light; the polarization is detected by changing the orientation of the receiving antenna at the focus of the parabolic reflecting dish of the radio telescope). The evidence suggests, therefore, that there are a vast number of charged atomic particles circulating around Jupiter, spiraling through the lines of force of a magnetic field associated with the planet. The phenomenon is reminiscent of the Van Allen belt around the earth (Section 13.5b).

14.5 SATURN

Saturn is the solar system's second largest planet. Its unique *ring* system makes it one of the most impressive of telescopic objects. It is named for the Titan god of seed sowing, the father of Jupiter. It is the most remote of the planets known before the invention of the telescope.

(a) Orbit and Satellites

The orbit of Saturn has an eccentricity of 0.056, about the same as that of Jupiter. The planet's distance from the sun varies from about 840 to 935 million mi and has a median value of 887,100,000 mi. The orbit is inclined 2½° to the ecliptic plane. At a mean orbital speed of 6 mi/sec, Saturn completes one sidereal revolution in 29½ years.

In addition to its ring system, Saturn has 10 known satellites. Several of them are easily visible with small telescopes. The largest, Titan, is larger than our moon and is the only satellite in the solar system that is known definitely to possess an atmosphere; Kuiper identified methane in its spectrum in 1944. The second largest of Saturn's satellites is about 1000 mi in diameter. The most distant satellite is Phoebe, with a distance and period of about 8 million miles and 550 days, respectively. The last satellite to be discovered, and the innermost one, is Janus, found by Dollfus in France. A small gap in Saturn's ring system led Dollfus to suspect the existence of the satellite very close to the outer rim of the ring; the gravitational influence of the satellite would produce the gap. The new object was found during a successful search in December 1966, when the rings were turned edge on to the earth and thus appeared very faint. Janus is less than 100,000 mi from the center of Saturn and revolves about it in 18 hours.

(b) The Ring System

Galileo first saw the rings of Saturn, but he was unable to discern them clearly with his crude telescopes. It was not until half a century later, in 1655, that Huygens described their true form. The rings have approximately the appearance of the brim of a straw hat surrounding the planet in its equatorial plane. There are three concentric portions of the

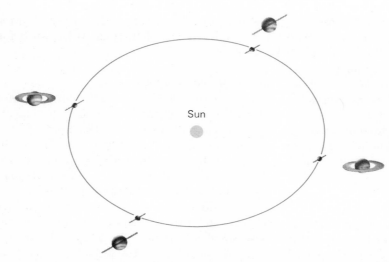

FIG. 14-8   Orientation of the rings as seen from the sun with Saturn at different places in its orbit.

rings. The brighest and broadest is the central, or *bright* ring. Surrounding this is the *outer* ring with an outside diameter of 171,000 mi. The faintest is the inner or *crape* ring, whose inside diameter of about 88,000 mi allows only a 8000-mi gap between it and the ball of the planet itself.

The rings do not form a solid sheet, as is deduced by the following:

(1) Background stars can be seen shining through them.

(2) The Doppler shift in the spectrum of sunlight reflected from them shows the inner portions of the rings to be revolving about the planet in a shorter period than the outer portions, in accordance with Kepler's third law (the periods would be the same if the rings were a solid sheet).

(3) All parts of the rings are within Roche's limit, where no extended solid body could withstand the destructive tidal forces produced by the planet (Section 6.4). The intensity of the reflected light from the rings, and its spectrum, show that they cannot be gaseous. There is no question that the rings are composed of many billions of minute solid particles, probably gravel-sized or smaller. Sunlight reflected from

the countless myriads of particles gives the illusion of solid rings.

The rings lie in the equatorial plane of Saturn, whose orientation remains constant during the planet's revolution about the sun. This plane is inclined at about 28° to the ecliptic, so that during part of Saturn's orbital revolution we see one face of the rings, and when the planet is on the opposite side of the sun, we see the other face. At intermediate points, the rings may appear edge on to our line of sight; in that condition they virtually disappear (Figure 14-8), which shows that they are very thin. The thickness of the rings is too small to measure; observations show it not to exceed 10 or 15 mi. There are some theoretical reasons to suspect that the rings might be extremely thin — perhaps only a few inches or feet.

Between the outer two rings is a gap about 1000 mi wide, known as *Cassini's division*. A particle at this distance from Saturn would have a period of $11^h17\frac{1}{2}^m$, just half of the period of Mimas, the second innermost satellite. Such a particle would be nearest Mimas at the same place in its orbit every second time around and at that location would feel the maximum gravitational "tug" of the satellite. These perturbations, occurring repeatedly in the same direction, would gradually accumulate until they removed any particle from the region, leaving a gap. When one revolving body has a period that is a simple integral multiple of the period of another, those periods are said to be *commensurable* with each other. Many observers have reported other, less conspicuous divisions in the rings; these could occur at places where particles would have periods commensurable with those of other satellites. One such gap, as stated above, led to the discovery of Janus in 1966.

(c) Mass, Size, and Structure

Saturn has a mass equal to 95 earth masses, and the mean diameter of the planet proper is just under 72,000 mi. These figures give it an average density (in terms of water) of only 0.71 — the lowest of any planet in the solar system. In fact, Saturn would be light enough to float, if an ocean existed large enough to launch it. Its surface gravity is only 17 percent greater than the earth's; nevertheless, it has a velocity of escape of 22 mi/sec, great enough for Saturn to hold the lightest gases.

The internal structure of Saturn is very similar to that of Jupiter. DeMarcus calculates a central pressure of about 50 million times sea-level atmospheric pressure and a central density of about 16 times that of water. According to DeMarcus' model, Saturn must be at least 63 percent hydrogen.

(d) Telescopic Appearance and Rotation

Except for its rings, Saturn looks very much like Jupiter. It, too, has parallel alternately dark and light cloud bands, although the details, color, and irregularities in them are much less distinct than in Jupiter's clouds. Small light spots are occasionally seen on the planet.

The rotation period of the planet, as determined both from the Doppler shift in its spectrum and from the apparent motions of the spots on its disk, is just over 10 hours at the equator. Like Jupiter, however, Saturn rotates more slowly at

latitudes away from the equatorial regions. The mean rotation period for most of the planet is near 10^h38^m. Because of its rapid rotation, Saturn is the most oblate of all the planets; its equatorial diameter is about 10 percent greater than that through its poles.

(e) Atmosphere

Measures of the radiant heat from Saturn indicate a temperature of 120 to 130°K, or near −230°F. This is some 30 to 40°F higher than might be expected for a planet of Saturn's albedo and distance from the sun. It therefore may be appropriate for a depth somewhat below the visible cloud layer. The only gases detected spectrographically are hydrogen and methane. Evidently, because of the somewhat lower temperature of Saturn (compared to Jupiter), the ammonia has "snowed" out of the atmosphere. The structure of Saturn's atmosphere is thought to be similar to Jupiter's. The gaseous region probably merges into a hydrogen "ocean" a few hundred miles below the cloud layers. The ocean may extend a little deeper than in Jupiter, but probably, at a depth of not more than 200 or 300 mi., it solidifies into the icelike interior of the planet.

14.6 URANUS

Uranus is a planet that must surely have been seen by the ancients and yet was unknown to them. It was discovered on March 13, 1781, by the German-English astronomer, William Herschel, who was making a routine telescopic survey of the sky in the constellation of Gemini. Herschel noted that through his telescope the planet did not appear as a stellar point but seemed to present a small disk. He believed it to be a comet and followed its motion for some weeks. Several months later, a preliminary solution for its orbit was computed and it was found to be a nearly circular one, lying beyond that of Saturn; the object was unquestionably a new planet.

Uranus can be seen by the unaided eye on a dark clear night, but is near enough to the limit of visibility so that it is indistinguishable from a very faint star. It is so inconspicuous that its motion escaped notice until after its telescopic discovery. However, it turned out that Uranus had been plotted as a star on charts of the sky on at least 20 previous occasions since the year 1690. These earlier observations were later of use in the determination of how perturbations were altering the planet's orbit.

Herschel proposed to name the newly discovered planet *Georgium Sidus*, in honor of George III, England's reigning king. Others suggested the name Herschel; the name finally adopted, in keeping with the tradition of naming planets for gods of Greek mythology, was Uranus, father of the Titans, and grandfather of Jupiter.

(a) Orbit and Satellites

The orbit of Uranus lies more nearly in the plane of the ecliptic than that of any of the other planets, its inclination being only 46′. Its eccentricity is 0.047, and it varies by a little more than 80 million mi from its median distance of 1783 million mi from the sun. Being so far from the sun, its orbital speed is low (4½ mi/sec), and its period is long (84 years).

Uranus has five known satellites; the last to be discovered, and the faintest, was found by Kuiper in 1948. Herschel himself found the two brightest moons of the planet. None of the satellites is probably much over 1000 mi in diameter. Their distances range from 81,000 to 364,000 mi from the center of Uranus.

(b) Gross Characteristics

The mass of Uranus is found from the motions of its satellites and from its perturbative effects upon the motion of Saturn; it is 14½ times the mass of

FIG. 14.9   Uranus and its satellites, photographed with the 120-inch telescope. *(Lick Observatory.)*

the earth. It presents a small angular diameter, just under 4″, because of its great distance from the sun. Its linear diameter, therefore, is difficult to measure accurately, but it is about 30,000 mi. With these values for its mass and size, its mean density is calculated to be 1½ times that of water, its surface gravity about the same as on the earth, and its velocity of escape about 20 km/sec. Micrometric measures and also studies of the perturbations on its satellites both show Uranus to be somewhat oblate. The accepted value for the difference in equatorial and polar diameters is 1 part in 14, but this figure is quite uncertain.

Uranus has a high albedo (0.66), so probably it is surrounded by a cloud layer that reflects sunlight. The spectrum of the planet reveals a strong concentration of methane, and also a series of infrared absorption lines attributed to molecular hydrogen. The identification of hydrogen was achieved in 1952 by Kuiper and Herzberg. Uranus is the first planet in which hydrogen was identified spectrographically. Ammonia does not appear in the spectrum, probably because it is frozen out of the atmosphere; the temperatures to be expected on Uranus are below −300°F.

The internal structure of Uranus is probably similar to that of Jupiter and Saturn. Its mean density is high enough, however, to show that it must contain a far smaller proportion of hydrogen than do the two giant planets.

(c) Appearance and Rotation

Uranus appears as a greenish disk when seen through the telescope. The green color is probably due to its atmospheric methane. A few observers report faint markings, but these are too indefinite to indicate the rotation. The rotation period, therefore, must be obtained from the Doppler shift in the spectrum of light from different parts of its disk. The accepted value, somewhat uncertain, is 10^h45^m.

A unique feature about Uranus is that its axis of rotation lies almost in the plane of its orbit; during some parts of its revolution, it is so oriented that we look almost directly at one or the other of its poles. The actual inclination of its equatorial plane to that of its orbit is 82°. Its direction of rotation is the same as that of the revolution of its satellites, and both are in the *reverse* direction from the rotation of the other planets. Its direction of orbital revolution, however, is normal — that is, from west to east.

14.7 NEPTUNE

Whereas the discovery of Uranus was quite unexpected, Neptune was found as the result of mathematical prediction. The discoveries of the two planets could hardly have been made under more different circumstances, yet in other respects Uranus and Neptune are more alike than any two other worlds in the solar system.

(a) Discovery

By 1790, an orbit had been calculated for Uranus (first by Delambre) on the basis of observations of its motion in the decades following its discovery. Even after allowance was made for the perturbative effects of Jupiter and Saturn, however, it was found that Uranus did not move on an orbit that fitted exactly the earlier observations of it made since 1690, before it was known as a planet (see Section 14.6). By 1840, the discrepancy between the positions observed for Uranus and those predicted from its computed orbit amounted to about 2′ — an angle barely discernible to the unaided eye but still much larger than the probable errors in the orbital calculations. In other words, Uranus did not seem to move on an orbit that would have been predicted from Newtonian theory.

In 1843 John Couch Adams, a young Englishman who had just completed his work at Cambridge, began an analysis of the irregularities in the motion of Uranus to see whether they could be produced by the perturbative action of an unknown planet. His calculation indicated the existence of a planet more distant than Uranus from the sun. This unknown planet was pulling on Uranus, accelerating it slightly ahead in its orbit until 1822, when Uranus passed between the planet and the sun. After 1822, the motion of Uranus was being retarded slightly. In October 1845 Adams sent his results to Sir George Airy, the Astronomer Royal, informing him where in the sky he should look to find the new planet. We know today that Adams' predicted position for the unknown body was correct to within 2°. Airy, however, apparently had little faith in the ability of so young and unknown a mathematician, and posed a simple problem to him as a test. When Adams did not bother to respond, Airy let the whole matter drop.

Meanwhile, Leverrier, a French mathematician, unaware of Adams or his work, attacked the same problem, and published its solution in June 1846. Airy, noting that Leverrier's predicted position for the unknown planet agreed to within 1° with Adams', posed the same test problem to the Frenchman and received a prompt and correct reply.

FIG. 14-10 Neptune and a satellite, photographed with the 120-inch telescope. *(Lick Observatory.)*

Airy then suggested to Challis, director of the Cambridge Observatory, that he begin a search for the new object. The Cambridge astronomer, having no up-to-date star charts of the region of the sky in *Aquarius* where the planet was predicted to be, proceeded by plotting all the faint stars he could observe with his telescope in that location. It was Challis' plan to repeat such plots at intervals of several days, in the hope that the planet would reveal its presence and distinguish itself from a star by its motion. Unfortunately, he was negligent in examining his observations; although he had actually seen the planet, he did not recognize it.

About 1 month later, Leverrier suggested to Galle, an astronomer at the Berlin Observatory, that he look for the planet. Galle received Leverrier's letter on September 23, 1846, and, pos-

sessing new charts of the *Aquarius* region, he found and identified the planet that very night. It was only 52′ from Leverrier's predicted position.

The discovery of the eighth planet, now known as Neptune (named for the god of the sea), was a major triumph for gravitational theory and ranks as one of the great scientific achievements. The honor for the discovery is quite properly shared by the two mathematicians, Adams and Leverrier.

Although a great triumph, the discovery of Neptune was not a complete surprise to astronomers, who had long suspected the existence of the planet. On September 10, 1846, Sir John Herschel, son of the discoverer of Uranus, remarked: "We see it as Columbus saw America from the shores of Spain. Its movements have been felt trembling along the far-reaching line of our analysis with a certainty hardly inferior to ocular demonstration."

(b) Orbit, Satellites, and Rotation

Neptune is almost exactly 30 AU from the sun. Its orbit is very nearly circular, having an eccentricity of only 0.009; its distance from the sun varies by only 24 million mi on each side of its median value of 2797 million mi. Traveling with an orbital speed of only 3⅓ mi/sec, it requires 165 years to complete one revolution around the sun. It will not have completed its first revolution since its discovery until 2011.

Its axial rotation period, determined from the Doppler shift in its spectrum, is 15.8 hours; the value is uncertain by 1 hour or so. Its equatorial plane is inclined about 29° to that of its orbit. Its oblateness is small, the equatorial diameter exceeding the polar diameter by only about 1 part in 40.

Neptune has two satellites. *Triton,* the larger, is somewhat greater in diameter than our moon. Its distance from the center of Neptune is only about 220,000 mi, and its motion is backward — that is, east to west. Neptune's other satellite, discovered by Kuiper in 1949, is very much smaller, and has the most eccentric orbit of any satellite in the solar system; its distance from Neptune varies from 1 to 6 million mi.

(c) Other Properties

Neptune's mass is 17.2 times that of the earth. Since it subtends an angle of only about 2″, its diameter is difficult to measure, but it is approximately 28,000 mi. Its mean density is somewhat more than twice that of water — higher than the density of Uranus. Its internal structure is believed to resemble that of Uranus.

Neptune appears telescopically as a small greenish disk. There are no conspicuous markings. Its high albedo of 0.5 to 0.6, however, suggests that light is reflected from a gaseous atmosphere. Both methane and hydrogen have been detected spectrographically, as in Uranus. Helium is probably also present but cannot be observed in the spectrum. Ammonia appears to be absent in the gaseous state, as we would expect, since the temperature of Neptune must be near −350°F.

14.8 PLUTO

After the perturbative effects of Neptune upon Uranus had been taken into account, the discrepancies between the observed and predicted positions of Uranus were reduced to only $\frac{1}{60}$ of what they had been. Today those remaining discrepancies are realized to be smaller than the errors of the original observations and thus are almost certainly not real. Yet, several investigators made attempts to account for these remaining deviations by a gravitational influence of a ninth planet, beyond the orbit of Neptune. Among them were Gaillot, W. H. Pickering, and Percival Lowell. It was Lowell's solution to the problem that led to the discovery of Pluto.

The orbital elements calculated by Lowell, Pickering, and Gaillot, along with the actual elements of Pluto, are shown in Table 14.1.

TABLE 14.1 Orbital Elements of Pluto

	GAILLOT 1909	LOWELL 1915	PICKERING 1928	ACTUAL
Semimajor axis (AU)	66	43.0	30	39.5
Period (years)	536	282	165	248
Eccentricity	(0)	0.20	0.20	0.25
Perihelion date	—	1991	1974	1990
Magnitude (see Chapter 20)	—	12–13	12.2	15
Mass (earth = 1)	24	6.6	0.8	$<\frac{1}{5}$

(a) Discovery

By the beginning of the twentieth century, when Lowell made his calculations, Neptune had moved such a short distance in its orbit that a knowledge of the perturbations upon it was not yet available. Therefore Lowell, as did the others, based his calculations entirely on the minute remaining irregularities of the motion of Uranus. His computa-

FIG. 14-11 Percival Lowell. *(Yerkes Observatory.)*

FIG. 14-12   Two photographs of Pluto, showing its motion among the stars in a 24-hour period; photographed with the 200-inch telescope. *(Mount Wilson and Palomar Observatories.)*

tions indicated two places where a perturbing planet could be, the more likely of the two being in the constellation of *Gemini*. Lowell predicted a mass for the planet intermediate between that of the earth and that of Neptune (his calculations gave about 6.6 earth masses), and also an angular diameter in excess of 1″. Lowell searched for the unknown planet at his Arizona Observatory from 1906 until his death in 1916, without success. Subsequently, Lowell's brother donated to the observatory a 13-in. photographic telescope that could record a 12° by 14° area of the sky on a single photograph. The new camera went into operation in 1929, and the search was continued for the ninth planet.

Unfortunately, Gemini lies near the Milky Way, and is a region in which some 300,000 star images were recorded on each exposure. It was an immense task to compare all the star images on each of two or more photographs of the same field in the hope of finding one image that changed position with respect to the rest, revealing itself as the new planet. The job was facilitated by the invention of the *blink microscope,* a device in which are placed two different photographs of the same region of the sky. The operator's vision is automatically shifted back and forth between corresponding parts of the two photographs. If the star patterns are the same on the two plates, the

observer sees a constant, although flickering picture. However, if one object has moved slightly in the interval between the times the two plates were taken, the image of that object appears to jump back and forth as the view is transferred from one photograph to the other. In this way, moving objects can quickly be picked out from among the many thousands of star images. The blink microscope is also used as a means of locating stars that vary in brightness.

In February 1930, Clyde Tombaugh, comparing photographs made on January 23 and 29 of that year, found an object whose motion appeared to be about right for a planet far beyond the orbit of Neptune. It was within 6° of Lowell's predicted position for the unknown planet; subsequent investigation of the object showed its orbit to have elements very similar to those Lowell had calculated. Announcement of the discovery was made on March 13, 1930. The new planet was named for Pluto, the god of the underworld. (Appropriately, the first two letters of Pluto are the initials of Percival Lowell.)

(b) Orbit of Pluto

Pluto's orbit has the highest inclination to the ecliptic (17°9′) of any planet and also the largest eccentricity (0.249). Its median distance from the sun is 39.52 AU, or 3,675,000,000 mi, but its aphelion distance is over 4½ billion mi, and its perihelion distance under 2.8 billion mi. Part of its orbit is closer to the sun than the orbit of Neptune. There is no danger of collision between the two planets, however; because of its high inclination, Pluto's orbit clears that of Neptune by 240,000,000 mi. Moreover, according to calculations by Cohen and Hubbard, dynamical affects between Pluto and Neptune, due to their gravitational interactions, will never allow them to approach closer to each other than 18 AU's. Pluto completes its orbital revolution in a period of 248.4 years, poking along at a mean velocity of only 2.9 mi/sec.

(c) Nature of Pluto

Following Pluto's discovery, estimates were made of its probable mass from a study of the motion of Neptune. A recent estimate indicates that the mass of the planet is about 0.18 times that of the earth. The diameter of Pluto was measured by Kuiper at the 200-in. telescope; he found the angular diameter to be about ¼″, which at Pluto's distance corresponds to a linear diameter of about 6000 km (3700 mi). If the mass and size are both correct, Pluto's density would have to be about one and two-fifths times the density of the earth. Such a high density may not be possible for a planet of such a small mass as Pluto's, and it has been suggested that either the mass or diameter determinations (or both) are in error.

An opportunity for a check was provided in the Spring of 1965, when Pluto was expected to occult a star. However, careful observations of the star from several observatories in the United States revealed no trace of the star's light being dimmed. Excellent measures of the positions of the star and Pluto were made at Flagstaff. They showed that the center of Pluto must have passed only ⅛″ from the star, as seen from the McDonald Observatory in Texas, but actual observations made at McDonald indicated that Pluto did *not* cover the star at any time. Thus the radius of Pluto must be smaller than ⅛″, and its corresponding linear diameter, after allowances for all suspected errors are made, cannot be more than 6000 km — the size found by Kuiper. A lower limit to the size of the planet is found by assuming that it is perfectly reflecting — that is, has an albedo of 1.0. Even then, it would have to be at least 2000 km in diameter to account for its observed brightness. Thus Pluto's diameter lies between 2000 and 6000 km, and is probably near the larger value, because its albedo is believed to be small.

Kuiper and Harris reported that they thought Pluto might be variable in its brightness. Walker and Hardie, therefore, carefully observed the planet with a photoelectric photometer attached to a telescope and found that it does, in fact, vary in light with a period of 6.3867 days, with an uncertainty of about 0.0003 day. They interpret the variation as a rotation of Pluto, alternately turning to our view hemispheres of greater and lesser albedo. Evidently, the surface is not uniform.

Pluto is several thousand times too faint to see with the unaided eye. Even under the best conditions a telescope of aperture 6 in. or more is required to see it, and even then it appears only as a star. No gases are revealed in its spectrum, but at its expected temperature of only about 40°K (nearly 400°F below zero) all common gases except neon, hydrogen, and helium would be frozen, and of these all but neon would probably escape the planet, and neon is not a gas that one would expect to find there in appreciable abundance. In any case, none of the gases that would remain unfrozen would be easy to observe spectrographically. Probably Pluto has little or no atmosphere.

(d) The Problem of Pluto's Discovery

Pluto's discovery was based on alleged residuals in the positions of Uranus after account was taken of the perturbations that could have been produced on Uranus' motion by Neptune and other known planets. These residuals, only about 4″ to 5″, were in observations of the positions of Uranus made in the early eighteenth century before it had been recognized as a planet. They were, in fact, smaller than the errors expected in observations made that long ago, and have now been generally discredited by the experts as having any significance. In view of this, one may well wonder whether there can have been any real validity in the calculations of the predicted orbit of Pluto.

Even if the remaining residuals in the positions of Uranus *were* real, to produce them with perturbations by Pluto, Pluto would need a mass substantially greater than that of the earth (note that Lowell predicted a mass of 6.6 earth masses). On the other hand, the theory of planetary structure indicates that Pluto's mean density should not exceed that of the earth and Mercury. This requires a mass for the planet even lower than the recent estimate of less than one fifth that of the earth. We must conclude, therefore, that Pluto's mass is too small to have produced perturbations in the amount required, and that the prediction of Pluto's position, although mathematically correct, was based on invalid data; Pluto's discovery was accidental!

The fact that Pluto *was* discovered, with so nearly the orbital elements that Lowell predicted, is one of the startling coincidences in the history of astronomy. Be that as it may, it was Lowell's faith and enthusiasm that led to our knowledge of Pluto, and he has justly earned the honor for its discovery.

(e) The Solar System from Pluto

From Pluto, the solar system must appear a bleak and empty place. The sun would appear as a bright star, although it would still provide Pluto with 250 times more light than we on earth receive from our full moon. The earth as seen from Pluto would scarcely ever be at more than 1° angular separation from the sun and would be slightly too faint to be seen with the unaided eye. Even Jupiter could only be 7° from the sun and would appear only as a medium-bright star.

14.9 ARE THERE UNKNOWN PLANETS?

The possibility of undiscovered planets exists. However, a careful search by Tombaugh after his discovery of Pluto failed to reveal any other trans-Neptunian planet. He should have picked up any object as large as Neptune within a distance of 270 AU.

(a) The Orbit of Mercury; Vulcan

The major axis of the orbit of a planet is called the *line of apsides*. One of the effects of perturbations by other planets is to produce a slow rotation of the line of apsides in the orbital plane, thereby gradually shifting the point of perihelion of a planet. It has long been known that the line of apsides of the orbit of Mercury is changing orientation by the amount of 574″ per century. However, when Newtonian gravitational theory is applied, and due account is taken of perturbations by known bodies, it is found that Mercury's advance of perihelion should amount to only 531″, a discrepancy of 43″.

At one time it was thought that an unknown intramercurial planet was perturbing Mercury, accounting for the extra 43″. The hypothetical body was given the name *Vulcan* (the Roman god of fire). Vulcan was even reported as having been observed once, during a total solar eclipse. However, the observation has never been confirmed, and it is now generally discredited.

(b) Explanation of Relativistic Theory

We recall that Einstein's general theory of relativity (Section 4.7) introduced a new way to look on gravitational phenomena. Relativity predicts motions of material bodies that are so nearly identical to the motions predicted by Newtonian theory that only in the most exceptional circumstances is the distinction between the predictions of the two theories observable. Einstein, however, called attention to three astronomical phenomena in which the two theories predicted results that are just different enough so that they should be detectable by observation, and which could, therefore, serve as tests for general relativity. These are

(1) The bending of starlight in passing near a massive body, like the sun, the deflection being observable during a solar eclipse (Section 9.5c).

(2) A gravitational *redshift* produced in the spectra of stars of very high surface gravity (Chapter 23).

(3) A rotation of the line of apsides of Mercury in the amount of 574″ per century, 43″ more than predicted by Newtonian theory.

The strange behavior of the orbit of Mercury is thus completely explained by general relativity, and the "need" for Vulcan has vanished.

14.10 LIFE ON OTHER WORLDS

Realizing that the other planets are worlds, we naturally wonder whether any of them might be abodes of life. We find, however, that only the planets Venus, Earth, and Mars possess conditions under which we can even remotely imagine forms of living organisms to exist. Of these three, only earth offers conditions compatible with the existence of higher forms of plant or animal life as we know them. We cannot rule out the possibility of life processes unknown to us, but neither have we evidence for their existence. The best we can say is that we have no knowledge of any life on other planets of the solar system.

It is often pointed out that the sun is a typical star, and that there must be many other suns with planetary systems similar to our own. To be sure, among the countless billions of stars known to exist in the universe, we have no reason to expect that there are not many with planets, and that some of these planets might have conditions approximating those on earth. On the other hand, we do not know how life began on our own planet and have no objective means of assessing the probability of life beginning elsewhere, much less of evaluating the likelihood of the critical steps of evolution which may lead to intelligent beings. Life may abound in the universe, or it may be unique to the earth; on this topic we can do no more than speculate.

Exercises

1. Explain why Mercury is visible in the west after sunset when it is at eastern elongation, and in the east before sunrise when it is at western elongation. Draw a diagram.

2. At what seasons of the year would Mercury, when at eastern elongation, be above the horizon longest after sunset? Explain.

3. From the range of Mercury's distances from the sun, find the range of its orbital velocities. (*Hint:* Use Kepler's second law.)

   Answer: 24 to 36 mi/sec

4. Show that the mean period from noon to noon at a place on Mercury is 176 days.

5. Venus requires 440 days to move from western to eastern elongation but only 144 to move from eastern to western elongation. Explain why. A diagram will help.

6. At its nearest, Venus comes within about 25 million mi of the earth. How distant is it at its farthest?

7. Why isn't Venus always *exactly behind* the sun at superior conjunction?

8. On what occasion *can* Venus be observed at inferior conjunction? Explain.

9. Would astronomers be likely to learn more about the earth from observatories on Venus or Mars? Explain.

10. Why does Mars have the longest synodic period of any planet but a sidereal period of only 687 days?

11. Show that the satellites of Mars obey Kepler's third law. Use the data in this chapter or in Appendix 11.

12. Why is it not easy for us to get a good look at the southern polar cap of Mars?

13. Which satellite would have the greatest period, one 1 million mi from the center of Jupiter or one 1 million mi from the center of earth? Why?

14. The satellite *Io* of Jupiter is 262,000 mi from the center of the planet and has a nearly perfectly circular orbit and a period of $1^d18^h28^m$. It requires 3.6 minutes from the time it starts to enter Jupiter's shadow until it is completely eclipsed. What is its diameter? Show your reasoning.

    Answer: About 2300 mi

15. How often are Saturn's rings turned edge on to us on the earth? Assume, for this exercise, that Saturn's orbit is in the plane of the ecliptic. What is the relevance of this assumption?

16. Saturn's rings extend from a distance of 40,000 to 85,000 mi from the center of the planet. What is the approximate variation factor for the periods of time required for various parts of the rings to revolve about the planet?

    Answer: About 3:1

17. If Saturn's mass is 95 times the earth's, how does it happen that its surface gravity is so nearly the same as the earth's?

18. Describe the seasons of Uranus.

CHAPTER 15

The Minor Planets

Of the many thousands of minor planets, the vast majority appear telescopically like stars — that is, as small points of light. The term *asteroid* (meaning "starlike") is therefore often applied to these tiny worlds. Another synonym is *planetoid*. The term *minor planet*, preferred by many, is used in this text.

The orbits of most of the minor planets, including that of Ceres, the largest and first to be discovered, lie between the orbits of Mars and Jupiter. From the time of Kepler it was recognized that this region of the solar system constituted a considerable gap in the spacing of the planetary orbits; consequently, when Ceres was found there in 1801 it occasioned no great surprise. Moreover, the median distance of Ceres from the sun fitted well into Bode's law, a sequence of numbers that gives the approximate distances of the planets from the sun in astronomical units.

15.1 "BODE'S LAW"

The term "Bode's law" is a complete misnomer, for it was not discovered by Bode and is not a law. A law in science is a statement of an invariable order or relation of phenomena. Bode's law can be described more accurately as a scheme for remembering the distances of the planets from the sun. It was discovered by Titius of Wittenberg in 1766, and was brought into prominence by Bode, director of the Berlin Observatory, in 1772.

Titius' progression can be obtained easily by writing down the numbers: 0, 3, 6, 12, . . . , each succeeding number in the sequence (after the first two) being obtained by doubling the preceding one. If now 4 is added to each of the numbers, and

in each case the sum is divided by 10, the numbers obtained give the approximate distances from the sun (in astronomical units) of the planets known in 1766. The progression is illustrated in Table 15.1.

TABLE 15.1 Bode's Law

TITIUS' PROGRESSION	PLANET	PLANET'S ACTUAL DISTANCE (AU)
(0 + 4)/10 = 0.4	Mercury	0.387
(3 + 4)/10 = 0.7	Venus	0.723
(6 + 4)/10 = 1.0	Earth	1.000
(12 + 4)/10 = 1.6	Mars	1.524
(24 + 4)/10 = 2.8		
(48 + 4)/10 = 5.2	Jupiter	5.203
(96 + 4)/10 = 10.0	Saturn	9.539
(192 + 4)/10 = 19.6	Uranus	19.18
(384 + 4)/10 = 38.8	Neptune	30.6
(768 + 4)/10 = 77.2	Pluto	39.4

When Uranus was discovered in 1781, it was found to fit fairly well into Bode's law. There is no major planet at a distance corresponding to the 2.8 position, and a search for the "missing planet" was actually organized. The discovery of the minor planets was independent of this search, but they fell nicely into the gap at 2.8 AU; the orbit of Ceres has a semimajor axis of 2.767 AU. Bode's law, however, breaks down entirely for Neptune and Pluto.

Bode's law seems to be a favorite starting point for those who delight in working out theories for the origin of the solar system, but it is not, today, regarded as a fundamental relationship. On the other hand, the surprisingly regular, although not precise, geometrical spacing of the planetary orbits can hardly be accidental. The accuracy of Titius' progression must result somehow from the distribution of matter and motions in the material from which the principal planets were formed, and satisfactory theories of the origin of the solar system must account for the observed spacing of the planets.

15.2 DISCOVERY OF THE MINOR PLANETS

The Sicilian astronomer Giuseppe Piazzi (1746–1826) was engaged in mapping a region of the sky in Taurus, when on January 1, 1801 (the first night of the nineteenth century) he observed an uncharted "star." During the next two nights, he noted that the new object had shifted its position slightly. At first, he suspected that it was a comet, but on January 14 it changed its direction of motion from westward to eastward, characteristic of a superior planet completing its retrograde motion when near opposition (Section 2.3d). Piazzi continued to observe the object until February 11, after which he was forced to interrupt his work because of illness.

In mid-January Piazzi wrote of his discovery to Bode, at the Berlin Observatory. Unfortunately, his letter did not reach Bode until March 20, when the object was too nearly in the direction of the sun to be observed. Bode suspected that Piazzi's object might be the "missing planet," but since it had only been observed for the brief span of 6 weeks, astronomers were not able to calculate an orbit for it.

(a) Rediscovery of Ceres

The discovery of the new planet was rescued by the brilliant young German mathematician, Karl Friedrich Gauss (1777–1855), who had devised new mathematical techniques that could be applied to orbit calculations. By November, Gauss had successfully calculated the orbit of the object from Piazzi's short series of observations. The object was found very near Gauss' predicted position in the constellation of Virgo by von Zach, on

FIG. 15-1   Gauss. *(Yerkes Observatory.)*

New Year's Eve, the last night of the year of its first discovery. At Piazzi's request, the new planet was named for *Ceres,* the protecting goddess of Sicily.

(b) Subsequent Discoveries

Ceres was widely assumed to be the missing planet predicted by "Bode's law." It came as a complete surprise, therefore, when in March 1802 Heinrich Olbers discovered a second moving starlike object — the minor planet to be named *Pallas.* It was a natural speculation that if there were room for *two* minor planets, there could be room for others as well, and a search for such objects began in earnest. The discovery of *Juno* followed in 1804 and of

FIG. 15-2   Time exposure showing trails left by two minor planets (marked by arrows). *(Yerkes Observatory.)*

Vesta in 1807. It was 1845 before Karl Hencke discovered the fifth minor planet after 15 years of search. Subsequently, new ones were found with increasing frequency, until by 1890 more than 300 were known.

In 1891 Max Wolf, of Heidelberg, introduced the technique of astronomical photography as a means of searching for minor planets. The angular motion of a minor planet is large enough (especially if it is near opposition) so that during a long time exposure its image will form a trail on the emulsion. The object appears on the photograph, therefore, as a short dash rather than a starlike point image. *Brucia,* the three-hundred twenty-third minor planet to be discovered, was the first to be found photographically.

Today, discoveries of minor planets are usually accidental; they most often occur when the tiny objects leave their trails on photographs that are taken for other purposes. Literally thousands of minor planet trails appear on the photographs taken for the National Geographic Society–Palomar

Observatory Sky Survey. The majority of these trails are of objects that have never been catalogued. Most of them have been ignored, because at least three separate observations of a minor planet are required, preferably separated by intervals of several weeks, before its orbital elements can be determined.

15.3 ORBITS OF THE MINOR PLANETS

A standard reference in which the orbital elements and ephemerides of minor planets are published is the Soviet *Minor Planet Ephemeris.* The 1969 edition gives data for 1735 minor planets.

The minor planets all revolve about the sun in the same direction as the principal planets (from west to east), and most of them have orbits that lie nearly in the plane of the earth's orbit; the mean inclination of their orbits to the plane of the ecliptic is $9°.5$. About two dozen, however, have inclinations greater than $25°$; the orbit of Betulia is the most inclined ($52°$) to the ecliptic.

The orbits of most of the minor planets have semimajor axes that lie in the range 2.3 to 3.3 AU. The sidereal periods of most of them are in the range 3.5 to 6 years. Icarus has the smallest orbit — with a semimajor axis equal to 1.0777 AU. It also has the orbit of highest eccentricity among the known minor planets (0.83), and its perihelion distance from the sun is only 17 million mi. It is the only known object in the solar system other than comets and meteorites that passes within the orbit of Mercury. Hildago is the minor planet with the largest known orbit (its semimajor axis = 5.79 AU), and second largest eccentricity ($e = 0.66$). At aphelion it is farther from the sun than the orbit of Jupiter. The mean value of the eccentricities of the minor planet orbits is 0.15, not much greater than the average for the orbits of the planets.

Three minor planets are known to have passed within less than 3 million mi of the earth. They are Apollo (1932), Adonis (1936), and Hermes

(1937). None of the three, however, was observed long enough to allow accumulation of sufficient data to compute its orbit, and each has been lost. The two minor planets that most closely approach the earth and that are "in captivity," that is, whose orbits are known and whose returns can be predicted, are Icarus, which on June 14, 1968, passed within about 4 million mi of earth, and Geographos, discovered on the National Geographic Society–Palomar Observatory Sky Survey and accordingly named for "the geographer." In 1969 Geographos passed within about 6 million mi of earth, but in the most favorable approaches it can pass as close as Icarus did in 1968.

(a) Kirkwood's Gaps

An interesting characteristic in the distribution of minor planet orbits is the existence of several relatively clear areas, or gaps. These were explained in 1866 by Daniel Kirkwood as being due to the perturbative effects of Jupiter. The situation is exactly analogous to the divisions in the rings of Saturn, which are due to perturbations produced by the satellite Mimas (Section 14.5b). The gaps occur in regions where minor planets would have periods commensurable with that of Jupiter. For example, a minor planet at about five eighths Jupiter's distance from the sun would have a period exactly half that of Jupiter, and so every two times around the sun it would be near Jupiter. The repeated attractions by Jupiter, always in the same direction, would eventually force any minor planet out of that region, leaving a gap. Similiar Kirkwood gaps occur where minor planets would have periods of one third, two fifths, three fifths, and so on, that of Jupiter.

(b) The Trojans

In 1772 the mathematician Joseph Louis Lagrange found a particular solution to the problem of three bodies which showed that there should be two points in the orbit of Jupiter at or near which a minor planet could remain almost indefinitely. These are the two points which, with Jupiter and the sun, make equilateral triangles (see Figure 15-3). It seemed possible, therefore, that at either or both of the two points there could be located minor planets circling the sun in the same orbit as Jupiter, and with the same period. In the interval from 1906 to 1908, four such minor planets were

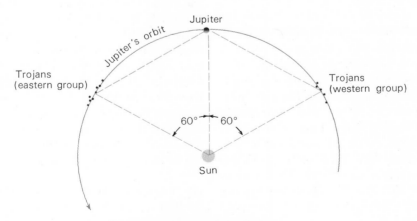

FIG. 15-3   Locations of the Trojans.

found. The number had increased to 14 by 1959. These minor planets are called the *Trojans;* they are named for the Homeric heroes. The convention today is to name those to the east of Jupiter for the Greek warriors and those to the west for the Trojans. (The custom was not established, however, until after the first three had already been named. As a result, there is one Trojan "spy" among the Greeks, and one Greek "spy" among the Trojans.) To a first approximation, the Trojan minor planets circle the sun with Jupiter's period of 12 years, only one sixth of a cycle ahead of or behind the Jovian planet. Their detailed motion, however, is very complicated; they slowly oscillate back and forth around the points of stability found by Lagrange, some of their oscillations taking as long as 140 years. At present there are five known Trojans in the west (or Trojan) side, and nine in the east (or Greek) side. It is possible that some of Jupiter's outer satellites may once have been Trojan minor planets that approached Jupiter too closely and were captured.

At the relatively great distance of Jupiter, only minor planets of fair size can be discovered; it is entirely possible that there are many undiscovered members of the Trojan group of minor planets too faint to be seen.

15.4 PHYSICAL NATURE OF THE MINOR PLANETS

On those occasions when the minor planet Vesta is at opposition while it is also at perihelion, it is faintly visible to the unaided eye. With this single exception, the minor planets are visible only telescopically. The four brightest, and presumably largest, show measurable disks. Even these, however, can be measured only with considerable difficulty; the diameters that have been found for them are therefore only approximate. Those values are given in Table 15.2, but the diameters are not as precisely known as the figures in the table would indicate.

TABLE 15.2 Sizes of Minor Planets

OBJECT	DIAMETER	
Ceres	785 km	488 mi
Pallas	489	304
Vesta	399	248
Juno	190	118

The albedo, or light-reflecting power, can be calculated for each of the minor planets of known diameter, because once we know its size, we can calculate how much sunlight falls upon it. From measurements of its brightness, we can then deduce how much light it reflects into space. By comparing the two quantities, we find the reflecting power. The albedos of the four minor planets of known size are found to average about 0.1; that is, they reflect about 10 percent of the light incident upon them. If we assume that the other minor planets have similar albedos, we can calculate the sizes they must have to account for their observed brightness.

It is thus found that there are probably fewer than a dozen minor planets with diameters greater than 100 mi. There may be a few hundred that are more than 25 mi across. Most observable minor planets must have diameters not much greater than 1 mi or so.

If it is assumed that the minor planets have mean densities comparable to that of the crust of the earth, or to that of the moon, their masses can be estimated from their volumes. The mass of Ceres, the largest, is probably only about $\frac{1}{8000}$ that of the earth. The velocity of escape from its surface must be only about $\frac{1}{3}$ mi/sec. Even Ceres, therefore, could not retain an atmosphere. From the smaller minor planets, a good pitcher could easily pitch a baseball into space. These small ones cannot even have enough gravity to pull themselves into a spherical shape.

(a) Rotation of Minor Planets

In 1901 it was observed that the minor planet Eros varies in brightness by about a factor of five in a period of 5^h16^m. These changes could be explained easily by the hypothesis that Eros has an elongated shape, like that of a brick, and that it rotates in a period of 5^h16^m about an axis roughly perpendicular to its long dimension. During a close approach of Eros to the earth in 1931, its elongated shape was actually observed telescopically, and its rotation in the above period was confirmed. It is probably about 15 to 20 mi long and about 5 mi thick and wide. Several other minor planets show similar changes in brightness over periods of a few hours and are also assumed to be irregular in shape and in rotation. Gehrels, for example, studying light variations of Vesta, has found that it rotates from west to east in a period of 5 hours 20 minutes, and that its equator is inclined at an angle of about 25° to the ecliptic plane.

(b) Combined Mass of Minor Planets

The total combined mass of all minor planets is not known. A guess is $\frac{1}{20}$ that of the moon, or about $\frac{1}{1600}$ that of the earth. Of this total, the two largest, Ceres and Pallas, probably account for an appreciable fraction. There are doubtless many thousands of unknown minor planets, but they must be small in size, and their combined mass may not contribute substantially to change the above total. Thus the whole aggregate might not make up even a first-rate moon, much less a "missing planet" as predicted by Bode's law.

15.5 TOTAL NUMBER OF MINOR PLANETS

It would be a formidable task to discover, determine orbits for, and catalogue all the minor planets bright enough to be observed with modern telescopes. Nevertheless, the total number of such objects can be estimated by systematically sampling selected regions of the sky.

Walter Baade, of the Mount Wilson Observatory, made such a sampling with the 100-in. telescope. He photographed a few selected regions of the zodiac, within which most minor planets are found. From the numbers recorded on his photographs, he calculated that in all there must be about 44,000 minor planets within reach of the 100-in. telescope. Similar results have been obtained in a more recent investigation by Kuiper and his associates.

15.6 NAMING MINOR PLANETS

By modern custom, after a newly found minor planet has had its orbit calculated, and has been observed again after having completed at least one circuit of the sun since its first discovery, it is given a name and a number. The number is a running index that indicates the order of discovery among the minor planets. The discoverer is customarily given the honor of supplying the name. The full designation of the minor planet contains both number and name, with the number encircled and preceding the name, thus: (1) Ceres, (2) Pallas, (433) Eros, (1566) Icarus, and so on.

Originally, the names were chosen from those of gods in Greek and Roman mythology. These, however, were soon used up. After exhausting the names of heroines from Wagnerian operas, discoverers chose names of wives, friends, flowers, cities, colleges, pets, and even favorite desserts. One is named for a computer (NORC). The thousandth minor planet to be discovered was named Piazzia, and number 1001 is Gaussia. Other minor planets bear the names Washingtonia, Hooveria, and Rockefellia.

Today, feminine names are generally given to "run of the mill" minor planets and masculine names to those of unusual interest. For example, Geographos has a masculine name because it can approach so near to the earth.

15.7 MINOR PLANETS AND CELESTIAL MECHANICAL PROBLEMS

Certain minor planets can be useful for the determination of other planetary data. If, for example, a minor planet passes near one of the larger planets, the gravitational influence of the latter on the motion of the former can be determined from the change produced in the orbit of the minor planet, enabling us to calculate the mass of the planet responsible for the perturbations. We have seen (Section 14.4c) how a recent determination of Jupiter's mass was made possible from Jupiter's perturbations on the minor planet (65) Cybele. Until the space age, minor planets (especially Icarus) provided the best means of improving our knowledge of the masses of Mercury and Venus; now we can determine the masses of those planets from their perturbations on space vehicles.

Also, until the space age, minor planets played a major role in the determination of the length of the astronomical unit. The orbits of minor planets can be learned with high precision, and hence their distances from the earth can be determined with great accuracy but only in astronomical units, for (as we shall see in Chapter 18) we use the earth's orbit (or some part of it) to survey the distances of solar system objects. Needed, in other words, is a "scale of miles" for the solar system — that is, the length of the astronomical unit in miles or kilometers. If, however, a minor planet passes close to the earth, its distance in miles can be determined by direct triangulation from two stations on earth. Its distance at that instant is then known both in miles and astronomical units; this information gives the actual length of the astronomical unit. Observations of Eros were used for such an investigation when it passed within 16 million mi of the earth in 1931.

Close approaches of minor planets afforded a second method of determining the value of the astronomical unit. Perturbations produced on the orbit of the minor planet by the earth during its close approach to us can be observed. Since its distance from the earth relative to that from the sun is known, the extent of the perturbations produced upon its orbit by the earth enables us to

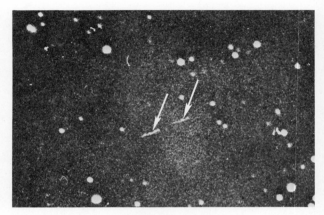

FIG. 15-4   Arrows show two trails left by the minor planet Icarus. This photograph was made by superimposing star images on two plates; each had two 5-minute exposures separated by an interval of 3 minutes. The uneven character of the trails is probably due to seeing or motion of the telescope. *(Griffith Observatory.)*

calculate the relative gravitational influence of the earth and sun upon the object. This information, in turn, enables us to calculate the relative masses of the earth and sun. The earth's mass is measured by laboratory experiments (Section 4.3e); thus we can calculate the mass of the sun. Because the magnitude of the acceleration of the earth toward the sun depends on the mass and distance of the sun, a knowledge of the sun's mass makes it possible for us to calculate its distance — and the value of the astronomical unit.

It is now possible to determine the length of the astronomical unit by direct radar observations of the nearer planets (Chapter 18), and planetary perturbations on space probes also yield information not only on the planets' masses but on the astronomical unit. Nevertheless, the investigations of minor planets played an important historical role and enabled us to develop the techniques that could later be applied in space technology. After all, interplanetary probes are minor planets in a sense.

15.8 ORIGIN OF THE MINOR PLANETS

Most authorities regard it as probable that the minor planets were formed from the same material that formed the principal planets, and at about the same time. It has been speculated that the minor planets, and perhaps the meteorites as well, may have originated from the breakup of a larger body, presumably a planet that once revolved about the sun at a distance of 2.8 AU. It must be remembered, however, that the combined mass of all the minor planets may well be less than a thousandth of the mass of the earth, itself a relatively small planet.

On the other hand, there is some evidence that many of the minor planets may have originated from the breakup of several somewhat larger bodies. In 1917 the Japanese astronomer K. Hirayama found that a number of the minor planets fall into "families" or groups of similar orbital characteristics. He hypothesized that each family may have resulted from an explosion of a larger body, or by the collision of two bodies.

Slight differences in the initial velocities given the fragments of the explosion or collision would have resulted in the relatively small differences now observed among the orbits of the different minor planets in a given family.

In 1950 Dirk Brouwer extended Hirayama's investigation to the study of 1537 minor planets. He found 29 families of objects, each family containing from 4 to 62 members. He accounted for the differences among the orbital elements of members of the same family by the effect of perturbations of Jupiter and Saturn. Some of these families can be subdivided further into two subgroups, which suggests that such families may have resulted from a collision of two bodies. Collisions among a relatively few larger objects would be extremely rare. If they once occurred, however, subsequent collisions would be more likely among the larger number of smaller bodies that resulted. If this collision and fragmentation hypothesis for the origin of the minor planets should be shown to be correct, it would be inferred that the fragmentation process is still going on.

Exercises

1. Minor planets can be discovered by the trails they leave on astronomical photographs. Fainter objects could be recorded, however, if their images were points rather than trails. Explain how you might plan a photographic search for faint minor planets.

2. What would be the period of a minor planet whose distance ranges from 2.5 to 3.5 AU from the sun?

    Answer: About 5.2 years

3. How far from the sun would a minor planet be if it had a period one third as long as that of Jupiter? Would such an object long remain in that orbit? Why?

    Answer: About 2.5 AU

4. Two minor planets are observed. One is 64 times brighter than the other. Both are at the same distance from the sun and from the earth. The brighter has a measured diameter of 240 mi. What is the diameter of the other?

5. What would be your answer to Exercise 4 if the two minor planets were equally far from the sun, but if the fainter were twice as far from the earth as the brighter?

6. If you weigh 200 lb on the earth, what would you weigh on the surface of a minor planet that has a mass about $\frac{1}{10,000}$ that of the earth and a diameter $\frac{1}{20}$ that of the earth? Can you name a minor planet that approximately fits this description?

    Answer: 8 lb

7. Assume that a minor planet rotates on its axis from west to east and that it is irregular in shape, so that its brightness varies. Assume, further, that its period of rotation is precisely constant. Will the *observed* light fluctuations be at a constant rate? Can they ever be at a greater or lesser rate than the sidereal period of rotation of the object? Explain.

8. At times Eros fluctuates in brightness by a factor of about five. At other times its light fluctuates by a much smaller amount. Can you offer an explanation for this phenomenon?

CHAPTER 16

Comets

Comets have been observed from the earliest times; accounts of spectacular comets are found in the histories of virtually all ancient civilizations. Yet, until comparatively recently, comets were not generally regarded as celestial objects.

A typical comet that is bright enough to be conspicuous to the unaided eye has the appearance of a rather faint, diffuse spot of light, somewhat smaller than the full moon and many times less brilliant. There may be a very faint nebulous tail, extending for a length of several degrees away from the main body of the comet. Like the moon and planets, comets slowly shift their positions in the sky from night to night, remaining visible for periods that range from a few days to a few months. Unlike the planets, however, most comets appear at unpredictable times. In medieval Europe, comets were usually regarded as poisonous vapors in the earth's atmosphere and as bad omens. More fear and superstition have been attached to comets than to any other astronomical objects.

16.1 EARLY INVESTIGATIONS

Perhaps the first scientific investigation of a comet, based on careful observation, was Tycho Brahe's diligent study of the brilliant comet that appeared in 1577 (Section 3.3a). Had the comet been inside the earth's atmosphere, as was then generally supposed, changes in its apparent direction would easily have been detectable to an observer who changed his position by several miles. Brahe, not being able to detect any such changes, realized that the comet was a celestial object. Moreover, his failure to detect any diurnal parallax of the comet led him to the conclusion that it was at least three times as distant as the moon and that it probably revolved about the sun.

Kepler described in detail the comet of 1607 (later known as Halley's comet). He held the view that this and other comets are celestial bodies that travel in straight lines through the solar system. Two comprehensive treatises were published in 1654 and 1668 by John Hevel. These works contained systematic references to all known comets, and proved highly valuable to later investigators.

When Newton applied his law of gravitation to the motions of the planets, he wondered whether comets might similarly be gravitationally accelerated by the sun. If so, their orbits should be conic sections. If comets, like planets, had nearly circular orbits, they should be visible at regular and frequent intervals. On the other hand, if a comet moved in an elongated elliptical orbit of large size, it would necessarily be visible during only that relatively brief period of time during which it passed near the sun (perihelion); over most of its orbit the comet would be so far from the sun as to be invisible. Furthermore, the periods of comets, moving in such orbits, would be very great. A comet that had been seen for a short time, and then was seen again many tens or hundreds of years later when it next appeared near perihelion, would naturally be mistaken for a new object. One end of an ellipse, of very great length and of eccentricity near unity, is nearly indistinguishable from a parabola; thus the motion of a comet moving on a parabolic orbit would closely resemble that of a comet whose orbit is a very long ellipse. Newton concluded that comets are gravitationally attracted about the sun in orbits that are either very long ellipses or parabolas.

(a) Halley's Comet

Edmund Halley greatly extended Newton's studies of the motions of comets. In 1705 he published calculations relating to 24 cometary orbits. In particular, he noted that the elements of the orbits of the bright comets of 1531, 1607, and 1682 were so similar that the three could well be the same comet, returning to perihelion at intervals of about 75 or 76 years. If so, he predicted that the object should return about 1758.

Alexis Clairaut calculated the perturbations that this comet should experience in passing near Jupiter and Saturn and predicted that it should first make its appearance late in 1758 and should pass perihelion within about 30 days of April 13, 1759. The comet was first sighted by an amateur astronomer, George Palitzsch, on Christmas night, 1758. It passed perihelion on March 12, 1759, 31 days earlier than Clairaut's calculations predicted. The comet has been named *Halley's comet,* in honor of the man who first recognized it to be a permanent member of the solar system. Subsequent investigation has shown that Halley's comet has been observed and recorded on every passage near the sun at intervals of from 74 to 79 years since 240 B.C. The period varies somewhat because of perturbations upon its orbit produced by the Jovian planets. It last appeared in 1910, and is due again about 1986.

16.2 DISCOVERY AND DESIGNATION

Observational records exist for about a thousand comets. Today, new comets are discovered at an average rate of about a dozen per year. Some of these are discovered accidentally on astronomical photographs taken for other purposes. Many are discovered by amateur astronomers. At one time medals and even hundred-dollar prizes were awarded to the discoverers of new comets. With the development of high-speed, wide-angle photo-

graphic telescopes, however, such as the 48-inch Schmidt telescope of the Palomar Observatory, comet discoveries have become almost a "mass production" business; more than a dozen were discovered on the photographs taken for the National Geographic Society–Palomar Observatory Sky Survey, alone. Consequently, the prizes and medals have been discontinued.

Most of the new comets found each year never become conspicuous, and are visible only on photographs made with large telescopes. Every few years, however, a comet may appear that is bright enough to be seen with the unaided eye. In 1957 there were two conspicuous naked-eye comets (Comet Arend-Roland and Comet Mrkos). In the past, about two or three times each century, there have appeared spectacular comets that reached naked-eye visibility even in daylight. The first "daylight comet" to appear since the brilliant comet of 1910 (that preceded the more famous Halley's comet by a few months), was comet Ikeya-Seki in 1965.

A newly found comet is named for its discoverer or discoverers (there are often more than one). It is also given a temporary designation consisting of the year of its discovery followed by a

FIG. 16-1   Halley's comet in 1066, as depicted on the Bayeux tapestry. *(Yerkes Observatory.)*

FIG. 16-2   Halley's comet. *(Yerkes Observatory.)*

lowercase letter indicating its order of discovery in that year. For example, Comet 1969a is the first comet discovered in 1969, 1969b is the second, and so on. Later, when the orbits of all recently observed comets have been calculated, each of the comets is given a permanent designation consisting of the year in which it passed nearest the sun, followed by a Roman numeral designating its order among the comets that passed perihelion during that particular year. For example, Comet 1969 I would be the first comet to pass perihelion in 1969; 1969 II, the second, and so on.

16.3 ORBITS OF COMETS

The orbit of a comet, like that of a planet, can be determined from three or more fairly well spaced observations of its position in the sky among the stars. Most comet orbits are indistinguishable from parabolas. Some are definitely elliptical; a very few comets have been observed whose orbits appear to be hyperbolas.

(a) Membership in the Solar System

Because the orbits of most comets appear to be parabolic, and a few even hyperbolic, the question arises whether all comets are members of the solar system or whether some might be accidental intruders from interstellar space. The evidence available, however, strongly suggests that comets have been members of the solar system for at least a long time.

If comets were intruders from interstellar space, their orbits should nearly all be hyperbolic. An elliptical orbit is possible only for a body permanently revolving about the sun, not for one coming from outside the solar system. Even a parabolic orbit would be possible only for a body that, before being attracted toward the sun, was moving through space in almost exactly the same direction and at the same speed as the sun, that is, which was *at rest with respect to the sun*. Such motion would be extremely improbable for an interstellar visitor.

Those few comets that do appear to have moved in hyerbolic orbits are believed to have approached the sun on nearly parabolic orbits that were perturbed when the comets passed near Jupiter or another planet. E. Strömgren has investigated the motions of 20 such "hyperbolic" comets and has found that in every case the gravitational influences of the planets had changed the orbit from a nearly parabolic one to a hyperbola. Although a comet, through planetary perturbations, may thus escape from the solar system, there is no case where a comet is known to have *approached* the sun on a hyperbolic orbit.

Moreover, the orbits of comets, unlike those of planets, are oriented at random in space. As many comets appear to approach the sun from one direction as from another. If, however, comets were interstellar objects, there should be a preponderance of them approaching from the direction of Hercules, toward which the sun is moving with a speed of about 12 mi/sec (see Chapter 19).

It is believed, therefore, that most or all comets are members of the solar system, and that they travel on elliptical orbits, most of which are extremely long and have eccentricity near unity. A comet can be observed from the earth only when it traverses that end of its orbit where is passes close to the sun. If the orbit has an eccentricity greater than about 0.99, that part of it which can be observed is indistinguishable from a parabola. Thus "parabolic" orbits observed for comets are probably only the ends of very long ellipses that carry comets tens of thousands of astronomical units from the sun to the "outposts" of the solar system. Such comets have periods that probably range into millions of years and thus may have been visible only once during man's existence on the earth.

A minority of comets have orbits of low enough eccentricities that their semimajor axes can be determined from the relatively small portions of the orbits that are actually observed. These comets, whose periods can be determined, are called *periodic comets*.

(b) Periodic Comets

Technically, all observed comets could be regarded as periodic, inasmuch as they are all presumed to revolve about the sun in elliptical orbits. The diffuse appearance of a comet, however, makes it difficult to measure its position with great precision; consequently, only those comets whose orbits differ rather substantially from parabolas can be definitely classed as periodic. The comets of longest definitely established periods are Pons-Brooks (71 years), Halley (76 years), and Rigollet (151 years). The comet of the shortest known period is Comet Encke (3.3 years).

There are about 45 known comets whose orbits are inclined at less than 45° to the ecliptic, that travel from west to east (like the planets), and whose aphelion distances are near Jupiter's mean distance from the sun. They comprise *Jupiter's family of comets*. They have periods that range from 5 to 10 years. They are believed to be comets whose orbits were originally very long (like those of most comets) but were changed into relatively small orbits as a result of perturbations that occurred when the comets passed close to Jupiter. There is also evidence of other less marked families of comets associated with the more distant planets.

(c) Disintegration of Short-Period Comets

Comets are such flimsy objects that they often suffer severe damage as the result of tidal forces produced by the sun when they pass perihelion. It is estimated that most periodic comets are completely disintegrated after a hundred or fewer perihelion passages.

Comet Biela, for example, with a period of 7 years, was discovered in 1772. During its approach to the sun in 1846, it was observed to break into two separate comets. Both comets returned on about the same orbit in 1852, but neither has been seen since. However, spectacular meteor showers (see Chapter 17) were observed on November 27, 1872, and on November 27, 1885, on which occasions the earth passed through the orbit of Comet Biela and encountered swarms of particles traveling in the path of the disintegrated comet. The orbit of the swarm has subsequently been altered by Jupiter; consequently, spectacular displays of Bielid meteors have not been observed in the present century.

Because of their high mortality rates, periodic comets cannot have been periodic for long but must originally have been comets of very long periods having nearly parabolic orbits. Within the recent past (perhaps the last few thousand years) their orbits have been drastically altered to their present relatively small size by perturbations produced by the planets, especially Jupiter. Those occasional comets that are highly spectacular, and hence cannot have suffered appreciable disintegration, almost invariably have nearly parabolic orbits, and also they have not been seen before in recorded history.

(d) Comets of Nearly Circular Orbits

There are two known comets whose orbits are nearly circular, like those of the planets, and which are under more or less continuous observation. They are Comet Schwassmann-Wachmann, which travels in a 16-year period in an orbit of eccentricity 0.14, lying between the orbits of Jupiter and Saturn; and Comet Oterma, which travels with an 8-year period in an orbit lying between those of Mars and Jupiter. Comet Schwassmann-Wachmann is particularly interesting. It is normally quite faint, but occasionally it displays sudden unexplained bursts of brightness in which it may brighten by more than 100 times. These intermittent flareups are believed to be associated with violent emissions of charged particles from the sun.

16.4 PHYSICAL NATURE OF COMETS

No two comets are alike, but they all have one characteristic in common — the *coma*, which appears as a small, round, diffuse, nebulous glow. As a comet approaches the sun, the coma usually grows in brightness, but sometimes it becomes so diffuse that it disappears. At distances of less than about 1 AU from the sun, the coma tends to contract. Often, but not always, there is a small bright *nucleus* in the center of the coma. Together, the coma and the nucleus (if any) comprise the *head* of the comet. Many comets, as they approach the sun, develop tails of luminous material that extend for millions of miles away from the head.

The coma of a typical comet is very large, of the order of a hundred thousand miles across, although individual comets differ enormously from one another in size. The coma of a comet is composed of extremely tenuous material; stars can easily be seen shining through it. Only a few miles of the terrestrial atmosphere substantially dims starlight; yet stars shone in full brilliance through more than 40,000 miles of the head of Halley's comet in 1910.

(a) Light and Spectra of Comets

Comets are seldom observed when they are much farther away from the sun than the planets Jupiter or Saturn. At that distance, they are usually very faint, and their spectra resemble that of the sun, a fact which shows them to be composed of solid particles, reflecting sunlight.

When a comet is within about 3 AU of the sun, its spectrum changes. Superimposed upon the spectrum of reflected sunlight, there are bright emission lines (or series of many closely spaced lines called *bands*) due to the chemical combinations C_2, OH, CN, NH, and NH_2. These chemical *radicals* are probably dissociated from the molecules CH_4 (methane), NH_3 (ammonia), and H_2O (water) and shine by the process of *fluorescence;* that is, they have been set aglow by ultraviolet radiation from the sun.

Some comets, at perihelion, pass within a few hundred thousand miles of the solar surface. Such a comet was Ikeya-Seki, which passed perihelion in October 1965. In such proximity to the sun, comets can become as hot as 4500°K. If a comet approaches very close to the sun, its spectrum changes again. Bright emission lines produced by atoms of various metals appear: sodium, iron, silicon, magnesium, and others.

(b) *Variation of Brightness of a Comet*

The apparent brightness *B* of a comet is proportional to the amount of light *L* leaving the comet and inversely proportional to the square of the distance Δ of the comet from the earth (because of the inverse-square law of light propagation — see Section 10.1a); that is,

$$B \propto \frac{L}{\Delta^2}.$$

When a comet is far from the sun, it shines only by reflected sunlight. In this case, L is inversely proportional to the square of the distance R of the comet from the sun, and

$$B \propto \frac{1}{R^2\Delta^2}.$$

On the other hand, when a comet is within a few astronomical units of the sun, it shines partly by fluorescence. At a distance of 1 AU from the sun, a comet shines almost entirely by fluorescence. As a comet approaches the sun, therefore, it brightens up far more rapidly than it would if it merely reflected sunlight. The amount of light L that leaves a comet depends on the kinds of gases that compose it and on the effectiveness of the solar radiation in exciting those gases to glow; L depends, therefore, in a rather complicated manner, on R. Empirically, it is found that

$$B \propto \frac{1}{R^n\Delta^2},$$

where the exponent n is usually near 4, but may be as low as 3 for some comets and as high as 6 for others, and may even vary with time for a single comet.

(c) Heads of Comets

Far from the sun, where a typical comet spends most of its time, all its material must be frozen into the solid state. There, the comet is believed to consist only of a nucleus, which may be a swarm of solid particles or may even be a single mass. Fred Whipple envisions a comet, far from the sun, as resembling a "dirty iceberg," about one third of its material consisting of rocky or metallic materials, the remainder being made up of substances that vaporize more easily. The latter sublime (change from solid to gaseous state without going through the liquid state) when the approaching

FIG. 16-3 Head of Halley's comet, photographed on June 5, 1910, by the 60-inch reflector with an exposure of 9 minutes. *(Mount Wilson and Palomar Observatories.)*

comet attains a distance of several astronomical units from the sun, thus forming the coma.

As the volatile materials vaporize from the nucleus, it is left as a swarm of discrete particles surrounded by the coma. Most of these particles must be in the form of microscopic dust grains. It is not known how large the largest may be. The evidence suggests, however, that they are probably not over a few miles in diameter. At least two comets, the great "daylight" comet of 1882 and Halley's comet in 1910, passed directly between the earth and the sun. In each case efforts were made to observe the nucleus silhouetted against the disk of the sun, but no trace whatever of any opaque body was visible. The largest particles in the nucleus must have been less than 50 mi across.

(d) Tails of Comets

Many comets develop tails as they approach the sun. Some, however, especially certain short-period

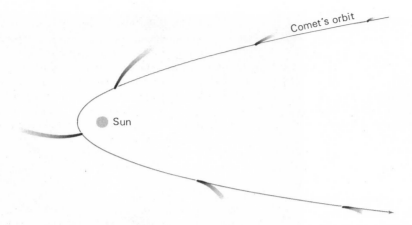

FIG. 16-4 Shape of a typical comet tail as the comet passes perihelion.

comets, have either lost the ability to form a tail or never possessed it. In the sixteenth century, the astronomers Fracastor and Apian described many comets, and remarked that their tails generally point *away* from the sun. This observation was later confirmed by Kepler. Newton attempted to account for the tails of comets as being produced by a repulsive force of sunlight driving particles away from the comets' heads, an idea close to our modern view.

Actually, the material forming the tails of comets is subject to three forces:

(1) A radial ejection force from the nucleus of the comet, which results from the thermal activity produced when the comet nears the sun and heats up.

(2) A force of repulsion directed radially away from the sun, due, in part at least, to pressure exerted on the gas molecules and on the solid particles by the sun's radiation.

(3) The sun's gravitational attraction, which causes each particle or molecule in a comet's tail to move in an orbit.

A steady flow of cometary material is ejected outward by the first force, away from the head of a comet; it is picked up by the solar repulsive force and driven radially away from the sun. Thereafter, it describes a new orbit, generally hyperbolic, about the sun. The description of the orbit followed by the tail material depends on the relative magnitudes of the forces acting on it, but in any case is different from that of the parent comet; thus this material is lost permanently to the comet and, usually, to the solar system.

Tails generally grow in size as comets near the sun; some comet tails have reached lengths of more than 100 million miles. Once the tail material has left the vicinity of the comet's head, the only forces acting upon it are radial forces toward and away from the sun (gravity and the force of repulsion). As the material recedes farther from the sun, its *angular velocity* about the sun decreases, and the material lags behind the comet. In general, therefore, comet tails lie in the plane of the comet's orbit, point more or less away from the sun, but curve somewhat backward, away from the direction of the comet's motion (see Figure 16-4). We shall see that the acceleration produced by the force of repulsion from the sun varies according to the size of the particles it acts upon and according to whether the particles are gas molecules or solid grains. Material of different types, therefore, is accelerated away from the sun in different amounts. If the repulsive force exceeds the gravitational force on a certain kind of particle by 20 or 30 times, particles of that type are driven outward from the sun so rapidly that they form a tail that is nearly straight. On the other hand, if the repulsive force is only 2 or 3 times the gravitational force, the particles move outward more slowly; the difference between the orbital motions of the comet and tail material becomes apparent before the latter has receded very far from the comet's head, and the tail appears noticeably curved. Many comets have two or more tails showing different curvatures; presumably these tails consist of different kinds of particles.

Several comets have displayed a phenomenon similar to that displayed by Comet Arend-Roland in April 1957, just after it passed perihelion. In addition to a normal tail pointing away from the sun, the comet had a short spikelike tail that appeared to be pointed almost directly toward the sun. The explanation that seems most plausible is that the comet was followed by a fan-shaped sheet of debris ejected from it and moving in its orbital plane. This material was too diffuse to be visible when viewed broadside, but just after the comet passed perihelion, the earth happened to lie in the plane of the comet's orbit so that we saw the material edge on. Of course, the material was not really between the comet and the sun; the effect was an illusion that resulted from our angle of sight (see Figure 16-6).

(e) The Force of Repulsion*

The first known force of repulsion from the sun was radiation pressure. Each photon of electromagnetic radiation has associated with it a small amount of momentum, which is numerically equal to the energy of the photon (Planck's constant times the frequency — see Section 10.1f) divided by the velocity of light. If a photon strikes a particle, or is absorbed by it, the momentum of the photon is transferred to the particle, thereby accelerating the latter. On particles of moderate size, the acceleration is entirely negligible in comparison to that produced by the gravitational forces acting on them. On very small particles in interplanetary space, however, the force of solar radiation can be greater than the sun's gravitational

attraction. Radiation pressure is most effective on particles whose sizes are near that of the average wavelength of sunlight. If they have a density near unity (like ice crystals), the force of radiation on such particles can exceed that of gravitation by about five times. Particles several times larger than the wavelength of the incident light have masses great enough so that the force of gravitational attraction between them and the sun exceeds the outward force of radiation pressure. The effect of radiation pressure on smaller particles depends on their chemical composition; some types of smaller particles can be repulsed. Many kinds of molecules can also absorb radiation and be driven away from the sun.

Radiation pressure, incidentally, can produce measurable perturbations on the motion of a balloon satellite revolving around the earth — for example, the Echo satellite, launched in the early 1960s. There is even a possibility of using radiation pressure for space travel. It has been shown that if large lightweight sails can be attached to space vehicles, "solar sailing" is a feasible mode of locomotion through interplanetary space. It would not, of

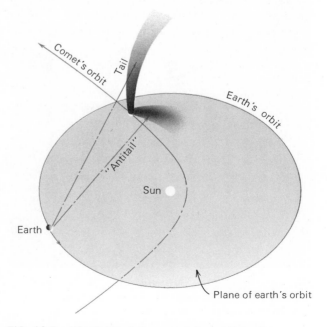

FIG. 16-5   "Antitail" of the comet Arend-Roland.

FIG. 16-6   Comet Arend-Roland photographed with the 48-inch Schmidt telescope on April 26, 1957, with an exposure of 15 minutes. The "antitail" was a prominent feature at this time. *(Mount Wilson and Palomar Observatories.)*

course, be a sufficient means of propulsion in the strong gravitational fields near the surfaces of planets.

Some comet tails have bright spots or knots or other recognizable features that persist for several days. The motions of such irregularities can be measured to determine the acceleration with which the material in the tail is moving away from the sun. In most cases, these observed accelerations are only a few times greater than the acceleration produced by solar gravitation, and radiation pressure can easily account for the formation of the tails. Sometimes, however, the accelerations are found to exceed the gravitational acceleration by many tens or even hundreds of times.

L. Biermann first proposed an explanation to account for the occasional very high acceleration observed in comet tails and also for the various abrupt changes that are sometimes observed in irregular features of tails. It is now known that the sun more or less continually ejects charged atomic particles into space, producing a solar wind of ionized gas (see Chapter 24). From time to time, in sudden outbursts, these particles stream outward from the sun in larger numbers than usual. As they impinge upon a comet, they interact with the gases in the coma; and charged, or ionized, gases are driven by electrostatic repulsion from the coma into the tail. Biermann's mechanism is believed to be responsible for those comparatively straight comet tails that must result from the ejection of matter with very high accelerations. Spectra of such tails have confirmed that they are composed largely of ionized gases, whereas spectra of the more curved tails show them to be composed of solid particles reflecting sunlight, and neutral molecules.

(f) Masses of Comets

We do not know the mass of a single comet. There are a few instances, however, that enable us to place *upper limits* on the masses of certain comets. In 1770 Lexell's comet passed close enough to the earth so that measurable perturbations on the earth's orbital motion would have been produced if the comet's mass had been as much as $\frac{1}{110,000}$ that of the earth; no effect, however, was observed. When an orbit had been calculated for Comet Brooks II, discovered in 1889, it was found that in 1886 the comet had passed through the satellite

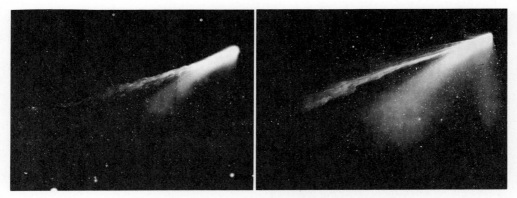

FIG. 16-7   Two photographs of the comet Mrkos obtained on successive nights with the 48-inch Schmidt telescope. *(Mount Wilson and Palomar Observatories.)*

system of Jupiter. It produced no observable perturbations on the motions of Jupiter's satellites, but the comet's own period was shortened by the encounter, from 27 to 7 years.

William Liller studied the scattering of sunlight from the tails of the two bright comets that appeared in 1957. He found that the observed reflection of sunlight could be accounted for if each tail contained about one or two hundred million tons of microscopic iron particles. From the rate at which material was leaving the comets to form their tails, Liller found that each of the comets must have been losing 100 to 1000 tons of iron per second while it passed near the sun. If it is assumed that such comets can survive 100 perihelion passages before losing the ability to form tails, they must contain something like 10^{11} to 10^{12} tons of iron. Even if iron constitutes only $\frac{1}{20}$ the total mass of a comet, the two comets studied by Liller still have masses less than one billionth that of the earth.

Many experts estimate that a typical comet may have only a trillionth (10^{-12}) the mass of the earth. Thus, despite the large sizes of comets, they are mostly "empty space." A comet has been described as the nearest thing to nothing that anything can be and still be something.

16.5 SOURCE OF SUPPLY OF COMETS

Unlike the planets, which all travel from west to east in orbits that lie in nearly the same plane, comets approach the sun from all possible directions along orbits that are inclined to the ecliptic at all possible angles. Comets travel to such great distances from the sun that most have passed perihelion only once during recorded history. Some pass so close to the sun that they would most certainly be torn apart after a few such visits; yet the solar system is at least 5 billion years old, and still comets are plentiful. What, then, can be their means of replenishment, in view of the fact that we still find new ones each year?

The most popular hypothesis for the origin of comets is probably that of the Dutch astronomer, Jan H. Oort. He suggests that there might be a vast "cloud" of comets revolving about the sun at distances of from 50,000 to 150,000 AU. Their orbits, ordinarily, cannot bring them close to the sun, because, as has been shown by A. J. J. van Woerkom, a comet that comes within 2 AU of the sun can survive, at most, only a few million years before suffering disintegration or being lost to the solar system. If the comets we now observe had come so close as this to the sun since the formation of the solar system, they would long ago have been

FIG. 16-8   Comet Seki-Lines photographed from Frazier Mountain, California, August 9, 1962. (*Alan McClure, Los Angeles.*)

destroyed by solar tidal forces or, alternatively, would have been deflected by planetary perturbations away from the solar system.

On the other hand, the outer fringe of the cloud of comets cannot extend beyond about 60 percent of the distance of the nearest star (Alpha Centauri) because beyond that point their orbits would be unstable. Even so, perturbations produced by those stars that pass, every few million years, within a few light-years of the sun must alter some of the primeval cometary orbits. Some comets might be lost to the solar system as the result of such stellar perturbations, but others, retarded in their motion by the star's attraction, would fall in toward the sun. Thus, Oort accounts for the steady influx of comets into the central part of the solar system.

Those comets that are so deflected toward the sun move subsequently on orbits of very high eccentricity, until, after one or a few perihelion passages, further perturbations by the planets either convert them into "periodic" comets, or deflect them into interstellar space on hyperbolic orbits. In the former case, they are eventually decomposed by tidal forces; in the latter, they are permanently lost to the solar system.

Although the supply of comets in the "cloud" must be gradually diminishing, the remaining number of comets may still be very large. Oort estimates that there are at least 100 billion, and that their combined mass is between $\frac{1}{100}$ and $\frac{1}{10}$ that of the earth.

16.6 COLLISIONS OF THE EARTH AND COMETS
The probability of a collision between the earth and a comet is small. Yet, over the billions of years of existence of the solar system, the event may well have occurred — even more than once. In all likelihood, the earth would pass on through the comet with virtually no gross effect. If large particles of the nucleus struck the earth, however,

FIG. 16-9   Comet Wilson, on July 25, 1961. *(Photograph by Alan McClure.)*

severe local damage could result. Some investigators take the view that the great 1908 meteorite fall (see Chapter 17) was a cometary collision. The earth very probably did pass through the tail of Halley's comet in 1910. There was no detectable evidence of the encounter, although it is reported that at least one enterprising person made his fortune selling "comet pills."

Exercises

1. Comet 1955b is also known as 1954 V. Why is it denoted by two different years?

2. Find the period of a comet which at perihelion just grazes the sun, and whose aphelion distance from the sun is:

(a) 200 AU

(b) 2000 AU

(c) 20,000 AU

(d) 200,000 AU

3. On the assumption that a comet can survive 100 perihelion passages, find the lifetime of each of the comets in Exercise 2.

*4. Suppose that the brightness of a comet is given by the law

$$B \propto \frac{1}{R^4 \Delta^2}.$$

When first discovered it is exactly at opposition and at a distance of 4 AU from the earth. How much brighter would it appear when it is 1 AU both from the sun and from the earth? Compare this ratio with that by which a body would brighten if it merely reflected sunlight.

5. With the aid of a diagram, explain how you could derive the true length of the tail of a comet from the observed angular length. Assume that the tail points directly away from the sun and that the comet is 1 AU from both the earth and sun.

6. Show by Kepler's laws that the tail of a comet should always curve back slightly away from the direction of motion of the comet.

7. The force due to radiation pressure on a particle is proportional to the amount of radiation that the particle intercepts. Show that the ratio of the force of radiation pressure to the solar gravitational force on a spherical particle is independent of the distance of the particle from the sun.

8. If Oort's hypothesis for the origin of comets is correct, why do comets that come close to the sun for the first time always have aphelion points in the region of the cloud from which they originate?

9. Why does the material in the tail of a comet follow a hyperbolic orbit about the sun?

CHAPTER 17

Meteoroids, Meteorites, and Meteors

Although the layman often confuses comets and meteors, these two phenomena could hardly be more different. Comets, when seen from the earth, are swarms of particles and glowing gases that revolve about the sun. They can be seen when they are many millions of miles away from the earth, and may be visible in the sky for weeks, or even months, slowly shifting their positions from day to day. They rise and set with the stars, and during a single night appear motionless to the casual glance. Meteors, on the other hand, appear only when small solid particles enter the earth's atmosphere from interplanetary space. Since they are moving at speeds of many miles per second, they vaporize as a result of the high friction they encounter with the air. The spot of light caused by the luminous vapors formed in such an encounter appears to move rapidly across the sky, and it fades out within a few seconds. Meteors are commonly called "shooting stars."

On rare occasions, an exceptionally large particle may survive its flight through the earth's atmosphere and land on the ground. Such an occurrence is called a *meteorite fall,* and if the particle is later recovered, it is known as a fallen *meteorite.* Frequently, the term "meteorite" is reserved for the fallen particle, and the particle, when still in space, is called a *meteor.* In this book, however, we shall use the terminology recommended by the International Astronomical Union in 1961. The particle, when it is in space, is called a *meteoroid;* the luminous phenomenon caused when the particle vaporizes in the earth's atmosphere is a *meteor;* and if it survives and lands on the ground, the particle is a *meteorite.*

17.1 PHENOMENON OF A METEOR

On a typical dark moonless night an alert observer can see half a dozen or more meteors per hour. However, to be visible, a meteor must be within 100 to 150 mi of the observer; over the entire earth, the total number of meteors bright enough to be visible must total about 25 million per day. Faint meteors are far more numerous than bright ones; the number of meteors that are potentially visible with binoculars or through telescopes, and that range down to a hundred times fainter than naked-eye visibility, must number near 8 billion per day.

More meteors can generally be seen in the hours after midnight than in the hours before. Moreover, there are certain times when meteors can be seen with much greater than average frequency — up to 60 or more per hour. These unusual meteor displays are called *showers.* Meteor showers occur when the earth collides with swarms of many meteoroids moving together in space. A distinction is made between shower meteors and nonshower, or *sporadic,* meteors.

(a) Fireballs

Occasionally an exceptionally bright meteor is seen, attracts much attention, and is reported by many observers. These bright meteors are called *fireballs,* or *bolides.* It is estimated that a few thousand fireballs appear every day over the entire earth, most of them over oceans or uninhabited regions. A few fireballs are visible in broad daylight; some have even reached the brilliance of the full moon. Sometimes they break up in midair with explosions that are audible from the ground.

FIG. 17-1   Photograph of the trail of a bright
meteor that happened to cross the field of view
of the telescope while photographing the
Andromeda galaxy. *(Photograph by F. Klepesta.)*

Fireballs occasionally leave luminous trails or
trains behind them, which may persist for periods
ranging from 1 second to ½ hour. The velocities
of upper atmospheric winds are sometimes revealed by the twisting and distortion of meteor
trains.

Bright fireballs have been recorded by very
ancient peoples. There are references to them, for
example, in the Book of Joshua, and in records kept
by the Chinese centuries before Christ.

(b) Observations of Meteors

Bright fireballs are rare at any one place, and
astronomers must rely upon lay observations for
the data required to analyze their paths through
the atmosphere. Unfortunately, the layman usually
provides very unreliable accounts of what he saw.
Fireballs often seem many times closer than they
actually are; also, their angular altitudes and
speeds are frequently grossly exaggerated. Nevertheless, the path of a spectacular fireball can
sometimes be derived by comparing reports from
many observers scattered over an area of hundreds
of miles.

On the other hand, the paths of many fainter,
more common meteors have been determined accurately from routine photographic surveys. Two
specially designed, high-speed, wide-angle *meteor
cameras* are placed many miles apart, and are
directed to the same region of the sky. When a
meteor chances to pass through the common field
of view, it is recorded by both cameras. At each
camera, the exposure is interrupted by a rotating
propeller-type shutter, so that the trail of the
meteor on the film consists of a series of dashes
rather than a continuous streak. When trails of
the same meteor are identified on photographs
obtained simultaneously by the two cameras, the
meteor's elevation and direction of motion through
the atmosphere can be computed by triangulation.
Since the rotation rate of the propeller shutters is
known, the spacing of the dashes that comprise
the meteor's trailed image indicates the speed with
which the image moved across the emulsion and
hence the velocity of the meteor.

Recently, meteors have been detected by radio
and radar as well as visually and photographically.
In the early 1940s it was found that meteors caused
brief interruptions in high-frequency broadcasting
reception; these interruptions took the form of
"whistles," which usually fell quickly in pitch.
Since then, the rate at which meteors occur has
been determined by counting the occurrences of
such whistles.

Meteoroids themselves are too small to reflect
radar waves back to the ground. These waves are
reflected, instead, by the ionized gases that are
formed when meteoroids vaporize in the air. Radar
waves travel with the speed of light, an accurately
known quantity; consequently, the time required

FIG. 17-2   Photograph *(left)* obtained with a Baker Super-Schmidt camera *(right)*. Trails of three meteors are visible; they are interrupted by a timing device in the telescope. *(Harvard Observatory.)*

for the waves to travel from the ground to the meteor and back again indicates the distance or *range* of the meteor from the radar station. The motion of the meteoroid toward or away from the radar station can be detected from the rate at which the range changes (*range rate*), which in turn gives the speed of the meteoroid in the line of sight. Simultaneous observations from two radar stations give the height, speed, and direction of the meteoroid.

It is found that meteoroids produce meteors at an average height of about 60 mi (95 km). The highest meteors form at heights of 80 mi. Nearly all meteoroids completely disintegrate, and their meteors disappear, by the time they reach altitudes of 50 mi. A few meteoroids doubtless skim out of the atmosphere, returning to space before they are completely burned up.

The vast majority of meteoroids have speeds between 8 and 45 mi/sec. The faster ones tend to be brighter and are more likely to leave trains, because of their greater energy upon entry into the earth's atmosphere. On the other hand, the faster-moving particles "burn up" at higher altitudes and are less likely to survive their flights and reach the ground.

Spectra of some meteors have been obtained by photographing them through cameras equipped with objective prisms (Section 11.6f). It is found that more than 99 percent of the light of meteors comes from emission lines, which indicates that meteoric light is produced by glowing gases rather than by incandescent solid particles. The spectral lines present show that the luminous vapors are at high temperatures — in the thousands of degrees Kelvin. Emission lines due to calcium and iron are usually present. Lines of the elements manganese, silicon, aluminum, magnesium, sodium, nitrogen, hydrogen, ionized iron, and calcium are also found.

17.2 ORBITS OF SPORADIC METEOROIDS

A knowledge of the speeds and directions of meteors in the earth's atmosphere enables us to calculate the orbits of the meteoroids in space before they encountered the earth. To determine the *heliocentric* orbit of a meteoroid (that is, its motion with respect to the sun), it is necessary to correct for (1) that component of its velocity relative to the earth that is produced by the earth's orbital motion of about 18½ mi/sec, (2) the acceleration of the meteoroid toward the earth that is produced by the earth's gravitational attraction, and (3) that component of its relative velocity that arises because of the earth's rotation. The latter two contributions to the meteoroid's velocity can range from 0.5 to 2.7 mi/sec.

(a) Orbits of Fireballs

Meteoroids that produce fireballs are relatively rare. Nevertheless, more than 100 have been photographed by the Prairie Network Meteoroid Patrol of the Smithsonian Astrophysical Observatory. The orbits of those meteoroids producing fireballs are found to lie close to the plane of the ecliptic and have small eccentricities, and the revolution of the particles is *direct* — that is, from west to east, as is the revolution of the planets. These meteoroids are thought to have orbits similar to those of the minor planets.

Fireball-producing meteoroids usually collide with the earth after catching up with it from the "rear," for they are moving in the same direction about the sun as the earth, but at greater speeds. Some, however, have smaller heliocentric speeds than the earth's, and we overtake them. Their speeds with respect to the earth, in any case, are relatively small, and they survive their passage through the earth's atmosphere down to comparatively low altitudes before becoming hot enough to vaporize. The fact that fireballs are formed at low altitudes partially accounts for their unusual apparent brilliance.

(b) Orbits of Typical Sporadic Meteoroids

For the vast majority of typical meteoroids — virtually all the thousands of nonshower objects whose orbits have been determined by photographic or radar methods — there seems to be no preference for the ecliptic plane. Like the comets, they approach the earth from all directions. Also, like the comets, most of them are found to be moving at speeds very near the velocity of escape from the solar system; that is, they travel on near-parabolic heliocentric orbits.

At the earth's distance from the sun, the parabolic velocity, or velocity of escape from the sun, is about 26 mi/sec (see Section 5.4). The velocity with *respect to the earth* of a particle moving on a parabolic orbit will be the vector sum of its velocity of 26 mi/sec and the earth's orbital velocity of 18½ mi/sec. If a meteoroid overtakes the earth from the west, it approaches the earth at a speed of 26 *minus* 18½, or about 7½ mi/sec; if it meets the earth from the east, it approaches us at 26 *plus* 18½, or about 44½ mi/sec. The circumstance that most meteors have geocentric speeds in the range 8 to 45 mi/sec is a consequence of the fact that the particles producing them have near-parabolic orbits.

The reason meteors are more frequently seen, and tend to appear brighter, in the hours after midnight than in the hours before is illustrated in Figure 17-3. In the morning hours we are on the side of the earth which is turned in the direction of its orbital motion; consequently, the meteors we see then are due either to particles that run into us from the east at high relative speeds, or that we overtake. In the evening hours we are on the side of the earth which is turned away from the direction of its orbital motion; consequently, the meteors we see then arise from particles that overtake us at low relative speeds. Because the brightness of a meteor depends strongly on the speed of its meteoroid, we are more likely to see meteors during the morning hours.

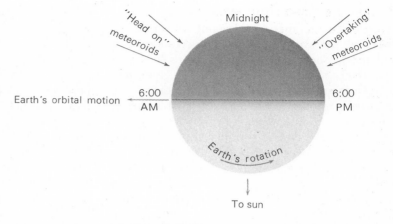

FIG. 17-3 Meteors seen after midnight are produced by particles approaching the earth from the east.

(c) Membership in the Solar System

The vast majority of meteoroids are believed to be permanent members of the solar system. The argument is exactly the same as that for the solar-system membership of comets (Section 16.3a). If meteoroids were chance intruders from interstellar space, they would have hyperbolic orbits. Only a very few meteoroid orbits have been observed to be slightly hyperbolic, and these orbits may have been altered from parabolic ones as a result of perturbations produced by other planets. Like the orbits of comets, the apparent parabolic orbits that we observe for meteoroids must actually be the ends of very long ellipses. At most, less than 1 percent of all observed meteoroids can be interstellar visitors.

17.3 METEOR SHOWERS

Several times each year meteors can be seen with much greater than average frequency. These *meteor showers* occur when the earth encounters swarms of particles moving together through space. Many such swarms of particles that the earth intercepts at regular intervals are known. Showers produced by them are predictable. On rare occasions the earth unexpectedly encounters swarms of particles that produce spectacular meteor displays.

(a) Shower Radiants

Unlike sporadic meteors, which seem to come from any direction, meteors belonging to a shower all seem to radiate or diverge away from a single point on the celestial sphere; that point is called the *radiant* of the shower. Recurrent showers are named for the constellation within which the radiant lies or for a bright star near the radiant.

The phenomenon that shower meteors seem to diverge away from a common point is easily explained. The meteoroids producing a meteor shower are members of a swarm; they are all traveling together in closely spaced parallel orbits about the sun. When the earth passes through such a swarm, it is struck by many meteoroids, all approaching it from the same direction. As we, on the ground, look toward the direction from which the particles are coming, they all seem to diverge from it. Similarly, if we look along railway tracks, those tracks, although parallel to each other, seem to diverge away from a point in the distance (see Figure 17-4).

Actually, a meteor radiant is not a perfectly sharp point. The meteors seem to radiate away from a small region of the celestial sphere. In some showers, that region is very small — as little as 3′ in diameter. In other showers, the radiant may be as large as 1° across. The size of the radiant

rn, photographed with the 200-inch telescope.
unt Wilson and Palomar Observatories.)

Comet Humason (1961a), photographed with
the 48-inch Schmidt telescope.
(Mount Wilson and Palomar Observatories.)

Comet Ikeya-Seki, photographed by
J. B. Irwin in Chile, October, 1965.

Section through Neenach, California stoney meteorite, showing
flecks of nickel-iron metal. *(From the collection of Ronald Oriti, Griffith
Observatory; photograph by Ivan Dryer.)*

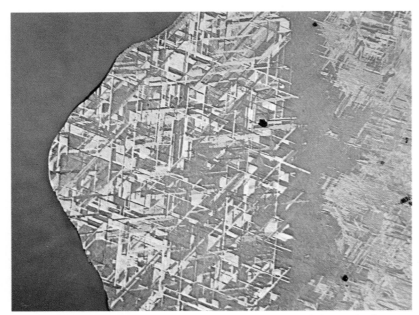

Slice of Kamkas iron meteorite which has been polished and then etched
with dilute nitric acid to show the criss-cross Widmanstätten figures. *(From the
collection of Ronald Oriti, Griffith Observatory; photograph by Ivan Dryer.*

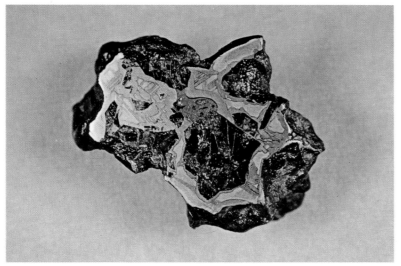

Stoney-iron meteorite, Glorieta Mountains, N.M. The specimen has been polished and etched to show the metallic structure. *(From the collection of Ronald Oriti, Griffith Observatory; photograph by Ivan Dryer.)*

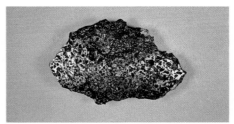

Albin, Wyoming stoney meteorite. *(From the collection of Ronald Oriti, Griffith Observatory; photograph by Ivan Dryer.)*

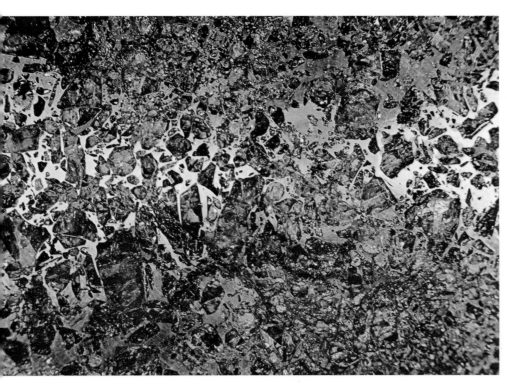

Polished slice of Albin, Wyoming stoney meteorite. This type of meteorite consists of nickel-iron metal with inclusions of the green mineral olivine. *(From the collection of Ronald Oriti, Griffith Observatory; photograph by Ivan Dryer.)*

A thin section of the Lucerne Valley stoney meteorite, polished to a thickness of 0.001 inch.
photograph was made in polarized light transmitted by the section. Note the rounded grains, or *chondrules*, found only in st
meteorites. *(From the collection of Ronald Oriti, Griffith Observatory; thin section cut by John DeGrosse, UCI*

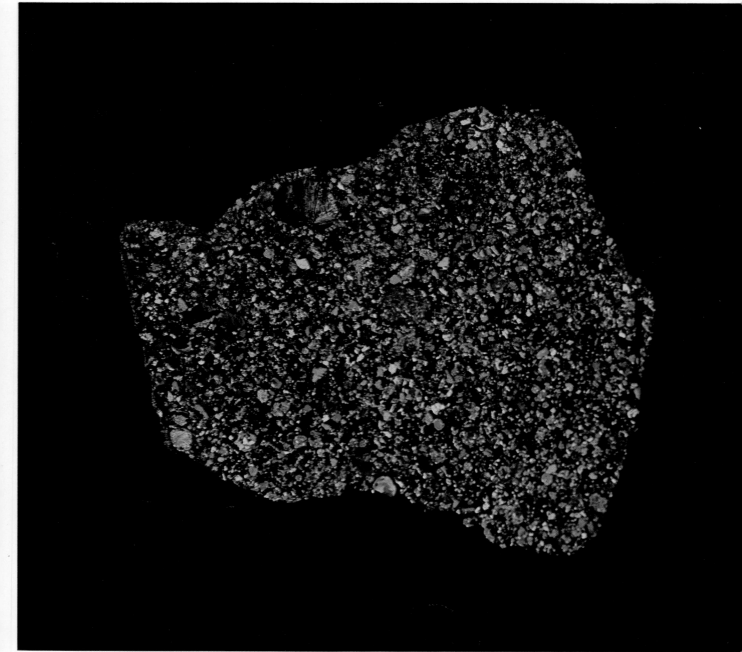

depends on how closely packed the swarm of particles is. The denser, or more closely packed the swarm, the more spectacular the meteor shower it produces. The spectacle also depends on how closely the earth approaches the densest part of the swarm.

Many meteors classed as sporadic may actually belong to showers that are so inconspicuous that they go unnoticed. Scarcely a night passes that a patient observer cannot see three or more meteors moving in the sky in directions away from a common point. Many of these may be members of minor showers.

(b) Association of Showers and Comets

The direction of the radiant of a meteor shower indicates the direction in which the swarm of particles that produces it is moving through space with respect to the earth. The velocity of the swarm is found, as explained in Section 17.1, from the velocity of the meteors. These are enough data to specify completely the orbit of the swarm.

On about August 11 of each year, the earth passes through a swarm of particles that approach from the direction of Perseus. In 1866 it was observed that the particles producing this *Perseid shower* travel in an orbit that is almost identical to that of Comet 1862 III. It was then realized that those meteoroids encountered each August are debris from the comet that has spread out along the comet's orbit.

Subsequently, it has been found that the elements of the orbits of many other meteoroid swarms are similar to those of the orbits of known

FIG. 17-4   A meteor shower radiant (a) and the apparent divergence or convergence of parallel lines (b).

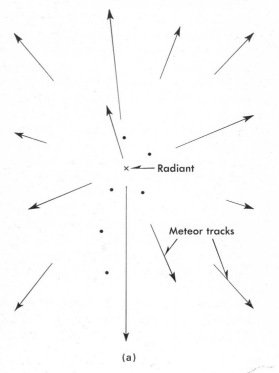

(a)

(b)

FIG. 17-5 Photograph showing trails of many meteors during the shower of October, 1946. Note the divergence of the trails from a radiant (off the field to upper right). Other streaks are star trails.

the Perseid shower. Consequently, the earth meets about the same number of particles each time it passes through the orbit of the swarm, and the Perseid shower is of about equal intensity each year. More often, a swarm will be bunched up [Figure 17-6(b)], and we will experience a spectacular display only on those occasions when the earth passes through the orbit of the swarm at just the right time to complete a rendezvous with the meteoroids as they reach that same point in their revolution about the sun. A good example is provided by the Leonid meteors in the last century, when the earth met the debris of Comet 1866 I in 1833 and again in 1866 (after an interval of 33 years — the period of the comet). Those were among the most spectacular showers ever recorded. As many as 200,000 meteors could be seen from one place within a span of a few hours. The last good Leonid shower was on November 17, 1966, where in some southwestern states, up to 140 meteors could be observed per second. Even in such dense swarms the individual particles of the swarm are separated by distances of 20 mi or more; in most meteoroid swarms, the particles are more than 100 mi apart.

The best meteor shower that can be depended on at present is the Perseid shower, which appears for about 3 days near August 11 each year. In the absence of bright moonlight, meteors can be seen with a frequency of about one per minute during a typical Perseid shower. It is estimated that the total combined mass of the particles in the Perseid swarm is near 5×10^8 tons; this gives at least a lower limit for the original mass of Comet 1862 III. The orbit of this comet (and hence of the Perseids) was nearly perpendicular to the ecliptic plane; thus the meteoroids are but little perturbed by the planets, accounting for the reliability of the Perseid shower.

One spectacular shower of recent decades was the *Draconid* shower that reached maximum display on October 9, 1946. On that date the earth

comets. Not all meteor showers have yet been identified with individual comets, but it is presumed that all showers have had a cometary origin. These swarms of debris, provided by the gradual disintegration of comets, give further evidence of the flimsy nature of comets.

Twice each year the earth passes through a swarm of particles moving in the orbit of Halley's comet. Debris from that famous comet gives us the *Eta Aquarids* in May and the *Orionids* in October.

Sometimes, if a meteoroid swarm is old enough, its particles are strewn more or less uniformly along their entire orbit [Figure 17-6(a)]; an example is the swarm of particles that produces

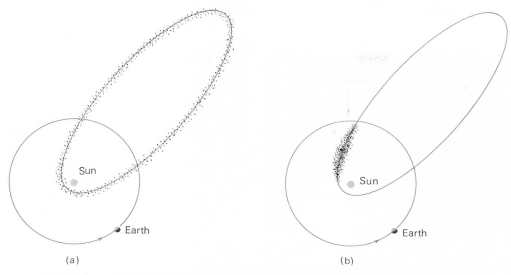

FIG. 17-6 Meteoroids in a swarm may be strewn more or less uniformly along their orbit (a), or bunched up (b).

reached the point that Comet Giacobini-Zinner had passed 15 days earlier. Debris from the comet produced meteors that could be counted from points in the southwestern United States at a rate of two per second, even though the moon was full at the time.

The characteristics of some of the more famous meteor showers are summarized in Table 17.1. Other, spectacular meteor showers can occur, however, at almost any time, just as some bright comets appear unexpectedly.

(c) Daytime Showers

We have seen that radar waves are reflected by the column of ionized gases left in the wake of a meteor. Now, some wavelengths of radio waves, those that lie in the range 4 to 5 m, are reflected efficiently only if they strike the ionized gas column nearly broadside (waves shorter than about 4 m usually are not reflected, except by the very densest ion clouds). Thus, meteors that belong to a shower can be detected by radar operating at a wavelength of 4 to 5 m only if the radar beam is

TABLE 17.1 Characteristics of Some Meteor Showers

SHOWER	DATE OF MAXIMUM DISPLAY	VELOCITY (MI/SEC)	ASSOCIATED COMET	PERIOD OF COMET (YR)
Quadrantid	Jan 3	27	—	7.0
Lyrid	Apr 21	30	1861 I	415
Eta Aquarid	May 4	37	Halley	76
Delta Aquarid	Jul 30	27	—	3.6
Perseid	Aug 11	38	1862 III	105
Draconid	Oct 9	15	Giacobini-Zinner	6.6
Orionid	Oct 20	41	Halley	76
Taurid	Oct 31	19	Encke	3.3
Andromedid	Nov 14	10	Biela	6.6
Leonid	Nov 16	45	1866 I	33
Geminid	Dec 13	23	—	1.6

directed at nearly right angles to the direction of the shower radiant (the direction from which the meteoroids are coming). If the sky is scanned by radar, the maximum amount of energy will be reflected back from meteors along a great circle in the sky that is 90° from the shower radiant. Radar observations of shower meteors, therefore, can locate the radiant, even though the shower occurs in daylight when the meteors cannot be observed visually. Four daylight meteor showers have been found, thus far, which, if they were visible at night, would be at least as spectacular as those listed in Table 17.1.

17.4 FORMATION OF A METEOR

As a meteoroid plunges into the earth's atmosphere at a speed of many miles per second, it undergoes many collisions with air molecules. The impinging molecules penetrate the meteoroid and chip off pieces of it. The surface of the meteoroid heats up, and the dislodged particles vaporize. The whole process is called *ablation.*

The gases from the meteoroid and the air immediately around it heat sufficiently so that the atoms comprising the gases emit light in the form of bright emission lines. The *meteor* is the region of glowing gas; it may be anywhere from a few feet to several hundred yards in diameter. A meteoroid in space, before entering the atmosphere, must be fairly cold; by the methods outlined in Section 12.2, we find its temperature should be about 40°F. On the other hand, the emission lines observed in the light of meteors show that the luminous gases involved must have temperatures of thousands of degrees.

Air resistance and the ablation process rob the meteoroid of energy and slow it down. The rate at which it decelerates depends on its original speed, its mass, and the density of air. Thus, as a by-product of the study of meteors, we learn something about the density of the upper atmosphere.

(a) Masses of Meteoroids

Minimum masses of meteoroids are estimated from the amount of light they produce in their meteors. From the brightness of the photographic image of a meteor trail, the amount of light energy received from the meteor is measured. Then, from the height and distance of the meteor, the amount of luminous energy it radiated during its brief existence can be calculated. The original source of this energy is the kinetic energy (the energy of motion) of the meteoroid as it plunged into the atmosphere.

The kinetic energy of a moving particle is one half the product of its mass and the square of its speed, that is,

$$KE = \tfrac{1}{2}mv^2.$$

FIG. 17-7 Full-scale model of the Goose Lake Meteorite. The original weighs 2573 lb. The white card in the photograph is 9 inches long. *(Griffith Observatory.)*

We have already discussed how the velocities of meteoroids are determined (Section 17.1b). Thus, if we assume that the kinetic energy of the meteoroid was completely converted into luminous energy of the meteor, we can use the above formula to compute the mass of the particle before it entered the earth's atmosphere. Because some of the particle's kinetic energy will have been dissipated by other means than conversion to visible light, its actual mass must be greater than that calculated for it.

The mass of a meteoroid producing a typical bright meteor is about ¼ gm (about 1/100 ounce). A very few meteoroids, however, have masses greater than a few grams. Many visible meteors are produced by particles with masses of only a few milligrams. Most meteoroids, therefore, are much smaller than pebbles.

Meteoroids of low density are more easily slowed down in the air and may survive to relatively lower altitudes before becoming meteors. The lower than average heights of some meteors, especially those in showers, indicate that the corresponding meteoroids may have densities much less than that of water. Meteoroids that originate from cometary debris, therefore, may be very porous.

The rate at which the earth accumulates meteoritic material is not accurately known. However, the total mass of meteor-producing meteoroids that enters the atmosphere each day is probably a few tons.

17.5 FALLEN METEORITES

Occasionally, a meteoroid survives its flight through the atmosphere and lands on the ground; this happens with extreme rarity in any one locality, but over the entire earth probably more than a thousand meteorites fall each year. Fallen meteorites have been found for centuries; it was 1800,

FIG. 17-8   Tektites from Thailand. The largest is about 3 inches long. *(Ronald Oriti, Griffith Observatory.)*

however, before their association with "shooting stars" was appreciated.

Today, fallen meteorites are found in two ways:

(1) Sometimes bright fireballs are observed to penetrate the atmosphere to very low altitudes. A search of the area beneath the point where the fireball was observed to burn out may reveal one or more remnants of the meteoroid. *Observed falls,* in other words, may lead to discoveries of fallen meteorites.

(2) Unusual-looking "rocks" are occasionally discovered that turn out to be meteoritic. These are termed "finds." Now that the public has become "meteorite conscious," many suspected meteorites are sent to experts each year. The late F. C. Leonard, a specialist in the field, referred to these objects as "meteorites" and "meteorwrongs." Genuine meteorites are turned up at an average rate of about 25 per year.

(a) Types and Compositions of Meteorites

Meteorites can be classified in three general groups†:

(1) *Irons* (*siderites*) — alloys of metals. From 85 to 95 percent of the mass of siderites is iron; the rest is mostly nickel.

(2) *Stony irons* (*siderolites*) — relatively rare meteorites that are about half iron and half "stony" silicates. The stony materials generally decompose more rapidly than the iron, leaving the latter in separated fragments.

(3) *Stones* (*aerolites*) — composed mostly of silicates and other stony materials. Aerolites contain about 10 to 15 percent iron and nickel in metallic flakes.

Stones are probably the most common kind of fallen meteorites. Ninety-three percent of those meteorites obtained from observed falls are stones. However, the aerolites gradually break up under erosive action. Moreover, they resemble terrestrial rocks, and generally only experts can ascertain their meteoritic nature from the iron pieces embedded in them. Consequently, about two thirds of the total number of recovered meteorites are irons. (Irons are seldom found, however, in those civilized regions of the world where for centuries man has made use of iron for tools.) When irons are cut, polished, and etched, about 80 percent show a characteristic crystalline structure (*Widmanstätten figures*) that makes their identification as meteorites certain.

FIG. 17-9 An iron meteorite cut and polished to show the Widmanstätten figures. *(Griffith Observatory.)*

So far as is known, all recovered meteorites are remnants of *sporadic* meteors. There has never been an observed fall from a *shower* meteor. The spectra of shower meteors, however, indicate that they contain some of the same elements as stones.

The chemical composition of many meteorites has been determined by laboratory analysis. It is found that meteorites contain the same common chemical elements that we find in the crust of the earth. When allowance is made for the fact that stones are at least 4 to 9 times more common than irons, the following abundances (by weight) of elements are derived: iron, 30 to 40 percent; oxygen, 30 percent; silicon, 15 percent; magnesium, 12 percent; sulfur and nickel, 2 percent each; calcium, 1 percent; and small amounts of carbon, sodium, aluminum, lead, chlorine, potassium, titanium, chromium, manganese, cobalt, copper, and other elements.

Meteorites often show evidence of ablation that occurred during their atmospheric flights. A skin-deep layer on the surface of an iron meteorite may show evidence that the surface has been molten. Meteoroids pass through the air too quickly, however, for their interiors to heat up. By the time a meteoroid is slowed, by air friction, to 2 mi/sec, the molten surface of the iron hardens;

---

†*Tektites,* somewhat rounded glassy bodies of indefinite shape, have been found in Indonesia, Australia, and elsewhere; some investigators suspect them to be of extraterrestrial origin, but their identity as true meteorites has never been established. They show evidence of having been molten at high pressure, and having undergone rapid cooling. They contain small spheres of iron and nickel, and schreibersite, a mineral known only in meteorites. The latter is believed to be produced by melting of the surface layer of crater-producing meteorites.

when its speed has slowed to 1 mi/sec, it stops glowing altogether. Popular opinion to the contrary, meteorites are not "red hot" when they strike the ground; meteorites that have been picked up just after they have fallen are generally cool enough to handle. The surfaces of the stones may also become molten during their atmospheric passage and turn glassy. Both stones and irons are sometimes ablated to conical shapes.

(b) Ages of Meteorites

Natural radioactivity provides a means of determining the ages of meteorites just as it provides a means of determining the ages of rocks in the earth's crust (Section 13.6d). The heaviest kinds of naturally occurring atoms are unstable and spontaneously split up into lighter atoms. In a sample of the most common kind of uranium,† for example, one half of all the uranium atoms will decay in a period of 4.5 billion years. In a second interval of 4.5 billion years, half of the remaining uranium atoms will break up, leaving only one fourth of the original number; after another 4.5 billion years, only one eighth of the uranium is left; and so on. The period of 4.5 billion years is called the half-life of uranium. The half-lives of the most abundant forms of thorium and radium are 13.9 billion years and 1620 years, respectively.

The end products of this natural radioactivity are helium (which is usually trapped in the ore containing the original radioactive element) and forms of lead. Thus, the relative concentrations in a meteorite of, say, uranium, helium, and the appropriate kind of lead indicate how long the radioactive decay process has been going on, and hence the age of the meteorite. The derived age, of course, applies only to the age of the solid particle in its present mineral state. The ages of meteorites are found to range from 3 to 4.7 billion years — comparable to the age of the earth.

---

†The isotope U^{238}.

(c) Largest Recovered Meteorites

The largest meteorite ever found on the earth is *Hoba West,* near Grootfontein, South West Africa. It has a volume of about 9 cubic yards and an estimated mass of more than 50 tons. The largest meteorite on display in a museum has a mass of 34 tons; it was found by Peary in Greenland in 1897 and is now at the American Museum of Natural History in New York. The largest fallen meteorite found in a single piece in the United States was discovered in a forest near Willamette, Oregon, in 1902; it has a mass of 14 tons. (The discoverer spent about 3 months hauling the meteorite to his own property, where he put it on display for an admission price. Among those interested in the new exhibit were the attorneys of the Oregon Iron and Steel Company, owner of the land on which the meteorite was found. After litigation, it was decided that the company was the rightful owner of the object.)

17.6 METEORITE FALLS

A total of 1800 meteorite falls are listed in the *Hey Catalogue of Meteorites;* of these, 785 are falls, and 1015 are finds.

(a) Recent Spectacular Falls

The two most spectacular meteorite falls in recent times took place in Siberia. On June 30, 1908, a brilliant fireball occurred that was seen in broad daylight. Remnants of its meteoroid landed in a forest near the Tunguska River in central Siberia. The impact produced shock waves that were registered on seismographs in distant Europe, and also set up an air-compression wave that was reported detected in Europe. Trees were seared of their branches and were felled radially from the impact over an area more than 20 mi in radius. About 1500 reindeer were killed, and a man standing on the porch of his home 50 mi away was knocked unconscious. No pieces of the original particle have been recovered, but its original mass before it entered the earth's atmosphere is estimated to have been about 10^5 tons.

A second spectacular Siberian fall occurred on February 12, 1947, near Vladivostok. The approaching fireball was described as "bright as the sun." The impact produced 106 craters ranging in size up to 30 yards across and 10 yards deep. Trees were felled radially around each of the large craters. The entire region covers nearly 2 square miles. More than 5 tons of iron meteorite fragments have been recovered in the area.

There is no authenticated case of the killing of a human being by a meteorite fall; however, there have been some close calls. On September 29, 1938, a woman in an Illinois town heard a crash in the back yard. Later, it was found that a meteorite had pierced the roof of a garage 50 ft away. A car was in the garage at the time, and the meteorite was found buried in the cushion of the car seat. In 1954 an Alabama woman was struck and injured by a falling meteorite — the only modern case of injury.

(b) Meteorite Craters

It is estimated that in each century several meteorites strike the earth with enough force to produce craters more than 10 yards across. To produce a crater of this diameter requires the fall of a meteorite with a mass of about 10 tons; such a body would be a yard or so in diameter and could be photographed with telescopes when it was as far away as the moon. It would differ from a minor planet only in size. A large meteorite produces a crater larger than itself, for when it strikes the ground, its kinetic energy is dissipated with explosive violence.

The first crater to be discovered, and the most famous definitely known to be meteoritic, is the Barringer Meteorite Crater, near Winslow, Arizona. The crater was shown to be meteoritic as a result of research instigated by the Barringers (owners

FIG. 17-10   The Barringer meteorite crater in Arizona. *(American Meteorite Museum.)*

of the land) early in the twentieth century. The crater is 4200 ft across and 600 ft deep, and its rim rises 150 ft above the level of the surrounding ground. In the area over 30 tons of iron meteorite fragments have been found, some buried near the crater and many more scattered at distances up to 4 miles around. Sizable chunks of iron can still be found shallowly buried in the vicinity of the crater. Drillings have been made into the earth beneath the crater floor in search of a main body of the meteorite, but none has been recovered. Whether a large mass of the original meteorite remains intact is controversial. Some experts believe that a large mass is buried beneath the south rim of the crater; others are of the opinion that the meteorite blew up completely into small pieces when it exploded, forming the crater. The age of the crater is not accurately known. Estimates run between 5000 and 75,000 years; it cannot be much older and have survived erosion.

A larger crater is the New Quebec Crater (formerly Chubb Crater) in Quebec. It was discovered in 1950 on aerial photographs of the region. It is similar in appearance to the Barringer Crater but is about twice as large, being more than

FIG. 17-11   Region of the lunar crater Goclenius, photographed by
the crew of Apollo 8 in December, 1968. *(NASA.)*

2 mi in diameter. It is currently filled with water and forms a lake in solid granite. No trace of meteoritic fragments has been found in the area, and it is not absolutely certain that the crater is meteoritic; however, the absence of volcanic activity in the region suggests that a meteoritic origin of the crater is likely.

In Algeria there is a large crater, somewhat over a mile across, that resembles meteoritic craters. Again no fragments of meteorites have been found, but there is no evidence of volcanism, either. There are more than a dozen other craters about the world that are believed to be meteoritic. Those craters that were formed more than a few tens of thousands of years ago, of course, have long since eroded away.

Older "fossil" craters must exist. The largest suspected fossil meteorite crater is in South Africa and is about 20 mi across. In size it is reminiscent of the lunar craters. If it is meteoritic, it has lost much of its original appearance. We do not yet know enough about meteorite craters to identify such fossils with certainty.

17.7 MICROMETEORITES

Many very tiny meteorites strike the earth each day. Those that are only a few microns in diameter (1 micron is 10^{-4} cm, or about 4×10^{-5} in.) will be slowed in the air before they have a chance to heat up; they will eventually settle to the ground. These particles, too small to make meteors, are called *micrometeorites*.

There are two kinds of evidence for the existence of micrometeorites:

(1) Many small particles, rich in iron and also containing silicon and magnesium, can be collected from the ground, from roofs, from rainwater, and even from the ocean floors. Although the fact has not yet been proved, these particles may be meteoritic; that is, they may be micrometeorites.

(2) Impacts of small particles have been detected by space vehicles that are equipped with microphones or other types of micrometeorite detectors. These particles must be meteoritic, for they are encountered in space.

In the vicinity of the earth, micrometeorites may number from 100 to 1000 per cubic mile, but this estimate is very uncertain.

The total daily accretion of micrometeorites by the earth is estimated at 50 to 100 tons.

17.8 THE INTERPLANETARY MATERIAL

We see, then, that the space between the planets contains a vast number of micrometeorites. These particles comprise a distribution of *interplanetary dust*. Micrometeorites are a few microns or more in diameter. Particles less than 1 micron in size are "blown" out of the solar system by radiation pressure from the sun, just as small particles are blown out of comets in the form of comet tails (Section 16.4e). Particles the size of micrometeorites are not blown away but revolve about the sun like tiny planets, which, indeed, they are. Their orbits are gradually diminishing in size, however, and they consequently spiral inward toward the sun, because of a perturbative effect known as the *Poynting-Robertson effect*.

(a) *The Poynting–Robertson Effect*[*]

For particles the size of micrometeorites, the sun's attractive gravitational force exceeds the repulsive force of the sun's radiation pressure. Like the planets, therefore, the micrometeorites revolve about the sun in Keplerian orbits. If the force of radiation pressure acting on such a particle were *exactly radial* (away from the sun), its only effect would be to reduce slightly the sun's attraction, resulting in a somewhat increased period of orbital revolution. However, the direction of the force of radiation pressure is not exactly radial but rather has a small component in a direction opposite to that of the particle's motion.

Suppose one runs through the rain. Even if the rain is falling vertically, it appears to strike the runner's face

FIG. 17-12 Zodiacal light behind telescope dome. The time exposure also shows star trails. *(Yerkes Observatory.)*

obliquely, because *relative to him* the raindrops have a horizontal component of motion — namely, a velocity equal and opposite to his velocity with respect to the ground. Similarly, to objects revolving about the sun, photons do not appear to come from an exactly radial direction, away from the sun, but have a slight component of velocity in a direction opposite to that of the objects' own motions. Thus, radiation pressure produces a slight "backward" force upon them. The effect of radiation pressure is negligible on particles of large mass (like the planets), but small particles, such as micrometeorites, are appreciably perturbed. The "backward" force acts like a "drag" on the orbital motion of such a particle, first making its orbit more and more circular and then gradually causing the orbit to diminish in size until the particle ultimately spirals into the sun. A particle 1 mm in diam-

eter that originates in the region of the minor planet belt (between the orbits of Mars and Jupiter) spirals into the sun in only 60 million years. A particle even as large as 10 cm in diameter at the earth's distance from the sun will spiral into the sun in less than 1 billion years. The fact that we find small particles around the earth is evidence that they are either newly formed or have newly arrived in our part of the solar system.

Attention was first directed to the effect by J. H. Poynting in 1903; it was confirmed by a rigorous application of relativistic theory by H. P. Robertson in 1937. It is thus called the *Poynting-Robertson effect.*

(b) Further Evidence of Interplanetary Material

In addition to micrometeorites, or interplanetary dust, space vehicles have found evidence for interplanetary gas. Large numbers of charged atoms (ions) have been encountered. It is now well established that these ions have been expelled from the sun, and this outflow of gas is known as the *solar wind.* In the neighborhood of the earth, the ion density is from 1 to 10 particles per cubic centimeter. Further evidence of the solar wind is provided by its effect on comets in producing the straight tails.

There is also additional evidence for the interplanetary dust (micrometeoritic material): the *zodiacal light* and the *gegenschein.*

(c) Zodiacal Light

The zodiacal light is a faint glow of light along the zodiac (or ecliptic). It is brightest along those parts of the ecliptic nearest the sun and is best seen in the west in the few hours after sunset or in the east in the hours before sunrise. Under the most favorable circumstances, the zodiacal light rivals the Milky Way in brilliance. It is sometimes called the "false dawn" because of its visibility in the morning hours before twilight actually begins.

The zodiacal light has the same spectrum as the sun, which shows it to be reflected sunlight. Gas molecules and atoms cannot be numerous enough in space to scatter enough sunlight to contribute appreciably to the zodiacal light. Not only are molecules inefficient scatterers of light, but what light they do scatter (that is, reflect helter-skelter) is mostly blue and violet light of short wavelength — the blue daylight sky comes from the scattering of sunlight by air molecules. The present interpretation of the zodiacal light, then, is that the interplanetary dust is concentrated most heavily in the plane of the ecliptic. The dust reflects enough sunlight to produce the faint glow along the zodiac.

The outer part of the sun's corona — the Fraunhofer or F corona — has been observed as far as 9° from the sun (see Chapter 24). The F corona has the same color and spectrum as the zodiacal light. It seems likely, therefore, that the zodiacal light is simply an outer extension of the F corona and that this outer corona is sunlight scattered by particles.

(d) The Gegenschein

The *gegenschein,* which means "counterglow," is a faint glow of light, centered on the ecliptic, which is exactly opposite the sun in the sky. Its angular size is from 8° to 10° by 5° to 7°. It is much more difficult to see than the zodiacal light, but it can be measured photoelectrically, and it has also been photographed. Like the zodiacal light, the gegenschein appears to be reflected sunlight.

Several explanations have been advanced for the gegenschein. Three of the more widely mentioned hypotheses follow:

(1) It has been shown as a special solution to the three-body problem that a body, under the mutual attraction of the sun and earth, could have an unstable orbit about the sun in a position exactly 930,000 mi in a straight line beyond the earth from the sun (Section 6.1b). It has been suggested that the gegenschein is light reflected from a collection of small particles in such an orbit.

(2) Solar radiation pressure can carry some of the gases of the earth's outer atmosphere radially away from the sun, forming a very diffuse "tail" of the earth, not unlike the tail of a comet. There is no definite evidence for a tail on any planet, but such diffuse gases would be very difficult to observe except when viewed "end on" through the length of the tail. It has been suggested that gases streaming away from the earth in this way might reflect enough sunlight to account for the gegenschein.

(3) Those solid particles opposite the sun in the sky would be viewed at the "full" phase, like the full moon, and would, therefore, reflect more light to us than particles in other directions that are viewed at gibbous or crescent phases. It is suggested that the gegenschein is simply a portion of the zodiacal light where the particles reflecting sunlight are seen at a more favorable illumination angle.

Any of the above hypotheses for the gegenschein requires the existence of interplanetary material.

FIG. 17-13 The Gegenschein (patch of light just to the left and below center) and the Milky Way, photographed by Osterbrock and Shapeless with the wide-angle Greenstein-Henyey camera. *(Yerkes Observatory.)*

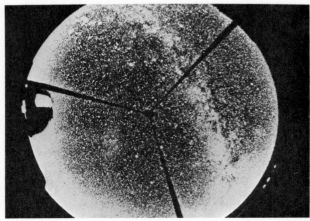

17.9 THE ORIGIN OF METEORITIC MATERIAL

There is growing evidence that many meteoroids are products of the breakup of larger bodies. For example, as we have seen, small particles cannot remain in the solar system indefinitely without being drawn into the sun by the Poynting-Robertson effect.

In size, the lesser bodies of the solar system seem to range continuously from minor planets through crater-producing meteorites, fireball-producing meteorites, ordinary sporadic meteoroids, on down to micrometeorites. The numbers of particles of various sizes appear to fit the distribution we would expect to result from some breakup process like rock crushing. Collisions between a few larger bodies would be extremely rare and unlikely; however, once a collision occurred and formed a larger number of smaller bodies, subsequent collisions between them would be more likely. As the breakup continued, collisions would become increasingly frequent, accelerating the fragmentation.

Studies of the chemical composition of various meteorites by J. Wasson and his associates at UCLA have shown that abundance of, for example, gallium and germanium in iron meteorites are sometimes found to be strikingly identical in meteoritic samples collected in very different parts of the world. Here is strong evidence that several different meteorites can have originated from the same body — that is, that fragmentation is indeed taking place.

It is tempting to hypothesize that the iron meteorites come from the ironlike core and the stones from the rocklike mantle of a former planet. It is difficult, however, to understand how such a planet could be shattered. Moreover, the present total mass of minor planets and meteoroids combined is very small compared to the mass of any known planet. On the other hand, it is possible that several bodies intermediate in size between minor planets and planets produced the small fragments we observe today.

Whereas the minor planets, the larger meteoroids, and possibly the micrometeorites may have originated from a few larger subplanetary bodies, we recall that the majority of the particles that produce observable meteors have orbits more characteristic of those of the comets and, moreover, that shower meteors are definitely associated with the debris from comets. Much remains to be learned concerning the origin of these small bodies. The best we can say is that some meteoroids seem to have an origin associated with that of comets, whereas others may have an origin in common with that of the minor planets.

Exercises

1. Two meteor cameras are located exactly 120 mi apart on an east-west line. A particular meteor is recorded by both cameras. The westernmost camera showed its trail to begin exactly 45° above the east point on the horizon, while the easternmost camera showed its trail to begin exactly 45° above the west point on the horizon. At what altitude did the meteor trail begin?

2. Suppose meteoroids all moved in nearly circular orbits. What then would be the range of velocities we should observe for them, relative to the earth?

3. Comets that have been associated with meteor showers are all periodic comets. Why do you suppose showers have not been identified with comets having near parabolic orbits?

4. Why must the perihelion of the orbit of a meteoroid swarm never be greater than 1 AU from the sun for us to see a shower produced by that swarm, and why are the perihelia of such swarms usually closer to the sun than 1 AU?

5. Why is it that a particular shower may be seen only at certain hours of the night, for example, early evening, or the hours before sunrise?

6. Suppose a meteor shower were detected in December, and that the radiant of the shower is in the constellation of Scorpio. How would the shower have been discovered?

7. From the data in Table 17.1 find the semimajor axis of the orbit of the Draconid swarm.

8. Show by a diagram how the earth can encounter a swarm left by Halley's comet twice each year.

9. Two meteoroids of the same mass enter the earth's atmosphere at the same instant. Both produce observable meteors. One is moving 40 mi/sec, and the other is moving 20 mi/sec. Which meteor gives out the more light, and give an estimate of how many times more light it produces than the other?

10. Two meteoroids enter the atmosphere side by side, moving at the same speed. An observer notes that one of the meteors produced is three times as bright as the other. Estimate the relative masses of the two objects.

11. Suppose a micrometeorite has a cubical shape and measures 4 microns on a side. If its density is five times that of water, what is its mass in grams? What is its mass in ounces (there are about 28 gm/ounce)? How many such particles would be required to make up a mass equal to that of the earth?

12. The F corona is much brighter, since it is nearer the sun, than the zodiacal light. Why is it, then, that the F corona is generally observed only at times of total solar eclipses, whereas the zodiacal light is visible on many clear evenings?

CHAPTER 18

Triangulation of Space

Nearly all measurements of astronomical distances, as well as of distances on the earth, depend directly or indirectly on the principle of triangulation. Some of the concepts of triangulation were known to the Egyptians; the art was developed by the Greeks.

18.1 TRIANGULATION

Six qualities describe the dimensions of a triangle: the lengths of the three sides and the values of the three angles. It is a well-known theorem in elementary geometry that any three of these quantities in succession around the perimeter of the triangle (for example, two sides and an included angle or two angles and an included side) determine the triangle uniquely.

As an example, suppose that in the triangle ABC (in Figure 18-1) the side AB and the angles A and B are all known. The triangle can then be constructed without ambiguity, for the side AB can be laid out and the lines AE and BD can be drawn at angles A and B, respectively, to the line AB. The two lines intersect at C, which completes the construction.

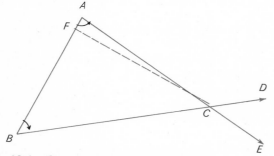

FIG. 18-1   Construction of a triangle with two angles and an included side given.

(a) Application to Surveying

Suppose, in the triangle in Figure 18-1 that the point C represents an inaccessible object — say, a remote mountain peak, or an island in a large river, or the moon. The distance to C can be determined by setting up two observation stations at A and B, separated by the known distance AB. This known distance (AB) is called a *base line*. At station A the angle A is observed between the directions to B and C. At station B, the angle B is observed between the directions to A and C. Enough information is now available to construct a scale drawing of the triangle ABC. The base line (AB) is first laid out on the drawing at some convenient scale. Lines AE and BD are then constructed at angles A and B to the base line, and their intersection is at C. The distance of C from any point on the base line (for example, A, or B, or F) can now be measured in the drawing; since the scale of the drawing is known, the real distances to C are determined.

The preceding two paragraphs described a *geometrical* method for solving a triangle. In practice the triangle can be solved more simply and more accurately by numerical calculation. The solution of triangles by calculation rather than by geometrical construction is the subject of *trigonometry*.

(b) The "Skinny" Triangle

In astronomy we frequently have to measure distances that are very large in proportion to the length of the available base line, and the triangle to be solved is thus long and "skinny." Suppose (Figure 18-2) that it is desired to measure the distance to the moon or a minor planet, located at O.

Two observation stations, *A* and *B*, are set up on the earth. The base line *AB* is known, since the size of the earth and the geographical locations of *A* and *B* are known. From *A*, *O* appears in direction *AS*, in line with a distant point, such as some very remote star. Because of the great distance of that star, observers at *A* and *B* would look along parallel lines to see it; thus from *B* the same star is in direction *BS'*. Observer *B* sees *O*, on the other hand, in direction *BT*, at an angle *p* away from direction *BS'*. This angle, *p*, is the same as the angle at *O* between the lines *AO* and *BO* (because the line *BO* intercepts the parallel lines *AS* and *BS'*). Note that *p* is the difference in the directions of *O* as seen from *A* and *B*; this is the apparent displacement (or parallax) of an object as seen from two different points, (Section 2.2d).

Now the problem is to find the distance to *O*, which lies somewhere along the line *AS*. This can be done in two ways. One way is to determine the angles *OAB* and *OBA* and then determine the distance to *O* by solving the triangle *OAB* as described in Section 18.1a. However, because the sides *AO* and *BO* are so nearly parallel, the two angles would have to be measured with very great precision. It would be very difficult to obtain adequate precision, because at each station (*A* and *B*) it would be necessary to compare the direction of *O* to that of the other terrestrial observation point. An alternative procedure for finding the distance *AO* is to measure the parallax *p* and then calculate how distant *O* must be in order for the base line *AB* to subtend that angle. Such a calculation is illustrated in Section 18.1c.

Because observers at *A* and *B* both see *O* projected against remote stars on the celestial sphere, each can measure *O*'s direction among those stars with great accuracy; the difference between those directions, the parallax, can be found, therefore, with considerable precision. In astronomical practice the greatest emphasis is thus placed on measuring the parallax, rather than the angles at the end

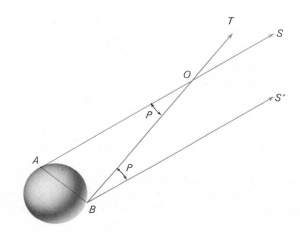

FIG. 18-2   "Skinny" triangle.

of the base line. Even the parallax, however, can be determined only as accurately as the direction *O* can be measured with respect to the background of stars on the celestial sphere. Under the most favorable circumstances parallaxes can be measured to within a few thousandths of a second of arc. The more distant the object, the smaller its parallax and the greater the uncertainty of its value in comparison to its size — that is, the greater its *percentage error,* and the greater the uncertainty in the distance to the object that is derived from its parallax. It is advantageous, therefore, that the base line be made as long as possible, so that the parallax will be as large as possible in comparison to the error of its measurement.

(c) An Example of the Solution of the "Skinny" Triangle*

Suppose it is found that the displacement in direction of an object (at *O*, in Figure 18-3), as viewed from opposite sides of the earth, is the angle *p; p,* then, is the angle at *O* subtended by the diameter of the earth. Imagine a circle, centered on *O*, that passes through points *A* and *B* on opposite ends of a diameter of the earth. If the distance of *O* is very large compared to the size of the earth, then the length of the chord *AB* is very nearly the same as the

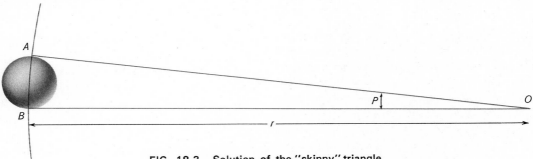

FIG. 18-3 Solution of the "skinny" triangle.

distance along the arc of the circle from A to B. This arc is in the same ratio to the circumference of the entire circle as the angle p is to $360°$. Since the circumference of a circle of radius r is $2\pi r$, we have

$$\frac{AB}{2\pi r} = \frac{p}{360°}.$$

By solving the above equation for r, the distance to O, we find

$$r = \frac{360°}{2\pi}\frac{AB}{p}.$$

If p is measured in seconds of arc, rather than in degrees, it must be divided by 3600 (the number of seconds in $1°$) before its value is inserted in the above equation. After such arithmetic, the formula for r becomes

$$r = 206{,}265\frac{AB}{p(\text{in seconds})}.$$

As an example, suppose p is 18 seconds of arc (about what would be observed for the sun). Since AB, the earth's diameter, is 7927 mi,

$$r = 206{,}265\frac{7927}{18} = 9.1 \times 10^7 \text{ mi.}$$

In Exercise 2 the moon's distance is to be calculated by the same procedure.

18.2 RELATIVE DISTANCES IN THE SOLAR SYSTEM

The planets and other bodies that revolve about the sun in the solar system are so far away that at any time each one is seen in almost the same direction by all terrestrial observers. The parallaxes of these bodies are, therefore, very small and are difficult to measure accurately; a larger base line than the diameter of the earth is required. As the earth moves about the sun, however, it carries us across a base line that can be as large as the diameter of the earth's orbit — nearly 200 million mi. Thus, we can determine accurate distances to the other members of the solar system by observing them at times when the earth is in two different places in its orbit. Kepler arranged that the different observations of the same body were made at intervals of its sidereal period so that it was at the same place in its orbit during each observation (Section 3.4a). This is not necessary, however, for we can take into account the motion of the object during the interval between the sightings of it. It is convenient, of course, not to have to restrict our observations of a body to intervals of its sidereal period. There are various mathematical techniques for unscrambling the effects of the combined motions of the body and the earth (which will not be gone into here).

Unfortunately, measures described above — measures within the solar system — are not obtained directly in miles or kilometers; they are rather found in terms of the *astronomical unit*, the semimajor axis of the earth's orbit. It is a different problem to determine the length of the astronomical unit in miles or kilometers. The foregoing procedure for surveying the distances to the planets provides us with an accurate map of the solar system, but does not give us the absolute scale of the map. On a map of the United States that is correct proportionally, the relative distances between cities, for example, can be determined simply by measuring distances on the drawing itself. Distances so obtained cannot be converted to miles or kilometers, however, until the scale of the map is established — that is, one inch equals so many miles, kilometers, or other units of distance. The scale can be determined if the actual distance in miles or kilometers between two places (perhaps New York and Philadelphia) is known. Similarly, to find the scale of the solar system — that is, to evaluate the astronomical unit — we must find the distance to some object that revolves about the sun, both in astronomical units and in miles or other units.

18.3 DETERMINATION OF THE LENGTH OF THE ASTRONOMICAL UNIT

The earliest known attempt to find the length of the astronomical unit was made by Aristarchus in the third century B.C. (Section 2.3a). His value of 200 earth diameters for the distance to the sun (the earth's size was measured by Aristarchus' contemporary, Eratosthenes — Section 2.3b) was later corrected to 600 earth diameters by Hipparchus (Section 2.3c). The latter value survived until the seventeenth century, when Kepler estimated that it was at least three times too small because of the absence of a diurnal parallax of Mars (Section 3.4c). Today there are several procedures for evaluating the length of the astronomical unit.

(a) Direct Triangulation of Some Other Planet

The most obvious method of finding the length of the astronomical unit is to observe directly the parallax of the sun and then compute its distance by the procedure of Section 18.1c. One difficulty is that the maximum displacement of the sun as seen from opposite sides of the earth is less than 18″, a very small angle and consequently hard to measure accurately. Further, the sun's brilliance and large angular size make the determination of its exact direction among the stars somewhat uncertain. On the other hand, the distances to the planets are accurately known in astronomical units, so it is just as satisfactory to triangulate directly the distance to Mars, for example, thus establishing one distance which will give the scale of the whole solar system map. Mars, at its closest (at the time of a favorable opposition), passes within about ⅓ AU from the earth. At such a time its parallax is three times as great as the sun's. Moreover, the planet is above the horizon all night and its direction among the stars can be ascertained with much greater ease than the sun's.

Even more useful than Mars for the determination of the length of the astronomical unit are certain close-approaching minor planets, which come nearer the earth than any other body except the moon. In 1932 the minor planet Eros passed within about 14 million mi of the earth; at that time its total displacement in direction as seen from opposite sides of the earth was about 2′, and a worldwide effort, under the direction of Harold Spencer Jones, was made to determine this parallactic angle as accurately as possible.

Alert students often ask why the moon's parallax cannot be used to evaluate the astronomical unit; the moon is closer than any other natural astronomical body, and its parallax, and distance in miles, are known with the greatest accuracy of any body. The moon cannot be used to *determine* the astronomical unit because its distance is determined directly in terrestrial units (such as miles) but not in astronomical units. The reason is that if we attempt to observe the moon from different places in the earth's orbit's, we find that the moon has followed right along with the earth in its orbital revolution.

(b) Perturbations of Minor Planets

When a minor planet passes close to a planet, the gravitational perturbations produced by the latter can substantially alter, or perturb, the orbit of the minor planet. The magnitude of the perturbing force can be found accurately by comparing the orbital elements of the minor planet before and after the encounter. The perturbing force (being due to gravity) depends on the distance between the two bodies at the time of their close approach to each other, so the amount of the perturbation produced indicates the actual separation of the planet and minor planet. On the other hand, that separation is also known in astronomical units, since an accurate "scale map" of the solar system, in astronomical units, is available. Comparison of the numbers of astronomical units and miles that correspond to this distance gives the value of the astronomical unit.

(c) The Earth's Orbital Velocity

A knowledge of the earth's orbital velocity gives the value of the astronomical unit, because the earth's speed, say, in miles per second, multiplied by the length of the sidereal year, in seconds, gives the number of miles in the circumference of its orbit; since the circumference is also known in astronomical units, the number of miles per astronomical unit is readily determined.

The simplest way to find the earth's orbital velocity is to determine the radial (line of sight) velocity of a star on the ecliptic from the Doppler shift of the lines in its spectrum (Section 10.3b). Since the star itself, in general, is approaching or receding from the sun, two observations of its radial velocity are necessary — one when the earth is approaching the star and one when the earth is moving directly away from the star. The mean of the two measures gives the star's radial velocity with respect to the sun; the difference between the two is twice the earth's orbital speed.

The aberration of starlight (Section 7.3a) also reveals the earth's orbital velocity. The 20″.5 angle by which the direction of a star is displaced because of aberration shows that the velocity of the earth is about $\frac{1}{10,000}$ that of light, or about 18½ mi/sec.

These methods of finding the earth's orbital velocity require a knowledge of the speed of light. Any method for determining distances that involves the speed of light (including those described in Sections 18.3d and 18.3e) gives those distances in terrestrial units, such as miles, because, just as in triangulation methods, the method is based on the measurement of a terrestrial distance. In this case, that measurement is the distance over which the velocity of light has been measured.

FIG. 18-4 Two spectograms of Arcturus, taken 6 months apart. On July 1, 1939, *(top)* the measured radial velocity was +18 km/sec; on January 19, 1940, it was −32 km/sec. The difference of 50 km/sec is due to the orbital motion of the earth. *(Mount Wilson and Palomar Observatories.)*

(d) *Interplanetary Rockets**

The Doppler shift in the wavelength of radio waves received from a transmitter carried by a space rocket indicates how fast that rocket recedes from the earth. By keeping track of the recession velocity of a rocket and the time since it was launched, the rocket's distance can be accurately found at all times. If such a space vehicle is put into a solar orbit, its distance can also be found, just as the distance to a planet is found, in astronomical units, and the length of the astronomical unit is determined. Such analyses have been carried out for several of the interplanetary probes, and the values found for the length of the astronomical unit are in good agreement with those found from other methods.

(e) Radar Measures

The most accurate technique for evaluating the astronomical unit at present is by means of radar. We saw (Section 9.2a) how the distance to the moon is accurately found from the time required for radar waves to make the round trip to the moon and back. In February 1958 the Lincoln Laboratories of the Massachusetts Institute of Technology sent radar waves of frequency 440 Mc to Venus and observed what they believed to be a return echo. The distance in miles traversed by the beam was found from the observed travel time of the signal and the accurately known velocity of light. Comparison of this distance with the known distance of Venus in astronomical units determines the length of the unit. The experiment has now been repeated many times with better equipment and higher accuracy.

(f) Best Value to Date

By convention, the *solar parallax* is defined as the angle subtended by the radius of the earth at a distance of 1 AU. The best determination of the solar parallax (or astronomical unit), obtained from radar observations of Venus, gives 8″.7941; the corresponding value of the astronomical unit is 149,597,893 km (or about 92,960,000 mi) with an uncertainty of only a few kilometers. Before 1961

the best value of the solar parallax was that obtained by Rabe from the perturbations of Eros; the value he obtained was 8″.798.

18.4 SURVEYING DISTANCES TO STARS

The measurement of the parallax of a star enables us to calculate its distance in terms of the earth-sun distance — that is, in astronomical units. However, the stars are so remote that their parallaxes are imperceptible to the unaided eye, even when the stars are observed from both ends of a diameter of the earth's orbit — a base line of 2 AU. When it was conceded that the earth is in orbital motion, therefore, it was realized that the stars are thousands of times more distant than the planets. This fact alone shows that the stars are self-luminous, like the sun, and do not shine merely by reflected sunlight. The illumination that sunlight produces upon a body drops off as the square of that body's distance from the sun (Section 10.1a). A typical naked-eye star would receive so little illumination from the sun that to reflect enough light back to us to account for its apparent brightness, it would have to be extremely large and would therefore show an observable disk; no star shows an observable disk.

With telescopes, the parallaxes of the nearest stars actually can be measured by observing their apparent displacements against the background of more distant stars as the earth traverses its orbit. Even the nearest star, however, shows a total displacement of only about 1″.5; small wonder that Tycho Brahe was unable to observe the stellar parallaxes and thus concluded that the earth was not in motion. The first successful measurements of stellar parallaxes took place in about the year 1838, when Friedrich Bessel (Germany), Thomas Henderson (Cape of Good Hope), and Friedrich Struve (Russia) detected the parallaxes of the stars 61 Cygni, Alpha Centauri, and Vega, respectively.

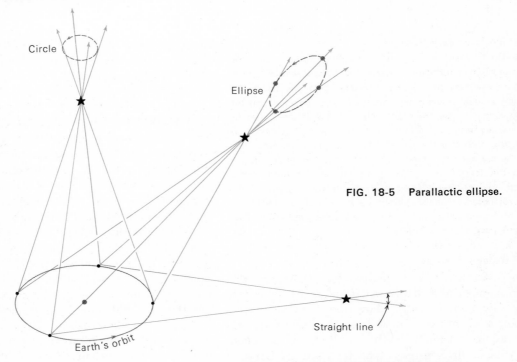

FIG. 18-5   Parallactic ellipse.

(a) The Parallactic Ellipse

As the earth moves about its orbit, the place from which we observe the stars is continually changing. Consequently, the positions of the comparatively near stars, projected against the more remote ones, are also continually changing. If a star is in the direction of the ecliptic, it seems merely to shift back and forth in a straight line as the earth passes from one side of the sun to the other. A star that is at the pole of the ecliptic (90° from the ecliptic) seems to move about in a small circle against the background of more distant stars, as we view it from different positions in our nearly circular orbit. A star whose direction is intermediate between the ecliptic and the ecliptic pole seems to shift its position along a small elliptical path during the year. The eccentricity of the ellipse ranges from that of the earth's orbit (nearly a circle) for a star at the ecliptic pole to unity (a straight line) for a star on the ecliptic (see Figure 18-5). This small ellipse is called the *parallactic ellipse*.

(b) Stellar Parallax and Stellar Distances

The angular semimajor axis of the parallactic ellipse is called the *stellar parallax* of the star. Since the major axis of the ellipse is the maximum apparent angular deflection of the star as viewed from opposite ends of a diameter of the earth's orbit, the stellar parallax, the semimajor axis of this ellipse, is the angle, at the star's distance, subtended by 1 AU perpendicular to the line of sight. The angular displacements in the direction of a star are always measured with respect to more distant stars; the stellar parallax derived, therefore, is, strictly speaking, a *relative parallax*. The *absolute parallax* is determined from the relative parallax by applying corrections — always exceedingly small — for the parallaxes of the background stars. The latter must be derived by indirect methods (see Section 18.5).

Until now we have implied that the star whose parallax is observed is motionless with respect to the sun. Actually, the stars are all moving, at many kilometers per second (Chapter 19). The effects of the relative motion of the star and the sun can be separated from the effects of the earth's motion (the star's parallax) by observing the star's direction at the same time of the year several years apart. Any observed change in its direction then must be due to its own motion relative to the sun. The change indicates the corrections that must be applied to observed total changes in the direction of the star in order to obtain its true stellar parallax.

(c) Units of Stellar Distance

If a line segment of length D subtends an angle of p seconds of arc as seen from a distant object, the distance r of that object is given by the formula derived in Section 18.1c:

$$r = 206,265 \frac{D}{p}.$$

Since the parallax of a star is the angle, in seconds, subtended by 1 AU at the star's distance, the distance of a star, in astronomical units, is $206,265/p$. The length 206,265 AU is defined as a *parsec* (abbreviated pc). One parsec is, therefore, the distance of a star, in astronomical units, is $206,265/p$. The distance of any star, in parsecs, is thus given by

$$r = \frac{1}{p},$$

where p is the parallax of the star. A star with a parallax of ½″, for example, has a distance of 2 pc; one with a parallax of $\frac{1}{10}$″ has a distance of 10 pc. Parallaxes of stars are usually measured in seconds, so the parsec is a convenient unit of distance — a star's distance in parsecs is simply the reciprocal of its parallax. One parsec is 1.92×10^{13} mi, or 3.08×10^{13} km.

Since 1 pc is 206,265 AU, the sun's distance is $\frac{1}{206.265}$ pc, and its stellar parallax is 206,265 sec. [This should not be confused with the *solar parallax*, which is defined as the angle subtended by the earth's radius at 1 AU; the latter has the value of approximately 8″.8 (see Section 18.3f)].

Another unit of stellar distance is the *light-year*, which is the distance traversed by light in 1 year at the rate of 186,000 mi/sec. One light-year (abbreviated LY) is 5.88×10^{12} mi or 9.46×10^{12} km; 1 pc contains 3.26 LY.

(d) The Nearest Stars

The nearest stellar neighbors to the sun are three stars that make up a multiple system. To the naked eye the system appears as a single bright star, Alpha Centauri, which is only 30° from the south celestial pole and hence is not visible from the United States. Alpha Centauri itself is a double star — two stars revolving about each other that are too close together to be separated by the naked eye. Nearby is the third member of the system, a faint star known as *Proxima Centauri*. Proxima is believed to be slightly closer to us than the other two stars of the system. All three have a parallax of about 0″.76, and a distance of $\frac{1}{0.76}$, or about 1.3 pc (4.3 LY). The nearest star visible to the naked eye from most parts of the United States is the brightest appearing of all the stars, *Sirius*. Sirius has a distance of 2.6 pc, or about 8 LY. It is interesting to note that light reaches us from the sun in 8 minutes and from Sirius in 8 years.

Parallaxes have been measured for thousands of stars. Only for about 700 stars, however, are the parallaxes large enough (about 0″.05 or more) to be measured with a precision of 10 percent or better. Of those 700 or so stars within about 20 pc, most are invisible to the unaided eye and actually are intrinsically less luminous than the sun. Most of the stars visible to the unaided eye, on the other hand, have distances of hundreds or even thousands of parsecs and are visible not because they are relatively close, but because they are intrinsically very luminous. The nearer stars are described more fully in Chapter 23.

18.5 OTHER METHODS OF MEASURING STELLAR DISTANCES

The vast majority of all known stars are too distant for their parallaxes to be measured, and we must resort to other methods to determine their distances. Most of these methods are either statistical or indirect; they are discussed in later chapters. For completeness, however, some of the more important procedures, other than parallax measurement, for determining stellar distances may be listed:

(1) **Stellar motions** (Chapter 19). All stars are in motion, but only for those comparatively nearby ones are the angular motions perceptible. Statistically, therefore, the stars that have large apparent motions are the nearer ones; we can estimate the average distance to stars in a large sample from the average angular motions of those stars.

(2) **Moving clusters** (Chapter 19). In a few cases the direction of motion through space of a cluster or swarm of stars can be determined from the apparent convergence or divergence of the directions of motions of the individual stars in that cluster. In such a case, an analysis of the apparent motions and radial velocities of the member stars gives the distance to the cluster.

(3) **Inverse-square law of light** (Chapter 20). The apparent brightness of a star depends on both its intrinsic luminosity and its distance (through the inverse-square law of light). Very often it is possible to infer the intrinsic brightness of a star from its spectrum (Chapter 23) or because it is a recognizable type of variable star (Chapter 25). Then its distance can be calculated from a knowledge of its observed brightness.

(4) **Dynamical parallaxes** (Chapter 22). The time required for the two stars in a binary (or double) system to revolve about each other depends, through Newton's formulation of Kepler's third law, on their combined mass and mean separation. If the masses of the stars can be assumed, or if an intelligent guess is made, their separation, in astronomical units, can be calculated from their period of mutual revolution. A comparison of their linear and angular separations then gives the distance to the system.

(5) **Interstellar lines** (Chapter 26). The space between the stars throughout much of space contains a sparse distribution of gas. Sometimes this interstellar gas leaves absorption lines superposed upon the spectrum of a star whose light must shine through the gas to reach us. The amount of the star's light absorbed by these interstellar lines indicates the total mass of the gas that must lie in the light path from the star. If we can estimate the approximate density of the gas in space (as we often can) we can tell what the total path length of the star's light must be, and hence the distance of the star.

(6) **Galactic rotation** (Chapter 27). The sun and its neighboring stars are part of a vast system of stars — our Galaxy. The Galaxy is rotating; the stars that compose it all revolve about its center much as the planets revolve about the sun. The speeds with which distant stars in the Galaxy approach us or recede from us as a result of this galactic rotation depends on the directions and distances of these stars. Observations of their radial velocities, therefore, can lead to an estimate of their distances.

Exercises

1. At a distance of 3500 ft from the base of a vertical cliff, the top of the cliff is observed to have an angular altitude of 45°. What is its height?

*2. The moon's position among the stars was observed at moonrise. Six hours and 12 minutes later the moon happened to be exactly at the zenith. Its position among the stars by then had shifted to the east by 148'. On the other hand, because of the moon's orbital motion, it is known to move eastward on the celestial sphere at the rate of 33' per hour. Find the distance to the moon. Compare your result with the value given elsewhere in the text.

3. At a particular opposition of Mars, the planet was simultaneously observed from two points on the equator — one where it was rising and one where it was setting. Its directions among the stars as found at the two observing stations differed by 41". What was the distance to Mars in astronomical units?

 Answer: 0.43 AU

4. Why would observations such as those described in Exercise 3 be very difficult to make?

5. When Geographos is only 6 million miles away, how large an angular displacement in direction can be observed for it from different places on earth?

6. Describe how you might organize a program to determine the length of the astronomical unit from observations of Geographos. Explain what kinds of observations would be needed, how many would be desired, and what use you would make of them.

7. A star lying on the ecliptic and exactly 90° to the west of the sun is observed to have a radial velocity of exactly 30 mi/sec toward the earth. Six months later the star's radial velocity is 7 mi/sec away from us. What is the star's radial velocity with respect to the sun? What is the earth's orbital velocity? Show how this gives the value of the astronomical unit.

8. Suppose that when Mars is exactly 35 million miles away, radar waves are beamed toward it. Exactly 4½ minutes later a return "beep" is picked up that the radar operator thinks is a radar echo reflected from Mars. Do you agree with him? Why?

9. Show that a light year contains 5.88×10^{12} mi.

10. Show that 1 pc equals 1.92×10^{13} mi and that it *also* equals 3.26 LY. Show your reasoning.

11. Make up a table relating the following units of astronomical distance: mile, kilometer, earth radius, astronomical unit, light-year, parsec.

Following page: View looking northwest into Mare Tranquillitatis, photographed from the Apollo 8 spacecraft in December, 1968. *(NASA.)*

CHAPTER 19

Motions of Stars

The ancients distinguished between the "wandering stars" (planets) and the "fixed stars," which seemed to maintain permanent patterns with one another in the sky. The stars are, indeed, so nearly fixed on the celestial sphere that the apparent groupings they seem to form — the constellations — look today much as they did when they were first named, more than 2000 years ago. Yet the stars are moving with respect to the sun, most of them with speeds of many kilometers per second. Their motions are not apparent to the unaided eye in the course of a single human lifetime, but if an ancient observer who knew the sky well — Hipparchus, for example — could return to life today he would find that several of the stars had noticeably changed their positions relative to the others. After some 50,000 years or so, terrestrial observers will find the handle of the Big Dipper unmistakably "bent" more than it is now. Apparent changes in the positions of the nearer stars can be measured with telescopes after an interval of only a few years.

19.1 ELEMENTS OF STELLAR MOTIONS

(a) Proper Motion

The *proper motion* of a star is the rate at which its direction in the sky changes, and is usually expressed in seconds of arc per year. It is almost always an angle that is too small to measure with much precision in a single year; in an interval of 20 to 50 years, on the other hand, many stars change their directions by easily detectable amounts. The modern procedure for determining proper motions is to compare the positions of the star images on two different photographs of the same region of the sky taken at least two decades apart. Most of the star images on such photographs do not appear to have changed their positions measurably; these are, statistically, the more distant stars that are relatively "fixed," even over the time interval separating the two photographs. With respect to these "background" stars, the motions of a few comparatively nearby stars can be observed. Modern practice is to measure proper motions of stars with respect to remote galaxies, which can show no measurable proper motions themselves (Chapter 32).

The star of largest known proper motion is *Barnard's star,* which changes its direction by $10''.25$ each year. This large proper motion is partially due to the star's relatively high velocity with respect to the sun but is mostly the result of its relative proximity. Barnard's star is the nearest known star beyond the triple system containing Alpha Centauri; its distance is only 1.8 pc, but it emits less than $\frac{1}{1000}$ as much light as the sun and is nearly a hundred times too faint to see with the unaided eye. There are about 330 other stars with proper motions as great as $1''.0$. The mean proper motion for all naked-eye stars is less than $0''.1$; nevertheless, the proper motions of most stars are larger than their stellar parallaxes.

The compete specification of the proper motion of a star includes not only its angular rate of motion, but also its *direction* of motion in the sky.

(b) Radial Velocity

The *radial* velocity (or line-of-sight velocity) of a star is the speed with which it approaches or recedes from the sun. This can be determined from the Doppler shift of the lines in its spectrum (Section 10.3b). Unlike the proper motion, which is observable only for the comparatively nearby stars, the radial velocity can be measured for any star that is bright enough for its spectrum to be photographed. The radial velocity of a star, of course, is only that component of its actual velocity that is projected along the line of sight — that is, that carries the star toward or away from the sun. Radial velocity is usually expressed in kilometers per second, and is counted as *positive* if the star is moving *away* from the sun, and *negative* if the star is moving *toward* the sun. Since motion of either the star or the observer (or both) produces a Doppler shift in the spectral lines, a knowledge of the radial velocity alone does not enable us to decide whether it is the star or the sun that is "doing the moving" (indeed, as we saw in Section 7.5, it does not even make sense to ask "which" is moving). What we really measure, therefore, is the speed with which the distance between the star and sun is increasing or decreasing — that is, the star's radial velocity *with respect to the sun.*

Actually, the radial velocity of a star with respect to the sun is not obtained directly from the measured shift of its spectral lines, because we must observe the star from the earth, whose rotational and orbital motions contribute to the Doppler shift. Since the direction of the star is known, however, as are the speed and direction in which the moving earth carries the telescope at the time of observation, it is merely a problem in geometry (albeit a slightly complicated one) to correct the observed radial velocity to the value that would have been found if the star had been observed from the sun.

(c) Tangential Velocity

Radial velocity is a motion of a star along the line of sight, while proper motion is produced by the star's motion *across,* or at right angles to, the line of sight. Whereas the radial velocity is known in kilometers per second, the proper motion of a star does not, by itself, give the star's actual *speed* at right angles to the line of sight. The latter is called the *tangential* or *transverse* velocity. To find the tangential velocity of a star, we must know both its proper motion and *distance.* A star with a proper motion of $1''.0$, for example, might have a relatively low tangential velocity and be nearby, or a high tangential velocity and be far away.

The relation between tangential velocity and proper motion is illustrated in Figure 19-2. As seen from the sun S, a star A, at distance r, is in direction SA. During 1 year it moves from A to B and then appears in direction SD, at an angle μ (the proper motion) from SA. The star's radial motion is AC — it has moved a distance AC farther away

FIG. 19-1   Two photographs of Barnard's Star, showing its motion over a period of 22 years. *(Yerkes Observatory.)*

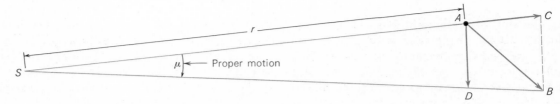

FIG. 19-2   Calculation of tangential velocity.

during the year; its *tangential* motion is AD — it has moved that distance across the line of sight. The motions shown in the figure are grossly exaggerated; actual stars do not move enough to change their distances by an appreciable percentage, even in 100 years (see Exercise 3).

The tangential motion AD can be approximated very accurately by a small arc of a circle of radius r centered on the sun. The arc AD is the same fraction of the circumference of the circle, $2\pi r$, as the proper motion is of $360°$. The proper motion is expressed in seconds of arc, so we have

$$\frac{\mu}{1,296,000} = \frac{AD}{2\pi r}$$

(there are $1,296,000''$ in $360°$). If we solve for AD and note that AD is the product of the star's tangential velocity in miles per second and the number of seconds in a year (3.16×10^7), we obtain

$$T = \frac{\mu r}{6.52 \times 10^{12}} \text{ mi/sec,}$$

where r must be in miles. If r is to be expressed in parsecs, we must multiply the right-hand side of the above equation by 1.92×10^{13}, the number of miles per parsec. The formula for the tangential velocity then becomes

$$T = 2.95\mu r = 2.95\frac{\mu}{p} \text{ mi/sec,}$$

where p is the stellar parallax, which is merely the reciprocal of the distance when the latter is expressed in parsecs (Section 18.4) In practice, T is

usually expressed in kilometers per second rather than in miles per second. In those metric units, T is given by

$$T = 4.74\frac{\mu}{p} \text{ km/sec.}$$

The above equations show how the tangential velocity of a star is related to its proper motion and distance (or parallax).

(d) Space Velocity

The *space velocity* of a star is its total velocity, in miles or kilometers per second, with respect to the sun. The radial velocity is the distance the star moves toward or away from the sun in 1 second; the tangential velocity is the distance it moves at right angles to the line of sight in 1 second. The space velocity, therefore, being the total distance the star moves in 1 second, is simply the hypotenuse of the right triangle whose sides are the radial and tangential velocities (Figure 19-3). It is found immediately by the theorem of Pythagoras,

$$V^2 = V_r^2 + T^2,$$

where V and V_r are the space and radial velocities, respectively.

19.2 THE SOLAR MOTION AND PECULIAR VELOCITIES OF STARS

It might be expected that the sun, a typical star, is in motion, just as the other stars are. We turn now to the motion of the sun, and how it influences the apparent motions of the stars.

(a) The Local Standard of Rest

Before we can speak of the motion of any body, we must first define a reference system with respect to which we can refer that motion. As described earlier (Section 4.4), near the end of the last century Michelson and Morley attempted to measure the *absolute velocity* of the earth in space, but failed completely. Einstein adopted, as one of the postulates of his *special theory of relativity,* the concept that all motion is relative, that is, the motion of one body can only be referred to some other body or bodies. Before we can discuss the motion of the sun, therefore, we must decide upon what reference frame to use.

As we know, the sun is a member of our Galaxy, a system of a hundred billion or so stars. The Galaxy is flat, like a pancake, and is rotating. The sun, partaking of this general rotation of the Galaxy, moves with a speed of about 250 km/sec or 150 mi/sec to complete its orbit about the galactic center in a period of about 200 milllion years. At first thought it might seem that the galactic center is the natural reference point with which to refer the stellar motions. However, our observations of the proper motions and radial velocities of the stars that surround the sun in space, all in our own so-called local "neighborhood" of the Galaxy, do not give us directly the motions of these stars about the galactic center. The reason is that the stars' orbits around the galactic center

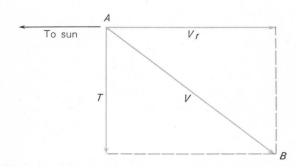

FIG. 19-3   Space velocity; the star moves from *A* to *B* in one second.

and their orbital velocities are both nearly the same as those of the sun. The motions we observe are merely small "residual" or *differential* motions of these stars with respect to the sun. These small residual motions arise because our neighboring stars' orbits about the galactic center are not absolutely identical to our own. We are overtaking and passing some stars, while others are passing us; the slightly different eccentricities and inclinations of our respective orbits bring us closer to some stars and carry us farther from others. We can study these residual motions without knowing anything about the actual motions of stars around the center of the Galaxy. Our situation is analogous to that of a man driving an automobile on a busy highway. All the cars around him are going the same direction and at roughly the same speed, but some are changing lanes and others are passing each other. More or less like the highway traffic, the residual motions of the stars around us seem to be helter skelter.

Astronomers have defined a reference system (that is, a coordinate system) within which the motions of the stars in the solar neighborhood — within a hundred parsecs or so — average out to zero. In other words, those stars in our neighborhood appear, on the average, to be at rest in this system; it is thus called the *local standard of rest.* The local standard of rest is not really at "rest," of course, but shares the average motion of the sun and its neighboring stars around the center of the Galaxy. We would not realize, however, that we are part of a huge rotating galaxy if we observed only stars near us. In fact, for many purposes it is useful to forget about the rotation of the Galaxy as a whole and to consider only the motions of stars, and of the sun, with respect to the local standard of rest, which we can pretend, for these purposes, to really be at "rest."

(b) The Solar Motion and Solar Apex

We deduce the motion of the sun with respect to the local standard of rest by analyzing the proper motions and radial velocities of the stars around us. The easiest way to understand how the sun's motion can be found is to consider the effect it has on the apparent motions of the other stars.

First, consider the radial velocities of stars with respect to the sun. Note that we have defined the local standard of rest as being stationary with respect to the average of the motions of the stars in the solar neighborhood. Therefore, if we could correct the observed space motions of the stars to those values they would have if the sun were not moving in the local standard of rest, they would then average out to zero. Now it is clear that in a direction that is at right angles to the direction in which the sun is actually moving, the solar motion cannot affect the observed radial velocities of the stars. In those directions, indeed, we find as many stars approaching us as receding from us — their radial velocities *do* average to zero. On the other hand, if we look in the direction *toward* which the sun is moving we find that most of the stars are approaching us, because, of course, we are moving forward to meet them. The only stars in that direction that have radial velocities of recession are those that are moving in the same direction we are going, but at a faster rate, so that they are pulling away from us. The observed radial velocities of all the stars in the direction toward which the sun is moving do not average to zero, but to −20 km/sec, (about 12 mi/sec) shows that we are moving toward them at about 20 km/sec. Similarly, stars in the opposite direction have an average radial velocity of +20 km/sec, because we are pulling away from them at that speed.

Now consider the proper motions of stars. Part of a star's proper motion, in general, is due to its own motion and part is due to the sun's motion. However, the sun's motion can contribute nothing to a star's proper motion if the star happens to lie in the direction toward which the sun is moving. Therefore, if we look at stars that lie in a path along the direction of the solar motion, as many of

their proper motions should be in one direction as any other; the average of the motions of many stars in those directions, therefore, should be zero. On the other hand, the maximum effect on the proper motions of stars should occur in directions that are at right angles to the direction of the solar motion. If the stars were at rest, they would all show a backward drift due to our forward motion. As it is, the stars have motions of their own, but only those moving in the same direction we are, but at a faster rate than we, appear to have "forward" proper motions — the rest, by far the majority, *do* appear to drift backward.

William Herschel was the first to attempt to detect the direction of the solar motion from the proper motions of stars. In 1783 he analyzed the proper motions of 14 stars and deduced that the sun was moving in a direction toward the constellation Hercules — a nearly correct result.

Modern analysis of the proper motions and radial velocities of the stars around the sun has shown that the sun is moving approximately toward the direction now occupied by the bright star Vega in the constellation of Lyra at a speed of 20 km/sec (about 12 mi/sec or 4.2 AU/year), with an uncertainty of about 0.5 km/sec. The direction in the sky toward which the sun is moving is called the *apex* of solar motion,† and the opposite direction, away from which the sun is moving, is called the *antapex*.

(c) Peculiar Velocities of Stars

The velocity of a star with respect to the local standard of rest is called its *peculiar velocity*. The *space velocity* of a star, being its motion with respect to the sun, is made up of both the star's peculiar velocity and a component due to reflection of the solar motion. Since the solar motion is known, the peculiar velocity can be calculated for a star of known space velocity.

---

†The equatorial coordinates of the *solar apex* are (1950) $\alpha = 18^h 4^m \pm 7^m$, $\delta = +30° \pm 1°$.

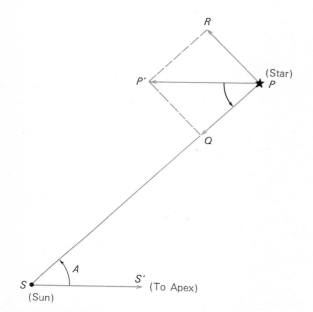

FIG. 19-4   How the solar motion affects a
star's space motion.

Consider, for example, the special case of a
star that is not moving with respect to the local
standard of rest and which therefore has zero
peculiar velocity. The entire space motion of that
star, then, is merely a reflection of the solar motion;
its space velocity is a vector of magnitude exactly
equal to the solar velocity but directed toward the
solar antapex. In Figure 19-4, the vectors SS' and
PP' represent the solar velocity and that star's space
velocity, respectively; A is the angle at the sun
between the directions to the star and to the solar
apex. Vector PQ, the projection of PP' onto the
line of sight from the sun to the star, is the star's
radial velocity. Vector PR, the component of PP'
that is perpendicular to the line of sight, is the
star's tangential velocity. Since this hypothetical
star has no peculiar velocity, PQ and PR are due
entirely to the solar motion and are, therefore, the
corrections that must be applied to the observed
radial and tangential velocities of *any* star in that
same direction to obtain, respectively, its *peculiar*

radial velocity and *peculiar tangential velocity*. The
angle A can always be observed; consequently, the
desired corrections to the radial and tangential
velocities can always be found geometrically or
calculated trigonometrically.† The correction to
the star's observed proper motion that is required
to obtain its *peculiar proper motion* can be found
from the relation between proper motion and tangential velocity given by the formulas in Section
19.1c. The peculiar proper motion can be found
only for stars of known distance.

19.3 DISTANCES FROM STELLAR MOTIONS

The proper motions of stars can be expected to be
largest, statistically, for the nearest stars. If, for
example, a star is only a few parsecs away, its
proper motion will almost certainly be observable
after a few years (but see Exercise 14). The proper
motion of a very distant star, on the other hand,
may only be detectable after a long time, and then
only if the star has a very great space velocity.
Searches for nearby stars, therefore, are usually
conducted by searching for stars of large proper
motion. Conversely, remote stars can be identified
by their lack of observable proper motions; the
latter are needed to serve as standards against
which we can measure the parallaxes of the nearby
stars to determine their distances. The proper motion of an individual star does not uniquely indicate its distance. However, proper motions and
distances of stars are inversely correlated, and investigations of such motions do give some information about stellar distances.

(a) *Secular Parallaxes* *

Trigonometric parallaxes of stars are based on triangulation with a base line of only 2 AU, the diameter of the
earth's orbit. The solar motion, however, amounts to

---

†The corrections to the observed radial and tangential
velocities are $V_0 \cos A$ and $V_0 \sin A$, respectively, where
V_0 is the solar velocity (about 19.4 km/sec).

4.09 AU/year. If we wait long enough between observations, the motion of the sun will carry us over as long a base line as we desire. For example, after 20 years we move 81.8 AU; with such a base line we should be able to detect parallaxes of stars 40 times as distant as we can with a base line of 2 AU. By this method, accurate distances to many stars could be determined if they did not have peculiar velocities of their own. Unfortunately, without knowing a star's distance in advance, we have no way of knowing how much of its proper motion is due to the solar motion and how much to its own peculiar velocity. On the other hand, the peculiar velocity of a large number of stars, spaced well about the sky, should average out to zero, for peculiar velocities are measured with respect to the local standard of rest, which is *defined* to have zero average velocity with respect to stars in the sun's neighborhood. Consequently, the *average* of the space velocities for a large sample of stars must be due entirely to the solar motion. The proper motion corresponding to that average space velocity depends upon the average of the distances to those stars. This motion is an average angular drift of the stars in the sample toward the solar antapex; it results from the fact that the sun carries us over a long base line. Since we know how far the sun carries us in any interval of time, we can calculate the average of the distances for the stars in question (more accurately, we calculate the *mean parallax*). The procedure is called the method of *secular parallaxes.*

At first thought, it might not seem particularly useful to know only the average of the distances to a large number of stars, because some of those stars will be much more distant than others and we have no way of knowing which are which. Suppose, however, that we have chosen the sample in such a way that we have reason to believe that all the stars in it have more or less equal luminosities (as determined, say, from their spectra). In that case, the brighter-appearing stars would be the nearer and the fainter-appearing stars the more distant. The inverse-square law of light, in other words, allows us to calculate the relative distances of stars of similar luminosities. A knowledge of their mean distance, found from the method of secular parallaxes, then enables us to find both their actual distances and luminosities. (The relation between the apparent brightness, luminosity, and distance of a star will be further explained in Chapter 20.) The method of

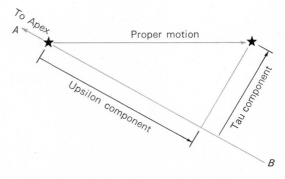

FIG. 19-5 Components of proper motion; the figure is in the plane of the sky, perpendicular to the line of sight. The line *AB* is part of a great circle running through the star and the solar apex and antapex.

secular parallaxes has proved very useful for determining the luminosities of relatively rare stars of certain types which are not represented by examples near enough for direct parallax measurement.

(b) Statistical Parallaxes*

The method of secular parallaxes is based on the detection of the average backward drift, due to solar motion, in a large sample of stars. Now, because the sun moves toward the apex, that part of a star's motion due to the solar motion is a drift toward the antapex. However, the solar motion cannot affect the component of a star's proper motion that is in a direction *perpendicular* to a line in the apex-antapex direction. It is convenient, therefore, to regard the proper motion of a star as having two components: one — the *upsilon component* — along a great circle on the celestial sphere from the star through the antapex, that is, parallel to the direction to the solar apex, and another — the *tau component* — perpendicular to the direction of the apex direction (see Fig. 19-5). The upsilon component of the proper motion of a star can be augmented or diminished by the solar motion, but the tau component can involve only the star's peculiar motion — in fact, only that part of it that contributes to the star's tangential velocity perpendicular to the direction of the solar apex. If we had some way of knowing the value of this perpendicular component of the star's tangential velocity, $T_\perp$, we could find its parallax from the tau component, τ, of its proper motion, by applying the formulas of Section 19.1c:

$$T_\perp = 4.74 \frac{\tau}{p} \,\text{km/sec.}$$

There is no way, in general, of knowing $T_\perp$ for an individual star. The stars in a large sample, however, should be moving at random in the local standard of rest — as many are going in any one direction as in any other. We should expect, therefore, that in such a sample the average value of $T_\perp$ should be the same as the average value of the *peculiar radial velocities*, and so $T_\perp$ can be replaced by the latter (without regard to sign) in the above equation. If the quantity τ is then replaced by the average value of the many individual values of τ for the stars in the sample, the average value of the parallaxes of those stars can be computed. This procedure is called the method of *statistical parallaxes.*

Like secular parallaxes, statistical parallaxes do not indicate distances of individual stars but only the average distances for large numbers of stars. The method of statistical parallaxes, however, like that of secular parallaxes, is very useful if applied to a sample of stars that can all be expected to have the same intrinsic brightness. It gives greater accuracy than does the method of secular parallaxes, when applied to stars whose peculiar velocities are, statistically, larger than the sun's velocity.

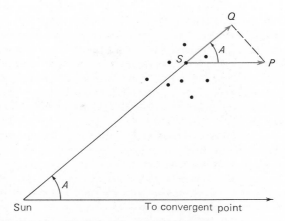

FIG. 19-7   Space velocities of the stars in a moving cluster.

(c) *Moving Clusters**

The stars that belong to a group or cluster, moving more or less together through space, all have about the same space velocity. If such a cluster is close enough to the sun, however, the proper motions of its member stars may not all be exactly parallel. If, for example, the cluster is approaching us, its stars appear to radiate away from a distant point in the direction from which the cluster is coming. Conversely, if the cluster recedes from us, the stars appear to converge toward the direction in which the cluster is moving. In either case, the proper motions

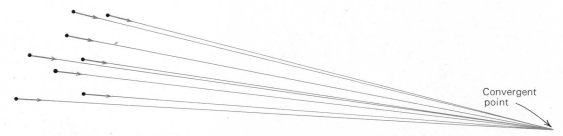

FIG. 19-6   Proper motions of the stars in a moving cluster. Note that here we see the cluster projected on the plane of the sky. Although the stars are moving parallel to one another in space, their proper motions appear to converge toward a point infinitely far away. The direction from the earth to that convergent point is also parallel to the space motions of the cluster stars.

of the member stars indicate the direction in which the cluster is traveling. The situation for a hypothetical cluster is illustrated in Figure 19-6, which shows the apparent convergence of the proper motions of its stars toward a distant "convergent point," and in Figure 19-7, which shows the actual space velocities of the stars. In these diagrams the cluster is viewed from two different directions.

In studying a moving cluster for which the direction of convergence or divergence can be observed, we have the great advantage (see Fig. 19-7) of knowing the angle A between that direction of motion and the line of sight from the sun to the cluster. Now consider any star in the cluster, such as star S. Its space velocity, radial velocity, and tangential velocity are SP, SQ, and QP, respectively. Angle SQP is a right angle, and angle A and the star's radial velocity are both observed. Thus the tangential velocity QP can be calculated by solving the triangle.†

---

†$QP = SQ \tan A$.

Then the parallax (or distance) of the star S, and thus that of the cluster, can be found from the observed proper motion of the star and from the formula relating tangential velocity and proper motion:

$$QP = 4.74\mu r = 4.74 \frac{\mu}{p} \text{ km/sec.}$$

There are several groups and clusters that are near enough for us to observe the convergence or divergence of the proper motions of their member stars. The three best-known examples are the *Hyades* in Taurus, the *Ursa Major group* (which contains most of the naked-eye stars in the Big Dipper), and the *Scorpio-Centaurus group.* These three clusters contain, in all, more than 500 stars. This is almost as great a number of stars as the number whose parallaxes can be directly observed to an accuracy of 10 percent or better. Thus, this method of using moving clusters to obtain stellar distances is very important and fruitful. It has contributed greatly to our knowledge of fundamental data about stars.

Exercises

1. In 50 years a star is seen to change its direction by 1'40''. What is its proper motion?

2. What factors must be considered in converting the observed radial velocity of a star to the value that would be observed from the sun
(a) At the ecliptic pole?
(b) At the celestial pole?

3. Suppose a star at a distance of 10 pc has a radial velocity of 150 mi/sec. By what percentage does its distance change in 100 years?

4. Find the tangential velocities of the following stars:
(a) Proper motion = 1''.5; distance = 20 pc.
(b) Proper motion = 0''.01; distance = 1000 pc.
(c) Proper motion = 0''.01; distance = 20 pc.
(d) Proper motion = 0''.01; parallax = 0''.001.

5. The wavelength of a particular spectral line is normally 6563 Å; in a certain star, the line appears at 6565 Å. How fast is that star moving in our line of sight? Is its motion one of approach or recession?

6. On June 22 a star is observed that is exactly in the direction of the autumnal equinox; its radial velocity appears to be +22 mi/sec. On December 22, the same star appears to have a radial velocity of −14 mi/sec. What is the radial velocity of the star as seen from the sun?

7. Show by a diagram how two stars can have the same radial velocity and proper motion but different space motions.

8. Suppose a star has a parallax of 0''.001 but zero peculiar velocity. How long must we wait to see the star apparently change its direction by 4''.09?

9. Do we need to correct the observed proper motion of a star for the earth's motion around the sun, to obtain the star's proper motion as seen from the sun? If so, explain how we might do so.

10. If a star moves 150 mi/sec, how many astronomical units does it move per year?

11. A star has a proper motion of 3''.00, a parallax of 0''.295, and a radial velocity of 40 mi/sec. What is its space motion?

Answer: 50 mi/sec

*12. Under what conditions would the proper motion of a nearby star be zero:
(a) If it were in the direction of the solar apex?
(b) If it were *not* at the solar apex?

Following page: View of earth from Apollo 8 spacecraft. South America is conspicuous near the center. *(NASA.)*

CHAPTER 20

The Light from Stars

The most casual glance at the sky shows that stars differ from one another in apparent brightness. They differ not only because of actual differences in their output of luminous energy, but also because they are at widely varying distances. Measures of the amounts of light energy, or *luminous flux,* received from stars are among the most important and fundamental observational data of astronomy; they are used in estimating both the distances and the actual output of energy of stars.

20.1 STELLAR MAGNITUDES

In the second century B.C., Hipparchus compiled a catalog of about a thousand stars (Section 2.3). He classified these stars into six categories of brightness, which are now called *magnitudes.* The brightest-appearing stars were placed by him in the *first magnitude;* the faintest naked-eye stars were of the *sixth magnitude.* The other stars were assigned intermediate magnitudes. This system of stellar magnitudes, which began in ancient Greece, has survived to the present time, with the improvement that today magnitudes are based on precise measurements of apparent or total luminosity rather than arbitrary and uncertain eye estimates of star brightnesses.

(a) Measuring Starlight — Photometry

That branch of observational astronomy which deals with the measurement of the intensity of starlight is called *photometry.* In the latter part of the eighteenth century, William Herschel devised a simple and direct, although only approximate, method of stellar photometry. The principle of his method depends on the fact that the light-gather-

ing power of a telescope is proportional to the area of its aperture (lens or mirror). Thus, if the same star is viewed through two telescopes, identical in construction but different in size, more light energy will be received through the larger telescope than through the smaller one, and the ratio of these two energies will be in the same proportion as the areas of the apertures of the two telescopes. Conversely, if two stars appear of the same brightness even though each is viewed with a separate telescope, then the earth must really receive *less* light from the star that is viewed with the larger of the telescopes. Actually, it is not necessary to use more than one telescope. The aperture size of a telescope can be varied at will by the use of circular diaphragms or an iris diaphragm, like the one on a camera. As an example, consider two stars that appear of different brightness in the field of view of a telescope. Suppose, now, that the dimmer of the stars becomes just barely visible when the telescope is closed down to an 8-in. diameter, while the brighter one becomes barely visible after the aperture is stopped down to 4 in. The two stars, when made just barely visible, appear equally bright. However, since only one fourth as much light enters the telescope with the 4-in. diaphragm as with the 8-in. one, we must, in fact, actually be receiving four times as much light from the brighter star as from the fainter. By such reasoning, Herschel estimated the relative amounts of luminous flux received from different stars. He determined, for example, that the average first-magnitude star delivers to the earth somewhat more than 100 times as much light as a star which is just barely visible on a dark night — that is, a star of approximately sixth magnitude.

A more accurate method of stellar photometry developed later employs the *visual photometer,* a device, attached to a telescope, which produces an artificial star image. The astronomer, looking through the telescope, sees the actual star image that he wishes to measure and the artificial "star" side by side. The artificial star can be varied in brightness and color until it matches the real star in appearance; the amount of energy provided to the artificial star image to accomplish this match is a measure of the luminous flux from the real star.

A widely used modern method of comparing the amounts of light received from stars is that of *photographic photometry,* in which the degrees of blackness and the sizes of star images on photographic negatives are measured. The most common of several techniques in use is to pass a collimated light beam through the negative, with the star image to be measured centered in the beam. The blacker, larger image of a brighter star attenuates the beam more than the smaller, less black image of a fainter star; the attenuation of the beam, measured with a photoelectric cell, thus serves as a measure of the star's light.

Actually, even large telescopes do not resolve the stars into disks. In principle, stars make point-like images; the photographic images of bright stars, however, are larger than those of faint ones because of the *turbidity* of the photographic emulsions — the fact that during the time exposure the light from the brighter star images is scattered about within the emulsion, exposing a larger area than that occupied by the true image. Bright star images, in other words, are very much overexposed in long time exposures. Atmospheric turbulence (seeing) also enlarges star images. To a lesser extent (at least in a good-quality telescope) optical imperfections (aberrations) and diffraction account for some image enlargement. Finally, images of the faintest stars, those which are just barely recorded photographically, would not be geometrical points even if these other effects were

absent because of the finite resolution of the photographic emulsions used. On the most common types of photographic plates and films used in telescopic photography the smallest images attainable, in good seeing, are usually about 0.03 mm in diameter.

The most modern and precise method of stellar photometry, *photoelectric photometry,* has come into widespread use since World War II. The light from a star, coming to a focus at the

FIG. 20-1   Effect of telescope aperture and exposure time. *(Top left)* 14 in., 1 min. Stars to 12th magnitude. *(Top right)* 60 in., 1 min. Stars to 15th magnitude. *(Lower left)* 60 in., 27 min. Stars to 18th magnitude. *(Lower right)* 60 in., 4 hr. Stars to 20th magnitude. *(Mount Wilson and Palomar Observatories.)*

focal plane of the telescope, is allowed to pass through a small hole in a metal plate and thence onto the photosensitive surface of a photomultiplier (Section 11.6g). The electric current generated in the photomultiplier is amplified and recorded and provides an accurate measure of the light passing through the hole. Because even the darkest night sky is not completely dark, however, not all the light striking the photomultiplier is due to the star, but some is provided by the light of the night sky, which is, of course, also gathered by the telescope and passed through the hole. Consequently, the hole is next moved to one side of the star image, so that only light from the sky passes through and is recorded; the difference between the first and second readings is a measure of the star's light entering the telescope.

(b) The Magnitude Scale

In 1856, after early methods of stellar photometry had been developed, Norman R. Pogson proposed the quantitative scale of stellar magnitudes that is now generally adopted. He noted, as did Herschel, that we receive about 100 times as much light from a star of the first magnitude as from one of the sixth, and that, therefore, a difference of five magnitudes corresponds to a ratio in luminous flux of 100:1. It is a widely accepted assumption in the physiology of sense perception that what appear to be equal intervals of brightness are really equal *ratios* of luminous energy. Pogson suggested, therefore, that the ratio of light flux corresponding to a step of one magnitude be the fifth root of 100, which is about 2.512. Thus, a fifth-magnitude star gives us 2.512 times as much light as one of sixth magnitude, and a fourth-magnitude star, 2.512 times as much light as a fifth, or 2.512 $\times$ 2.512 times as much as a sixth-magnitude star. From stars of third, second, and first magnitude, we receive 2.512^3, 2.512^4, and 2.512^5 ($= 100$) times as much light as from a sixth-magnitude star. By assigning a magnitude of 1.0 to the bright stars Aldebaran and Altair, Pogson's new scale gave

magnitudes that agreed roughly with those in current use at the time.

TABLE 20.1 Magnitude Differences and Light Ratios

DIFFERENCE IN MAGNITUDE	RATIO OF LIGHT
0.0	1:1
0.5	1.6:1
0.75	2:1
1.0	2.5:1
1.5	4:1
2.0	6.3:1
2.5	10:1
3.0	16:1
4.0	40:1
5.0	100:1
6.0	251:1
10.0	10,000:1
15.0	1,000,000:1
20.0	100,000,000:1
25.0	10,000,000,000:1

Table 20.1 gives the approximate ratios of light flux corresponding to several selected magnitude differences. Note that a given ratio of light flux, whether between bright or faint stars, always corresponds to the same magnitude interval. Further, note that the numerically *smaller* magnitudes are associated with the *brighter* stars; a numerically *large* magnitude, therefore, refers to a faint star.†

With optical aid, stars can be seen that are beyond the reach of the naked eye. The *limiting magnitude* of a telescope is the magnitude of the faintest stars that can be seen with that telescope under ideal conditions. A 6-in. telescope, for example, has a limiting magnitude of about 13. The *photographic limiting magnitude* of a telescope is the magnitude of the faintest stars that can be photographed with it. The photographic limiting magnitude of the 200-inch telescope on Palomar Mountain is about 23.5.

---

†If m_1 and m_2 are the magnitudes corresponding to stars from which we receive light flux in the amounts l_1 and l_2, the difference between m_1 and m_2 is defined by

$$m_1 - m_2 = 2.5 \log \frac{l_2}{l_1}.$$

The so-called first-magnitude stars are not all of the same apparent brightness. The brightest-appearing star, Sirius, sends us about 10 times as much light as the average star of first magnitude, and so has a magnitude of $1.0 - 2.5$, or of about -1.5. Several of the planets appear even brighter; Venus, at its brightest, is of magnitude -4. The sun has a magnitude of -26.5. Some magnitude data are given in Table 20.2. It is of interest to note

TABLE 20.2 Some Magnitude Data

OBJECT	MAGNITUDE
Sun	-26.5
Full moon	-12.5
Venus (at brightest)	-4
Jupiter; Mars (at brightest)	-2
Sirius	-1.4
Aldebaran: Altair	1.0
Naked-eye limit	6.5
Binocular limit	10
6-in. telescope limit	13
200-inch (visual) limit	20
200-inch photographic limit	23.5

that the brightness of the sun and Sirius differ by 25 magnitudes — a factor of 10 billion (10^{10}) in light energy, or flux — and that we also receive 10 billion times as much light from Sirius as from the faintest stars that can be photographed with the 200-inch telescope. The entire range of light flux represented in Table 20.2 covers a ratio of about 100 quintillion (10^{20}) to 1.

(c) Magnitude Standards

Magnitudes, as explained above, are defined in terms of ratios of the light received from stars (or other celestial objects); thus magnitude differences between objects indicate the *relative* amounts of luminous flux received from them. To set the scale unambiguously, however, it is necessary to pick out some standard, or standards, with respect to which all other stars are to be compared. Originally, it was planned to define the brightest stars as having first magnitude. The brightest stars, on the other hand, are not always conveniently disposed for observation. At one time, by agreement among

astronomers, it was decided to define as standards several stars in the vicinity of the north celestial pole, which is always visible from northern observatories. This group of stars was called the *North Polar Sequence.* The magnitudes of these stars were intended to serve as a permanent basis of the magnitude system. The actual magnitude values were assigned so that the brightest-appearing stars in the sky, in accordance with previous convention, would still be reasonably near the first magnitude. At the Mount Wilson Observatory, 139 secondary standard regions, known as *selected areas,* were set up over most of the sky by comparing stars in those areas with the North Polar Sequence. Standard regions were also set up by the Harvard College Observatory, which included regions in the southern sky observed by the Southern Stations of the Observatory in South America and South Africa.

Since the development of photoelectric photometry, however, it became clear that the North Polar Sequence and the selected areas, which had been based largely upon photographic photometry, were inadequate for calibrating the precise measures that are obtained today. Consequently, the old standards have been superseded by accurate photoelectric measures of a large number of new standard stars distributed over the sky.

20.2 THE "REAL" BRIGHTNESSES OF STARS

Even if all stars were identical, and if interstellar space were entirely free of opaque matter, stars would not all appear to have the same brightness, because they are at different distances from us, and the light that we receive from a star is *inversely proportional to the square of its distance* (Section 10.1a). The apparent brightnesses of stars therefore do not provide a basis for comparing the amounts of light that they actually emit into space. To make such a comparison we would first have to calculate how much light we would receive from each star if all stars were at the same distance from us. Fortunately, if we know a star's distance we can make such a calculation.

(a) Absolute Magnitudes

The sun gives us billions of times as much light as any of the other stars, but on the other hand, it is hundreds of thousands of times closer to us than any other star is. To compare the intrinsic luminous outputs of the sun and of other stars, we first have to determine what magnitude the sun would have if it were at a specified distance, typical of the distances of the other stars. Suppose we choose 10 pc as a more or less representative distance of the nearest stars. Since 1 parsec is about 200,000 AU, the sun would be 2,000,000 times as distant as it is now if it were removed to a distance of 10 pc; consequently, it would deliver to us $(\frac{1}{2,000,000})^2$ or $1/(4 \times 10^{12})$ of the light it now sends. A factor of 4×10^{12} corresponds to about 31½ magnitudes (which can be verified by raising 2.512 to the 31.5 power). The sun, therefore, if removed to a distance of 10 pc, would appear fainter by some 31½ magnitudes than its present magnitude of −26.5; that is, the sun would then appear as a fifth-magnitude star.

Similarly, we can use the inverse-square law of light to calculate how luminous all other stars of known distance would appear if they were 10 pc away. Suppose, for example, that a tenth-magnitude star has a distance of 100 pc. If it were only 10 pc away, it would be only one tenth as far away, and hence 100 times as bright — a difference of five magnitudes. At 10 pc, therefore, it too would appear as a fifth-magnitude star.

We now define the *absolute magnitude* of a star as the magnitude that star would have if it were at the standard distance of 10 pc (about 32.6 LY). The absolute magnitude of the sun is about +5. Most stars have absolute magnitudes that lie in the range 0 to +15. The extreme range of absolute magnitudes observed for normal stars is −9 to +20, a range of a factor of nearly 10^{12} in

intrinsic light output. The absolute magnitudes of stars are measures of how bright they "really" are; they provide a basis for comparing the actual amount of light emitted by stars.

(b) Distance Modulus

The absolute magnitude of a star is the magnitude that the star *would* have if it were at a distance of 10 pc, and therefore is a measure of the actual rate of emission of visible light energy by the star, which, of course, is independent of the star's actual distance. On the other hand, the magnitude of a star (sometimes called the *apparent magnitude* to avoid confusion with *absolute magnitude*) is a measure of how bright the star *appears* to be, and this depends on both the star's actual rate of light output and its distance. The *difference* between the star's apparent magnitude, symbolized m, and absolute magnitude, symbolized M, can, however, be calculated from the inverse-square law of light and from a knowledge of how much greater or less than 10 pc the star's distance actually is. The difference $m - M$ therefore depends only on the distance of the star and is called its *distance modulus*. The distance modulus of a star, then, is a measure of its distance, from which the actual distance, in parsecs, can be calculated.†

---

†Let $l(r)$ be the observed light of a star at its actual distance, r, and $l(10)$ the amount of light we would receive from it if it were a distance of 10 pc. From the definition of magnitudes, we have

$$m - M = 2.5 \log \frac{l(10)}{l(r)},$$

and from the inverse-square law of light,

$$\frac{l(10)}{l(r)} = \left(\frac{r}{10}\right)^2.$$

Combining the above equations, we obtain

$$m - M = 5 \log \frac{r}{10}.$$

The quantity, $5 \log (r/10)$, then, is the distance modulus.

For example, suppose the distance modulus of a star is 10 magnitudes. Ten magnitudes (see Table 20.1) corresponds to a ratio of 10,000:1 in light. Thus, we actually receive from the star $\frac{1}{10,000}$ of the light that we would receive if it were 10 pc away; it must, therefore, be 100 times as distant as 10 pc, or at a distance of 1000 pc. In succeeding chapters we shall see that the absolute magnitude of a star can often be inferred from its spectrum or from some other observable characteristic. Since the apparent magnitude of a star can be observed, a knowledge of its absolute magnitude is equivalent to a knowledge of its distance.

Most of us make use of this same principle, subconsciously, in everyday life. Every experienced motorist, for example, has an intuitive notion of the actual brightness of a stop light. If, while driving down the highway at night, he sees a stop light, he judges its distance from its apparent faintness. In other words, the difference between the light's *apparent* and *real* brightness indicates its distance. The computation of a star's distance from its distance modulus is analogous.

20.3 COLORS OF STARS

Until now, this discussion has ignored the fact that stars have different colors and that all colors do not produce an equal response in the human eye. The apparent brightness of a star can depend to some extent, therefore, upon its color.

(a) Color Response of the Eye and Other Detecting Devices

Every device for detecting light has a particular color or spectral sensitivity. The human eye, for example, is most sensitive to green and yellow light; it has a lower sensitivity to the shorter wavelengths of blue and violet light and to the longer wavelengths of orange and red light. It does not respond at all to ultraviolet or to infrared radiation. The eye, in fact, responds roughly to the same kind of light that the sun emits most intensely; this coincidence is probably not accidental — the eye may have evolved to respond to the kind of light most available on earth.

Another detecting device is the photographic plate (or film). The early photographic emulsions, before the development of yellow- and red-sensitive and panchromatic emulsions, were sensitive only to violet and blue light and did not respond to light of wavelengths longer than about 5000 Å (in the blue-green). The basic photographic emulsion is still sensitive to violet and blue; dyes must be added to the basic emulsion to make it sensitive to longer wavelengths.

Suppose, now, that the total amount of light energy entering a telescope from each of two stars is exactly the same if light of all wavelengths is considered, but that one star emits most of its light in the blue spectral region and the other in the yellow spectral region. If these stars are observed visually (that is, by looking at them through the telescope) the yellow one will appear brighter, that is, will have a numerically smaller magnitude, because the eye is less sensitive to most of the light emitted by the blue star. If the stars are photographed on a blue-sensitive photographic plate, however, the blue star will produce the more conspicuous image; measures of the photographic images will show the blue star appearing brighter and having the smaller magnitude. Consequently, when a magnitude system is defined, it is necessary also to specify how the magnitudes are to be measured — that is, what detecting device is to be used.

Magnitudes, whether apparent or absolute, that are based on stellar brightness as they are observed with the human eye are called *visual magnitudes* or *absolute visual magnitudes* and are symbolized m_v and M_v, respectively. Magnitudes based on measures of star images on standard violet- and blue-sensitive photographic emulsions are called *photographic magnitudes* and *absolute*

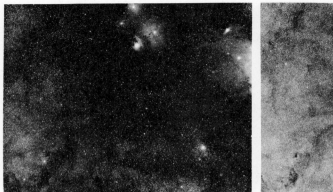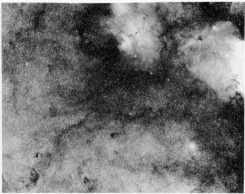

FIG. 20-2   Two photographs of the same region of the Milky Way taken with the 48-inch Schmidt telescope; *(left)* in blue light, *(right)* in red light. Note the difference with which the stars and nebulae show up in the different colors. *(National Geographic Society-Palomar Observatory Sky Survey.)*

photographic magnitudes and are symbolized m_{pg} and M_{pg}. When green- and yellow-sensitive photographic emulsions were perfected, it was found that they could be used in conjunction with a suitable color filter to approximate closely the spectral response of the human eye; stellar photographs so obtained provide *photovisual magnitudes* (m_{pv} and M_{pv}). Today photographic plates or photomultipliers can be used in conjunction with many kinds of color filters to produce a great variety of different magnitude systems — for example, red magnitudes, ultraviolet magnitudes, infrared magnitudes, and so on. A certain few spectral bands have become more or less standard, however, and are now widely used to define magnitudes.

(b) Color Indices

Until about 1955, the most commonly used magnitudes were photographic magnitudes and photovisual magnitudes. The *difference* between the photographic and photovisual magnitudes of a star is called its *color index, CI:*

$$CI = m_{pg} - m_{pv}.$$

Since the inverse-square law of light applies equally to all wavelengths, the color index of a star would not change if the star's distance were changed. The color index of a star, therefore, can be defined in terms of either its apparent or absolute photographic and photovisual magnitudes:

$$M_{pg} - M_{pv} = m_{pg} - m_{pv} = CI.$$

The system of photographic and photovisual magnitudes and color index is called the international system. (Actually, space is not completely transparent; microscopic particles of "cosmic dust" not only dim the light of stars in some parts of the Galaxy, but also *redden* their light. Chapter 26 includes a discussion of the effect of this interstellar absorption of starlight and how it can be allowed for. In this chapter, however, space is assumed to be perfectly transparent.)

A very blue star will appear brighter on a blue-sensitive photograph than on a yellow-sensitive one; its photographic magnitude, therefore, will be algebraically *less* than its photovisual magnitude, and its color index will be *negative*. A yellow or red star, on the other hand, will have a brighter (smaller) photovisual magnitude, and a *positive* color index. Color indices, therefore, provide measures of the *colors* of stars. Colors, in turn, indicate the temperatures of stars. Photographic and photovisual magnitudes are adjusted to be the same, so as to give a color index of zero to a star with a temperature of about 10,000°K (a spectral–type A star — see Chapter 21). Color indices of stars range from −0.6 for the bluest to more than 2.0 for the reddest.

If a star field is photographed on separate blue-sensitive and yellow-sensitive plates, the photographic and photovisual magnitudes and color indices so obtained comprise a powerful tool for determining the temperatures of all the stars appearing on the photographs. Since the star images on these two black and white photographs can be accurately measured, the color-index method is a more precise means of obtaining colors and temperatures of stars than the method of color photography would be. Moreover, color films generally require excessively long exposure times. (Astronomical photographs in color can and have been made, however.)

Since the development of the more precise methods of photoelectric photometry, it has become customary to measure the light of stars in three or more spectral regions instead of only two, and thereby to increase the amount of information obtained about their light. In the early 1950s, H. L. Johnson and W. W. Morgan defined the U,B,V magnitude system, which is now rather generally adopted. The U,B,V system utilizes measures in the ultraviolet (U), blue (B), and green-yellow or visual (V) spectral regions. It provides two independent color indices, B-V and U-B. There are many other more or less standardized magnitude systems.

The maximum information is obtained in photoelectric spectral scanning. In this technique, a photomultiplier replaces the photographic plate in a spectrograph. The photomultiplier scans the spectrum (usually by slowly rotating the grating), and the output of the tube is recorded, yielding a point by point record of the distribution of energy according to wavelength in the star.

20.4 BOLOMETRIC MAGNITUDES AND LUMINOSITIES

All the magnitude systems so far discussed refer only to certain spectral regions. A magnitude system based on *all* the electromagnetic energy reaching the earth from the stars would seem to be more fundamental. Magnitudes so based are called *bolometric magnitudes, m_{bol}*, and the bolometric magnitudes that stars would have at a distance of 10 pc are *absolute bolometric magnitudes, M_{bol}*. Unfortunately, bolometric magnitudes cannot be observed directly because some wavelengths of electromagnetic energy do not penetrate the earth's atmosphere. It is true that most of the radiation from stars such as the sun does reach the earth's surface, and for such stars bolometric magnitudes can be determined approximately. A large part of the energy from stars that are substantially hotter or cooler than the sun, however, lies in the far ultraviolet or in the infrared, and is either blocked by the earth's atmosphere or cannot be observed easily; for those stars, bolometric magnitudes can only be estimated or calculated from theoretical considerations. Recently, observations from rockets and artificial satellites have begun to provide measures in spectral regions to which the earth's atmosphere is opaque, and have been of much value in improving our knowledge of bolometric magnitudes of stars — especially of those stars of extreme temperatures.

The difference between the photovisual and bolometric magnitudes of a star is called its *bolometric correction, BC:*

$$m_{pv} - m_{bol} = M_{pv} - M_{bol} = BC.†$$

A bolometric correction, then, is a sort of color index, except that for stars of most temperatures it is not obtained from direct observation but from theoretical calculation. Bolometric corrections are

---

†Sometimes bolometric corrections are defined as $m_{bol} - m_{pv}$, in which case they are always negative, rather than positive, as in the convention used in this book.

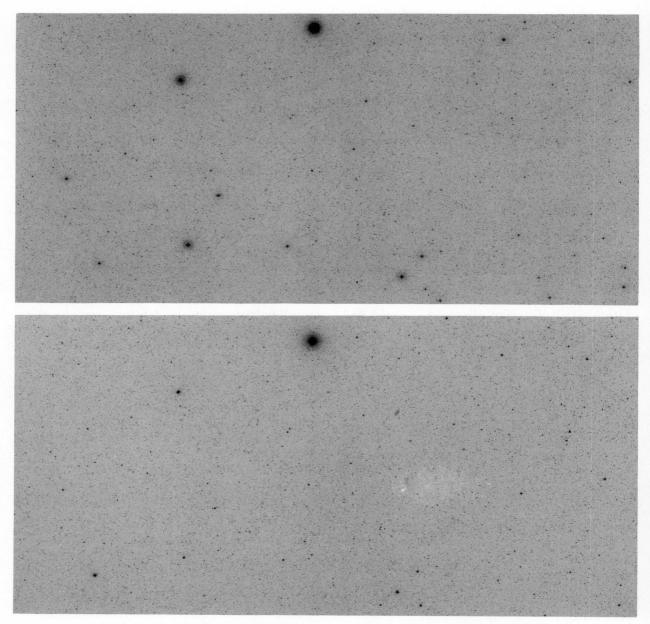

FIG. 20-3   Two photographs of the same region of the sky (in Coma Berenices). The upper is in blue light, and the lower in red; note how the colors of the stars are apparent from comparison of the two. These are negative prints, the stars showing as black dots on a white sky. *(National Geographic Society-Palomar Observatory Sky Survey.)*

adjusted to be nearly zero for stars like the sun. They are positive quantities for stars that are much hotter or cooler than the sun because these stars would appear brighter (would have numerically smaller bolometric than photovisual magnitudes) if all of their radiant energy could be observed. For the hottest stars, calculated bolometric corrections reach as much as eight magnitudes.

(a) Stellar Luminosities

The absolute bolometric magnitude of a star is a measure of the rate of its entire output of radiant energy. The rate at which a star pours radiant energy into space, usually expressed in ergs per second, is called its *luminosity*. Absolute bolometric magnitudes of stars bear the same relation to their luminosities as their visual magnitudes do to the amount of light we receive from them.† The steps required to determine the luminosity of a star are the following:

(1) Its apparent photovisual magnitude is observed.

(2) The bolometric correction is calculated from theory and from a knowledge of the star's temperature. The latter can be estimated, say, from the spectrum or the color of the star. The bolometric correction is applied to the star's photovisual magnitude to obtain its bolometric magnitude.

(3) The absolute bolometric magnitude is calculated from the star's apparent bolometric magnitude and its distance.

(4) The luminosity is found by comparing the star's absolute bolometric magnitude to that of some other star (say, the sun) whose luminosity is already known.

(b) The Luminosity of the Sun

The first step in determining the luminosity of the sun is to measure the rate at which its radia-

---

†If two stars of absolute bolometric magnitudes, $M_{bol}(1)$ and $M_{bol}(2)$, have luminosities L_1 and L_2,

$$M_{bol}(1) - M_{bol}(2) = 2.5 \log \frac{L_2}{L_1}.$$

tion falls upon the earth. There are various devices for measuring this quantity. In one method, the sun's radiation is allowed to fall upon the blackened surface of some metal (for example, platinum or silver), which absorbs this radiant energy and rises in temperature. The rate of the temperature rise can be measured by the increased resistance that the metal offers to an electric current. A simpler device measures the rate at which the temperature of water rises as the sun's radiation is absorbed by it. The measures obtained of the sun's radiation must, in any case, be corrected for the attenuation of the sunlight passing through the earth's atmosphere. (Measures of stellar magnitudes must similarly be corrected.)

It is found that a surface of 1 cm² just outside the atmosphere and oriented perpendicular to the direction of the sun receives from the sun 1.36 × 10⁶ ergs/sec, or 1.95 cal/min. This value is known as the *solar constant*.

The total energy that leaves the sun during an interval of 1 second diverges outward, away from the sun, in all directions. Since one astronomical unit is 1.49 × 10¹³ cm, the area of the spherical surface over which the solar radiation has spread by the time it reaches the earth's distance from the sun (about 8 minutes later) is 2.8 × 10²⁷ cm². The solar constant of 1.36 × 10⁶ ergs/sec/cm² is the energy that crosses just one of those square centimeters. The total energy that leaves the sun in 1 second — its luminosity — is thus 1.36 × 10⁶ × 2.8 × 10²⁷ = 3.8 × 10³³ ergs/sec. The corresponding absolute bolometric magnitude of the sun is +4.6.

The sun's luminosity is equivalent to 5 × 10²³ horsepower. A dramatic illustration of the magnitude of that amount of energy is obtained by imagining a bridge of ice 2 mi wide and 1 mi thick and extending over the nearly 100-million-mile span from the earth to the sun. If all the sun's radiation could be directed along the bridge, it would be enough to melt the entire column of ice in 1 second.

Exercises

1. Suppose that star A is just barely visible through a 6-in. telescope and that star B is just barely visible through a 24-in. telescope. Which star gives us more light and by what factor? What is the approximate difference in magnitudes of the stars?

2. Show that if all stars were of the same intrinsic brightness, the "depth" in space that could be penetrated by a telescope would be proportional to its aperture.

3. What magnitude would be assigned to an object from which we receive
(a) 100 times as much light as from Venus when it is at its brightest?
(b) $\frac{1}{10}$ as much light as a star at the naked-eye limit?
(c) 10,000 times as much light as from stars at the photographic limit of the 200-inch telescope?

4. Find the absolute magnitude of each of the following stars:
(a) apparent magnitude $m = 7$; distance $r = 10$ pc
(b) $m = 20$; $r = 100$ pc
(c) $m = 0$; $r = 100$ pc
(d) $m = 3\frac{1}{2}$; $r = 5$ pc
(e) $m = 17\frac{1}{2}$; $r = 20$ pc
(f) $m = -5$; $r = \frac{1}{10}$ pc (there is no such star as this)

5. What are the distances of stars of the following apparent magnitudes m and absolute magnitudes M?
(a) $m = 10$; $M = 5$
(b) $m = 5$; $M = 10$
(c) $m = 13.5$; $M = 15$
(d) $m = 20$; $M = 10$
(e) $m = -26.5$; $M = 5$

6. Draw a diagram showing how two stars of equal bolometric magnitude, but one of which is blue and the other red, would appear on photographs sensitive to blue and to yellow light.

7. What is the distance of the following star:

$$m_{pg} = 12.4; \quad M_{pv} = 6.8; \quad CI = 0.6?$$

8. If a star has a bolometric correction of six magnitudes, what fraction of its radiant energy is observed in its visual magnitude?

9. The apparent photovisual magnitude of a star is 10.4. Its bolometric correction is 0.8. Its parallax is $0\rlap{.}''001$. What is its luminosity?

    Answer: 3.8×10^{35} ergs/sec

10. By what steps do you suppose the absolute bolometric magnitude of the sun is found?

CHAPTER 21

Spectra of Stars

About 1665 Newton showed (perhaps not for the first time) that white sunlight is really a composite of all colors of the rainbow (Section 10.2c), and that the various colors, or wavelengths, of light could be separated by passing the light through a glass prism. William Wollaston (Section 10.3) first observed dark lines in the solar spectrum, and Joseph Fraunhofer catalogued about 600 such dark lines. As early as 1823, Fraunhofer observed that stars, like the sun, also have spectra that are characterized by dark lines crossing a continuous band of color. Sir William Huggins, in 1864, first identified some of the lines in stellar spectra with those of known terrestrial elements.

21.1 CLASSIFICATION OF STELLAR SPECTRA

When the spectra of different stars were observed, it was found that they differed greatly among themselves. In 1863 the Jesuit astronomer Angelo Secchi classified stars into four groups according to the general arrangement of the dark lines in their spectra. Secchi's scheme subsequently was modified and augmented, until today we recognize seven such principal *spectral classes*.

(a) The Spectral Sequence

As we have seen (Sections 10.3 and 10.5), each dark line in a stellar spectrum is due to the presence of a particular chemical element in the atmosphere of the star observed. It might seem, therefore, that stellar spectra differ from each other because of differences in the chemical makeup of the stars. Actually, the differences in stellar spectra are due mostly to the widely differing temperatures in the outer layers of the various stars.

Hydrogen, for example, is by far the most abundant element in all stars, except probably in those at an advanced stage of evolution (Chapter 30). In the atmospheres of the very hottest stars, however, hydrogen is completely ionized, and can, therefore, produce no absorption lines. In the atmospheres of the coolest stars, on the other hand, hydrogen is neutral and can produce absorption lines. In these stars, however, since practically all the hydrogen atoms are in the lowest energy state (unexcited), they can absorb only those photons that can lift them from that first energy level to higher ones; the photons so absorbed produce the *Lyman series* of absorption lines (Section 10.5b), which lies in the unobservable ultraviolet part of the spectrum. In a stellar atmosphere with a temperature of about 10,000°K, many hydrogen atoms are not ionized; nevertheless, an appreciable number of them are excited to the second energy level, from which they can absorb additional photons and rise to still higher levels of excitation. These photons correspond to the wavelengths of the *Balmer series,* which is in the part of the spectrum that is readily observable. Absorption lines due to hydrogen, therefore, are strongest in the spectra of stars whose atmospheres have temperatures near 10,000°K, and they are less conspicuous in the spectra of both hotter and cooler stars, even though hydrogen is, roughly, equally abundant in all the stars. Similarly, every other chemical element, in each of its possible stages of ionization, has a characteristic temperature at which it is most effective in producing absorption lines in the observable part of the spectrum.

Once we have ascertained how the temperature of a star can determine the physical states of the gases in its outer layers, and thus their ability to produce absorption lines, we need only to observe what patterns of absorption lines are present in the spectrum of a star to learn its temperature. We can therefore arrange the seven classes of stellar spectra in a continuous sequence in order of decreasing temperature. In the hottest stars (temperatures over 25,000°K) only lines of ionized helium and highly ionized atoms of other elements are conspicuous. Hydrogen lines are strongest in stars with atmospheric temperatures of about 10,000°K. Ionized metals provide the most conspicuous lines in stars with temperatures from 6000° to 8000°K. Lines of neutral metals are the strongest in somewhat cooler stars. In the coolest stars (below 4000°K), bands of some molecules are very strong. The most important among the molecular bands are those due to titanium oxide, a tenacious chemical compound which can exist at the temperatures of the cooler stars. The sequence of spectral types is summarized in Table 21.1 and Figure 21-1. [Hot stars (types O, B, A) are sometimes referred to as having *early* spectral types, and cool stars (G, K, M) as having *late* spectral types. This jargon derived from old ideas of stellar evolution that are now discarded, but the terminology is still in wide use.]

The spectral classes of stars listed in Table 21.1 can be subdivided into tenths; thus a star of spectral class A5 is midway in the range of A-type

TABLE 21.1 Spectral Sequence

SPECTRAL CLASS	COLOR	APPROXIMATE TEMPERATURE (°K)	PRINCIPAL FEATURES	STELLAR EXAMPLES
O	Blue	>25,000	Relatively few absorption lines in observable spectrum. Lines of ionized helium, doubly ionized nitrogen, triply ionized silicon, and other lines of highly ionized atoms. Hydrogen lines appear only weakly.	10 Lacertae
B	Blue	11,000–25,000	Lines of neutral helium, singly and doubly ionized silicon, singly ionized oxygen and magnesium. Hydrogen lines more pronounced than in O-type stars.	Rigel Spica
A	Blue	7,500–11,000	Strong lines of hydrogen. Also lines of singly ionized magnesium, silicon, iron, titanium, calcium, and others. Lines of some neutral metals show weakly.	Sirius Vega
F	Blue to white	6,000–7,500	Hydrogen lines are weaker than in A-type stars but are still conspicuous. Lines of singly ionized calcium, iron, and chromium, and also lines of neutral iron and chromium are present, as are lines of other neutral metals.	Canopus Procyon
G	White to yellow	5,000–6,000	Lines of ionized calcium are the most conspicuous spectral features. Many lines of ionized and neutral metals are present. Hydrogen lines are weaker even than in F-type stars. Bands of CH, the hydrocarbon radical, are strong.	Sun Capella
K	Orange to red	3,500–5,000	Lines of neutral metals predominate. The CH bands are still present.	Arcturus Aldebaran
M	Red	<3,500	Strong lines of neutral metals and molecular bands of titanium oxide dominate.	Betelgeuse Antares

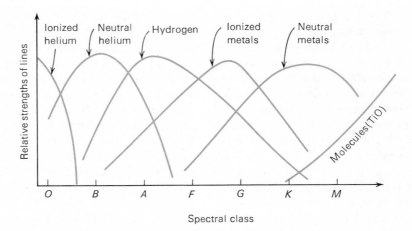

FIG. 21-1 Relative intensities of different absorption lines in stars at various places in the spectral sequence.

stars — that is, halfway between stars of type A0 and F0. The sun is of spectral class G2 — two tenths of the way from class G0 to K0.

The famous American astronomer Henry Norris Russell proposed a scheme by which every student can remember the order of classes in the spectral sequence: The class letters are the first letters in the words, "Oh, Be A Fine Girl, Kiss Me!"

(b) The Role of Pressure*

Although temperature is the most important factor that determines the characteristics of a stellar spectrum, other factors are present. Two stars whose atmospheres have the same temperatures but different pressures, for example, will have somewhat different spectra. The degree to which atoms of a particular kind are ionized in a gas depends on the rate at which those atoms can recapture electrons, and this, in turn, depends on how closely the atoms, ions, and electrons in the gas are packed together. If the density of the gas is high, the particles are close together; it is easier for ions to capture electrons, and the fraction of atoms that are ionized at any instant of time is then lower than if the gas density is low. At any given temperature, the density of a gas is proportional to its pressure (*Boyle's law*); consequently, in a star whose atmospheric gases are at a high pressure, a smaller fraction of each kind of atom will be ionized than in a star

of the same temperature whose atmospheric gases are at a low pressure.

Atmospheric pressure, either on a planet or a star, results from the *weight* of the atmosphere — that is, from the gravitational attraction exerted on the atmosphere. The weight of a unit mass upon the surface of a spherical body is called the *surface* gravity of that body; it is proportional to the mass of the body and is inversely proportional to the square of its radius (Section 4.3c). We shall see in Chapter 22 that stars differ enormously from one another in radius, but only moderately in mass. Two stars that differ very greatly in radii may thus have masses that are only slightly different, and consequently these stars must have very different surface gravities (see Exercise 4). The photosphere, where most of the spectral lines are formed, may extend in depth through a different total mass of material in a large star than in a small one, but the difference is generally small compared to the difference in surface gravity. Consequently, the weight of the atmosphere of the larger star, and hence its pressure, is less than in the smaller star. Even if the two stars are at the same temperature, therefore, the gases in the larger one must be at a higher degree of ionization than those in the smaller one. The spectrum of a larger star thus resembles that of a smaller star of greater temperature.

The resemblance is not, however, perfect. The lower

atmospheric pressure of the larger star partially compensates for its lower temperature, but the compensation is not exactly the same for all elements; there are subtle differences in the fractions of the atoms of the various elements that are ionized in the two stars. Spectroscopic evidence for these differences was first noted in 1913 by Adams and Kohlschütter at the Mount Wilson Observatory. A trained spectroscopist can tell from its spectrum whether a star is a giant with a tenuous atmosphere, or a smaller, more compact star of somewhat higher temperature. The importance of the subtle spectroscopic differences between large and small stars will become clear when we consider the method of spectroscopic parallaxes in Chapter 23

(c) Special Spectral Classes*

The vast majority of all known stars fit into the spectral sequence outlined in Section 21.1. Some relatively rare stars, however, require special classification. Among the most important of these groups are the following:

The Wolf-Rayet Stars   The Wolf-Rayet stars are O-type stars that have broad emission lines in their spectra. These emission lines are presumed to originate from material that has been ejected from the star at a high velocity. The ejected gas absorbs light from the star and reemits it as emission lines; the different Doppler shifts of the light coming from different parts of the expanding envelope of gas smear each emission line into a broad "band," as observed. Some of the more prominent broad emission lines, or bands, are due to carbon in some stars and to nitrogen in others. Wolf-Rayet stars probably have abnormally high abundances of carbon or nitrogen.

Early-Type Emission-Line Stars   The Of, Be, and Ae stars are stars whose spectra display bright emission lines of hydrogen. These lines, like those of the Wolf-Rayet stars, are presumed to come from extended gaseous envelopes surrounding the stars.

Peculiar A Stars   The *peculiar A stars* are stars of spectral type A that show abnormally strong lines of certain ionized metals. The intensities of these absorption lines sometimes vary periodically with time. These effects may be due to magnetic fields in the stars, or possibly to unequal distribution of the chemical elements over the stars' surfaces.

R Stars   R-type stars are stars with spectral characteristics of K-type stars, except that molecular bands of C_2 and CN (carbon and cyanogen, respectively) are present.

FIG. 21-2   Spectra of several stars of representative spectral classes. *(UCLA Observatory.)*

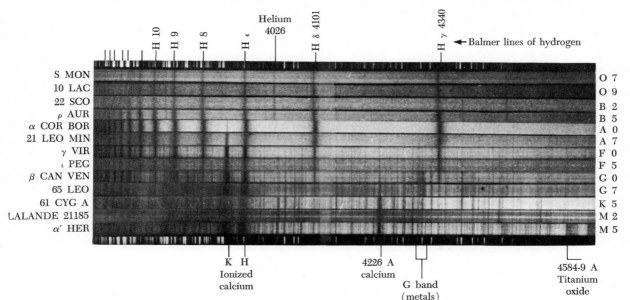

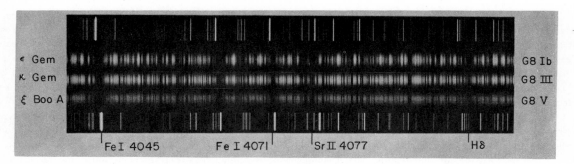

FIG. 21-3   Spectra of three G8 stars; *from top to bottom:* a supergiant, a giant, and a main-sequence star. Note the subtle differences in spectra due to the lower surface gravities of the larger stars. *(Lick Observatory.)*

N Stars   The N stars are like M-type stars, except that bands of C_2, CN, and CH are strong, rather than those of titanium oxide.

S Stars   The S stars are like M–type stars except that molecular bands of zirconium oxide and lanthanum oxide are present in addition to, or instead of, bands of titanium oxide.

Unusual stars are discussed more exhaustively in Chapter 25.

21.2 SPECTRUM ANALYSIS AND THE STUDY OF THE STELLAR ATMOSPHERES

At least half of the observing time of many large telescopes is assigned to the study of stellar spectra. Analyses of stellar spectra yield an enormous amount of information about the stars. Unfortunately, spectrum analysis is a very complicated subject; in the following section the approach is only briefly outlined. Some of the more important data that can be gleaned from the detailed study of stellar spectra will be summarized at the end of this section.

(a) *How Stellar Spectra Are Observed**

The design of a spectroscope or spectrograph is described in Section 11.6e. Stellar spectra were first observed by means of a spectroscope placed at the focus of a telescope. With a magnifying glass (eyepiece) the observer viewed the spectrum of the light from a star gathered by the telescope and passed through the apparatus. Today, however, stellar spectra are usually observed photographically. With a long time exposure we can photograph a spectrum that is far too faint to be seen visually; moreover, a photograph is a permanent record that can be measured and reproduced at will.

In practice, the spectra of the star and a laboratory source are both photographed on the same negative. The laboratory source is usually iron vaporized in an electric arc; the light emitted from the glowing iron vapor is passed through portions of the slit of the spectrograph adjacent to the portion through which the starlight is passed. The spectrum of the iron arc consists of many bright emission lines whose wavelengths have been measured accurately in the laboratory. They serve as a convenient standard with which the absorption (or emission) lines in the star's spectrum can be compared, so

FIG. 21-4   A typical stellar spectrogram. The bright streak along the middle, crossed by dark lines, is the spectrum of the star ξ Boötis. The bright lines on either side form the comparison (emission) spectrum of iron.

that the wavelengths of the stellar lines can be determined. In this way, the chemical elements that give rise to the lines can be identified, and the Doppler shift of those lines (due to the radial velocity of the star) can be measured.

Spectra are also observed photoelectrically. The observer, instead of focusing the spectrum of the starlight directly upon a photographic plate, allows part of the spectrum to fall upon the photosensitive surface of a photomultiplier tube. The current generated in the tube is a measure of the total energy in that wavelength range of the stellar spectrum. If the photomultiplier tube is then slowly moved along the spectrum, the current generated fluctuates in step with the variation of intensity with wavelength in the spectrum. A continuous recording of the current output from the tube provides a permanent and accurate measure of the star's spectral energy distribution.

If a stellar spectrum has been photographed, the blackness of the image of the spectrum at various wavelengths indicates the intensity of the starlight at those wavelengths. The intensity of the light drops down, for example, where there is an absorption line. These variations in intensity along a stellar spectrum, and through its various absorption lines, can be exhibited conveniently if the spectrogram (the photograph of the spectrum) is scanned with a *microphotometer* (or *microdensitometer*). In this device a narrow beam of light is passed through

the spectrogram. The intensity of the transmitted beam is a measure of the blackness of the image; it can be recorded as various parts of the spectrogram are passed along the light beam so that the beam scans along different wavelengths. Figure 21-5 illustrates the appearance of a recording of a portion of a spectrogram made with a microphotometer.

We note that the spectral lines are not perfectly sharp but have a finite width (the reasons will be explained below. The *profile* of a spectral line is the exact shape the line would have on the recording of a microphotometer scan that was calibrated in such a way as to indicate directly the intensity of the light in the spectrum at various wavelengths. Because the profiles of different lines differ in shape, it is convenient to define some measurable quantity that can be used to calculate the total amount of light energy that is abstracted (or subtracted) from the spectrum by a line. The most widely used measure is the *equivalent width*, the width of a hypothetical line with rectangular profile of zero intensity along its entire width. The equivalent width represents the same subtraction of light from the stellar spectrum as is removed by the actual line (see Figure 21-6).

(b) *Analysis of the Continuous Spectrum**
In Chapter 20 we saw that the continuous spectrum of a star consists of radiation of a wide range of wavelengths that escapes from a relatively shallow layer of

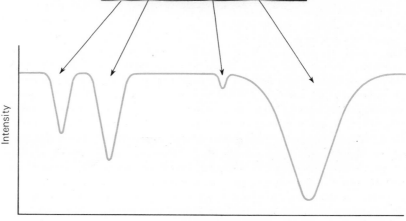

FIG. 21-5   Appearance of the recording of a microphotometer scan of a portion of the negative of a spectrograph.

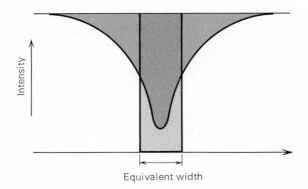

FIG. 21-6   Equivalent width of a line. The
shaded and striped regions have the same area.

gases in the outer part of the star called the photosphere.
One of the problems of astrophysics is to understand the
atomic processes that give rise to this emission of light.

In any given region in a stellar photosphere the gas
atoms must emit light at the same rate that they absorb
it. If the atoms emitted less light than they absorbed,
they would heat up indefinitely, and stellar temperatures
would continually rise; if they emitted more light than
they absorbed, stellar temperatures would continually
drop — both cases are contrary to observation. The problem of understanding the emission of light in stellar
photospheres is equivalent to that of understanding how
the photospheric gases absorb light — that is, what their
source of opacity is.

Calculations show that only hydrogen is sufficiently
abundant in the atmospheres of stars of most spectral
classes to contribute appreciably to the opacity of their
outer layers. We have seen (Section 10.5b) that hydrogen
atoms can absorb light over a range of wavelengths in
the process of becoming ionized. The ionization of hydrogen atoms by the absorption of photons, then, is a possible source of opacity in a stellar photosphere. On the
other hand, in stars as cool as the sun, or cooler, most
atoms of hydrogen are in their lowest energy states. They
can be ionized only if they absorb the invisible radiation
of wavelengths less than 912 Å (which possesses energy
greater than the ionization energy of hydrogen). Opacity
at visible wavelengths can result only from the ionization
of hydrogen atoms that are already excited to their third
or higher energy levels, so that less energetic photons, of

longer wavelengths, can be absorbed. Those excited hydrogen atoms are so rare in the solar photosphere that if they
were the only source of opacity, the photosphere should
be far more transparent than it actually is; we should be
able to see into far hotter and deeper layers of the sun
than we do.

The problem was solved by the realization that in
stars of temperatures near that of the sun and lower,
an occasional hydrogen atom can capture and temporarily
hold a passing electron. Such a hydrogen atom, possessing
two electrons, is called a *negative hydrogen ion*. The
negative hydrogen ion can absorb visible light and dissociate into an ordinary hydrogen atom and a free electron again. Only about one hydrogen atom in 1 hundred
million in the solar photosphere has an extra electron at
any given time. This number is enough, however, to
account for the opacity of the sun's atmospheric gases
in the visible spectrum. Negative hydrogen ions re-form
at the same rate that they dissociate, and in doing· so
emit the visible continuous spectrum that we observe.

The hypothesis that negative hydrogen ions are responsible for most of the opacity of the solar photosphere
was proposed in the 1940s. Strong confirming evidence
for the theory is provided by data obtained from the
limb darkening of the sun (see also Chapter 24). The
limb of the sun (or of any celestial body) is its apparent
edge as it is seen in the sky. The limb of the sun appears
darker than the center of the disk, because when we
look at the center of the disk we see into deeper, hotter
layers of the sun's photosphere than when we look at
the limb. The explanation for this phenomenon is seen
in Figure 21-7. Point A is relatively deep in the photosphere and is beneath the exact center of the sun's disk.
Suppose photons that are emitted at point A travel, on
the average, a distance d_a before being absorbed. The
only photons that have an even chance of escaping from
the photosphere are those that are moving radially outward, a direction that happens to be toward the earth.
We see, then, sunlight emerging from points as deep as
point A only at the center of the sun's disk. Suppose
that·at point B, photons travel, on the average, a distance
d_b before being absorbed. To get into the line of sight
to the earth they must travel outward obliquely, rather
than radially, and hence can reach the earth only if point
B is closer to the surface of the sun than point A. To
reach the earth from point C, photons would have to

traverse an extremely long path through the solar photosphere unless point *C* were practically at the outermost photospheric level. The depths to which we see into the sun at various distances from the center of its disk are indicated (although not to scale) by the dashed line.

The gases of the sun are not equally opaque at all wavelengths. Light from the limb of the sun of *any* wavelength reaches the earth only if it originates from the *highest photospheric layers;* on the other hand, we find that at short wavelengths (blue and violet) the distance we can see into the photosphere at the center of the disk is greater than at long wavelengths (at least over the visible spectrum). The detailed manner in which the limb darkening varies with wavelength enables us to calculate from observations how the opacity of the photospheric gases varies with wavelength. It is found that the propensity of the atoms in the sun's outer layers to absorb light at various wavelengths is just what we would expect if negative hydrogen ions were the main source of opacity. The fact that negative hydrogen ions in the sun account for its continuous spectrum, therefore, is quite well established.

Once we know the source of opacity in a stellar photosphere, we can calculate how the temperature, pressure, and density of its gases increase with depth. One clue is provided by the condition that the pressure of the gases in a photosphere must be just great enough to balance the weight of the overlying layers of gases (the condition of *hydrostatic equilibrium* — see Chapter 29). The laws of gases (also see Chapter 29) tell us how density, pressure, and temperature are related. Finally, the known opacity of the gases tells us how they impede the flow of radiation through them. It is the opacity of the gases, blocking in the radiation, that maintains the increase of temperature inward through the stellar photosphere. Thus, a knowledge of the opacity enables us to calculate the temperature at each depth in the photosphere. When he puts the clues together, the astrophysicist can calculate what is called a *model photosphere* or *model atmosphere* for a star. In practice, the calculations are difficult and the results obtained for the march of pressure, temperature, and density through a photosphere are somewhat uncertain. Model photospheres do indicate, however, at least roughly, the physical nature of the outer layers of a star. Table 24.2 gives the data for a model photosphere of the sun.

FIG. 21-7   Illustration of limb darkening.

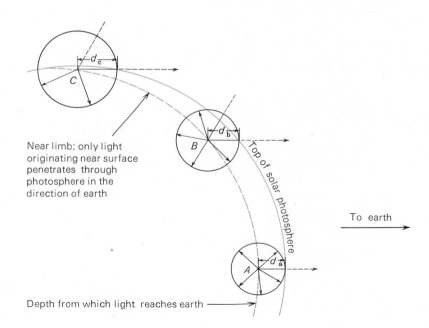

Near limb; only light originating near surface penetrates through photosphere in the direction of earth

Top of solar photosphere

To earth

Depth from which light reaches earth ⟶

(c) *Analysis of the Line Spectrum* *

The analysis of the absorption-line spectrum (sometimes called the *Fraunhofer spectrum)* of a star is even more difficult than that of the continuous spectrum, but it yields a great deal of additional information about the star. The object of the analysis is to account in detail for the appearance of each spectral line, and to extract such information as is possible concerning the physical and chemical characteristics of the star.

Before we can account for the shape, or profile, of a spectral line, we must consider those processes that *broaden* it — that is, that keep it from being an infinitely sharp line of one precise wavelength only. The principal sources of *line broadening* are the following.

Natural Broadening   We saw in Section 10.5 that atoms have discrete energy levels. The levels are not *perfectly* sharp, however. The nominal energy levels of an atom, and the wavelengths of the photons it can absorb, are actually only the most probable energies (or wavelengths). The energy levels of an atom are somewhat "fuzzy" in that there is a small range of energy about each nominal energy that the atom can have.† There is, therefore, a small range of wavelengths that can be absorbed or emitted. The resulting (usually) slight range of wavelengths over which an atom can absorb radiation in the

---

†According to the Heisenberg uncertainty principle, the mean range in energy of an energy level, ΔE, and the average time an atom spends in that level, Δt, are related by the formula

$$\Delta E\,\Delta t = h,$$

where h is Planck's constant (6.6×10^{-27} erg · sec). A typical value of Δt is 10^{-8} second; the corresponding value of ΔE is 10^{-18} erg.

vicinity of a line is called the *natural width* of the line.

Doppler Broadening   Photospheric atoms are in rapid motion because of the high temperatures of the outer layers of stars. Atoms, moving at various speeds with respect to the earth, "see" the wavelengths of the photons they encounter as different from the lengths we would measure on earth, because of the Doppler Shift. At any instant of time, some atoms approach us and others recede from us. Some, therefore, absorb photons of wavelengths that we measure to be a little shorter than the nominal value for an absorption line, and others absorb photons that we measure to have slightly longer wavelengths. The result is that the line is smeared out into a finite width. This *Doppler broadening* is the main source of broadening of most "weak" lines — that is, lines in which only a little energy is subtracted from the spectrum.

Collisional Broadening   The energy levels of an atom can be *perturbed* by the presence of other atoms and ions that pass near it or that collide with it. These perturbations of the energy levels shift them to slightly different energies, so that atoms then absorb at wavelengths slightly different from their usual ones. As in Doppler broadening, the effect is to smear out each absorption line. *Collisional broadening* is the most important source of line broadening for the majority of "strong" lines.

Zeeman Effect   Other effects can also perturb energy levels; in particular, magnetic fields cause energy levels to split into two or more separate levels, which are shifted from the nominal energies by amounts that depend on the strength of the field. Although this *Zeeman effect* splits each line into two or more lines, the several components are not usually resolved in stellar spectra, so the effect merely appears to broaden a line.

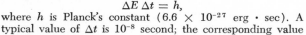

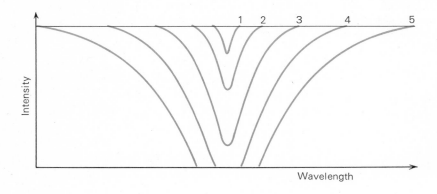

FIG. 21-8 Profiles of spectral lines of different equivalent widths. The profiles of larger numbers correspond to larger numbers of atoms that can produce the line being present in the stellar photosphere.

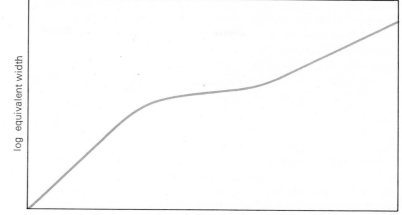

FIG. 21-9   Curve of growth —
a graph that shows how the
equivalent width of a line
depends on the number of atoms
in the photosphere.

log equivalent width

log number of atoms

Were it not for line broadening, very little energy could be removed from the continuous spectrum by line absorption. Since, however, atoms can absorb radiation over a small range of wavelengths near each line, they can subtract enough energy from the continuous spectrum so that the lines can be observed easily and analyzed in considerable detail.

The strength of an absorption line depends not only on the total abundance of the relevant atomic species, but also on the fraction of those atoms that are in the right state of ionization and excitation to produce the line. From a model photosphere for a star, we have the temperatures and pressures as functions of depth in its atmosphere. For each depth in the photosphere, therefore, we can calculate what fraction of the atoms are in the relevant ionization and excitation state (Sections 10.5d and 10.5e), and also how that particular absorption line is broadened at each depth. Since an absorption line arises from atoms at all depths throughout the photosphere,† we must consider how the contributions from atoms at all layers combine to predict the observed profile or equiva-

lent width of the line. We can then predict how the equivalent width will vary with the total number of atoms of the element in question. It is found that for a relatively low abundance, the equivalent width of a line is approximately proportional to the total number of atoms of the element present. For a great enough number of atoms, the line "saturates" and, with the addition of a still larger number of atoms, its strength increases only moderately. If the abundance of the element is very great, however, broad "wings" appear on the line as the result of collisional broadening; as the abundance of the element is increased further, the equivalent width of the line increases roughly as the square root of the numbers of atoms. A graph that shows how the equivalent width of a line depends on the number of atoms that produce it is called a *curve of growth.* Figure 21-8 shows how the profiles of successively stronger lines appear. Figure 21-9 shows the appearance of a typical curve of growth.

(d) What Is Learned from Stellar Spectra

Having described some of the procedures by which stellar spectra are analyzed, we now summarize some of the more important kinds of data that can be obtained from spectrum analysis.

TEMPERATURE

The kind of temperature measured by an ordinary thermometer is called *kinetic temperature;* it is a

---

†At one time, stellar photospheres were regarded as discrete *surfaces* in the atmospheres of stars, and absorption lines were described as being produced in a *reversing layer* floating immediately above the photosphere. The concept of a reversing layer, in the literal sense, has now generally been discarded, although the concept is still useful for the mathematical representation of some spectral lines.

measure of the kinetic energies of the atoms or molecules. Since we cannot place a thermometer in the photosphere of a star, we cannot measure its kinetic temperature directly, but must infer its temperature from the quality of the radiation that we receive from the star. Because that radiation emerges from a variety of depths in the stellar photosphere, it can not be expected to indicate a unique temperature that corresponds to any one layer in the star. The temperature that we derive is a representative temperature, corresponding to some representative depth in the photosphere. What that depth is depends on the manner in which the temperature is measured. If the star's continuous spectrum, or a portion of it, is fitted to the spectral energy distribution of a black body (perfect radiator), for example, the temperature of that black body is called the star's *color temperature*. A temperature that is estimated from the extent to which lines of ionized atoms appear in the spectrum is called an *ionization temperature*. A temperature that is estimated from the extent to which lines of certain atoms in certain states of excitation are present is called an *excitation temperature*. The temperature of a perfect radiator that would emit the same total amount of radiant energy per unit area as the star does is the *effective temperature* of the star; the effective temperature cannot be determined directly from the spectrum alone, but is found, rather, from the luminosity and radius of a star if they are known (see Chapter 22). If a star were a perfect radiator, and were in complete thermodynamic equilibrium, all these temperatures would agree. As it is, the different kinds of temperatures usually do agree with each other fairly well — a circumstance that results from the fact that the observable radiation from most stars emerges from a relatively small range of depths in their photospheres.

PRESSURE

We have already discussed how the pressure in a stellar photosphere can affect its spectrum and how stars of high photospheric pressure and high temperature can be distinguished from those of low photospheric pressure and a somewhat lower temperature. It is generally possible, therefore, to make a rough estimate of the pressures of the gases in a stellar photosphere.

CHEMICAL COMPOSITION

Dark lines of a majority of the known chemical elements have now been identified in the spectra of the sun and stars. Of course, the lines of *all* elements are not observable in the spectrum of a star, nor are lines of any one element visible in the spectra of *all* stars. As we have seen, because of variations among the stars in temperature and pressure of the photosphere, only certain of the prevailing kinds of atoms are able to produce absorption lines in any one star. The absence of the lines of a particular element, therefore, does not necessarily imply that that element is not present. Only if the physical conditions in the photosphere of a particular star are such that lines of an element *should* be visible, were the element present in reasonable abundance in that star, can we conclude that the absence of observable spectral lines implies low abundance of the element. On the other hand, spectral lines of an element, in the neutral state or in one of its ionized states, certainly imply the presence of that element in the star.

RELATIVE ABUNDANCES OF THE ELEMENTS

Once due allowance has been made for the prevailing conditions of temperature and pressure in a star's photosphere, analyses of the strengths of absorption lines in its spectrum can yield information regarding the relative abundances of the various chemical elements whose lines appear. In practice, abundances of elements in stars are determined by comparison of observed line strengths with theoretical curves of growth. It is found that the relative abundances of the different chemical elements in the sun and in most stars

FIG. 21-10 A portion of the solar spectrum and a comparison spectrum of iron photographed with the same spectrograph. Notice that many of the dark lines in the solar spectrum are matched by the bright lines in the comparison spectrum, showing that iron is present in the sun. *(Lick Observatory.)*

(as well as in most other regions of space that have been investigated) are approximately the same. Hydrogen seems to comprise from 50 to 80 percent of the mass of most stars. Hydrogen and helium together comprise from 96 to 99 percent of the mass; in some stars they comprise more than 99.9 percent. Among the 4 percent or less of "heavy elements," neon, oxygen, nitrogen, carbon, magnesium, argon, silicon, sulfur, iron, and chlorine are among the most abundant. Generally, but not invariably, the elements of lower atomic weight are more abundant than those of higher atomic weight.

RADIAL VELOCITY

The radial or "line-of-sight" velocity of a star can be determined from the Doppler shift of the lines in its spectrum (Section 10.3b). Huggins made the first radial-velocity determination of a star in 1868. He observed the Doppler shift in one of the hydrogen lines in the spectrum of Sirius and found that the star was receding from the solar system.

ROTATION

If a star is rotating, unless its axis of rotation happens to be directed exactly toward the sun, one of its limbs approaches us and the other recedes from us, relative to the star as a whole. For the sun or a planet, we can observe the light from one limb or the other and measure directly the Doppler shifts that arise from the rotation. A star, however, appears as a point of light, and we are obliged to analyze the light from its entire disk at once. Nevertheless, if the star is rotating, part of the light from it, including the spectral lines, is shifted to shorter wavelengths and part is shifted to longer

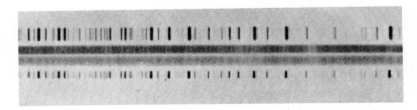

FIG. 21-11 Rotational broadening of spectral lines: *(top)* spectrum of WZ Ophiuchi, which does not rotate rapidly and whose lines are relatively sharp; *(bottom)* spectrum of UV Leonis, a rapidly rotating star whose lines are very broadened and have a washed out appearance. These spectrograms are negative prints.

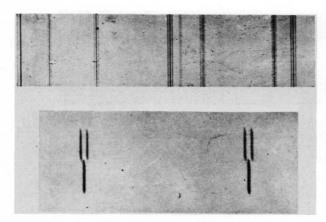

FIG. 21-12 The Zeeman effect. *(Above)* the splitting of several iron lines in a magnetic field; *(below)* two yellow mercury lines in the presence of *(upper half)* and in the absence of *(lower half)* a magnetic field.
(Yerkes Observatory.)

wavelengths. Each spectral line of the star is a composite of spectral lines originating from different parts of the star's disk, all of which are moving at different speeds with respect to us. The effect produced by a rapidly rotating star is that all its spectral lines are broadened so that their profiles have a characteristic "dish" shape (see Exercise 8). Fortunately, this dish shape is highly characteristic and usually can be distinguished from broadening produced by other sources. The amount of rotational broadening of the spectral lines, if observable, can be measured, and a lower limit to the rate of rotation of the star can be calculated (see Exercise 9).

TURBULENCE

If large masses of a star's photospheric layers have vertical turbulent motion, this motion, like rotation, causes a broadening of the spectral lines. Anomalously, high equivalent widths of the lines in the spectra of some stars have been attributed to such turbulence in their photospheres.

MAGNETIC FIELDS

The Zeeman effect splits or broadens the spectral lines produced by gases in a strong magnetic field. The actual splitting of the spectral lines, for example, can be observed in light coming from the magnetic fields associated with spots on the sun (Chapter 24). However, in the spectra of those stars that possess appreciable magnetic fields, the lines are usually only broadened, because the separate components of a line are not resolved. Zeeman broadening can still be differentiated, however, because the light that is not absorbed in the components of the split lines is polarized. The amount by which a spectral line is broadened by a magnetic field can be detected, therefore, by photographing the star's spectrum through a polarizing filter. A few stars have been found to have very strong magnetic fields. In some cases, these field strengths are variable. One possible explanation is that the field on such a star is localized, and rotation of the star alternately carries the face with the magnetic field into and out of our view.

SHELLS AND EJECTED GASES

In the spectra of some stars, absorption lines are observed that do not appear to originate in the photospheres. Sometimes such lines can be associated with shells or rings of material ejected by the star. If enough material is ejected, and if the star has radiation of high enough energy to excite or ionize the gas, the latter produces emission lines instead of absorption lines, superposed on the stellar spectrum. Examples are Wolf-Rayet stars, and early type emission-line stars. Stars that are ejecting matter are further discussed in Chapter 25.

Exercises

1. What are the probable approximate spectral classes for stars whose wavelengths of maximum light have the following values (see Section 10.4c):
(a) 2.9×10^{-5} cm
(b) 0.5×10^{-5} cm
(c) 6.0×10^{-5} cm
(d) 12.0×10^{-5} cm
(e) 15×10^{-5} cm

2. What are the probable approximate spectral classes of stars described as follows:
(a) Balmer lines of hydrogen are very strong; some lines of ionized metals are present.
(b) Strongest lines are those of ionized helium.
(c) Lines of ionized calcium are the strongest in the spectrum; hydrogen lines show with only moderate strength; lines of neutral and ionized metals are present.
(d) Strongest features are lines C_2, CN, and CH; titanium oxide is absent.

3. The spectrum of a star shows lines of ionized helium and also molecular bands of titanium oxide. What is strange about this spectrum? Can you suggest an explanation?

4. Most stars have masses in the range $\frac{1}{10}$ to 10 times the sun's mass. Stellar radii range, however, from 0.01 to 1000 times that of the sun. What is the ratio of the surface gravities of two stars, each of 1 solar mass, but one with a radius of 0.01 times the sun's radius, and one with a radius of 1000 times the sun's radius?

*5. Sketch how the microphotometer tracing of a spectrogram might appear if the spectrum showed *both* absorption and emission lines.

*6. At what part of the solar disk would the temperature of the visible gases correspond to the "boundary temperature" of the sun? How could you define a "boundary"?

*7. Explain why extremely little light could be subtracted from a continuous spectrum if there were no line broadening.

8. Explain (with a diagram) how stellar rotation broadens spectral lines.

9. Why is it that only a lower limit to the rate of stellar rotation can be determined from the rotational broadening, rather than the actual rotation rate?

Following page: Looking south on the lunar farside from the Apollo 8 spacecraft. *(NASA.)*

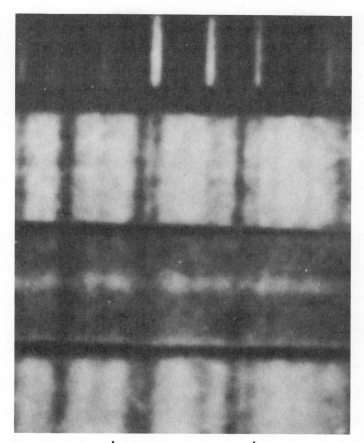

CHAPTER 22

Weighing and Measuring the Stars— Binary Stars

Chapters 18 through 21 have dealt with the methods by which we determine the distances, motions, luminosities, temperatures, and atmospheric characteristics of stars. Now we shall consider the methods by which their masses and radii are determined. The observational problems involved in "weighing" and "measuring" the stars directly are very difficult to overcome, and we have accurate measures for only a handful of stars. This handful, however, makes it possible for us to devise and calibrate indirect methods that we can use to estimate, at least, the masses and radii of many other stars. Most of the data on the masses and sizes of stars come from analysis of *binary* or *double stars:* in this chapter, therefore, we shall be concerned especially with a discussion of binary stars.

22.1 DETERMINATION OF THE SUN'S MASS

The masses of stars must be inferred from their gravitational influences on other objects. We can measure the force exerted by the sun, for example, on the earth and other planets; and the mass of the sun can be determined more reliably than that of any other star.

The most direct way to calculate the mass of the sun is from the acceleration of the earth. For the sake of illustration, let us assume that the orbit of the earth is circular. Then the acceleration required to keep the earth in its orbit is the centripetal acceleration (Section 4.2):

$$a = \frac{v^2}{R},$$

where v is the earth's orbital speed and R is the radius of its orbit. This acceleration must be provided by the gravitational attraction of the sun on a unit mass at the earth's distance; that is,

$$a = \frac{v^2}{R} = \frac{GM_s}{R^2},$$

where G is the universal gravitational constant and M_s is the desired mass of the sun. Solving the above equation for the sun's mass, we obtain

$$M_s = \frac{v^2 R}{G}.$$

Both v and R are known from observation and G is determined from laboratory measurements (Section 4.3e). In metric units, $v = 3 \times 10^6$ cm/sec, $R = 1.49 \times 10^{13}$ cm, and $G = 6.67 \times 10^{-8}$. Substituting these values into the above formula, we find for the mass of the sun,

$$M_s = 2 \times 10^{33} \text{ gm} = 2 \times 10^{27} \text{ metric tons.}$$

Because the earth's orbit is nearly circular, the value thus found for the mass of the sun is very nearly the correct one. Of course, the mass of the sun can be found more accurately by using the instantaneously correct acceleration of the earth and its exact distance from the sun at the given moment in its elliptical orbit.

A simple way to calculate the mass of the sun in terms of that of the earth is to apply Kepler's third law, as it was refined by Newton (Section 5.3). The sun and earth can be considered a pair of mutually revolving bodies. Since the earth's mass is negligible compared to the sun's, Kepler's third law, when applied to the earth-sun system, becomes

$$M_s P_{es}^2 = K a_{es}^3,$$

where P_{es} is the period of revolution of the earth about the sun (sidereal year), a_{es} is the semimajor axis of the relative orbit of the earth and sun (the length of the astronomical unit), and K is a known constant of proportionality whose value depends on the units used to measure mass, time, and length. The same formula may be applied, then, to another pair of mutually revolving bodies, the earth-moon system:

$$(M_e + M_m)P_{em}^2 = \frac{82.3}{81.3} M_e P_{em}^2 = K a_{em}^3,$$

where M_e is the earth's mass, which is 81.3 times the mass of the moon, M_m; P_{em} is the period of revolution of the moon about the earth (the sidereal month); and a_{em} is the semimajor axis of the relative orbit of the earth-moon system. If the first of the above equations is divided by the second, the result is,

$$\frac{M_s}{M_e} = \frac{82.3}{81.3}\left(\frac{P_{em}}{P_{es}}\right)^2\left(\frac{a_{es}}{a_{em}}\right)^3.$$

The sidereal month is about 27.32 days, or about $\frac{1}{13.4}$ sidereal years, and the astronomical unit is about 389 times the moon's distance. If these figures are inserted in the above equation,

$$\frac{M_s}{M_e} = 333,000.$$

That is, the sun is about one third of a million times as massive as the earth. Since the earth's mass is 6×10^{27} gm (Section 4.3e), the sun's mass is 2×10^{33} gm (2.2×10^{27} English tons), as found above.

22.2 BINARY STARS

It would be helpful if we could observe directly planets (if any exist) in orbital revolution about stars other than the sun, but unfortunately we are unable to do so. We do observe, however, many cases of double stars — systems of two stars that revolve about each other under the influence of their mutual gravitational attraction. Frequently, we even find cases of three or more stars that belong to the same dynamical system.

(a) Discovery of Binary Stars

In 1650, less than half a century after Galileo turned a telescope to the sky, the Italian astronomer Jean Baptiste Riccioli observed that the star Mizar, in the middle of the handle of the Big Dipper, appeared through his telescope as *two* stars; Mizar was the first *double star* to be discovered. In the century and a half that followed, many other closely separated pairs of stars were discovered telescopically.

Usually, one star of a pair is brighter than the other. William Herschel assumed that the fainter star was the more distant, and that the brighter one was a foreground object. It occurred to him that it might be possible to measure the parallax (Section 18.4) of the nearer star with respect to the more distant one, which was so nearly in line with it. Accordingly, he began a search for such pairs, and between 1782 and 1821 he published three catalogues, listing more than 800 double stars. Actually, only rarely does a double star consist of one nearby and one distant star; the vast majority of these systems found by Herschel are *physical pairs* of stars, *revolving about each other*. This had been suggested as early as 1767 by John Michell, who pointed out the extreme improbability of there being more than a very few close pairs of stars that are really only *optical doubles*, that is, lined up in projection.

One famous double star is Castor, in Gemini. The telescope reveals Castor to be two stars separated by an angle of about 5″. By 1804 Herschel had noted that the fainter component of Castor had changed, slightly, its direction from the

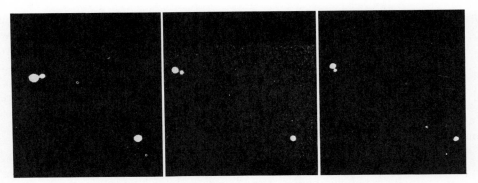

FIG. 22-1   Three photographs, covering a period of about 12 years, which show the mutual revolution of the components of the double star Kruger 60. *(Yerkes Observatory.)*

brighter component. Here, finally, was observational evidence that one star was moving about another; it was the first evidence that gravitational influences exist outside the solar system. Herschel had failed in his program to facilitate parallax determinations but had found something of far greater interest. As he put it, he was like Saul, who had gone out to seek his father's asses and had found a kingdom. His son, John Herschel, continued the search for double stars, and prepared a catalogue (published posthumously) of more than 10,000 systems of two, three, or more stars.

If the gravitational forces between stars are like those in the solar system, the orbit of one star about the other must be an *ellipse*. The first to show that such is the case was Felix Savary, who in 1827 showed that the relative orbit of the two stars in the double system ξ *Ursae Majoris* is an ellipse, the stars completing one mutual revolution in a period of 60 years.

Another class of double stars was discovered by E. C. Pickering, at Harvard, in 1889. He found that the lines in the spectrum of the brighter component of Mizar (the first double star to be discovered) are usually *double,* but that the spacing of the components of the lines varies periodically, and at times the lines even become

single. He correctly deduced that the brighter component of Mizar (Mizar A) itself, is really *two* stars that revolve about each other in a period of 104 days. When one star is approaching us, relative to the center of mass of the two, the other star is receding from us; the radial velocities of the two stars, and therefore the Doppler shifts of their spectral lines, are different, so that when the composite spectrum of the two stars is observed, each line appears double. When the two stars are both moving across our line of sight, however, they both have the same *radial* velocity (that of the center of mass of the pair), and hence the spectral lines of the two stars coalesce.

Stars like Mizar A, which appear as single stars when photographed or observed visually through the telescope, but which the spectroscope shows really to be double stars, are called *spectroscopic binaries;* systems that can be observed visually as double stars are called *visual binaries.* In 1908 Frost found that the fainter component of Mizar, Mizar B, is also a spectroscopic binary.

Almost immediately following Pickering's discovery of the duplicity of Mizar A, Vogel discovered that the star Algol, in Perseus, is a spectroscopic binary. The spectral lines of Algol were not observed to be double, because the fainter star of the pair gives off too little light compared to the brighter for its lines to be conspicuous in the composite spectrum. Nevertheless, the periodic shifting back and forth of the lines of the brighter star gave evidence that it was revolving about an unseen companion; the lines of both components need not be visible in order for a star to be recognized as a spectroscopic binary.

The proof that Algol is a double star is significant for another reason. In 1669 Montonari had noted that the star varied in brightness; in 1783 John Goodricke established the nature of the variation. Normally, Algol is a second-magnitude star, but at intervals of $2^d20^h49^m$ it fades to one third of its regular brightness; after a few hours, it brightens to normal again. Goodricke suggested that the variations might be due to large dark spots on the star, turned to our view periodically by its rotation, or that the star might be eclipsed regularly by an invisible companion. Vogel's discovery that Algol is a spectroscopic binary verified the latter hypothesis. The plane in which the stars revolve is turned nearly edgewise to our line of sight, and each star is eclipsed once by the other during every revolution. The eclipse of the fainter star is not very noticeable because the part of it that is covered contributes little to the total light of the system; the second eclipse can, however, be observed. A binary such as Algol, in which the orbit is nearly edge on to the earth so that the stars eclipse each other, is called an *eclipsing binary*.

Binary stars are now known to be very common; they may be the rule — not the exception. In the stellar neighborhood of the sun, somewhere between one half and two thirds of all stars are members of binary or multiple star systems.

We may summarize the types of binary stars before proceeding to explain how they are used to determine stellar masses.

OPTICAL BINARIES

An optical binary consists of two stars, in nearly the same line of sight, of which one is far more distant than the other; they are not true binary stars, and they are relatively rare.

VISUAL BINARIES

A visual binary is a gravitationally associated pair of stars; the members are either so near the sun, or so widely separated from each other (usually, both), that they can be observed visually (in a telescope) as two stars. Typical separations for the two stars in a visual binary system are hundreds of astronomical units; thus the orbital speeds of the stars are usually quite small and their orbital

FIG. 22-2 Two spectra of the spectroscopic binary κ Arietis. When the components are moving at right angles to the line of sight *(bottom)*, the lines are single. When one star is approaching us and the other receding *(top)*, the spectral lines of the two stars are separated by the Doppler effect. *(Lick Observatory.)*

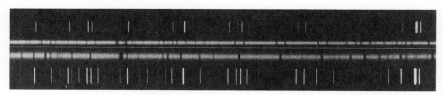

motion may not be apparent over a few decades of observation. Nevertheless, two closely separated stars are generally assumed to comprise a visual binary system if there is no reason to doubt that they are at the same distance from us and if they have the same proper motion and radial velocity, indicating that they are moving together through space. Over 64,000 such systems are catalogued.

ASTROMETRIC BINARIES

Sometimes one member of what would otherwise be a visual binary system is too faint to be observed; its presence may be detected, however, by the "wavy" motion of its companion, revolving about the center of mass of the two stars as they move through space. In 1844 Bessel discovered that the bright star Sirius displays such a sinusoidal motion with a period of 50 years. Sirius remained an *astrometric binary* until 1862, when Alvan Clark found its faint companion — a member of the class of stars known as *white dwarfs* (discussed in Chapter 23).

SPECTROSCOPIC BINARIES

When the binary nature of a star is known only from the variations of its radial velocity (or of both radial velocities if the spectral lines of both stars are visible), it is said to be a spectroscopic binary. Over 700 systems have been analyzed.

SPECTRUM BINARIES

If the orbit of what would otherwise be a spectroscopic binary is turned nearly "face on" to us (that is, perpendicular to our line of sight), or if the masses of the member stars are so low that they have very small orbital velocities, we can see no radial-velocity variations. It may still be obvious, however, that there are two stars if the composite spectrum contains lines that are characteristic of both hot and cool stars and which would not be expected to occur in the spectrum of a single star. Such systems are called spectrum binaries.

ECLIPSING BINARIES

If the orbit of a binary system is turned nearly edge on to us, so that the stars eclipse each other, it is called an eclipsing binary.

The different kinds of binaries are not mutually exclusive. An eclipsing binary, for example, may *also* be a spectroscopic binary, if it is bright enough that its spectrum can be photographed, and if its radial-velocity variations have been observed.

(b) Mass Determinations of Visual Binary Stars

The two stars of a binary pair revolve mutually about their common center of mass (or barycenter), which in turn moves in a straight line among the neighboring stars. Each star, therefore, describes a wavy path around the course followed by the barycenter. With careful observations it is possible to determine these individual motions of the member stars in a visual binary system. It is far more convenient, however, simply to observe the motion of one star (by convention the fainter), about the other (the brighter). The observed motion shows the *apparent relative orbit.* The periods of mutual revolution for visual binaries range from a few years to thousands of years, but only for those systems with periods less than a few hundred years can the apparent relative orbits be determined with much precision; even then, a long series of observations covering a number of decades is usually necessary. The data observed are the angular separation of the stars and the *position angle,* which is the direction, reckoned from the north, around toward the east, of the fainter star from the brighter one. These data can be measured on a photograph if the separation of the stars is not too small; in any case, they can be measured directly at the telescope with a filar micrometer (Section 11.6). A typical apparent relative orbit, assembled from many observations of separation and position angle, is shown in Figure 22-3.

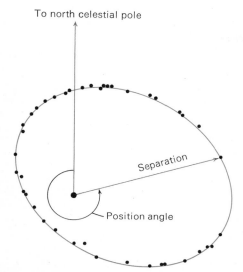

To north celestial pole

Separation

Position angle

FIG. 22-3 Apparent relative orbit of a hypothetical visual binary system.

this geometry problem, so that the shape and angular size of the true orbit can be found. The period of mutual revolution, of course, is observed directly if the system has completed one revolution during the interval it has been under observation; otherwise, the period must be calculated from the rate at which the fainter star moves in the relative orbit.

If the semimajor axis of the true relative orbit (that is, the semimajor axis the orbit would appear to have if it were seen face on) has an angular length, in seconds of arc, of a'', and if the system is at a distance of r pc, the semimajor axis, in astronomical units, is $r \times a''$ (see Exercise 7). The sum of the masses of the two stars, in solar units, is given by Kepler's third law:

$$m_1 + m_2 = \frac{(r \cdot a'')^3}{P^2}.$$

To find what share of the total mass belongs to each star, it is necessary to investigate the indi-

The true orbit of the binary system will not, in general, happen to lie exactly in the plane of the sky (that is, perpendicular to the line of sight). Consequently, the apparent relative orbit is merely the *projection* of the *true relative orbit* into the plane of the sky. Now it is easy to show that when an ellipse in one plane is projected onto another plane (that is, is viewed obliquely), the projected curve is also an ellipse, although one of different eccentricity. However, the foci of the original ellipse do *not* project into the foci of the projected ellipse. Therefore, the brighter star, although it is located at one focus of the *true* relative orbit, is not at a focus of the *apparent* relative orbit. This circumstance makes it possible to determine the inclination of the true orbit to the plane of the sky. The problem is simply one of finding the angle at which the ellipse of the true relative orbit must be projected in order to account for the amount of displacement of the brighter star from the focus of the apparent relative orbit (see Figure 22-4). There are several techniques for solving

FIG. 22-4 Relation between a true and apparent relative orbit.

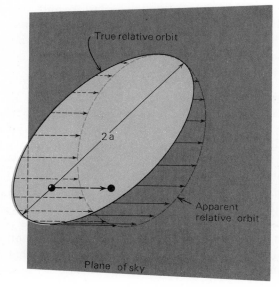

True relative orbit

2 a

Apparent relative orbit

Plane of sky

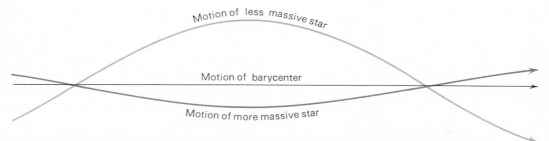

FIG. 22-5   Possible paths of two stars of a visual binary about their barycenter.

vidual motions of the stars with respect to the center of mass of the system. The distance of each star from the barycenter is inversely proportional to its own mass (Figure 22-5).

As an example, consider the visual binary, Sirius A and Sirius B. The semimajor axis of the true relative orbit is about 7½″ and the distance from the sun to Sirius is 2.67 pc. The period of the binary system is 50 years. The sum of the masses of the stars, then, is

$$m_1 + m_2 = \frac{(2.67 \times 7.5)^3}{(50)^2} = 3.2 \text{ solar masses.}$$

Sirius B, the fainter component, is about twice as far from the barycenter as Sirius A, and so has only about half the mass of Sirius A. The masses of Sirius A and Sirius B, therefore, are about 2 and 1 solar masses, respectively.

(c) Barnard's Star—An Interesting Astrometric Binary

Often, interesting information can be learned from the study of an astrometric binary. A very interesting example of a probable astrometric binary is Barnard's star — the star of largest known proper motion (Section 19.1). Careful observations of the motion of Barnard's star over 51 years were analyzed in 1968 by P. van de Kamp of the Sproul Observatory. Van de Kamp's study shows that the star has a very slight wave in its proper motion with a period of 25 years, which suggests that it revolves about the barycenter of it and an unseen

companion with this period. Barnard's star, however, has a very small mean distance from the barycenter — only 0.″028, which at its distance of 1.83 pc corresponds to 0.051 AU. Barnard's star is an M5 main-sequence star; such stars have masses of only about 0.15 solar mass. If we denote the masses of Barnard's star and its companion, in solar units, by m_B and m_c, respectively, their distances from the barycenter by a_B and a_c AU, and their period of mutual revolution by P years, we have the following two equations that apply to them:

$$\frac{m_B}{m_c} = \frac{a_c}{a_B}$$

and

$$m_B + m_c = \frac{(a_B + a_c)^3}{P^2}.$$

If we now substitute the known quantities $P = 25$, $m_B = 0.15$, and $a_B = 0.051$, and solve the two equations for the two unknowns m_c and a_c, we find the surprising result

$$a_c = 4.5 \text{ AU}$$

and

$$m_c = 0.0017 \text{ solar mass;}$$

that is, the companion of Barnard's star has a mass only 80 percent greater than that of Jupiter, and must therefore be a planet.

This is the only object of planetary mass known beyond the solar system. It is, of course, far too faint to observe directly with any telescope (its apparent magnitude is estimated to be about 30). It was discovered only because Barnard's star is so near the sun and because it has a small enough mass itself that its own motion about the barycenter could be detected. A planet of appreciably smaller mass, let alone anything as small as the earth, would never have been discovered.

(d) Spectroscopic Binaries

If the two stars of a binary system have a small linear separation, that is, if their relative orbit is small, there is little chance that they will be resolved as a visual binary pair. On the other hand, they have a shorter period, and their orbital velocities are relatively high, as compared with the stars of a visual binary system; unless the plane of orbital revolution is almost face on to our line of sight, there is a good chance that we will be able to observe radial-velocity variations of the

stars due to their orbital motions. In other words, they comprise a spectroscopic binary system.

Most spectroscopic binaries have periods in the range from a few days to a few months; the mean separations of their member stars are usually less than 1 AU. If the two stars of a spectroscopic binary are not too different in luminosity, the spectrum of the system will display the lines of both stars, each set of lines oscillating back and forth in the period of mutual revolution. More often, lines of only one star will be observed. A graph showing the radial velocity of a member of a binary star system plotted against time is called a *radial-velocity curve* or, simply, a *velocity curve*.

In general, the orbit of each star about the barycenter is an ellipse, and the radial-velocity curve is skewed (Figure 22-6). Both stars of the system, of course, have the same period, for they stay on opposite sides of the barycenter as they revolve about it. Also, the orbits of both stars have the same eccentricity (see Exercise 10) but differ in size by a ratio that is in inverse proportion to

FIG. 22-6   Radial-velocity curves for the spectroscopic binary system φ Cygni. (Adapted from Rach and Herbig.)

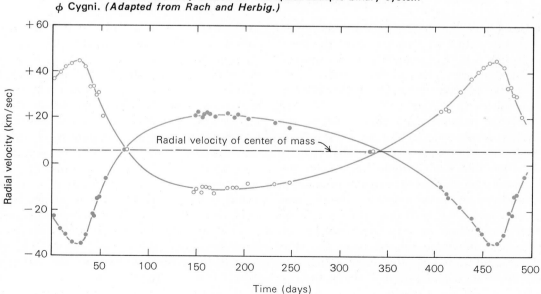

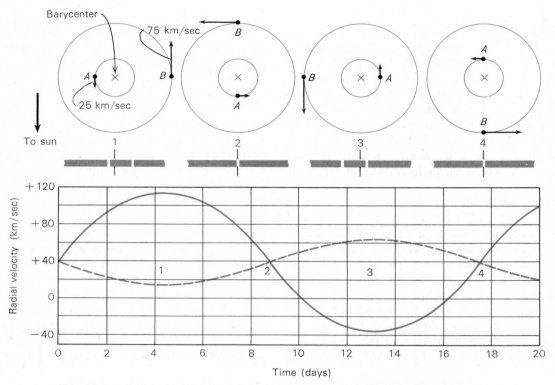

FIG. 22-7 Hypothetical spectroscopic binary system with circular orbits.

their masses (since the masses of the stars are in inverse proportion to their distances from the barycenter — Section 5.1). Consequently, if the spectral lines of both stars are visible, so that radial-velocity curves can be plotted for each, those velocity curves are mirror images of each other, differing only in scale (or in *amplitude*); this is true for the binary system whose velocity curves are shown in Figure 22-6. From the shape of each radial velocity curve, it is possible to determine the eccentricity of the orbits of the stars, as well as the orientation of the major axis of the orbit in the orbital plane.

Here, however, we shall only consider, as an example of the analysis of a spectroscopic binary, the simple case in which the orbits of the stars are circular, and in which the spectral lines of both stars are observed. Figure 22-7 shows the two stars in their orbits, the Doppler shifts of a hypo-

thetical spectral line, and the two radial-velocity curves. Because the orbits of the stars in this example are circular, the velocity curves are symmetrical; they have shapes that are known, technically, as *sine curves*. A complete cycle of variation of radial velocity — the period of the system — is seen to be 17.5 days. At position 1, star A has its maximum possible component of velocity toward the sun, and star B has its maximum possible component of velocity away from the sun. The conditions are reversed at position 3. At positions 2 and 4, both stars are moving across our line of sight and neither has a component of velocity in our line of sight due to orbital motion; both have the same radial velocity as that of the center of mass of the system, 40 km/sec. The radial velocity of star B ranges from 115 to −35 km/sec, a range of 150 km/sec. The maximum difference between the radial velocities of star *B* and that of the bary-center, then, is 75 km/sec. The corresponding value for star *A* is only 25 km/sec. Because both stars have the same period and since star *A* moves only one third as fast as star *B*, with respect to the barycenter, star *A* must have only one third as far to go to get around its orbit; its orbit must be one third the size of that of star *B*, and its mass, there-fore, must be three times as great.

Stars *A* and *B* are moving in opposite direc-tions with respect to their center of mass; thus the maximum radial velocity (as observed from the solar system) of star *B* with respect to star *A* must be the sum of the radial velocities of the two stars with respect to the barycenter, or 100 km/sec. If the orbital plane of the system were in our line of sight, this would be the relative orbital velocity of one star with respect to the other. As it is, however, the orbital plane is tilted at some un-known angle to our line of sight, and the 100 km/sec is only the maximum *radial component* of the relative velocity; the actual relative velocity will be greater by an unknown factor. Now, the distance around the relative orbit — its circum-ference — is the relative orbital velocity multiplied by the period — the time one star takes to get around the other. The distance between the stars, *a*, is the circumference of the orbit divided by 2π; that is,

$$a = \frac{V \times P}{2\pi},$$

where *V* is the relative velocity. If we substitute the observed lower limit to the relative velocity — in our example, 100 km/sec — for *V* in the above equation, we will obtain a lower limit to the separation of the stars. If this lower limit to *a* is applied to Kepler's third law, we obtain a lower limit to the sum of the masses of the stars:

$$m_1 + m_2 = \frac{a^3}{P^2}.$$

The calculation for the numerical example

given in Figure 22-7 may be illustrated as follows: A velocity of 100 km/sec is equivalent to about 20 AU/year, and a period of 17.5 days is about 0.048 year; thus a lower limit to *a* is given by

$$a = \frac{20 \times 0.048}{2\pi} = 0.153 \text{ AU.}$$

A lower limit to the sum of the masses, then, is

$$m_1 + m_2 = \frac{0.153^3}{0.048^2} = 1.6 \text{ solar masses.}$$

Since star *A* is three times as massive as star *B*, it has 75 percent of the total mass; lower limits to the individual masses of the stars are therefore

$$m_A \geq 1.2 \text{ solar masses}$$

and

$$m_B \geq 0.4 \text{ solar mass,}$$

where the symbol "$\geq$" means "greater than or equal to."

The analysis is more difficult in the general case of elliptical, rather than circular, orbits, but it can, nevertheless, be carried out. The results are similar except that *a* is then the semimajor axis of the relative orbit instead of the constant separation between the stars that exists if the orbits are circular. In either case, individual masses of the stars are not found, only lower limits to their masses.* If the spectral lines of only one star are visible, we do not even find lower limits to the masses of the individual stars but only a relation between their masses, known as the *mass function*.†

---

*What is actually found for each star is $m \sin^3 i$, where *i* is the inclination angle of the plane of orbital revolution to the plane of the sky.

†The *mass function* is

$$\frac{m_2^3 \sin^3 i}{(m_1 + m_2)^2},$$

where m_2 is the mass of the star whose spectrum is not observed.

The analyses of spectroscopic binaries, therefore, yield only lower limits to stellar masses. If large numbers of spectroscopic binary systems are investigated, and if it is assumed that their planes of orbital revolution are oriented at random in space, we can apply statistical corrections to the lower limits to find the *average* masses of the stars in the pairs that are studied. To find the masses of *individual* binary systems, however, we must have some way of determining the angle at which the orbital plane of the system is inclined to our line of sight (or to the plane of the sky); we can find this angle only when the orbit is almost edge on, so that the spectroscopic binary is also an eclipsing binary.

(e) Eclipsing Binaries

To simplify our discussion of eclipsing binaries we shall assume that the two stars in such a system revolve about each other in circular orbits. If the orbits were highly eccentric, some of the remarks in this section would have to be qualified. Such complicating qualifications, however, would serve mostly to confuse the student and divert his attention from the simple principles that make the analysis so beautiful. Keeping in mind, therefore, that in practice far more sophisticated mathematical techniques are employed in the analysis of eclipsing binaries — techniques that take account of the eccentricities of the orbits — we pretend that the orbits are all circular. Actually the assumption is not so bad. Most stars in binary systems that are seen to eclipse each other are relatively close to each other. Their mutual proximity enhances the effects of their perturbations on each other's motions, and these perturbations tend to produce circular, or nearly circular, orbits.

During the period of revolution of an eclipsing

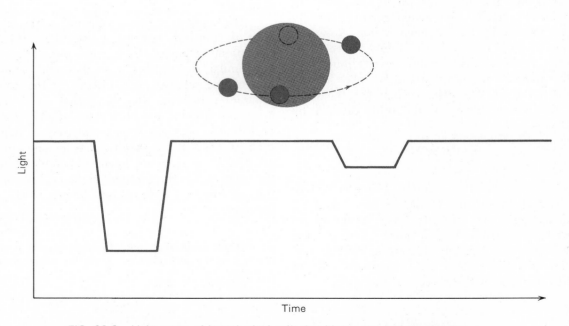

FIG. 22-8   Light curve of hypothetical eclipsing binary star with total eclipses.

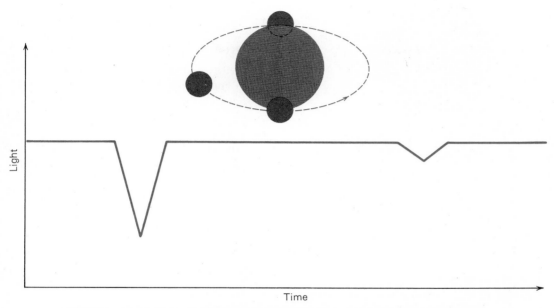

FIG. 22-9   Light curve of hypothetical eclipsing binary star with partial eclipses.

binary, there are two times when the light from the system diminishes — once when the smaller star passes behind the larger one and is eclipsed, and once when the smaller star passes in front of the larger one and eclipses part of it. If the smaller star goes completely behind the larger one, that eclipse is *total* and the other eclipse half a period later is *annular* (see Figure 22-8). If the smaller star is never completely hidden behind the larger star, both eclipses are partial (Figure 22-9). Each interval during an eclipse when the light from the system is farthest below normal is called a *minimum*. Both minima will not, in general, be equally low in light. The same area of each star is covered during the time it is eclipsed; for example, if the eclipses are total or annular, an area equal to the total cross-sectional area of the smaller star is eclipsed in each minimum. The relative amount of light drop at each minimum, however, depends on the relative surface brightnesses of the two stars, and hence on their temperatures. *Primary mini-*

mum occurs when the hotter star is eclipsed (whether it is a total, an annular, or a partial eclipse), and *secondary minimum* occurs when the cooler star is eclipsed. A graph of the light from an eclipsing binary system, plotted against time through a complete period, is called a *light curve*.

The most important data that are obtained from the analysis of the light curve of an eclipsing binary system are the sizes of the stars, relative to their separation, and the inclination of the orbit to our line of sight. To illustrate how the sizes of the stars are related to the light curve, we may consider a hypothetical eclipsing binary in which the stars are very different in size, and in which the orbit is exactly edge on, so that the eclipses are *central* (Figure 22-10). When the small star is at point *a* (*first contact*), and is just beginning to pass behind the large star, the light curve begins to drop. At point *b* (*second contact*), it has gone entirely behind the large star and the total phase of the eclipse begins. At *c* (*third contact*) it be-

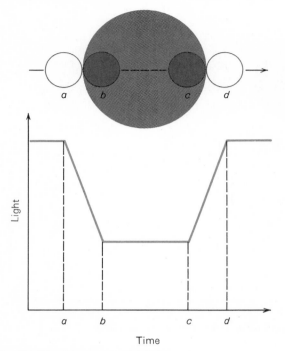

FIG. 22-10   Contacts in the light curve of a hypothetical eclipsing binary with central eclipses.

gins to emerge, and when it has reached *d* (*last contact*) the eclipse is over. During the time interval between first and second contact (or between third and last contacts) the small star has moved a distance equal to its own diameter. That time interval is in the same ratio to the period of the system as the diameter of the small star is to the circumference of the relative orbit. During the time interval from first to third contacts (or from second to last contacts) the small star has moved a distance equal to the diameter of the large star; that time interval is to the period as the diameter of the large star is to the circumference of the relative orbit. We see, therefore, that the light curve alone gives the sizes of the stars in terms of the size of their orbit. If the lines of both stars are visible in the composite spectrum of the binary, both radial velocity curves can be found.

Then the size of the relative orbit can be found, and we can determine the actual (linear) radii of the stars. In other words, the velocity of the small star with respect to the large one is known, and, when multiplied by the time intervals from first to second contacts and from first to third contacts, gives, respectively, the diameters of the small and large stars.

In actuality the orbits are not, generally, exactly edge on, and the eclipses are not central. To see how the inclination affects the light curve, consider the system shown in Figure 22-11. Here the smaller star does not pass behind the larger one in a direction perpendicular to its limb, but instead passes obliquely, and thus takes longer to be eclipsed. Moreover, it does not pass behind a full diameter of the disk of the large star but

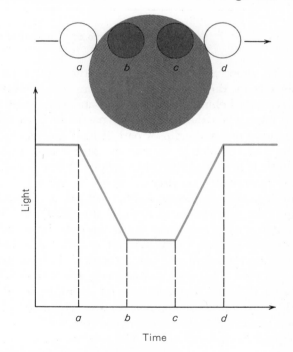

FIG. 22-11   Contacts in the light curve of a hypothetical eclipsing binary with noncentral eclipses.

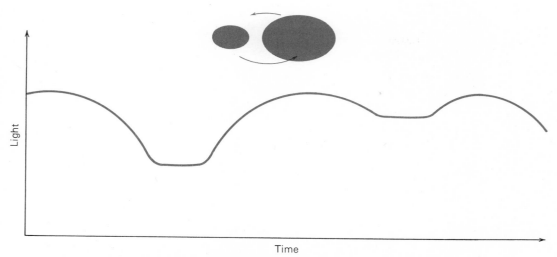

FIG. 22-12   Effect of tidal distortion in a binary system on the light curve.

along a shorter chord, so that the entire eclipse, especially the total phase, is of shorter duration than when the eclipses are central. The effects of the sizes of the stars and of the inclination of the orbit are interdependent. However, it is a relatively simple geometry problem, at least in principle, to separate these effects, and from the depths of the minima and the exact instants of the various contacts to calculate both the inclination of the orbit and the sizes of the stars, relative to their separation.

The foregoing discussion applies only to eclipses that are total and annular. If they are partial, the analysis is far more difficult, although it can still be accomplished.

There are various other complications which have been ignored here. For example, stars such as the sun exhibit limb darkening (Section 21.2b) which affects the rate of the drop of light during eclipse. Also, frequently, the two stars in an eclipsing binary are so close together that they suffer severe tidal distortion and have shapes more like footballs than spheres. The light from such systems is not constant, even outside of eclipse, but is great-

est when the stars' longest dimensions are turned "broadside" to us, and is less just before and just after a minimum (Figure 22-12). Such close binaries are even found to lose or exchange matter as a result of their interactions (see Chapter 25). Still another complication arises when the two stars are fairly close together and are very different in temperature. The hotter star can heat up the portion of the cooler star that is nearest it, which causes that portion of the cool star to radiate more intensely than the rest of its surface. The light curve of such a system shows small maxima just before and after secondary minimum, when the "hot spot" is turned toward us (Figure 22-13). Finally, even when the eclipses are total and annular, and all contacts are well defined, it is not always possible to tell whether it is the primary or secondary eclipse that is total. Often, fortunately, spectroscopic evidence enables us to deduce whether the larger or smaller star is the cooler one, and hence is the one which is in front during the primary minimum.

To summarize: From the analysis of the light curve of an eclipsing binary we can find the in-

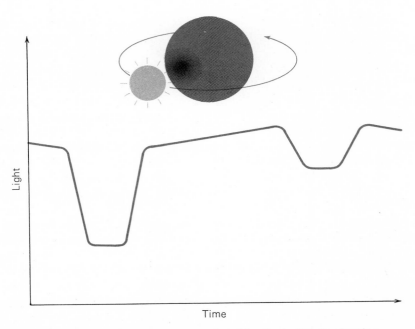

FIG. 22-13   Effect on the light curve produced by the hotter star heating a region of the cooler star in an eclipsing binary system.

clination of the orbit and the sizes of the stars, relative to their separation. If, in addition, we can measure the Doppler shifts of the spectral lines of both stars during their period of revolution, we can obtain their velocity curves. The analysis of the velocity curves, as described in the preceding section, leads to a determination of lower limits to the masses of the stars. The knowledge of the inclination of the orbit now allows us to convert these minimum values for the masses to actual masses for the individual stars. We can also convert the lower limit to the separation of the stars (or the semimajor axis of their relative orbit) to the actual value when the inclination is known; since the sizes of the stars relative to this separation are found from the light curve, we find their actual diameters, or radii. Finally, from the relative depths of the primary and secondary minima, we can calculate the relative surface brightnesses of the stars

and hence their effective temperatures (the surface brightness of a star is proportional to the fourth power of its effective temperature — Sections 10.4d and 21.2d). Among the thousands of known eclipsing binaries, however, there are only a few dozen that are so favorably disposed for observation that all the necessary data can be obtained. Only for these few binaries have complete analyses led to fairly reliable values of the masses and radii of their member stars.

22.3 THE MASS-LUMINOSITY RELATION

Studies of binary stars have provided a fairly accurate knowledge of the masses of a few dozen individual stars. When the masses and luminosities of those stars for which both of these quantities are well determined are compared, it is found that, in general, the more massive stars are also the more luminous. This relation, known as the *mass-luminosity relation,* is shown graphically in Figure 22-14. Each point represents a star of known mass and luminosity; its horizontal position (abscissa) indicates its mass, given in units of the sun's mass, and its vertical position (ordinate) indicates its luminosity, in units of the sun's luminosity.

It is seen that most stars fall along a narrow sequence running from the lower left (low mass, low luminosity) corner of the diagram to the upper right (high mass, high luminosity) corner. The relation between the mass and luminosity of a star is not accidental or mysterious but results from the fundamental laws that govern the internal structures of stars; the explanation was supplied by the British astrophysicist, Arthur S. Eddington, in 1924 and will be discussed in Chapter 29. It is estimated that about 90 percent of all stars obey the mass-luminosity relation. These, as we shall see later, are the so-called *main sequence stars* (Chapter 23). Stars that are still contracting after recent forma-

tion from interstellar matter (Chapter 30), giants, and white dwarfs (Chapter 23) are the most numerous among those stars to which the relation does *not* necessarily apply. We can usually identify those nonconforming stars by their spectral characteristics; for most other stars the mass-luminosity relation provides a useful means of estimating the masses of stars of known luminosity that do not happen to be members of visual or eclipsing binary systems.

In particular, it should be noted how very much greater the range of stellar luminosities is than the range of stellar masses. Luminosities of stars are roughly proportional to their masses raised to the 3.5 power. Most stars have masses between $\frac{1}{10}$ and 50 times that of the sun; according to the mass-luminosity relation, however, the corresponding luminosities of stars at either end of the range are respectively less than 0.01 and about 10^6 solar luminosities. The intrinsically faintest known star has a luminosity of 10^{-6} that of the sun; its mass

FIG. 22-14   Mass-luminosity relation. The three points at the bottom represent white dwarf stars, which do not conform to the relation.

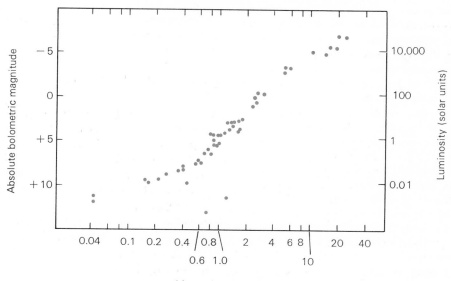

Mass (solar masses)

is probably greater than 0.01 of the sun's. If two stars differ in mass by a factor of 2, their luminosities would then be expected to differ by a factor of 10.

(a) *Dynamical Parallaxes*[*]

If the mass-luminosity relation is assumed to hold for the stars in a visual binary system of unknown distance, it provides a means of calculating the distance to that system. In Section 22.2b Kepler's third law was given in the form

$$m_1 + m_2 = \frac{(r \times a'')^3}{P^2}.$$

The above equation may now be solved for *r*, the distance to the visual binary. The result is

$$r = \frac{[P^2(m_1 + m_2)]^{1/3}}{a''}.$$

The period, *P*, and the angular semimajor axis of the relative orbit, *a''*, are both obtained from observation. The sum of the masses of the stars, of course, is not known if the distance to the system is not known in advance. On the other hand, most stars have masses that are not so very different from the sun's, so a good guess for the sum of the masses of the two stars would be 2. Even if this guess is considerably in error, the percentage error in the calculated distance will be less, because the masses occur in the equation only to the one-third power.

Now, with the calculated value of *r* and the observed apparent magnitudes of the two stars, their absolute magnitudes can be calculated. With these absolute magnitudes, the mass-luminosity relation is applied to obtain a better estimate for the mass of each star. These new mass values are put into the above equation and an improved value of *r* is calculated. The procedure can be repeated as many times as desired — a process known as *iteration*. With each step a closer approximation to the distance is found. After two or three iterations, the maximum accuracy is reached that is allowed by the uncertainties in the mass-luminosity relation and in the observed values of the period, apparent magnitudes, and angular size of the orbit.

The above method of calculating distances makes use of principles of mechanics (through Kepler's third law), and so is a *dynamical* procedure. Because stellar distances (largely for historical reasons) are often expressed in terms of their reciprocals, stellar parallaxes, the method is called that of *dynamical parallaxes*.

The visual binary star 70 Ophiuchi furnishes an example of the calculation of a dynamical parallax. The period of the system is 87.7 years and the angular semimajor axis of the relative orbit is 4″.5. For the first approximation to *r*, $m_1 + m_2$ is set equal to 2, and thus

$$r = \frac{[(87.7)^2 \times 2]^{1/3}}{4.5} = 5.5 \text{ pc.}$$

With this provisional value for the distance, absolute magnitudes are calculated for the two stars and then the mass-luminosity relation is applied. For the masses, the results are 0.85 and 0.6. The second approximation to *r* is, then,

$$r = \frac{[(87.7)^2 \times 1.45]^{1/3}}{4.5} = 5.0 \text{ pc.}$$

It happens that the trigonometric parallax can be measured for 70 Ophiuchi; it leads also to a value for the distance of 5.0 pc.

Unfortunately, dynamical parallaxes can only be determined for visual binaries whose periods are short enough to be determined accurately. Such systems are usually relatively nearby, and their parallaxes can often be measured directly, by triangulation.

22.4 DIAMETERS OF STARS

The eclipsing binary systems, as we have seen, provide one means of determining the diameters of stars, but of course only of those stars that happen to be members of eclipsing systems for which the necessary analysis can be carried out. It would be convenient if the angular sizes could be measured directly for many stars of known distances; then their linear diameters could be calculated just as they are for the moon or planets. The sun is unfortunately the only star whose angular size can be resolved optically and whose diameter can be calculated simply. There are a few other stars, however, whose angular sizes are only slightly beyond the limit of resolution of the largest telescopes and which can be measured with a device known as the *stellar interferometer*.

(a) *The Stellar Interferometer**

The stellar interferometer, invented by the American physicist, A. E. Michelson, can be employed to increase, effectively, the resolving power of a telescope. It is beyond the scope of this book to describe in detail how the stellar interferometer works; we can only describe briefly how it is used.

The instrument consists of a long beam or track upon which two mirrors are mounted on opposite sides of, and equidistant from, the center of the beam. The two mirrors can be separated by varying amounts by moving each inward or outward from the center along the track. In 1920 Michelson and Pease mounted an interferometer with a 20-ft beam on the front of the 100-inch telescope at Mount Wilson. Light from a star was gathered by each of the movable mirrors, which reflected the starlight, by means of auxiliary stationary mirrors, to the 100-inch mirror of the telescope. The 100-inch mirror then focused this incoming light. Now if the star were a true point source, its image would not be a perfect point, but a small disk, called a *diffraction disk* (Section 11.2c), crossed with alternate dark and bright bands called *fringes.* (The term "fringe" as used here does not refer to the "edge" of the image.) The fringes are produced because the light waves from the two movable mirrors interfere with each other; the spacing of the fringes decreases with increasing separation between the movable mirrors.

Now, if a double star is observed with the instrument, the image of *each* star consists of such a fringe pattern. If the angular separation of the stars is small, the fringe patterns of their images overlap; if the spacing of the movable mirrors is a certain critical amount, the dark fringes of one star image fall on top of the bright fringes of the other image, thus "washing out" the composite fringe pattern. The spacing of the movable mirrors that is required for such a disappearance of the fringe pattern provides a way to calculate the angular separation of the stars. The stellar interferometer, therefore, can be used to measure the angular separations of close binary stars.

An interesting application of the stellar interferometer, and the one to which Michelson and Pease directed most of their energy, is the measurement of the angular sizes of single stars. Each point on the apparent disk of a star can be thought of as producing its own fringe pattern. If the two mirrors are separated far enough, the patterns produced by the different parts of the star's disk will cancel each other out. The greater the separation of the mirrors that is required to make the fringes disappear, the smaller is the angular size of the star. Michelson and Pease were able to measure the angular sizes of seven giant stars that are large enough and near enough so that their disks, although too small to photograph or to see directly, are still just large enough to measure with the interferometer. The data for these stars are given in Table 22.1.

TABLE 22.1 Stars Measured with the Stellar Interferometer

STAR	ANGULAR DIAMETER	DISTANCE (IN PC)	LINEAR DIAMETER (IN TERMS OF SUN'S)
Betelgeuse	0."034†	150	500
(α Orionis)	0.042		750
Aldebaran			
(α Tauri)	0.020	21	45
Arcturus			
(α Bootis)	0.020	11	23
Antares			
(α Scorpii)	0.040	150	640
Scheat			
(β Pegasi)	0.021	50	110
Ras Algethi			
(α Herculis)	0.030	150	500
Mira			
(ο Ceti)	0.056	70	420

†Variable in size.

Later, Michelson built a 50-ft beam interferometer that had its own built-in telescope. Unfortunately, the larger instrument did not perform to expectations and no additional stars could be measured.

An electronic analogue of the stellar interferometer has been applied more recently to the measurement of star diameters. Separate optical telescopes, placed up to hundreds of feet apart, and equipped with photomultiplier tubes, are used to observe the same star simultaneously. The electric currents generated in the two tubes are

STELLARUM
INERRANTIUM
CATALOGUS BRITANNICUS,
Ad Annum Chrifti Completum, 1689.

Ab Obfervationibus GRENOVICI in OBSERVATORIO Regio habitis.

Affiduis Vigilijs, Cura, & Studio

JOHANNIS FLAMSTEEDII, Aftronom. Reg.

Deductus & Supputatus.

In Conftellatione ARIETIS.							
ORDO	STELLARUM Denominatio.	Bayer Char.	Afcenfio Recta. ° ′ ″	Diftantia à Polo B. ° ′ ″	Longitudo. s ° ′ ″	Latitudo. ° ′ ″	Varia. Afc.R. ′ ″	Varia. D. à P. ′ ″	Magnitud. ′ ″

Ptol.	Tych.	STELLARUM Denominatio.	Bayer	Afcenfio Recta	Diftantia à Polo B.	Longitudo.	Latitudo.	Varia. Afc.R.	Varia. D.àP.	Magnitud.
				20 46 0	69 17 15	♈26 58 25	11 4 58 B	58 7	22 25	7. 6
				21 25 45	71 15 25	26 48 15	9 1 26 B	57 52	22 19	7. 6
				22 26 15	74 10 55	26 36 18	5 57 3 B	57 31	22 12	6
1	1	Quæ in Cornu duarum præcedens	γ	22 51 15	74 36 55	26 49 4	5 23 59 B	57 28	22 07	7. 6
				24 8 30	72 14 45	28 51 0	7 8 58 B	58 02	21 55	4
2	2	Sequens & Borea eft	β	24 23 30	70 43 55	29 37 59	8 28 16 B	58 22	21 50	3
5	6	In Cervice	ι	24 39 45	67 57 45	♉ 0 54 20	10 57 12 B	58 57	21 49	tel.
		In Vertice	λ	25 7 0	73 13 15	♈29 10 57	5 26 12 B	57 54	21 44	6
				25 11 0	67 56 -5	♉ 1 22 15	10 47 47 B	59 00	21 44	5
				26 32 3	65 35 5	3 26 14	12 31 52 B	59 45	21 28	6. 7
	17	*Infra Lucidam*	κ	27 19 30	65 48 15	4 2 12	12 4 2 B	59 48	21 20	6
				27 19 30	68 51 5	2 55 8	9 13 29 B	59 16	21 21	6. 5
Inf.1	3	Infor.fup.Caput, *Lucida* ♈tis	α	27 26 30	68 1 45	3 19 18	9 57 12 B	59 22	21 20	2
				27 57 30	65 33 15	4 40 46	12 5 32 B	59 56	21 12	6
				28 22 30	71 58 45	2 43 49	5 56 58 B	58 36	21 08	6
				28 24 45	65 32 40	5 4 35	11 57 0 B	60 4	21 06	8
3	4	In Roftro duarum Borea	η	28 52 30	70 16 25	3 46 50	7 22 45 B	59 2	21 02	6
				28 58 45	71 33 15	3 25 14	6 8 45 B	58 46	21 00	7

A

FIG. 22-15   A page from Flamsteed's star catalogue, published in 1689. (*Yerkes Observatory.*)

brought together to a single amplifier. The electrical impulses carry information that can be made to interfere in a way somewhat analogous to the production of fringes in the optical interferometer. The radio astronomers Brown and Twiss first used the device to measure the angular diameter of the star Sirius.

The Brown-Twiss type of interferometer is known as an *intensity interferometer.* The largest model, assembled in Australia, uses two telescopes that can be separated up to 200 m. Each uses a mosaic mirror 6.5 m in diameter, which in turn consists of 251 separate small mirrors, all mounted to reflect the light from a star to one point. These special-purpose mirrors are not of high optical quality and could not be used for ordinary telescopic observations. They are adequate, however, to focus starlight on the light-sensitive surface of a photomultiplier. This large intensity interferometer can achieve a resolution down to 5×10^{-4} seconds of arc and (at the time of writing) has been used to measure the angular diameters of 15 stars. These range from Sirius, whose angular diameter is 0."00585, to Epsilon Orionis, with an angular diameter of only 0."00070.

The vast majority of stars, however, are too remote for even sophisticated interferometers to be able to measure their angular sizes directly.

(b) Stellar Radii from Radiation Laws

For most stars we must use an indirect method by which we can calculate their radii from theory. The theory involved is the Stefan-Boltzmann law (Section 10.4d); we calculate the radius of a spherical perfect radiator that has the same luminosity and effective temperature that a star does.

The luminosity of a star can be obtained by the procedure discussed in Chapter 20, and the temperature of a star can be obtained in various ways, as from its color or its spectrum (Section 21.2d). Now the energy emitted per unit area of a star (given by Stefan's law), multiplied by its entire surface area, gives the star's total output of radiant energy — that is, its luminosity. Since the surface area of a sphere of radius R is $4\pi R^2$, the luminosity of a star is given by

$$L = 4\pi R^2 \times \sigma T^4.$$

The above equation can be solved for the radius of the star.

Note that the temperature appearing in the above formula is raised to the fourth power; if it is in error, therefore, the computed value of the star's radius can be substantially incorrect. In particular, because stars are *not* perfect radiators, values of stellar temperatures as determined by different methods do not all agree precisely. There is no assurance that the color temperature, or a temperature estimated from the star's spectrum, is the most appropriate value to use in the computation of stellar radii from radiation laws. In the relatively few cases (described above) in which both the radius and luminosity of a star can be determined independently, the radius and luminosity of the star can be used with the above relation to derive the star's temperature. A temperature so determined, the *effective temperature* (T_{eff}), is the temperature of a black body that emits the same amount of energy per unit area as the star does, and thus is the appropriate kind of temperature to use in the calculation of stellar radii. Those effective temperatures which have been determined for stars enable us to see what small corrections must be applied to temperatures derived by other means (for example, from colors) to make the determination of stellar radii more reliable. Different kinds of stellar temperatures were discussed more fully in Section 21.2d.

We shall illustrate the use of Stefan's law for the computation of stellar radii with two examples. Consider, first, a star whose red color indicates that it has a temperature of about 3000°K, roughly half the temperature of the sun. Each square centimeter of the star, therefore, emits only $\frac{1}{16}$ as much light as the sun (for the light emitted is proportional to the fourth power of the temperature). Suppose, however, that the star is, nevertheless, 400 times as luminous as the sun. It must be many times larger than the sun to emit more light despite its much lower surface brightness. We can find its radius, in terms of the sun's, by noting that $L \propto R^2T^4$, and thus (since the constants of proportionality cancel in each of the ratios),

$$\frac{R_*}{R_\odot} = \sqrt{\frac{L_*}{L_\odot}}\left(\frac{T_\odot}{T_*}\right)^2 = \sqrt{400} \times 4 = 80.$$

(The subscripts $*$ and $\odot$ refer to the star and sun, respectively.) This star has 80 times the sun's radius; if the sun were placed at its center the star's surface would reach past the orbit of Mercury.

Next, consider a star whose blue color indicates a temperature of about 12,000°K — twice the sun's temperature. Suppose, however, that this star has a luminosity of only $\frac{1}{100}$ that of the sun. Now we find, for the star's radius,

$$\frac{R_*}{R_\odot} = \sqrt{\frac{L_*}{L_\odot}}\left(\frac{T_\odot}{T_*}\right)^2 = \sqrt{\frac{1}{100}}\left(\frac{1}{2}\right)^2 = \frac{1}{40}.$$

The star has only $\frac{1}{40}$ the sun's radius — less than three times the radius of the earth. These two examples are by no means extreme cases, but are more or less typical of stars known as *red giants* and *white dwarfs,* respectively (Chapter 23).

(c) Summary of Stellar Diameters

The few dozen good geometrical determinations of stellar radii come from (1) direct measure of the sun's angular diameter, (2) measures of the angular diameters of seven nearby giant stars with the stellar interferometer, (3) measures of the angular sizes of 15 other stars with an electronic adaptation of the stellar interferometer (the method may well have been applied to more stars by the time this is read), and (4) analyses of the light curves and radial-velocity curves of eclipsing binary systems. All other determinations of stellar radii make use of the radiation laws; the validity of this indirect method is verified by noting that it gives the correct radii for those stars whose sizes can also be measured by geometrical means.

The temperatures obtained from observations of the colors or spectra of stars are not found to be exactly the same as the effective temperatures which must be used with Stefan's law to compute stellar radii. The necessary corrections to color or spectral temperatures that are required to obtain effective temperatures are known for those stars whose radii are determined geometrically. Unfortunately, many classes of stars do not contain objects whose sizes have been measured by direct or geometrical methods, and effective temperatures for them must be predicted from theoretical studies of stellar atmospheres. Although radii computed for those stars are somewhat uncertain, we can use Stefan's law to find the sizes of most stars with a precision of 10 or 20 percent.

Exercises

1. Many eclipsing binaries can be observed which are *not observed* as spectroscopic binaries. Can you suggest an explanation?

2. A few stars are both visual binaries *and* spectroscopic binaries (in that their radial-velocity variations can be detected). Why do you suppose such stars are rare? Can you suggest a way of determining the distance to such a system? (*Hint:* Consider the method of determining the parallax to a moving cluster, Section 19.3c.)

3. Describe the apparent relative orbit of a visual binary whose true orbital plane is edge on to the line of sight. Describe the apparent motions of the individual stars of the system among the background stars in the sky.

*4. What, approximately, would be the periods of revolution of binary star systems in which each star had the same mass as the sun, and in which the semimajor axes of the relative orbits had the values:

(a) 1 AU?   (c) 6 AU?   (e) 60 AU?
(b) 2 AU?   (d) 20 AU?   (f) 100 AU?

*5. In each of the binary systems in Exercise 4, at what distance would the two stars appear to have an angular separation of 1″? (Assume circular orbits.)

6. Why do you suppose most visual binaries have long periods and most spectroscopic binaries shorter periods?

7. Show that the semimajor axis of the true relative orbit of a visual binary system, in astronomical units, is equal to its angular value, in seconds of arc, times the distance of the system, in parsecs.

8. The true relative orbit of ξ Ursae Majoris has a semimajor axis of 2″5, and the parallax of the system is 0″127. The period is 60 years. What is the sum of the masses of the two stars in units of the solar mass?

    Answer: 2.1 solar masses

9. Explain why it is unlikely that a binary star system with a period of more than a few months would be observed as a spectroscopic binary. Under what conditions could a binary with a relatively long period be observed as a spectroscopic binary?

10. Explain why the two stars in a binary system must have orbits of identical shape (eccentricity), differing only in size.

11. A hypothetical spectroscopic-eclipsing binary star is observed. The period of the system is 3 years. The maximum radial velocities, with respect to the center of mass of the system, are as follows:

$$\text{Star } A: \quad \tfrac{4}{3}\pi \text{ AU/year}$$
$$\text{Star } B: \quad \tfrac{2}{3}\pi \text{ AU/year}$$

(a) What is the ratio of the masses of the stars?

(b) Find the mass of each star (in solar units). Assume that the eclipses are central.

12. Describe the analysis of the light curve of a hypothetical eclipsing binary system in which both stars are *cube-shaped* and in which one face of each star

is in the plane of the sky. Include a discussion of the analysis of both total and annular and of partial eclipses. Sketch the light curve for each case. Ignore limb darkening and other complications.

13. In an eclipsing binary in which the eclipses are exactly central, and in which a small star revolves about a considerably larger one, the interval from first to second contacts is 1 hour and from first to third contacts is 4 hours. The entire period is 3 days. The centers of the stars are separated by 11,460,000 mi. What are the diameters of the stars?

   Answer: 1,000,000 and 4,000,000 mi

14. Although the periods of known eclipsing binaries range from 4^h39^m to 27 years, the average of their periods is less than the average period of all known spectroscopic binaries. Can you suggest an explanation?

15. How many times as massive as the sun would you expect a star to be that is 1000 times as luminous? What if it were 10,000 times as luminous? (Assume that the mass-luminosity relation holds for these stars.)

16. Most eclipsing binary stars are giant stars of high luminosity. Why do you suppose this is the case?

17. What is the radius of a star (in terms of the sun's radius) with the following characteristics:

(a) Twice the sun's temperature and four times its luminosity?

(b) Eighty-one times the sun's luminosity and three times its temperature?

18. Assume the wavelength of maximum light of the sun to be exactly 5000 Å, its temperature exactly 6000°K, and its absolute bolometric magnitude exactly 5.0. Another star has its wavelength of maximum light at 10,000 Å. Its apparent photovisual magnitude is 15.5, its bolometric correction is 0.5, and its parallax is 0".01. What is its radius in terms of the sun's?

   Answer: $R/R_\odot = 0.4$

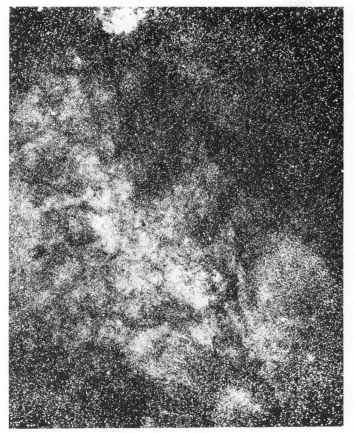

CHAPTER 23

The Stellar Population

Chapters 18 through 22 describe the methods by which we are able to obtain basic information about individual stars — their distances, masses, luminosities, radii, colors, temperatures, motions, and so on. This chapter summarizes the data that have been gathered about the normal stars and the stellar population.

23.1 THE NEAREST AND THE BRIGHTEST STARS

Let us consider first our most conspicuous stellar neighbors, the brightest-appearing stars in the sky. Appendix 13 lists some of the properties of the 20 brightest stars. Many of these are double- or triple-star systems; data are given, in such cases, for each component.

(a) The Brightest Stars

The most striking thing about the brightest-appearing stars is that they are bright not because they are nearby, but because they are actually of high intrinsic luminosity. Of the 20 brightest stars listed in Appendix 13, only 6 are within 10 pc of the sun. The absolute magnitude of a star (Section 20.2a) is the apparent magnitude that it would have if it were at a distance of 10 pc. Since the 20 brightest stars are of apparent magnitude 1.5 or brighter, the 14 of them that are more distant than 10 pc must have absolute magnitudes *less* (that is, brighter) than 1.5 Even among the approximately 3000 stars with apparent magnitudes less than 6.0, only about 60 are within 10 pc. Most naked-eye stars are tens or even hundreds of parsecs away and are many times more luminous than the sun. Figure 23-1 is a histogram showing the distribution among various absolute visual magnitudes of the 30 brightest-appearing stars (the absolute visual magnitude of the sun is 4.7).

From an inspection of Appendix 13 or Figure 23-1 we might gain the impression that the sun is far below average among stars in luminosity. This is not so. Most stars, as we shall see in the next subsection, are much less luminous than the sun is. They are too faint, in fact, to be conspicuous unless they are nearby. Stars of high luminosity are rare — so rare that the chance of finding one within a small volume of space, say, within 10 pc of the sun, is very slight. Why, then, are the most common, intrinsically faint stars not among the most common naked-eye stars, while rare, highly luminous stars are?

The question is best answered with the help of some numerical examples. The sun, whose absolute visual magnitude is +4.7, would appear as a very faint star to the naked eye if it were 10 pc away. Stars much less luminous than the sun would not be visible at all at that distance. Stars with absolute magnitudes in the range +10 to +15 are very common, but a star of absolute magnitude +10 would have to be within 1.6 pc to be visible to the naked eye. Only Alpha Centauri is closer than this. The intrinsically faintest star observed has an absolute magnitude of about +20. For this star to be visible to the naked eye, it would have to be within 0.016 pc, or 3200 AU. The star could not even be photographed with the 200-inch telescope if it were more distant than 50 pc. It is clear, then, that the vast majority of nearby stars, those less luminous than the sun, do not send enough light across the span of interstellar distances to be seen without optical aid.

In contrast, consider the highly luminous stars. Stars with absolute magnitudes of 0 have luminosities of about 100 times that of the sun. They are far less common than stars less luminous than the sun, but they are visible to the naked eye even out to a distance of 160 pc. A star with an absolute magnitude of −5 (10,000 times the sun's luminosity) can be seen without a telescope to a distance of 1600 pc (if there is no dimming of light by interstellar dust — see Chapter 26). Such stars are very rare, and we would not expect to find one within a distance of only 10 pc; the volume of space included within a distance of 1600 pc, however, is about 4 million times that included within a distance of only 10 pc. Hence many stars of high luminosity are visible to the unaided eye.

(b) The Nearest Stars

Evidently, the brightest naked-eye stars do not provide a very representative sample of the stellar population. Let us turn then to the nearest known stars. Appendix 12 lists data for the 37 known stars within 5 pc of the sun (some are double or multiple systems). The table must not be taken as the "final word," for additional nearby stars are discovered from time to time, and the measurements of distances, luminosities, and so on, for known nearby stars are being continually refined. On the other hand, the table does indicate some general characteristics of the sun's nearest stellar neighbors.

First, the table shows that only three of the 37 stars are among the 20 brightest stars: Sirius, Alpha Centauri, and Procyon. This fact is further confirmation that the nearest stars are not the brightest-appearing stars. The nearby stars also tend to have large proper motions, as would be expected (Section 19.3). In fact, the large proper motions of many of these stars led to the discovery that they are located nearby. Another interesting observation is that 14 of the 37 stars are really binary- or multiple-star systems; the table contains, actually, 54 rather than 37 stars. Thirty-one of these 54 stars are members of systems containing more than one star.

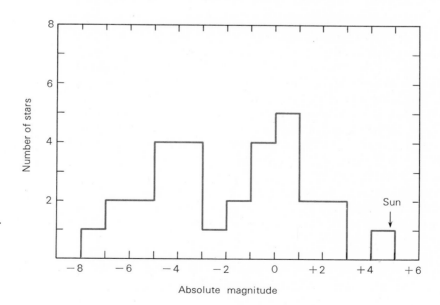

FIG. 23-1 Distribution among absolute magnitudes of the 30 brightest appearing stars.

The most important datum concerning the nearest stars is that most of them are intrinsically faint. Only eight of the nearest stars are visible to the unaided eye. Only three are as intrinsically luminous as the sun; 26 have absolute magnitudes fainter than +10. If the stars in our immediate stellar neighborhood are representative of the stellar population in general, we must conclude that the most numerous stars are those of low luminosity.

An estimated lower limit can be established for the mean density of stars in space in the solar neighborhood. There are at least 54 stars within 5 pc (counting the members of binary- and multiple-star systems). We would expect eight times as many stars within 10 pc (for that distance includes a volume of space eight times larger), and about 64 times as many stars within 20 pc. W. Luyten estimates there to be more than 500 stars within 10 pc and some of the fainter stars within that distance must have escaped discovery. Within 20 pc there must be at least 4000 stars. A sphere of radius 5 pc has a volume of $\frac{4}{3}\pi(5)^3$, or about 520 pc^3. Since this volume of space contains at least

54 stars, the density of stars in space is at least one star for every 10 pc^3; the actual stellar density, of course, can be greater than this figure if there are undiscovered stars within 5 pc. The mean separation between stars is the cube root of 10, or about 2.1 pc. If the matter contained in stars could be spread out evenly over space, and if a typical star has a mass 0.4 times that of the sun, the mean density of matter in the solar neighborhood would be about 3×10^{-24} gm/cm^3.

The nearest stars comprise a much more nearly representative sample of the stellar population than do the brightest stars. We are still not sure, however, that we have identified all of the faintest stars in the solar neighborhood. Moreover, there do not happen to be any stars of high luminosity in this "tiny" volume of space. Yet we can identify all the luminous stars, with a reasonable degree of completeness, out to a much greater distance. If we make allowance for the different volumes of space that we must survey to catalogue large samples of stars of different intrinsic luminosities, we can gain some indication of their relative abundances. For example, within 10 pc

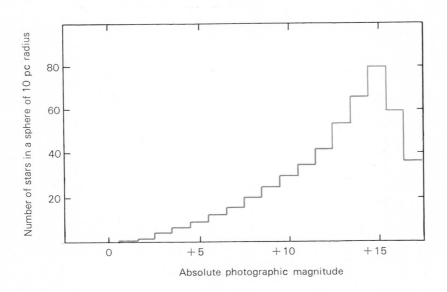

FIG. 23-2 Luminosity function of stars in the solar neighborhood.

there are about 12 known stars brighter than absolute magnitude +4, while within 5 pc there are about 35 known stars fainter than absolute magnitude +4. We would expect, however, some 8 × 35, or 280, stars fainter than absolute magnitude +4 within 10 pc; therefore, the ratio of stars with absolute magnitude greater (fainter) than +4 to the number of more luminous stars is about 280 to 12, or 23:1. This calculation is only an example and may not indicate the precise ratio that exists in the stellar population.

(c) The Luminosity Function

Once the numbers of stars of various intrinsic luminosities have been found, the relative numbers of stars in successive intervals of absolute magnitude within any given volume of space can be established. This constitutes what is called the *luminosity function.* Figure 23-2 shows the luminosity function for stars in the solar neighborhood, as it has been determined by W. J. Luyten. Compare Figure 23-2 with Figure 23-1.

The sun, we see, is more luminous than the vast majority of stars. Most of the stellar mass is contributed by stars that are fainter than the sun. On the other hand, the relatively few stars of higher luminosity than the sun compensate for their small numbers by their high rate of energy output. It takes only 10 stars of absolute magnitude 0 to outshine 1000 stars fainter than the sun, and only one star of absolute magnitude −5 to outshine 10,000 stars fainter than the sun. Most of the starlight from our part of space, it turns out, comes from the relatively few stars that are more luminous than the sun.

23.2 THE HERTZSPRUNG–RUSSELL DIAGRAM

In 1911 the Danish astronomer E. Hertzsprung compared the colors and luminosities of stars within several clusters by plotting their magnitudes against their colors. In 1913 the American astronomer Henry Norris Russell undertook a similar investigation of stars in the solar neighborhood by plotting the absolute magnitudes of stars of known distance against their spectral classes. These investigations, by Hertzsprung and by Russell, led to an extremely important discovery concerning the relation between the luminosities and surface temperatures of stars. The discovery is exhibited graphically on a diagram named in honor of the two astronomers — the *Hertzsprung-Russell,* or *H-R, diagram.*

(a) Features of the H-R Diagram

Two easily derived characteristics of stars of known distances are their absolute magnitudes (or luminosities) and their surface temperatures. The absolute magnitudes can be found from the known distances and the observed apparent magnitudes. The surface temperature of a star is indicated either by its color or its spectral class. Before the development of yellow- and red-sensitive photographic emulsions and, of course, photoelectric techniques, spectral classes of stars were usually used to indicate their temperatures. Now that stellar colors can be measured with precision, the color index is more often employed, even though the use of spectral classes is still of great value.

If the absolute photographic magnitudes of stars are plotted against their temperatures (or spectral classes, or color indices), an H-R diagram like that of Figure 23-3 is obtained. The most significant feature of the H-R diagram is that the stars are not distributed over it at random, exhibiting all combinations of absolute magnitude and temperature, but rather cluster into certain parts of the diagram. The majority of stars are aligned along a narrow sequence running from the upper left (hot, highly luminous) part of the diagram to the lower right (cool, less luminous) part. This band of points is called the *main sequence.* A substantial number of stars, however, lie above the main sequence on the H-R diagram, in the upper right (cool, high luminosity) region.

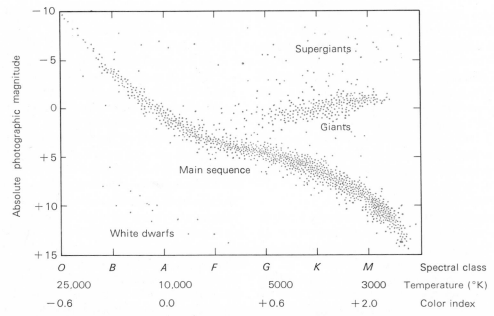

FIG. 23-3   Hertzsprung-Russell diagram for stars of known distance.

These are called *giants*. At the top part of the diagram are stars of even higher luminosity, called *supergiants*. Finally, there are stars in the lower left (hot, low luminosity) corner known as *white dwarfs*. To say that a star lies "on," or "off," the main sequence does not refer to its position in space, but only to the point that represents its luminosity and temperature on the H-R diagram.

An H-R diagram, such as Figure 23-3, that is plotted for stars of known distance does not show the relative proportions of various kinds of stars, because only the nearest of the intrinsically faint stars can be observed. To be truly representative of the stellar population, an H-R diagram should be plotted for all stars within a certain distance (see Exercise 9). Unfortunately, our knowledge is reasonably complete only for stars within a few parsecs of the sun, among which there are no giants or supergiants. It is estimated that about 90 percent of the stars in our part of space are main-sequence stars and about 10 percent are white

dwarfs. Less than 1 percent are giants or supergiants.

Chapters 29 and 30 deal with the theoretical interpretation of the distribution of stars on the H-R diagram.

(b) Method of Spectroscopic Parallaxes

One of the most important applications of the H-R diagram is in the determination of stellar distances. Suppose, for example, that a star is known to be a spectral class G2 star on the main sequence. Its absolute magnitude could then be read off the H-R diagram at once; it would be about +5. From this absolute magnitude and the star's apparent magnitude, its distance can be calculated (Section 20.2b).

In general, however, the spectral class alone is not enough to fix, unambiguously, the absolute magnitude of a star. The G2 star described in the last paragraph could have been, for example, a main-sequence star of absolute magnitude +5, a giant of absolute magnitude 0, or a supergiant of still higher luminosity. We recall, however (Section 21.1b), that pressure differences in the atmospheres of stars of different sizes result in slightly different degrees of ionization for a given temperature. It will be seen in the next subsection that giant stars are larger than main-sequence stars of the same spectral class and that supergiants are larger still. In 1913 Adams and Kohlschütter, at the Mount Wilson Observatory, first observed the slight differences in the degrees to which different elements are ionized in stars of the same spectral class but different luminosities (and therefore different sizes). It is now possible to classify a star by its spectrum, not only according to its temperature (spectral class) but also according to whether it is a main-sequence star, a giant, or a supergiant.

The most widely used system of classifying stars according to their luminosities is that of W. W. Morgan and his associates at the Yerkes Observatory. In favorable cases it has been found possible to divide stars of a given spectral class into as many as six categories, called *luminosity classes*, that depend on their luminosities (or sizes).

These luminosity classes are:

Ia Brightest supergiants
Ib Less luminous supergiants
II Bright giants
III Giants
IV Subgiants (intermediate between giants and main-sequence stars)
V Main-sequence stars

A small number of stars that may lie below the normal main sequence are called *subdwarfs* (Sd). The white dwarfs are much fainter. Main-sequence stars (luminosity class V) are often termed "dwarfs" to distinguish them from giants. The term "dwarf" is even applied to main-sequence stars of high luminosity, which may have diameters several times as great as the sun's. The term "dwarf," when applied to a main-sequence star, should not be confused with its use as applied to a white dwarf. The full specification of a star, including its luminosity class, would be, for example, for a spectral class F3 main-sequence star, F3 V. For a spectral class M2 giant, the specification would be M2 III. Figure 23-4 illustrates the approximate mean positions of stars of various luminosity classes on the

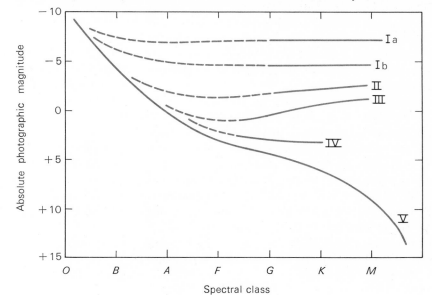

FIG. 23-4 Luminosity classes on the Hertzsprung-Russell diagram.

H-R diagram. The dashed portions of the lines represent those spectral classes (for a given luminosity class) for which there are very few stars.

With both its spectral class and luminosity class known, a star's position on the H-R diagram is uniquely determined. Its absolute magnitude, therefore, is also known, and its distance can be calculated. Distances determined this way, from the spectral and luminosity classes, are said to be obtained from the *method of spectroscopic parallaxes*.

There are some stars that do not fit into the standard classification scheme (see, especially, Chapter 25). The method of spectroscopic parallaxes does not apply to them.

(c) Extremes of Stellar Luminosities, Radii, and Densities

Let us now investigate the extremes in size, luminosity, and density found for stars. The most massive stars are the most luminous ones, at least for main-sequence stars. These stars have absolute magnitudes of −6 to −8. A few stars are known that have absolute bolometric magnitudes of −10; they are a million times as luminous as the sun. These super-luminous stars, most of which are at the upper left on the H-R diagram, are very hot spectral-type O and B stars, and are very blue. These are the stars that would be the most conspicuous at very great distances in space.

Consider now the stars at the upper right corner of the H-R diagram, both giants and super-giants. Some have surface temperatures less than half that of the sun, so each unit area of the surface of such a star must emit only $\frac{1}{16}$ or less as much light as the sun does (Section 22.4b). Yet they are at least a few hundred times as luminous as the sun (if they are giants) or some thousands of times as luminous (if they are supergiants). We see, then, that the cooler stars of high luminosity are called giants or supergiants because they are truly large stars.

Let us consider an illustration that is rather extreme but uses actually known stars. Suppose a red, cool supergiant has a surface temperature of 3000°K and an absolute bolometric magnitude of −5. This star has 10,000 times the sun's luminosity but only half its surface temperature. Since each unit area of the star emits only $\frac{1}{16}$ as much light as a unit area of the sun, its total surface area must exceed the sun's by 160,000 times. Its radius, therefore, is 400 times the sun's radius. If the sun could be placed in the center of such a star, the star's surface would lie beyond the orbit of Mars.

In the past several years special telescopes with infrared detectors have been built — especially by R. Leighton and his associates at the Mount Wilson and Palomar Observatories. These telescopes have been used to search for very cool stars that emit most of their energy in the invisible infrared. Many such stars have been discovered. These extreme red giants, or *infrared giants,* can have sizes even more extreme than the supergiant described in the last paragraph. It is difficult, however, to determine distances to them, so their absolute magnitudes are not well known.

Red giant stars have extremely low mean densities. The volume of the star described in the example in the paragraph before last would be 64 million times the volume of the sun. The masses of such giant stars, however, are probably at most only 50 solar masses, and very likely much less. Plaskett's Star, a spectral-type O star, has a mass of at least 50 solar masses but is one of the most massive stars known. If we assume that the supergiant star with 64 million times the sun's volume has only 10 times its mass, we find that it has just over 1 ten-millionth the sun's mean density, or only about 2 ten-millionths the density of water; the outer parts of such a star would probably constitute an excellent laboratory vacuum.

In contrast, the very common red, cool stars of low luminosity at the lower end of the main sequence are much smaller and more compact than the sun. As an example, consider such a "red dwarf," the star Ross 614B, which has a surface temperature of 2700°K and an absolute bolometric magnitude of about +13 ($\frac{1}{2300}$ of the sun's luminosity). Each unit area of this star emits only $\frac{1}{20}$ as much light as a unit area of the sun, but to have only $\frac{1}{2300}$ the sun's luminosity, the star must have only about $\frac{1}{115}$ the sun's surface area, or $\frac{1}{11}$ its radius. A star with such a low luminosity also has a low mass (Ross 614B has a mass of about $\frac{1}{12}$ that of the sun), but still would have a mean density about 100 times that of the sun's. Its density must be higher, in fact, than that of any known solid found on the surface of the earth.

The faint red main-sequence stars are not the stars of the most extreme densities, however. The white dwarfs, at the lower left corner of the H-R diagram, have the highest densities known to exist in nature.

(d) The White Dwarfs

The first white dwarf stars to be discovered were the companions to the stars 40 Eridani, Sirius, and Van Maanen's star. Sirius, the brightest-appearing star in the sky, is the most conspicuous star in the constellation of Canis Major (the big dog). It is a rather interesting coincidence that Procyon, the brightest star in the constellation of Canis Minor, (the little dog) also has a white-dwarf companion. Both Sirius and Procyon are visual binaries within 5 pc of the sun. One other star within 5 pc, 40 Eridani, is a multiple-star system that contains a white dwarf.

The white-dwarf companion of 40 Eridani, 40 Eridani B, is a good example of a typical white dwarf. Its absolute magnitude is 10.7 and its temperature is about 12,000°K; it has, therefore, 2.1 times the sun's surface temperature and $\frac{1}{200}$ its luminosity. Its surface area is only $1/(200 \times 2.1^4)$, or $\frac{1}{3900}$, of the sun's, which gives it a radius of

0.016 and a volume of 4.1×10^{-6} the sun's. Its mass, however, is 0.43 times that of the sun, so its mean density is $0.43/(4.1 \times 10^{-6})$, or about 100,000, times the mean density of the sun, and nearly 200,000 times the density of water.

One of the tests that Einstein suggested for the general theory of relativity is that light waves should show a slight shift to longer (redder) wavelengths in the presence of a strong gravitational field (Section 4.4a). Light leaving the surface of the sun should suffer such a red shift, but the effect is so small compared to other effects that it cannot be observed with certainty. At the surfaces of white dwarfs, on the other hand, there exist very strong gravitational fields (see Exercise 12). The theory of relativity predicts that the red shift of light leaving the white-dwarf companion of 40 Eridani should be equivalent to that produced by a radial velocity of 17 km/sec. Now the actual radial velocity of 40 Eridani B is known from the velocities of its nonwhite-dwarf companions. In 1954, D. M. Popper observed a red shift for this star that corresponds to a velocity 21 km/sec greater than that which should be produced by its known radial velocity. Popper's observed red shift agrees within the uncertainty of the observations with that predicted by the theory of relativity. Since then, gravitational red shifts have been observed for other white dwarf stars as well.

Since white dwarfs are intrinsically faint stars, they must be relatively nearby to be observed. Today they are usually discovered by searching for faint stars of large proper motion. Some hundreds of white dwarfs have now been found, largely due to the efforts of the astronomer W. J. Luyten of the University of Minnesota.

The theory of white dwarfs (Chapter 30) predicts that there should be a relation between their masses and radii. Unfortunately, masses can be observed directly for only the three nearby

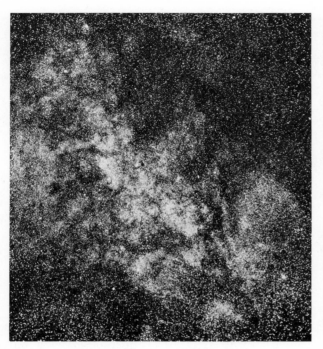

FIG. 23-5 The Great Star Cloud in Sagittarius.
(Yerkes Observatory.)

white dwarfs that are members of visual multiple-star systems; at least the masses and radii of these three stars are not inconsistent with the theoretical predictions. The masses of all white dwarfs should range from 0.1 to 1.2 solar masses, and their corresponding radii should range from 4 times to less than half that of the earth (the more massive white dwarfs being the smaller). The predicted mean densities of these stars range from about 50,000 to over 1,000,000 times that of water. At such densities, matter can not exist in its usual state. Although it is still gaseous, its atoms are completely stripped of their electrons, and the latter are obliged to move according to certain restrictive laws. The matter in white dwarfs is said to be *degenerate*. The white dwarfs are believed to be the final state of stellar evolution (see Chapter 30).

23.3 THE DISTRIBUTION OF THE STARS IN SPACE

In the immediate neighborhood of the sun, the stars seem to be distributed more or less at random (except for their tendency to form small clusters). The larger the volume of space we survey, the more stars we find, and if allowance is made for the fact that the faintest stars become invisible at larger distances, it is found that the number of stars we can count is roughly proportional to the cube of the distance to which we look. Eventually, however, the stars do thin out more rapidly in some directions than in others. The way they thin out is a clue to the nature of the stellar system to which the sun belongs. The idea that the sun is a part of a large system of stars was suggested as early as 1750 by Thomas Wright in his *Theory of the Universe*. Immanuel Kant, the great German philosopher, suggested the same hypothesis 5 years later. It was the German-English astronomer William Herschel, however, who first demonstrated the nature of the stellar system.

(a) Herschel's Star Gauging

Herschel sampled the distribution of stars about the sky by a procedure he called *star gauging*. He observed that in some directions he could count more stars through his telescope than in other directions. In 1785 he published the results of gauges or counts of stars that he was able to observe in 683 selected regions scattered over the sky. While in some of these fields he could see only a single star, in others he was able to count nearly 600. Herschel reasoned that in those directions in which he saw the greatest numbers of faint stars, the stars extended the farthest, and in other directions they thinned out at relatively short distances. As a result of his star gauging, Herschel arrived at the conclusion (only partially correct, as we shall see) that the sun is, indeed, inside a great sidereal system, and that the system is disk-shaped, roughly like a grindstone, with the sun near the center.

(b) The Phenomenon of the Milky Way

All of us who have looked at the sky on a moonless night away from the glare of city lights are aware of the Milky Way, a faint, luminous band of light that completely encircles the sky. Galileo solved the first mystery of the Milky Way when he turned his telescope on it and saw that it really consists of myriads of faint stars. Herschel's grindstone hypothesis solved the second mystery by explaining why the Milky Way should appear as a band all the way around the sky.

It must be recalled that we view our sidereal system from the inside. Figure 23-6 shows a portion of the "grindstone," viewed edge on. The sun's position is at *O*. If we look from *O* toward either face of the wheel, that is, in directions *a* or *b*, we see only those stars that lie between us and the nearest boundary of the stellar system. In these directions in the sky, therefore, we see only scattered stars. On the other hand, if we look edge on through the wheel, say in directions *c* or *d*, we encounter so many stars along our line of sight that we get the illusion of a continuous band of light. Since the greatest dimensions of the grindstone extend in all directions along its flat plane, the band of light extends completely around the sky. This band of light is the Milky Way; it is simply the light from the many distant stars that appear lined up in projection when we look edge on through our own flattened stellar system.

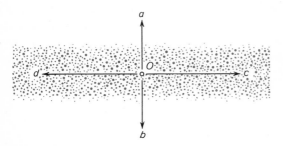

FIG. 23-6   Explanation of the Milky Way.

FIG. 23-7   NGC 5897, open star cluster in Libra. Photographed with the 200-inch telescope. *(Mount Wilson and Palomar Observatories.)*

(c) The Galaxy

We call our stellar system the *Galaxy,* or sometimes, the *Milky Way Galaxy.* In our modern view, the Galaxy is more complicated than Herschel's image of the grindstone. It is a vast, wheel-shaped system of some 100 billion stars, with a diameter that probably exceeds 30,000 pc (100,000 LY). The flattened shape of the Galaxy is a consequence of its rotation. The sun, about two-thirds of the way from the center out to the rim of the wheel, moves at a speed of about 250 km/sec (about 150 mi/sec) to complete its orbital revolution about the galactic center in some 200 million years. (The rotation of the Galaxy is not quite like that of a wheel, for it does not rotate as a solid body — see Chapter 27.)

At the center of the Galaxy is a huge hub or

FIG. 23-8   Globular star cluster, Omega Centauri.
(Harvard Observatory.)

spiral arms. The spiral arms consist of vast clouds of gas and cosmic dust — the so-called interstellar medium. Associated with these gas and dust clouds are many young stars, a few of which are very hot and luminous. It is in the interstellar gas and dust clouds of the spiral arms that star formation is believed to be still taking place. The sun is thought to be located in or near a spiral arm.

In addition to individual stars and clouds of interstellar matter, the Galaxy contains many *star clusters* — groups of stars having a common origin and probably a common age. The most common star clusters, numbering in the thousands, are the *open* or *galactic clusters.* Typically, an open cluster consists of a few hundred stars, loosely held together by their mutual gravitation, and moving together through space. The open clusters are located in the main disk of the Galaxy and are usually in or near spiral arms. Besides the open clusters, there are over a hundred *globular clusters* — beautiful, spherically symmetrical clusters, each containing hundreds of thousands of member stars. Most of the globular clusters are scattered in a roughly spherical distribution about the main wheel of the Galaxy, grouped around it rather like bees around a flower. They form a more or less spherical *halo* or *corona* surrounding the main body of the Galaxy.

Our Galaxy is not alone in space. Today we know that far beyond its borders there are countless millions of other galaxies, extending as far as we can see in all directions in space. Our Galaxy, interstellar matter, star clusters, and other galaxies are the subjects of some of the later chapters.

nucleus of stars, where the stars are somewhat closer together than they are in the solar neighborhood (although still light-months or light-years apart). Extending outward from the nucleus, and winding through the disk of the Galaxy, like the spirals of light in a gigantic pinwheel, are the

Exercises

1. (a) At what distance would a star of absolute magnitude +15 appear as a fifth magnitude star?

(b) At what distances would a star of absolute magnitude −10 and one of absolute magnitude +15 appear brighter than the fifth apparent magnitude?

2. Describe an everyday situation that is analogous to the fact that most naked-eye stars are of far more than average stellar luminosity.

3. If stars of all kinds were uniformly distributed through space, what would the approximate luminosity function have to be in order for intrinsically faint stars to be the most common among naked-eye stars?

4. At the actual luminosity function, how would stars have to be distributed in space in order for the intrinsically faint ones to be most common among naked-eye stars?

*5. Verify that the mean density of stellar matter in the solar neighborhood is about 3×10^{-24} gm/cm³.

6. Suppose that within 10 pc there were 11 stars of absolute magnitude +5, and that within 30 pc there were 11 stars of absolute magnitude 0. Estimate the true ratio of the number of stars of absolute magnitude +5 to the number of absolute magnitude 0 in a given volume of space.

7. From the data in Appendix 12 (the nearest stars), plot the luminosity function for stars nearer than 5 pc. Compare your plot with Figure 23-2.

8. Why are most faint-appearing stars blue?

9. Plot a Hertzsprung-Russell diagram for the stars within 5 pc of the sun. Use the data of Appendix 12. How does this H-R diagram differ from the one in Figure 23-3? Explain the reasons for these differences.

10. Find the distances to the following stars (See Figure 23-4):

(a) $m = +10$; spectral designation A0 Ib

(b) $m = +5$; spectral designation K5 III

(c) $m = 0$; spectral designation G2 V

11. Consider the following data on five stars:

STAR	m	SPECTRUM
1	15	G2 V
2	20	M3 Ia
3	10	M3 V
4	15	B9 V
5	15	M5 V

(a) Which is hottest?

(b) Which is coolest?

(c) Which is most luminous?

(d) Which is least luminous?

(e) Which is nearest?

(f) Which is most distant?

In each case, give your reasoning.

12. Suppose you weigh 150 lb on the earth. How much would you weigh on the surface of a white-dwarf star, the same size as the earth, but having a mass of 300,000 times the earth's (nearly the mass of the sun)?

13. Why can the test for the relativistic red shift only be made on a white-dwarf star that is a member of a binary- or multiple-star system?

14. Why do you suppose that most visual binaries are stars of low luminosity?

15. Sometimes our Galaxy is called, simply, the "Milky Way." Why is this poor terminology? Where, exactly, *is* the Milky Way? Does the question make sense? Why?

16. Suppose the Milky Way were a band of light extending only halfway around the sky (that is, in a semicircle). What then would you conclude about the sun's location in the Galaxy?

CHAPTER 24

A Typical Star—
The Sun

The preceding chapters have described the properties of stars in general; here a typical star is examined in detail. Fortunately, our own sun is a typical "garden-variety" star and serves as a good example.

This chapter is concerned only with those properties of the sun that are learned of directly from observation — its gross characteristics, the nature of its outer layers, solar activity, and the terrestrial disturbances associated with solar activity. It is also important for us to understand the nature of the sun's interior, where the temperatures of the gases range up to more than 10 million degrees Kelvin. There the thermonuclear conversion of hydrogen to helium gives rise to the sun's enormous energy output. However, we must derive the internal conditions of the sun almost entirely from theoretical arguments. Therefore, discussion of the solar interior is postponed to a later chapter.

24.1 GROSS PROPERTIES

The gross properties of the sun have been determined by procedures described earlier, mostly in Chapters 18 through 22. They are summarized in Table 24.1.

24.2 OUTER LAYERS OF THE SUN

The only parts of the sun that can be observed directly are its outer layers, collectively known as the sun's *atmosphere*. The solar atmosphere is not stratified into physically distinct layers with sharp boundaries. Yet there are three general regions, each having substantially different properties, even though there is a gradual transition from one region to the next. These are the *photosphere*, the *chromosphere*, and the *corona*.

(a) The Solar Photosphere

What we see when we look at the sun is the solar photosphere. As stated in Section 21.2b, the photosphere is not a discrete surface but covers the range of depths from which the solar radiation escapes. Most of the absorption of visible light in the solar photosphere, as we have seen, is done by negative hydrogen ions. As one looks toward the limb of the sun, his line of sight enters the photosphere at a grazing angle, and the depth below the outer surface of the photosphere to which he can see is even less than at the center of the sun's disk. The light from the limb of the sun, therefore, comes from higher and cooler regions of the photosphere. Analysis of this *limb darkening* can be used to determine the variation of temperature with depth in the photosphere, as described in Section 21.2. From the data obtained from observations of limb darkening, using our knowledge of the physics of gases and the way in which atoms absorb and emit light, we can calculate a *model solar photosphere*. Such a model solar photosphere is given in Table 24.2.

It is evident from Table 24.2 that within a depth of about 160 mi the pressure and density increase by a factor of 10, while the temperature climbs from 4500 to 6800°K. The corresponding range of temperature on the Fahrenheit scale is

TABLE 24.1 Solar Data

DATUM	HOW FOUND	TEXT REFERENCE (SECTION)	VALUE
Solar parallax	Radar reflection from Venus	18.3	8″.794
Mean distance	Radar reflection from Venus	18.3	1 AU 92,960,000 mi 149,597,893 km
Maximum distance			94,500,000 mi
Minimum distance			91,500,000 mi
Mass	Acceleration of earth	22.1	333,400 earth masses 2.2×10^{27} English tons 1.99×10^{33} gm
Mean angular diameter	Direct measure	12.3	31′59″.3
Diameter of photosphere	From angular size	22.4	109.3 times earth diameter 864,400 mi 1.39×10^{11} cm
Mean density	Mass/volume	4.1	1.41 gm/cm³
Gravitational acceleration at photosphere (surface gravity)	$\dfrac{GM}{R^2}$	4.3	27.9 times earth surface gravity 27,300 cm/sec² 900 ft/sec²
Solar constant	Measure with instrument such as bolometer	20.4b	1.92 cal/min/cm² 1.36×10^6 ergs/sec/cm²
Luminosity	Solar constant times area of spherical surface 1 AU in radius	20.4b	3.8×10^{33} ergs/sec 5×10^{23} horsepower
Spectral class	Spectrum	21.1	G2V
Visual magnitude			
Apparent	Received visible flux	20.1	−26.5
Absolute	Apparent magnitude and distance	20.2	+4.7
Rotation period at equator	Sunspots, and Doppler shift in limb spectra	24.4	$24^d 16^h$
Inclination of pole to ecliptic	Motions of sunspots	24.4	7°10′.5

TABLE 24.2 Model Solar Photosphere

DEPTH BELOW SURFACE (MI)	PERCENT OF LIGHT THAT EMERGES FROM THAT DEPTH	TEMPERATURE (°K)	PRESSURE (EARTH ATMOSPHERES)	DENSITY (GM/CM³)
0	100	4500	1.0×10^{-2}	2.8×10^{-8}
30	95	4800	1.7×10^{-2}	4.2×10^{-8}
60	91	5000	2.6×10^{-2}	6.2×10^{-8}
85	82	5300	3.8×10^{-2}	8.7×10^{-8}
105	67	5600	5.4×10^{-2}	11.5×10^{-8}
140	37	6200	8.3×10^{-2}	16.0×10^{-8}
160	13	6800	11.2×10^{-2}	$20 \quad \times 10^{-8}$

7600 to 11,700°. At a typical point in the photosphere, the pressure is only a few hundredths of sea-level pressure on the earth, and the density is about 1 ten-thousandth of the earth's atmospheric density at sea level.

(b) Chemical Composition

More than 60 of the elements known on the earth have now been identified in the solar spectrum. Those that have not been identified in the sun either do not produce lines in the observable spectrum or are so rare on the earth that they cannot be expected to produce lines of observable strength on the sun unless, proportionately, they are far more abundant there. Most of the elements found in the sun are in the atomic form, but more than 18 types of molecules have been identified. Most of the molecular spectra are observed only in the light from the cooler regions of the sun, such as the sunspots.

The relative abundances of the chemical elements in the sun are similar to the relative abundances found for other stars. About 60 to 80 percent of the sun (by weight) is hydrogen; 96 to 99 percent is hydrogen and helium. The remaining few percent is made up of the other chemical elements, in approximately the amounts that are described in Section 21.2.

(c) The Chromosphere

There is a change in the physical state of the gases just above the photosphere. The photosphere ends at about the place where the density of negative hydrogen ions has dropped to a value too low to result in appreciable opacity. Gases extend far beyond the photosphere, but they are transparent to most radiation. The region of the sun's atmosphere that lies immediately above the photosphere is the chromosphere.

Until this century the chromosphere could be observed only when the photosphere was occulted by the moon during a total solar eclipse. In the

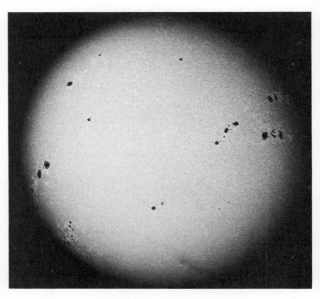

FIG. 24-1   The sun, photographed under excellent conditions, showing a large number of sunspots; September 15, 1957. *(Mount Wilson and Palomar Observatories.)*

seventeenth century several observers described what appeared to them as a narrow red "streak" or "fringe" around one limb of the moon during a brief instant after the sun's photosphere had been covered. Not until the careful observations of the solar eclipses of 1842, 1851, and 1860, however, was much attention paid to the chromosphere. Some of these observations were made by photographic methods and established beyond doubt the existence of the chromosphere. In 1868 the spectrum of the chromosphere was first observed; it was found to be made up of bright lines, which showed that the chromosphere consists of hot gases that are emitting light in the form of emission lines. These bright lines are difficult to observe against the bright light of the photosphere but appear in the spectrum of the light from the extreme limb of the sun just after the moon has eclipsed the photosphere. They disappear within a few seconds, when the moon has covered the chromosphere as well. Because of the brief instant during which the chromospheric spectrum can be photographed during an eclipse, its spectrum, when so observed, is called the *flash spectrum*. The element *helium* (from *helios,* the Greek word for "sun") was discovered in the chromospheric spectrum before its discovery on earth in 1895.

Today it is possible to photograph both the chromosphere and its spectrum outside of eclipse. One instrument used for this purpose is the *coronagraph,* a telescope in which a black disk at the focal plane occults the photosphere, producing an artificial eclipse. The chromosphere can also be photographed in limited narrow spectral regions

through monochromatic filters or with a *spectro-heliograph* (Section 24.3).

The chromosphere is usually considered to be a zone of the solar atmosphere about 1000 to 2000 mi thick, but in its upper region it breaks up into a forest of jets (called *spicules*) and other irregular motions, so the position of the upper boundary of the chromosphere is somewhat arbitrary. Its reddish color (from whence comes its name) arises from the fact that one of the strongest emission lines in the visible part of its spectrum is the bright red line due to hydrogen (the Hα line — first line in the Balmer series — Section 10.5). The density of the chromospheric gases decreases upward above the photosphere, but spectrographic studies show that

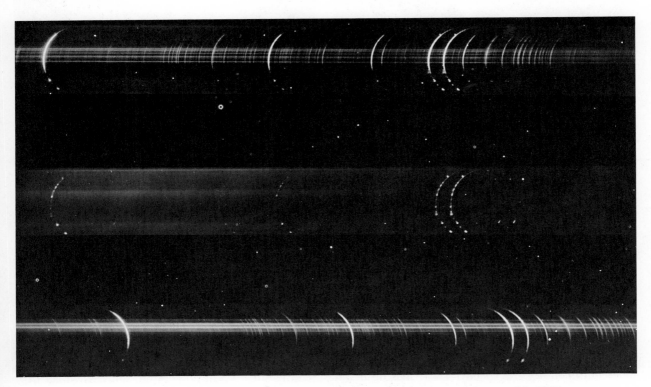

FIG. 24-2   Flash spectrum, photographed during the total eclipse of January 24, 1925, Middletown, Conn. *(Mount Wilson and Palomar Observatories.)*

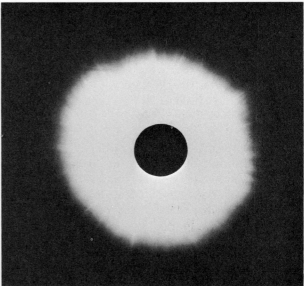

FIG. 24-3.   The solar corona, photographed during total eclipses. Note the difference in the shape of the corona during sunspot minimum *(left)* and sunspot maximum *(right). (Yerkes Observatory.)*

the temperature *increases* through the chromosphere, from 4500°K at the photosphere to 100,000°K or so at the upper chromospheric levels. The processes that heat the corona (see below) evidently also heat the upper chromosphere.

(d) The Corona

The chromosphere merges into the outermost part of the sun's atmosphere, the corona. Like the chromosphere, the corona was first observed only during total eclipses, but unlike the chromosphere, the corona has been known for many centuries; it is referred to by Plutarch and was discussed in some detail by Kepler. Many of the early investigators regarded the corona as an optical illusion, but photography proved its actual existence in the nineteenth century. Its spectrum was first observed in 1869 by the American astronomers Harkness and Young. The corona extends millions of miles above

the photosphere and emits half as much light as the full moon. Its invisibility, under ordinary circumstances, is due to the overpowering brilliance of the photosphere. Like the chromosphere, the corona can now be photographed, with the coronagraph and other instruments, under other than eclipse conditions. The first successful photograph made with a coronagraph was taken by the inventor of the instrument, Bernard Lyot, at Pic du Midi, France, in 1930.

Although transparent to visible light, the corona is opaque to radio waves of length greater than about 5 m. Thus radio observations of the sun provide information about conditions high in the corona. Not only are radio waves originating in the corona observed, but the corona produces scintillations in distant radio sources when they are observed through its outer part. The phenomenon is somewhat analogous to the scintillation of the light from stars caused by the earth's atmosphere. Scintillations of remote radio sources observed 90° away from the sun in the sky show that the corona actually reaches out beyond the earth. Coronal atoms have also been detected in the vicinity of the earth with space vehicles.

Two parts of the corona are recognized: the inner, "real" part, or K corona, and the outer, "false" part, or F (Fraunhofer) corona. The F corona has a solar-type spectrum that shows it to be sunlight, apparently reflected by tiny particles; presumably the F corona is an inner extension of the zodiacal light. It is superimposed upon and overlaps the K corona, which has a continuous spectrum of sunlight scattered by free electrons. The continuum is broken by about two dozen bright emission lines that arise from ions in the hot corona.

The emission lines in the corona were not identified with known chemical elements until 1942, when the Swedish physicist B. Edlén showed them to be "forbidden" lines of calcium, iron, and nickel. When atoms become excited (Section 10.5d) they ordinarily deexcite themselves by emitting a photon of light within a period of time which is typically 10^{-8} sec. There are some excited levels, however, in which an atom will spend, on the average, several seconds (or, in some cases, hours, days, or even years) before jumping to a lower level with the emission of a photon. The vast majority of atoms both enter and leave these so-called *metastable* levels by collisions with other atoms; an extremely small fraction of them happen to remain in such levels long enough to leave them by radiating photons. These photons are emitted so infrequently that they are said to give rise to *forbidden* lines — lines not usually observed. Normally they are overwhelmed by the ordinary lines and continuous radiation. In a very hot rarefied gas like the corona, however, the ordinary lines are in the ultraviolet, and the continuous radiation is very weak, so the gas is quite transparent, and

we observe the forbidden radiation coming from the great number of atoms along the line of sight.

Studies of the solar corona reveal two important facts. First, the gas density is extremely low. The corona is very tenuous; despite its large size, it cannot contribute appreciably to the total mass of the sun. Second, and even more interesting, the lines of iron arise from atoms that have lost 13 or more electrons, and, very rarely, are observed lines of calcium from atoms that have lost 14 or more electrons. Lines of highly ionized iron, nickel, argon, and calcium are frequently observed; much energy is required to ionize these atoms so highly (Section 10.5e). Such a high degree of ionization requires a temperature of at least 1 million degrees Kelvin. The corona, in other words, is many times hotter than the photosphere. It must be noted, however, that because of the low density of the corona, it does not contain much actual heat, despite its high temperature. One heating mechanism of the corona is believed to be shock waves originating from convective currents in the photosphere. Because of its high temperature, the bulk of the spectral lines emitted by the corona lies in the far ultraviolet. In fact, observations of the sun made from rockets show that an appreciable fraction of the far-ultraviolet solar spectrum is light from the corona and the chromosphere. Rockets and satellites have permitted observations of these lines, ranging up to those of iron 16 times ionized.

24.3 PHENOMENA OF THE SOLAR ATMOSPHERE
In its gross characteristics the sun is very stable, but the detailed features of its atmosphere are constantly changing.

(a) Photospheric Granulation
Either direct telescopic observation or photography shows that the photosphere is not perfectly smooth

but has a mottled appearance resembling rice grains — this structure of the photosphere is now generally called *granulation*. Among the best photographs of the granulation are those obtained with a telescope carried to an altitude of 80,000 ft in a balloon. These observations, under the direction of Martin Schwarzschild and J. B. Rogerson of Princeton University, are equal to the best of those made from ground-based observatories because they were obtained far above most of the atmospheric disturbances that cause bad "seeing." Typically, granules are about 1000 km in diameter; the smallest observed are about 300 km across. They appear as bright spots surrounded by narrow darker regions.

The motions of the granules can be studied by the Doppler shifts in the spectra of gases just above them. It is found that the granules themselves are columns of hotter gases arising from below the photosphere. As the rising gas reaches the photosphere it spreads out and sinks down again. The darker intergranular regions are the cooler gases sinking back. The centers of the granules are hotter than the intergranular regions by 50 to 100°K. The vertical motions of the granules have speeds of about 2 or 3 km/sec. Individual granules persist for about 8 minutes. The granules, then, appear to be the tops of convection currents of gases seen through the photosphere.

Detailed studies of the motions of photospheric gases show that the granules themselves form part of a structure of still larger scale called *supergranulation*. Supergranules are cells about 30,000 km in diameter, within which there is a flow of gases from center to edge. In addition to the vertical currents of gases in the granules and the center-toward-edge flow in the supergranules, in each region of the sun (from 3500 to 7000 km across) the gases rhythmically pulse up and down with speeds of about ⅓ km/sec, taking about 5 minutes for a complete cycle. The surface gases of the sun, then, comprise a seething, sloshing sea.

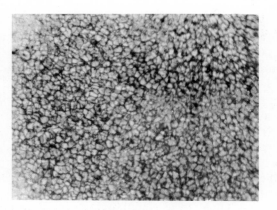

FIG. 24-4   A portion of the solar photosphere photographed from a balloon at an altitude of 80,000 feet. At this altitude, excellent resolution of the solar granulation is possible because disturbances produced by the earth's atmosphere ("seeing") are greatly reduced. *(Project Stratoscope, Princeton University.)*

(b) Sunspots

The most conspicuous of the photospheric features are the *sunspots*. Occasionally, spots on the sun are large enough to be visible to the naked eye, and such have been observed for many centuries. It was Galileo, however, who first showed that sunspots are actually on the surface of the sun itself, rather than being opaque patches in the earth's atmosphere or the silhouettes of planets between the sun and earth (Section 3.5b).

In 1774 the Scot Alexander Wilson suggested that the spots were "holes" through which we could see past the photosphere into a cooler interior of the sun. William Herschel held a similar view; he imagined that the sun had a cool, probably inhabited, interior that was surrounded by two cloud layers. The outermost layer, he thought, was the intensely luminous photosphere; beneath the photosphere was a cool cloud layer that acted as a "fire screen" and protected the interior. Although Herschel's hypothesis seems wild to us today, no observational or theoretical arguments

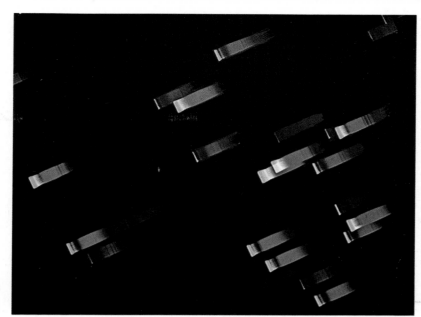

Stellar spectra of Pleiades
photographed with an objective prism.
(Warner and Swasey Observatory.)

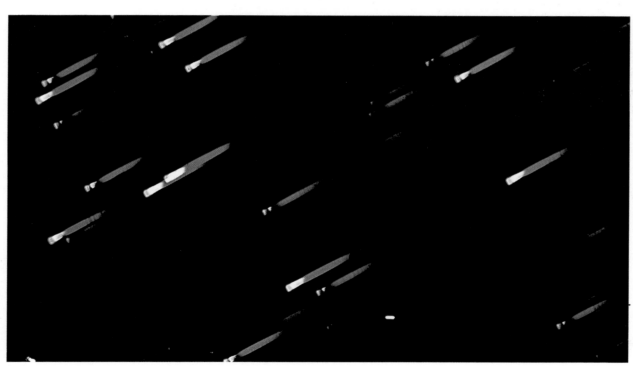

Stellar spectra of Hyades photographed with an objective prism.
(Warner and Swasey Observatory.)

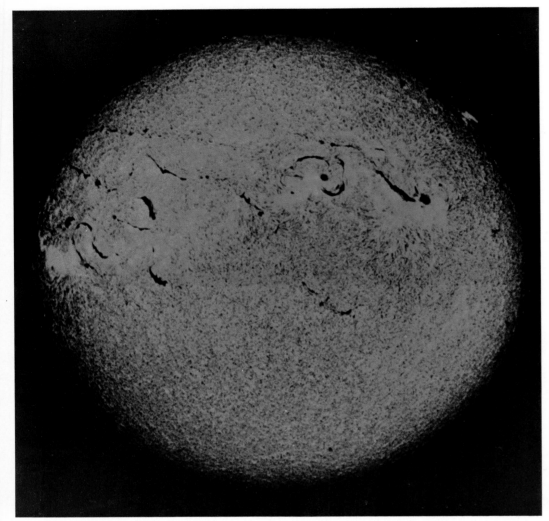

The Sun in the red light of the first
Balmer line of hydrogen. *(Carl Zeiss.)*

Six photographs, over a 40-minute interval, of an
active region of the sun showing a spot group
and a filament. Each is a composite of two black
and white photographs made through filters that
pass light of wavelength slightly less than and
greater than that of the first Balmer line of
hydrogen. The former is printed through a green
filter and the latter through a red one. When the
light from the two photographs is mixed, objects
not moving in the line of sight appear yellow.
Because of the Doppler shift, regions approaching
us (rising in the solar chromosphere) appear
red in the above prints, and those receding
(falling) appear green. Thus we can follow
the motions of the filament by the color
changes. *(Lockheed Solar Observatory.)*

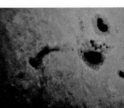
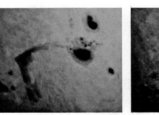
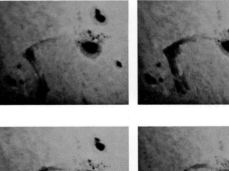
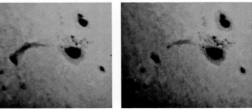
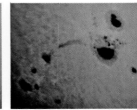

were raised at the time to refute it. Indeed, sunspots were generally thought to be depressions and could well have been "holes" through which an inner, cool, dark cloud layer could be seen.

We know today that the spots are not actually depressions in the photosphere but are, rather, regions where the gases are cooler than those of the surrounding regions. Their spectra show them to be up to 1500°K cooler than the photosphere itself. Sunspots are nevertheless hotter than the surfaces of many stars. If they could be removed from the sun, they would be seen to shine brightly; they appear dark only by contrast with the hotter, brighter surrounding photosphere.

Individual sunspots have lifetimes that range from a few hours to a few months. They are first seen as small dark "pores" somewhat over 1000 mi in diameter. Most of them disappear within a day, but a few persist and may last for a week or, occasionally, much longer. If a spot lasts and develops, it is usually seen to consist of two parts: an inner darker core, the *umbra,* and a surrounding less dark region, the *penumbra.* Many spots become much larger than the earth, and a few have reached diameters of 100,000 mi. Frequently spots occur in groups of from two to twenty or more. If a group contains many spots, it is likely to include two large ones, oriented approximately east-west, and many smaller spots clustered around the two principal ones. The leading principal spot (in the direction toward which the sun is rotating) is most often the largest one of the group. The largest groups are very complex and may have over a hundred spots. Like storms on the earth, sunspots may move slowly on the surface of the sun, but their individual motions are slow when compared to the solar rotation, which carries them across the disk of the sun.

The motions of the gases around a sunspot can be studied by the Doppler shifts in the spectra of their light. It is found that close to the photospheric surface, gases are flowing horizontally outward away from the spot, while at higher elevations (in the chromosphere) gases are flowing horizontally inward, toward the center of the sunspot. The velocities of these gas motions range from ½ to 1 km/sec.

(c) The Sunspot Cycle

In 1851 a German apothecary and amateur astronomer, Heinrich Schwabe, published an important conclusion he had reached as a result of his observations of the sun over the previous decade. He found that the number of sunspots visible, on the average, varied with a period of about 10 years. Since Schwabe's work, the *sunspot cycle* has been clearly established. Although individual spots are short-lived, the total number of spots visible on the sun at any one time is likely to be very much greater during certain periods, the periods of *sunspot maximum,* than at other times, the periods of *sunspot minimum.* Sunspot maxima have occurred at an average interval of 11.1 years, but

FIG. 24-5   A large spot group on the sun, photographed on May 17, 1951. *(Mount Wilson and Palomar Observatories.)*

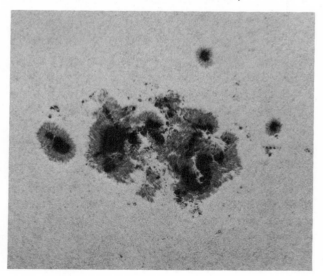

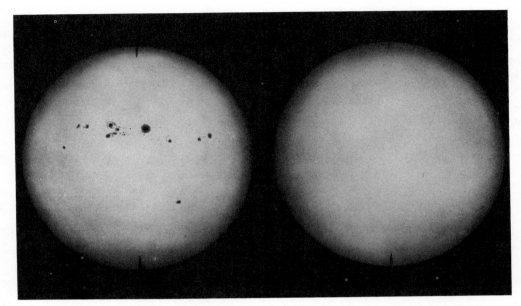

FIG. 24-6   Direct photographs of the sun near the time of sunspot maximum *(left)* and near sunspot minimum *(right). (Mount Wilson and Palomar Observatories.)*

the intervals between successive maxima have ranged from as little as 8 years (from 1830 to 1838) to as long as 16 years (from 1888 to 1904). During sunspot maxima, more than 100 spots can often be seen on the sun at once. During sunspot minima, the sun sometimes has no visible spots.

At the beginning of a cycle, just after a minimum, a few spots or groups of spots appear at latitudes of about 30° on the sun. As the cycle progresses, the successive spots occur at lower and lower latitudes, until, at the maximum of the cycle, their average latitude is about 15°. Near minimum, the last few spots of a cycle appear at about 8° latitude. About the same time, the next cycle begins with a few spots occurring simultaneously at higher latitudes. Sunspots almost never appear at latitudes greater than 40° or less than 5°. The locations of sunspots on the sun in both the Northern and Southern Hemispheres are related to the sunspot cycle in the same way; however, sunspot activity in one hemisphere may dominate for long periods.

The occurrence of the sunspot cycle, as well as of the spots themselves, is still unexplained. There is no general agreement on any of the many theories that have been proposed to explain them.

(d) Magnetic Fields on the Sun

As stated in Section 21.2c, each spectral line is split up into several components in the presence of a magnetic field, the phenomenon known as the Zeeman effect. Because the light absorbed in the shifted components of a split line is circularly polarized, each component can be isolated if the spectrum is photographed through a polarizing filter. The measured displacement of a Zeeman component of a spectral line gives a measure of the strength of the magnetic field that caused the line to split, and also the polarity of the field. In 1908 the American astronomer George E. Hale observed the Zeeman effect in the spectrum of sunspots and found them to possess strong magnetic fields. Magnetic fields are usually measured in terms of a unit known as the *gauss*. For example, the magnetic field of the earth (which aligns our compass needles) has a strength of about 1 gauss. The magnetic fields observed in sunspots range from 100 to nearly 4000 gauss. This is 10 times as great as the field of a good alnico magnet, and moreover, is spread over a region tens of thousands of miles across. The magnetic field is present in the region surrounding a spot as well as in the spot itself, and persists even after the spot has disappeared.

Whenever sunspots are observed in pairs or in groups containing two principal spots, one of the spots usually has the magnetic polarity of a north-seeking magnetic pole, and the other has the opposite polarity. Moreover, during a given cycle, the leading spots of pairs (or leading principal spots of groups) in the Northern Hemisphere all tend to have the same polarity, while those in the Southern Hemisphere all tend to have the opposite polarity. During the next sunspot cycle, however, the polarity of the leading spots is reversed in each hemisphere. For example, if during one cycle the leading spots in the Northern Hemisphere all had the polarity of a north-seeking pole, the leading spots in the Southern Hemisphere would have the polarity of a south-seeking pole; during the next cycle, the leading spots in the Northern Hemisphere would have south-seeking polarity and those of the Southern Hemisphere would have north-seeking polarity. We see, therefore, that the sunspot cycle does not repeat itself as regards magnetic polarity until *two* 11-year maxima have passed. The sunspot cycle is therefore sometimes said to last 22 years, rather than 11.

FIG. 24-7   Solar magnetograms showing the intensity and polarity of magnetic fields in the photosphere of the sun on August 5 *(left)* and August 6 *(right)*, 1968. Successive contours show regions with field strengths of 5, 10, 20, 40, and 80 gauss.

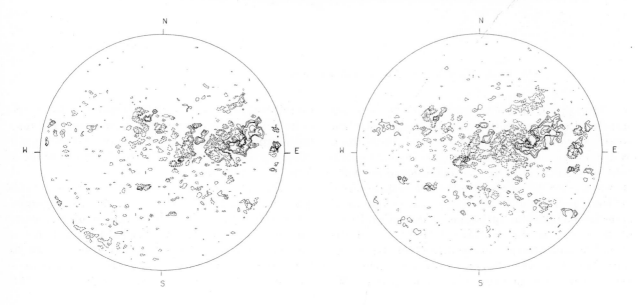

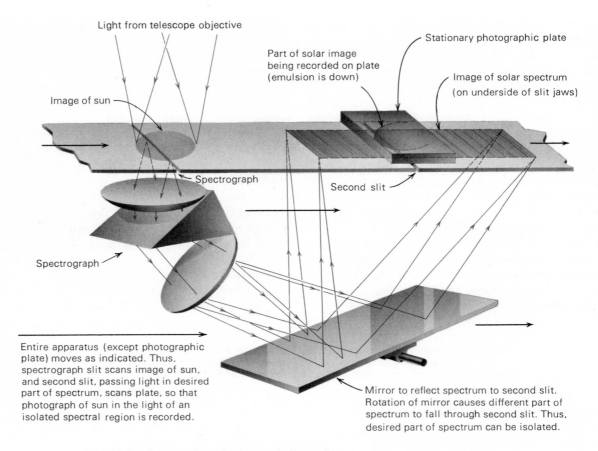

Light from telescope objective

Part of solar image being recorded on plate (emulsion is down)

Stationary photographic plate

Image of solar spectrum (on underside of slit jaws)

Image of sun

Spectrograph

Second slit

Spectrograph

Entire apparatus (except photographic plate) moves as indicated. Thus, spectrograph slit scans image of sun, and second slit, passing light in desired part of spectrum, scans plate, so that photograph of sun in the light of an isolated spectral region is recorded.

Mirror to reflect spectrum to second slit. Rotation of mirror causes different part of spectrum to fall through second slit. Thus, desired part of spectrum can be isolated.

FIG. 24-8   Construction of a spectroheliograph.

The strong magnetic fields on the sun are associated with sunspots. When the spots die, their fields are spread out towards the poles, contributing to a more general, but far weaker, solar magnetic field, with the polarity of the following spots of the previous cycle. Until recently there was little direct evidence indicating the existence of a general magnetic field of the sun analogous to the general field of the earth. H. W. and H. D. Babcock, of the Mount Wilson Observatory and the Hale Solar Observatory, however, have investigated magnetic fields on the sun for many years, and in the 1950s established the existence of a general polar magnetic field with a strength of a few gauss. Because the sunspot polarity reverses each cycle, the general magnetic field of the sun also reverses, but in a rather irregular way.

The solar magnetic field is not uniform. The motions of gas in the supergranules, for example, cause the field strength to build up at the inter-supergranule boundaries. Magnetism is still not a dominant force in the photospheric layers, but in the chromosphere and corona, where the gas density and pressure have dropped enormously, the magnetic fields are still strong and play an important role in influencing the motions of ionized gases. Far out in the corona magnetic lines of force manifest themselves by organizing ionized coronal gases into streamers, which are easily seen and photographed during total solar eclipses. Low in the corona, magnetic fields guide the motions of ions in solar prominences (Section 24.3h). The solar magnetic field even extends into interplanetary space and is measured in the vicinity of the earth and other planets with magnetometers carried on space probes.

(e) The Spectroheliograph

In order to see regions of the sun that lie directly above the photosphere, we may observe in spectral regions to which the photospheric gases are especially opaque — at the centers of strong absorption lines such as those of hydrogen and calcium.

In 1892 Hale and, independently, Deslandres in France, invented the *spectroheliograph,* with which the sun can be photographed in light comprising only a very narrow spectral region, generally that corresponding to a small part of a particular line of some element. The instrument produces a spectrum of the sun and then isolates the particular line or spectral region by masking off all other light in the spectrum with a second slit. The entire apparatus, spectrograph and slit, is then scanned across the telescopic image of the sun. The light that passes through the slit isolating a small part of the spectrum simultaneously scans across a photographic plate, leaving an image of the sun as it appears in that narrow spectral region (Figure 24-8). The photograph so obtained is called a *spectroheliogram.* If the photographic plate is replaced by an eyepiece, and if the scanning is done very quickly, an image of the sun in the same part of the spectrum can be observed visually; then the instrument is called a *spectrohelioscope.*

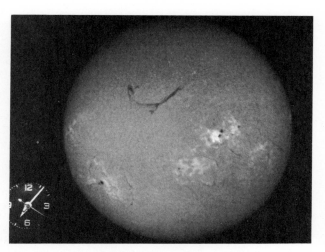

FIG. 24-9   Filtergram of the sun in the light of the first Balmer line of hydrogen. *(Lockheed Solar Observatory.)*

It is most common in practice to place the slit in the spectrum so that it isolates one of the absorption lines of ionized calcium in the ultraviolet (the K line of ionized calcium) or the Hα line of hydrogen in the red. These spectral lines appear dark when viewed against the rest of the solar spectrum, but they are not completely dark; what light remains in the centers of the lines is light that is emitted from atoms of calcium or hydrogen, respectively, in the chromosphere. Thus, spectroheliograms reveal the appearance of the chromosphere in the light of calcium, or hydrogen. Special filters have also been devised which pass only light in narrow spectral regions. In particular, considerable success has been attained in photographing the sun in the light of hydrogen through such *monochromatic* filters. These photographs are called *filtergrams.* Even filters that pass only the Lα line (the first line in the Lyman series of hydrogen, Section 10.5b) have been used in cameras carried above most of the atmosphere by rockets to photograph the sun in this ultraviolet light of hydrogen.

Spectacular motion pictures have been taken with spectroheliographs. Time-lapse photographs, in which frames are exposed every few seconds or every few minutes and then run through a projector at normal speed, show in a dramatic way changes that occur in the solar chromosphere. The technique is referred to under the impressive title of *spectroheliokinetography*.

(f) Plages (Flocculi) and Faculae

Spectroheliograms in the light of calcium and hydrogen show bright "clouds" in the chromosphere in the magnetic-field regions around sunspots. These bright regions were formerly called *flocculi* ("tufts of wool") but are now more commonly known as *plages*. Calcium and hydrogen plages are also sometimes seen in regions where there are no visible sunspots, but these regions are generally those of higher-than-average magnetic fields.

Plages are often referred to as "clouds of calcium" or "clouds of hydrogen" (depending on the wavelength region used to make the spectroheliogram). The description is misleading, however. All the chemical elements, including hydrogen and calcium, probably maintain uniform relative abundance distributions over the entire chromosphere. The plages, rather, are regions where calcium and hydrogen happen to be emitting more light at the observed wavelengths. These elements are partially ionized throughout most of the visible chromosphere, and some of the atoms emit light as they capture electrons and become neutral (or less ionized) or as those atoms (or ions) cascade down through the various excited energy levels. The plages, then, are not clouds of any particular element, but are regions where some of the atoms of the element observed are changing their states of ionization or excitation and are emitting more light than other regions. Plages of hydrogen and calcium usually occur in approximately the same projected regions at the same time, but they display different small-scale structure.

Light is sometimes emitted at many wavelengths and can be seen in white light, that is, in the direct image of the sun. These "white-light" plages are called *faculae* ("little torches"), and were first described by Galileo's contemporary Christopher Scheiner. Faculae are seen best near the limb of the sun where the photosphere is not so bright and the contrast is more favorable for their visibility.

(g) Spicules

The chromosphere also contains many small jet-like spikes of gas rising vertically through it. These features, called *spicules*, occur at the edges of supergranule cells, but when viewed near the limb of the sun so many are seen in projection that they give the effect of a forest of them. They show up best when the chromosphere is viewed in the light of hydrogen. They consist of gas jets moving upward at about 30 km/sec and rising to heights of from 500 to 20,000 km above the photo-

FIG. 24-10   A solar prominence 205,000 miles high, photographed in violet light of the calcium K line, July 2, 1957. *(Mount Wilson and Palomar Observatories.)*

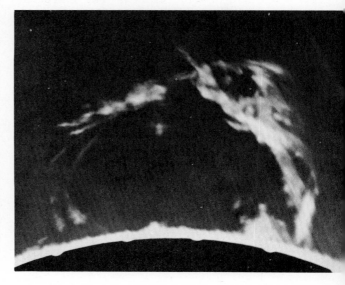

sphere. Individual spicules last only ten minutes or so. Through the spicules matter continually flows into the corona.

(h) Prominences

Among the more spectacular of coronal phenomena are the *prominences*. Prominences have been viewed telescopically during solar eclipses for centuries. They appear as red flamelike protuberances rising above the limb of the sun. Prominences can now be viewed at any time on spectroheliograms and filtergrams. Motions of prominences are exhibited in motion pictures. The gross features of some, the *quiescent* prominences, may remain nearly stable for many hours, or even days, and may extend to heights of tens of thousands of miles above the solar surface. Others, the more active prominences, move upward or have arches that surge slowly back and forth. The relatively rare *eruptive* prominences appear to send matter upward into the corona at speeds up to 700 km/sec, and the most active *surge* prominences may move upward at speeds up to 1300 km/sec. Some eruptive prominences have reached heights of over 1 million miles above the photosphere. When seen silhouetted on the disk of the sun, prominences have the appearance of irregular dark filaments.

Superficially, prominences appear to be material ejected upward away from the sun, but the motion pictures show that whereas a prominence may grow in size and rise higher and higher above the photosphere, the actual material in the prominence most often appears to move downward in graceful arcs, evidently along lines of magnetic force. Apparently, most prominences form from coronal material that cools and moves downward, even though the disturbance that characterizes the prominence may move upward. Prominences are cool and dense regions in the corona where atoms and ions are capturing electrons and emitting light. Their origin is unknown, but it is significant that they usually originate near regions of sunspot activity and lie

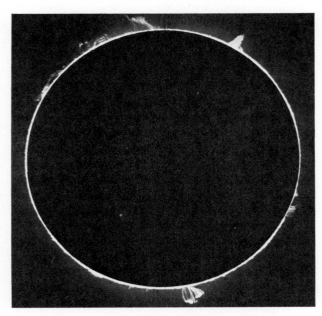

FIG. 24-11   Entire limb of the sun photographed in the light of the calcium K line, showing several prominences. *(Mount Wilson and Palomar Observatories.)*

on the boundary between regions of opposite magnetic polarity. Quiescent prominences are supported by coronal magnetic fields, and eruptive prominences evidently result from sudden changes in the magnetic fields. Prominences seem to be further symptoms of the same general disturbances that produce spots and plages, that is, magnetic-field regions.

(i) Flares

Occasionally, the chromospheric emission lines (the Hα and ionized calcium lines in particular) in a small region of the sun brighten up to unusually high intensity. Such an occurrence is called a *flare*. A flare is usually discovered on a spectroheliogram that is being made in the light of one of the spectral lines that brighten. It appears as an intensely bright spot on the photograph a few thousand or tens of thousands of kilometers in

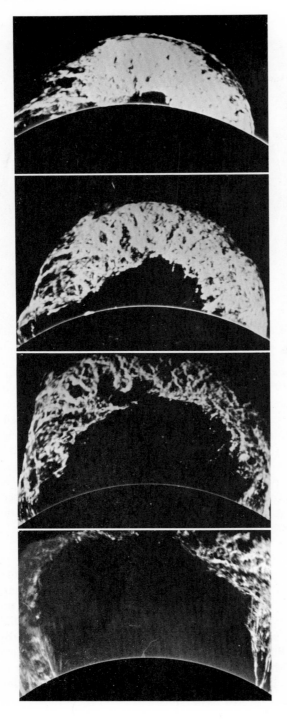

diameter. Very rarely, the continuous spectrum of that part of the solar surface affected also brightens during a flare. These white-light flares are among the most intense observed. A flare usually reaches maximum intensity a few minutes after its onset. It fades out slowly, and after an interval ranging from a few minutes to a few hours, it disappears.

During major flares, an enormous amount of energy is released. The visible light emitted is, of course, very small compared to that from the entire sun, but a flare covering only a thousandth of the solar surface can actually outshine the sun in the ultraviolet. X-rays are emitted as well and are observed from space probes and satellites. In addition, matter is thrown out at speeds of 500 to 1000 km/sec. X-rays and atomic nuclei ejected from the sun by flares not only disturb the ionosphere in the earth's atmosphere (producing radio "fadeouts") but comprise a hazard to space travelers as well. Major, potentially hazardous, flares are fortunately rare, a few occurring each year, but smaller flares occur every day.

Flares are most frequent in the regions of complex sunspot groups with complicated magnetic field structure. Material thrown into the corona from a flare expands and cools and, as ions and electrons recombine, it may emit light and be seen as a loop prominence over the active region of the sun. Loop prominences are the hottest of all prominences and are indicative of rather violent activity in the sun's atmospheres. As is the case with almost all solar phenomena, we do not know the source of energy of flares or the mechanism that triggers them into existence, but we feel that they are somehow connected with the magnetic fields of sunspot regions.

FIG. 24-12   Four successive photographs (spectroheliograms) of the great explosive prominence of June 4, 1946, taken with a coronagraph. The total elapsed time between the first *(top)* and the last *(bottom)* picture was one hour. *(Harvard Observatory.)*

(j) Summary

There are many solar phenomena that seem to be associated with a particular kind of disturbance in the solar atmosphere. The disturbances are accompanied by strong magnetic fields, sunspots, plages, prominences, and solar flares. All these phenomena are related to the semiregular sunspot cycle of 11 (or perhaps 22) years. We shall see in Sections 24.5 and 24.6 that radio radiation from the sun, as well as solar-terrestrial effects, seem also to be associated with these centers of solar activity. Even the shape of the corona varies with the sunspot cycle. During sunspot maximum it is nearly spherical and distended, while at sunspot minimum it is contracted but extends to relatively greater distances near the plane of the solar equator.

24.4 SOLAR ROTATION

Galileo first demonstrated that the sun rotates on its axis when he observed the apparent motions of the sunspots as the turning sun carried them across its disk. He found from his spot observations that the rotation period of the sun is a little less than 1 month. The period was determined more accurately by subsequent observations of others. In 1859 Richard Carrington found that in the equatorial regions the sun rotates in about 25 days, but that at a latitude of 30° the period is about 2½ days longer. The sun's rotation rate can be determined also from the difference in the Doppler shifts of the light coming from the receding and approaching limbs (Section 21.2d). From 1887 to 1889, Duner, at Uppsala, determined the solar rotation by spectroscopic methods at latitudes beyond those at which spots are found. The solar rotation determined from the Doppler method confirms that the sun rotates most rapidly at low latitudes, and with a longer period at high latitudes. At a latitude of 75°, the rotation period is about 33 days, and at the poles it may be near 35 days. Its *direction* of rotation is from west to east, like the orbital revolution of the planets and

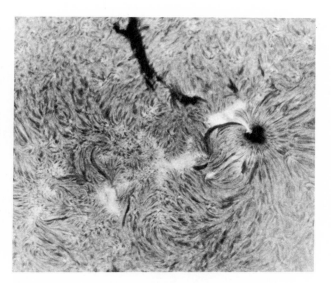

FIG. 24-13   An Hα filtergram showing a small flare (brightest region) near a sunspot. *(J. Harvey, Lockheed Solar Observatory.)*

(except for Uranus and Venus) like their axial rotations. It is possible for different parts of the sun to rotate at different rates because, of course, it is fluid rather than solid like the planets. The details of its differential rotation, however, are not yet completely explained.

The apparent motions of sunspots are not usually straight lines across the sun's disk but rather slightly curved arcs, because the axis of rotation of the sun is not exactly perpendicular to the plane of the ecliptic. The angle of inclination of the solar equator to the ecliptic is about 7°.

It has been suggested that the sun may have a different rotation period in its interior. R. H. Dicke, at Princeton, has proposed a modification of the general theory of relativity, but in order for his theory to be compatible with the exact motion of Mercury the sun would have to have a rapidly rotating core, which, because of its rotation, is significantly flattened. Now an oblate inner core of the sun would produce a slight oblateness in the photospheric layer. The expected flattening

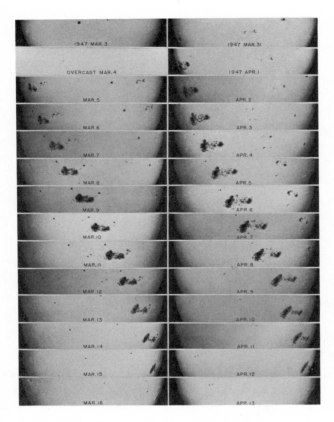

FIG. 24-14   Series of photographs showing the motions of sunspots, indicating the solar rotation. *(Mount Wilson and Palomar Observatories.)*

(if Dicke is right) should be very little, the sun's polar diameter being smaller than an equatorial diameter by less than 1 part in 10,000. Dicke has constructed a very ingenious attachment to a solar telescope, with which he has attempted to measure the solar oblateness. Provisional results indicate that the sun has, in fact, the slight flattening predicted. On the other hand, some theorists regard it as very doubtful, and probably impossible, that the sun can have a rapidly rotating core, and feel that Dicke's observations — or at least his interpretation of them — are in need of confirmation. It has been suggested, for example, that the observed apparent flattening might be due to white-light plages (faculae) at the sun's equatorial limb.

24.5 THE RADIO SUN

In Great Britain in 1942, unexpected noise was picked up on radar receivers. It was subsequently learned that the source of this noise was the sun, which emits electromagnetic radiation at radio as well as optical wavelengths. Since World War II, radio observations of the sun have been made regularly at many radio astronomical observatories. Shortwave radio energy (near 1 cm wavelength) can escape the sun from the lower chromosphere. The corona is more and more opaque, however, to longer and longer radio wavelengths. Those of 15 m escape the sun only if they originate high in the corona. Thus, by observing the sun at different radio wavelengths, we observe to different depths in the corona and chromosphere and can determine the heights in the solar atmosphere at which various disturbances giving rise to radio emission occur.

When the sun is comparatively inactive (that is, has few spots, plages, and so on) its background radio radiation is about as intense as that which would be expected from a body of the sun's temperature; this is called the *quiet sun*. Even at these times there are occasional very intense short bursts. When the sun is active, however, these bursts of unusual radio energy are frequently superposed upon that of the quiet sun. Solar flares are often accompanied by exceptionally strong surges of radio energy. Initially the energy of bursts is received only at short wavelengths, but at successively greater intervals of time after the start of a burst, energy is received over longer and longer wavelengths, which shows that the source of the radio energy is rising higher and higher in the corona; the outward velocity of such a source may reach 1500 km/sec.

If the sun is typical among stars as a radio emitter, we would expect radio emission from other stars, because of their distances, to be too feeble to detect. The various cosmic radio sources known are not ordinary stars; we shall discuss them in later chapters.

24.6 SOLAR-TERRESTRIAL EFFECTS

In 1852, 1 year after Schwabe announced his discovery of the solar sunspot cycle, three investigators, Edward Sabine in England, and Rudolf Wolf and Alfred Gautier in Switzerland, independently discovered that sunspot activity was correlated with magnetic storms on the earth. During geomagnetic storms, the earth's magnetic field is disturbed, and the direction of the compass needle shows fluctuations. Today we know that long-range shortwave radio interference and displays of the aurora are also correlated with geomagnetic storms and the sunspot cycle. Some investigators are of the opinion that even changes in climate that occur with time may be connected with solar activity.

(a) Solar Corpuscular Radiation

In addition to electromagnetic radiation (X-rays, ultraviolet radiation, visible light, infrared light, and radio waves), the sun emits *corpuscular radiation*. The corpuscular radiation is in the form of charged atomic particles, mostly protons and electrons. There is a more or less continuous emission of these particles from the sun, in the form of what is called the *solar wind*. Particles in the solar wind have been recorded on instruments carried on artificial earth satellites and cosmic space probes. It has also been found that corpuscular radiation from the sun is far from constant in intensity. Following instances of strong solar activity, particularly solar flares, the flux of charged particles is greatly enhanced. Following a strong solar flare, almost invariably there is an extra intense rain of particles at the earth about 1 day later. The fastest moving of these are cosmic-ray particles (Chapter 31). To reach the earth in 1 day, these particles must

escape the sun with velocities of millions of miles per hour.

At the surface of the earth, we are protected from these particles by the atmosphere and by the earth's magnetic field. Some of the particles may be trapped temporarily in the earth's magnetic field. Many others are deflected toward the earth's magnetic poles. Only the most energetic of the particles strike the upper atmosphere of the earth above the equatorial regions. In any case, they interact with molecules of the air and do not penetrate directly to the earth's surface.

These charged particles, along with electromagnetic radiation from the sun interacting with the upper atmosphere of the earth, do, however, disrupt the ionized layers of gas in the *ionosphere* (Section 13.3). It is these ionized layers that reflect shortwave radio waves back to the earth, making long-range radio communication between distant stations on the earth possible. When the ionospheric layers are disturbed, they may disrupt the reflection of radio waves transmitted from the ground; this results in radio "fadeouts."

The rain of particles impinging upon the upper atmosphere and on the radiation belts of the earth is also responsible for the aurora (northern and southern lights). As the particles strike atoms and molecules in the upper atmosphere they excite them. Radiation from the ions and atoms in the atmosphere gives rise to the auroral emission of light. The most spectacular auroras occur at elevations of 50 to 100 mi. Even in the absence of conspicuous auroras, however, atoms in the earth's atmosphere are being bombarded by particles from the sun continually and emit light in the form of the *airglow* (also called the *night-sky glow* or *permanent aurora*). The airglow is strongest at heights of 60 to 120 mi and is the major source of illumination of the night sky in regions away from large cities and in the absence of moonlight. Like auroras, radio fadeouts, and other solar-terrestrial effects, the airglow is more intense at times of sunspot maximum.

Exercises

1. How might you convince an ignorant friend that the sun is not hollow?

2. Give at least three good arguments that refute the view proposed by Herschel that the sun has a cool interior which is inhabited.

3. Suppose an eruptive prominence is seen to rise at 100 mi/sec. If it did not change speed, how far from the photosphere would it extend in 3 hours?

4. Would the material in the prominence in Exercise 3 escape the sun? Why? (See Section 4.4a.)

5. From the Doppler shifts of the spectral lines in the light coming from the east and west limbs of the sun, it is found that the radial velocities of the two limbs differ by about 4 km/sec. Find the approximate period of rotation of the sun.

6. If the rotation period of the sun is determined by observing the apparent motions of sunspots, must any correction be made for the orbital motion of the earth? If so, explain what the correction is and how it arises. If not, explain why the earth's orbital revolution does not affect the observations.

7. If the corona, which is outside the photosphere, has a temperature of 1,000,000°K, why do we measure a temperature near 6000°K, for the surface of the sun?

CHAPTER 25

Unusual Stars

Most stars are stable and fit into the general scheme of classification discussed in the preceding chapters. A minority, however, deviate from the usual pattern.

25.1 STARS THAT VARY IN LIGHT

The standard international index of stars that vary in light is the Soviet *General Catalogue of Variable Stars*. The 1958 edition of this catalogue (the most recent at the time of writing) lists 14,711 known variable stars in our Galaxy. The first and second supplements of this catalogue bring the total number of catalogued variables to 18,791.

(a) Designation

Variable stars are designated in order of time of discovery in the constellation in which they occur. If a star that is discovered to vary in light already has a proper name or a Greek-letter designation, it retains that name; examples are Polaris, Betelgeuse (α Orionis), Algol, and δ Cephei. Otherwise, the first star to be recognized as variable in a constellation is designated by the capital letter R, followed by the possessive of the Latin name of the constellation (for example, R Coronae Borealis). Subsequently discovered variables in the same constellation are designated with the letters S, T, ... , Z, RR, RS, ... , RZ, SS, ST, ... , SZ, and so on, until ZZ is reached. Then the letters AA, AB, ... , AZ, BB, BC, ... , BZ, and so on, are used up to QZ (except that the letter J is omitted). This designation takes care of the first 334 variable stars in any one constellation. Thereafter, the letter V followed by a number is used, beginning with V 335. Examples are V 335 Herculis and V 969 Ophiuchi.

(b) The Light Curve

A variable star is studied by analyzing its spectrum and by measuring the variation of its light with lapse of time. Some stars show light variations that are apparent to the unaided eye. Generally, however, the apparent brightness of a variable star is

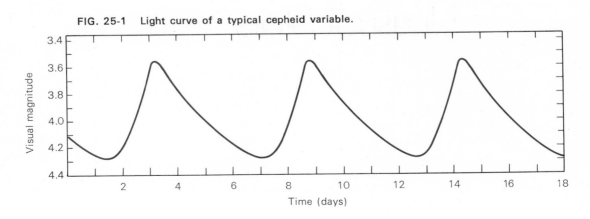

FIG. 25-1 Light curve of a typical cepheid variable.

determined by telescopic observation. The three techniques most commonly employed are the following:

(1) The magnitude of the variable is estimated by visual observation through the telescope, by comparing its brightness with the brightnesses of neighboring stars of known magnitudes.

(2) The magnitude of the variable star is measured by comparing its image with the images of comparison stars on a telescopic photograph (photographic photometry).

(3) The magnitude of the variable is determined by photoelectric photometry.

A graph that shows how the magnitude of a variable star changes with time is called a *light curve* of that star. An example is given in Figure 25-1. The *maximum* is the point on the light curve where the maximum amount of light is received from the star; the *minimum* is the point where the least amount of light is received. If the light variations of a variable star repeat themselves periodically, the interval between successive maxima is called the *period* of the star. The *amplitude* is the difference in light (usually expressed in magnitudes) between the maximum and the minimum.

(c) Types of Variable Stars

The *General Catalogue of Variable Stars* lists three types of variable stars: (1) *pulsating variables,* (2) *eruptive variables,* and (3) *eclipsing variables.* Pulsating variables are stars that periodically expand and contract, pulsating in size as well as in light. Eruptive variables are stars that show sudden, usually unpredictable, outbursts of light, or, in some cases, diminutions of light. Eclipsing variables

TABLE 25.1 Numbers of Variable Stars

TYPE	NUMBER
Pulsating	9855
Eruptive	959
Eclipsing	2763
Unclassified or unstudied	1134
All kinds	14711

are binary stars whose orbits of mutual revolution lie nearly edge-on to our line of sight and which periodically eclipse each other (eclipsing binaries). Eclipsing variables are not, of course, true variable stars; they have been discussed in Chapter 22 and will not be considered further here. The known numbers of different kinds of variable stars (in 1958) are summarized in Table 25.1.

25.2 PULSATING STARS

Pulsating variable stars are included among all spectral classes from B through M. Most of these stars are giants or supergiants. The largest group of pulsating stars consists of the red giant variables known as *Mira*-type stars; these are named for their prototype, Mira, in the constellation of Cetus. Other large groups of pulsating stars are the *RR Lyrae variables,* the *semiregular variables,* and the *irregular variables.* We shall open our discussion of pulsating stars, however, with an account of the relatively rare, but historically very important, *cepheid variables.*

(a) Cepheid Variables

The cepheid variables are yellow supergiants, named for the prototype and first known star of the group, δ Cephei. The variability of δ Cephei was discovered in 1784 by the young English astronomer John Goodricke just 2 years before his death at the age of 21. The magnitude of δ Cephei varies between 3.6 and 4.3 in a period of 5.4 days. The star rises rather rapidly to maximum light and then falls more slowly to minimum light (see Figure 25-1).

More than 600 cepheid variables are known in our galaxy. Most cepheids have periods in the range 3 to 50 days and absolute magnitudes (at median light†) from −1.5 to −5. The amplitudes

---

†The *median light* of a variable star is the amount of light it emits when it is halfway between its maximum and minimum brightness.

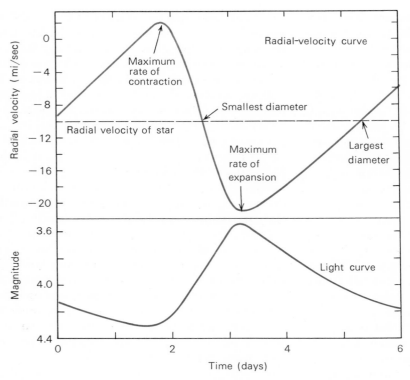

FIG. 25-2 Radial-velocity curve *(top)* and light curve *(bottom)* of the star δ Cephei.

of cepheids range from 0.1 to 2 magnitudes. Polaris, the *north star,* is a small amplitude cepheid variable that varies betwen magnitudes 2.5 and 2.6 in a period of just under 4 days.

(b) Nature of Cepheid Variation

In the latter part of the last century, the light variations of Algol were shown to be caused by eclipses of one star by another in a binary system (Section 22.2a). At one time it was suspected that the light variations of many other stars, including the cepheids, might similarly be explained, but today the evidence is conclusive that the variability of the cepheids is intrinsic.

A hint of the physical nature of the variability of these stars comes from their spectra. At maximum light, their spectral classes correspond to higher surface temperatures than at minimum light. The light variations of a cepheid, therefore, are due in part to variations in the temperature of its radiating surface. Further spectroscopic evidence reveals that the temperature fluctuations are accompanied by actual pulsations in the sizes of these stars. The lines in the spectrum of a cepheid show Doppler shifts that vary in exactly the same period as that of the star's light fluctuations. Evidently, the changes in light are associated with a periodic rise and fall of the cepheid's radiating surface.

A graph that displays changes in the Doppler shifts of the spectral lines of a cepheid, with lapse of time is called a *radial-velocity curve* (see Figure 25-2). It is like the radial-velocity curve of a spectroscopic binary star, except that in a cepheid the Doppler shifts are due to the periodic rising and falling of its radiating surface rather than to the orbital motion of the star as a whole. The mean value of the apparent radial velocity corresponds to the line-of-sight motion of the star itself; the photospheric pulsations cause variations about this mean value. When the photosphere expands, it is approaching us with respect to the rest of the star, and each spectral line is shifted to slightly shorter wavelengths than that of its mean position; when the photosphere contracts, the lines are shifted to slightly longer wavelengths. When the photosphere reaches its highest or lowest point — that is, when the star is at its largest or smallest size — the position of the spectral lines corresponds to the radial velocity of the star itself.

We can calculate the total distance through which the cepheid's photosphere rises or falls by multiplying the velocity of its rise at each point in the pulsation cycle by the time it spends at that velocity, and then adding up the products.† For δ Cephei, for example, we find by such a calculation that the photosphere pulsates up and down over a distance of somewhat under 2 million miles. On the other hand, the mean diameter of δ Cephei as calculated from Stefan's law (Section 22.4b) is about 25 million miles. During a pulsation cycle, therefore, the radius of the star changes by about 7 or 8 percent.

We might expect a pulsating star to be hottest when it is at its smallest size and is most compressed. δ Cephei, however, like other cepheid variables, is hottest and brightest at about the time when its radiating surface is rushing outward at its maximum speed. Evidently, the greatest compression of the star as a whole does not correspond to the maximum temperature at its surface; the explanation is related to the mechanism by which energy is transferred outward through the outer layers of the star.

_____
†Or, technically, by *integrating* the velocity curve over the time of rise or fall.

In a stable star the weights of the constituent layers, bearing toward the center of the star under the influence of gravity, are just balanced by the pressure of the hot gases within (Chapter 29). A pulsating star, on the other hand, is something like a spring: as the star contracts, its internal pressures build up until they surpass the weights of its outer layers. Eventually, these pressures start the star pulsing outward, but because of their inertia, the outward-moving layers overshoot the equilibrium point where their weights will just balance the internal pressures. As the star expands further, the weights of the overlying layers decrease, but the internal pressure decreases faster. Hence, the overlying layers are not supported adequately, and the star begins to contract. As it does so, it overshoots again, and this time it becomes too highly contracted. Once more the inner pressures cause the star to expand — and so the pulsation continues. The time required for a complete cycle of pulsation is greater for a giant star of low density than for a smaller compact star of higher density — just as a long piano string vibrates more slowly than a short one. It can be shown that for pulsating stars of any one type, the period of pulsation of a particular star is inversely proportional to the square root of its mean density.

(c) The Period-Luminosity Relation
Many other stars, besides cepheids, vary in light because of physical pulsations; indeed, cepheids are a comparatively rare class of pulsating variables. The great importance of cepheid variables lies in the fact that a relation exists between their periods of pulsation (or light variation) and their median luminosities, that is, their average absolute magnitudes. The relation was discovered in 1912 by Henrietta Leavitt, an astronomer of the Harvard College Observatory. Some hundreds of cepheid variables had been discovered in the Large and Small Magellanic Clouds, two great stellar systems that are actually neighboring galaxies (although

they were not known to be galaxies in 1912 — see Chapter 32).

Many photographs of the Magellanic Clouds had been taken at the Southern Station of the Harvard College Observatory in South America. (These stellar systems are too far south in the sky to be observed from the United States.) On some of the photographs, Miss Leavitt identified 25 cepheids in the smaller cloud and plotted light curves for them. She found that the periods of the stars were related to their relative brightnesses, in the sense that the brighter-appearing ones always had the longer periods of light variation. Since the 25 stars are all in the same stellar system, they are all at about the same distance — that is, the distance to the Small Magellanic Cloud — and their relative apparent brightnesses, therefore, indicate their actual relative luminosities. Miss Leavitt, in other words, had found that the luminosities, or absolute magnitudes of the cepheid variables, were correlated with their periods. Subsequent investigation showed that this relation exists between all the cepheids in the Large and Small Magellanic Clouds.

Now the cepheid variables known in our Galaxy are intrinsically very luminous stars. If those cepheids in the Magellanic Clouds are the same kind of object, they must also be highly luminous. They appear very faint, however, of fifteenth or sixteenth apparent magnitude, which indicates that these stellar systems must be very remote. In 1912 the actual distances to the Clouds of Magellan were not known. While it was known then that the periods of cepheids are correlated with their absolute magnitudes, it was not known what those absolute magnitudes are; only *differences* between absolute magnitudes of cepheids of different periods could be determined. For example, if one cepheid had a period of 3 days and another of 30 days, it was known that the one of longer period was about 2 magnitudes, or six times, brighter than the star of shorter period. Unfortunately, however, the actual absolute magnitude of

neither star was known, beyond the fact that they were both very luminous; hence, the true distances to the stars remained inaccessible.

If the distances to only one or a few cepheid variables could be determined by independent means, however, the absolute magnitudes of those few stars could be found (Section 20.2b), and if it is assumed that the period-luminosity relation applies to cepheids in the Galaxy, as well as in the clouds, the scale of the relation would be set. Thereafter, the distance could be determined to any system or cluster of stars in which a cepheid variable had been identified. Harlow Shapley was one of the astronomers who recognized the importance of cepheids as distance indicators, and he pioneered the work of determining distances to some of them in our galaxy. His work was extended by others, and by 1939 it appeared that the distances to these stars were well determined. Unfortunately, however, the determination of distances to cepheid variables is a very difficult observational problem. In the first place, there is not a single cepheid near enough so that its trigonometric parallax can be measured (Section 18.4b); therefore, statistical methods involving the proper motions and radial velocities of these stars must be employed (Section 19.3). In the second place, most of the cepheid variables in our Galaxy lie close to the plane of the galactic system, where clouds of interstellar dust heavily obscure their light (see Chapter 26). Corrections must be applied, therefore, to the measured apparent magnitudes of these cepheids; these corrections are difficult to determine accurately and are thus rather uncertain.

In the three decades prior to about 1950, the best available observational evidence indicated that the cepheids of shortest period (about 3 days) had absolute magnitudes between 0 and -1.0, while those of longest period (about 50 days) had absolute magnitudes of about -3.5. It will be seen in Section 25.2f that these figures have recently been revised and are still not precisely known.

(d) Cepheids in Globular Clusters

In a very few globular star clusters, and also outside of clusters in the galactic corona (Section 23.3c), there are variable stars with periods in the range 10 to 30 days. The light curves of these stars are similar to those of the cepheid variables except that they fall from maximum light to minimum more slowly, and the stars are somewhat bluer in color. These objects are sometimes called *W Virginis* stars, after the prototype W Virginis, or sometimes *type II cepheids,* to distinguish them from ordinary or "classical" cepheids, which are called *type I cepheids.* Although only a few dozen type II cepheids are known, a period-luminosity relation also seems to hold for them, and it resembles that for type I cepheids, in that stars of longer period are the more luminous. Because of the similar characteristics of type I and type II cepheids, they were originally thought to be the same kind of star, and only since about 1950 has it been realized that type I and type II cepheids are actually different types of pulsating stars.

(e) RR Lyrae Stars

Next to the long-period variables (Section 25.2g), the most common variable stars are the *RR Lyrae* stars, named for RR Lyrae, best known member of the group. Nearly 2500 of these variables are known in our Galaxy. Almost all of them are found in the nucleus or the corona of our Galaxy or in globular clusters. In fact, nearly all globular clusters contain at least a few RR Lyrae variables, and some contain hundreds; these stars, therefore, are sometimes called *cluster-type* variables.

The periods of RR Lyrae stars are less than 1 day; most periods fall in the range 0.3 to 0.7 day. Their amplitudes never exceed 2 magnitudes, and most RR Lyrae stars have amplitudes less than 1 magnitude. They are of spectral classes A or F. Several subclasses of RR Lyrae stars are recognized, but the differences between these subclasses are small and need not be considered here.

It is observed that the RR Lyrae stars occurring in any particular globular cluster all have about the same median apparent magnitude. Since they are all at approximately the same distance, it follows that they must also have nearly the same absolute magnitude. Because the RR Lyrae stars in different clusters are all similar to each other in observable characteristics, it is reasonable to assume that *all* RR Lyrae stars have about the same absolute magnitude. If we could learn what that absolute magnitude is, we could immediately calculate the distances to all globular clusters that contain these stars.

Unfortunately, as is true for cepheids, not a single RR Lyrae star is near enough to measure its parallax by direct triangulation. Like the cepheids, distances to RR Lyrae stars have also had to be determined by statistical means, that is, by analyzing their proper motions and radial velocities. The early distance investigations, carried out from 1917 to about 1940, seemed to indicate an absolute magnitude of about 0 for these important stars — they are, therefore, about 100 times as luminous as the sun, whose absolute magnitude is about +5. More recent work shows that different RR Lyrae stars may differ from each other slightly in median absolute magnitude, and that the average value may be a little fainter than was once thought — somewhere between 0 and +1. These recent revisions do not vitiate the important results derived from the study of RR Lyrae stars concerning the nature of our Galaxy (Chapter 27).

(f) *Relation of Cepheids, W Virginis Stars, and RR Lyrae Stars to the Period-Luminosity Relation**

Until the early 1950s, the evidence seemed to indicate a common period-luminosity relation for cepheids, W Virginis stars, and RR Lyrae stars. If the period-luminosity relation that was found for the cepheids was extrapolated to shorter periods than any actually observed for those stars, it indicated that if there were cepheids of periods less than 1 day, those hypothetical stars should have absolute magnitudes of about 0. Since this was the absolute magnitude found for the RR Lyrae stars, it was natural to assume that they were actually short-period cepheids that lay along a continuation of the same period-luminosity relation. RR Lyrae stars, therefore, were called "cluster-type cepheids" or "short-period cepheids." The few W Virginis stars known in globular clusters, also thought to be cepheids, lay approximately along the same relation, and apparently completely confirmed the assumption.

Several independent investigations in the late 1940s and early 1950s showed, however, that these three kinds of variable stars do not really all lie on the same period-luminosity relation. With the 100-in. telescope on Mount Wilson, for example, W. Baade found that he could photograph many cepheids, but not RR Lyrae stars, in the Andromeda galaxy (Chapter 32). When he determined the distance to the galaxy on the basis of the accepted period-luminosity relation for the cepheids, however, he calculated that its RR Lyrae stars should be visible on photographs obtained with Palomar's 200-inch telescope. To his surprise, they were not.

It is now realized that the cepheids are between 1 and 2 magnitudes brighter than was thought. The absolute magnitudes found for them originally had been based upon statistical studies of proper motions of the cepheids, which require very critical measurements, and upon rather uncertain corrections for the dimming of their light by interstellar dust. Neither was accurately known. A much better technique in use today to find the absolute magnitudes of cepheids — one due to J. B. Irwin — to identify them in certain open star clusters whose distances can be found by fitting the main sequences in their color-magnitude diagrams to the well-established main sequence on the Hertzsprung-Russell diagram (Section 28.3).

As we shall see (Chapter 32), the distances to other galaxies are based primarily on the apparent brightness, or magnitude, of the cepheids visible in the nearest of them. As soon as those cepheids were realized to be substantially brighter than had been thought, it meant that those galaxies were also farther away than was originally supposed. The revision of about 1½ magnitudes in the absolute magnitudes of the cepheids corresponded to a factor of 4 in light. In other words, all cepheid variables observed in other galaxies are intrinsically four times brighter than was previously thought and, therefore, twice

as far away. This is the origin of the spectacular announcement in the early 1950s that the extragalactic distance scale had been suddenly doubled.

Subsequent investigations have shown that those cepheids in globular clusters — W Virginis stars — that appeared to confirm the uniqueness of the period-luminosity relation are entirely different kinds of stars from type I cepheids, and average about 1 to 2 magnitudes fainter than type I cepheids of the same period. Moreover, there is considerable scatter about the period-luminosity relation for each kind of cepheid.

Figure 25-3 shows the form of the period-luminosity relation for all three kinds of variable stars.

(g) Long-Period Variables

Most common among the pulsating stars are the *long-period variables*. These stars, all red giants or red supergiants, are sometimes called "red variables." A typical example, and the best known star of this group, is Mira (o Ceti). Ordinarily, Mira is at apparent magnitude 8 to 10, too faint to be visible to the unaided eye. About every 11 months, however, it brightens to naked-eye visibility and averages magnitude 2.5 at maximum. (Once it reached magnitude 1.2.)

The spectral type of Mira varies from M6 to M9, which corresponds to a variation in surface temperature of 2600 to 1900°K. With this moderate range of surface temperature, one would expect (from Stefan's law) a variation in luminosity of only about a factor of 3½. Yet, in the visible spectral region, Mira varies in light by about 100 times or more. One reason is that most of the light emitted by the star lies in the invisible infrared spectral region. Although the *total* energy emitted by the star at maximum light may be only a few times greater than at minimum, the higher temperature at maximum causes that energy to be emitted at shorter wavelengths (Section 10.4c). A far greater fraction of it is thus in the visible part of the spectrum than at minimum. Another reason

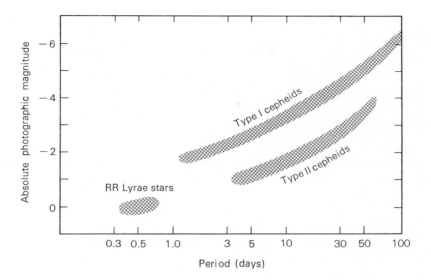

FIG. 25-3 Period-luminosity relation for type I cepheids *(upper curve)*, type II cepheids *(lower curve)*, and RR Lyrae stars *(left)*.

for the enormous light variations of Mira may be that when it is at minimum, more molecules can form in its outer atmosphere, and much of its visible light may be veiled by increased absorption in molecular bands, or even by temporary condensations of solid or liquid particles.

Mira is one of the few stars whose angular diameters have been measured with the stellar interferometer (Section 22.4a). Its angular size was observed to be 0″.056, which, for an assumed distance of 70 pc, corresponds to a linear diameter of 420 times that of the sun. The radius of Mira changes by 20 percent during a pulsation cycle. The mean density of the star is less than one millionth that of the sun.

Other long-period variables have periods in the range 80 to 600 days, and amplitudes ranging from 2.5 to more than 7 magnitudes. The periods of these red-giant variables tend to be somewhat unstable; Mira's period, for example, *averages* 330 days, but the interval between successive maxima has varied from this period by as much as several weeks. In general, the longer the period of a red variable, the less regular is its pulsation.

(h) Irregular and Semiregular Variables

Many red variables show very marked irregularity, and some exhibit no trace whatever of regularity in their light variations. The latter are designated *irregular variables.* Those stars which usually exhibit periodic variability, but which are at times unpredictably disturbed, are classed as *semiregular variables.* Generally, semiregular and irregular variables have light amplitudes that are less than 2 magnitudes — that is, a smaller range than the range of ordinary long-period variables. To the class of semiregular variables belongs the famous star Betelgeuse, which marks Orion's right shoulder.

The long-period, semiregular, and irregular red variables must be constantly watched to gather information about their light fluctuations. It would require far too much of the time of professional astronomers to keep constant vigil on all these variables. It is here that amateur astronomers provide a real service to astronomy; the society known as the *American Association of Variable Star Observers* has organized a careful surveillance of red variables and has been gathering data on them for years.

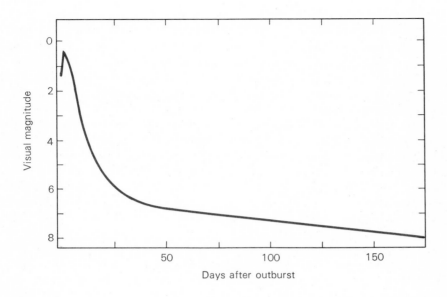

FIG. 25-4   Light curve of Nova Puppis, 1942.

(i) "Spectrum" Variables*

Many peculiar main-sequence spectral-type A stars are known that comprise a class called *spectrum variables*. They show unusually strong lines of the ions of strontium, silicon, and chromium, or of ionized rare-earth elements. These lines vary in strength, however, in periods of from 1 to 25 days. Often one set of lines (such as those of rare-earth elements) weakens, while another set (such as those of chromium) strengthens. At the same time the brightness of the star varies, but with an amplitude of only about 0.1 magnitude. Moreover, these stars have strong magnetic fields, which vary in the same period.

An example of the group, α^2 Canum Venaticorum, has a period of 5.5 days. H. W. Babcock, of the Mount Wilson and Palomar Observatories, has observed a magnetic field associated with this star which varies from −4000 to +5000 gauss. (In comparison, the magnetic field of the earth has a strength of less than 1 gauss.) One hypothesis which explains some of these phenomena, suggested by A. J. Deutsch, also of Mount Wilson and Palomar, is that the star (like the earth) has a magnetic axis that is inclined to its axis of rotation and that the stellar rotation, in a period of 5.5 days, alternately carries first the north, then the south magnetic pole into the view of earth, causing the observed variation of the magnetic field. According to Deutsch's hypothesis, the lines of certain elements vary in intensity because these elements tend to congregate in the vicinity of one or the other magnetic pole.

25.3 ERUPTIVE VARIABLES

There are many types of eruptive variables; they range from the *R W Aurigae* (or *T Tauri*) types, which display rapid, irregular increases in brightness, through the *novae*, to the spectacular *supernovae*. Also to the eruptive variables belong the *R Coronae Borealis variables*, stars which show sudden, unpredicted decreases in brightness.

(a) Novae

The most famous of the eruptive variables are the *novae*. Nova literally means "new." Actually, a nova is an existing star that suddenly emits an outburst of light. In ancient times, when such an outburst brought a star's luminosity up to naked-eye visibility, it seemed like a new star. Novae remain bright for only a few days or weeks and then gradually fade. They seldom remain visible to the unaided eye for more than a few months. The Chinese, whose annals record novae from centuries before Christ, called them "guest stars." Only occasionally are novae visible to the naked eye, but, on the average, two or three are found telescopically each year. Many must escape detection; altogether there may be as many as two or three dozen nova outbursts per year in our galaxy.

Novae, before and after their outbursts, are hot, subdwarf stars — stars of very high temperature but small size, so that despite their high temperatures they have lower luminosities than equally hot main-sequence stars. A typical nova, however, flares up during its outburst to thousands or even tens of thousands of times its normal luminosity, and may reach an absolute magnitude of from −6 to −9. The rise to maximum light is very rapid, often requiring less than 1 day. The subsequent decline in light is much slower; the star requires years, or even decades, to return to normal. A typical nova light curve is shown in Figure 25-4. Different types of novae, however, decline at different rates. Some (possibly all) show variability during their gradual fading.

The mechanism that gives rise to a nova outburst is not known with certainty, but it is known that at the time of the outburst, an outer layer of the star is ejected in the form of a shell of gas. The rapidly growing shell gives the effect of an enlarging photosphere, thus accounting for most of the increase of light. Evidence for the expanding shell comes first from the fact that the star's spectral absorption lines are shifted suddenly to the violet, indicating that the photosphere is rising, that is, approaching us. Very soon after the outburst, however, the shell has grown large enough so that it has thinned out and no longer absorbs all the light striking it from within. When the shell thus becomes partially transparent, it no longer emits a continuous spectrum. It still does absorb some of the star's light, however, especially in the far ultraviolet; it reemits this light in the form of bright emission lines, which appear in the spectrum of the nova soon after maximum light. Now, because the shell has become partially transparent, we simultaneously observe light coming from all parts of it, those parts moving out across our line of sight, some other parts moving away from us,

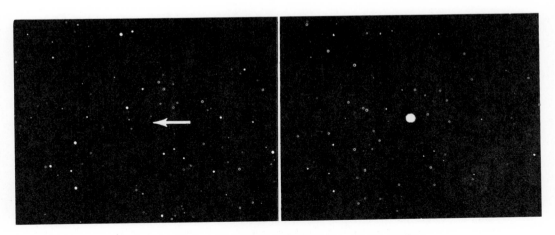

FIG. 25-5   Nova Herculis, as it appeared before and after its outburst in 1934. *(Yerkes Observatory.)*

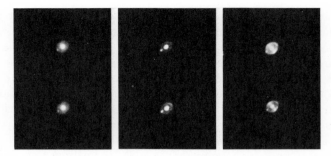

FIG. 25-6   200-inch telescope photographs of Nova Herculis in ultraviolet *(left)*, green *(center)*, and red *(right)* light 17 years after its outburst in 1934. Note the surrounding shell of gas. *(Mount Wilson and Palomar Observatories.)*

and those parts coming toward us. Thus the light from the shell that we observe is displaced in wavelength by the Doppler shift, in differing amounts (see Figure 25-8). Light reemitted from parts A and B of the shell, moving away from the earth, is shifted to longer wavelengths; light from parts C and D is not shifted; light from parts E and F is shifted to the violet. Only part G of the shell, directly between the earth and the star itself, produces absorption lines in the observed spectrum of the star; these are shifted farthest of all to the violet. The emitted light from the shell, therefore, does not appear *to us* in the form of sharp bright lines but as broadened lines, or emission bands. At the violet end of each emission band is the sharp absorption line produced by the part of the shell directly between us and the star.

From the widths of the emission bands, or

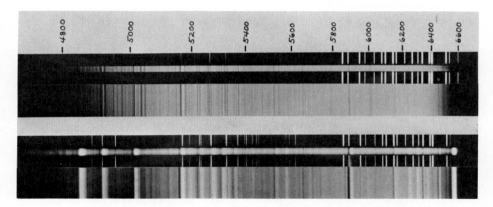

FIG. 25-7   Spectra of Nova Herculis in two stages of its expansion. At maximum light *(top)*, dark absorption lines predominate. As the star begins to fade *(bottom)*, broad emission lines appear. In both cases, the absorption lines are shifted toward the violet *(left)* end of the spectrum, which indicates that an envelope of gases is expanding about the star. In each case, the narrow spectrum between the comparison spectra is the actual one while the lower spectrum has been widened artificially to show more detail. *(Lick Observatory.)*

from the displacement of the absorption lines at their violet ends, we can calculate the velocity with which the shell is expanding. Velocities of ejection of up to 1000 km/sec or more are found. From the total light emitted by the shell, we can calculate the amount of material it contains. A typical nova shell is found to contain from 10^{-5} to 10^{-4} times the solar mass. The star, therefore, ejects only a small part of its total mass. Some months after the outburst, the expanding envelope may become visible on telescopic photographs, as in the case of Nova Aquilae (Figure 25-9).

Some novae have undergone more than one outburst over the period they have been under observation; these are called *recurrent novae*. There is no way of knowing, of course, whether ordinary novae may have had previous flareups before recorded history, or whether they will produce outbursts again. It is possible, in other words, that such novae may be recurrent but show more spectacular outbursts, with longer intervals between them.

There is now rapidly accumulating evidence that novae always occur in stars that are members of binary-star systems in which the two stars, the one that undergoes the outburst and its companion, are very close together. A popular theory for the nova phenomenon involves the transfer of mass from the companion star to the nova star, and the possibility that this mass exchange triggers the outburst. The theory is described a little more fully in Chapter 30.

(b) Supernovae

Among the more spectacular of the cataclysms of nature is the *supernova*. In contrast to an ordinary nova, which increases in luminosity a paltry few thousands or at most tens of thousands of times, a supernova is a star that flares up to hundreds of millions of times its former brightness. At maximum light, supernovae reach absolute magnitude -14 to -18, or possibly even -20. The three most

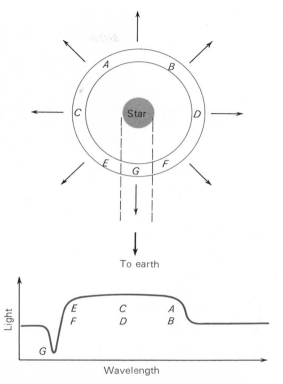

FIG. 25-8   Diagram of an expanding nova shell. Below is a profile of an emission and absorption line formed by the shell.

famous supernovae to have been observed during the last 10 centuries in our Galaxy are (1) the Supernova of 1054 in Taurus (described in the *Chinese Annals*), (2) Tycho's "star" of 1572 (Section 3.3) in the constellation Cassiopeia, and (3) the supernova of 1604 in Serpens, described by both Kepler and Galileo. Supernovae are now commonly observed in other galaxies. In a typical galaxy, supernovae appear to occur at the rate of at least one every few hundred years.

The light curve of a supernova is similar to that of an ordinary nova, except for the far greater luminosity of the supernova. It now appears that there may be several different kinds of supernovae, but they all rise to maximum light extremely quickly (in a few days or less), and for a brief time one may outshine the entire galaxy in which it appears. Just after maximum, the gradual decline sets in, and the star fades in light until it disappears from telescopic visibility within a few months or years after its outburst. Bright emission lines in the spectra of supernovae indicate that they, like ordinary novae, eject material at the time of their outbursts. The velocities of ejection may be substantially greater than in ordinary novae, however, since speeds of up to 4000 km/sec have been observed. Moreover, a much larger amount of material is ejected; in fact, a large fraction of the original star may go off in the expanding envelope.

(c) The Crab Nebula

An example of the remnant of a supernova explosion is the Crab nebula in Taurus, a chaotic, expanding mass of gas, visible telescopically, and spectacular on telescopic photographs.

The outer parts of this gas cloud are observed to be moving away from the center at the rate of about 0''.21 per year. The Doppler shift of light from the center of the nebula shows the gases there to be approaching us at the rate of 1100 km/sec. If we assume that the nebula expands at the same rate in all directions, we conclude that at the distance of the Crab, 1100 km/sec produces an annual proper motion of 0''.21. In section 19.1c we developed the formula

$$T = 4.74\mu r \text{ km/sec},$$

which relates tangential velocity, T, proper motion, μ, and distance, r. Upon substituting 1100 for T and 0.21 for μ, we find, for the distance of the Crab nebula, 1100 pc. If we now assume that the rate of expansion of the nebula has not changed since the initial supernova explosion, we can calculate how long ago that explosion occurred. We

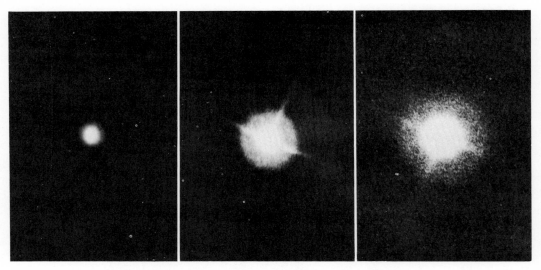

FIG. 25-9 Expanding nebulosity around Nova Aquilae, Nov. 3, 1918. *Left*, July 20, 1922; *center*, September 3, 1926; *right*, August 14, 1931. *(Mount Wilson and Palomar Observatories.)*

FIG. 25-10 The Crab Nebula, photographed in red light with the 200-inch telescope. *(Mount Wilson and Palomar Observatories.)*

divide the present angular radius of the nebula, 180″, by 0″21 per year, and find that the cloud of gas has required about 860 years to reach its present size, which places the event in the eleventh century. Both the location and computed time of formation of the Crab nebula are in good agreement with the occurrence of the supernova of 1054. The Crab nebula is believed, therefore, to be the material ejected during that stellar explosion.

The Crab nebula is a strong source of radio waves and X-rays. The radio spectrum (variation of radio energy with wavelength) resembles that expected from electrons which are spiraling at nearly the velocity of light in a magnetic field — a phenomenon known as *synchrotron radiation*. When the nature of the radio radiation from the Crab nebula was suspected, it was suggested that some of its visible light might similarly originate from the synchrotron mechanism. If so, theory

predicted that the light should be polarized. The polarization of some of the light from the Crab nebula was observed at the 200-inch telescope by W. Baade in 1956. There is little doubt, therefore, that the radio energy, as well as much of the visible light from the Crab nebula, is emitted from electrons accelerated in magnetic fields; the origins of the electrons and magnetic fields remain a mystery. It is also possible to observe synchrotron radiation, incidentally, from electrons that are being accelerated in a laboratory synchrotron.

(d) Other Galactic Supernovae

Radio sources are now known that correspond to positions of other supernovae, including those observed by Tycho (1572) and Kepler (1604). If the position of a temporarily appearing bright star, as given in historical records, agrees with that of

FIG. 25-11 Filamentary nebula in Cygnus, photographed with the 48-inch Schmidt telescope. The nebula is believed to be the remnant of an ancient supernova. *(Mount Wilson and Palomar Observatories.)*

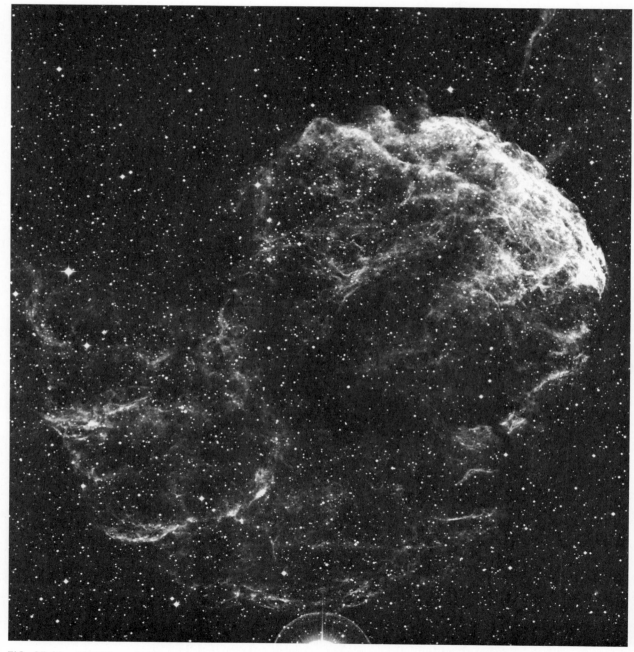

FIG. 25-12   IC 443, probable supernova remnant in Gemini. *(Mount Wilson and Palomar Observatories.)*

an observed radio source, and if the characteristics of the radio radiation suggest synchrotron emission, the star is generally assumed to have been a supernova. The identification is further strengthened if gas filaments or nebulosity can be observed at that same position. Provisional identifications of 14 supernovae in our galaxy within the past 2000 years have so been made. Descriptions of the temporary or "guest" stars themselves appear in the annals of the Chinese (especially), the Japanese, Koreans, Arabs, and Europeans. Several of these were observed with the naked eye for more than 1 year, and one was seen for 2 years. The latter reached an estimated apparent magnitude (at brightest) of -8, many times as bright as Venus; several were easily visible in daylight. Still other radio sources have been identified with filaments of gas that are believed to be the remnants of prehistoric supernovae. An example of these is the filamentary "loop" or Veil nebula in Cygnus.

The 14 supernovae observed in our galaxy in the past 2000 years give an average rate of one supernovae per 140 years. Many supernovae must have been missed, however, especially those in remote parts of our galaxy, hidden by interstellar dust. Shklovsky estimates that the frequency of supernovae in our galaxy may be as high as one every 30 years. This is greater than the rate estimated for other galaxies, but the data are still so few that the figures cannot be regarded irreconcilable.

(e) Flare Stars

Flare stars (or *UV Ceti stars*) are red main-sequence stars, mostly of spectral types M3 to M6. These stars show rare flareups in brightness. The total increase in light may be from 0.1 to several magnitudes. Their rise to maximum is very rapid — a few seconds — and the decline lasts for from a few minutes to 1 hour or so. They behave rather like miniature novae.

The color of the light associated with a flare is blue, appropriate to that from a star with a temperature of 10,000 to 20,000°K. A large area at such a temperature would radiate far more energy than is observed to come from flares. Consequently, a flare must be a localized disturbance on a small part of a star. In total energy output, a typical flare observed on a UV Ceti-type star resembles the more spectacular of solar flares (Section 24.3i).

Flares cause proportionately larger increases in the light of intrinsically faint stars than of brighter ones. Suppose a flare causes an M2 star to brighten by 1 magnitude — a factor of 2½ in light. The flare itself, then, gives off (during the brief period of its existence) 1½ times as much light per second as does its parent star. A flare of exactly the same energy would increase the light of a star 2 magnitudes brighter than the M2 star by only about 20 percent, and would increase the sun's light by only 1 percent, an amount too little to be noticed if the sun were as distant as the other stars. We cannot be sure, in other words, that flares do not occur on intrinsically bright, as well as intrinsically faint, stars; they are obvious only on the latter. On stars of intermediate luminosity, flares might be observed spectrographically. At least one flare was observed on a K star by means of bright emission lines that happened to appear and be recorded while the star's spectrum was in the process of being photographed.

25.4 SUMMARY OF VARIABLE STARS

It is neither feasible nor appropriate to describe in detail the many other kinds of variable stars in a survey of astronomy such as this book. A tabular summary of the more important types of variables is presented in Tables 25.2 and 25.3.

TABLE 25.2 Pulsating Variables

TYPE OF VARIABLE	SPECTRA	PERIOD (DAYS)	MEDIAN MAGNITUDE (ABSOLUTE)	AMPLITUDE (MAGNITUDES)	DESCRIPTION	EXAMPLE	NUMBER KNOWN IN GALAXY†
Cepheids (type I)	F to G supergiants	3 to 50	-1.5 to -5	0.1 to 2	Regular pulsation; period-luminosity relation exists	δ Cep	610
Cepheids (type II)	F to G supergiants	5 to 30	0 to -3.5	0.1 to 2	Regular pulsation; period-luminosity relation exists	W Vir	(About 50; included with type I)
RV Tauri	G to K yellow and red bright giants	30 to 150	-2 to -3	Up to 3	Alternate large and small maxima	RV Tau	92
Long-period (Mira-type)	M red giants	80 to 600	$+2$ to -2	>2.5	Brighten more or less periodically	o Cet (Mira)	3657
Semiregular	M giants and supergiants	30 to 2000	0 to -3	1 to 2	Periodicity not dependable; often interrupted	α Ori	1675
Irregular	All types	Irregular	<0	Up to several magnitudes	No known periodicity; many may be semiregular, but too little data exist to classify them as such	π Gru	1370
RR Lyrae or cluster-type	A to F blue giants	<1	0 to $+1$	<1 to 2	Very regular pulsations	RR Lyr	2426
β Cephei or β Canis Majoris	B blue giants	0.1 to 0.3	-2 to -4	0.1	Maximum light occurs at time of highest compression	β Cep	11
δ Scuti	F subgiants	<1	0 to $+2$	<0.25	Similar to, and possibly related to, RR Lyrae variables	δ Sct	5
Spectrum variables	A main sequence	1 to 25	0 to $+1$	0.1	Anomalously intense lines of Si, Sr, and Cr vary in intensity with same period as light; most have strong variable magnetic fields	α^2C Vn	9

†According to the 1958 edition of the Soviet *General Catalogue of Variable Stars.*

TABLE 25.3 Eruptive Variables

TYPE OF VARIABLE	SPECTRA	DURATION OF INCREASED BRIGHTNESS	NORMAL ABSOLUTE MAGNITUDE	AMPLITUDE (MAGNITUDES)	DESCRIPTION	EXAMPLE	NUMBER KNOWN IN GALAXY†
Novae	O to A hot subdwarfs	Months to years	>0	7 to 16	Rapid rise to maximum; slow decline; ejection of gas shell	GK Per	146
Novalike variables or P Cygni stars	Hot B stars	Erratic	−3 to −6	Several magnitudes	Slow, erratic, and novalike variations in light; may be unrelated to novae. Gas shell ejected	P Cyg	35
Supernovae	?	Months to years	?	15 or more	Sudden, violent flareup, followed by decline and ejection of gas shell	CM Tau (Crab nebula)	7
R Coronae Borealis	F to K supergiant	10 to several hundred days	−5	1 to 9	Sudden and irregular drops in brightness. Low in hydrogen abundance, but high in carbon abundance	R CrB	39
T Tauri or RW Aurigae	B to M main sequence and subgiants	Rapid and erratic	0 to +8	Up to a few magnitudes	Rapid and irregular light variations. Generally associated with interstellar material. Subtypes from G to M are called T Tauri variables	RW Aur, T Tau	590
U Geminorum or SS Cygni or "dwarf novae"	A to F hot subdwarfs	Few days to few weeks	>0	2 to 6	Novalike outbursts at mean intervals which range from 20 to 600 days. Those with longer intervals between outbursts tend to have greater amplitudes. Many, if not all, are members of binary-star systems	SS Cyg, U Gem	112
Flare stars	M main sequence	Few minutes	>8	Up to 6	Sudden flareups in light; probably localized flares on surface of star	UV Cet	15
Z Camelopardalis variables	A to F hot subdwarfs	Few days	>0	2 to 5	Similar to U Geminorium, except that variations are sometimes interrupted by constant light for several cycles. Intervals between outbursts normally range from 10 to 40 days	Z Cam	15

†According to the 1958 edition of the Soviet *General Catalogue of Variable Stars.*

25.5 STARS WITH EXTENDED ATMOSPHERES

We have already seen that novae, supernovae, and novalike stars eject gas shells at the times of their outbursts. In addition, many other stars, for which no outbursts have been observed, are known to be surrounded by extended atmospheres of expanding gas shells. The extended atmosphere is usually revealed by the presence of emission lines or bands superposed on the continuous spectrum of the star, just as in nova shells (Section 25.3a). Sometimes the light absorbed and reemitted by the gas shell is too feeble to be observed, but the shell may still reveal its presence by producing absorption lines that, because of their wavelengths or sharpness, cannot originate in the stellar photosphere. In a few cases, a gaseous envelope surrounding a star can be seen or photographed telescopically.

Many red giants are examples of stars with extended atmospheres, whose spectra show that the gaseous shells have been and are being ejected from them, and are now expanding about them. The mechanism by which this material is ejected is unknown, but nevertheless it appears that some of these stars may lose a substantial fraction of their original mass over a period of a few hundred million years.

About 4000 spectral class B stars are known whose spectra show emission lines, usually of hydrogen, and sometimes of other elements as well. These stars, known as *B emission,* or *Be, stars,* have evidently ejected material from their outer layers. Most of them are rapidly rotating and the mass ejection may be related in some manner to their rapid rotation.

(a) Shell Stars

A good example of the class of stars known as *shell stars* is the B5 star Pleione, one of the brighter members of the Pleiades cluster. In 1888, bright emission lines of hydrogen appeared in the spectrum of Pleione, but by 1905 they had disappeared. They reappeared, however, in 1938, and this time sharp dark absorption lines appeared in the centers of the emission lines. The bright lines originated from gases, apparently ejected from and surrounding the star. The dark lines were formed by the gases directly between the star and us. The significant thing about the absorption lines produced by the ejected gas is that they were sharp, while the dark lines produced in the photosphere of the star are greatly broadened by the star's rapid rotation, which produces a linear velocity of about 350 km/sec at its surface. This is more than 100 times greater than the linear rotational velocity at the surface of the sun.

In such a rapidly spinning star, the equatorial regions, where the speed is greatest, are moving at nearly the circular satellite velocity (Section 5.7). A small ejection force, therefore, can cause gases in the equatorial region to move out, spiraling into ever-larger orbits. Such ejected material would produce a slowly expanding ring. It is generally supposed that Pleione, and other shell stars, are stars surrounded by *rings* of gaseous material, which have been ejected from the equatorial parts of the stars. Several dozen shell stars, all more or less resembling Pleione, have been studied.

(b) Wolf–Rayet Stars*

Nearly 200 *Wolf-Rayet stars* (also called *W stars*) are known. The brightest of them is the second magnitude star, γ Velorum. Wolf-Rayet stars are named for the two astronomers who discovered the first ones in 1867.

W stars are spectral class O stars of moderately high luminosity (absolute magnitudes of about −5) and very high surface temperatures, ranging up to estimated values of 100,000°K. In size they are more modest, however, averaging about only twice the radius of the sun.

Bright, broad emission lines appear in the spectra of Wolf-Rayet stars. These lines are usually of the elements helium, nitrogen, oxygen, silicon, or carbon. At the violet edge of each emission line is a sharp absorption line. The widths of the emission lines show that material must be ejected at speeds of from 1000 to 3000 km/sec. The gas shells in Wolf-Rayet stars, then, are similar to the envelopes ejected from novae and P Cygni stars (see Table 25.3), except that they are not known to accompany a sudden outburst of light. Some investigators feel that W

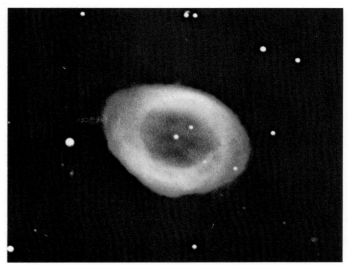

The ring nebula in Lyra, M57; 200-inch telescope photograph. *(Mount Wilson and Palomar Observatories.)*

7293 in Aquarius; 200-inch telescope photograph. *nt Wilson and Palomar Observatories.)*

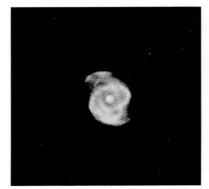

NGC 6543. *(Lick Observatory.)*

NGC 6781 in Aquila; 48-inch Schmidt telescope photograph. *(Mount Wilson and Palomar Observatories.)*

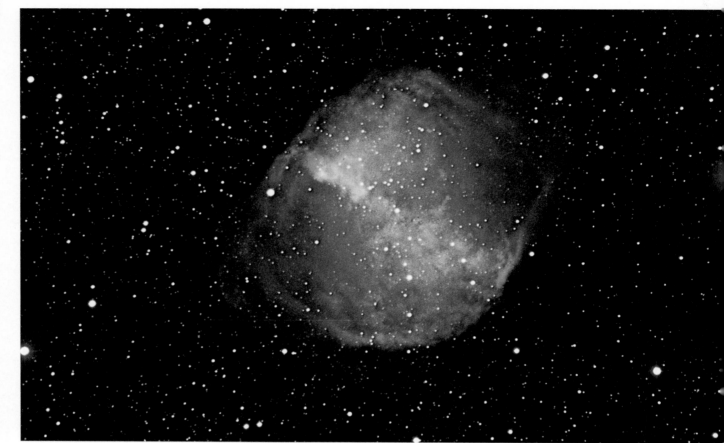

The Dumbell nebula, NGC 6853, M27; photographed by the 200-inch teles
(*Mount Wilson and Palomar Observator*

stars may be related to P Cygni stars, differing mainly in that the former are hotter and, of course, are not known to be eruptive variables.

Many Wolf-Rayet stars are members of spectroscopic binary systems; a few are even eclipsing stars. The binary condition may be related to the other unusual properties of W stars, but how is not known. It is also suspected that U Geminorum stars (Table 25.3) are usually, if not always, members of binary systems, and, as we have seen, at least some old novae have been shown to have close stellar companions.

(c) Planetary Nebulae

Planetary nebulae are shells of gas ejected from and expanding about certain extremely hot stars. They derive their name from the fact that a few bear a superficial telescopic resemblance to planets; actually they are thousands of times larger than the entire solar system, and have nothing whatever to do with planets.

Planetary nebulae are identified in two ways. Often they appear large enough to see or to photograph with a telescope. The most famous example is the ring nebula, in Lyra. It is typical of many planetaries in that, although actually a hollow shell of material emitting light, it appears as a ring. The explanation is that we are looking through the *thin* dimensions of the front and rear parts of the shell, while along its periphery our line of sight encounters a long path through the glowing material. Similarly, a soap bubble often appears to be a thin ring. The other way in which planetary nebulae are identified is by their spectra. Those that are very distant from us are unresolved and appear stellar, but their spectra show emission lines which indicate the existence of luminous shells of gas surrounding stars, as do spectra of nova shells. Altogether, there are about 1000 planetary nebulae known. Doubtless there are many distant ones that have escaped detection, so there might be some thousands or even tens of thousands in the Galaxy. Nevertheless, among the tens of billions of stars in the system, planetary nebulae must be classed as rare objects.

Planetary nebulae differ from the other kinds of stars that have ejected gas shells that we have discussed — novae, Be stars, W stars, and so on — in an important respect: an appreciable amount of material is ejected in the shell of a planetary nebula. From the light emitted by the shells, we calculate that they must have masses of 10 to 20 percent that of the sun. The shells, typically, expand about their parent stars at speeds of 20 to 30 km/sec.

It is an interesting problem to determine the distances to planetary nebulae. One is in a globular cluster of known distance (NGC 7078), and one (NGC 246) belongs to a star that appears to be a member of a binary-star system. The distance to the companion star (and thus to the planetary) can be estimated from the companion's spectrum by the method of spectroscopic parallaxes (Section 23.2b). The central stars of planetary nebulae themselves, unfortunately, do not fit into the usual pattern of spectral and luminosity classes, and the method of spectroscopic parallaxes cannot be applied to them. Most planetary nebulae are too distant for their parallaxes to be measured directly; therefore, their distances are estimated by indirect means. The most widely applied technique is to assume that all planetary nebula shells have about the same mass — say $\frac{1}{5}$ that of the sun. Then the distance to an individual nebula is estimated by calculating the distance it must have, if the mass assumed for it is correct, in order that it have its observed angular size and its observed brightness.

The linear diameters of planetary-nebula shells can be calculated from their angular diameters and distances (although the latter are known only with considerable uncertainty). A typical planetary appears to have a diameter of about ½ LY to 1 LY. If it is assumed that the gas shell has always expanded at the speed with which it is now enlarging about its parent star, its age can be calculated. Most of the gas shells have been ejected within the past 50,000 years; an age of 20,000 years is more or less typical. After about 100,000 years, the shell is so enlarged that it is too thin and tenuous to be seen. The rarity of planetary nebulae, therefore, is due entirely to the fact that they cannot be seen for very long; they are temporary phenomena. When we take account of the relatively short time over which a planetary nebula exists as such, we find that they are actually very common; indeed, an appreciable fraction of all stars must sometime evolve through the planetary nebula phase (see Chapter 30).

The gas shells of planetary nebulae shine by the process of fluorescence. They absorb ultraviolet radiation from their central stars and reemit this energy as visible light (the process will be explained more fully in Chapter 26 in connection with gaseous nebulae). During the first few tens of thousands of years after the ejection of the shell, the gas in it is dense and thick enough to prevent the star's ultraviolet radiation from penetrating all the way through it. This energy is completely absorbed, therefore, within the inner part of the shell, and only that inner part is luminous. The outer portion of the gaseous shell is dark and cold; it is transparent, however, so we see all the re-radiated, longer-wavelength energy emitted from within.

The mechanism of the fluorescence process enables us to calculate, or at least to estimate, the temperatures of the central stars of planetary nebulae. All the visible light emitted by the gas shell is converted from ultraviolet energy originally emitted by the star. Knowing the details of the atomic processes of absorption and emission of light that are involved in the fluorescence phenomenon, we can calculate the rate at which ultraviolet radiation must be leaving the star to account for the visible light coming from the gas shell; it turns out to be a far greater amount of energy than the star radiates in its observable, visible spectrum. Most of the sun's radiant energy, on the other hand, is in the form of visible light. The central star of a planetary nebula, therefore, must be many times hotter than the sun for so large a fraction of its luminosity to be in the ultraviolet (see Section 10.4). Nearly all these stars are hotter than 20,000°K, and some have temperatures well in excess of 100,000°K, which makes them the hottest known stars.

Despite their high temperatures, the central stars of planetary nebulae do not have exceedingly high luminosities — some emit little more total energy than does the sun. They must, therefore, be stars of small size; some, in fact, appear to have the dimensions of white dwarfs (see Exercise 10). In Chapter 30 it will be seen that we have good reason to regard white dwarfs as stars near the end of their life spans.

The central stars of planetary nebulae comprise a class of very hot, and usually very small, dense stars. It does not follow, however, that the stars were small and hot when the gas shells were ejected. The velocity of escape from such a small dense star is extremely high — up to thousands of kilometers per second. It is very difficult to imagine an ejection mechanism whereby a shell of gas can be shot off at such speeds, and a few thousand years later be expanding at a leisurely 20 to 30 km/sec. It is more likely that the nebulae were ejected from their parent stars when the latter were, at an earlier stage in their evolution, large red giants, from which the escape velocity would be under 100 km/sec.

(d) Gaseous Envelopes of Binary Stars*

It has been pointed out (Chapter 23) that some stars are of enormous size. It is not uncommon to find large stars in binary-star systems, and there is evidence that some of these stars are losing material, which either flows toward the companion star or forms a gaseous envelope surrounding the entire binary system. This form of mass loss may not always have been occurring in a particular binary system. As will be shown in Chapter 30, stars are believed to increase in size during a particular period in their evolution. We must now investigate why such stars may suddenly begin to shed material.

On the line between the centers of the two stars of a binary system there is a point where the gravitational attraction between a small body and one of the stars would be equal to that between the body and the other star. If the two stars were of equal mass, for example, that point would lie halfway between them. There are, in fact, entire surfaces that partially or completely surround either or both stars. A small body can move freely along one of these surfaces under the combined effects of the gravitational forces of the two stars and the mutual revolution system (see Section 6.1b).

Now suppose that one of the stars of a binary system increases in size, so that its outer surface extends through one of these regions between the stars. The stellar material along such a surface "doesn't know which star it belongs to" and is no longer gravitationally bound to the original star. An exchange of material from one star to the other may result. Also, some material may flow away from both stars and form a ring encircling both, or even a common envelope. Such a gaseous envelope surrounding a binary system can reveal itself by producing absorption lines superposed upon the spectra of both stars, lines which do not show large variations in Doppler shift, since the material in the outer envelope is not moving as rapidly as the two stars are in their mutual revolution.

Of the many binary systems in which the member stars are exchanging or losing matter, the best known is the eclipsing binary, β Lyrae. The spectrum of the system shows a set of absorption lines that appear to belong to a star of spectral type B8, and which show variable Doppler shifts that yield a radial-velocity curve with a period of just under 13 days. Also in the spectrum, however, appear both dark and bright lines that do not exhibit the Doppler shifts of the lines of the B star. The model suggested for the system consists of a B8 star and an F star in mutual revolution. The F star is too faint for its spectrum to be observed, but its spectral type is surmised from the relative amounts of light lost during the two eclipses of the system. Material is presumed to be flowing from each star and forming a ring surrounding the system; this ejected material produces the absorption and emission lines that do not show large variations in Doppler shift. As a result of the mass loss, the period of mutual revolution of the two stars is increasing by about 10 sec/year. At the present rate at which they are ejecting material, the stars would be depleted of matter in about 300,000 years. The existing state of affairs for β Lyrae, however, is almost certainly a very temporary one; the system will doubtless change its character in a very much shorter time.

There is a large group of eclipsing binary stars, known as *W Ursae Majoris stars*, in which the two members of each system are so close together that they are nearly in contact and revolve about each other in a period of less than 1 day. These stars are greatly distorted by their mutual tidal forces, and material from each of them forms a gaseous envelope surrounding both stars.

25.6 X-RAY STARS

X-rays are electromagnetic radiation of wavelengths in the range 0.1 to 100 Å, to which the earth's atmosphere is opaque. Since 1962, however, we have been scanning the sky with X-ray detectors that are carried by rockets above practically all the atmosphere. Thus a whole new window in the electromagnetic spectrum has been opened to astronomical investigation. The field is still new; at the time of writing the sum total of all astronomical observations at X-ray wavelengths, involving many rocket flights, has covered just over an hour of time. Yet, important and exciting new discoveries have been made.

The first object from which X-rays were detected is the sun. However, if the sun is typical of other stars as an X-ray emitter, with existing equipment we could not observe X-rays from any other single star, because the other stars are so much farther away. Nevertheless, an effort was made to scan the sky with X-ray detectors of improved design carried by later rockets. On June 18, 1962, the first such systematic scan was made. A strong source of X-rays was detected in the constellation Scorpio. That first X-ray source discovered in Scorpio is now called Sco X-1. Subsequently another source was discovered — the Crab nebula (Section 25.3c).

The Crab nebula radiation was soon found to be coming from (essentially) the entire region of the sky occupied by the visible nebula. Sco X-1, on the other hand, was found to be of small angular size. The best guess was that the X-rays were coming from a star, and it was predicted that the star should be at least as bright as of the thirteenth magnitude. When an accurate position of Sco X-1 was derived, it was, indeed, found to correspond to that of a star of magnitude 13. The star, investigated both in Japan and in the United States, was found to be very blue. From the strength of the lines of interstellar gas in its spectrum, caused by gas between us and it, and also from the lack of any observable proper motion, the star's distance is estimated to be a few hundred parsecs. It emits a thousand times as much energy at X-ray wavelengths as it does in visible light, and its X-ray energy is about equal to the total radiation that the sun puts out at all wavelengths combined. Old astronomical photographs show that the star has been about the same for at least the past 80 years. Over short intervals, however, the star is variable, both in visible light and in X-rays. One hypothesis is that the star is an old nova.

By early 1968 about 30 X-ray sources had been found, and, in addition, a faint more or less uniform background of X-ray emission in the sky had been detected. The only positive identifications with optically visible objects are Sco X-1 and the Crab nebula. Another source in Cygnus, Cyg X-2, probably corresponds to a faint star that has properties similar to the one identified with Sco X-1. Cyg X-2, however, may be a binary star. There are three other possible identifications; two correspond to supernova remnants and one to the galaxy M87 (see Chapter 32). Good positions exist for three other X-ray sources, but no objects visible in the optical range are seen at those positions. The remaining 20 or so X-ray sources still have poorly determined positions, and attempts to identify them with optical objects are not yet possible.

Although X-ray astronomy is still in its infancy,

it is a rapidly expanding field, and it is anticipated that future editions of this book will have a far more complete story to tell.

25.7 THE PULSARS

Something has to be written about *pulsars,* even though it is virtually certain to be wrong, or at least outdated, when it is read.

Pulsars, or *pulsating radio stars,* are radio sources emitting sharp, intense, rapid, and extremely regular pulses. The first to be discovered, in mid-1967, is in the constellation Vulpecula. It emits a radio pulse every 1.33728 seconds. Detailed studies show each pulse to be complex and to consist of at least three much shorter pulses, each lasting only a few milliseconds, and the entire pulse sequence only a few tens of milliseconds. Although extremely regular in period, the pulses vary considerably from one to the next in intensity. The radiation in all pulses is polarized.

The pulses are emitted over a wide range of radio frequencies, presumably simultaneously, but they are received at the earth first at high frequency, with successively lower frequencies being received with successively longer delay times. The delay in the arrival of the radiation is believed to be due to the retardation of radio waves by free electrons in interstellar space. The interstellar electrons, in effect, give interstellar space an index of refraction (at radio wavelengths) greater than unity, so that the speed of radio radiation through space is reduced below its value in a perfect vacuum; the reduction in speed is greater the greater the wavelength, accounting for the observed relation between frequency and delay time of the received pulse.

By early 1969 more than 20 other pulsars were discovered, all more or less similar to the first one discovered in Vulpecula. They have pulse periods ranging from 1/30 to 2 seconds. If we knew the density of free electrons in space we could calculate the distances to the pulsars from the relation between frequency and the delay of reception time of the pulses. At present we can only guess the electron density, and our guesses indicate distances for the pulsars of (typically) a few hundred parsecs. A few pulsars have radio (21 cm) absorption lines (Section 26.2i), which show them to lie at least as far as certain instellar hydrogen clouds producing those lines. Some pulsars, we find, are more than 1000 pc away. At such distances, the radio energy emitted in each pulse must be enormous. Moreover, because of the sharpness of the pulses, that radio energy must be coming from a region at most a few hundred kilometers in diameter; otherwise, the light-travel time across the emitting region would result in lengthening the pulse time.

One pulsar is in the middle of the Crab nebula (Section 25.3c). It has the shortest pulse period known (at this writing) — 0.033 sec, and the period is observed to be very slowly increasing, suggesting that pulsars evolve, pulsing gradually more slowly as they age. Most remarkable, this pulsar is also observed to emit optical (visible light) pulses with that same 0.033-second period. It has tentatively been identified with what appears to be a star of about sixteenth magnitude.

So much is known about the pulsars. What they really are and what the origin of the pulses is are among the most perplexing astronomical puzzles of our time. Of the various theories proposed to account for them — pulsating white dwarfs, pulsating, rotating, or binary neutron stars (see Chapter 30 for a brief discussion of neutron stars), oscillating plasmas in the atmospheres of stars, even signals from advanced civilizations — none has enough promise to have gained general acceptance.

Doubtless many more pulsars will be found, and hopefully a reasonable explanation for them will be forthcoming.

Exercises

1. Sketch a Hertzsprung-Russell diagram, and indicate upon it the positions of as many of the kinds of variable stars mentioned in this chapter as you can.

*2. Sketch the period-luminosity relation for cepheids, W Virginis stars, and RR Lyrae stars that was assumed to hold prior to 1950. Indicate each of the three kinds of variables on the sketch.

3. Verify that for the temperature range mentioned in the chapter, the luminosity of Mira can be expected to vary by only about a factor of 3½.

4. Sketch the spectral energy distribution of Mira, both at maximum and at minimum, and show how a greater fraction of its total energy is in the visible spectral region when it is at maximum than when it is at minimum.

5. Since supernovae occur so rarely in any one galaxy, how might a search for them be conducted? (It may help to glance at Chapter 32.)

*6. Explain why a rotating ring surrounding a shell star produces a sharp absorption line at the *center* of each emission line rather than at the violet edge, as in the case of an expanding shell of gas.

*7. Calculate the diameter of a Wolf-Rayet star, in terms of the sun's diameter, if the star has 10,000 times the sun's luminosity and 7 times its temperature.

8. Compare and contrast: nova shells; supernova shells; planetary nebulae.

9. Suppose the luminous shell of a planetary nebula is easily resolved with a telescope. Now suppose the spectrum of the nebula is photographed, with the

slit of the spectrograph extending completely along one diameter of the shell. Sketch the appearance of a typical emission line in the spectrogram, and explain your sketch. It may help to look over the description of the construction of a spectrograph in Chapter 11.

10. Suppose the central star of a planetary nebula is 16 times as luminous as the sun, and 20 times as hot (about 110,000°K). Find its radius, in terms of the sun's. Does this star have the dimensions of a white dwarf?

11. The gas shell of a particular planetary nebula is expanding at the rate of 10 mi/sec. Its diameter is 1 LY. Find its age. For this calculation, assume that there are 3×10^7 seconds/year and 6×10^{12} mi/LY.

12. Assume that a pulsar is 100 pc away. Suppose that no star brighter than apparent magnitude 23 shows up in that position of the sky. What is the brighter limit to the absolute magnitude that a star associated with the pulsar could have?

CHAPTER 26

The Interstellar Medium

By earthly standards the space between the stars is empty, for in no laboratory on earth can so complete a vacuum be produced. Yet, throughout large regions of space this "emptiness" consists of vast clouds of gas and tiny solid particles. Most of this interstellar material is found between the stars in the spiral arms of our own and other galaxies. It exists, for example, in the neighborhood of the sun. The gas and dust is not distributed uniformly, however, but has a patchy, irregular distribution, being denser in some areas than in others, hence forming "clouds."

Sometimes, these tenuous clouds are visible, or partially so, in the form of *nebulae* (Latin for "clouds"). More often, they are invisible, and their presence must be deduced. In the spiral arms, on the average, there is about one atom of gas per cubic centimeter in interstellar space, and about 25 or 50 tiny particles or "grains," each less than a thousandth of a millimeter in diameter per cubic kilometer. In some of the denser clouds, the densities of gas and dust may exceed the average by as much as a thousand times, but even this is more nearly a vacuum than any attainable on earth. In air, for contrast, the number of molecules per cubic centimeter at sea level is of the order 10^{19}.

26.1 COSMIC "DUST"

The tiny solid grains in interstellar space, commonly called interstellar dust, are manifested in the following ways: (1) dark nebulae, (2) general obscuration, (3) reddening of starlight, (4) reflection of starlight, and (5) polarization of starlight.

(a) Dark Nebulae

Relatively dense clouds of the solid grains produce the *dark nebulae,* the opaque-appearing clouds that are conspicuous on any photograph of the Milky Way. Even in the densest clouds, the particles are very sparse, but the clouds extend over such vast regions (measured in parsecs) that they absorb or scatter a considerable portion of the starlight passing through them. Such concentrations of dust often have the appearance of dark curtains, greatly dimming or completely obscuring the light of stars behind them.

The "dark rift," running lengthwise down a long part of the Milky Way and appearing to split it in two, is an excellent example of such an obscuring cloud. The obstruction of light from the stars located behind it is so great that less than a century ago astronomers thought that it was a sort of "tunnel" through which they could see beyond the Milky Way, into extragalactic space. Today, we know that the dark rift is not such a tunnel; the Galaxy extends far beyond such observable dark nebulae (Chapter 27).

The conspicuous obscuring dust clouds are relatively close to us, within 1000 pc. The more distant opaque clouds would be difficult to discover because the large number of stars lying between them and us reduces the contrast produced by their own obscuration of starlight. The contrast produced by remote absorbing clouds is further reduced by the general obscuration caused by the foreground dust itself, which pervades the entire spiral-arm region of the Galaxy.

In addition to the large dark clouds, many

very small dark patches can be seen on Milky Way photographs, silhouetted against bright backgrounds of star fields or glowing gas clouds. Many of these patches, called *globules,* are round or oval and have angular diameters of only a few seconds of arc. The sharp contrast of their dark boundaries shows that they cannot be more distant than about 1000 pc, which allows us to calculate upper limits for their linear diameters. They probably range from a few thousand to a hundred thousand astronomical units across. The high opacity of the globules (they dim background objects by 5 magnitudes or more) implies that they must be very dense compared to the usual interstellar material. It has been suggested that the globules may be condensations of matter that may ultimately form into stars.

(b) The General Obscuration

Although the distribution of the interstellar dust is spotty, and dense clouds produce conspicuous dark nebulae, some of the dust is thinly scattered more or less generally throughout the spiral arms of the Galaxy. As a result, some absorption of starlight occurs even in regions where dark clouds are not apparent. Unfortunately, the presence of such sparse absorbing matter is not obvious, and it has been the cause of considerable difficulty in the determination of stellar distances.

It has been shown (Section 20.2b) that the distance to a star can be calculated from comparison of its apparent and absolute magnitudes. If light from a star has had to pass through interstellar dust to reach us, however, it is dimmed, much as a traffic light is dimmed by fog. We therefore underestimate the apparent brightness of the star — that is, we assign to it too large (that is, faint) an apparent magnitude, and the distance we calculate for the star, corresponding to its known (or assumed) absolute magnitude, is too large. Analogously, a motorist is likely to over-

FIG. 26-1   A portion of the Milky Way in Cygnus, showing the "dark rift." *(Mount Wilson and Palomar Observatories.)*

estimate the distance to a stoplight that he views through fog.

Astronomers once tried to calculate the extent of the Galaxy by counting the numbers of visible stars in various directions and calculating the distances required to account for their observed apparent magnitudes. Because of the general obscuration, however, they overestimated the distances of the stars, and the error increased with increasing actual distances (and hence increasing obscuration). If stars are placed too far away, they are spread out too thinly in space; in other words, interstellar absorption of light produces an *apparent* thinning out of stars with distance. The early investigators arrived at the erroneous conclusion that the Galaxy was centered on the sun, and thinned out (that is, its "edge" was reached) at a distance of only about 10,000 LY. In actual fact, we do not even see (in visible light) as far as the Galaxy's brilliant central nucleus. Were it not for the obscuring dust in space, we would be able to read at night by the light of the Milky Way.

Turning the problem around, we find that if we try to determine the intrinsic luminosity (or absolute magnitude) of a star of known distance and observed apparent magnitude, we underestimate the luminosity. Section 25.2f described how just such an error was made in the determination of the absolute magnitudes of cepheid variables whose distances were found from statistical studies of their motions.

The presence of the general interstellar obscuration can be demonstrated in several simple ways. We now have independent knowledge that the Galaxy does not thin out a few thousand parsecs from the sun (Chapter 27). The numbers of stars counted in any given field (that is, any given region of the sky) would be expected to increase, therefore, as the limiting brightness to which stars are counted is decreased (see Exercise 2). As we count to fainter and fainter magnitudes in any direction along the Milky Way, we count more and more stars, but their numbers do not increase as rapidly as we would expect for a more or less uniform stellar distribution because of the dimming of the stars by the dust. It is possible, in fact, to estimate the average absorbing power of the interstellar material by making such star counts in directions at different angles to the plane of the Galaxy.

Interstellar obscuration also shows up in the apparent distribution of external galaxies. Because of the disklike shape of our Galaxy, we would expect to encounter more absorption in the direction of the Milky Way (in the plane of the system) than at right angles to it (out either face of the disk). Throughout a region near the Milky Way, the absorption is so heavy that practically no external galaxies show through. Hubble called this region the "zone of avoidance." However, more

and more galaxies can be observed as one turns away from the Milky Way, the maximum occurring at roughly 90° from the plane of the Galaxy (see Figure 32-6). Light from the latter direction is probably dimmed by less than 30 percent. However, the varying numbers of galaxies in various directions also indicate that there is a spotty, irregular distribution of the absorbing material.

(c) Interstellar Reddening

It is a fortunate circumstance for observational astronomy that the interstellar obscuration is *selective;* that is, light of short wavelengths is obscured more readily than that of long wavelengths. Fifty years ago, astronomers were puzzled by the existence of stars whose spectra indicate that they are intrinsically hot and blue, of spectral type B, although they actually appear as red as cool stars of spectral type G. We know today that the light from these stars has been reddened by the interstellar absorbing material; most of their violet, blue, and green light has been obscured, leaving a greater percentage of their orange and red light, of longer wavelengths, which penetrates through the obscuring dust. This *reddening* of starlight by interstellar dust not only shows that the stars are dimmed, but also provides a means of estimating the amount of obscuration they have suffered.

The manner in which the absorption depends on wavelength can be evaluated by comparing, at various wavelengths, the relative brightnesses of two appropriate stars. Stars are chosen whose spectra show them to be approximately identical. One, however, is dimmed and reddened by interstellar dust, while the other is not, being in a direction in the sky that is relatively free of interstellar obscuration. As an illustration, suppose the nearer of the two stars, in the absence of obscuration, would be brighter than the other by ½ magnitude at all wavelengths. If this nearer star were dimmed by dust, then, at a wavelength of 10,000 Å it might be, say, only 0.2 magnitude brighter than the unobscured star; at 5000 Å it would appear about ½ magnitude fainter, and at 3300 Å it would be about 1 magnitude fainter than the more distant, unobscured star. From a study of several pairs of such obscured and unobscured stars, the absorbing power of the interstellar material at various wavelengths has been determined. Over the visible spectral region, it turns out that the absorption, expressed in magnitudes, is roughly inversely proportional to wavelength. In other words, the interstellar dust is twice as effective at obscuring starlight of 5000-Å wavelength as it is at obscuring starlight of 10,000-Å wavelength.

The law of interstellar obscuration has been found to be usually about the same in different directions in the Galaxy. Thus it is possible to estimate the total amount by which a star is dimmed from the amount that it is reddened. The reddening of the light from a star increases its apparent color index (the redder the star, the greater the color index — see Section 20.3b). The difference between the *observed* color index and the color index that the star *would have* in the absence of obscuration and reddening is called the *color excess*. The $m_{pg} - m_{pv}$ color excess, for example, is the amount by which the difference between the photographic and visual magnitudes of a star is increased by reddening. In most directions in the Galaxy, the total absorption, in visual magnitudes, is found empirically to be about three times the $m_{pg} - m_{pv}$ color excess.

The calculation of the distance to a star that is dimmed by obscuration may be illustrated with a numerical example.† Suppose a spectral type G8

---

†The distance modulus of a star dimmed by interstellar obscuration is given by

$$m_v - M_v = 5 \log \frac{r}{10} + 3CE,$$

where m_v and M_v are its apparent and absolute photovisual magnitudes, r its distance in parsecs, and CE its photographic minus photovisual color excess.

star is observed to have an apparent photographic magnitude of 14.6 and an apparent photovisual magnitude of 13.0. Its *observed* color index, therefore, is +1.6. Now it is known that this type of star has an absolute photovisual magnitude of +5.0, an absolute photographic magnitude of +5.6, and therefore, an intrinsic color index of +0.6. Its color excess is 1.6 − 0.6 = 1.0, and its obscuration, in photovisual magnitudes, is about three times its color excess, or about 3.0 magnitudes. In the absence of obscuration, therefore, we estimate that its apparent photovisual magnitude would be 13.0 − 3.0 = 10.0, just 5 magnitudes fainter than its absolute magnitude (its distance modulus is +5 — see Section 20.2b). In the absence of interstellar dimming, in other words, the star would appear 5 magnitudes, or 100 times, fainter than it would appear at a distance of 10 pc; we estimate the true distance of the star, therefore, at 100 pc.

(d) Reflection Nebulae

Until now, the term "absorption" has been used loosely in this discussion. The tiny interstellar grains actually absorb some of the starlight they intercept, but most of it they merely scatter — that is, they redirect it in helter skelter directions. Since the starlight that is scattered, as well as that which is truly absorbed, does not reach us, the effect is the same as if the loss were all due to actual absorption. The whole process is more correctly termed *interstellar extinction*. The scattered or reflected light, on the other hand, illuminates the dust itself. Consequently, even the darkest dark nebulae are not completely dark but are illuminated by a faint glow of scattered starlight that can actually be measured. It is estimated that about one third of the light of the Milky Way is diffused starlight, scattered by interstellar dust.

The light scattered by a particularly dense cloud of dust around a luminous star may be bright enough to be seen or photographed telescopically. Such a cloud of dust, illuminated by starlight, is called a *reflection nebula*. One of the

FIG. 26-2 Reflection nebula about the star Merope in the Pleiades. *(Mount Wilson and Palomar Observatories.)*

best known examples is the nebulosity around each of the brightest stars in the Pleiades cluster.

Blue light is scattered more than red by the dust. A reflection nebula, therefore, always appears bluer than its illuminating star. A reflection nebula could be red only if the star which is the source of its light were very red. However, it takes a bright star to illuminate a dust cloud sufficiently for it to be visible to us, and in the regions of the Galaxy where dust clouds are found, the brightest stars are usually blue main-sequence stars. Most of the reflection nebulae that are conspicuous, therefore, are blue since they are illuminated by blue stars. Sometimes, of course, an intrinsically blue star illuminating a reflection nebula may appear much redder than it actually is because of interstellar absorption and reddening.

From the measured brightness and size of a reflection nebula we can calculate how much light it actually reflects. We find that in general this quantity is almost identical with the amount of starlight intercepted by the dust. Thus, practically all the obscured starlight is scattered (or reflected) by the dust and is not truly absorbed. The interstellar dust, in other words, has a very high reflecting power or *albedo* — about like that of snow. The small fraction of the light that is truly absorbed by the dust eventually may be reradiated in longer wavelengths or may be converted into energy of motion of the particles.

(e) Interstellar Polarization

Molecules of gas scatter light and polarize it at the same time. The light of the blue daylight sky, for example, is highly polarized. The dust particles of interstellar space also polarize light, but not so much. The light from reflection nebulae is observed to be polarized by as much as 20 percent, which means that the brightness of a reflection nebula as observed through a polaroid filter varies by that percentage as the filter is rotated through various angles. The light of stars dimmed by interstellar dust is also slightly polarized. Evidently, the dust grains preferentially scatter light in a particular plane of vibration.

(f) Nature of the Interstellar Grains

The preceding paragraphs have described various phenomena associated with the interstellar dust. These phenomena reveal something of the nature of the solid grains themselves.

First, it is found that the absorption of light is accomplished by *solid particles,* and not by interstellar gas. Atomic or molecular gas is almost transparent. Consider the earth's atmosphere; despite its incredibly high density, as compared to interstellar gas, it is so transparent as to be practically invisible. The absorbing power of the interstellar medium exceeds that of an equal mass of gas by more than 100,000 times. The quantity of gas that would be required to produce the observed absorption in space would have to be many thousands of times the amount that can possibly exist. The gravitational attraction of so great a mass of gas would produce effects upon the motions of stars that would be easily detected; such effects, however, are not observed.

Moreover, molecules or atoms (of a gas) scatter light quite differently from the interstellar material. Scattering by gas molecules, as happens in the earth's atmosphere, is called *Rayleigh scattering.* Rayleigh scattering discriminates very strongly among colors, blue light being scattered very efficiently, as we see from the brilliant blue of the daytime sky; the scattering efficiency is inversely proportional to the fourth power of the wavelength. Interstellar particles, too, scatter selectively, but not so strongly as the earth's atmosphere. Interstellar extinction, as we have seen, is approximately inversely proportional to the first power of the wavelength. Light scattered by molecules is also far more strongly polarized than that scattered by interstellar dust.

Whereas gas can contribute only negligibly to absorption of light, we know from our everyday experience that tiny particles can be very efficient absorbers. Water vapor in the gaseous state in the air is quite invisible. When some of that vapor condenses into tiny water droplets, however, the resulting cloud is opaque. Dust storms, smoke, and smog furnish other familiar examples of the opacity of solid particles.

Calculations confirm that widely scattered particles in interstellar space must be responsible for the observed dimming of starlight. The next problem is to learn the size and composition of particles that can produce just the observed scattering, absorption, and polarization. The physical theory that deals with the scattering of light by solid particles is very complicated; here we can only summarize the results of calculations that have been performed.

It is found that metallic grains, for example, iron, could produce the observed effects only if the grains were all close to a critical size. It seems more likely, however, that a variety of sizes of particles exists. Moreover, since hydrogen has been found to be by far the most abundant chemical element in the universe, one might expect that the interstellar particles are composed mostly of frozen compounds of hydrogen. Calculations show that such particles can exist in interstellar space. It is, therefore, quite probable that the interstellar "dust" is a mixture of different-sized frozen particles of nonmetallic substances, those most responsible for the spatial extinction having diameters in the range 0.0002 to 0.0005 mm.

Finally, the interstellar polarization can be explained only if the particles are presumed to be elongated in shape and at least partially aligned with each other. The mechanism by which the elongated particles become aligned is not understood in detail but it is undoubtedly associated with the presence of interstellar magnetic fields (see also Chapter 27).

26.2 INTERSTELLAR GAS

Although interstellar gas is possibly 100 hundred times as dense, on the average, as the dust, because of its high transparency it is not visible by reflected starlight, nor does it contribute to the general interstellar absorption. It does manifest itself, however, in several other ways. Because of the process of *fluorescence,* clouds of gas near hot stars often shine brightly. The gas also produces narrow absorption lines superposed upon the spectra of stars whose light passes through the gas. Finally, the gas is responsible for the emission of radio waves over a broad range of wavelengths and for certain emission and absorption lines at radio wavelengths, including the important line at 21 cm.

(a) General Gas in Space

Interstellar gas is distributed generally throughout the regions of the spiral arms of the galaxy. Hydrogen comprises about three quarters of the gas, and hydrogen and helium together comprise from 96 to 99 percent of it (by mass). Most of the gas is cold and nonluminous. Near very hot stars, however, it is ionized by the ultraviolet radiation from those stars. Since hydrogen is the main constituent of the gas, we often characterize a region of interstellar space according to whether its hydrogen is neutral — an "H I region" — or ionized — an "H II region."

The gas in the H II regions glows by the process of fluorescence. The light emitted from these regions of ionized gas consists largely of emission lines, so they are also called *emission nebulae*. Those emission nebulae in which the gas happens to be much denser than average (it occasionally reaches densities of 10^3 or 10^4 atoms per cubic centimeter — still an extremely high vacuum on earth) are especially conspicuous. The best-known example is the Orion nebula, which is barely visible to the unaided eye, but easily seen with binoculars, in the middle of the sword of the hunter. Other famous emission nebulae are the North America nebula in Cygnus and the Lagoon nebula in Sagittarius. (These three nebulae are shown in accompanying figures.)

The cold gases in H I regions are invisible on direct photographs, but they can be observed by their radio emission (Sections 26.2h and 26.2i). Also, some of the gases in H I regions produce absorption lines on stellar spectra (Section 26.2g).

(b) Fluorescence in H II Regions

All ultraviolet radiation of wavelength 912 Å or less can be absorbed by neutral hydrogen, and in the process the hydrogen is ionized (Section 10.5). An appreciable fraction of the energy emitted by the hottest stars lies at wavelengths shorter than 912 Å. If such a star is embedded in a cloud of interstellar gas, the ultraviolet radiation from that star ionizes the hydrogen in the gas, converting it into positive hydrogen ions (protons) and free electrons. These detached protons and electrons are then a part of the gas, each of them acting like an individual molecule, that is, a free particle. Protons in the gas are continually colliding with electrons and capturing them, becoming neutral hydrogen again. As the electrons cascade down through the various energy levels of the hydrogen atoms on their way to the ground states, they emit light in the form of emission lines. Lines belonging to all the series of hydrogen (Section 10.5b) are emitted — the Lyman series, Balmer series, Paschen series, and so on — but only the lines of the Balmer series are easily observed from the surface of the earth because of the opacity of our atmosphere to most wavelengths. Part of the invisible ultraviolet light from the star is thus transformed into visible light in the Balmer emission lines of hydrogen. After an atom has captured an electron and emitted light, it loses that electron again almost immediately by the subsequent absorption of another ultraviolet photon from the star. Thus, although neutral hydrogen absorbs and emits light in H II regions, almost all the hydrogen, at any given time, is in the ionized state.

The interstellar gas, of course, contains other elements besides hydrogen. Many of them are also ionized in the vicinity of hot stars and are capturing electrons and emitting light, just as the hydrogen does. Of these, only helium is abundant enough to contribute an appreciable amount of light to an emission nebula by the process of electron capture that we have described. On the other hand, there are a few other mechanisms by which some of the less abundant elements do contribute substantial light to the H II regions. These processes are described in the next subsections.

(c) The Bowen Mechanism*

Certain blue and violet emission lines due to doubly ionized atoms of oxygen and nitrogen are so bright as to indicate a relatively high abundance of these elements, while their other lines in the nebular spectra are weak and indicate low abundances. An explanation for the anomalously high strength of those oxygen nitrogen lines has been given by the American physicist I. S. Bowen. A strong emission line of ionized helium consists of photons that by coincidence happen to be of just the right energy (304 Å) to raise doubly ionized oxygen atoms to one of their excited levels. The large number of photons of 304 Å emitted by helium excites a large percentage of the doubly ionized oxygen atoms; as the latter de-excite themselves by dropping through various lower energy levels to the ground states of those ions, strong emission lines are produced. Some of these are the observed lines in the blue and violet; one ultraviolet line of wavelength 374 Å is of special interest.

Radiation of 374 Å happens to have just the right energy to excite atoms of doubly ionized nitrogen to one of their excited levels. Thus, a much greater number of ionized nitrogen atoms are excited than would otherwise be expected. As these ions drop back to their ground states, they emit the other observed lines in the blue and violet region of the spectrum. The surprising strength of the lines of ionized oxygen and nitrogen are, therefore, fully explained by this double coincidence.

(d) Forbidden Radiation*

It was explained in Section 10.5d that an atom or ion can be excited in either of two ways, by absorbing radiation or by collision with another particle. Atoms of singly ionized nitrogen and singly and doubly ionized oxygen all contain energy levels that correspond to low energies above their ground states. The ions are easily excited to these "low-energy" levels by collisions with free electrons in the H II regions (most of these electrons have been freed from hydrogen atoms by ionization). Ordinarily, observed emission lines originate from atoms or ions that have

FIG. 26-3   The Lagoon Nebula in Sagittarius, photographed in red light with the 200-inch telescope. *(Mount Wilson and Palomar Observatories.)*

remained excited for only a very brief period — of the order of a hundred millionth to a ten millionth of a second — before becoming de-excited with the emission of radiation. The levels to which oxygen and nitrogen ions are excited by collision, however, are said to be metastable levels, because the ions will normally remain in them for periods of hours before radiating energy and dropping to their ground states.

When the gas is at moderate or high pressure, as in stars or in laboratory experiments, ions that have been collisionally excited to these levels are normally de-excited by subsequent collisions long before the radiation process can occur. Consequently, emission lines corresponding to such atomic transitions are exceedingly weak compared to other lines and are not usually observed in the laboratory; they are known as *forbidden lines.*

Collisions between particles are infrequent in the rarefied gases of a nebula, but at the temperatures of H II regions many of the free electrons have just the right kinetic energy that when they do collide with ions of oxygen and nitrogen they often excite those ions to their low-lying metastable levels. Although transitions from these levels are slow to occur, so many ions are excited to them at any time that many such "forbidden" transitions occur in an H II region. Now the gas is transparent to visible light, so the photons emitted through the entire depth of an H II region contribute to visible emission lines; indeed, the forbidden radiation often comprises half or more of the observable light from H II regions. We might say that the cards are stacked in favor of it. The most important forbidden lines in the spectra of emission nebula are two green lines (5007 and 4959 Å) due to doubly ionized oxygen. Other important forbidden lines are two ultraviolet lines (near 3727 Å) due to singly ionized oxygen, two red lines (6584 and 6548 Å) due to singly ionized nitrogen and two ultraviolet lines (3867 and 3968 Å) due to doubly ionized neon.

When the green forbidden oxygen lines were first observed in the spectra of emission nebulae, their origin was a mystery. For a time, they were ascribed to an unknown element, *nebulium*, named for its apparent prevalence in gaseous nebulae. The correct explanation of the "nebulium lines" was provided by I. S. Bowen in 1927.

(e) *Size and Brightness of an H II Region**

If a cloud of gas surrounding a hot star is not very extensive, some of the ultraviolet radiation emitted by the star that is capable of ionizing hydrogen may leak out through the gas, and the apparent boundary of the emission nebula will be the actual edge of the gas cloud. Such a nebula is said to be *optically thin.* Usually, however, an H II region is *optically thick*, which means that all of the star's ultraviolet radiation (of wavelengths less than 912 Å) is absorbed within the gas. In this case, the boundary of the H II region is merely the limiting distance through the gas to which the star's ultraviolet radiation penetrates.

Near the star, the hydrogen is virtually completely ionized. Since the hydrogen ions have no electrons, they cannot absorb photons and thus become further ionized or excited. A few hydrogen atoms, to be sure, will have just captured electrons and become temporarily neutral (or we would not see emission lines of hydrogen in the spectra of nebulae), but are almost immediately re-ionized by the absorption of ultraviolet photons. At any given time, these neutral atoms are so few that the amount of ultraviolet radiation they absorb is relatively negligible. In the H II region, therefore, the gas is practically trans-

parent to all ultraviolet radiation that can ionize hydrogen.

Radiation thins out, however, as the square of the distance from its source (Section 10.1a). Eventually a distance is reached where the ultraviolet radiation from the central exciting star is so feeble that ions of hydrogen recapture electrons just as quickly as neutral hydrogen atoms are being ionized by photons. Beyond this distance the star's radiation can no longer keep all of the gas ionized, and the many neutral atoms absorb all of its ultraviolet energy; the outer boundary of the emission nebula is thus determined. Detailed calculations show that the transition from completely ionized hydrogen (the H II region) to completely neutral hydrogen (the H I region) occurs over a relatively short distance; the H II region, in other words, is rather sharply bounded.

If the interstellar gas were distributed with absolute uniformity, every emission nebula would be a spherical H II region exactly centered on a hot star. Because the distribution of the gas is patchy and irregular, actual H II regions are only approximately spherical, with irregular boundaries corresponding to the irregularities in the gas density; where the gas is denser, the ionizing radiation is consumed, and the H II region ends, closer to the star. Further irregularities may result when two or more stars are responsible for the radiation, and their H II regions overlap. The theory of H I and H II regions was worked out in detail by the astronomer B. Strömgren. The more or less spherical emission regions are, therefore, sometimes called *Strömgren spheres.*

The linear size of an H II region depends on two things: (1) the "ultraviolet" luminosity of the central star, that is, how much energy it emits per second in wavelengths less than 912 Å; and (2) the density of the gas. The higher the density, the more hydrogen per unit volume there is to ionize, and the shorter is the distance through the gas that the ultraviolet energy can penetrate before it is completely absorbed. If the gas density is very low, and if the star is very hot and luminous, the H II region can be very large; if the density is one atom per cubic centimeter, a main-sequence spectral-type O6 star can ionize a region more than 100 pc in diameter. Main-sequence stars of types B0 and A0 would produce in the same gas H II regions having diameters of 40 and 1 pc, respectively.

The amount of light emitted by a unit volume of gas in an H II region depends on the rate at which the ionized

The Trifid nebula, M20, in
Sagittarius. The entire region is
immersed in gas and dust. The
blue region on the left is a reflection
nebula. The region on the right
is an emission nebula; 200-inch
telescope photograph. *(Mount Wilson
and Palomar Observatories.)*

ega nebula Sagittarius; 48-inch Schmidt telescope photograph.
unt Wilson and Palomar Observatories.)

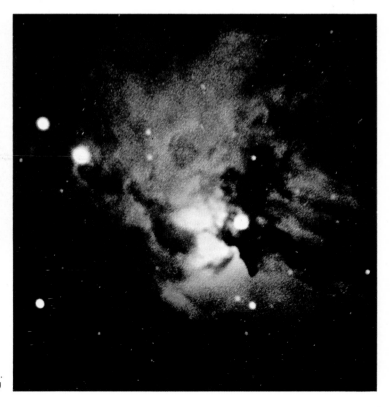

Central region of the Lagoon nebula (M8).
(Lick Observatory.)

Rosette nebula in Monoceros, NGC 2237; photographed by the 48-inch Schmidt telescope. *(Mount Wilson and Palomar Observatories.)*

Orion nebula.
(Lick Observatory.)

al portion of Orion nebula. *(Lick Observatory.)*

The North American nebula; photographed by the
48-inch Schmidt telescope. *(Mount Wilson
and Palomar Observatories.)*

The Horsehead nebula in Orion; 48-inch Schmidt telescope photograph. *(Mount Wilson and Palomar Observatories.)*

The 300-foot Radio Telescope. *(National Radio Astronomy Observatory.)*

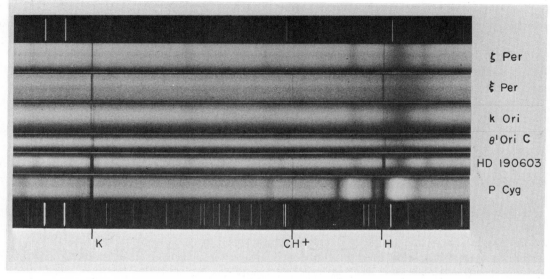

FIG. 26-4 Interstellar H and K lines of ionized calcium and a line of the CH-radical showing the spectra of several stars. *(Lick Observatory.)*

atoms capture electrons, which depends in turn on the gas density; the higher the density, the greater the rate of recombinations of ions and electrons, and the more light that is emitted. Since visible light travels freely through interstellar gas, the observed surface brightness of a glowing H II region is proportional both to the amount of light emitted per unit volume within the region, and to the length of the column along the observer's line of sight intercepted by the region. The surface brightness of an H II region, therefore, depends on both the density of the gas within the region and its size.

(f) Intermixture of Interstellar Gas and Dust

Gas and dust are generally intermixed in space, although the proportion of one to the other is not everywhere exactly the same. The presence of dust is apparent on many photographs of emission nebulae. Clouds of dark material can be seen silhouetted on the Orion nebula, actually hiding a large part of the H II region from our view. Foreground dust clouds produce the lagoon in the

Lagoon nebula, and the Atlantic Ocean and Atlantic coastline in the North America nebula. Although the dust is most conspicuous when it is in front of an emission nebula and is silhouetted against it, the dust is also intermixed with the gas. Spectra of H II regions often reveal the faint continuous spectrum (with absorption lines) of the central star, whose light is reflected to us by the dust associated with the gas. In other words, emission nebulae are generally superimposed upon *reflection nebulae*.

Both the emission component (due to the gas) and the reflection component (due to the dust) are brighter the farther the central star is up the main sequence. The brightness of an emission nebula, however, is far more sensitive to the kind of central, exciting star than is that of a reflection nebula. Stars cooler than about 25,000°K have so little ultraviolet radiation of wavelengths shorter than 912 Å (that is, which can ionize hydrogen) that the reflection nebulae around such stars outshine the emission nebulae. The dust around a star with a surface temperature of only 10,000°K scatters more than 5000 times as much visible light as is emitted by the gas around the same star, although even the reflection component would probably not be conspicuous unless the star were a supergiant. Stars hotter than 25,000°K emit enough ultraviolet energy so that the emission nebulae produced around them generally outshine the reflection nebulae. The dust around a star of 50,000°K reflects less than one tenth the amount of light emitted by the gas that is probably also present.

(g) Interstellar Absorption Lines

The cold interstellar gas — that in the H I regions — is not visible by reflected or emitted light, nor does it appreciably dim the light of stars shining through it. Yet it often reveals its presence by leaving dark absorption lines superposed upon the spectra of stars that lie beyond it. There are several ways of knowing that the interstellar lines seen in the spectrum of a star do not originate in the star itself. Since the interstellar gas is cold, most of its atoms are neutral and are practically all in the state of lowest energy; the lines they produce, therefore, are generally not the same as the ones that are produced by the atoms of the hot gases in stellar photospheres. Moreover, the lines of the cold interstellar gases are very sharp, while those formed in stellar photospheres show the characteristic broadening associated with the spectra of hot gases at relatively high pressure (Section 21.2c). Finally, an interstellar gas cloud does not, in general, move with the same radial velocity as the star whose spectrum is observed, and the Doppler shifts of the interstellar lines are thus different from that of the stellar lines.

Interstellar lines have been found of most of those elements for which observable lines would be expected. The most conspicuous interstellar lines are produced by sodium and calcium. Lines are also observed of titanium, iron, and potassium, as well as bands of the radicals CN and CH+. The strengths of interstellar lines lead to estimates of the relative abundances of the elements that produce them. Such estimates do not differ markedly from the relative abundances of the same elements in stellar photospheres.

Sometimes the strength of an interstellar line seen in the spectrum of a star provides an indication of the distance to that star. A very strong interstellar line, for example, indicates that the starlight has traversed a considerable amount of interstellar gas. Because the density of gas in space is very low, the starlight would have to travel a long distance to encounter that much material; that is, the star would have to be very far away. Sometimes the interstellar lines are double or even multiple, which indicates that the starlight has traversed two or more gas clouds, moving with respect to each other, whose different radial velocities produce different Doppler shifts.

(h) Radio Emission From Interstellar Gas

We receive from the Milky Way a large amount of radio energy in the frequency range 10 to 300 Mc (or wavelength range 1 to 30 m). This emission from the Milky Way was, in fact, the first radio radiation of astronomical origin to be observed (Section 11.8a). Similar radiation at radio frequencies is received from nearby galaxies. The radio energy from the Milky Way does not appear to originate in stars. The sun, to be sure, is an apparently strong source of radio waves, but the sun is very close to us. If its radio emission is typical of that of the other stars, all the stars in the Galaxy would not emit a billionth of the radio energy actually observed. It is concluded, therefore, that the radio energy originates in the interstellar gas.

Many strong sources of radio energy have been identified with individual gaseous nebulae. One example is the Crab nebula (Section 25.3c); another is a group of faint gaseous filaments in Cassiopeia (known as the Cassiopeia A source).

FIG. 26-5   Orion Nebula. *(Lick Observatory.)*

The spectra of the different filaments in the Cassiopeia A source show Doppler shifts that indicate that they are moving with respect to one another with speeds in excess of 4000 km/sec. Several other discrete sources of radio radiation have been identified with supernovae remnants (in addition to the Crab nebula), and a few have been identified with the more conspicuous nebulae (such as the Orion nebula). The majority of discrete radio sources scattered along the Milky Way have not yet been associated with objects visible optically. Superposed upon them all is the general background of radio radiation from those regions where interstellar material prevails.

(Thousands of discrete radio sources have been catalogued that are in directions in the sky away from the Milky Way. Most of these are extragalactic sources, and will be discussed in more detail in Chapters 32 and 33.)

One cause of galactic radio emission is free-free transitions (bremsstrahlung — Section 10.5e) in H II regions. An electron, passing near an ionized hydrogen atom (proton), can change from its original hyperbolic orbit to another hyperbola, emitting or absorbing in the process a photon of a definite wavelength. The many photons of different wavelengths arising from such free-free transitions in the interstellar gas comprise continuous radiation, a good part of which lies in the radio part of the spectrum. On the other hand, some galactic radio sources are apparently due to synchrotron radiation (Section 25.3c), which results when electrons spiral around magnetic lines of force with speeds near that of light. In particular, the radio sources associated with old supernovae (such as the Crab nebula) appear to be produced by the synchrotron mechanism.

(i) Radio Lines; The 21-cm Line of Hydrogen

A number of emission (and absorption) lines at radio wavelengths can now be observed. The first radio line to be detected is the 21-cm line of neutral hydrogen.

Some of the most important radio observations carried out in the interstellar medium are of the spectral line of hydrogen at the radio wavelength of 21.11 cm. A hydrogen atom possesses a tiny amount of angular momentum by virtue of the axial spin of its electron and the electron's orbital motion about the nucleus (proton). In addition, the proton has an axial spin of its own, and the angular momentum associated with this spin may either add to or subtract from that of the electron, depending on the direction of spin of the nucleus, with respect to that of the electron. If the spins of the two particles oppose each other, the atom as a whole has a very slightly lower energy than if the two spins are aligned. Ordinarily, an atom of hydrogen is in the state of lower energy. If the requisite minute amount of energy is imparted to the atom, however, the spins of the proton and electron can be aligned, leaving the atom in a slightly *excited state*. If it loses that same amount of energy again, the atom returns to its ground state. The amount of energy involved is that associated with a photon of 21-cm wavelength. An atomic transition of this type, which involves the spin of the nucleus of an atom, is called a *hyperfine transition*.

Neutral hydrogen atoms in H I regions can be excited to this 21-cm level by collisions with electrons and other atoms. Such collisions are extremely rare in the sparse gases of interstellar space; a typical atom may have to wait millions of years before such an encounter reverses its nuclear spin, that is, aligns it with the electron spin. Nevertheless, over many millions of years a good fraction of the hydrogen atoms are so excited. An excited atom can then lose its excess energy either by a subsequent collision or by radiating a photon of 21-cm wavelength. It happens, however, that hydrogen atoms are extremely loath to do the latter; an excited atom will wait, on the average, about 10 million years before emitting a photon and returning to its state of lowest energy (it is a highly forbidden transition — see Section 26.2d). On the other hand, collisions between particles in the interstellar vacuum are rare, too, and there is a definite chance that the atom will radiate before a second collision carries away its energy of excitation. In 1944 the Dutch astronomer H. C. van de Hulst predicted that enough atoms of interstellar hydrogen would be radiating photons of 21-cm wavelength to make this radio emission line observable. Equipment sensitive enough to detect the line was not available until 1951. Since that time, "21-cm astronomy" has been a very active field of astronomical research.

Observations at 21 cm show that the neutral hydrogen in the Galaxy is confined to an extremely flat layer, most of it being in a sheet less than 100 pc thick, extending throughout the plane of the Milky Way. The strength of the line indicates that neutral interstellar hydrogen makes up from 1 to 2 percent of the mass of the Galaxy. One important use of 21-cm radiation is the measurement of its Doppler shifts in various directions, which helps us to map out the spiral structure of our galactic system (Chapter 27).

Since the discovery of the 21-cm line, other radio spectral lines have been observed. Especially interesting is a multiplet of 4 closely spaced lines due to the OH radical — a bond of one oxygen and one hydrogen atom. Emission lines of OH are at times associated with H II regions containing dust, and may bear some relation to the dissociation of solid grains in hot H II regions. Also, very high level hydrogen emission lines are now observed. The various series of hydrogen lines are described in Section 10.5b. Transitions from energy level 110 to 109 of the hydrogen atom, for example, give rise to a radio line of 6-cm wavelength; those from level 157 to 156 produce a line at 17 cm.

(j) Temperatures of H I and H II Regions*

When we speak of the temperature of a gas, we usually mean its *kinetic temperature*, which is a measure of the energy with which typical particles in the gas are moving. To understand the temperature of the interstellar medium,

we must first consider the processes that heat the gas and then those that cool it. Finally, we calculate the equilibrium temperature that exists when these two processes — the heating and the cooling of the gas — exactly balance each other.

In H II regions the heating is principally by ionization of hydrogen. The neutral hydrogen atoms, in becoming ionized, can absorb *any* photon of wavelength shorter than 912 Å. The energy absorbed by the atom, in excess of that required for its ionization, is converted to kinetic energy of the freed electron. By the process of collisions, the electrons gradually share their energy with the other particles of an H II region. Excess energy absorbed in the process of ionization of hydrogen, therefore, is slowly converted into heat in the gas.

Electrons can lose energy by collisions with ions of oxygen and nitrogen, because these ions can be excited by such collisions and can eventually radiate the energy away in the form of "forbidden" emission lines (Section 26.2d). This mechanism, in other words, takes energy *from* the gas, and therefore *cools* it. Calculations show that the heating (by ionization) and cooling (by the emission of forbidden lines) should balance each other at an equilibrium temperature somewhere in the range of 7000° to 20,000°K. If the temperature should drop much below 10,000°K, the rate of collisional excitations of oxygen and nitrogen would decrease (because the electrons would be moving more slowly) and the gas would heat up. If the temperature rose much above 20,000°K, the collisional excitations would become so numerous that the gas would cool rapidly. The heating and cooling mechanisms, therefore, act like a thermostat, and keep the gas in the H II region at a relatively even temperature.

We may measure the temperature of H II regions by comparing the intensities of two or more lines originating from different collisionally excited levels of the same kind of atom or ion. Measures of intensities of the forbidden lines of doubly ionized oxygen (the "nebulium" lines) are especially useful for this purpose. Temperatures of H II regions can also be estimated from the intensities of their radio emission at different wavelengths. Most such measurements suggest that H II regions have temperatures between 8000° and 10,000°K.

In the H I regions heating can occur by the ionization of certain heavier atoms (such as carbon and silicon) that can be ionized by the lower-energy photons of visible

FIG. 26-6   NGC 6611; Nebula in Scutum Sobiesky. Photographed in red light with the 200-inch telescope. *(Mount Wilson and Palomar Observatories.)*

starlight. Collisional excitation of carbon and other atoms is an efficient cooling process and would be expected to keep the equilibrium temperature down near 20°K. The strength of the 21-cm line of neutral hydrogen, however, indicates a temperature of the gas in H I regions near 125°K, so there is apparently another, unknown, mechanism for heating the cool gas. It seems possible that the heating might be provided by cosmic rays (Chapter 31).

In any case we can conclude safely that H II regions are hot, with temperatures of the order of 10,000°K, and that H I regions are cold, with temperatures of the order of 100°K.

(k) *Expansion of H II Regions**

It will be seen in Chapter 30 that the very hot, luminous stars — the type that are generally responsible for producing H II regions — are relatively short-lived. H II regions, therefore, are temporary phenomena; most existing emission nebulae have been formed within the past few million years. Consider, now, what happens when a very hot, luminous star is first formed, and rather suddenly ionizes the gas surrounding it in space, thereby producing an emission nebula. The ionized gas, although initially of the same density as the un-ionized gas in the adjacent H I region, is about 100 times hotter than the cool gas. Since the pressure in a gas is proportional to the product of its density and temperature (see Chapter 29), the H II region has a pressure that is also about 100 times greater than that of the H I region. Consequently, the H II region begins to expand, decreasing its own density, and pushing outward, away from the central star, against the cold gas in the H I region. At the same time, the radiation pressure exerted by the light from the star upon the interstellar dust particles pushes them outward, away from the star, much as sunlight pushes out the tail of a comet.

The expanding front of an emission nebula — the interface between the H I and H II regions — may move at first very rapidly. As a result, there is often a buildup in the density of the gas and dust at the boundary of the nebulae. The bright edges visible in many emission nebulae constitute observational evidence for the higher densities of the gas at the expanding front.

Eventually, the increasing density of the material in the H I region surrounding the hot, expanding emission nebula offers sufficient resistance to slow down the expansion. Sometimes points of higher-than-average density in the H I region break through the expanding front of hot gas and appear as "intrusions" into the emission nebula. Such intrusions generally look like dark cones with bright edges pointing toward the star at the center of the nebula; the emission nebula Messier 16 in Scutum is an excellent

example. The dark intrusions are sometimes referred to as "comet-tail" or "elephant-trunk" structures. They are not, of course, actually moving inward; the hot gas expands *outward*, around them. Dark globules are often found near, or as dissociated extensions of, such elephant-trunk structures.

The fact that stars exist which we have reason to believe are very young (Chapter 29) implies that stars must be continually forming. The only material for them to form from is the interstellar gas and dust. Regions within the interstellar medium, therefore, must be the "birthplaces" of new stars. Places where the densities of the interstellar gas are unusually high may be where stellar condensations begin.

Exercises

1. Identify several dark nebulae on photographs of the Milky Way in this book. Give the figure numbers of the photographs, and specify where on them the dark nebulae are to be found.

2. Suppose all stars had the same absolute magnitude, and that they were distributed uniformly through space. Show that the number of stars from which we receive an amount of light b is inversely proportional to b raised to the $3/2$ power. Now suppose that stars of many absolute magnitudes exist, but that the numbers of stars of various absolute magnitudes exist in the same relative proportions everywhere in space. Is the proportionality between the number of stars appearing brighter than b and $b^{-3/2}$ changed? Explain.

*3. Suppose stars are counted to increasingly fainter limiting brightnesses. Make a sketch showing how the numbers of stars counted will increase as b is decreased, both in the absence and in the presence of general interstellar obscuration.

4. A spectral type A star normally has a color index of 0.0. The photographic and photovisual magnitudes of an A star are observed to be:

$$m_{pg} = 11.6,$$
$$m_{pv} = 10.8.$$

(a) What is the color excess of the star?
(b) What is the total absorption of its light in visual magnitudes?
(c) What is the total absorption of its light in photographic magnitudes?

5. The sun is observed from a distant star to have an apparent photographic magnitude of 12.4 and an apparent photovisual magnitude of 10.8. How far away, approximately, is the star? (See Chapter 20 for the absolute magnitude of the sun.) Assume that the sun's color index is $m_{pg} - m_{pv} = 0.4$.

6. Suppose a bright reflection nebula appeared yellow. What kind of star probably is producing it?

7. The red color of the sun, when seen close to the horizon, and the blue color of the daytime sky provide analogies to the reddening of starlight and the blue color of reflection nebulae. Discuss this analogy more fully, and also explain how it breaks down.

***8.** The amount of light emitted in the hydrogen emission lines of an emission nebula depends on the amount of ionized hydrogen. This in turn depends on the amount of ultraviolet radiation, capable of ionizing hydrogen, that is emitted by the central star. The star's brightness in the visible part of the spectrum can be observed and compared to the amount of light in the hydrogen lines of the nebula. Does this suggest a way by which the temperature of the star might be estimated? Explain. (See Chapters 10 and 22.)

***9.** Explain why "nebulium" lines are not observed in the solar spectrum. Might they be possible in the spectrum of the very rarefied upper atmosphere of the earth?

***10.** Describe in detail the appearance of the spectrum of an emission nebula such as the Orion nebula. Would you expect any continuous spectrum to be present? Explain.

***11.** Explain in detail why emission nebulae are brighter in regions where the gas is denser, and why their brightness depends primarily upon the gas density.

***12.** Suppose you examined the spectrum of some nebulosity surrounding a main-sequence spectral-type O star and found that it contained no emission lines, only the continuous spectrum of the star. What conclusions could you draw about the nature of the interstellar material around that star?

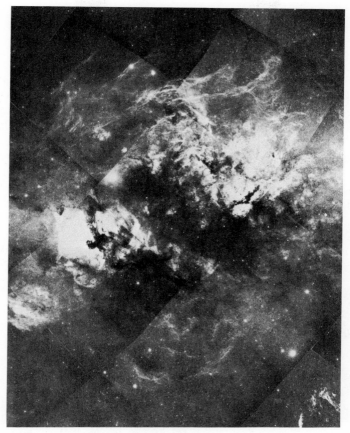

CHAPTER 27

The Galaxy

In 1610 Galileo described his telescopic observations of the Milky Way, which showed it to be a multitude of individual stars. In 1750 Thomas Wright published a speculative explanation, which turned out to be substantially correct — that the sun is part of a disk-shaped system of stars, and that the Milky Way is the light from the surrounding stars that lie more or less in the *plane* of the disk. The disk-shape of the stellar system to which the sun belongs — the *Galaxy* — was demonstrated quantitatively in 1785 by Herschel's "star gauging." We have already described the results of Herschel's investigations (Section 23.3), as well as our modern view of the Galaxy (Figure 27-2).

It was the second decade of the twentieth century before astronomers had deduced, approximately, the true size of the Galaxy and the fact that the sun is located eccentrically in the disk-shaped system. In 1924 the existence of other, similar, stellar systems was proved; the "pinwheel" nature of many of these other galaxies suggested that our own system might also contain spiral arms winding outward from a central nucleus. Only within the last two decades, however, has substan-

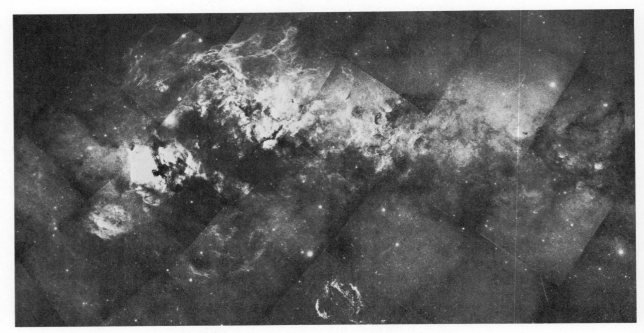

FIG. 27-1   Mosaic of the Milky Way in Cygnus, photographed with the 48-inch Schmidt telescope. *(National Geographic Society-Palomar Observatory Sky Survey.)*

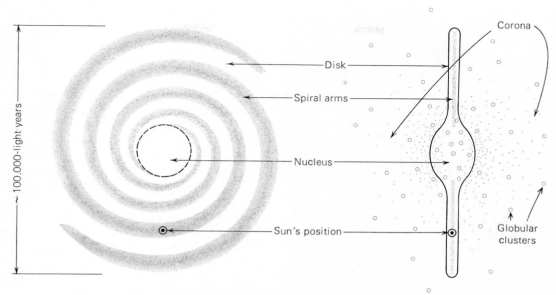

FIG. 27-2   Schematic representation of the Galaxy.

tial progress been made in identifying the spiral arms and in mapping them. This chapter deals with twentieth-century investigations of the structure of our Galaxy.

27.1 SIZE OF THE GALAXY, AND OUR POSITION IN IT

Until early in the twentieth century, the Galaxy was generally believed to be centered approximately at the sun and to extend only a few thousand light years from it. The shift from the "heliocentric" to the "galactocentric" view of our system, as well as the first knowledge of its true size, came about largely through the efforts of Harlow Shapley and his investigation of the distribution of the globular clusters. Globular clusters (Section 23.3c) are great symmetrical star clusters, containing tens of thousands to hundreds of thousands of member stars each. Because of their brilliance, and the fact that they are not confined to the central

plane of the Galaxy where they would otherwise be largely obscured by interstellar dust, they can be observed (with telescopes) to very large distances.

(a) Distribution of the Globular Clusters

Most globular clusters contain at least a few RR Lyrae variable stars (or cluster-type variables—Section 25.2e), whose absolute magnitudes are known to be about 0.[†] The distance to an RR Lyrae star in a globular cluster, and hence to the cluster itself, can therefore be calculated from its observed apparent magnitude (Section 20.2b). Shapley was able to determine the distances to the closer globular clusters which contained RR Lyrae stars that could be observed individually. Distance estimates to globular clusters that do not contain RR Lyrae

---

[†]Recent evidence suggests that the RR Lyrae stars may be fainter than this value by a few tenths of a magnitude. We have not attempted to incorporate this correction, still very uncertain, in the data presented here.

FIG. 27-3   Harlow Shapley. *(Yerkes Observatory.)*

stars, or which are too far away to permit resolution of the variables, were obtained indirectly. Shapley measured angular diameters of globular clusters of known distance, thus obtaining their true diameters. Assuming a statistical average for the true diameter of the clusters, he was then able to obtain distance estimates for the remote ones from their observed angular diameters.

From their directions and derived distances, Shapley, in 1917, mapped out the three-dimensional distribution in space of the 93 globular clusters then known. He found that the clusters formed a spheroidal system with the highest concentration of clusters at its own center. That center was not at the sun, however, but at a point in the middle of the Milky Way in the direction of Sagittarius, and at a distance of some 25,000 to 30,000 LY. Shapley then made the bold — and correct — assumption

that the system of globular clusters represented the "bony frame" of the entire Galaxy; that not only is the distribution of clusters centered upon the center of the Galaxy, but, moreover, that the extent of the galactic system is indicated by the cluster distribution. Today the assumption has been verified, not only by the layout of the spiral structure in our own galaxy, as deduced from the 21-cm radiation from neutral hydrogen, but by the observed distributions of globular clusters in other spiral galaxies. Although the sun lies far from the galactic center, the main disk of the Galaxy probably extends a nearly equal distance beyond the sun and comprises a gigantic system 100,000 LY (more or less) across. The exact size of the disk is not known, however, nor is the exact distance of the sun from its center. Today, the center of the galactic nucleus is believed to be about 10,000 pc from the sun (or about 30,000 LY), with an uncertainty of 2000 pc.

(b) The Galactic Corona

It is to be noted that the main body of the Galaxy is confined to a relatively flat disk, while the globular clusters define a more or less spheroidal system superimposed upon the disk. A sparse "haze" of individual stars — not members of clusters — is now known to exist also in the region outlined by the globular clusters. This haze of stars and clusters forms the galactic *corona*, or "halo," a region whose volume exceeds that of the main disk of the Galaxy by many times. The presence of stars in the corona was first suspected when RR Lyrae stars not belonging to clusters were found lying in such directions and at such distances as to place them far from the galactic plane. These variables, of course, represent only an extremely minute fraction of all coronal stars, but they serve as markers, indicating the distribution of stars in the corona, not only because their distances can be determined easily, but because they have fairly high luminosities and can be seen to relatively large distances. Some of the galactic radio emission also appears to originate in the corona, which suggests that there must be some gas there as well as stars. The corona contains, however, no conspicuous or obvious H II regions.

The spatial density of stars and clusters in the coronal haze of the Galaxy increases toward the Milky Way plane, particularly toward the galactic nucleus. When we look to either side of the Milky Way (that is, slightly above or below its plane) in the directions of Scorpius, Ophiuchus, and Sagittarius, our line of sight skims near the nuclear "bulge" in the middle of the disk of the galaxy. In those directions we find the greatest numbers of globular clusters and stars in the corona. The largest number of RR Lyrae stars seen in these directions have apparent magnitudes near 15, which means that they must be at distances of about 10,000 pc (a slightly smaller distance is found when correction is made for interstellar obscuration).

Individual RR Lyrae stars have been found as far away as 30,000 to 50,000 LY on either side of the galactic plane, which shows that the corona must have an over-all thickness of up to 100,000 LY. A few globular clusters discovered on the Palomar Sky Survey appear to have distances from the sun of more than $\frac{1}{4}$ million light years. We are not sure whether they are true members of the Galaxy or are intergalactic objects. In any event, the possibility remains that the corona extends to very great distances in some directions. The corona is either a spherical or spheroidal system at least 100,000 LY thick, and in the direction of the galactic plane may have a diameter of two or three times this figure, extending far beyond the "rim" of the main disk of the Galaxy. Coronas of some other galaxies have been traced to similar distances.

27.2 REVOLUTION OF THE SUN IN THE GALAXY

Like a gigantic "solar system," the entire Galaxy is rotating. The sun, partaking of the galactic rotation, moves with a speed of 200 to 300 km/sec in a nearly circular path about the nucleus. The method has been described (Chapter 19) by which we determine the motions of stars with respect to the sun (*space motions*), or with respect to the local standard of rest (*peculiar velocities*). Proper motions, however, and thus space motions and peculiar velocities, can be detected only for those stars that are relatively near the sun and that occupy a volume of space which is very small compared to the size of the Galaxy. All the stars with observable space motions, therefore, are in *our part* of the Galaxy (within a few thousand light years), and most of them are moving in galactic orbits that differ only minutely from the sun's. These stars, in other words, are moving along with us about the center of the Galaxy; their space motions, most of which are less than a few tens of miles per second, result from the slight differences between the inclinations, eccentricities, and sizes of their galactic orbits and the sun's. Only by observing far more distant objects can we determine our own true motion in the Galaxy.

(a) The Sun's Galactic Orbit

The motion of the sun in the Galaxy is deduced from the apparent motions of objects surrounding us that do not share in the general galactic rotation. The globular clusters are the most convenient of such objects. These clusters are moving, to be sure, but the fact that they are found in a spheroidal distribution, rather than being confined to the flat plane of the Galaxy, is evidence that the system of globular clusters as a whole is not rotating with the disk of the Galaxy. By analyzing the radial velocities of the globular clusters in various directions, we can determine the motion of the sun with respect to them very much in the same way that we determine the solar motion with respect to the local standard of rest (Section 19.2b). In one direction, the globular clusters, on the average, seem to be approaching us, while in the opposite direction they seem to recede from us.

The motion of the sun in the Galaxy can also be deduced from the radial velocities of the nearby external galaxies. The Large Magellanic Cloud, for example, has a radial velocity of recession of about 170 mi/sec, while the Andromeda galaxy has a velocity of approach of about 190 mi/sec (see Chapter 32). Studies of the radial velocities of other nearby galaxies show that if they are moving at random, their individual velocities with respect to each other and to the center of the Galaxy are relatively small, and that most of the high velocities observed for them are actually due to the orbital motion of the sun about the center of our own galaxy.† We conclude, therefore, that the sun is moving in the general direction away from the Large Magellanic Cloud and toward the Andromeda galaxy.

When the data from various sources are combined, they indicate that the sun is moving in the direction of the constellation Cygnus, with a speed that, although somewhat uncertain, probably lies in the range 200 to 300 km/sec. This direction lies in the Milky Way, and is about 90° from the direction of the galactic center, which shows that the sun's orbit is probably nearly circular and lies, approximately, in the main plane of the Galaxy. As viewed from the north side of the galactic plane, the orbital motion of the sun is in a clockwise direction. The period of the sun's revolution about the nucleus, the *galactic year,* can be found by dividing the circumference of the sun's orbit by its speed; it comes out *roughly* 200 million (2×10^8) of our terrestrial years. We can observe, therefore, only a "snapshot" of the Galaxy in rotation; we do not actually see stars traverse appreciable portions of their orbits.

---

†This statement applies only to the very nearest external galaxies — those that belong to the Local Group; see Chapters 32 and 33.

(b) High- and Low-Velocity Stars

As already mentioned, the majority of the stars near the sun move nearly parallel to the sun's path about the galactic nucleus, and their speeds with respect to the sun are generally less than 40 or 50 km/sec. These are said to be *low-velocity stars.* The radial velocities of nearby gas clouds, as indicated by the Doppler displacements of the interstellar absorption lines and the bright lines of emission nebulae (Section 26.2), are low and show that they, like the sun, move in roughly circular orbits about the galactic nucleus. In other words, the interstellar material in our part of the Galaxy also belongs to the class of low-velocity objects.

Some stars, on the other hand, have speeds relative to the sun in excess of 80 km/sec, and are called *high-velocity stars.* They move along orbits of rather high eccentricity that cross the sun's orbit in the plane of the Galaxy at rather large angles (Figure 27-4). Nearby stars moving on such orbits are passing through the solar neighborhood and are only temporarily near us. Stars in the galactic corona and globular clusters also have orbits very different from the sun's and are high-velocity objects.

It should be noted that the term "high velocity" or "low velocity" refers to the speed of an object *with respect to the sun* and has nothing to do with its motion in the Galaxy. Most high-velocity stars are actually revolving about the Galaxy with speeds less than those of the low-velocity stars near the sun.

The component of velocity of an individual star in a direction perpendicular to the plane of the Galaxy (sometimes called the z component of its velocity) cannot, in general, be determined from its radial velocity alone. However, the average of the z-velocity components for stars of a given class or group can be found from a statistical analysis of their radial velocities. It is learned that low-velocity stars usually have lower velocity com-

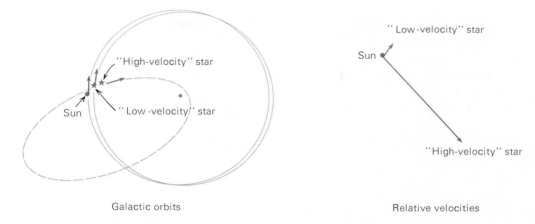

FIG. 27-4   Orbits of high- and low-velocity stars.

ponents perpendicular to the galactic plane than do high-velocity stars. Consequently, high-velocity stars tend to be less strongly concentrated to the plane of the Galaxy than low-velocity stars. Stars in the corona are extreme examples and usually have very high z-velocity components. It is, of course, the high velocities of coronal objects perpendicular to the galactic plane that accounts for their spheroidal distribution in the Galaxy. Globular clusters, in particular, are believed to revolve about the nucleus of the Galaxy in orbits of high eccentricity and inclination to the galactic plane, perhaps rather like the comets revolving about the sun in the solar system. A globular cluster must pass through the plane of the Galaxy twice during each revolution. The large distances between stars, both within the cluster and within the Galaxy itself, make stellar collisions during the cluster's penetration of the galactic disk exceedingly improbable.

The different kinds of motion of stars in the Galaxy may be related to the times and places of formation of those stars; the high-velocity objects are believed to be among the older members of the Galaxy (see Chapter 30).

27.3 THE MASS OF THE GALAXY

The mass of the Galaxy can be calculated from the gravitational influence of the system as a whole upon the sun. For purposes of illustration, a few simplifying assumptions will be made which enable us to estimate the mass with the use of Kepler's third law.

The greatest spatial density of stars occurs in the region of the nucleus of the Galaxy. To a first approximation, therefore, we can assume that it acts, gravitationally, as if its mass were concentrated at its center. Let us assume further that the sun moves on a strictly circular orbit of radius 10,000 pc and that it completes an orbital revolution about the galactic center in 2×10^8 years. Finally, we ignore the contribution of matter farther from the galactic nucleus than the orbit of the sun and assume that all of the Galaxy's mass is interior to the sun's orbit. With these assumptions, we can regard the sun and the Galaxy as a whole as two mutually revolving bodies.

Since there are 2×10^5 AU per parsec (Section 18.4c), the radius of the sun's orbit is 2×10^9 AU. Applying Kepler's third law (as corrected by

Newton) and ignoring the mass of the sun in comparison to that of the Galaxy, we find

$$M_{\text{Galaxy}} = \frac{(2 \times 10^9)^3}{(2 \times 10^8)^2}$$
$$= \frac{8 \times 10^{27}}{4 \times 10^{16}},$$

or that the mass of the Galaxy is 2×10^{11} (200 billion) times that of the sun.

The above calculation is crude because the Galaxy cannot be expected to act, gravitationally, like a point mass. On the other hand, more sophisticated calculations that take into account the actual distribution of matter in the Galaxy (as well as it is known) give very nearly the same result. Moreover, the radius of the sun's galactic orbit is rather uncertain, as is the amount of material *outside* the sun's orbit. In the opinion of some investigators, the corona of the Galaxy may extend so far that despite its very low spatial density it might still contribute so much to the mass of the system that more than half of its mass lies farther than the sun from the nucleus. It appears, therefore, that the result of the above simple calculation is about as good a value for the mass of the Galaxy as can be obtained with our present knowledge; it is probably correct to within a factor of 2 or 3.

Presumably, most of the material of the Galaxy is in the form of stars. If the mass of the sun is taken as average, we find that the Galaxy contains some hundreds of billions of stars.

27.4 SPIRAL STRUCTURE OF THE GALAXY

Observations have been made of many other spiral galaxies which contain interstellar matter and in which individual stars can be resolved (Chapter 32). In those systems both the interstellar matter and the most luminous resolved stars are generally confined to the spiral arms. The association of the brightest stars and the gas and dust is not mysterious; highly luminous stars are almost certainly

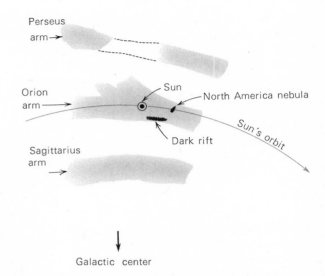

FIG. 27-5   Spiral structure of the Galaxy as deduced from optical observations.

relatively young objects and, in all probability, have recently formed from clouds of interstellar gas and dust (Chapter 30). In mapping out the spiral structure of our own galaxy, therefore, we can use the gaseous nebulae and the very luminous main-sequence O and B stars as "tracers" to identify spiral arms.

(a) The Spiral Structure from Optical Observations

Because of our position in the disk of the Galaxy, and because we are surrounded by interstellar dust, it is difficult for us to identify even the nearby spiral arms optically. Nevertheless, short pieces of three nearby arms have been detected from the observed directions and distances of a large number of O and B stars and from the distribution of emission nebulae. The fragmentary data on the spiral structure of the Galaxy, as deduced from such optical observations, are summarized in Figure 27-5.

The sun appears to be near the inner edge of an arm which contains such conspicuous features as the North America nebula, the Coalsack (near the

Southern Cross), the Cygnus Rift (great dark nebula in the summer Milky Way), and the Orion nebula. More distant, and therefore less conspicuous, emission nebulae can be identified in the *Sagittarius* and *Perseus arms,* located, respectively, about 2000 pc inside and outside the sun's position with respect to the galactic nucleus.

There are more powerful methods for determining the locations of the spiral arms of the Galaxy, but before we can describe them, we shall have to discuss the galactic rotation in more detail.

(b) *Differential Galactic Rotation**

The Galaxy does not rotate like a solid wheel. In our part of the Galaxy, stars farther from the nucleus move more slowly in their orbits, just as planets farther from the sun

have lower orbital speeds than those which are closer. This effect produces a sort of "shearing" motion in the plane of the Galaxy, called *differential galactic rotation.* We shall now consider how the differential rotation affects the radial velocities of stars which are moving in circular orbits concentric to the sun's but which are observed in different directions from the sun in the plane of the Galaxy.

It is useful here to define the coordinate known as *galactic longitude.* The galactic longitude, *l*, is the angle at the sun, measured eastward in the plane of the Galaxy, from the direction of the galactic center. (Galactic coordinates are described more fully and precisely in Appendix 7.)

Stars in the direction of the galactic center, and in the direction opposite the galactic center (*l* = 0° and 180°, respectively — see Figure 27-6), have no component of motion toward or away from the sun and thus have zero radial velocities. Stars on orbits very close to

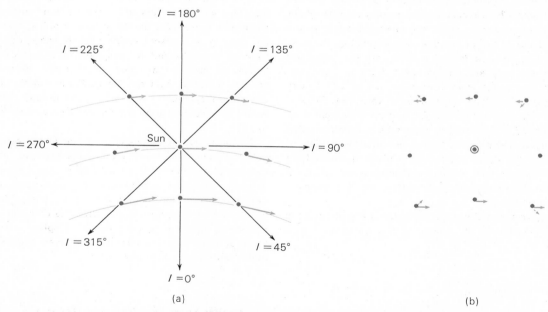

(a)

(b)

FIG. 27-6   Effect of differential galactic rotation on radial velocities of the stars. (a) Velocities of stars relative to the galactic center; (b) velocities of stars relative to the sun. Dashed arrows are the observed radial velocities.

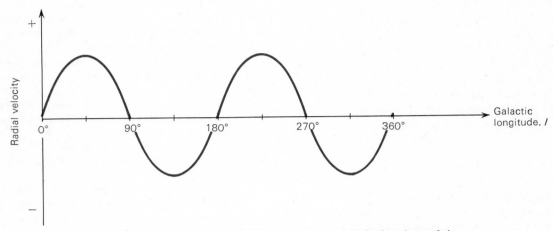

FIG. 27-7   Plot of the observed radial velocities of stars in the plane of the galactic equator at a certain distance from the sun against their galactic longitudes.

the sun's orbit, but which are ahead of or behind the sun ($l = 90°$ and $270°$, respectively) are moving with about the same speed and in about the same direction as the sun and have no radial velocity with respect to the sun. Stars at $l = 45°$, on the other hand, are moving faster than the sun and are pulling away from us, and we, similarly, are pulling away from stars at $l = 225°$; stars in these directions have positive radial velocities. Stars at $l = 315°$ are catching up with us, and we are catching up with stars at $l = 135°$; these objects have negative radial velocities. If we plot the observed radial velocities of all stars in the galactic plane at a given distance from the sun against their galactic longitudes, we obtain a graph like that in Figure 27-7. Such a curve is called a *double sine curve.*[†]

The actual radial velocity of a star at a given galactic longitude — that is, the height of the waves in Figure

27-7 — depends on the distance of the star, being greater for stars of greater distance. This circumstance makes it possible to estimate the distance of a rather remote star that lies in the galactic plane from a knowledge of its galactic longitude and its observed radial velocity. Distances so obtained are not very precise, because the galactic orbits of stars are not generally exactly circular; therefore, their radial velocities are not exactly those that would be predicted from the simple theory of differential rotation.

The clouds of interstellar gas are also presumed to move in roughly circular orbits, like the sun, and the differential galactic rotation applies to them as well as to stars. The radial velocities of the gas clouds can be determined from the Doppler shifts of the interstellar lines in the spectra of more distant stars. The distances of the interstellar gas clouds (and hence of spiral arms), like those of stars, can be estimated from their galactic longitudes and radial velocities.

[†]The mathematical equation for this curve is

$$V_r = Ar \sin 2l,$$

where V_r is the observed radial velocity, r the distance of the stars, and A a constant of proportionality known as *Oort's A constant.* A is about 15 to 20 km/sec per 1000 pc.

(c) *Orbital Speeds of Stars and Gas in the Galaxy*[*]
In most directions, interstellar dust limits our view through the plane of the Galaxy to only a few thousand light-years

when we make our observations in visible or ultraviolet light. Radio waves, on the other hand, are long compared to the size of the interstellar dust particles and penetrate easily through the interstellar medium. In particular, the emission line at 21 cm due to neutral hydrogen (Section 26.2i) can be observed from gas clouds in all parts of the Galaxy. As with spectral lines in the optical part of the spectrum, we can measure the Doppler shift of this radio emission line and determine the radial velocity of its source. If we knew in advance the galactic orbital velocities of the gas clouds at varying distances from the galactic center, we could tell from their observed radial velocities how far they were from the sun, and hence their locations in the Galaxy. Since this is, potentially, a powerful means of mapping the spiral structure of the system, it is important to examine in detail how the Galaxy rotates as a function of distance from its center.

Stars beyond the sun are observed to revolve about the galactic center more slowly than the sun. The mass of the Galaxy is probably concentrated enough toward its nucleus so that we can use Kepler's third law to calculate the orbital speeds (in terms of the sun's speed) of objects substantially *farther* than we are from the center. (See Exercise 7.) On the other hand, the orbital speeds of objects whose orbits are *smaller* than the sun's depend in a critical way on the distribution of mass in the central parts of the Galaxy, and we cannot calculate these speeds with so simple an approach.

Suppose we point a radio telescope at galactic longitude *l* in the plane of the Galaxy (Figure 27-8). If *l* is between 0° and 90° (as in the figure), all gas clouds along the line of sight that are within the sun's orbit must be moving *away* from the sun, and the 21-cm line from each will be shifted to a slightly longer wavelength. We actually receive 21-cm radiation simultaneously from many clouds at various distances, moving away from us at various speeds. That radiation, therefore, is received as a *band* covering a small range of wavelength, each part of the band being 21-cm radiation from a different cloud which is along our line of sight, and is therefore shifted by a different amount. The long-wavelength edge of the band corresponds to radiation coming from the interstellar hydrogen moving away from us with the greatest radial velocity. It can be shown (if all galactic

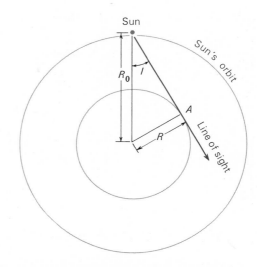

FIG. 27-8 Determination of the orbital velocities of objects with smaller orbits than that of the sun.

orbits are assumed to be circular) that at a given galactic longitude, the gas with the greatest radial velocity will be at position *A* (in Figure 27-8), where our line of sight passes closest to the galactic center. The distance, *R*, of *A* from the galactic center is a simple function of the radius of the sun's orbit, R_0, and the galactic longitude, *l*.[†] Since our line of sight is tangent to the orbit of a cloud at *A*, the maximum observed radial velocity, after correction for the sun's motion, is the actual orbital speed of a cloud moving on a circular orbit of radius *R*. By directing the radio telescope at various other galactic longitudes, we find the circular orbital speeds at other distances from the galactic center.

It has been found from such investigations that the inner part of the Galaxy rotates almost like a solid wheel; over these inner regions, orbital speeds increase with distance from the galactic center. Farther out (before the sun's orbit is reached) the trend reverses, and orbital speeds begin to decrease with distance. At still greater distances there is a gradual transition to "Keplerian"

[†] $R = R_0 \sin l$.

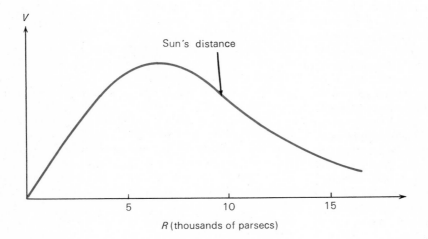

FIG. 27-9 Orbital velocity at various distances from the galactic center.

rotation, like that of the planets about the sun. Figure 27-9 is a schematic plot of rotation speed at various distances from the galactic center.

(d) Spiral Structure as Found From 21-cm Observations*

Once the rotation of the Galaxy at various distances from its center is known, we can use 21-cm observations to map out the spiral structure of our stellar system. The problem is actually very complicated because the neutral hydrogen clouds probably do not revolve about the galactic center in perfectly circular orbits. In principle, however, the procedure is straightforward. Suppose, for example, a radio telescope is pointed in the direction shown in Figure 27-10; note that our line of sight (in this example) passes through two spiral arms, arm (1) at point A and arm (2) at point B. If we knew the positions of these arms in advance, we would know how far points A and B are from the galactic center, and thus their orbital speeds. Then, with simple geometry (or trigonometry) we could calculate their speeds, relative to the sun, projected onto our line of sight — that is, we could predict their radial velocities. We have, of course, no way of knowing the locations of the spiral arms in advance, but by turning the problem around, we can calculate how far away points A and B must be to account for their observed radial velocities, and thus where spiral arms (1) and (2) cross our line of sight.

As has already been mentioned, many hydrogen clouds of differing radial velocities lie along a given line of sight, and consequently the 21-cm radiation received from these clouds covers a short band of wavelengths. Most of the interstellar gas, however, is concentrated at points A and B (in the example shown in Figure 27-10), and thus its radiation will be Doppler-shifted by amounts corresponding to the radial velocities of those points. The energy gathered by a radio telescope is generally passed into a radio receiver not unlike that of an ordinary home radio. Just as the frequency of a home radio can be changed to tune in different stations, so the receiver used with a radio telescope can be swept in frequency, or wavelength. As the receiver is tuned through wavelengths near 21 cm, the signal strength increases whenever radiation is received from neutral hydrogen. In the example illustrated in Figure 27-10, the strongest radiation, coming from points A and B, is received at wavelengths λ_1 and λ_2, respectively; λ_1 and λ_2 are both slightly less than 21 cm because the sun is overtaking A and B in their galactic revolution, giving them *negative* radial velocities. A plot of observed signal strength versus wavelength would look something like Figure 27-11. The difference between λ_1 (or λ_2) and the nominal wavelength of the 21-cm line (actually 21.11 cm) gives the radial velocity of point A (or B); these data, plus a knowledge of the manner in which the Galaxy rotates in general, enables the distances to A and B to be determined and thus

The foregoing discussion has been somewhat over-simplified in many respects. One of the difficulties glossed over is that in directions near galactic longitude 0° and 180° the gas clouds are moving more or less parallel to the sun about the center of the Galaxy and no radial velocities are observed; this is the reason for the triangular "gaps" in Figure 27-12. Another problem is that when the line of sight is between galactic longitudes 270° and 90° it passes through two points in the plane of the Galaxy that have the same radial velocity with respect to the sun; these points are located symmetrically on the far and near side of the point where our line of sight passes nearest the galactic center. There is an ambiguity, therefore, whether a given signal is from a nearby or distant portion of a spiral arm. The ambiguity can usually be resolved by the total strength of the radiation or the angular size of its source. We have also ignored the effects of the random motions of the gas clouds, which are superimposed upon their orbital motions about the galactic center. Such noncircular motion can result in spurious spiral arms, and many features in maps, such as that shown in Figure 27-12, are doubtless wrong. Moreover, in the vicinity of the galactic nucleus, gas seems to be streaming outward in the galactic plane, away from the center. The reason for this radial motion of the gas in the central parts of the Galaxy is unknown, as is the origin of the material itself; it is conjectured that it may be fed into the nucleus, somehow, from the galactic corona.

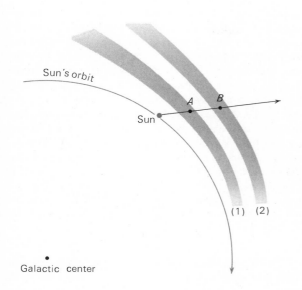

FIG. 27-10   Observation of spiral arms at a wavelength of 21 cm.

locates them in the galactic plane. Similar observations made at other galactic longitudes reveal the distances of spiral arms in other directions. Thus, a map of the spiral structure of the Galaxy is built up. Such a map is shown in Figure 27-12.

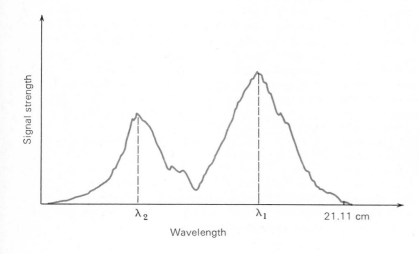

FIG. 27-11   Strength of received radio signals versus wavelength for the observation illustrated in Figure 27-10.

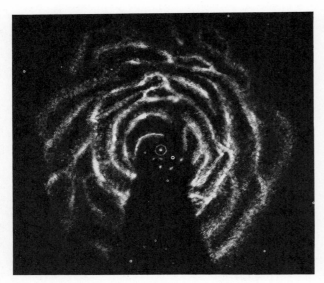

FIG. 27-12   A drawing of the spiral structure of the Galaxy, as deduced from 21-cm observations made at Leiden (in The Netherlands) and at Sydney, Australia. The large and small dots show the location of the galactic center and the sun, respectively. *(Courtesy of Gart Westerhout.)*

In any case, the distribution of neutral hydrogen, as deduced from 21-cm observations, has definitely established the existence of spiral arms in the Galaxy. Although future research can be expected to modify greatly the picture shown in Figure 27-12, at least the gross features of the spiral structure are now revealed.

(e) *Formation and Permanence of Spiral Structure**

It is not surprising that the interstellar material is concentrated into spiral arms. No matter what the original distribution of the matter might be, the differential rotation of the Galaxy would be expected to form it into spirals. Figure 27-13 shows the development of spiral arms from two irregular blobs of interstellar matter, as the portions of the blobs closest to the galactic center move fastest, while those farther away trail behind.

It is difficult to understand, however, why the arms are not wound tighter than they are. At the sun's distance from the center of the Galaxy, the Galaxy rotates once

in about 2×10^8 years. Its total age, however, is believed to be at least 5×10^9 years, in which case the sun has made at least 25 revolutions, and probably more. With so many turns, we would expect the spiral arms to be wound very tightly and to lie very much closer together than they do. Apparently, there are effects that tend to disrupt the spiral structure and to prevent the arms from winding indefinitely. The random motions of the interstellar clouds may be responsible. Perhaps the spiral arms are dissolving as they form, and the spiral structure that we observe is an equilibrium state between the development and dissipation of the arms. One theory, due to C. C. Lin of the Massachusetts Institute of Technology, is that the spiral arms are higher-than-average-density standing waves in the Galaxy. Many problems of galactic structure remain to be solved.

27.5 DIFFERENT STELLAR POPULATIONS IN THE GALAXY

Striking correlations are found between the characteristics of stars and other objects and their locations in the Galaxy. Some classes of objects, for example, are found only in regions of interstellar matter, that is, in the spiral arms of the Galaxy. Examples are bright supergiants, main-sequence stars of high luminosity (spectral classes O and B), Wolf-Rayet stars, type I cepheid variables, and young open star clusters.

The distributions of some other classes of objects show no correlation with the location of spiral arms. These objects are found throughout the disk of the Galaxy, with greatest concentration toward the nucleus. They also extend into the sparse galactic corona. Examples are planetary nebulae, novae, type II cepheids, RR Lyrae variables, and Mira-type variables of periods less than 250 days. The globular clusters, which also belong to this group of objects, are found almost entirely in the corona and nucleus of the Galaxy.

Main-sequence stars of spectral types F through M exist in all parts of the Galaxy, as do red giants and, probably, white dwarfs (observations of white dwarfs, of course, are limited to our immediate neighborhood because of their low luminosities).

There are also differences in the chemical compositions of stars in different parts of the Galaxy. Nearly all stars appear to be composed mostly of hydrogen and helium, but the residual abundance of the heavier elements seems to spread over a large range for different stars. In the sun and in other stars associated with the interstellar matter of the spiral arms, the heavy elements (elements other than hydrogen and helium) account for about 1 to 4 percent of the total stellar mass. Stars in the galactic corona and in globular clusters, however, have much lower abundances of the heavy elements — often less than one tenth, or even one hundredth that of the sun. There is also some evidence that many stars in the disk of the Galaxy, whose galactic orbits show that they are not associated with interstellar matter, may have lower heavy-element abundances than the sun.

The stars associated with the spiral arms are sometimes said to belong to *population I*, while those found elsewhere in the Galaxy are said to belong to *population II*. The terms "population I" and "population II" were first applied to different classes of stars by W. Baade, late astronomer at the Mount Wilson and Palomar Observatories. During World War II, Baade was impressed by the similarity of the stars in the nucleus of M31—the Andromeda galaxy — to those in the globular clusters in the corona of our own galaxy and concluded that the stars situated in spiral arms must, collectively, display different properties from those located elsewhere in the Galaxy. Note that high-velocity stars in the disk of our galaxy (Section 27.2b) have orbits such that they cannot be permanently associated with any given portion of a spiral arm, and hence are population II stars.

Today we can interpret the phenomenon of different stellar populations in the light of stellar evolution. As has been mentioned, and as will be discussed further (Chapter 30), only in the interstellar matter of spiral arms is star formation expected to take place. Thus, population I is comprised of stars of many different ages, including some that were recently formed or are still forming. Population II, on the other hand, consists entirely of old stars, formed, probably, early in the history of the Galaxy.

It is clear today that two stellar populations are insufficient to account completely for the distribution of all the different kinds of stars in the Galaxy. Modern investigators often speak of several stellar populations, ranging from "extreme popula-

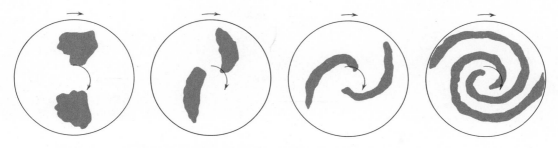

FIG. 27-13   Hypothetical formulation of two spiral arms from irregular clouds of interstellar material.

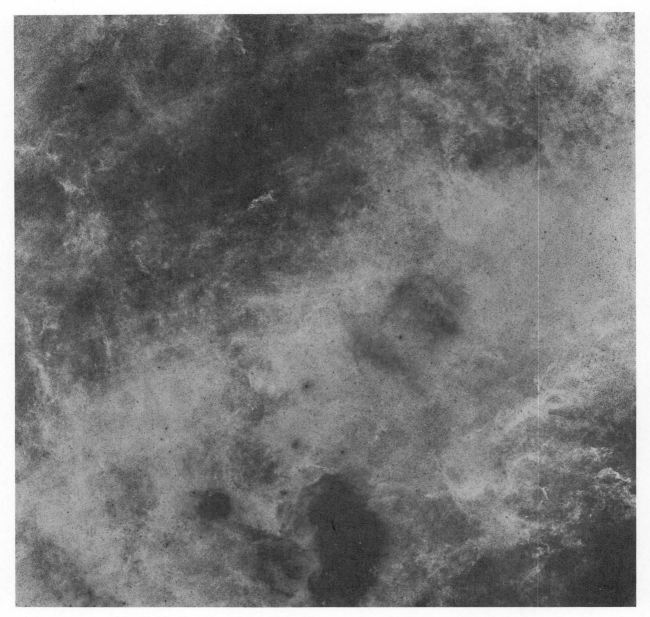

FIG. 27-14   A region of the Milky Way near the galactic center. Negative print, photographed in red light with the 48-inch Schmidt telescope. *(Copyright, National Geographic Society-Palomar Observatory Sky Survey.)*

tion I" (spiral-arm objects), through "disk population" objects, to "extreme population II" (corona objects). The concept of two stellar populations is still a useful one, however, in interpreting many of the general characteristics of our own and other galaxies. In this book we shall use the term "population I" to apply to spiral-arm objects, and "population II" to apply to all others.

27.6 MAGNETIC FIELDS IN THE GALAXY*

It is fashionable, these days, to explain away nasty problems as due to magnetic fields. Charged particles (such as protons and electrons) moving in a magnetic field are accelerated, and accelerated charged particles radiate electromagnetic energy. Thus magnetic fields, properly introduced, can account for a host of otherwise unexplained phenomena. The trouble is that it is hard to measure weak magnetism beyond the solar system, and it is difficult to test our hypotheses of its existence.

Yet, there is strong circumstantial evidence to support the existence of weak magnetic fields in the spiral arms of the Galaxy. The polarization of light from reflection nebulae, and of starlight dimmed by interstellar dust, can be understood only if we assume that the dust particles are elongated and aligned, and the only suitable aligning force that we can imagine is a magnetic field. Moreover, the acceleration of electrons by interstellar magnetic fields could account for much of the galactic continuous radio energy. Finally, acceleration of atomic nuclei in interstellar magnetic fields may be responsible for a large fraction or even most of the cosmic rays that continually bombard the earth (Chapter 31).

Exercises

1. Sketch the distribution of globular clusters about the Galaxy, and show the sun's position. Show how they would appear on a Mercator-type map of the sky, with the central line of the Milky Way chosen as the "equator."

2. The globular clusters probably have highly eccentric orbits, and either oscillate back and forth through the plane of the Galaxy or revolve about its nucleus. Suppose the latter is the case; where would the clusters spend most of their time? (Think of Kepler's second law.) At any given time, would you expect most globular clusters to be moving at high or low speeds with respect to the center of the Galaxy? Why?

3. The period of the sun's revolution about the center of the Galaxy was calculated from its measured speed and distance from the center of the Galaxy. How would the period be changed if the sun's distance from the galactic center were 20 percent greater than the figure assumed?

4. What would the mass of the Galaxy be if the sun's distance from the center were 10,000 pc but its period of revolution about the nucleus were only 100,000,000 years?

5. Suppose we correctly knew the sun's distance from the center of the Galaxy but had derived a value for the speed of the sun in its orbit that is too high by 10 percent. How much would our calculated mass of the Galaxy be in error, and in which direction would the error be?

6. Suppose the mean mass of a star in the Galaxy were only ⅓ solar mass. Using the value for the mass of the Galaxy found in the text, find how many stars the system contains. What did you assume about the total mass of interstellar matter in finding your answer?

7. According to Kepler's third law, what would be the period and orbital speed of a gas cloud moving on a circular orbit of radius 20,000 pc? Assume that the sun moves on a circular orbit of 10,000 pc radius with a period of 200,000,000 years.

*8. Why are we not able to map out the spiral structure of the Galaxy in directions $l = 0°$ and 180° from 21-cm observations? Why do you suppose we *are* able to map out its spiral structure in directions $l = 90°$ and 270°?

9. Describe the details of an experiment in which you witness the formation of "spiral arms" of cream in a cup of coffee that you have stirred vigorously before putting in the cream.

10. Distinguish clearly between the orbital motion of the sun, toward galactic longitude 90°, and the *solar motion*, toward the *solar apex*, which was described in Chapter 19.

CHAPTER 28

Star Clusters

A star cluster is a congregation of stars that have a stronger gravitational attraction for each other (because of their proximity) than do stars of the general field (that is, those *not* in clusters). Clusters range from rich aggregates of many thousands of stars to loose associations of only a few stars. The mutual gravitation of the stars in the larger clusters may hold them together more or less permanently; the small clusters may be held together so weakly that they must be gradually dissipating into the field.

The study of star clusters is very important, because the stars in a single cluster are all at about the same distance (the distance to the cluster); consequently, they can be intercompared with regard to their luminosities, colors, and so on. Moreover, the stars of a cluster probably have had a common origin, being formed at about the same time from the same prestellar material. Investigations of clusters of different ages are a great aid to the study of stellar evolution (Chapter 30).

A number of clusters bear proper names. Some of these are names of mythological characters (the Pleiades); other clusters bear the names of the constellations in which they appear (the Double Cluster in Perseus). Most of the conspicuous clusters are listed in the early catalogues of star clusters and nebulae, especially in Sir John Herschel's *General Catalogue* (1864). The catalogue designations of clusters most often referred to today are those in Messier's (1781) catalogue (for example, M13 — the famous globular cluster in Hercules), and in Dreyer's revisions of Herschel's catalogue, the *New General Catalogue* (*NGC*) and the *Index Catalogue* (*IC*), published between 1888 and 1908. For example, in Dreyer's catalogue, M13 is known as NGC 6205.

28.1 DESCRIPTIONS OF STAR CLUSTERS

The different types of star clusters have already been mentioned briefly (Section 23.3c); here we may describe them more fully. Clusters that contain a great many stars are said to be *rich;* those that contain comparatively few are said to be *poor.* Rich clusters are likely to be conspicuous objects, and their identification as genuine stellar systems is certain. Poor clusters, on the other hand, are much more difficult to pick out against the background of the general star field. Sometimes a real cluster may not be identifiable as such against the background. Other groups of stars that appear to be real systems may actually be stars at different distances seen close together in projection. Most of the clusters that are catalogued, however, contain a high enough density of stars to stand out against the background so that there is virtually no chance of their being accidental superpositions of stars at different distances. Even so, it is often difficult or impossible to say with certainty whether a given individual star is a member of the cluster or not. In general, therefore, a few of the stars studied as cluster members will actually be stars in the foreground or background — that is, stars which belong to the field.

(a) Globular Clusters

As we have already mentioned, about a hundred globular clusters are known, most of them in the corona and nucleus of our Galaxy. All are very far from the sun, and some are found at distances of 60,000 LY or more from the galactic plane. A few, nevertheless, are bright enough to be seen with the naked eye; they appear as faint, fuzzy stars. One of the most famous naked-eye globular clusters is M13, in the constellation of Hercules, which passes nearly overhead on a summer evening at most places in the United States. Through a good pair of binoculars the more conspicuous globular clusters resemble tiny moth balls. A 4- to 6-in. telescope reveals their brightest stars, while a large telescope shows them to be beautiful, globe-shaped systems of stars. Visual observation, however, even through the largest telescope, does not reveal the multitude of fainter stars in globular clusters that can be recorded on telescopic photographs of long exposure.

A good photograph of a typical globular cluster shows it to be a nearly circularly-symmetrical system of stars, with the highest concentration of stars near its own center (a few globular clusters, such as Omega Centauri, appear slightly flattened). Most of the stars in the central regions of the cluster are not resolved as individual points of light but appear as a nebulous glow. Two photographs of a globular cluster made on emulsions sensitive to two different colors of light, say red and blue, show that the brightest stars are red. These stars are 2 or 3 magnitudes brighter than the RR Lyrae variable stars that are almost always found in globular clusters (also called cluster variables — see Section 25.2e); since RR Lyrae stars average about absolute magnitude 0, the brighter red stars must be red giants. Other kinds of variables sometimes found in globular clusters include type II cepheids and RV Tauri stars. One cluster (NGC 7078) contains a planetary nebula.

Distances to globular clusters are usually calculated from the apparent magnitudes of the RR Lyrae stars they contain (Section 27.1a). From their angular sizes (typically a few minutes of arc) their actual linear diameters are found to be from 20 to 100 pc or more. In one of the nearer globular clusters more than 30,000 stars have been counted, but if those stars too faint to be observed are considered, most clusters must contain hundreds of thousands of member stars. The combined light from all these stars gives a typical globular cluster

FIG. 28-1   The globular cluster, M3.
(Lick Observatory.)

an absolute magnitude somewhere in the range −5 to −9.

The average star density in a globular cluster is about 0.4 star per cubic parsec. In the dense center of the cluster the star density may be as high as a 100 or even 1000 per cubic parsec. There is plenty of space between the stars, however, even in the center of a cluster. The "solid" photographic appearance of the central regions of a globular cluster results from the finite resolution of the telescope, seeing effects of the earth's atmosphere, and the scattering of light in the photographic emulsion. A bullet fired on a straight line through the center of a cluster would have far less than one chance in 100 billion of striking a star. If the earth revolved not about the sun, but about a star in the densest part of a globular cluster, the nearest neighboring stars, light-months away, would appear as points of light. Thousands of stars, however, would be scattered uniformly over the sky. The Milky Way would be hard, if not impossible, to see, and even on the darkest of nights the brightness of the sky would be comparable to what it is on earth in bright moonlight.

The motions of globular clusters were described in Chapter 27. They do not partake of the general galactic rotation and are high-velocity objects. They are believed to revolve about the nucleus of the Galaxy on orbits of high eccentricity and high inclination to the galactic plane (rather like the orbits of comets in the solar system). Obeying Kepler's second law, a cluster spends most of its time far from the nucleus; a typical cluster probably has a period of revolution of the order of 10^8 years.

Because most globular clusters lie outside the plane of the Milky Way, probably nearly all have been discovered. It is doubtful that more than a few dozen, hidden by the obscuring dust clouds, remain undiscovered in the disk and nucleus of the Galaxy.

(b) Open Clusters

In contrast to the rich, partially unresolved globular clusters, *open clusters* appear comparatively loose and "open" (hence their name). They contain far fewer stars than globular clusters and show little or no strong concentration of stars toward their own centers. Although open clusters are usually more or less round in appearance, they lack the high degree of spherical symmetry that characterizes a globular cluster; some open clusters actually appear irregular. On photographs made with long-focus telescopes, the stars in these clusters are usually fully resolved, even in the central regions.

Open clusters, as mentioned previously, are found in the disk of the Galaxy, often associated with interstellar matter. Because of their locations, they are sometimes called *galactic clusters* rather than open clusters (this term should not be confused, however, with *clusters of galaxies* — Chapters 32 and 33). They are low-velocity objects and belong to stellar population I; they are presumed to originate in or near spiral arms. Nearly 900 open clusters are catalogued (at the time of writing), but probably well over 1000 are identifiable on good search photographs such as those of the Palomar Sky Survey. Yet only the nearest open clusters can be observed, because of interstellar obscuration in the Milky Way plane. We conclude, therefore, that we see only a small fraction of the open clusters that actually exist in the Galaxy; possibly tens or even hundreds of thousands of them escape detection.

Several open clusters are visible to the unaided eye. Most famous among them is the *Pleiades,* which appears as a tiny group of six stars (some people see more than six) arranged like a tiny dipper in the constellation of *Taurus;* a good pair of binoculars shows dozens of stars in the cluster, and a telescope reveals hundreds. (The Pleiades is *not* the Little Dipper; the latter is part of the constellation of *Ursa Minor,* which also contains the north star.) The *Hyades* is another famous open cluster in Taurus. To the naked eye, the cluster appears as a V-shaped group of faint stars, marking the face of the bull. Telescopes show that the Hyades actually contains more than 200 stars. The naked-eye appearance of the *Praesepe,* in Cancer, is that of a barely distinguishable patch of light;

FIG. 28-2 The open star cluster M67 (NGC 2682) in Cancer. Photographed with the 200-inch telescope. *(Mount Wilson and Palomar Observatories.)*

this group is often called the "Beehive" cluster, because its many stars, when viewed through a telescope, appear like a swarm of bees.

Typical open clusters contain several dozen to several hundred member stars, although a few, such as M67, contain more than a thousand. Compared to globular clusters, open clusters are small, usually having diameters of less than 10 pc. Bright supergiant stars of high luminosity in some open clusters, however, may cause them to outshine the far richer globular clusters. The RR Lyrae stars are never found in open clusters, but other kinds of variable stars, such as type I cepheids, are sometimes present.

(c) Associations

For more than 40 years it has been known that the most luminous main sequence stars — those of spectral types O and B — are not distributed at random in the sky but tend to be grouped into what are now called *associations,* lying along the spiral arms of our Galaxy. In the decade following World War II, interest was greatly revived in these groups of hot stars, especially by the Soviet astronomer V. A. Ambartsumian, who called attention to many of them and pointed out that they must be very young groups of stars. Because the stars of an association lie in the galactic plane and are spread over tens of parsecs, each will revolve about the galactic center with a slightly different orbital speed; Ambartsumian showed that the different orbital speeds of the different members of an association would completely disrupt the group after a few million years. The fact that associations exist at all, therefore, shows that they must be very young objects on the astronomical time scale.

We now distinguish between two kinds of associations: those containing O and B stars are called *O-associations;* the others contain T Tauri stars and are called *T-associations.* T Tauri, or RW Aurigae stars, are red eruptive variables of intermediate to low luminosity that show rapid and irregular increases in brightness (Table 25.3). They are always associated with interstellar matter and are believed to be very young stars that have just formed from it. They are often present in extremely young star clusters (Chapter 30). Sometimes, as in Orion, T-associations and O-associations coexist.

FIG. 28-3   A wide angle photograph of the constellation Orion. An association of young stars is centered in the "sword" (lower part of the picture). *(Mount Wilson and Palomar Observatories.)*

An O-association appears as a group of several (say, 5 to 50) O stars and B stars, and sometimes Wolf-Rayet stars, scattered over a region of space some 30 to 200 pc in diameter. Because these stars are rare, it would be very unlikely for so many of them to exist by chance in so relatively small a volume of space. It is assumed, therefore, that the stars in an association are either physically associated or at least have had a common origin. Stars of other spectral types may also belong to associations, but these more common stars are not conspicuous against the general star field and do not attract attention to themselves as belonging to any particular group.

Often a small open cluster is found near the center of an association. It is presumed, in such cases, that the stars in the association are the outlying members of the cluster, and that they probably had a common origin with the cluster stars. It is probable that some associations may be expanding — their member stars moving radially outward from the association center. An example of such an association is one in Perseus, known as the *Zeta Persei association*. More than a dozen of its member stars are bright enough so that their positions have been measured accurately for more than a century. Most of the stars seem to be moving outward, from the center of the group. If we extrap-

olate their motions backward, we find an age for the expansion of about 1½ million years. Another famous association is in *Orion*, centered on a group of stars in the middle of the Orion Nebula. Rather far from the main association are three luminous O and B stars that seem to be moving away from it at speeds of from 70 to 130 km/sec. It is tempting to postulate that these stars originated with those of the association, but there is as yet no satisfactory hypothesis to explain their extremely high speeds. Analysis of the statistics of associations and O and B stars, however, have led some investigators to conclude that all such stars have originated in associations.

About 80 associations are now catalogued. Like ordinary open clusters, however, they lie in regions occupied by interstellar matter, and many others must be obscured. There are probably several thousand undiscovered associations in our galaxy.

(d) Summary of Clusters

The foregoing descriptions of star clusters are summarized in Table 28.1. The numbers of globular clusters, open clusters, and associations are taken from the Czechoslovakian *Catalogue of Star Clusters and Associations,* which includes all objects reported up to October 7, 1961, and is the most comprehensive star cluster catalogue available. The

TABLE 28.1 Characteristics of Star Clusters

	GLOBULAR CLUSTERS	OPEN CLUSTERS	ASSOCIATIONS
Number known	119	867	82
Location in Galaxy	Corona and nucleus	Disk (especially spiral arms)	Spiral arms
Diameter (pc)	20 to 100	<10	30 to 200
Mass (solar masses)	10^4 to 10^5	10^2 to 10^3	10^2 to 10^3 ?
Number of stars	10^4 to 10^5	50 to 10^3	10 to 100 ?
Color of brightest stars	Red	Red or blue	Blue
Integrated absolute visual magnitude of cluster	-5 to -9	0 to -10	-6 to -10
Density of stars (solar masses per cubic parsec)	0.5 to 1000	0.1 to 10	<0.01
Examples	Hercules Cluster (M13)	Hyades, Pleiades, h and χ Persei, Praesepe	Zeta Persei, Orion

sizes, absolute magnitudes, and numbers of stars listed for each type of cluster are approximate only and are intended as representative values.

28.2 DYNAMICS OF STAR CLUSTERS

The problem of two bodies moving under the influence of their mutual gravitational attraction was solved by Newton. We have seen (Chapter 5) that each member of such a two-body system moves on a path (a conic section) that can be described by a simple algebraic equation. A star cluster, on the other hand, is a system of many bodies. Each star in a cluster moves under the influence of gravitational forces exerted upon it simultaneously by all the other stars. Since all the stars in a cluster are moving, the gravitational forces they exert on each other are constantly changing, and it is an enormously complicated problem to predict the future path of any one.

In principle, the motion of an individual star in a cluster could be computed in detail. On a much smaller scale, such calculations are actually performed today with electronic computers, especially in space science applications — for example, to compute the motion of a space vehicle moving in the combined gravitational fields of the earth, moon, and sun. The particular wanderings of an individual star in a cluster, however, are not of much interest. Far more significant is the way all the cluster members move on the average, for their average motion depends upon certain fundamental characteristics of the cluster — its mass, size, and structure. This section will describe, briefly, the statistical methods that enable us to learn some of the properties of clusters from the average motions of their member stars.

(a) The Virial Theorem and Masses of Clusters*

If a stone is raised high above the ground and released, the force of the earth's gravity upon it accelerates it downward; as it falls, it picks up more and more speed, or energy of motion. The energy associated with the motion of an object is called its *kinetic energy.* The potential ability of gravity to acclerate a body and give it kinetic energy is called *gravitational potential energy.* The greater the height the stone is dropped from, the farther it can fall, the more it will accelerate, and the more kinetic energy the stone will gain; thus, the higher the stone is held, the greater is its potential energy before it is released. As it falls, its potential energy is converted to its kinetic energy.

Similarly, a star in a cluster feels itself attracted generally toward the center of the cluster, and so has gravitational potential energy. The actual amount of its potential energy depends upon the strength of the total resultant gravitational force acting upon it and its distance from the cluster center. As a star moves about in a cluster, it sometimes decreases its distance from the center, and thus speeds up, converting some of its potential energy to kinetic energy. At other times the star increases its distance from the cluster center, pulling against gravity, and therefore slows down; it then converts kinetic to potential energy. The actual path of the star may be very complicated, because occasionally it may pass near another star, and its direction of motion will then be deflected by the gravitational attraction between the two. Such a deflection is called a *gravitational encounter.* An "encounter" in this case is not a real collision; because of the distances separating the stars in a cluster, actual stellar collisions are all but impossible.

There is associated with a star cluster a certain total kinetic energy, which is defined as the sum of the individual kinetic energies of its member stars. (The kinetic energy of a star of mass m and speed v is $\frac{1}{2}mv^2$.) By convention, the potential energy of a star in a cluster at a given instant is defined as the work that must be done on the star (that is, the energy that must be given it) to remove it from its location in the cluster at that instant to a point infinitely far away, pulling, in the process, against the gravitational attraction between the star and cluster. The potential energy of the entire cluster is the energy required to separate all the stars infinitely far apart. Actually, as defined, the potential energy of a star cluster is the potential energy its stars would have given up if they had all fallen together under their mutual gravitational attraction, from a configuration in which they were extremely widely separated to their present configuration in the cluster. Clearly, more potential energy would have been released if the cluster were smaller and more concentrated than it is, and less if it were more spread out.

After a cluster has existed for a long enough time (usually an interval of hundreds of millions or billions of years) there will have been enough encounters between stars to divide the total energy of the cluster (potential plus kinetic) approximately evenly among the stars. The energy will never be distributed *exactly* equally; there will always be a few stars that by virtue of recent encounters have more energy than average, and others that have less energy than their share. On the average, however, each star will have its share of the total energy of the cluster; no one *kind* of star would be expected to have more energy than any other kind. The cluster is then said to be in a state of *statistical equilibrium.*

Under the above conditions of statistical equilibrium it can be shown with the calculus that the total potential energy of a cluster (as defined above) is twice its total kinetic energy. Thus if we assume that a particular cluster is in equilibrium we can state this relation as an equation applying to the cluster. This statement, relating the kinetic and potential energies of a system of particles interacting under their mutual gravitation, is called the *virial theorem.* An important application of the virial theorem in astronomy is in the estimation of masses of systems of stars that are presumed to be in statistical equilibrium.

We can express the kinetic energy of a cluster in terms of its total mass and the mean of the speeds of its member stars, and the potential energy can be stated in terms of the mass of the cluster, its size, and how the stars are distributed throughout it. Now the stellar speeds are found from the Doppler shifts in the spectra of individual stars compared to the average Doppler shift of all the stars, which, of course, corresponds to the radial velocity of the cluster as a whole. We can observe directly the size of the cluster, the approximate position of its center of mass, and the internal distribution of its member stars. Thus the only unknown in the equation expressing the equality of twice the kinetic energy and the potential energy is the cluster mass, and the mass can be solved for. From such analyses we find that the mass of a typical globular cluster is about 10^5 solar masses. The

Pleiades, a more or less typical open cluster, is found to have a total mass of about 300 suns.

The above discussion of the application of the virial theorem has been somewhat oversimplified, but does illustrate an important method of mass determination. It will be referred to again in a later discussion of the masses of galaxies and clusters of galaxies. The underlying physical idea can be summarized briefly as follows: If we assume that a cluster is in equilibrium, its members are held together and their motions are determined by their mutual gravitation. The virial theorem tells us how much mass the cluster must have (and hence how much gravitation) in order that its stars are moving with their observed speeds, and in such paths that the cluster has its observed shape.

(b) Stability of Star Clusters*

A condition for the stability of an isolated cluster is that its total potential energy be greater than its kinetic energy. Otherwise, the average stellar speeds exceed the escape velocity from the cluster, in which case it dissipates into space. If the potential energy of the cluster exceeds its kinetic energy, it is gravitationally bound — that is, its member stars cannot all escape.

Clusters, however, are not completely isolated but move in various orbits in the Galaxy. Thus, an added condition for the stability or permanence of a cluster is that it be bound together with gravitational forces that are stronger than the disrupting tidal forces of the Galaxy, or other nearby stars, upon it. The more compact a cluster, the greater is its own gravitational binding force compared to the disrupting forces, and the better chance it has to survive to old age.

Globular clusters are highly compact systems and are, consequently, very stable. Most globular clusters can probably maintain their identity almost indefinitely. Even these clusters lose some stars, however — especially those of relatively small mass. A few stars in a cluster are always moving substantially faster than average. Every now and then one of them, through an encounter, will be given enough speed to escape the cluster. Some of the stars in the galactic corona are very likely stars that have, in the past, escaped in this way from globular clusters. The stellar escape rate from these rich clusters, however, is so slow that the clusters can survive for many billions of years. The situation is analogous to the evaporation of molecules from a more or less permanent planetary atmosphere.

Matters are very different for most open clusters. Those that we have discovered are relatively close to the sun; at our distance from the galactic center, the tidal force of the Galaxy will disrupt a cluster in short order if the star density within it is much less than about 1 star per cubic parsec. The Pleiades, for example, is probably just stable in its central regions, while its outer parts are probably dissipating. The Hyades is on the verge of instability. Typical open clusters have maximum lifetimes as clusters of only a few hundreds of millions of years; a handful of the richer, denser ones, like M67, can be expected to survive for billions of years. Most open clusters, in other words, are relatively young stellar groups.

The case for stellar associations, which have very low star densities, is even more extreme. These loose groups cannot possibly be permanent stellar systems. If they are actually expanding, as one or two appear to be, they would be highly unstable even in the absence of disruptive galactic tidal forces.

28.3 DETERMINATION OF DISTANCES OF CLUSTERS

It has already been explained (Chapter 27) how the distances to globular clusters are determined — from the RR Lyrae variable stars found in most of them and, at least statistically, from the angular diameters of the clusters. Open clusters, on the other hand, differ from each other in size too much for us to use their angular diameters to estimate their distances. Moreover, as we have seen, they do not contain RR Lyrae stars, although some have cepheid variables. In practice, the most dependable way to determine the distance to an open cluster is from its color-magnitude diagram (see below) rather than from any variable stars it may contain; in fact, one of the newest methods of calibrating the period-luminosity relation of cepheid variables of population I (Section 25.2c) is from the known distances of open clusters which possess these stars.

It was described in Section 19.3c how distances to three moving clusters, the Hyades, the Ursa Major Group, and the Scorpio-Centaurus Group, have been found from the apparent convergence or divergence of their member stars as they move across the sky. In addition, there are a few other examples of groups of stars that may have common motion in space (the Czechoslovakian *Catalogue of Star Clusters* lists five such groups as of October 1960). If it is found that the stars in each of these groups are really moving parallel to one another, we can find the distances to them with the same technique.

The most useful and powerful method of determining the distance to an open cluster is from a plot of the apparent magnitude versus the color index of its member stars as shown in Figure 28-4. The time required to photograph spectra of many individual cluster stars would be prohibitive, but their colors can be observed very quickly. It is to be recalled that the color index of a star de-

FIG. 28-4   Color-magnitude diagram for a typical (hypothetical) open star cluster.

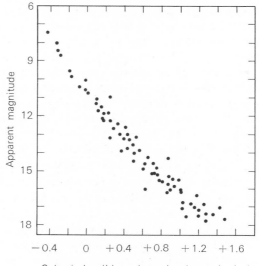

Color index (blue minus visual magnitudes)

pends upon the star's temperature and is thus related directly to its spectral class. Therefore, the plot of magnitude versus color index (called a *color-magnitude diagram*) is, with one exception, like a Hertzsprung-Russell diagram for the cluster (Section 23.2). The exception is that a normal H-R diagram is a plot of the *absolute* magnitudes (rather than apparent magnitudes) of stars against their spectral classes (or color indices). All the stars in a cluster, however, are at the same distance, and the difference between the apparent and absolute magnitudes is the same for every star — the *distance modulus* of the cluster (Section 20.2b). Most of the stars in a cluster generally lie along a *main sequence* similar to that defined by stars in the neighborhood of the sun. We find a main sequence, therefore, in the color-magnitude diagram of a cluster (see Figure 28-4), with the bluer, brighter-appearing stars (which are really the more luminous main-sequence stars in the cluster) farther up on the diagram than the redder, fainter-appearing stars. From the ordinary H-R diagram, we know what absolute magnitudes correspond to various color indices along the main sequence. The difference, at any given color index, between the apparent magnitude of the cluster stars and the absolute magnitude of known main-sequence stars of the same color is the distance modulus of the cluster.

As a numerical example, consider the hypothetical cluster whose color-magnitude diagram is shown in Figure 28-4. A main-sequence cluster star of color index $+0.6$ is seen to have an apparent magnitude of $+15$. But this is a star like the sun, whose absolute magnitude is $+5$. At a distance of 10 pc, therefore, this star would appear 10 magnitudes brighter than it does at the actual distance of the cluster. Since 10 magnitudes corresponds to a factor of 10,000 in light, the cluster must be 100 times as distant as 10 pc or must be 1000 pc away.

In the actual application of the procedure just described for obtaining cluster distances, the apparent magnitudes and color indices must first be corrected for effects of interstellar absorption and reddening (Chapter 26) — not always an easy task. We shall have more to say about the color-magnitude diagrams for star clusters in the next section.

28.4 STELLAR POPULATIONS OF STAR CLUSTERS

The Hertzsprung-Russell diagrams, or color-magnitude diagrams of star clusters, are extremely useful in the study of stellar evolution. Here we shall describe some of the properties of the color-magnitude diagrams for different kinds of clusters and some other properties of the stellar populations of clusters. The interpretation in terms of stellar evolution is discussed in Chapter 30.

(a) Color-Magnitude Diagrams of Globular Clusters

Globular clusters nearly all have very similar appearing color-magnitude diagrams. Figure 28-5 shows, semischematically, the color-magnitude diagram for a typical globular cluster of known distance, and for which the apparent magnitudes have been converted to absolute magnitudes. The region from *a* to *b* is the main sequence. Presumably, the main sequence would extend farther down than *a* if the cluster were near enough for us to observe its fainter stars. Above point *b*, however, the main sequence seems to terminate; in most globular clusters, this point occurs at about absolute magnitude, $M_v = +3.5$. From *b* to *c* there extends a sequence of stars that are yellow and red giants, the brightest and reddest of them (at *c*) being at $M_v = -3$; note that these latter stars are brighter than typical red giants in the solar neighborhood. A third sequence of stars extends from *d* to *f*; it is called the *horizontal branch* of the H-R diagram for a globular cluster. There is a gap in the horizontal branch at $M_v = 0$ (point *e*), where no star of constant light output is found. The stars observed in this gap are the RR Lyrae variables. The color-magnitude or H-R diagram of a globular

FIG. 28-5 Hertzsprung-Russell diagram for a typical (hypothetical) globular star cluster.

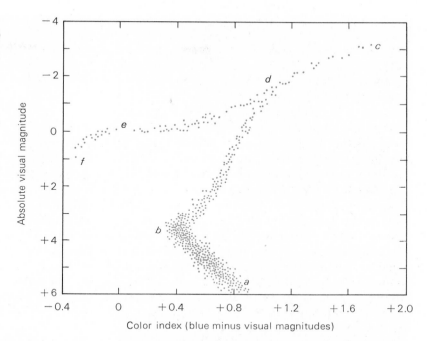

Absolute visual magnitude

Color index (blue minus visual magnitudes)

cluster is more or less similar to that of any system of stars which belongs to stellar population II.

(b) Color-Magnitude Diagrams of Open Clusters

Whereas globular clusters nearly all have very similar H-R diagrams, those of open clusters differ widely from one another. Figure 28-7 shows, schematically, superposed H-R diagrams for four hypothetical open clusters, which will serve as illustrative examples.

In cluster A the main sequence extends to high-luminosity stars, and the highly luminous cool stars in the cluster are red supergiants. The double cluster, h and χ Persei, has an H-R diagram similar to that of cluster A. The main sequence of cluster B (representative of the open clusters, M11 and M41) extends to less luminous stars, and the red giants of that cluster are less luminous than those in A as well. Cluster C (characteristic of the open cluster, M67) has a main sequence that terminates at about the same absolute magnitude as do those

of globular clusters. Cluster C has a sequence of yellow and red giants that resembles, but is less luminous than, that of the globular clusters (dashed line). The red giants in cluster C have about the same luminosity as typical red giants near the sun. At least one open cluster of the "C type" (M67) has a suggestion of a horizontal branch.

Note that in the three clusters discussed so far, the upper ends of the main sequences terminate at different absolute magnitudes and that the red giants (if present) usually lie off to the right of the top of the main sequence in each H-R diagram. In a few clusters (such as the Pleiades), no red giants are observed, but such clusters are not generally very rich. We note that there is a gap between the top of the main sequence and the red giants in the H-R diagrams of clusters A and B. This gap, called the *Hertzsprung gap*, is broadest in the color-magnitude diagrams of clusters whose main sequences extend to high luminosities, and narrows for clusters whose main

FIG. 28-6   The Pleiades. North is to the left. *(Lick Observatory.)*

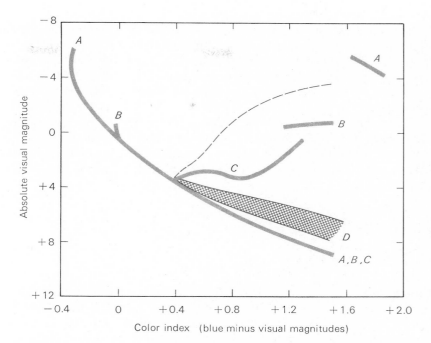

FIG. 28-7 Composite Hertzsprung-Russell diagram for four hypothetical open clusters *A, B, C,* and *D*. The dashed line shows the location of the giants in globular clusters.

sequences terminate at successively lower luminosities. The gap finally disappears in the color-magnitude diagrams of clusters whose main sequences extend only as far as that of cluster *C*. We shall see (Chapter 30) that this gap is probably due to relatively rapid stages of evolution in the "lives" of certain stars.

The least luminous stars observed in clusters *A, B,* and *C* all lie on the main sequence. In cluster *D*, however, the brighter stars are on the main sequence, while the faintest observed ones lie off to the *right*. The open cluster, NGC 2264, has such an H-R diagram. We shall see (Chapter 30) that the faintest stars in this cluster (and others like it) are believed to be still in the process of contraction from interstellar matter.

(c) Differences in Chemical Composition of Stars in Different Clusters

Hydrogen and helium, the most abundant elements in stars in the solar neighborhood, are also the most abundant constituents of the stars in all kinds of clusters. The exact abundances of the heavy elements (those heavier than helium), however, vary from cluster to cluster. In the sun, and most of its neighboring stars, the combined abun-

FIG. 28-8 The double open star cluster *h* and chi Persei. *(Yerkes Observatory.)*

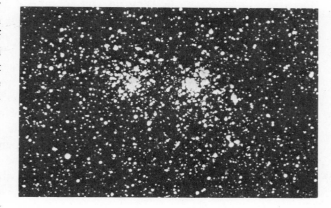

dance (by mass) of the heavy elements seems to be between 1 and 4 percent. The strengths of the lines of heavy elements in the spectra of stars in most open clusters show that they, too, have 1 to 4 percent of their matter in the form of heavy elements.

Globular clusters, however, are a different story. Spectra of their brightest stars often show extremely weak lines of the heavy elements. The heavy-element abundance of stars in typical glob-

ular clusters is found to range from only 0.1 to 0.01 percent, or even less. Other stars of population II also have spectra that often indicate low abundances of heavy elements, although the difference between them and the sun is not usually as extreme as for some of the globular cluster stars. Apparently, differences in chemical composition are related to differences in stellar population. Some possible explanations of these phenomena are discussed in Chapter 30.

Exercises

*1. The RR Lyrae stars in a particular globular cluster appear at apparent magnitude +15. How distant is the cluster? (Ignore interstellar absorption.)

2. Where in the Galaxy do you suppose undiscovered globular clusters may exist?

3. Suppose globular clusters have orbits about the galactic center with very high eccentricities — near unity. When a particular globular cluster is at its farthest from the center of the Galaxy, its distance from the center is 10^4 pc. What is its period of galactic revolution? (*Hint:* 1 pc $= 2 \times 10^5$ AU. Assume that the mass of the Galaxy is 10^{11} solar masses.)

4. In Table 28.1 it is indicated that stellar associations can emit even more light than a globular cluster. How is this possible if the associations have so few stars?

5. What color would a globular cluster appear? Why?

6. Why do you suppose it is sometimes said that the problem of dealing with the motions of more than two bodies (interacting gravitationally) has no solution? Is the statement true? Explain.

*7. At what time of year is the earth's potential energy greatest? At what time of year is its kinetic energy greatest?

*8. Could the virial theorem be used to compute the mass of a flock of birds flying together in formation? Why?

9. A main-sequence star of color index 0 has an absolute magnitude of about +1. In the color-magnitude diagram of a certain cluster, it is noted that stars of 0 color index have apparent magnitudes of about +6. How distant is the cluster? (Ignore interstellar absorption.)

10. It is often possible to observe fainter main-sequence stars in open clusters than in globular clusters. Why do you suppose this is the case?

CHAPTER 29

Structure and Energy of Stars

Except for the sun, the stars appear as unresolved points of light, even when viewed with the world's greatest telescopes. It might seem, therefore, that a knowledge of the internal structure of a star would be hopelessly inaccessible. Yet, in many respects we have a better understanding of the interiors of stars than we do of their outer, observable layers. In some respects, we know more about the interiors of stars than we do about the interior of our own earth. This chapter will summarize how we study the structure of stars and the energy sources that keep them shining.

29.1 EQUILIBRIUM IN STARS

The sun and the vast majority of other stars are *stable;* that is, they are in a *steady state* — neither expanding nor contracting. Such a star is said to be in a condition of *equilibrium;* all the forces within it are balanced, so that at each point within the star the temperature, pressure, density, and so on, are maintained at constant values. We shall see (Chapter 30) that even these stable stars, including the sun, are changing as they evolve, but such evolutionary changes are so gradual that to all intents and purposes the stars are still in a state of equilibrium. The way in which the conditions of equilibrium within a stable star enables us to learn about its internal structure are described below. A comparable discussion of unstable stars, such as novae and variables, would be beyond the scope of this text.

(a) Hydrostatic Equilibrium

The mutual gravitational attraction between the masses of various regions within a star produces tremendous forces that tend to collapse the star toward its center. Yet, since stars like the sun have remained more or less unchanged for billions of years, the gravitational force that tends to collapse a star must be exactly balanced by a force from within.

Even at the surfaces of stars, the temperatures are so high that all chemical elements are vaporized to gases. The internal temperatures of stars are many times higher than their surface temperatures; thus, stars are gaseous throughout. Most of the internal force that supports the outer layers of a star against the downward pull of gravitation upon them is the pressure of the gases themselves. In some very luminous stars the pressure of radiation (Section 16.4e) also contributes appreciably to support of the outer layers.

If the internal gas pressure in a star were not great enough to balance the weight of its outer parts, the star would collapse somewhat, contracting until the gas pressure inside built up to the point where it could support the star. Analogously, if the air in a tire or balloon cools and its pressure is thus reduced, the rubber contracts, further compressing the air until a balance is restored. On the other hand, if the gas pressure in a star were greater than the weight of the overlying layers, the star would expand, thus decreasing the internal pressure. Expansion would stop when the pressure at every point within the star again equaled the weight of the stellar layers above that point. Similarly, if the air in a balloon or tire heats up and increases in pressure, the balloon expands until the increased tension in the rubber is just enough to match the higher air pressure.

Thus, a star must adjust itself until it is in balance. For mathematical purposes, the star can be regarded as composed of a large number of thin spherical layers, or concentric spherical shells, piled one upon the other (rather like the skin layers in an onion); this is not to say that the star is stratified into physically distinct layers, but only that we can divide it into concentric shells for descriptive purposes, much as we speak of layers or depths in the ocean. The pressure within a star increases downward from the surface to the center; the increase of pressure through each concentric shell or layer of the star is exactly the right amount necessary to support the weight of that shell. This condition is called *hydrostatic equilibrium*. Stable stars are all in hydrostatic equilibrium; so are the oceans of the earth, as well as the earth's atmosphere. (Recall that the pressure of the air keeps the air from falling to the ground.)

(b) The Perfect Gas Law

The particles that comprise a gas (molecules or atoms) are in rapid motion, frequently colliding with each other and with the walls of the container of the gas. This constant bombardment is the *pressure* of the gas. The pressure is greater, the greater the number of particles within a given volume of the gas, for of course the impact carried by the moving particles increases with their number. The pressure, in other words, is proportional to the density of the gas. The pressure is also greater the faster the molecules or atoms are moving; since their rate of motion is determined by the temperature of the gas, the pressure is greater the higher the temperature.

Most students have run across these concepts in high school, in the form of Boyle's law, which states that the pressure of a gas at constant temperature is inversely proportional to the volume to which it is constrained (that is, in proportional to its density), and Charles' law, which states that the pressure (at constant volume) is proportional

to the temperature of the gas. These two laws (Boyle's and Charles') can be combined to give us what is called the *perfect gas law* (also called the *equation of state* for a perfect gas). The perfect gas law provides a mathematical relation between the pressure, density, and temperature of a perfect, or ideal, gas (one in which intermolecular or interatomic forces can be ignored), and states that the pressure is proportional to the product of the density and temperature of the gas. The gases in most stars closely approximate an ideal gas; thus, they must obey this law.

(c) Minimum Pressures and Temperatures in Stellar Interiors

Suppose we knew how the matter was distributed within a star, that is, what fraction of the star's mass was included within each concentric layer or shell. Since the weight of a shell is the gravitational attraction between it and all the underlying layers, we could then calculate the weight of each shell. From the condition of hydrostatic equilibrium, we could next calculate how the pressure must increase downward through each shell to support its weight. At the surface of the star, where there are no overlying layers of stellar matter, the pressure is zero. By simply adding up the increases of pressure through the successive layers inward, we would be able to find the pressure at each point within the star, in particular at its center. Using the pressures and densities thus determined at all points along the radius of the star, we could then find the corresponding temperatures from the perfect gas law. In other words, if we only knew how the material within a star was distributed, we would be able to calculate the density, pressure, and temperature at all its internal points.

It is not known in advance, of course, how the matter in a particular star is distributed. On the other hand, some ways that it is *not* distributed can be specified. Internal gravity, for example, must force the gases comprising a star into higher

FIG. 29-1   A. S. Eddington — a pioneer in the study of stellar structure. *(Yerkes Observatory.)*

and higher compression at deeper and deeper levels in the star's interior. In other words, the material of a star would be expected to show high central concentration; the densities of outer layers would certainly not exceed those of inner layers. To assume that the matter in a star is distributed with uniform density, therefore, would certainly be to underestimate its central compression, and the values calculated for its internal pressures and temperatures would certainly be lower than the true values. Here, then, is a method by which *lower limits* can be found for the pressures and temperatures in stellar interiors.

Applying this method to a typical star, the sun, we find that the mean pressure is at least 500 million times the sea-level pressure of the earth's atmosphere, that the central pressure is at least 1.3 billion times that of the earth's atmosphere, and that the mean temperature is at least 2.3 million degrees Kelvin. Since these pressures and temperatures would exist if the sun were uniform in density, the actual values must be much higher. Under such conditions all elements are in the gaseous form, and the atoms cannot be combined into molecules. Moreover, most of the atoms are almost completely ionized — that is, stripped of their electrons (Section 10.5). These electrons, freed from their parent atoms, become part of the gas itself, moving about as individual particles.

Thus, with only the assumption of hydrostatic equilibrium and a knowledge of the perfect gas law, it is possible to learn something of the conditions that prevail in stellar interiors.

(d) Thermal Equilibrium

From observation we know that energy (electromagnetic energy) flows from the surfaces of the sun and stars. According to the second law of thermodynamics, heat always tries to flow from hotter to cooler regions. Therefore, as energy filters outward toward the surface of a star, it must be flowing from inner hotter regions. The temperature can never *decrease* inward in a star, or energy would flow in and heat up those regions until they were at least as hot as the outer ones. We conclude that the highest temperature occurs at the center of a star and that temperatures drop to successively lower values toward the stellar surface.† The outward flow of energy through a star, however, robs it of its internal heat and would result in a cooling of the interior gases were that

---

†The high temperature of the sun's corona may therefore appear to be a paradox. The actual heat energy in the sun's corona is relatively small because the corona is a highly rarefied gas. Its high temperature is believed to be maintained by shock waves or by some other process that would not exist for a gas in thermodynamic equilibrium (Section 24.2d).

energy not replaced. There must therefore be a source of energy within each star.

If a star is in a steady state (that is, in hydrostatic equilibrium and shining with a steady luminosity), the temperature and pressure at each point within it must remain approximately constant. If the temperature were to change suddenly at some point, the pressure would similarly change, causing the star to contract suddenly, or to expand, or to otherwise deviate from hydrostatic equilibrium. Energy must be supplied, therefore, to each layer in the star at just the right rate to balance the loss of heat in that layer as it passes energy outward toward the surface. Moreover, the rate at which energy is supplied to the star as a whole must, at least on the average, exactly balance the rate at which the whole star loses energy by radiating it into space; that is, the rate of energy production in a star is equal to the luminosity. We call this balance of heat gain and heat loss for the star as a whole and at each point within it the condition of *thermal equilibrium*. A later section will deal with the source of stellar energy.

(e) Heat Transfer in a Star

There are three ways in which heat can be transported from one place to another: by *conduction,* by *convection,* and by *radiation.* The rate at which heat passes through gases by conduction, however, is so low that this mode of transfer can be ignored in stellar interiors. (An exception is the transfer of energy in white dwarf stars, but the matter in these stars is not in the form of an ordinary gas — see Chapter 30.)

Stellar convection occurs as actual currents of gas flow in and out through the star. While these convection currents travel at moderate speeds and do not upset the condition of hydrostatic equilibrium, they nevertheless carry heat outward through a star very efficiently. However, convection currents cannot be maintained unless the temperatures of successively deeper layers in a star

increase rapidly in relation to the rate at which the pressures increase inward. Convection does occur, nevertheless, in certain parts of many stars, and convection currents may travel completely through some of the least luminous stars.

Unless convection occurs, the only significant mode of energy transport through a star is by radiation, in which electromagnetic radiation gradually filters outward as it is passed from atom to atom. However, radiative transfer is not an efficient means of energy transport, because under the conditions that prevail in stellar interiors gases are very opaque — that is, a photon does not go far before it is absorbed by an atom (typically, in the sun, about 1 cm). The energy absorbed by atoms is always reemitted, to be sure, but most of it is reemitted in random directions. A photon that is traveling outward in a star when it is absorbed has almost as good a chance of being reemitted back toward the center of the star as toward its surface. A particular quantity of energy being passed from atom to atom, therefore, zigzags around in an almost random manner and takes a long time to work its way from the center of the star to the surface; in the sun, the time required is of the order of 1 million years.

The measure of the ability of a gas to absorb radiation is called its *opacity.* It should not surprise the reader that the gases in a star are opaque. If they were not, stars would be completely transparent. We would then, for example, be able to see all the way through the sun. We have already discussed most of the sources of opacity, that is, the ways in which atoms can absorb radiation, in Chapter 21. An atom that still retains one or more electrons can absorb radiation either by photoionization (losing another electron) or by excitation, wherein it jumps from a lower to a higher energy level. A relatively small number of atoms in stellar interiors are not completely ionized; yet there are enough of them so that photoionization is an important source of opacity. Even transitions from lower to higher energy levels occur often enough to contribute appreciably to stellar opacity. Other ways in which the flow of radiation can be interrupted or impeded are by free-free transitions and by scattering of photons ("reflection," helter-skelter) by free electrons. These atomic processes were described in Chapter 21.

If the temperature and density of a gas and its exact composition are known, its opacity can be calculated by the methods of quantum mechanics. These calculations are very involved but are now regularly performed with electronic computers. Once we know the opacity of a given layer in a star, we can determine how much the flow of energy through it is impeded. Of course, some energy *must* filter through the layer or the star would have no luminosity. From the actual energy flow through the layer it now becomes possible to calculate how much the temperature must increase across the layer to force the observed amount of radiation through it. In other words, knowledge of the opacity of the gases in a stellar interior and of the amount of energy flowing through them makes possible a calculation of the temperature variations throughout the star.

If the temperature difference across some regions of a star should be high enough to support convection, convection currents, rather than radiation, would carry most of the energy. Within the regions of a star where convection occurs, if it occurs at all, the variation of temperature with depth is determined by the expansion of outward-moving masses of gas and the contraction of inward-moving ones. Here again, knowledge of the energy transport mechanism within a star makes possible calculation of the temperature distribution.

This section has dealt with some of the physical principles that give us clues to the internal structure of a star in equilibrium. When enough of these clues are put together, a theoretical model of the star can be computed, one that describes the pressure, temperature, mass distribution, and

so on, throughout the star. Any discussion of the construction of stellar models, however, must follow a description of the sources of stellar energy.

29.2 STELLAR ENERGY

The rate at which the sun emits electromagnetic radiation into space, and thus the rate at which energy must be generated within it, is about 4×10^{33} ergs/sec (Section 20.4b). Moreover, the power output of the sun has been about the same throughout recorded history and, according to geological evidence, not very different since the formation of the earth billions of years ago. Our problem is now to find what sources can provide the gigantic amounts of energy required to keep the stars like the sun shining for so long.

(a) Thermal and Gravitational Energy

Two large stores of energy in a star are its internal heat, or *thermal energy*, and its *gravitational energy*. The heat stored in a gas is simply the energy of motion (kinetic energy) of the particles that comprise it. If the speeds of these particles can decrease, the loss in kinetic energy can be radiated away as heat and light. This is how a hot iron cools after it is withdrawn from a fire (except that the atoms in a solid vibrate within a crystalline structure, rather than moving freely, as in a gas).

Because a star is bound together by gravity, it has gravitational potential energy, as does a star cluster (Chapter 28). If the various parts of a star fall closer together, that is, if the star contracts, it converts part of its potential energy into heat, some of which can be radiated away. About the middle of the nineteenth century, the physicists Helmholtz and Kelvin postulated that the source of the sun's luminosity was indeed the conversion of part of its gravitational potential energy into radiant energy.

The sun cannot be infinitely old, of course, for no source of energy can last forever. Sometime in the past, therefore, the sun must have formed, presumably from interstellar gas or dust or both. As the presolar material gradually fell or gravitated together into a condensing mass, later to become the sun, it steadily gave up its potential energy. It can be shown by thermodynamics that about half the potential energy released by a contracting star goes into radiation (or luminosity) and the other half goes into heating up its interior. Thus the internal heat or thermal energy of a star is numerically equal to about half the potential energy it has given up in its contraction.

Helmholtz and Kelvin showed that because of its enormous mass, the sun need contract only extremely slowly to release enough gravitational potential energy to account for its present luminosity. In fact, over the time span of recorded history, the decrease in the sun's size resulting from its contraction would be so negligible as to escape detection. It seemed to these researchers, therefore, that the sun's gravitational and thermal energies were sufficient to keep it shining for an extremely long time, and were certainly the source of its power.

The amount of potential energy that has been released since the presolar cloud began to contract is of the order of 10^{49} ergs. This is the amount, according to the Helmholtz and Kelvin theory, that it could have converted to thermal energy and luminosity. Since the present luminosity of the sun is 4×10^{33} ergs/sec, or about 10^{41} ergs/year, its contraction can have kept it shining at its present rate for a period of the order of 100 million years. It is only within the present century that it was learned that the earth, and hence the sun, has an age of at least several billion years, and therefore that the sun's gravitational energy is grossly inadequate to account for the luminosity it has generated over its lifetime.

(b) Nuclear Energy

As stated in Section 10.5a the nuclei of atoms can be regarded as composed of still smaller particles called *nucleons*. The most common nucleons are protons, which carry a positive electric charge, and neutrons, which are electrically neutral. Although neutrons are slightly more massive than protons, both have masses that are approximately 1 atomic mass unit (amu); 1 amu = 1.66×10^{-24} gm.

The number of protons in the nucleus of a particular kind of atom determines the total positive charge on the nucleus, and the number of protons plus the number of neutrons is equal to the atomic mass number of the nucleus (see Chapter 10). The mass of an ordinary hydrogen nucleus is exactly that of a proton (1.00813 amu). The total mass of every other kind of nucleus, however, is slightly less than the sum of the masses of the nucleons that would be required to build it. This slight deficiency in mass, always only a tiny fraction of 1 amu, is called the *mass defect* of the nucleus. The mass defect is greatest for the nucleus of the iron atom, and is less for both more massive and less massive nuclei.

A *nuclear transformation* is a buildup of a heavier nucleus from lighter ones, or a breakup of a heavier nucleus into lighter ones. Now the total mass of all the nuclei entering a nuclear transformation (or reaction) is, in general, slightly different from the total mass of the nuclei that result from the transformation. This difference in mass is simply the difference between the sums of the mass defects of the initial and final nuclei involved in the reaction. It might seem at first sight that the principle of conservation of mass is violated in nuclear transformations. However, Einstein showed, as part of his special theory of relativity, that there is an equivalence between mass and energy, given by the famous equation

$$E = mc^2,$$

where E is the energy equivalent to the mass, m, and c is the velocity of light. The mass unaccounted for in a nuclear transformation is always balanced by the absorption or release of energy in the amount given by the above formula. In this case m is the mass discrepancy involved in the nuclear reaction. In point of fact, the equivalence of mass and energy enables us to think of the mass defect of a single nucleus as the energy binding that nucleus together.

Natural radioactivity provides an illustration of the release of nuclear energy by the disintegration of certain heavy nuclei. Nuclei of radium, for example, spontaneously change into nuclei of radon and helium, and electrons. Radon and helium are two of the elements produced in a long chain of radioactive transformations that begin with uranium and end with lead. The total mass of the radon, helium, and electrons produced from radium is less than that of the radium that disintegrates; the lost mass appears as energy in the form of kinetic energy of the disintegration products, especially of gamma rays. The energy released when nuclei of uranium or plutonium break into smaller nuclei (as in some kinds of nuclear bombs and reactors) is another example of energy produced by the breakup of heavy nuclei. In the latter case, when the nuclear products of the breakup are of comparable mass, the process is called *nuclear fission.*

The opposite of nuclear fission is nuclear *fusion,* in which lighter nuclei combine to form heavier ones. It was suggested about 1928 that the energy source in stars might be fusion of light elements into heavier ones. Since hydrogen and helium account for about 98 or 99 percent of the mass of most stars, we logically look first to these elements as the probable reactants in any such fusion reaction. Helium nuclei are about four times as massive as hydrogen nuclei, so it would take four nuclei of hydrogen to produce one of helium. The masses of hydrogen and helium nuclei are 1.00813 and 4.00389 amu, respectively. Let us compute the difference in initial and final mass:

$$4 \times 1.00813 = \quad 4.03252 \text{ amu (mass of initial hydrogen)}$$
$$- 4.00389 \text{ amu (mass of final helium)}$$
$$\overline{\quad 0.02862 \text{ amu (mass lost in the transaction).}}$$

The mass lost, 0.02862 amu, is 0.71 percent of the mass of the initial hydrogen. Thus if 1 gm of hydrogen turns into helium, 0.0071 gm of material is converted into energy. The velocity of light is 3×10^{10} cm/sec, so the actual energy released is

$$E = 0.0071 \times (3 \times 10^{10})^2$$
$$= 6.4 \times 10^{18} \text{ ergs.}$$

This 6×10^{18} ergs is enough energy to raise the 200-inch telescope nearly 100 mi above the ground.

To produce the sun's luminosity of 4×10^{33} ergs/sec, some 600 million tons of hydrogen must be converted to helium each second, with the simultaneous conversion of about 4 million tons of matter into energy. As large as these numbers are, the store of nuclear energy in the sun is still enormous. Suppose half of the sun's mass of 2×10^{33} gm is hydrogen that can ultimately be converted into helium; then the total store of nuclear energy would be 6×10^{51} ergs. Even at the sun's current rate of energy expenditure, 10^{41} ergs/year, the sun could survive for more than 10 billion years.

There is little doubt today that the principal source of energy in stars stems from thermonuclear reactions. Deep in the interiors of stars, where the temperatures range up to many millions of degrees, nuclei of atoms are changing from one kind to another with an accompanying release of energy. The most important of these changes is the conversion of hydrogen to helium. The discovery that some of the energy locked up in the nuclei of atoms can be released in the interiors of stars is perhaps the most significant contribution of astronomy in the twentieth century.

(c) Rate of Nuclear Reactions

As we have seen, the rate at which the sun actually converts hydrogen into helium is known. Let us now inquire whether, from the theory of nuclear physics, it is possible to predict this rate, that is, whether we can calculate the rate at which thermonuclear reactions will occur in a gas of given chemical composition under various physical conditions.

For a nuclear reaction to occur, the nuclei of the reacting atoms must first collide with each other. As a result of this collision, a temporary "compound" nucleus is formed which either turns into a single nucleus of a different kind of atom from those originally taking part in the collision, or breaks into two or more less massive nuclei, again of different atomic types from those of the original reactants. How rapidly such events occur depends on (1) how fast the nuclei in the gas are moving (which depends on the temperature), (2) what the effective sizes of the nuclei are (which is computed from quantum theory for the simplest kinds of nuclei and measured in the nuclear physics laboratory for the more complicated ones), (3) how close together the nuclei are (which depends on the density of the gas), and (4) the probability that a particular compound nucleus that is formed will break into the relevant kinds of new nuclei (which is also determined either from theory or laboratory experiment).

With a rigorous mathematical treatment of the factors discussed in the preceding paragraph, nuclear physicists combine theory and experimental data to derive formulas that predict the rate of nuclear-energy production. Such formulas enable astronomers to predict the rate of energy release in a given region of a star in terms of the chemical composition of that region, its temperature, and the gas density. Since the total rate of energy release from a star (its luminosity) is known, the formulas obtained from nuclear physics give new information about the physical conditions in the stellar interior, which, when combined with our knowledge of the conditions of hydrostatic and thermal equilibrium, provide additional clues about the structure of the star.

There are two different known series of nuclear reactions by which hydrogen can be converted to helium under the conditions prevailing in stellar interiors. One process, known as the carbon cycle, involves collisions between nuclei of hydrogen and of carbon. In successive collisions with hydrogen nuclei (protons), a carbon nucleus is built up into nitrogen and then into oxygen, which in turn disintegrates back into a carbon nucleus again, and a new helium nucleus.

The other mechanism is called the proton-proton chain. In this process, protons collide directly to form, first, deuterium nuclei (heavy hydrogen), which, after further collisions with protons, are transformed into nuclei of a light form of helium. These light helium nuclei finally collide with each other to produce ordinary helium. In the sun, and in less luminous stars, the proton-proton chain contributes most of the nuclear energy, while in more luminous stars the carbon cycle is more effective. The actual nuclear reactions involved in both of these processes (proton-proton chain and carbon cycle) are given in Appendix 8.

Most of the electromagnetic radiation released in these nuclear reactions is at very short wavelengths — in the form of X-rays and gamma rays. Nuclear reactions are important, however, only deep in the interior of a star. Before this released energy reaches the stellar surface, it is absorbed and reemitted by atoms a very great number of times. Photons of high energy (short wavelength) that are absorbed by atoms are often reemitted as two or more photons, each of lower energy. By the time the energy filters out to the surface of the star, therefore, it has been converted from a relatively small number of photons, each of very high energy, to a very much larger number of photons of lower energy and longer wavelength, which constitute the radiation we actually observe leaving the star.

29.3 MODEL STARS

To determine the internal structure of a star, we must now combine the principles we have described: hydrostatic equilibrium, the perfect gas law, thermal equilibrium, the mode of energy transport, the opacity of gases, and the rate of energy generation from nuclear processes. These physical ideas are formulated into mathematical equations which are solved to determine the march of temperature, pressure, density, and other physical variables throughout the stellar interior. The set of solutions so obtained, based upon a specific set of physical assumptions, is called a theoretical model for the interior of the star in question.

(a) Computation of a Stellar Model*

There are many ways to formulate mathematically the physical principles that govern the structure of a star and to solve the resulting equations to obtain a stellar model. Here we shall illustrate the construction of a model star by describing a particular procedure that has been used widely.

Four quantities are chosen to describe the physical conditions at any distance from the star's center: the pressure, $P(r)$, the temperature, $T(r)$, the mass, $M(r)$, contained within a sphere of radius r concentric with the star's center, and the contribution to the star's total luminosity, $L(r)$, that is generated within this sphere. Once the pressure and temperature are known, the density can be calculated at once from the perfect gas law. The opacity and rate of energy generation at each point in the star involve no new parameters, for they, like the density, can also be expressed as functions of the pressure, temperature, and chemical composition of the stellar material at that point.

The four quantities $P(r)$, $T(r)$, $M(r)$, and $L(r)$ are then combined into four equations that express the physical principles involved. Each equation describes how one of the quantities changes through a small radial distance within the star (say, across one of the imaginary layers or shells described in Section 29.1a). The change in pressure across such a layer is given by the condition of hydrostatic equilibrium. The mass of the shell is given by the density of the layer, which in turn is given by the pressure and temperature of the layer and the perfect gas law. The change in luminosity across the layer is given by the rate of energy generation within the shell. The change in temperature is governed by the mode of energy transport; if in that region of the star the energy is transported by radiation, the opacity determines the temperature variation, while if that region is in convection, the expansion and contraction of the gases determine the change in temperature. Since there are four equations that hold for each point within a star, their simultaneous solution determines the four desired quantities at that point.

The solution of the equations may begin at the surface of the star, where the physical conditions are known. The mass and luminosity included within the surface are, of course, the total mass and luminosity of the star. The pressure and temperature often can be considered zero to a sufficient approximation, since their actual values in the photospheric layers of a star are very small compared to those in the interior. If higher precision is desired, "surface" values of pressure and temperature can be taken from the solution of a model photosphere of the star (Section 21.2b).

The four equations are then used to calculate how the values of pressure, temperature, mass, and luminosity change over a short distance inward, beneath the surface of the star, thus yielding the values of these quantities at that new depth. Next, the equations are used to calculate the changes over the next short distance inward. So, step by step, the pressure, temperature, mass, and luminosity are found at successively deeper layers in the star, until the center is reached.

At the center, of course, the mass and luminosity should be zero, for no mass or luminosity can be contained within a point. This would be true if all the physical laws governing a star were precisely understood, and if the chemical composition were precisely known at each depth in the star. In actual practice, neither the physical details entering into the opacity and nuclear-energy-generation formulas nor the chemical composition are

FIG. 29-2   Modern computing center.
(Control Data Corporation.)

of the pressure, temperature, mass, and luminosity throughout the star are known and constitute a finished model for its interior.

Since a possible model is found only for a "correct" choice of chemical composition, something is learned of the distribution of the various chemical elements in a star as well as of the physical conditions in its interior. Hydrogen and helium are thus found to comprise (usually) more than 95 percent of the mass of a star, and more than two thirds of that 95 percent is hydrogen. Unfortunately, however, the physical parameters in the opacity and nuclear physics theory are not yet known accurately enough to determine the chemical composition within a star with high precision from such studies.

The solution of the equations of structure to obtain a stellar model is a difficult and tedious business. Until the 1950's it often took as long as a year to compute a stellar model, and such a computation was a satisfactory topic for the dissertation of a student earning his Ph.D. degree. Now, however, high-speed electronic computers enable the calculation of a model in a few minutes or, in some cases, even in a few seconds.

(b) A Model for the Sun

The sun is the most studied of all stars, and models of its interior have been calculated for several decades. Each new model of the sun represents a refinement resulting from an improvement in our knowledge of physics or of computing methods or both. The general run of the physical parameters

known with absolute accuracy. Consequently, the solution of the four equations of structure may not lead to zero values of mass and luminosity at the center. The physical laws, therefore, are expressed as accurately as knowledge permits, and trial adjustments to the chemical composition are made until a set of solutions is found for which $M(r)$ and $L(r)$ do equal zero at the center. Then the runs

TABLE 29.1 Model for the Structure of the Sun†

FRACTION OF RADIUS	FRACTION OF MASS	FRACTION OF LUMINOSITY	TEMPERATURE (MILLIONS OF DEGREES K)	DENSITY (GM/CM^3)	FRACTION HYDROGEN (BY WEIGHT)
0.00	0.000	0.00	15.4	158	0.38
0.05	0.014	0.14	14.8	128	0.47
0.10	0.095	0.55	12.6	84	0.59
0.15	0.20	0.79	11.0	59	0.67
0.20	0.35	0.94	9.4	36	0.71
0.30	0.62	1.00	6.6	13	0.73
0.40	0.80	1.00	5.0	3	0.73
0.60	0.95	1.00	3.1	2	0.73
0.80	0.99	1.00	1.5	0.2	0.73
1.00	1.00	1.00	0.0	0.0	0.73

†Adapted from Torres-Peimbert, Ulrich, and Simpson.

in the sun, however, was fairly well established even in the early approximate models. The temperature within the sun increases gradually toward its center and reaches a value somewhere between 12 and 18 million degrees at the center. The density (like the pressure), on the other hand, increases very sharply near the center of the sun (indicating a high degree of central concentration of its material) and reaches a maximum value over 100 times the density of water.

One of the uncertainties that affects the central temperature and density of the sun is its age, for as time goes on, the thermonuclear conversion of hydrogen to helium in its central regions gradually changes its chemical composition. The exact temperature and density of the material at the sun's center that are required to account for its observed size and luminosity depend on how much the composition of that material has been changed — that is, on the age of the sun.

Series of models that trace the past history of the sun, therefore, have been calculated (more of this in Chapter 30). Table 29.1 and Figures 29-3 and 29-4 exhibit a model appropriate to the present-day sun that results from one such set of calculations (by Torres-Peimbert, Ulrich, and Simpson). The model is based on the best physical data available in 1968, and on the assumption that the sun was originally 73 percent hydrogen and 24.5 percent helium. According to this model, the outer layers of the sun are in convection, while the inner parts transport energy by radiation. The hydrogen abundance at the very center has been reduced (by nuclear reactions) to only 38 percent, and the present age of the sun is about 4.5 billion years.

It must be borne in mind that all models of a stellar interior, such as this model for the sun, are only as accurate as the state of our knowledge of the relevant physical processes at the times the models are constructed. Even at the time of writing,

FIG. 29-3   (a) Distribution of mass within the sun, according to the model in Table 29.1; numbers show what percentage of the sun's mass is included within the radial zones shown. (b) Distribution of energy generation in the sun according to the same model. Successively smaller circles show the regions within which 100, 80, 60, 40, and 20 percent of the sun's energy is produced.

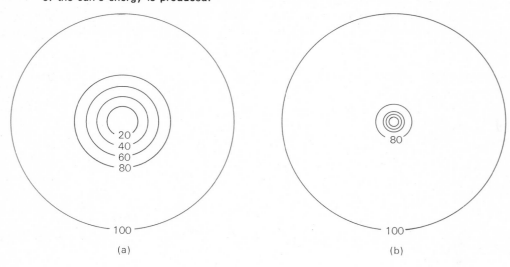

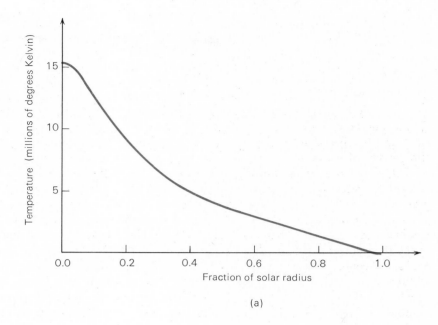

(a)

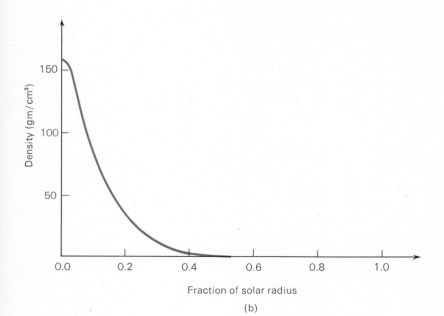

(b)

FIG. 29-4 (a) Temperature in millions of degrees Kelvin; (b) density in grams per cubic centimeter. Abcissas are fractions of the sun's radius. *(Adapted from a model by Torres-Peimbert, Ulrich, and Simpson.)*

new calculations are being performed (at Berkeley, Los Alamos, and elsewhere), using an improved theory of opacity. The model shown here, however, serves to illustrate the general nature of the internal structure of the sun.

(c) The Russell-Vogt Theorem and the Interpretation of the Main Sequence

A certain minimum amount of information, then, is required to determine the structure of a star, if it is assumed that all pertinent physical laws are perfectly understood and that infinitely accurate calculations can be performed.

It can be shown that if a star is in hydrostatic and thermal equilibrium, and if it derives all its energy from nuclear reactions, then its structure is completely and uniquely determined by its total mass and by the distribution of the various chemical elements throughout its interior. This is not to imply that we necessarily have the ability to compute perfect models for such stars; the stars themselves, nevertheless, must conform to the physical laws that govern their material and will adjust themselves to unique configurations. It should not, of course, surprise us that the very properties that a star is "born" with, and over which it "has no control," are just those that determine its structure. This important principle is known as the *Russell-Vogt theorem*.

Suppose a cluster of stars were to form from a cloud of interstellar material whose chemical composition was similar to the sun's. All condensations that become stars would then begin with the same chemical composition and would differ from each other only in mass. Suppose now that we were to compute a model for each of these stars for the time at which it became stable, and derived its energy from nuclear reactions, but before it had had time to alter its composition appreciably as a result of these reactions. The admixture of chemical elements would be the same, then, at all points within the star; such a composition is *homogeneous*.

The models we would calculate for these stars would indicate, among other things, their luminosities and radii. From Stefan's law we know that the luminosity of a star is proportional to the product of its surface area and the fourth power of its effective temperature (Section 10.4d). We can, therefore, calculate the temperature for each of the stars and plot it on the Hertzsprung-Russell diagram (Section 23.2). We would find that the most massive stars were the hottest and most luminous and would lie at the upper left corner of the diagram, while the least massive were coolest and least luminous and would lie at the lower right. The other stars would all lie along a line running diagonally across the diagram — the *main sequence*. The main sequence, then, is the locus of points on the H-R diagram representing stars of similar chemical composition but different mass. The observed fact that most stars in the Galaxy do lie along the main sequence is evidence that they have compositions similar to the sun's and are nearly chemically homogeneous. The observed scatter about the main sequence represents slight differences in the chemical compositions of actual stars. This explanation of the main sequence was first presented in the 1930s by the Danish astrophysicist B. Strömgren.

If we now plot the masses of the stars in our hypothetical cluster against their luminosities, we also find that the points lie along a line — the *mass-luminosity* relation. We have seen (Section 22.3) that most real stars do, indeed, obey a mass-luminosity relation; these stars are also of similar chemical composition. The locus of points on a plot of the masses of stars against their luminosities is simply the main sequence, plotted on a different kind of diagram. (Actually, it is possible for stars of the same mass but different chemical composition to have nearly the same luminosity — although they will differ in radius and temperature. Some stars, therefore, may obey the mass-luminosity relation even though they are *not* main-sequence stars, but this circumstance can be considered fortuitous.)

Those stars that do not lie on the main sequence in the Hertzsprung-Russell diagram (for example, red giants and white dwarfs) must differ somehow from the majority in their chemical compositions, or else they are not stable and are not shining by nuclear energy alone. We have seen, however, that as stars age they convert hydrogen to helium, and so change their compositions, especially near their centers. Chapter 30 will describe how most non-main-sequence stars can be interpreted either as stars that are still forming from interstellar matter and are not yet deriving all their energy from nuclear sources, or as stars that, by virtue of nuclear transformations, have altered their chemical compositions and hence their entire structures.

Exercises

1. Give some everyday examples of hydrostatic equilibrium. It is known that the pressure in a container of water increases with depth in the container. Is this a consequence of hydrostatic equilibrium? Explain. Compare the pressure-depth relation in water with that in the earth's atmosphere. Why is the case much simpler for water?

*2. Show that the combination of Boyle's law with Charles' law leads to the perfect gas law.

3. If the atmospheric pressure were the same on two different days, but if one day were much hotter than the other, what could you say about the relative density of the air on the two days?

4. If, in a vacuum chamber, the pressure is only one millionth of sea-level pressure, how does the density of the gas in the chamber compare with the average density of air at sea level?

5. Give everyday examples of convection and radiation of heat through a gas.

6. Verify that some 600 million tons of hydrogen are converted to helium in the sun each second.

7. Stars exist that are as much as a million times as luminous as the sun. Consider a star of mass 2×10^{35} gm and luminosity 4×10^{39} ergs/sec. Assume that the star is 100 percent hydrogen, all of which can be converted to helium, and calculate how long it can shine at its present luminosity. There are about 3×10^7 seconds in a year.

8. Perform a similar computation for a typical star less massive than the sun, such as one whose mass is 1×10^{33} gm and whose luminosity is 4×10^{32} ergs/sec.

9. Why do you suppose so great a fraction of the sun's energy comes from its centralmost regions? Within what fraction of the sun's radius does practically all of the sun's luminosity originate? (See Table 29.1.) Within what radius of the sun has its original hydrogen been partially used up? Discuss what relation the answers to these questions bear to each other.

Following page: Star cluster and nebula in Sagittarius, Messier 8. Note the small dark globules, possible sites of future star formation. *(Lick Observatory.)*

CHAPTER 30

Stellar Evolution

There are two approaches to the study of stellar evolution: the theoretical and the observational. In the theoretical approach, we calculate from the theory of stellar structure how stars should change as they contract gravitationally or age through changes in their chemical composition produced by nuclear reactions. In the observational approach, we observe stars or groups of stars that are at different stages in their evolution, and we check whether they actually exhibit the characteristics expected of them from the theoretical predictions.

The difficulty in the observational approach is that a star ages extremely slowly by human standards, and except for those brief periods in its existence where it may be variable or explosive (and the evolutionary significance of these phenomena is not yet well understood), we do not actually "see" a star evolving. A typical lifetime for a star might be something like 10 or 20 billion years, while the modern study of stellar evolution is barely a few decades old. Our problem of studying the evolution of a star by observing it for so short a time is like that of studying the evolution (or aging) of a man by observing him for only about 10 seconds. Effectively, therefore, we obtain only "snapshots" of stars of different ages.

Star clusters are the most useful objects to study in the observational approach to stellar evolution, because the stars within a cluster can usually be presumed to have a common origin and age, and to have all had, originally, similar chemical composition. From observations of cluster stars, we obtain their colors (or temperatures) and their relative brightnesses; if the distance to the cluster is known, we obtain the luminosities of the stars as well. For a cluster, therefore, we can plot a color-magnitude or Hertzsprung-Russell diagram (Section 28.4). Consequently, it is most convenient to compare theory and observations if we consider the tracks of evolution of stars on the H-R diagram. The reader must remember that the "position" of a star, or its "evolution," on the H-R diagram does not refer to its position or motion in space. Rather, these terms refer to the position and motion on a diagram of a point that represents the star's luminosity and surface temperature — they indicate changes in the structure of the star.

30.1 EARLY STAGES OF STELLAR EVOLUTION

No star that is shining today can be infinitely old, for eventually it exhausts its sources of energy. Moreover, the stars of highest known luminosity (100 thousand to 1 million times that of the sun) can continue to exist at the rate they are now expending energy for only a few million years. Had they been formed when the sun was formed, billions of years ago, they would long since have burned themselves out. At least some stars, therefore, have formed recently (in the cosmic time scale), and there is every reason to expect that stars are still forming today. The "birthplaces" of stars must be the clouds of interstellar material (gas and dust). That period in a star's existence during which it condenses from interstellar matter, and contracts into an "adult" star, may be considered as its "early stages."

(a) Theoretical Studies of Early Stages of Stellar Evolution

Here and there, in comparatively dense regions of interstellar matter, small condensations begin to form — atoms of gas and particles of dust slowly begin to collect under the influence of their mutual gravitation. As a condensing region grows, so does its gravitational influence, and more and more material is attracted to it. Eventually, material over a large region of space falls toward the central condensation.

Gas atoms and solid particles that are moving with speeds equal to the velocity of escape of the gravitating region can avoid becoming part of it and, unless collisions with other atoms or particles interfere, will not join the new star. Close to the center of condensation the velocity of escape is high. Far away, however, the velocity of escape is low, so that a small motion of a particle may give it enough speed to escape. At a large enough distance, just the random motions of the particles of dust or atoms of gas will suffice. The size of the contracting region is thus limited by the random motions of the particles.

As the material in the condensation falls together, gravitational potential energy is released (Section 28.2); much of this energy is radiated away, so the system loses energy. It is thus assured that the particles within the condensation will not, on the average, ever have high enough speed to carry them away from each other again. That is, the condensing mass is gravitationally bound and a protostar is born.

Eventually, the material in the condensation has a high enough density to become opaque to radiation. Thereafter the radiation can filter through the opaque matter only slowly. As the condensing mass continues to fall together, more and more of its potential energy is released, and since this energy is trapped in the interior rather than being radiated away at once, the interior heats up. Any solid grains vaporize and the gas becomes largely ionized. The pressures, especially, build up because of the increasing density and temperature. Finally the pressures are high enough to stop the further infall of material. Hydrostatic equilibrium

has been achieved (see Chapter 29) and a stable young star is formed.

It has been shown that as soon as the gravitational collapse of a protostar is halted (or at least very shortly thereafter) the material in the new star is in complete convection; that is, energy is carried from the center to the surface by leisurely convection currents (see Section 29.1e). We say that such a star is in convective equilibrium. The star's internal temperature is not yet high enough to support thermonuclear reactions, however, and its only sources of energy, so far, are gravitational and thermal. It tries to radiate away its internal heat, but this would cause the interior temperature and pressure to drop. To maintain hydrostatic equilibrium, then, the star must contract slightly, which builds up the temperature and pressure again. It turns out, in fact, that the pressure and temperature are even higher than before because the star has contracted slightly, so the pressure must support somewhat greater weights of the overlying material (remember that the weight of a layer of matter in a star is inversely proportional to the square of its distance from the center). Meanwhile, as the star contracts, more potential energy is released. Now that the star is in hydrostatic equilibrium, half of the gravitational energy released is radiated away and half is converted to internal heat in the star. Thus the star is gradually radiating energy into space and is deriving this energy from a very slow shrinking while its internal temperature and pressure continue to rise. The whole process is so gradual that hydrostatic equilibrium is never upset. This is exactly the process whereby Helmholtz and Kelvin attempted to explain the sun's source of energy more than a century ago (Section 29.2a). In these young stars, however, convection currents passing through their entire interiors are the principal mode of energy transport.

To follow theoretically the evolution of a star we compute a series of models, each successive

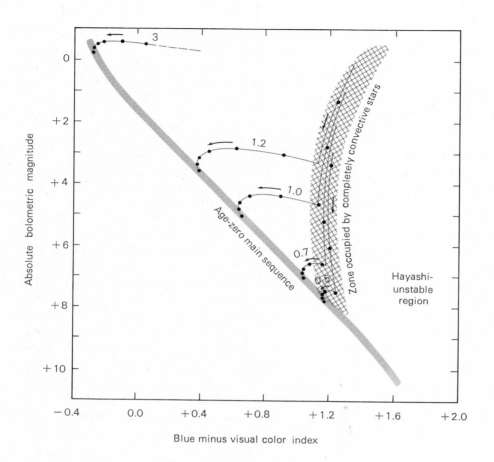

FIG. 30-1 Theoretical evolutionary tracks of contracting stars on the Hertzsprung-Russell diagram. Numbers labeling the tracks are the masses of the contracting stars in units of the sun's mass.

one representing a slightly later point in time. Given one (say, initial) model, we calculate the rate at which changes should occur due to, say, gravitational contraction or changing chemical composition. Then the total change that would occur during the time interval in question is found, after which a new model for the star is computed, appropriate to the new (changed) conditions that pertain at the end of the given time interval. With each new model, the luminosity and effective temperature of the hypothetical star are calculated so that its progress on the Hertzsprung-Russell (H-R) diagram is followed as it evolves.

Now we have seen that a star begins its existence in convective equilibrium and shines by virtue of its slow gravitational contraction. We now consider its theoretically predicted track of evolution on the H-R diagram (Figure 30-1). The Japanese astrophysicist C. Hayashi was the first to show that completely convective stars must lie in a zone on the H-R diagram extending nearly vertically from the lower main sequence to the right extreme of the regions occupied by red giants and red supergiants (crosshatched region in the figure). There can be no stable star such that the point representing it on the H-R diagram lies to the right of this zone. As convective stars contract in the initial stages of their evolution, they move (on the H-R diagram) downward in the zone along what are called *Hayashi lines*. The exact location of the Hayashi line for a particular star depends somewhat on its mass and chemical composition. More or less representative tracks for stars of several masses and of chemical composition more or less like the sun's are shown in Figure 30-1. We see that during its convective equilibrium stage a star decreases in luminosity without changing much in temperature.

Except for a star of low mass, after a period of some thousands or millions of years the convection currents cease at its center and there energy must be transported by radiation. The central zone in radiative equilibrium gradually grows in size, while the convection currents extend less and less deeply beneath the stellar surface. In this stage of its evolution, the star, still slowly shrinking and deriving its energy from gravitational contraction, turns sharply away from its Hayashi line on the H-R diagram and moves left, almost horizontally, toward the main sequence. Eventually, as the release of gravitational energy continues to heat up its interior, its central temperature becomes high enough to support nuclear reactions. Soon this new source of energy supplies heat to the interior of the star as fast as energy is radiated away. The central pressures and temperatures are thus maintained and the contraction of the star ceases; it is now on the main sequence. The small hooks in the evolutionary tracks of the stars shown in Figure 30-1, just before they reach the main sequence, are the points (according to theory) where the onset of nuclear-energy release occurs.

By the time stars of mass appreciably greater than the sun's have reached the main sequence, their outer convection zones have disappeared, but new cores of convection exist at their centers. Main-sequence stars of mass near that of the sun still have appreciable regions in their outer layers in convection, with their deep interiors in radiative equilibrium. Stars of rather low mass remain in complete convective equilibrium throughout and follow their Hayashi lines right down to the main sequence, where nuclear reactions finally stop their contraction. Stars of extreme low mass, on the other hand, never achieve high enough central temperature to ignite nuclear reactions. They continue to contract until (after an extremely long time) they are so dense that their matter becomes degenerate and they reach the white dwarf stage (Section 30.3). The lower end of the main sequence is considered to be that point at which stars have a mass just barely great enough to sustain nuclear reactions at a sufficient rate to stop gravitational contraction; this critical mass is believed to be near $\frac{1}{12}$ that of the sun.

At the other extreme the upper end of the main sequence terminates at the point where the mass of a star would be so high and the internal temperature so great that radiation pressure would dominate (Section 16.4e). The radiation produced from nuclear reactions would be so extreme that when absorbed by the stellar material it would impart to it a force greater than that produced by gravitation; hence, such a star could not be stable. The upper limit to stellar mass is believed to be in the range 60 to 100 solar masses.

In general, the pre-main-sequence evolution of a star slows down with time; the points on each evolution track in Figure 30-1 are intended to divide the track into segments over which the star spends roughly equal intervals of time. The time for the whole evolutionary process, however, is highly mass-dependent. Stars of mass much higher than the sun's reach the main sequence in a few million years or less; the sun must have required tens of millions of years; for stars to evolve to the lower main sequence requires hundreds of millions of years. For all stars, however, we should distinguish three evolutionary time scales:

(1) The initial gravitational collapse from interstellar matter is relatively quick. Once the condensation is, say, 1000 AU in diameter, the

time for it to reach hydrostatic equilibrium is measured in thousands of years.

(2) Pre-main-sequence gravitational contraction is much more gradual; from the onset of hydrostatic equilibrium to the main sequence requires, typically, millions of years.

(3) Subsequent evolution on the main sequence is very slow, for a star changes only as thermonuclear reactions alter its chemical composition. For a star of a solar mass, this gradual

FIG. 30-2 Nebulosity in Monoceros, situated in the south outer region of the young cluster NGC 2264. Photographed in red light with the 200-inch telescope. *(Mount Wilson and Palomar Observatories.)*

process requires billions of years. All evolutionary stages are relatively faster in stars of high mass and slower in those of low mass.

So much for the theoretical predictions. Now let us examine some of the observational data that give some hint that our theory must be at least partially correct.

(b) Observations of Very Young Star Clusters

The theoretical calculations described above enable us to predict what a cluster of stars that is now forming from interstellar matter should be like. Within a few million years, the most massive stars of the cluster should complete the contraction phase of their evolution and settle on the main sequence. As time goes on, more and more stars that are less and less massive should reach the main sequence. When the contraction phase is over, all the stars in the cluster should line up on the main sequence — just as we observe in the H-R diagrams of many clusters (but see Exercise 6).

On the other hand, since star formation is constantly going on, among the hundreds of known star clusters we might expect to find a few that are still in the process of formation, that is, with some of their stars still in the contraction phase. The more massive and luminous cluster stars might be expected to have reached the main sequence while their less massive companions would still be "on their way in." Within the last decade, several clusters have been observed that fit this description.

The first such cluster to be studied was NGC 2264, a small open cluster embedded in a cloud of gas and dust in the constellation Monoceros. The H-R diagram for this cluster (by M. Walker at the Lick Observatory) is shown in Figure 30-3. The brighter stars in the cluster are on the main sequence (solid line in the figure), while the less luminous ones (presumably also less massive) are off to the right. They are interpreted as being young stars, still contracting from the interstellar material associated with the cluster.

FIG. 30-3   Color-magnitude diagram of NGC 2264. The solid line is the position of the main sequence for "normal" stars. Stars indicated with crosses are T Tauri stars. *(Adapted from data by M. Walker.)*

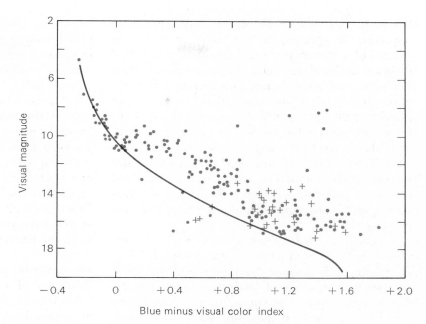

Today, several other such clusters, evidently very young ones, are known, all of which are associated with interstellar matter. One of them is the cluster in the central part of the Orion nebula. The probability seems very high that in these clusters we are witnessing the early evolution of stars.

Some of the stars in these young clusters are variable stars of the T Tauri type (Table 25.3). They show rapid and irregular light variations but are usually among the less luminous stars observed in the clusters. This type of variability may be associated with the gravitational contraction of stars of relatively low mass. The T Tauri stars in NGC 2264 are shown as crosses in Figure 30-3.

30.2 EVOLUTION FROM THE MAIN SEQUENCE TO GIANTS

As soon as a star has reached the main sequence, it derives its energy almost entirely from the thermonuclear conversion of hydrogen to helium. It re-

mains on the main sequence for most of its "life." Since only 0.7 percent of the hydrogen used up is converted to energy, the star does not change its mass appreciably, but in its central regions, where the nuclear reactions occur, the chemical composition gradually changes as hydrogen is depleted and helium accumulates. This change of composition forces the star to change its structure, including its luminosity and size. Eventually, the point that represents it on the H-R diagram evolves away from the main sequence. The original main sequence, corresponding to stars of homogeneous chemical composition, is called the *age-zero main sequence*.

(a) Evolution on the Main Sequence

As helium accumulates at the expense of hydrogen in the center of a star, calculations show that the temperature and density in that region must increase. Consequently, the rate of nuclear-energy generation increases, and the luminosity of the star slowly rises. A star, therefore, does not remain indefinitely *exactly* on the original age-zero main sequence. It appears, in fact, that the main sequence of a star cluster gradually rises in the H-R diagram as the cluster ages. The most massive and luminous stars alter their chemical composition the most quickly; thus, the main sequence rises most rapidly at the bright end, but scarcely at all at the faint end, even after billions of years. This stage of evolution does not cause the main sequence of a star cluster to deform appreciably, because a star increases its luminosity only by a moderate amount — probably less than a magnitude — before subsequent more rapid changes alter its structure enormously.

When the hydrogen has been depleted completely in the central part of a star, a core develops containing only helium, "contaminated" by whatever small percentage of heavier elements the star had to begin with. The energy source from hydrogen burning† is now used up, and with nothing more to supply heat to the helium core, it begins again to contract gravitationally. The star's energy is now partially supplied by potential energy released from the contracting core; the rest of its energy comes from hydrogen burning in the region immediately surrounding the core. These changes result in a substantial and rather rapid readjustment of the star's entire structure, so that the star "leaves" the vicinity of the main sequence altogether. Calculations suggest that a rather critical fraction of a star's mass must be depleted of hydrogen before the star evolves away from the main sequence in this way. That fraction is found to be about 10 percent. Clearly, however, the more luminous and massive a star, the sooner it ends its term on the main sequence. Because the total rate of energy production in a star must be equal to its

luminosity, the hydrogen is used up first in the very luminous stars. The most massive stars spend less than 1 million years on the main sequence; a star of 1 solar mass may remain there for more than 10 billion years.

Figure 30-4 summarizes the predicted evolution of the main sequence of a cluster of stars formed together. The dashed line is the original age-zero main sequence, and the solid line is the main sequence at some later time. The most luminous stars have evolved entirely away from the vicinity of the main sequence and are not shown. Other stars of successively lower luminosity (and mass) have evolved less and less far from their age-zero positions.

(b) Evolution to Red Giants

A number of investigators have calculated theoretical tracks of evolution for stars of various masses after they leave the main sequence entirely. These calculations necessarily involve assumptions about the appropriate physical formulas that apply in the interiors of these stars, and the uncertainty of these assumptions makes it impossible to predict exactly how a star will change. The general course of evolution away from the main sequence, however, is fairly well established.

The detailed calculations show that the central core of a star, where the hydrogen is exhausted, contracts, and as it does so the core increases in temperature, density, and pressure. The gravitational potential energy released from the contracting central regions of the star forces the outer parts of the star to greatly distend themselves. The star as a whole, therefore, expands to enormous proportions; all but its central parts acquire a very low density. The expansion of the outer layers causes them to cool, and the star becomes red. Meanwhile, some of the potential energy released from the contracting core heats up the hydrogen surrounding it to ever higher temperatures. In these hot regions the conversion of hydrogen to helium accelerates;

---

†The term "burning" is often used to describe the depletion of an element by nuclear reactions. This "nuclear burning" is not, of course, burning in the literal chemical sense.

FIG. 30-4   Predicted evolution of the main sequence of a star cluster.

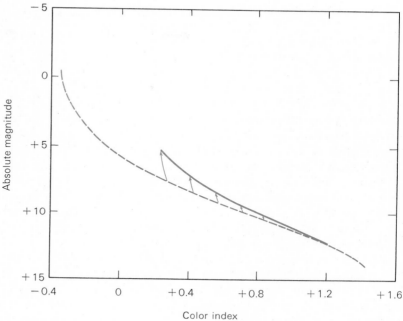

most stars actually increase in total luminosity. After leaving the main sequence, then, stars may be expected to move to the upper right portion of the H-R diagram; they become red giants.

(c) Observations of Star Clusters of Different Ages
The results of theoretical studies, outlined in the last two subsections, predict certain characteristics of the H-R diagrams of star clusters of different ages. In each cluster there should be some point along the main sequence where stars have just now reached the critical age where they rapidly evolve away from it; this point will be the upper termination of the main sequence for the cluster. In a young cluster, the main sequence should extend to stars of high luminosity; in successively older clusters, it will terminate at successively lower luminosities. As a cluster ages, its main sequence "burns down" like a candle.

Figure 30-5 shows a composite H-R diagram for several star clusters of different ages (compare with Figure 30-4). On the left side is shown the absolute visual magnitude scale. On the right side is a scale that gives the approximate ages of star clusters corresponding to the points where their main sequences terminate. These ages are based on computations of the times required for the cores of stars of various masses to become depleted of enough hydrogen to cause them to contract; the computations are almost certainly not precisely correct, but they do indicate the cluster ages to an order of magnitude. We see that the clusters shown range in age from only about 1 million years to several billion years.

As expected, most clusters have red giants. Those few which do not may just not happen to have any stars of the proper mass to be entering that stage of their evolution. In the younger clusters the red giants have magnitudes about the same as those of the brightest main-sequence stars; for stars of those masses and compositions, the tracks of evolution from the main sequence to the giant

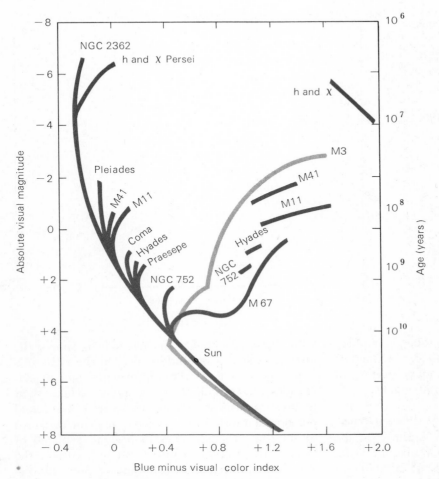

FIG. 30-5 Composite Hertzsprung-Russell diagram for several star clusters of different ages. *(Adapted from a diagram by Sandage.)*

stages are approximately to the right across the diagram. In the older clusters, however, the giants are brighter than the brightest main-sequence stars and must, therefore, have increased in luminosity during their evolution from the main sequence. Note the large gap in the center of the diagram where no cluster stars (actually, very few) appear (the Hertzsprung gap — Section 28.4b). There are theoretical reasons to expect stars of relatively high mass to become unstable as they leave the main sequence; they would be expected, therefore, to readjust themselves to the more stable red giant

configuration very rapidly. The observed absence of stars in the Hertzsprung gap does indeed suggest that stars of high mass evolve very quickly from the main sequence to the red giant domain.

In any case, the red giant stage must be a relatively brief part of a star's life. In the youngest clusters these stars are red *supergiants* of high luminosity. In the older clusters, stars *increase* their luminosities as they become red giants. In this stage of evolution, therefore, a star's nuclear fuel is consumed relatively quickly, and further evolutionary changes soon follow.

(d) The Oldest Star Clusters

The oldest assemblages of stars in the Galaxy are believed to be the globular clusters. The dashed line in Figure 30-5 outlines the H-R diagram for M3, a typical globular cluster; note that the main sequence terminates and branches into the giant sequence at a lower luminosity than for any of the other clusters. Ages for globular clusters are now variously estimated at from 5 to 20 billion years. A few open clusters (for example, NGC 188 and M67) may approach the globular clusters in age. In no cluster, however, does the main sequence terminate below the luminosity of the sun. Apparently, stars of solar mass in the Galaxy have not yet had time to evolve away from the main sequence.

Much study, both observational and theoretical, has gone into old star clusters. Figure 30-7 shows the appearance of the H-R diagram of a typical globular cluster, and Figure 30-8 shows the same for an old open cluster. It seems probable that the main sequences of at least some globular clusters lie lower on the H-R diagram (or more accurately, farther to the left) than do those of open clusters and most stars in the solar neighborhood. Moreover, the red giants in globular clusters are more luminous than are those in open clusters. These differences are believed to be due to differences in chemical composition; globular-cluster stars (population II) are, on the average, lower in abundance of heavier elements than the open-cluster stars of population I (Section 28.4c).

Analyses of counts of stars of various kinds in old clusters show that the number of giants is very small compared to the number of main-

FIG. 30-6   The globular cluster M13 in Hercules. *(Mount Wilson and Palomar Observatories.)*

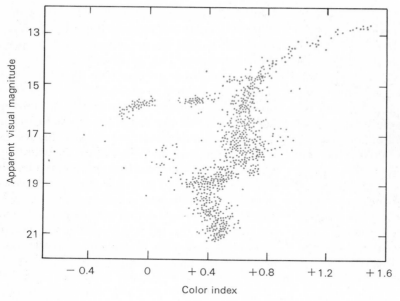

FIG. 30-7   Color-magnitude
diagram for the globular cluster
M3. *(Adapted from data by
Arp, Baum, and Sandage.)*

FIG. 30-8   Color-magnitude
diagram for the old, open
cluster M67. *(Adapted from data
by Sandage and Johnson.)*

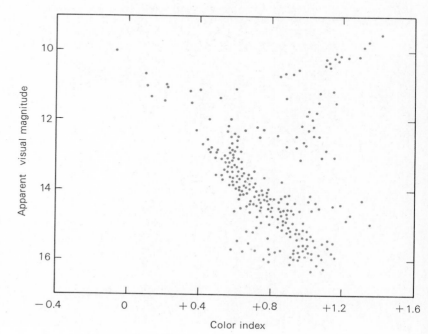

sequence stars, and that all the present giants in these clusters must have evolved from a very short segment of the original main sequence, just above its present termination point. In other words, all the giants in a single cluster are expected to have nearly the same mass, and also to have had almost the same luminosity when they were on the main sequence. Those stars at the top of the giant branch on a cluster H-R diagram are, to be sure, further evolved than those, say, only halfway up to the top, but they started from only very slightly greater luminosities on the main sequence, and so had only a slight "head start." We can conclude, therefore, that the sequence of stars forming the giant branch in the H-R diagram of an old cluster lies very nearly along the evolutionary tracks for the individual stars (see Figure 30-9).

The place on the main sequence from which the giant stars in an old cluster have evolved indicates that these stars have masses of about 1.1 to 1.2 solar masses. Several investigators have calculated theoretical tracks of evolution on the H-R diagram for stars of this mass after they leave the main sequence. The results of one such set of calculations are summarized in Figure 30-10. We can expect future work in the theory of stellar structure to force revisions of these approximate calculations, but the tracks shown in Figure 30-10 are probably correct in their general characteristics.

The shaded region in Figure 30-10 is the domain of main-sequence stars and giants in globular clusters. The solid line is the predicted track of a star of 1.2 solar masses and of negligible abundance of heavy elements (appropriate for some globular clusters). The dashed line is that for a star of the same mass but with chemical composition similar to the sun's. The theoretically predicted track for a star of essentially pure hydrogen and helium agrees fairly well with the observed sequence of giants in a typical globular cluster, while the star

FIG. 30-9   Presumed evolutionary tracks for individual giant stars on the Hertzsprung-Russell diagram for an old cluster. Shaded region is the observed domain of stars. *(Adapted from diagrams by Sandage.)*

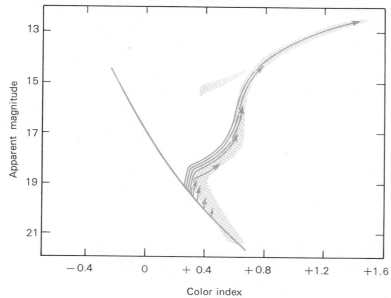

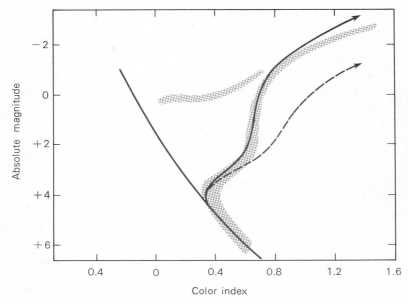

FIG. 30-10 Theoretical tracks of evolution for stars of 1.2 solar masses from main sequence to red giant. Solid line is track for a star of negligible heavy-element abundance; dashed line is for star of solar abundances of elements. Shaded region is observed main-sequence and giant domain for globular clusters. *(Adapted from a diagram by Schwarzschild and Hoyle.)*

with a "normal" abundance of heavy elements becomes less luminous as a giant, like the giants in old open clusters.

It seems, then, that the observed giant sequences in the H-R diagrams of old star clusters are quite well understood theoretically. We summarize by noting that globular clusters (and, presumably, other population II systems) have main sequences that terminate at about absolute magnitude 3.5; these systems are evidently old and have not experienced recent star formation. Open clusters, on the other hand, are often located in regions of interstellar matter, where star formation can still take place. Indeed, we find open clusters of all ages from less than 1 million to several billion years. The H-R diagrams of the oldest of them resemble those of globular clusters (at least as far as main-sequence and giant stars are concerned), except for differences that can be understood in terms of the differences in chemical composition of the two kinds of clusters.

30.3 LATER EVOLUTION STAGES

Our knowledge of the evolution of stars subsequent to their being red giants is less complete. We think we know, however, how stars end, and we can make some educated guesses about some of the intermediate steps.

(a) Termination of the Red Giant Stage

The most detailed models of red giant stars that have been constructed are for those of about 1.2 solar masses — presumably like the giants now observed in globular clusters. Part of the gravitational potential energy that is released in the contracting cores of these stars heats up those cores. By the time a star reaches the end of the red giant sequence in the H-R diagram, the temperature of its central regions must exceed 100 million degrees Kelvin.

At such a high temperature, nuclear processes other than the carbon cycle and proton-proton chain are possible. The most important of these is the formation of a carbon nucleus by three helium nuclei (the *triple-alpha process* — so named because the nucleus of a helium atom is also called an *alpha particle*). Successive bombardments of a carbon nucleus by helium nuclei can build up other, still heavier nuclei. One group of physicists and astronomers has found mechanisms whereby virtually all the chemical elements can be built up in the centers of red giant stars, in approximately the relative abundances that they occur in nature. It now seems quite probable that a gradual buildup of the heavy elements (heavier than helium) is continually going on in the hot centers of these stars.

It is thought that the triple-alpha process might begin explosively in the central core of a red giant. As the core evolves, not only does it get very hot but also very dense; in fact, the matter becomes electron-degenerate. One property of an electron-degenerate gas is that it conducts heat extremely well. (We shall describe an electron degenerate gas more fully in the next two subsections.) Now, as soon as the temperatures in one part of the core become high enough to start the triple-alpha process going, the extra energy released is transmitted through the entire degenerate core in a matter of seconds, producing a rapid heating of all the helium there. With the sudden rise in temperature, the helium burning accelerates like an explosion; the phenomenon is called the *helium flash.*

The new energy released expands the core rather quickly and reverses the growth of the outer parts of the red giant. The star then shrinks rapidly and increases in surface temperature. Although the calculations cannot be regarded as definitive, they indicate that the point representing the star on the H-R diagram moves more or less horizontally to the left. As soon as the helium is exhausted in a central region, however, the energy release from the triple-alpha process is shut off, and we have a situation analogous to a main-sequence star when its central hydrogen is used up and the hydrogen

burning ceases in its center. Now, however, we have a core of carbon (or perhaps heavier elements) surrounded by a shell where helium is still burning; farther out in the star we come to another shell where hydrogen is left and is still burning. The star now moves back to the right in the H-R diagram and returns to the red giant domain. The calculations indicate that a star may actually move first to the left across the H-R diagram, and then back to be a red giant several times, each time as a consequence of the onset of new nuclear reactions or of nuclear energy being released in new parts of the star. All these evolutionary stages occur in tens or hundreds of millions of years or less — a brief time compared to the stars' main-sequence lives. Some observational evidence supporting the theoretical calculations is the presence of the "horizontal branch" of stars on the H-R diagram of globular clusters and possibly in some open clusters (Section 28.4a). How the red giant phase of a star's evolution is finally terminated is not known. There is some thought that many, if not all, stars of about 1.2 solar masses eject planetary-nebula shells to end their careers as red giants (see below).

One stage of evolution of horizontal branch stars is through the "gap" occupied by the RR Lyrae variable stars, where no star is stable. As stars evolve across the horizontal branch they must some time pass through this stage of instability — as RR Lyrae variables. Other types of stellar variability must also represent stages of evolution of stars of other masses (compositions, or both), but ideas on how they fit into the picture are still rather speculative.

(b) White Dwarfs — A Final Stage of Evolution

Sooner or later a star must exhaust its store of nuclear energy. Then it can only contract and release more of its potential energy. Eventually, the shrinking star will obtain an enormous density. We observe such stars — the extremely compact white dwarf stars (Section 23.2d), whose mean densities range up to over 1 million times that of water.

White dwarf stars are far more dense, of course, than any solid substance. The high density of a white dwarf is possible because the atoms that comprise the gases in a stellar interior are almost completely ionized, that is, stripped of virtually all their electrons. Most of a neutral atom is empty space; once an atom is completely ionized, however, it and its freed electrons can occupy a volume many times smaller than when the electrons are still revolving about the nucleus.

It might seem that a star could collapse until the nuclei and freed electrons of its atoms were packed together like sardines. The contraction must cease, however, long before that. There is a theoretical limit to which certain kinds of particles, including electrons, can be packed. According to the quantum theory, no two electrons with almost the same velocity can exist in the same very small volume. There is a minimum volume, therefore, into which a given number of electrons can be crowded. How large that smallest possible volume is depends on the total range of velocities of the electrons; the *slower* they are moving, on the average, the *larger* is that minimum volume.

A star has a very great many electrons, each acting as a free particle in the stellar gas. Since some of the electrons in a white dwarf may have speeds near that of light, the total range of speeds of the electrons in the star is very large, but still it is not infinite. There exists, therefore, a minimum volume into which the electrons of a white dwarf can be compressed. As a star contracts, eventually the point is reached where its electrons have been packed into their smallest possible volume. They then exert an enormous pressure, which prevents the star from contracting any more. The electrons cannot slow down (that is, "cool") and allow further contraction, because if they lower their speeds they must occupy a *larger* volume, and there is no energy available to expand the star again.

A white dwarf is a star of such high density that its electrons are certainly packed to this limit, except in an extremely thin layer at the very surface of the star. In this state, the electrons are said to comprise a *degenerate* gas. White dwarfs are thus said to be degenerate stars. They have now reached their final size. There is still, however, a large amount of empty space between the gas particles; they are very far from "touching."

(c) Nature of a Degenerate Electron Gas
At usual stellar densities, the electrons in a gas behave as ordinary molecules and obey the usual gas laws (Section 29.1b). A degenerate electron gas is still a gas in that the particles (electrons) are still moving about at high speeds and exerting pressure in all directions, but it is not an ordinary gas. The electrons are not free to move in random directions, as are the particles in a perfect gas, for they would then encroach upon regions occupied by other electrons, already packed into their smallest possible space. Electrons can only change speeds or locations, therefore, as other electrons "get out of the way." The motions of all these particles are in a sense geared together.

The other particles that make up the gas in a white dwarf (the atomic nuclei) are still free to move as molecules of a perfect gas, and they exert their usual gas pressure. The pressure exerted by the degenerate electrons, however, is computed to far exceed that of the nuclei. The structure of white dwarfs, therefore, is dominated by this degenerate electron pressure.

(d) Structure of White Dwarfs
The structure of white dwarfs was first investigated by A. Eddington, the great British physicist and theoretical astronomer who was largely responsible for laying the foundations of the theory of stellar structure. Some of the electrons in some white dwarfs, however, have speeds near that of light, and Newtonian mechanics is not applicable to their

behavior. The first person to apply the theory of relativity to the construction of models of white dwarfs was the Indian (now American) astrophysicist, S. Chandrasekhar.

Chandrasekhar derived a relativistically accurate equation of state (gas law) that specifies the pressure of a degenerate electron gas in terms of the density of the stellar material. With this law and the condition of hydrostatic equilibrium (which still applies to white dwarfs), he was able to compute models for white dwarf stars of various masses.

It turns out that the entire structure of a white dwarf, including its radius, is specified uniquely by its *mass*. There is, therefore, a unique relation between the mass and radius of a white dwarf star. Chandrasekhar's mass-radius relation (with some small modifications that modern theory predicts) is shown in Figure 30-11. Note that the larger the mass of the star, the smaller is its radius. A white dwarf with a mass of about 1.2 solar masses would have a radius of zero! According to the theory, therefore, no white dwarf can have a mass quite as large as 1.2 times that of the sun.

Masses are known from direct observations for only three white dwarf stars that are members of nearby visual binary systems: the companions of Sirius, Procyon, and 40 Eridani. The radii of these three stars can be estimated from their rather uncertain temperatures and their luminosities (by Stefan's law — Section 22.4b). The observational data are not good enough to confirm definitely Chandrasekhar's mass-radius relation, but at least they are consistent with the theory.

The white dwarf upper limit of about 1.2 solar masses means that no stable electron-degenerate configuration of greater mass can exist. However, those stars that have had time to exhaust their nuclear fuel supply and evolve to the white dwarf stage must have had original masses greater than 1.2 solar masses, for those more massive stars are the very ones that use up their energy store most rapidly. At present we cannot rule out the possibil-

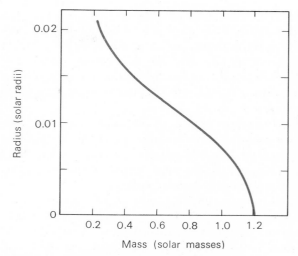

FIG. 30-11 Theoretical relation bewteen the masses and radii of white dwarf stars.

ity that a more massive star might continue to contract forever; such a star could not become electron-degenerate, so its electrons would have to avoid degeneracy by increasing their speeds tremendously as the star contracts. Eventually, the star would become so small that the velocity of escape from its surface would reach that of light, and photons could no longer escape. The star would, in a sense, "disappear" from the universe. There is no evidence, however, that any star has ever reached such a density, and there is some question whether it is possible. Stellar rotation, for example, might prevent it.

On the other hand, white dwarf stars are plentiful, and they must have come from somewhere. Moreover, the number of white dwarfs is high enough, as nearly as can be determined, to account for all evolved stars of original mass greater than 1.2 solar masses. It is widely accepted, therefore, that most, if not all stars eventually do, in fact, become white dwarfs. Consequently they must lower their masses somehow, before reaching that stage, by ejecting matter into space.

(e) Ejection of Matter by Stars

Shell stars, Wolf-Rayet stars, emission-line B stars, novae, supernovae, and various kinds of variable stars are all observed to be ejecting matter into space. It has even been shown that many red giants as well are steadily ejecting matter from their outer layers.

For stars of 1.2 or so solar masses — the most massive stars among red giants in globular clusters and in the galactic corona, nucleus, and disk — one of the most important mass-ejection mechanisms may be the planetary nebula phenomenon (Section 25.5c). Only about 1000 planetary nebulae are known, but when account is taken of the fact that they are identifiable as planetaries for less than 10^5 years, it is found that the number of planetaries is sufficient to be compatible with the hypothesis that all these red giants eventually undergo that form of mass loss. Moreover, there are compelling (although not conclusive) arguments that the planetary nebula shells are ejected from red giants and that their hot central stars are the inner parts of the red giants that are left after the mass ejection. The mass of a typical nebular shell is about 0.2 solar mass. Many of their central stars already have the dimensions of white dwarfs, and may well be evolving to that final state.

Stars of different mass probably undergo different means of mass loss. Although there are various theories for several mass-ejection mechanisms, we shall describe here only one — a modern theory for novae.

(f) A Nova Theory*

Evidence is rapidly accumulating that those stars that undergo nova outbursts are all members of close binary systems. A favored theory is that the prenova is a white dwarf — a star originally more massive than its non-white-dwarf companion but one that has already undergone its evolution, with mass loss, to reach that final stage. The companion is hypothesized to now be entering the red giant phase. At it distends its outer layers, they

pass out through the Lagrangian surface along which the matter is attracted equally to either star (see Sections 6.1b and 25.5d). Thus some matter from the new red giant flows onto the surface of the white dwarf. The surface gravity of a white dwarf is extremely high, and the added weight of this new matter produces a rise in pressure and temperature in the thin outer nondegenerate skin, where a little hydrogen may still be left. The rise of temperature and pressure, according to the theory, ignites the proton-proton cycle in that thin atmosphere of the white dwarf, and the energy released ejects some of that layer into space. The white dwarf then settles down to normal, but the nova can recur after a time as more matter flows from the giant to the white dwarf.

(g) Increase of Heavy Elements in the Universe

We have seen that stars convert hydrogen into helium and, moreover, that at least some stars in some stages of their evolution must be building up helium into carbon and heavier elements. Thus, inside stars, some of the lighter elements of the universe are gradually being converted into heavier ones. As these stars eject matter into the interstellar medium, that matter is richer in heavy elements than was the material from which the stars were formed. There is taking place, in other words, a gradual enrichment of the heavy-element abundance in interstellar matter. The heavy-element abundance in stars that are forming now should be higher than in those that formed in the past. The fact that the oldest known stars (those in globular clusters) are the stars with the lowest known abundance of heavy elements seems to support this idea.

Some astronomers have gone so far as to speculate that originally *all* stars in the Galaxy were formed of pure hydrogen, the lightest and simplest element, and that all other elements were formed in the hot centers of stars at advanced stages of evolution. Stars such as the sun, in whose outer layers heavy elements are observed spectroscopically, would then be of the second or third (or even higher) "generation," that is, have been formed of matter that was once part of other stars. The "original" pure hydrogen stars, much less massive than the sun, would not, however, be expected to leave the main sequence for many billions of years; while they remain on the main sequence, these stars would convert hydrogen to helium, but heavier elements would not yet be formed in them. Unless the Galaxy were at least 50 to 100 billion years old, we should see many of these "first-generation" stars without heavy elements; we do not. Moreover, there is now reason to believe that a substantial abundance of helium existed in the material from which the Galaxy formed — helium built up in the original "fireball" or "big bang" with which the universe began its present expansion (Chapter 33).

(h) Evolution of White Dwarfs

A white dwarf is presumed to have exhausted its available nuclear-energy sources. It cannot contract and release gravitational potential energy because of the great pressure of the degenerate electron gas. Thus the only source of energy is the thermal energy (that is, kinetic energy) of the nondegenerate nuclei of atoms, behaving as ordinary gas particles, scattered throughout the degenerate electrons. As these nuclei slow down (cool), the electron gas (which, unlike an ordinary gas, is highly conducting) conducts their thermal energy to the surface. At the boundary of the star, the very thin skinlike layer of nondegenerate gas radiates this energy into space. Only the opacity of this outer layer keeps the nuclei in the interior of the star from cooling off at once.

Gradually, however, a white dwarf does cool off, much like a hot iron when it is removed from a stove. The cooling is relatively rapid at first, but as the star's internal temperature drops, so does its cooling rate. Calculations indicate that its luminosity should drop to about 1 percent of the sun's in the first few hundred million years of its existence as a white dwarf. Several billion years more are required, though, before its luminosity has

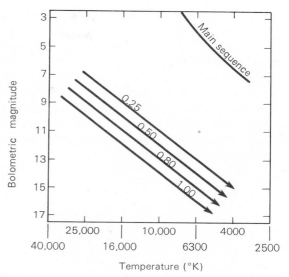

FIG. 30-12 Evolutionary tracks of white dwarf stars on the Hertzsprung-Russell diagram. Numbers give the masses of white dwarfs in units of the solar mass.

dropped to 1 ten-thousandth of the sun's. Since the radius of a white dwarf is constant, its surface temperature (by Stefan's law) is proportional to the fourth root of its luminosity. As it cools, its track on the H-R diagram is along a diagonal line, toward the lower right (low temperature and low luminosity). Figure 30-12 shows the evolutionary tracks of several white dwarf stars of different masses.

Eventually, a white dwarf will cease to shine at all. It will then be a *black dwarf*, a cold mass of degenerate gas, floating through space. An extremely long time is required, however, for a star to cool off to the black dwarf stage — many trillions of years. It is very doubtful that the Galaxy is old enough for any star to have had time yet to become a black dwarf.

(i) Neutron Stars

Another possible end product of stellar evolution is a neutron star. Neutron stars are hypothetical configurations composed entirely of neutrons. Ordinarily, a free neutron (one not bound in an atomic nucleus) survives only about 15 minutes before decaying into a proton and an electron. Under extremely high pressures, however, a neutron is stable. Suppose, somehow, that all the electrons in a star could be forced, under tremendous pressure, into the atomic nuclei. Since stars are electrically neutral there are just as many electrons as there are protons in the nuclei. Thus all the matter would become neutrons.

Neutrons behave as do electrons in the sense that they can become degenerate if crowded into a sufficiently small volume for a given velocity range. Thus, the structure of neutron stars is very analogous to that of white dwarfs, except that neutron stars are much smaller. A neutron star of 1 solar mass would have a radius of only about 10 km. There exists a mass-radius relation for neutron stars, and an upper mass limit as well, although some uncertainties in the calculations make

FIG. 30-13   Summary of evolutionary track on the Hertzsprung-Russell diagram for a star of 1.2 solar masses.

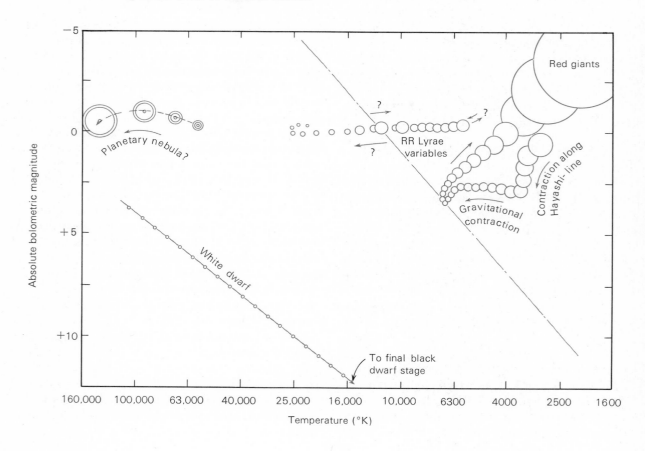

it difficult to say (at the time of writing) whether that upper mass limit is slightly less than or greater than the limiting mass for a white dwarf.

Theoretically, neutron stars could exist, but it is not known how a star would ever get into that state. If the upper mass limit exceeds that of white dwarfs, perhaps a star of mass greater than 1.2 solar masses could gravitationally contract to that state, missing the white dwarf configuration. However, if the limiting mass is less than that of white dwarfs, stars could not simply contract to that density, for they would stop on the way as white dwarfs. Possibly some mechanism within a star (say, a supernova explosion) could compress its inner regions to neutron-star densities and eject its outer layers, leaving a neutron star.

Neutron stars are discussed a great deal these days as possibly being associated with such phenomena as pulsars and quasars (Chapter 32). However, we do not know that they do, in fact, exist at all; there is no observational evidence — at least in late-1968 — for a single neutron star.

(j) Summary of Stellar Evolution

Figure 30-13 summarizes our current ideas on the evolution of a star of about 1.2 solar masses on the H-R diagram. In its early stages, the star contracts and moves to the left, reaching the main sequence with a size only slightly greater than that of the sun. In its subsequent evolution to the red giant stage, it grows to a radius of tens of millions of kilometers. The further evolution is uncertain. Perhaps the star goes through stages of variability, or emits material as a planetary nebula. Its final size as a white dwarf is only about that of the earth.

30.4 PAST AND FUTURE OF THE SUN AND SOLAR SYSTEM

The sun is a typical star, and much of the theory of stellar evolution that we have discussed applies to it. From theoretical calculations we can now form a fairly clear picture of the approximate past history of the sun, and we can make at least educated guesses about its future.

(a) Formation of the Solar System

A great many hypotheses have been proposed for the origin of the solar system. Most of these ideas fall into two general categories. First, Chamberlin, Moulton, Jeans, and Jeffreys are among those who have suggested that the planets were formed from material pulled from the sun by the tidal force of a passing star. Second, it was proposed in the mid-eighteenth century by Kant, and later by Laplace, that the planets and sun formed together from the same primeval matter — a large cloud of gas or dust or both. The facts that the planets all revolve in the same direction, and that their orbits lie in nearly the same plane, as well as the expected rarity of a close stellar encounter, have led modern investigators to lean to some variation of the second mechanism.

We can speculate that the original cloud of matter that formed the solar system must have been much larger than the orbits of the most distant planets. It must also have had at least some slight rotation (perhaps caused by the differential rotation of the Galaxy). As the cloud contracted, it rotated faster and faster to conserve its angular momentum. Most of the mass of the cloud found its way to the denser central condensation, ultimately to become the sun. A fraction of the material, however, was left behind and, because of its rotation, flattened into a disk. The planets and their satellites were formed from this disk.

The precise mechanism by which the planets formed is more speculative, as is the origin of the minor planets, meteoroids, and comets. One hypothesis, suggested first by C. F. Von Weizsäcker in the mid-1940s, has been modified by Kuiper and is now known as the protoplanet hypothesis. The protoplanets, according to the theory, were condensations that developed in the rotating disk. The planets themselves formed in the denser, central parts of these condensations and their satellites from smaller subcondensations. Only a small fraction of the material of these protoplanets, however, ended up in the planets and satellites; most of it was "blown" from the solar system by the force of electromagnetic and corpuscular radiation from the sun. The smaller planets (such as the earth), in fact, were not even able to retain the most abundant of the lighter elements.

Theories of the formation of the solar system, unfortunately, are less subject to observational verification than those of stellar evolution, and are less well founded. The reader who wishes to study this rather speculative subject will find references in Appendix 1.

(b) Early Solar Evolution

Several calculations of the sun's early evolution, analogous to those described in Section 30-1a, have been made. The exact track of evolution of the sun on the H-R diagram depends on its assumed initial chemical composition and on rather uncertain data involving the opacity of the outer layers of the young sun. The time required for it to contract to the main sequence is probably a few tens of millions of years.

Since it reached the main sequence, the sun has increased somewhat in luminosity, probably by about 30 to 50 percent. During that interval of from 4 to 5 billion years, it has depleted much of the hydrogen at its very center, but a pure helium core has not yet had time to form. It is not certain how much more time the sun has before starting to evolve to the red giant stage, but a good guess is that it has lived out about half of its main-sequence life. We can probably look forward to at least another 5 billion years before the sun's structure undergoes large changes.

(c) The Future of the Solar System

All available evidence leads us to expect that sometime in the future the sun will leave the main sequence and evolve to a red giant. We do not know exactly how large and luminous the sun will become, but it is possible that its photosphere will reach the orbit of Mercury and conceivably even that of the earth. Its luminosity will likely increase by 5 magnitudes or more. The oceans of the earth will boil. Life as we know it will end on the earth.

Eventually, the sun will leave the red giant stage and contract to a white dwarf. We do not know whether it will go through a stage of variability, eject mass in the form of a shell of gas, or otherwise. Ultimately, however, it must compress into a size comparable to that of one of the planets. If we could then come back to the cinder that was our earth, we would probably find it still revolving about the sun. What might conditions then be like?

Then oceans would have recondensed, but they would be frozen. The sun itself would appear as an unresolved point of light in the sky. At least for the first few hundred million years after becoming a white dwarf, however, the sun would deliver more than 100 times as much light to the earth as the full moon does now, and some of the nearer planets, shining feebly by reflected sunlight, would still be visible to the unaided eye. Most of the naked-eye stars, being more massive than the sun, would previously have become white dwarfs and would have faded from visibility. Main-sequence stars less massive than the sun would still be shining, but only those few of them passing temporarily through the solar neighborhood would be near enough to see with the unaided eye. It is doubtful, however, if by then *all* star formation in the Galaxy would have ceased. Luminous young stars might be shining in remote clouds of interstellar matter; a Milky Way might still stretch around the sky.

(d) Is the Solar System Unique?

In the solar neighborhood, at least, more than half the stars appear to be members of double- or multiple-star systems. Some of these systems contain invisible members whose presence is revealed only by their gravitational effects on the motions of the other stars in the systems. In a few cases, these unseen stars have very small masses; a companion of the star 61 Cygni, for example, has a mass that is only $\frac{1}{60}$ that of the sun, and Barnard's star (Section 22.2c) has a companion with a mass not much greater than Jupiter's. Some writers describe such objects as "planetlike." Actually, however, most cannot be solid bodies. Jupiter is nearly as massive as a solid object can be; if it had only a few times greater mass, it would be a feebly shining gaseous star.

Some astronomers have speculated that stars may form either in binary or multiple systems or else have planets. In a way, the sun and Jupiter can be considered a sort of "degenerate binary star." If the formation of the solar system were along the lines outlined above, it would not be surprising if there were many stars with planetary systems.

Unfortunately, our most powerful telescopes could not reveal an object as insignificant as the earth revolving about even the nearest other star. We may never know, then, whether worlds similar to the earth exist beyond the solar system.

At least we have no reason to believe that other planetary systems do *not* exist. They may be common, but at present we can only speculate on this possibility. Whether any of these unknown worlds possesses intelligent life, or even life at all, is a subject of further speculation which belongs in the field of biology, not astronomy. When we consider the many billions of other stars that could have planets, as does the sun, we might be tempted to believe that intelligent life on at least some of those hypothetical planets would be almost inevitable. Unfortunately, however, we have no basis whatsoever for guessing upon what fraction of those planets life may have developed, nor do we know the probability that simple life forms will ultimately develop into intelligent beings. If, for example, the chances of the latter are one in a million, intelligent life may abound elsewhere, but if they are only one in a trillion, we may be unique as "reasoning" beings.

Exercises

1. Where on the H-R diagram does a star *begin?*
2. Where on the H-R diagram does a star *end?*
°3. Suppose stars contracting from interstellar matter evolved exactly to the left across the H-R diagram (that is, at constant luminosity). The luminosities of main-sequence stars are approximately proportional to the cubes of their masses. Show that more massive stars would contract faster and reach the main sequence sooner than less massive stars.

4. The H-R diagram for field stars (that is, stars all around us in the sky) shows very luminous main-sequence stars and also various kinds of red giants and supergiants. We also find white dwarfs among the field stars. Explain these features, and interpret the H-R diagram for field stars.

5. Suppose a star spends 10 billion years on the main sequence and burns up 10 percent of its hydrogen. Then it quickly becomes a red giant with a luminosity 100 times as great as that it had while on the main sequence and remains a red giant until it burns up the rest of its hydrogen. How long a time would it be a red giant? Ignore helium burning and other nuclear reactions, and assume that the star brightens from main sequence to red giant almost instantaneously.

6. In the H-R diagrams for some young clusters, stars of very low and very high luminosity are off to the right of the main sequence, while those of intermediate luminosity are on the main sequence. Can you offer an explanation? Sketch an H-R diagram for such a cluster.

7. Why are all clusters that are known to be old also rich (that is, containing many stars)?

8. Why do you suppose masses are known from observation only for those few white dwarfs that are members of nearby visual-binary systems? (See Chapter 22.)

9. Show that the surface temperature of a particular, gradually cooling white dwarf is proportional to the fourth root of its luminosity. (See Chapter 10.)

10. Suppose the sun, while a red giant, were to become just 100 times as luminous as it is now. Show that the oceans would evaporate but that the water vapor would not escape from the earth. (*Hint:* See Chapter 12.)

11. At some future period when the sun is a white dwarf, it will be fainter than it is now in absolute visual magnitude by 6 magnitudes. At that time, which of the planets would be visible to the unaided eye from the earth? (Assume there is someone here to see them.)

12. By the time the sun becomes a white dwarf, the constellations familiar to us now could not be seen, even if their stars had never changed in luminosity. Why?

13. Do you expect that star formation in the Galaxy will eventually cease (a sort of Galactic Götterdämmerung)? Explain.

14. Calculate, very roughly, the density of a hypothetical neutron star of 1 solar mass.

The Pleiades, and associated
nebulosity in Taurus; 48-inch
Schmidt telescope photograph.
*(Mount Wilson and Palomar
Observatories.)*

cluster, M16, and nebulosity in Serpens;
nch telescope photograph. *(Mount Wilson
alomar Observatories.)*

The globular cluster
M13 in Hercules. *(U.S. Naval
Observatory; Flagstaff Station.)*

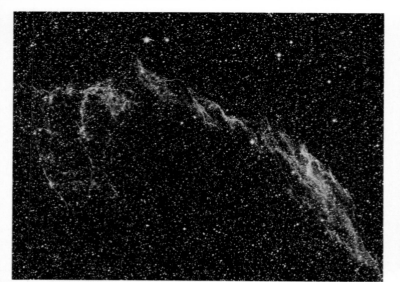

Veil nebula, or Cygnus "Loop," NGC 6992; photographed by the 48-inch Schmidt telescope. *(Mount Wilson and Palomar Observatories.)*

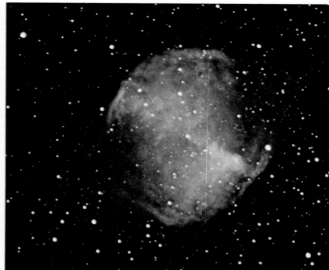

The Dumbbell nebula (M27) in Valpecula. *(U.S. Naval Observatory; Flagstaff Station.)*

The Crab nebula; 200-inch telescope photograph. *(Mount Wilson and Palomar Observatories.)*

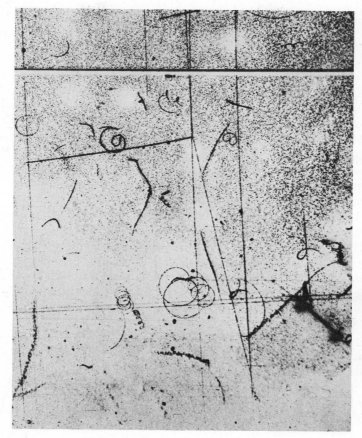

CHAPTER 31

Cosmic Rays

The foregoing chapters have described what has been learned of the sun and other stars, their distances, motions, compositions, structures, and evolution, as well as the structure and organization of our galaxy. All this knowledge has been acquired through the analysis of the electromagnetic radiation that we receive from these celestial objects. Now, before the analysis of the universe beyond our galaxy is considered, another important type of extraterrestrial radiation that we observe must be described briefly.

Our bodies are constantly being subjected to a rain of invisible high-energy particles passing through them. These bullets of radiation, undetected by our senses, result from the entrance of some 10^{18} atomic nuclei into the earth's atmosphere each second at speeds near that of light. The total rate of influx of energy to the earth from these particles is comparable to the rate at which the earth receives energy that comes from starlight.

The physicist has learned, through the study of this phenomenon, of kinds of subatomic particles (for example, *mesons* and *positrons*) hitherto unobserved. The incoming particles of extraterrestrial origin interest the astronomer because most are believed to come from beyond the solar system, and the discovery of the origin of these cosmic rays may provide us with new clues about the nature of the universe itself.

31.1 EARLY INVESTIGATIONS AND HISTORY OF COSMIC-RAY RESEARCH

For more than a century it has been known that the air has a slight electrical conductivity, because a charged body exposed to the air slowly loses its charge. Some of the atoms in the air, therefore, must be ionized. The electrons released from the atoms that have been ionized are attracted to a positively charged body, and the ions themselves are attracted to a negatively charged body; in either case, the charge on the body is reduced. The physicists Elster and Geitel investigated the conductivity of the air with an electroscope in 1899 and 1900 and found that the ionized particles in the air were continually being replenished.

Before proceeding, we must digress briefly to give a short description of some of the instruments with which charged particles can be detected, for the benefit of those unfamiliar with these detection techniques.

(a) Instruments for Detecting and Measuring Radiation*

A very simple device for detecting the presence of charged particles is an *electroscope.* An elementary "homemade" type of electroscope is illustrated in Figure 31-1. It operates on the principle that like charges repel. If the ball, A, at the top of the instrument is charged with electricity, part of that charge is conducted into the metal-foil leaves, B. Since the leaves have the same charge, they repel each other and separate; a measure of their separation is a measure of the total charge on the instrument. Now, if particles of the opposite charge are attracted to the ball, A, the charge on the electroscope is gradually neutralized, and the leaves come together. Thus, the rate of closing of the leaves measures the rate at which the electric charge leaks from the instrument as a result of ionization (that is, the presence of charged particles) in the air.

A radiation *counter* (for example, a *Geiger counter*) is a more sophisticated instrument than the electroscope.

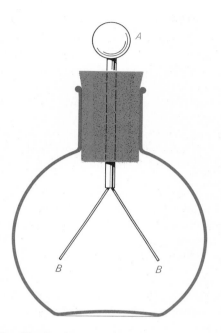

FIG. 31-1   Simple electroscope.

It consists of a gas-filled chamber across which an electric field is provided by oppositely charged electrodes on either side. When a high-energy charged particle enters the chamber, it ionizes some of the gas contained therein, so that the gas becomes momentarily conducting. This pulse of current flowing through the tube is amplified and recorded on a meter, or detected by means of a loudspeaker. A counter, therefore, can detect and record those individual charged particles that pass through it with enough energy to ionize the gas within the chamber.

A *cloud chamber* is a chamber filled with a gas that is saturated with the vapor of water or of some other liquid. The chamber is so designed that its volume can be enlarged suddenly, usually by moving a rubber diaphragm or a tightly fitting piston. As it is enlarged, the gas contained within it lowers in density and cools. If the chamber were enlarged sufficiently, the gas would cool enough for the liquid to precipitate. In actual use the cloud chamber is not enlarged quite enough for the precipitation to occur by itself. If, however, a charged particle should happen to pass through the chamber and

ionize some of the atoms of the gas just as it is cooling, those ions will serve as condensation nuclei, and a line of droplets will form, marking the path of the ionizing particle. The track can be observed visually or it can be photographed. Additional information is obtained if a magnetic field is superimposed upon the chamber because the path of a charged particle in a magnetic field is curved; the amount of curvature depends on the field strength, on the charge on the particle, and on its momentum. Cloud chambers with magnetic fields, therefore, not only show the tracks of the impinging particles but indicate something about their charges and momenta.

Tracks of charged particles can also be recorded in *photographic emulsions.* The grains in an emulsion are made capable of being developed not only by photons of light but also by ionization by charged particles. Ordinarily, a photographic emulsion is a very thin coating on a piece of celluloid or glass. For detecting particle radiation,

FIG. 31-2   Diffusion cloud chamber photograph. *(Brookhaven National Laboratory.)*

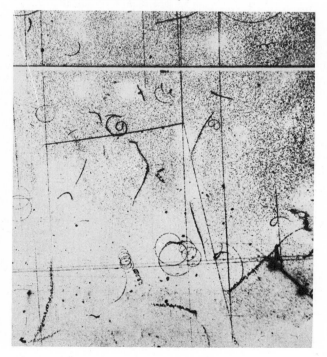

however, many layers of emulsion are often piled on top of one another, forming a thick emulsion *stack*. A charged particle passing through the stack leaves a track of developable grains behind it. After development the emulsion layers are separated and the track of the particle through each of them is measured; its course is thus determined.

Certain solid substances (for example, zinc sulphide) have the property that when bombarded by subatomic particles they emit flashes of light or *scintillations*. A scintillation counter is a device that utilizes such a scintillating phosphor to detect particles; the light flashes produced are amplified with a photomultiplier and recorded. Certain plastics fluoresce when hit by energetic particles, and so can also be used as scintillating materials. Large disks of such plastic spread out over a substantial area of the ground (for example, in the M.I.T. experiments at Volcano Ranch, New Mexico) have been used to detect great numbers of cosmic-ray particles.

One technique by which particles from a particular direction can be isolated is to arrange two or more counters in a line, and to record only pulses that occur almost simultaneously in all of them. Such coincidences are almost always the result of single particles that pass successively through each counter, and hence which approach along (or nearly along) the direction in which the counters are lined up. If sheets of lead (or other materials) of various thicknesses are placed between the counters, the penetrating power of a particle, and thus its energy, can be determined.

Electroscopes, cloud chambers, and geiger counters are no longer (or at best, rarely) used in cosmic-ray research, and emulsion stacks are used less and less in recent years, but these devices were important in the discovery and early investigation of cosmic rays. Space does not permit a discussion here of the many modern instruments which, in addition to scintillation counters, are in wide use today. At least, however, we have seen some ways in which fast-moving subatomic particles can be detected.

(b) Discovery of Cosmic Rays

In August 1912 the Austrian physicist Victor Hess carried an electroscope aloft in a balloon. He found that except near the ground the conduc-

tivity of the air *increased* with altitude. In 1914 this surprising result was confirmed by D. Kolhörster, who showed that the increase in conductivity with altitude continued to more than 25,000 ft. Apparently the radiation that ionized the air came either from high in the atmosphere or from beyond the earth.

Further evidence for extraterrestrial origin of the mysterious radiation was provided in 1928, when Millikan and Cameron lowered sealed electroscopes into two freshwater California lakes. They found that at successively greater depths under water, the radiation decreased, a result which would not be expected if it originated from within the earth's crust or atmosphere. Millikan gave the radiation the name *cosmic rays*.

(c) The Charged Nature of Cosmic Rays

At first, cosmic rays were believed to be very high energy photons, that is, electromagnetic energy of wavelengths even less than those of gamma rays (Section 10.1). In 1927, however, the Dutch physicist Clay found that the intensity of the ionizing radiation (cosmic rays) varies with latitude, being least near the geomagnetic equator (the circle halfway between the geomagnetic poles), and increasing as the geomagnetic poles are approached.† Clay's observations have been confirmed with many subsequent experiments. It is difficult to understand why photons, which have no electrical charge, should be in any way affected by the magnetic field of the earth as they approach it.

It is well known, on the other hand, that

---

†The *geomagnetic poles* are in line with the ends of a hypothetical ideal bar magnet whose magnetic field most nearly matches that of the earth. The actual field of the earth, however, is somewhat irregular, and the earth's *magnetic poles*, where the actual lines of force are perpendicular to the surface, deviate by some hundreds of miles from the idealized geomagnetic poles.

charged particles move on a curved path through a magnetic field. Charged particles entering the earth's magnetic field, therefore, would be deflected by the field; if their deflections, that is, the curvatures of their paths, are great enough, they will never strike the earth's atmosphere. Near the geomagnetic equator, only those particles of very high energy can reach the atmosphere. At higher geomagnetic latitudes, however, the lines of force of the earth's field curve toward its surface, and particles of lower energy can penetrate the field. At the magnetic poles, where the lines of force are perpendicular to the surface of the earth, particles of all energies reach the upper atmosphere. Thus, if the cosmic rays were charged particles, rather than photons, it would be easily understood why cosmic-ray intensity is greater nearer the magnetic poles. Today, it is well established that this explanation is, indeed, the correct one.

31.2 SOME PROPERTIES OF COSMIC RAYS

The realization only a few decades ago that the ionization noticed in the earth's atmosphere is due to charged particles striking the earth from space was one of the important discoveries of the twentieth century. Investigation of these particles became, in the 1940s, a major effort of modern physics. Although an exhaustive discussion of cosmic rays would not be appropriate here, some of their more interesting known properties may be summarized.

(a) Variation of Intensity with Altitude

The early observations of Hess and Kolhörster, as mentioned above, showed cosmic-ray intensity to increase with altitude. Modern investigations have shown, however, that the increase of intensity with altitude does not continue indefinitely. In Southern California it is found that at a height of 10,000 ft the cosmic-ray intensity is four times its sea-level value. The increase over sea-level intensity is 30-fold at 25,000 ft and 100-fold at 60,000 ft. At 100,000 ft, however, where the atmospheric pressure is only about 1 percent of its sea-level value, the cosmic-ray intensity is *less* than at 60,000 ft.

At first thought it might seem that the source of cosmic rays is near 60,000 ft above the earth's surface. Actually, however, it is found that at 60,000 ft, and below, most cosmic-ray particles are *secondary* particles, produced by collisions between the *primary*, extraterrestrial particles and molecules of air. At 100,000 ft, on the other hand, the atmospheric density is low enough that such collisions are relatively rare and most of the observed cosmic rays are the original primary particles from outer space.

(b) Primary and Secondary Particles

Analysis of the primary cosmic-ray particles shows that most of them are high-speed protons (nuclei of hydrogen atoms), that most of the rest are alpha particles (nuclei of helium atoms), and that a few are nuclei of the still heavier atoms. A primary particle traverses, on the average, only about one tenth of the earth's atmospheric gases, however, before colliding with the nucleus of an air molecule.

When such a collision occurs, the nucleus in the air molecule breaks into several smaller subatomic particles. If the primary particle has high energy, each of these secondary particles is also given considerable energy. It, in turn, collides with still another nucleus in an air molecule, producing more secondary particles. In this way an original primary particle moving with high speed dissipates its energy in a great many secondary particles, producing many of the particles that are recorded at intermediate and low altitudes in the atmosphere. A large proliferation of particles by successive collisions following the impingement of a primary particle of very high energy is called a *shower*.

Almost all the cosmic-ray energy observed at the earth's surface is due to secondary particles — direct or indirect products of collisions between the primary extraterrestrial particles and nuclei of air molecules. (Perhaps it is well that we do not see these primary and secondary particles or feel any sensation of them as they strike us. Otherwise, it might be discomforting to see repeated sky-rocketlike bursts high above our heads, and then observe many of the burst particles passing into one side of our bodies and out through the other.)

From their tracks in cloud chambers and through magnetic fields, physicists have learned the energies, charges, and masses of these secondary cosmic-ray objects. Most of them are found to be charged particles with masses intermediate between those of protons and electrons. Such particles of intermediate mass are called *mesons*. At sea level the most common mesons in secondary cosmic rays are called *mu mesons* (or *muons*); each carries either a positive or negative charge that is numerically equal to the charge on the electron and has a mass equal to 207 electron masses. Mu mesons are very unstable particles and disintegrate in periods (on the average) of only a few microseconds or less. When a mu meson disintegrates (or *decays*), for example, it forms an electron (if it had a negative charge) or a *positron* (if it had a positive charge). A positron is a particle equivalent to an electron but carrying a positive charge that is numerically equal to the negative charge on an ordinary electron. The excess mass of a meson, over that of an electron (or positron), is converted into energy and into *nutrinos* — particles that have energy but no mass. It is beyond the scope of this text to describe all these unusual subatomic particles; it is worth pointing out, however, that some of them were first observed in cosmic rays and that the study of cosmic rays has added much to our knowledge of the different kinds of fundamental particles that can exist, at least for brief periods.

(c) Energies of Primary Cosmic-Ray Particles

The unit in which we express the energies of cosmic-ray particles is a small unit of energy called the *electron volt*. One electron volt (eV) is the energy imparted to an electron that has been accelerated from rest through a voltage drop of 1 volt; it is equal to 1.602×10^{-12} erg. Ten million ergs of energy expended in 1 second is a *watt* of power. Typical electric light bulbs use about 100 watts of power.

The energies of most cosmic-ray primaries are found to be near 10^9 eV (a thousand million electron volts, abbreviated GeV), but a small percentage have energies in excess of 10^{18} eV. The relative number of particles of various energies is called the cosmic-ray *energy spectrum*. The energy spectrum is now known fairly well for cosmic-ray particles with energies in the range 10^8 to 10^{19} eV. In that range it is found that the numbers of primaries of successively higher energy drops rapidly; the total number of particles, $N(E)$, with total energies greater than E is represented, approximately, by the empirical formula

$$N(E) = \frac{\text{constant}}{E^{1.6}}.$$

The constant in the above equation is different for the different kinds of atomic nuclei that make up the primary cosmic-ray particles. The conclusion is that particles of higher energy are much less common than those of lower energy.

For energies above 10^{19} eV, the energy spectrum is much less well established. It is known, however, that a few primaries of extremely high energy exist. Spectacular showers have been detected for several decades with many techniques. For example, with the large fluorescent plastic disks at their Volcano Ranch site, the M.I.T. physicists have observed very large showers of secondary particles since about 1960. Since primary particles of the highest energy produce the largest showers, the energy of a primary particle can be inferred from the number of shower particles it produces. The M.I.T. group has found that for every 1 million showers of 1 million particles each, there are 1000 showers of 10 million particles each, and only 1 shower of 100 million particles.

Nevertheless, occasional showers of billions of particles have been observed. By 1962 many of the total number of showers observed at Volcano Ranch resulted from particles of energy greater than 10^{17} eV. Seven of the showers resulted from particles with energy above 2×10^{19} eV, three from particles with energy in excess of 5×10^{19} eV, and one from a particle with an energy of 10^{20} eV. A proton with an energy of 10^{19} eV has about 10^7 ergs of energy; this is an amount of energy equal to 10^{10} times that corresponding to the proton's mass — it could keep a 1-watt electric light burning for 1 second.

The energies observed for some primary cosmic-ray particles are many times those that can be obtained in laboratory accelerators (cyclotrons, synchrotrons, betatrons, and so on). The total energy density of cosmic rays in space is estimated at about one eV/cm³. Some of the possible origins of this energy will be considered later.

(d) Latitude Effects for Particles of Different Energies and the Isotropy of Cosmic Rays

Particles of low energy are deflected by the earth's magnetic field and can come close enough to the ground to interact with molecules of the earth's atmosphere only in the vicinity of the earth's magnetic poles. Particles that approach the earth near the geomagnetic equator cannot reach the denser parts of the atmosphere if they have energies much less than about 10 GeV (10^{10} eV). Particles of energy greater than about 60 GeV, on the other hand, always strike the atmosphere, even at the equator. Particles of intermediate energy may or may not reach the earth's dense atmosphere, depending on their direction of approach.

The *latitude effect*, that is, the variation of cosmic-ray intensity with latitude, should therefore be present only for cosmic-ray particles of relatively low energy. This prediction is confirmed by observation; for particles of energy above 60 GeV, the number striking the atmosphere is approximately the same at all latitudes.

Those high-energy particles from which the geomagnetic field does not successfully shield us are observed to approach the earth in approximately equal numbers from all directions. The high-energy cosmic-ray intensity, in other words, is *isotropic* about the earth. We presume that if it were not for the effects of the earth's magnetic field, the lower-energy particles would also appear to be isotropic. Further evidence for the isotropy of cosmic rays is that their diurnal (day to night) variation in intensity is less than ½ percent.

(e) Composition of the Primary Particles

As has been stated, the primary cosmic-ray particles are mostly atomic nuclei. The majority are protons (nuclei of hydrogen atoms). Most of the rest — about 15 percent — are alpha particles (nuclei of helium). The remaining 1 percent or so are nuclei of heavier elements; those of elements as heavy as iron are moderately abundant, and heavier ones have been observed. Very roughly, the relative abundances of the various atomic species in cosmic rays resemble those of elements elsewhere in the universe. There are exceptions, however. The cosmic-ray abundance of nuclei heavier than helium seems to be several times higher than the general cosmic abundance of those elements. The nuclei of the elements lithium, beryllium, and boron, in fact, are thousands of times as abundant in cosmic rays as in stars. Lithium, beryllium, and boron, however, are unstable at temperatures of a few million degrees, and in stars they would undergo nuclear transformations to other elements; their comparatively low abundances in stars, therefore, is not surprising. The high abundance of these nuclei in cosmic rays may result from the breakup of heavier nuclei. Finally, some recent evidence from balloon observations show that there is a small percentage (about 1 percent) of primary particles that are electrons.

31.3 ORIGIN OF COSMIC RAYS

The origin of cosmic rays is not yet fully understood and is a subject of considerable controversy. It is, in fact, one of the outstanding problems of modern astronomy and physics. We have learned enough of the nature of these strange particles, however, that we can at least *eliminate* some of the earlier hypotheses for their origin.

(a) The Sun as a Source of Cosmic Rays

Having determined that cosmic rays are extra-terrestrial, we look first, rather naturally, to the stars as possible sources of them. We have already seen that the sun is continually ejecting charged particles in the form of the "solar wind," and that many of these particles are responsible for magnetic storms, disturbances of the ionized layers in the earth's atmosphere, and auroras. The correlation between solar activity and these terrestrial disturbances, as well as satellite observations of increased corpuscular radiation approaching the earth at the times of increased activity on the sun, have shown conclusively that charged particles from the sun strike the earth. Can these particles be those we know as cosmic rays?

FIG. 31-3   Importance 2+ explosive solar flare, August 11, 1960. *(Lockheed Solar Observatory.)*

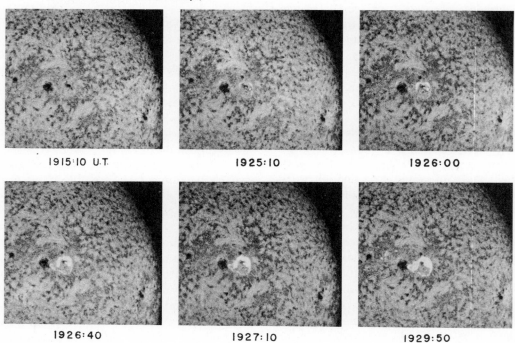

1915:10 U.T.   1925:10   1926:00

1926:40   1927:10   1929:50

Investigations have shown that occasionally the cosmic-ray intensity is indeed greater at times of unusual solar activity. In particular, the energy received in the form of cosmic rays has been observed to increase manyfold following violent solar flares.

We may suspect, therefore, that the sun is the source of cosmic rays. We know however, that on the long-time average the sun can contribute only a tiny fraction of the total high-energy cosmic radiation received at the earth.

(b) Why the Sun Cannot be Responsible for Most Cosmic Rays

There are several reasons why the particles ejected from the sun cannot account for most of the cosmic rays. First, the vast majority of the particles of solar origin have energies less than 0.5 GeV (that is, less than 5×10^8 eV), whereas cosmic-ray primaries have energies ranging from 1 GeV to more than 10^9 GeV. Only during violent flares has the sun been known to eject particles of energy as great as 10 GeV, and 20 GeV is about the highest energy ever observed for them. Moreover flares are operative only a small part of the time, whereas high-energy cosmic rays are observed all the time.

Second, the observed isotropy of the cosmic rays, and the extremely small diurnal variation of their intensity, argue against a solar origin for them. In other words, most cosmic rays approach the earth in equal numbers from all directions, not primarily from the sun. It has been suggested that clouds of ionized gases moving through the solar system from the sun might carry with them magnetic fields that could deflect particles ejected from the sun back toward the earth from any direction. Such a mechanism, however, could not deflect very high energy particles (those of energies greater than a few thousand GeV) even if such particles were ejected from the sun — and there is no evidence that they ever are.

Third, the primary cosmic-ray particles include nuclei of lithium, beryllium, and boron in relative abundances thousands of times higher than the relative abundances of these atoms in the solar atmosphere.

Finally, during the recent International Geophysical Year, it was found that there is actually an *anticorrelation* between solar activity and the cosmic-ray intensity observed at the earth. On the average, during times when there are many spots and active regions on the sun, we observe a slight *decrease* in the influx of primary cosmic-ray particles. Occasional solar flares, it is true, do contribute to the lower-energy cosmic rays (those of energy less than 10 GeV) received at the earth, and in rare cases these solar particles may be numerous enough to increase the total number of primary cosmic-ray particles by several times. But solar emission of this type is rare. More common are the many solar particles of energies less than 0.5 GeV that are emitted in great clouds when the sun is active. It is believed that these moving clouds of charged particles distort the magnetic field in the solar system in such a way as to shield the earth, slightly, from the cosmic-ray particles that come from beyond the solar system.

In summary, we find that the sun is a sporadic source of cosmic rays of energy less than 10 GeV (and especially of energy less than 1 GeV). Particles of higher energy, however, almost certainly cannot be coming from the sun. The conclusion is that most cosmic rays originate from outside the solar system.

(c) The Galaxy as a Source of Cosmic Rays

If cosmic rays do not originate from within the solar system, they must come either from the Galaxy or from beyond the Galaxy. An extragalactic origin for all cosmic rays seems unlikely, however; over the vast distances of intergalactic space, these particles would spread themselves so

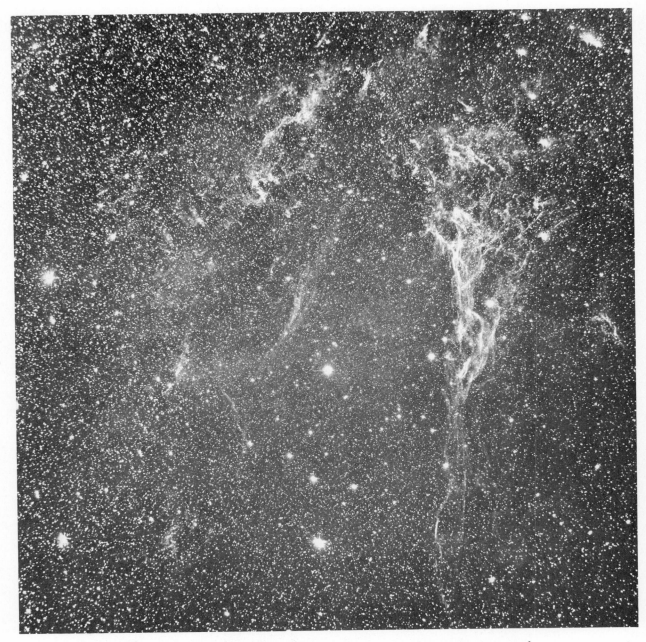

FIG. 31-4   Portion of the filamentary nebula in Cygnus, probably the remnant of
a prehistoric supernova, and possible source of cosmic rays. *(Lick Observatory.)*

thinly that truly fantastic sources of them would be required to account for the numbers that are observed to strike the earth. Most astronomers, therefore, lean to the hypothesis that most cosmic rays originate, somehow, within the Galaxy.

A satisfactory hypothesis for the origin of cosmic rays must account for the existence of the particles themselves, must explain how they obtain their high energies, and must explain how they can move about in the Galaxy in such a way as to approach the earth from all directions. We have already seen that there is strong evidence for magnetic fields in the Galaxy — at least in the regions occupied by interstellar matter (Section 27.6). It is strongly suspected that galactic magnetic fields are associated with the presence of cosmic-ray particles.

Calculations suggest that cosmic-ray particles of energy less than 10^{17} eV can be trapped by magnetic fields in the Galaxy and galactic corona for periods of 10 to 100 million years. Since these particles spiral around magnetic lines of force, they would strike the earth at all possible angles, and their directions of approach would have no relation to the direction from which they originated in the Galaxy. More energetic particles, on the other hand, and almost certainly those of energy greater than 10^{19} eV, could not be trapped within the Galaxy and should soon escape from it; those that are observed at the earth should be few in number and should not be isotropic if they had a galactic origin. Yet, many particles of energy 10^{17} eV and greater are observed. Moreover, the heavy nuclei are almost entirely absent among these energetic cosmic rays, which favors the view that the originally heavy nuclei may have been broken down to light ones by extremely long exposure to other radiation in space. It is likely therefore, that those primary particles of the highest energy have an extragalactic origin. The source of the particles and the mechanism by which they are accelerated to high energy are still to be explained.

(d) Origin of Cosmic-Ray Particles

We can, as yet, little more than speculate on the ultimate origin of cosmic-ray particles. The sun, we know, ejects particles of cosmic-ray energy at times, but only during rare violent flares are any of the solar particles ejected into interstellar space; all the stars in the Galaxy, if like the sun, would fall far short of providing the total cosmic-ray energy that is observed. Some stars, on the other hand, may eject particles more frequently and at higher energies than does the sun. Red giants, supergiants, T Tauri stars, and magnetic variables have all been suggested as possible sources. Collisions between magnetized gas clouds may also be responsible. Many astronomers, however, feel that the most promising candidates for the origin of at least galactic cosmic rays are the supernovae.

Supernovae are rare in any one galaxy (like our own), but particles ejected from them, and accelerated to high energies in galactic magnetic fields, could be stored in the Galaxy for many millions of years. The intense radio emission from expanding gas clouds about certain former supernovae appears to be due to very fast moving electrons spiraling about lines of force in magnetic fields surrounding those exploded stars (synchrotron emission). If electrons are ejected in the supernovae explosions, atomic nuclei must be poured into interstellar space as well. Possibly most of the primary cosmic-ray particles that now bombard the earth are tiny fragments of stars that exploded millions of years ago.

Exercises

1. The decay of naturally radioactive elements in the earth's crust would be expected to produce some ionization in the atmosphere. How did Hess's balloon experiment rule out the possibility that the total ionization of the air was so produced?

2. Describe how the observed cosmic-ray intensity at the surface of the earth would vary with the energy of the primary particle and with latitude if

(a) The earth had no magnetic field.

(b) The earth had a magnetic field millions of times stronger than its actual field.

3. Where on earth would you like to live to have the best chance of avoiding radiation resulting from primary cosmic-ray particles striking the earth at energies of 100 GeV? Why?

4. Discuss the relative abundances of the various kinds of atomic nuclei in cosmic rays in relation to the theory (discussed in Chapter 30) that heavy elements are "cooked" up in stellar interiors.

5. Verify the statement, given in the text, that if all stars in the Galaxy were like the sun, they would fall far short of being able to produce the observed amount of cosmic-ray energy. (*Hint:* Remember that the total cosmic-ray energy received at the earth is about the same as the energy received in the form of starlight.)

CHAPTER 32

Galaxies

The "analogy [of the nebulae] with the system of stars in which we find ourselves . . . is in perfect agreement with the concept that these elliptical objects are just [island] universes — in other words, Milky Ways . . . ."

So wrote Immanuel Kant (1724–1804) in 1755† concerning the faint patches of light which telescopes revealed in large numbers. Unlike the true gaseous nebulae that populate the Milky Way (Chapter 26), the nebulous-appearing luminous objects referred to by Kant are found in all directions in the sky *except* where obscuring clouds of interstellar dust intervene. Despite Kant's (and others') speculation that these patches of light are actually systems like our own Milky Way Galaxy, the weight of astronomical opinion rejected the hypothesis, and their true nature remained a subject of controversy until 1924. The realization, only a few decades ago, that our Galaxy is not unique and central in the universe ranks with the acceptance of the Copernican system as one of the great advances in cosmological thought.

32.1 GALACTIC OR EXTRAGALACTIC?

The discovery and cataloguing of nebulae had reached full swing by the close of the eighteenth century. A very significant contribution to our knowledge of these objects was provided by the work of the great German-English astronomer William Herschel (1738–1822) and his only son, John (1792–1871). William surveyed the northern

†*Universal Natural History and Theory of the Heavens.*

sky by scanning it visually with the world's first large reflecting telescope, instruments of his own design and manufacture. John took his father's telescopes to the Southern Hemisphere and extended the survey to the rest of the sky.

FIG. 32-1   William Herschel. *(Yerkes Observatory.)*

For a while, the elder Herschel himself, who had discovered thousands of "nebulae,"† regarded these objects as galaxies, like the Milky Way system; he was known to remark once that he had discovered more than 1500 "universes." He found, however, that many of the nebulae appeared as individual (although rather indistinct) "stars," surrounded by hazy glows of light. The fact that it is difficult to reconcile the appearance of such an object with that of a remote stellar system led Herschel, in 1791, to abandon the island-universe hypothesis. Still, the concept of the possibility of other galaxies never quite disappeared from astronomical thought.

(a) Catalogues of Nebulae

One of the earliest catalogues of nebulous-appearing objects was prepared in 1781 by the French astronomer Charles Messier (1730–1817). Messier was a comet hunter, and as an aid to himself and others in his field he placed on record 103 objects that might be mistaken for comets. Because Messier's list contains some of the most conspicuous star clusters, nebulae, and galaxies in the sky, these objects are often referred to by their numbers in his catalogue — for example, M31, the great galaxy in Andromeda (see also Chapter 28).

In the years from 1786 to 1802, William Herschel presented to the Royal Society three catalogues, containing a total of 2500 nebulae. The *General Catalogue of Nebulae*, published by John Herschel in 1864, contains 5079 objects, of which 4630 had been discovered by him and his father.

---

†*Nebula* (plural *nebulae*) literally means "cloud." Faint star clusters, glowing gas clouds, dust clouds reflecting starlight, and galaxies all appear as faint, unresolved luminous patches when viewed visually with telescopes of only moderate size. Since the true natures of these various objects were not known to the early observers, all of them were called "nebulae." Today, we usually reserve the word "nebula" for the true gas or dust clouds (Chapter 26), but some astronomers still refer to galaxies as nebulae or *extragalactic nebulae*.

FIG. 32-2   Herschel's 40-foot telescope. *(Yerkes Observatory.)*

The *General Catalogue* was revised and enlarged into a list of 7840 nebulae and clusters by J. L. E. Dreyer in 1888. Today most bright galaxies are known by their numbers in Dreyer's *New General Catalogue* — for example, NGC 224 = M31. Two supplements to the *New General Catalogue*, known as the first and second *Index Catalogues* (abbreviated "IC"), were published in 1895 and in 1908.

By 1908 nearly 15,000 nebulae had been catalogued and described. Some had been correctly identified as star clusters and others as gaseous nebulae (such as the Orion nebula). The nature of most of them, however, still remained unexplained. If they were nearby, with distances comparable to those of observable stars, they would have to be luminous clouds — probably of gas, and possibly with intermixed stars — within our Galaxy. If, on the other hand, they were very remote, far beyond the foreground stars of the Galaxy, they could be unresolved *systems* of billions of stars, galaxies in their own right — or as Kant had described them, "island universes." The resolution of the problem required the determination of the distances to at least some of the nebulae.

(b) Arguments For and Against the Island-Universe Hypothesis

By the early twentieth century the nebulae could be divided into two distinct groups: those largely irregular and amorphous, which are concentrated near the Milky Way, and those (either elliptical or wheel-shaped) showing symmetry, which are most numerous in parts of the sky far from the Milky Way. Those of the former type have bright-line spectra, and were clearly recognized as gaseous nebulae. The few available spectra of the brighter nebulae in the latter category, on the other hand, showed absorption lines like the spectra of stars. Moreover, most of their radial velocities were found to be very high (Chapter 33), and it seemed

FIG. 32-3   Two photographs of NGC 5457 made with the 200-inch telescope. Bright nova appears on the photograph taken February 7, 1951 *(right),* but not on the one taken June 9, 1950. *(Mount Wilson and Palomar Observatories.)*

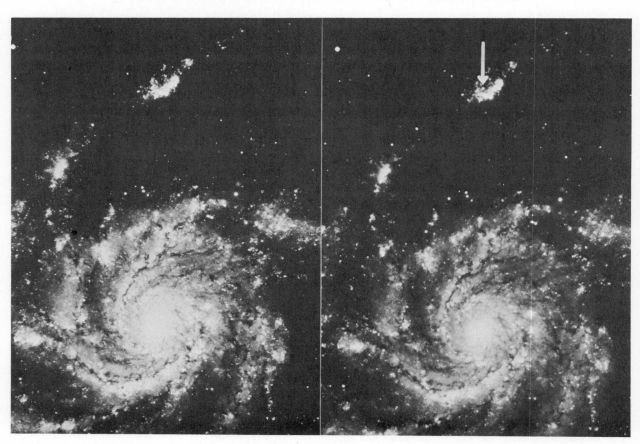

that such rapidly moving objects would escape the galaxy. These data supported the hypothesis that the nebulae of the second type are extragalactic.

In 1908 construction was completed of the 60-inch telescope on Mount Wilson, and within a decade the 100-inch telescope was also in operation. Photographs obtained with these instruments clearly resolved the brightest stars in some of the nearer "nebulae." By 1917 several novae (Section 25.3a) had been discovered in the more conspicuous nebulae. If those novae were as luminous as the 26 novae then known to have occurred in our own galaxy, they, and the nebulae in which they appeared, would have to be at distances of about 1 million LY — far beyond the limits of our galaxy. Similar distances were found for the nebulae if the brightest resolved stars in them were assumed to have the same intrinsic brightness as the most luminous stars in the Galaxy.

On the other hand, not all astronomers agreed that real stars had actually been resolved in any of the nebulae. Even the Mount Wilson astronomers who had photographed those stars were not convinced of their true stellar nature and had described them as "nebulous stars." Moreover, two of the novae that had been observed in the nebulae were far brighter than the rest. S Andromedae, for example, which appeared in the nebula M31 in 1885, reached magnitude 7.2. Today we recognize those two novae as *supernovae* (Section 25.3b), but supernovae were unknown in the early part of the century. If those two bright "novae" were assumed to be only as luminous as ordinary novae, the distances calculated for them turned out to be only a few thousand light-years, and the nebulae would not be extragalactic.

The most convincing evidence that seemed to prove that the nebulae were not galaxies was from measures in the early 1920s that appeared to show proper motions of brighter stars in a few of them. If those measures were valid, and the nebulae were remote enough to be extragalactic, those stars

FIG. 32-4   The Large and Small Magellanic Clouds (*top* and *lower left*). *(Sky and Telescope.)*

would have to have linear speeds near the speed of light or even greater. More recent and accurate measures have shown, of course, that those early results were in error, but at the time they seemed to give strong evidence for a small distance for the nebulae.

A remark should be made here concerning the Clouds of Magellan (Section 25.2c), two stellar systems that are not visible from as far north as the United States. The distances to the Clouds, determined from the cepheid variables in them, were thought (at that time) to be about 75,000 LY (they are actually at least twice that distance). In those years preceding 1924, however, the diameter of our own galaxy had been over-

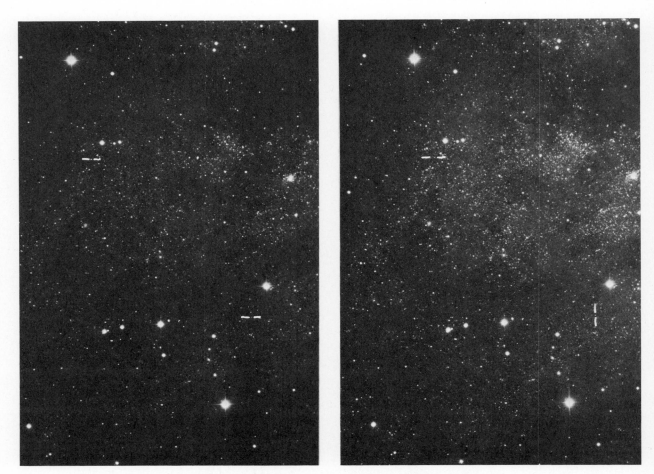

FIG. 32-5   A field of variable stars in the Andromeda Galaxy, with
two variables marked. Photographed with the 200-inch telescope.
(Mount Wilson and Palomar Observatories.)

estimated at about 300,000 LY (inadequate corrections having been made for the absorption of light
from the globular clusters by interstellar dust).
The Magellanic Clouds, therefore, although now
regarded as neighboring external galaxies, were
then considered to be outlying sections of our
galaxy.

Two of the major protagonists in the controversy over the nature of the nebulae were Harlow
Shapley, of the Mount Wilson Observatory, and
H. D. Curtis, of the Lick Observatory. Their
opposing views culminated on April 26, 1920, in
the famous Shapley-Curtis debate before the
National Academy of Sciences. Curtis supported
the island-universe theory, and Shapley opposed it.
Of course, the controversy was not settled by the
debate; according to A. R. Sandage, "Perhaps the
fairest statement that can be made is that Shapley

used many of the correct arguments but came to the wrong conclusion. Curtis, whose intuition was better in this case, gave rather weak and sometimes incorrect arguments from the facts, but reached the correct conclusion."[†]

(c) The Resolution of the Controversy

The final resolution of the controversy was brought about by the discovery of variable stars in some of the nearer "nebulae" in 1923 and 1924. Edwin Hubble, at the Mount Wilson Observatory, analyzed the light curves (Section 25.1b) of variables he had discovered in M31, M33, and NGC 6822 and found that they were cepheids (Section 25.2). Although cepheid variables are supergiant stars, the ones studied by Hubble appeared very faint — near magnitude 18. Those stars, therefore, and the systems in which they were found, must be very remote; the "nebulae" had been established as galaxies. Hubble's exciting results were presented to the American Astronomical Society at its thirty-third meeting, which began on December 30, 1924.

32.2 DISTANCES TO THE GALAXIES

When we realize how recently in the history of astronomy it was learned that galaxies are remote stellar systems, it may not seem strange to us that today, less than half a century later, their distances are known only roughly. Distances to even the nearest galaxies are not yet known with an accuracy of 1 part in 10. Most visible galaxies in the universe lie at distances that are uncertain by a factor of 2. The geometrical triangulation techniques that can be applied to yield many-place accuracy in the solar system do not work at all on galaxies; after many lifetimes of observation, the sun's motion in our Galaxy is still too little to pro-

vide us with a sufficient base line. We must, therefore, resort to indirect methods to "survey" the galaxies. The most important of these are based upon the photometry of individual stars and other objects contained within galaxies and upon the light of the entire galaxies themselves.

(a) Luminosity Methods of Estimating Distances of Galaxies

The most direct and reliable method of finding the distance to a galaxy is to identify in it an object or objects that are similar to known objects in our own Galaxy and which, presumably, have the same intrinsic luminosities (or absolute magnitudes). By comparing the apparent magnitude of such an object with its known absolute magnitude, we find its distance modulus, and hence its distance — which is, of course, the distance of the galaxy itself — by the method outlined in Section 20.2b. This was the procedure followed by Hubble when he derived distances to several nearby galaxies from the cepheid variables in them.

Cepheid variables are still our most important link to galaxian distances. That is why these relatively rare stars are so important and why so much effort is being expended today to ascertain as accurately as possible the absolute magnitudes of cepheids of various periods and other observable characteristics. We now think that we understand these stars well enough to estimate, with fair precision, distances to several neighboring galaxies in which cepheids can be observed. Unfortunately, however, there are only about 30 galaxies containing cepheids that are near enough so that we can observe those variables, even with the world's largest telescope. The nearest of these (the Clouds of Magellan) are only about 150,000 to 200,000 LY away from the sun, and the most distant are within 20 million LY.

There are many galaxies, closer to us than 20 million LY, which do not contain cepheid variables; these are systems that contain only population II

[†]A. R. Sandage, *The Hubble Atlas of Galaxies*. Carnegie Institution, Washington, D.C., 1961, p. 3.

stars (Section 27.5 and Chapter 30). In a few of them, RR Lyrae stars have been found, but these stars, with absolute magnitudes somewhere between 0 and +1, cannot be observed at a greater distance than about 1 million LY. A few population II galaxies (some of the dwarf ellipticals in the Local Group — see Table 32.2) are near enough that color-magnitude diagrams have been constructed for their brighter stars; their stellar content appears to resemble that of a typical globular cluster. It seems likely, therefore, that the brightest stars of such galaxies (always red stars) are red giants like those in globular clusters — at absolute magnitude $M_v = -3$. These stars can serve as distance indicators out to a distance of 4 or 5 million LY.

The brightest cepheids have an absolute magnitude of about -6. The most luminous stars in some kinds of galaxies are even brighter, and can be seen, therefore, to a greater distance. The brightest stars found in our own and other galaxies (usually blue supergiant stars) have absolute magnitudes of about -9. Bright novae, at maximum, reach about the same luminosity, and this is also about the absolute magnitude of the most luminous globular clusters. Thus the brightest supergiant stars in a galaxy, occasional nova outbursts that occur in it, and its brightest globular clusters all are distance indicators that can be observed to as far away as 80 million LY.

Some galaxies contain emission nebulae (H II regions — Section 26.2a) that exceed their brightest stars in luminosity by 3 magnitudes or more. Because of the uncertainty of how luminous such nebulae can be, however, they indicate distances at best only roughly. Promising as possible distance indicators of the future are the supernovae; some types of supernovae may reach, at maximum light, absolute magnitudes of -19 or even -20 and are visible at distances of billions of light years. Unfortunately, different supernovae cover a wide range in maximum brightness and we do not yet

know how to infer the intrinsic luminosities of various kinds of supernovae from their observable characteristics (such as their light variations, or spectra). At present, therefore, distances based upon identifiable objects *within* galaxies cannot be estimated with confidence for those galaxies too remote for their supergiant stars, novae, or globular clusters to be observed.

If all galaxies were identical, they would all emit the same total amount of light, and the magnitude of a galaxy as a whole would indicate its distance. We shall see that galaxies range enormously in total luminosity. Nevertheless, there is growing evidence that some types of galaxies (Section 32.4) have luminosities near certain *mean* luminosities; for those galaxies, rough distances can be estimated from their total apparent magnitudes. Of course, we can only establish what those "mean luminosities" are for the types of galaxies that are well represented among those whose distances can be determined from the magnitudes of individual stars contained in them. Unfortunately, many galaxies, probably the majority, do not have distinguishing characteristics that enable us to estimate their absolute magnitudes; we can only tell which ones are highly luminous and which ones are not if we see a collection of them, of various brightnesses, side by side in a cluster (Section 32.2c).

(b) Distribution of Galaxies

The distances that separate galaxies are hundreds of thousands to millions of light-years. There are less than 20 known galaxies within 2½ million LY, but there are many thousands within 50 million LY; galaxies extend in all directions as far as we can see. The more distant galaxies, of course, appear less conspicuous than the nearer ones. Tiny images of extremely remote galaxies are actually more numerous than images of the foreground stars of our own Galaxy on photographs obtained with large telescopes directed away from the

Milky Way. The images of galaxies at the limit of detection can be distinguished from those of faint stars only because the galaxian images are rather fuzzy or elongated and lack the sharp, point-like appearance of the star images.

In the years following his announcement of the true nature of the "nebulae," Hubble made an extensive study of galaxies and their distribution in space. He realized that it would take thousands of years to photograph the entire sky with the 100-inch telescope and to count all the galaxies that could be observed with that instrument. Instead, he photographed 1283 sample regions of the sky with the 100-inch telescope. The results of Hubble's survey are shown in Figure 32-6. Each symbol in the figure shows one of the sample regions, and

the size and type of the symbol shows how many galaxies were visible on the photograph of each region. From the 44,000-odd galaxies Hubble counted in these selected regions, he calculated that nearly 100 million galaxies must exist within the range of the 100-inch telescope. There may be 1 billion galaxies within the reach of the 200-inch instrument at Palomar.

Hubble found the largest number of galaxies in the regions near the galactic poles (90° from the great circle running through the Milky Way) and determined that, on the average, fewer and fewer galaxies could be seen in directions successively closer to the Milky Way. This, of course, is an effect of absorption of light by interstellar dust in our Galaxy (see Section 26.1b); from the

FIG. 32-6   Distribution of galaxies. The size of each dot plotted on the map of the sky indicates how many galaxies were counted on a photograph taken of that location with the 100-inch telescope. Dashes and open circles indicate few or no galaxies, and the irregular region across the center of the map where almost no galaxies are to be seen is the zone of avoidance. This zone coincides with the visible Milky Way, where interstellar dust obscures the exterior galaxies. The empty regions on the plot are those of the sky too far south to survey from Mount Wilson.

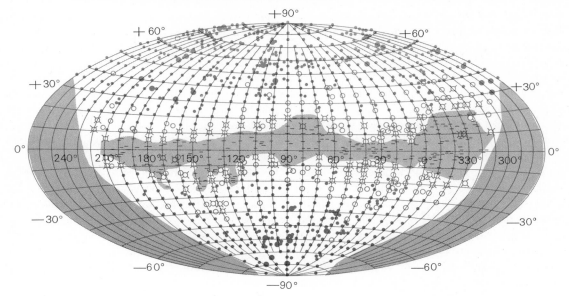

apparent thinning of galaxies toward the Milky Way, Hubble estimated that light is dimmed by about 0.25 magnitude (photographically) in traversing a half-thickness of the Galaxy at the sun's position. The obscuration of light by interstellar dust is so complete in a region along the Milky Way that no galaxy shows through (or nearly none); Hubble called this region the "zone of avoidance" (Section 26.1b).

Hubble found that when allowance is made for interstellar absorption, the distribution of galaxies, on the large scale, is the same in all directions, and that the mean density of galaxies in space seems to be about the same at all distances. On the small scale, however, galaxies tend to be grouped into *clusters of galaxies*. A discussion of these clusters is given in Section 32.5, but a description of how relative distances to them can be estimated is relevant here.

(c) Relative Distances to Clusters of Galaxies

All the galaxies in a cluster are at about the same distance — the distance to the cluster itself. We can intercompare those galaxies, therefore, and pick out those which are intrinsically bright and those which are faint. If a cluster is rich, that is, if it has many member galaxies, it is probable that it will contain a few highly luminous galaxies. We can estimate relative distances to different clusters, therefore, by comparing the apparent magnitudes of their brightest members. If the light received from the brightest galaxies in cluster *A*, for example, is four times less than from the brightest galaxies in cluster *B*, cluster *A* is probably twice as distant as *B*. It is assumed here that intergalactic space is perfectly clear and that correct allowance has been made for the interstellar absorption in our own Milky Way Galaxy. As we shall see (Section 32.8) there is no evidence for intergalactic absorption of photographic light.

Using the magnitudes of the brightest cluster members in this way works only for rich clusters, which can be expected to have similar galaxies for their brightest members. Unfortunately, there is no very rich cluster close enough to us so that we can determine its distance by independent means (say, from resolved stars in some of its galaxies). We are not certain, therefore, of the actual luminosities of the brightest galaxies in rich clusters. *Relative* distances to clusters are thus known more reliably than absolute distances.

Relative distances to clusters can also be estimated from their angular sizes and from the average apparent separations of their members.

(d) Summary

Distances to the nearest galaxies are found from the apparent faintness of stars and other objects of known luminosity that can be recognized in them. Although the distances found for these neighboring galaxies are very great compared to the largest astronomical distances known a half century ago, they are still very small compared to those of the overwhelming majority of galaxies that can be observed with large telescopes. Distances can be obtained for remote galaxies only when they happen to be members of rich clusters, and then only with considerable uncertainty.

We have not referred in this section to the well-known correlation between the distances of galaxies and their radial velocities, as indicated by the Doppler shifts of the lines in their spectra. Although distances are often estimated for galaxies from their radial velocities, the method is not really independent, for it depends on the calibration of the velocity-distance relation, which, in turn, must be based on distances measured by other means to selected galaxies with various observed Doppler shifts. The accurate calibration of this relation is one of the major problems of extragalactic astronomy. The velocity-distance relation, and its calibration, is so fundamental to the study of modern *cosmology* that discussion of it is postponed to Chapter 33.

32.3 DETERMINATION OF GROSS PROPERTIES OF GALAXIES

The linear size of that part of a galaxy that corresponds to an observed angular size can be calculated once the distance to the galaxy is known, just as we calculate the diameter of the sun or of a planet. Also, if we know its distance, we can apply the inverse-square law of light and calculate the total luminosity, or the absolute visual magnitude of a galaxy from the amount of light flux we receive from it or from its apparent visual magnitude. Thus a knowledge of galaxian radii and luminosities is provided by a knowledge of galaxian distances.

The determination of masses of galaxies, however, is more difficult, and, in fact, is possible (by present techniques) for only a small fraction of them. There are several techniques for measuring galaxian masses.

(a) Mass of Galaxies

We determine the masses of galaxies, like those of other astronomical bodies, by measuring their gravitational influences on other objects or on the

FIG. 32-7   The Andromeda Galaxy, M31, and its rotation curve. *(Yerkes Observatory.)*

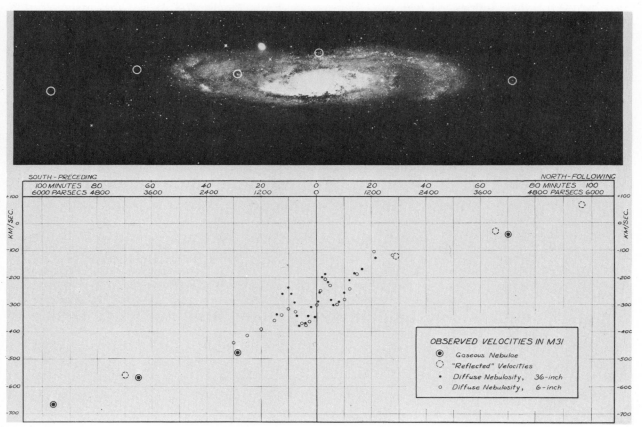

stars within them. We must assume, of course, that Newton's law of gravitation is valid over galaxian and extragalactic distances. The internal motions within galaxies and the structures of clusters of galaxies suggest that Newton's law does indeed apply, but we must, nevertheless, bear in mind the assumption on which our calculations are based.

Internal motions in galaxies provide the most reliable methods of measuring their masses but are observable only in comparatively nearby galaxies. The most important procedure is to observe the rotation of a galaxy and then to compute its mass with the help of Kepler's third law.

As an illustration, we shall consider the rotation of M31 — the Andromeda galaxy (Figure 32-7). Galaxy M31 is probably quite similar to our own galaxy; it has a brilliant central nucleus and conspicuous spiral arms that wind through a presumably circular disk. The galaxy is inclined at an angle of only about 15° to our line of sight, so we see it highly foreshortened. Figure 32-8 is a plot of the radial velocities measured at various points along the major axis of the projected image of

M31. The radial velocity of the brilliant nucleus shows that the galaxy as a whole approaches us at nearly 300 km/sec. The still more negative radial velocities of regions southwest of the nucleus indicate that that side of the galaxy is turning toward us and the northeast side away from us. We see that the maximum rotational speeds in M31 occur at about 70′ from its center. Parts farther from the center of the galaxy have smaller orbital speeds. They approach those speeds that would be predicted at various distances from the nucleus by Kepler's third law, on the assumption that those parts are well outside the bulk of the mass, or that the whole mass acts as if it were concentrated at the center (like the sun in the solar system).

From Figure 32-8 we see that a star 2° from the center of the Andromeda galaxy would have a speed of about 200 km/sec relative to the center of the galaxy. Since the distance to M31 is about 2.2×10^6 LY, that star would be about 2.4×10^4 pc, or 5×10^9 AU, from the nucleus, and the circumference of its orbit would be 3×10^{10} AU. A speed of 200 km/sec corresponds to 42 AU/year, so the period of the star would be 7×10^8 years.

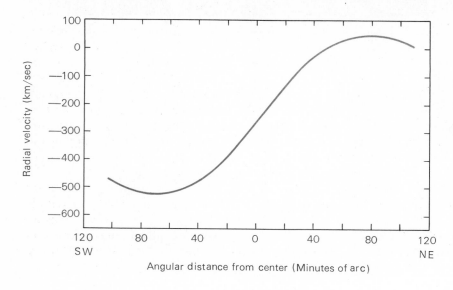

FIG. 32-8 Rotation curve of M31, based on measures of radial velocities of emission nebulae in the spiral arms by N. U. Mayall.

If we apply Kepler's third law to the mutual revolution of M31 and this hypothetical star, we find that the mass of the galaxy is given approximately by

$$\text{mass} = \frac{(5 \times 10^9)^3}{(7 \times 10^8)^2} = 2.5 \times 10^{11} \text{ solar masses,}$$

which is of the same order as the mass of our own Galaxy. The masses have now been determined of more than two dozen nearby galaxies from comparable analyses of their rotations. Some galaxies contain enough neutral hydrogen to be observed in 21-cm radiation (Section 27.4). For such galaxies the rotation curves can sometimes be obtained from the Doppler shifts of the 21-cm line.

The radio observations and long-exposure photographs show that galaxian material (stars, gas, and dust) extends farther from the bright nuclei than was first thought. In fact, the view one gets of a large galaxy, even through a large telescope, is a disappointing, hazy patch near the nucleus. Most of the extents and structural details of galaxies were not recognized until telescopic photographs were obtained.

Many galaxies are not highly flattened and rotating rapidly. Nevertheless, the velocities of the stars in such a galaxy depend upon its gravitational attraction for them, and hence upon its mass. The spectrum of a galaxy is a composite of the spectra of its many stars, whose different motions produce different Doppler shifts. The lines in the composite spectrum, therefore, are broadened, and the amount by which they are broadened indicates the range of speeds with which the stars are moving with respect to the center of mass of the galaxy. Application of the virial theorem, then, enables us to calculate the mass of the galaxy, just as we calculate the mass of a star cluster (Section 28.2a).

Like stars, galaxies are often observed in close pairs. Double galaxies are so common that no more than a small fraction can be "optical doubles," that is, two objects at different distances, merely lined up in projection. Most, in other words, are true binary systems of mutually revolving galaxies. However, we cannot "see" the motion of one galaxy about the other, as we can observe the mutual revolution of the members of a binary star system. Nevertheless, a study of double systems tells us something about galaxian masses.

Suppose the orbit of such a binary galaxy were exactly edge on to our line of sight, and that the two galaxies were at their maximum possible projected separation, so that one would be coming straight toward us and one going straight away. Then the difference in their radial velocities would be the actual orbital speed of one galaxy about the other. From the distance to the pair, we could calculate their true separation and the circumference of the relative orbit. Division of the latter by the orbital speed would give the period of mutual revolution. Application of Kepler's third law, then, would give the sum of the masses of the two galaxies. The problem is somewhat analogous to determining the masses of double stars.

We have, unfortunately, no way of knowing the inclination of the orbit plane to our line of sight, nor can we know the positions of the galaxies in their orbits. The observed difference in radial velocities of the two, therefore, is less than their true relative orbital speed by an unknown amount, and the masses we calculate are less than the true masses. If we assume that the orbits are oriented at random, however, and that the galaxies in the various pairs are distributed at random in their orbits, we can calculate by how much, *on the average*, we would be underestimating the masses of the galaxies. Analyses of many pairs of double galaxies, therefore, yield *average* galaxian masses. T. Page has investigated about 100 galaxies in binary systems and has derived average masses for different types of galaxies. His results are in-

FIG. 32-9   NGC 4565, a spiral galaxy in
Coma Berenices, seen edge on. Photographed
in red light with the 200-inch telescope.
(Mount Wilson and Palomar Observatories.)

corporated into the discussion of galaxies of various kinds in Section 32.4.

The masses of *clusters* of galaxies can be calculated by the same technique used to "weigh" star clusters. The radial velocities of many galaxies in a cluster are first measured. The average of these velocities is that of the center of mass of the cluster, and the differences between the velocities of individual galaxies and this mean value tells us how fast they are moving within the cluster. With the help of the virial theorem (Section 28.2a), we can then calculate the gravitational potential energy of the cluster, and hence its mass.

Here, however, we find a discrepancy which, at the time of writing, has not been completely resolved. The masses found by this method for clusters of galaxies are usually from 2 to 10 times as large as the masses we would expect from the numbers of observable galaxies in the clusters, and would indicate greater masses for galaxies than we find by other methods. Unless new, improved observations remove the discrepancy, many clusters must contain large amounts of invisible intergalactic matter or unseen galaxies whose combined mass contributes enough gravitational attraction to hold the cluster members together; otherwise, the rapid motions of these member galaxies would cause the clusters to gradually dissolve into intergalactic space. We cannot rule out completely the possibility that Newton's laws (and thus, the virial theorem) does not apply over the vast spaces occupied by clusters of galaxies, but such a conclusion would be reached only as a desperate last resort.

32.4 TYPES OF GALAXIES

We have seen how the fundamental properties of galaxies — their luminosities, radii, and masses — are estimated. Now let us consider briefly the different kinds of galaxies that are found. The vast majority of observed galaxies fall into two general classes: spirals and ellipticals. A small minority are classed as irregular.

(a) Spiral Galaxies

Our own galaxy and M31, which is believed to be much like it, are typical spiral galaxies. Like our galaxy (Chapter 27), a spiral consists of a nucleus, a disk, a corona, and spiral arms. Interstellar material is usually observed in the arms of spiral galaxies. Bright emission nebulae are present, and absorption of light by dust is also often apparent, especially in those systems turned almost edge on to our line of sight (see Figure 32-9). The spiral arms contain the young stars, which include luminous supergiants. These bright stars and the emission nebulae make the arms of spirals stand out like the arms of a fourth-of-July pinwheel. The individual interarm stars are usually not observable at all, save in the nearest galaxies, although their collective light may be appreciable as a uniform glow. Open star clusters can be seen in the arms of nearer spirals, and globular clusters are often visible in their coronas; in M31, for example, more than 200 globular clusters have been identified. The spiral arms of a galaxy are composed of population I stars, while the stars in the

FIG. 32-10   The Sc galaxy NGC 5194 (M51) and its irregular II companion, NGC 5195. 200-inch telescope. (Mount Wilson and Palomar Observatories.)

FIG. 32-11   NGC 598 (M33), a spiral galaxy in Triangulum, photographed with the 200-inch telescope. (Mount Wilson and Palomar Observatories.)

nucleus, corona, and disk (except for the arms themselves) are population II. Spiral galaxies, in other words, have mixed stellar population types.

Some famous spirals are illustrated in these pages. Galaxies M51 and M33 (Figures 32-10 and 32-11, respectively) are seen nearly face on; NGC 4565 (Figure 32-9) is nearly edge on. Note the absorbing dust lane in NGC 4565 — a thin slab in the central plane of the disk — which is silhou-

FIG. 32-12   NGC 3031 (M81), spiral galaxy in Ursa Major. Photographed with the 200-inch telescope. *(Mount Wilson and Palomar Observatories.)*

straight bar can persist, rather than winding up; the detailed structures and dynamics of barred spirals, however, are not yet understood.

In both normal and barred spirals we observe a gradual transition of morphological types. At one extreme, the nucleus is large and luminous, the arms are small and tightly coiled, and bright emission nebulae and supergiant stars are inconspicuous. At the other extreme are spirals in which the nuclei are small — almost lacking — and the arms are loosely wound, or even wide open. In these latter galaxies, there is a high degree of resolution of the arms into luminous stars, star clusters, and emission nebulae. Our Galaxy and M31 are both intermediate between these two extremes. Photographs of spiral galaxies, illustrating this transition of types,

FIG. 32-13   NGC 1300, barred spiral galaxy in Eridanus, photographed with the 200-inch telescope. *(Mount Wilson and Palomar Observatories.)*

etted against the nucleus. M81 (Figure 32-12), like M31 (Figure 32-7), is viewed obliquely.

A large minority (perhaps a fourth or more) of spiral galaxies display "bars" running through their nuclei; the spiral arms of such a system usually begin from the ends of the bar, rather than winding out directly from the nucleus. These are called *barred spirals*. A famous example is NGC 1300 (Figure 32-13). The bar in a barred spiral is in a sense a straight portion of spiral arm and usually contains interstellar matter and population I stars. Studies of the rotations of some barred spirals show that their inner parts (out to the ends of the bars) are rotating approximately as solid wheels. In the absence of differential shearing rotation, the

FIG. 32-14   Types of spiral galaxies. *(Mount Wilson and Palomar Observatories.)*

FIG. 32-15   Types of barred spirals. *(Mount Wilson and Palomar Observatories.)*

are shown in Figures 32-14 and 32-15. So far as is known, all spirals and barred spirals rotate in such a direction that their arms trail, as does our own Galaxy (Section 27.4).

Spiral galaxies range in diameter from about 20,000 to more than 100,000 LY. From the limited observational data available, their masses are estimated to range from 10^9 to 2×10^{11} times the mass of the sun. A useful datum is the ratio of the mass of a galaxy to its luminosity, M/L, where both the mass and luminosity are expressed in solar units (for example, if a galaxy were composed entirely of solar-type stars, its M/L ratio would be exactly 1). For spiral galaxies, M/L appears to be between about 1 and 20. The absolute magnitudes of most spirals fall in the range -16 to -21. Our Galaxy and M31 are probably relatively large and massive, as spirals go.

(b) Elliptical Galaxies

More than two thirds of the thousand most conspicuous galaxies in the sky are spirals. For this reason it is often said that most galaxies are spirals. Actually, however, the most numerous galaxies in any given volume of space are those of relatively low luminosity, which cannot be seen at large distances, and which, therefore, are not among the brightest-appearing galaxies. (Similarly, the most numerous stars are faint main-sequence stars, very few of which can be seen with the unaided eye — see Section 23.1.) Most of these dwarf galaxies fall into the class of *elliptical* galaxies. Moreover, the rich clusters, which contain a good fraction of all galaxies, are composed almost entirely of ellipticals. Elliptical galaxies, therefore, are really far more numerous than spirals.

Elliptical galaxies are spherical or ellipsoidal systems which consist entirely of population II stars and contain no trace of spiral arms. They resemble the nucleus and corona components of spiral galaxies. Although dust and conspicuous emission nebulae are not easily observed in elliptical galaxies, some do show evidence of sparse interstellar gas in their spectra. In the larger (nearby) ones, many globular clusters can be identified. The elliptical galaxies show various degrees of flattening and range from systems that are approximately spherical to those that approach the flatness of spirals (Figure 32-16). The distribution of light in a typical elliptical galaxy shows that while it has many stars concentrated toward its center, a

FIG. 32-16   Types of elliptical galaxies. *(Mount Wilson and Palomar Observatories.)*

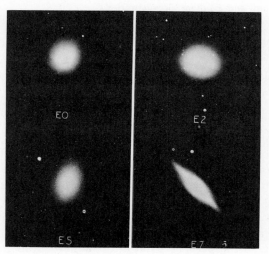

sparse scattering of stars extends for very great distances and merges imperceptibly into the void of intergalactic space (compare with the corona of our own galaxy — Section 27.1b). For this reason it is nearly impossible to define the total size of an elliptical galaxy. Similarly, it is not obvious how far the corona of a spiral galaxy extends.

The fact that elliptical galaxies are not disk-shaped shows that they are not rotating as rapidly as the spirals. It is hypothesized that they are systems that formed from pregalaxian material that had little angular momentum per unit mass — that is, that their original material had low net rotation. Consequently, as such a cloud of primeval material contracted, it did not flatten into a disk, and the density of the material was high enough that it completely (or nearly completely) condensed into stars. In a spiral, on the other hand, a considerable amount of gas (and/or dust) in the flat, rapidly rotating disk was not able to condense into stars at once. This material was presumably formed into spiral arms by the rotation of the galaxy, where it now still slowly condenses into stars at a gradual rate. Elliptical galaxies must consist entirely of old stars (at least if those galaxies are as old as our Galaxy), which would explain why they are pure population II systems.

Elliptical galaxies have a much greater range in size, mass, and luminosity than do the spirals. The rare giant ellipticals (for example, M87 — see Figure 32-17) are more luminous than any known spiral. The brightest elipticals in some rich clusters (for example, NGC 4886, in the Coma cluster of galaxies — see Section 32.5) probably reach absolute magnitudes that are brighter than −23 — more than 100 billion times the luminosity of the sun, and more than 10 times the luminosity of the Andromeda galaxy. The mass data from double galaxies and from clusters suggest that the M/L ratio for giant ellipticals is between 20 and 100. Some of these galaxies, therefore, have masses in excess of a trillion times that of the sun. While, as stated above, the diameters of these large galaxies are difficult to define, they certainly extend over at least several hundred thousand light-years. The often-made statement that the largest of the galaxies are spirals like our own is a misconception.

Elliptical galaxies range all the way from the giants, just described, to dwarfs, which are believed to be the most common kind of galaxy. An example

FIG. 32-17 NGC 4486 (M87), giant elliptical galaxy in Virgo, photographed with the 200-inch telescope. Note the many visible globular clusters in the galaxy. *(Mount Wilson and Palomar Observatories.)*

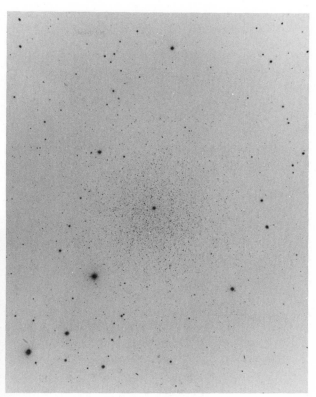

FIG. 32-18 Leo II, a dwarf elliptical galaxy (negative print). Photographed with the 200-inch telescope. *(Mount Wilson and Palomar Observatories.)*

is the Leo II system, shown in Figure 32-18. There are so few bright stars in this galaxy that even its central regions are transparent. The total number of stars, however (most too faint to show in Figure 32-18), is probably at least several million. The absolute magnitude of this typical dwarf is about −10; its luminosity is about 1 million times that of the sun. It is so near to us (about 750,000 LY) that its diameter (about 5000 LY) is probably limited by the tidal force exerted on it by our galaxy; this tidal force would pull more outlying stars away from the dwarf system.

FIG. 32-19   NGC 6822, an irregular galaxy in the Local Group. *(Mount Wilson and Palomar Observatories.)*

Whether still smaller galaxies than dwarfs like Leo II exist depends on how galaxies are defined. Several globular clusters are known that are more than 200,000 LY from the nucleus of our galaxy. It is not known whether objects like these are distributed through intergalactic space or whether they are outlying members of our galaxy. If the former is the case, they must be galaxies in their own right. Perhaps even individual stars exist in intergalactic space (Section 32.7).

Intermediate between the giant and dwarf elliptical galaxies are systems such as M32 and NGC 205, two near companions to M31. They can be seen in the photograph of M31 (Figure 32-34); NGC 205 is the one that is farther from M31.

(c) Irregular Galaxies

About 3 percent of the brightest appearing galaxies in the northern sky are classed as irregular. They show no trace of circular or rotational symmetry, but have an irregular or chaotic appearance. The irregular galaxies divide into two groups. The first group, denoted *Irr I* galaxies, consists of objects showing high resolution into O and B stars and emission nebulae. The best-known examples are the Large and Small Clouds of Magellan (Figure 32-4), our nearest galaxian neighbors. We find many star clusters in these galaxies, as well as variable stars, supergiants, and gaseous nebulae; they contain both population I and population II stars. The Small Cloud, however, is apparently lacking in dust, although it does contain interstellar gas. The lack of conspicuous dust clouds is common in this first kind of irregular galaxy.

Galaxies of the second irregular type (*Irr II*) resemble the Irr I objects in their lack of symmetry. These objects, however, display no resolution into stars or clusters, but are completely amorphous in

FIG. 32-20   The Large Magellanic Cloud. *(Lick Observatory.)*

Andromeda Galaxy; 48-inch Schmidt telescope photograph.
(Mount Wilson and Palomar Observatories.)

NGC 7331; 200-inch teles[cope]
photograph. *(Mount Wilson [and]
Palomar Observator[ies.)*

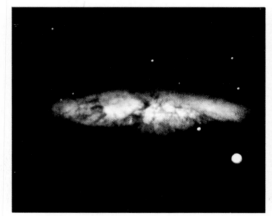

NGC 3034, M82; 200-inch telescope photograph.
(Mount Wilson and Palomar Observatories.)

NGC 253; photographed with the 48-inch
Schmidt telescope. *(Mount Wilson and
Palomar Observatories.)*

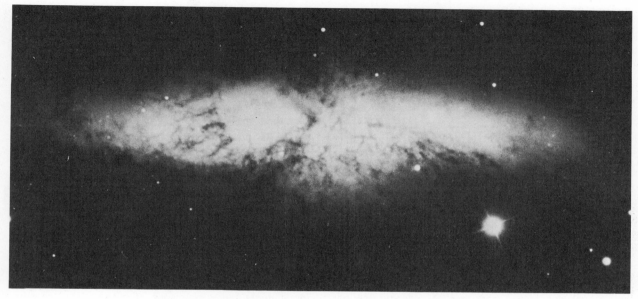

FIG. 32-21   NGC 3034 (M82), an irregular II galaxy in Ursa Major.
(Mount Wilson and Palomar Observatory.)

texture. Their spectra are continuous with absorption lines and resemble the spectra of type A5 stars, showing that stars not luminous enough to be resolved must exist in these galaxies. The Irr II galaxies generally also show conspicuous dark lanes of absorbing interstellar dust. Examples are M82 (Figure 32-21) and the companion to the spiral galaxy, M51 (Figure 32-10).

(d) Classification of Galaxies*

Of the several classification schemes that have been suggested for galaxies, one of the earliest and simplest, and the one most used today, was invented by Hubble during his study of galaxies in the 1920s. Hubble's scheme consists of three principal classification sequences: ellipticals, spirals, and barred spirals. The irregular galaxies (Irr I and Irr II) form a fourth class of objects in Hubble's classification.

The ellipticals are classified according to their degree of flattening or *ellipticity*. Hubble denoted the spherical galaxies by E0, and the most highly flattened by E7. The classes, E1, E2, . . ., E6, are used for galaxies of intermediate ellipticity.† Hubble's classification of elliptical galaxies is based on the appearance of their *images*, not upon their true shapes. An E7 galaxy, for example, must really be a relatively flat elliptical galaxy seen nearly edge on, but an E0 galaxy could be one of any degree of ellipticity, seen face on. A statistical analysis of the numbers of galaxies of various apparent flattening indicates, however, that all degrees of real flattening are about equally represented.

Hubble classed the normal spirals as S and the barred spirals as SB. Lowercase letters *a*, *b*, and *c* are added to denote the extent of the nucleus and the tightness with which the spiral arms are coiled. For example, S*a* and SB*a* galaxies are spirals and barred spirals in which the nuclei are large and the arms tightly wound.

---

†Each of the numbers 0 through 7 that describe the flattening of a galaxy is defined in terms of the major and minor axes of the image of the galaxy, a and b, respectively, by $10(a-b)/a$.

Sc and SBc are spirals of the opposite extreme. Our Galaxy and M31 are classed as Sb.

In rich clusters, galaxies are observed which have the disk shape of spirals but no trace of spiral arms. Hubble regarded these as galaxies of type intermediate between spirals and ellipticals and classed them S0. Their possible origin is discussed in one of the next subsections.

Hubble's classification scheme for galaxies is illustrated in Figure 32-22, in which the morphological forms are sketched and labeled, and with the three principal sequences joined at S0. The diagram is based on one by Hubble himself.

(e) Some Unusual Classes of Galaxies*

There are also some other classes of unusual galaxies that may be briefly mentioned:

cD Galaxies   cD galaxies are supergiant elliptical galaxies, usually E0 or E1, that are frequently found in (or near) the centers of clusters of galaxies. They are the largest galaxies known and tend to outshine the next brightest cluster galaxies by as much as a factor of 2. Often they are strong radio sources as well.

Compact Galaxies   The class of compact galaxies consists of a large number of galaxies of relatively small size and high surface brightness. They are usually elliptical or irregular.

N Galaxies   An N galaxy is an elliptical galaxy with a very bright, nearly stellar appearing nucleus. The rest of the galaxy appears as a sort of faint, extended haze.

Seyfert Galaxies   About a dozen galaxies of this class were first described by Seyfert, from whom the class derives its name. A Seyfert galaxy is a spiral that has a small bright region in its nucleus, whose spectrum shows broad bright emission lines, presumably arising from hot gases there. Seyfert galaxies also may be strong radio emitters. They, and certain other peculiar galaxies, have even been suggested as sources of very high energy cosmic rays.

(f) Evolution of Galaxies*

The continuity of the morphological forms of galaxies along classification sequences suggests that these different forms might represent stages of evolution for galaxies. It was speculated decades ago, for example, that elliptical galaxies gradually flatten, develop spiral arms, and be-

FIG. 32-22   Hubble's classification scheme for galaxies.

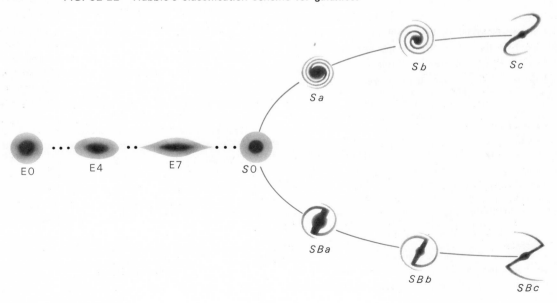

come spiral galaxies. Most of the modern investigators who suggest that galaxies evolve through all types, on the other hand, would reverse this direction of evolution. They envision galaxies beginning as irregulars, evolving through various stages of spiral forms, until finally their gas and dust is completely condensed into stars; then the galaxies become ellipticals.

There is much doubt, however, that galaxies evolve from one type to another at all. The fact that different kinds of galaxies are flattened by different amounts almost certainly results from their having different amounts of angular momentum — that is, from their different rotation rates. In other words, galaxies might always have had essentially their present forms (at least since their formation), the form of a particular galaxy depending mostly on its mass and angular momentum per unit mass.

Despite our lack of definite information concerning evolution of galaxies from one type to another, we do expect that *stars* within galaxies will evolve, as outlined in Chapter 30. Elliptical galaxies may always have been elliptical, but they may have had supergiant stars when they were young. Spirals may never become elliptical, but eventually their spiral arms may disappear when (and *if*) virtually all of their interstellar matter is converted into stars.

One fairly definite form of galaxian evolution that we suspect to occur is the conversion of spirals to S0 galaxies (Section 32.4d) in rich clusters. Most galaxies in a rich cluster may have suffered several collisions with other galaxies since the cluster was formed. The stars in the colliding galaxies are so widely spaced that they are virtually unperturbed by the encounter. The galaxies would pass right through each other, in fact, were it not for their interstellar matter, which *does* collide, and which can be swept completely out of the galaxies and left in the intracluster space. When there is no more interstellar gas or dust, star formation ceases, and these ex-spiral galaxies eventually acquire a stellar population like that of the elliptical galaxies.

(g) Summary

The gross features of the different kinds of galaxies are summarized in Table 32.1. Many of the figures given, especially for mass, luminosity, and diameter, are very rough and are intended only to illustrate orders of magnitude, not precise values.

32.5 CLUSTERS OF GALAXIES

Until World War II, only a few dozen clusters of galaxies were known; most of these had been discovered by accident on telescopic photographs. Recent photographic surveys of the sky have shown, however, that clusters are far more common than had previously been thought. It now appears that all or nearly all galaxies may be members of groups or clusters. Clusters of galaxies, in other words, are now regarded as fundamental condensations of matter in the universe.

(a) Types of Clusters

Clusters of galaxies can be roughly classified into two categories: *regular clusters* and *irregular clusters*. The regular clusters have spherical symmetry and show marked central concentration. They tend

TABLE 32.1 Gross Features of Galaxies of Different Types

	SPIRALS	ELLIPTICALS	IRREGULARS
Mass (solar masses)	10^9 to 2×10^{11}	10^6 to 10^{13}	10^8 to 3×10^{10}
Diameter (thousands of light-years)	20 to 150	2 to 500 (?)	5 to 30
Luminosity (solar units)	10^8 to 10^{10}	10^6 to 10^{11}	10^7 to 2×10^9
Absolute visual magnitude	-15 to -20	-9 to -23	-13 to -18
Population type	I and II	II	I and II
Composite spectral type	A to K	G to K	A to F
Interstellar matter	Both gas and dust	Almost no dust; little gas	Much gas; much dust in Irr II; little or no dust in Irr I

FIG. 32-23 Cluster of galaxies in Corona Borealis, photographed with the 200-inch telescope. *(Mount Wilson and Palomar Observatories.)*

to be very rich clusters, and most of them probably contain at least 1000 members brighter than absolute magnitude −16. The regular clusters have structures resembling those of globular star clusters; some investigators refer to them as *globular clusters of galaxies.* A typical example is the famous cluster in Corona Borealis (Figure 32-23). Regular clusters consist entirely, or nearly entirely, of elliptical and S0 galaxies.

The irregular clusters, sometimes called *open clusters,* have a more nearly amorphous appearance and possess little or no spherical symmetry nor central concentration. The irregular clusters sometimes, however, have several small subcondensations, and resemble loose swarms of small clusters. They contain all kinds of galaxies — spirals, ellipticals, and irregulars. Irregular clusters are more numerous than the regular clusters and range from rather rich aggregates of more than 1000 galaxies (for example, the Virgo cluster) to small groups of a few dozen members or less. An example of the latter is the *Local Group,* the small cluster of galaxies to which our Galaxy belongs.

(b) The Local Group

The Local Group contains 17 known members, spread over a region about 3 million LY in diameter. The two largest members are both spiral galaxies — our own system and M31. Altogether, there are 3 spirals, 4 irregulars, and 10 ellipticals, of which 6 are dwarf elliptical galaxies. In addition, there are a few other outlying irregular galaxies that some investigators consider members of the Local Group. There may well be a few undiscovered members, especially in regions hidden by the dust clouds of the Milky Way. It is also unknown how many "subdwarf" galaxies of very small luminosity may exist — objects like those few globular clusters that seem too remote to belong to our own Galaxy.

Table 32.2 lists the known members of the Local Group that all investigators agree are galaxies. Some of the data are very uncertain and are given in parentheses. Figure 32-25 is a plot of the Local Group; the galaxies have been projected onto an arbitrary plane centered on our Galaxy; then their distances from the center of the plot have been increased so that they are shown at the correct relative distances from us.

The radial velocities of some of the Local Group galaxies have been measured. If we assume that they are moving at random within the Local Group, we can find the motion of the sun with respect to the center of mass of the Group. The procedure is analogous to the determination of the motion of the sun compared to its neighboring stars in our part of our own Galaxy — that is, with respect to the local standard of rest (Section 19.2). The Andromeda galaxy, for example, approaches us at about 260 km/sec, while the Large Cloud of Magellan recedes from us at about 276 km/sec. Most of the apparent radial velocity of each of these objects is due to the revolution of the sun about the center of the Galaxy and the motion of the Galaxy in the Local Group. From an analysis of the radial velocities of all the other galaxies we find that the sun's orbital velocity is about 250 km/sec,

TABLE 32.2 The Local Group

GALAXY	TYPE	RIGHT ASCENSION 1950	DECLINATION 1950	VISUAL MAGNITUDE (m_V)	DISTANCE (KPC)	DISTANCE (1000 LY)	DIAMETER (KPC)	DIAMETER (1000 LY)	ABSOLUTE MAGNITUDE (M_V)	RADIAL VELOCITY (KM/SEC)	MASS (SOLAR MASSES)
Our Galaxy	Sb	— —	— —	— —	—	— —	30	100	(−21)	—	2×10^{11}
Large Magellanic Cloud	Irr I	5^h 26^m	−69°	0.9	48	160	10	30	−17.7	+276	2.5×10^{10}
Small Magellanic Cloud	Irr I	0 50	−73	2.5	56	180	8	25	−16.5	+168	
Ursa Minor system	E4 (dwarf)	15 8.2	+67 18′		70	220	1	3	(−9)		
Sculptor system	E3 (dwarf)	0 57.5	−33 58	8.0	83	270	2.2	7	−11.8		(2 to 4 × 10⁶)
Draco system	E2 (dwarf)	17 19.4	+57 58		100	330	1.4	4.5	(−10)		
Fornax system	E3 (dwarf)	2 37.7	−34 44	8.4	250	800	4.5	15	−13.6	+39	(1.2 to 2 × 10⁷)
Leo II system	E0 (dwarf)	11 10.8	+22 26	12.04	230	750	1.6	5.2	−10.0		(1.1 × 10⁶)
Leo I system	E4 (dwarf)	10 5.8	+12 33	12.0	280	900	1.5	5	−10.4		
NGC 6822	Irr I	19 42.1	−14 54	8.9	460	1500	2.7	9	−14.8	−32	
NGC 147	E6	0 30.4	+48 13	9.73	570	1900	3	10	−14.5		
NGC 185	E2	0 36.1	+48 4	9.43	570	1900	2.3	8	−14.8	−305	
NGC 205	E5	0 37.6	+41 25	8.17	680	2200	5	16	−16.5	−239	
NGC 221 (M32)	E3	0 40.0	+40 36	8.16	680	2200	2.4	8	−16.5	−214	
IC 1613	Irr I	1 00.6	+ 1 41	9.61	680	2200	5	16	−14.7	−238	
Andromeda galaxy (NGC 224; M31)	Sb	0 40.0	+41 0	3.47	680	2200	40	130	−21.2	−266	3×10^{11}
NGC 598 (M33)	Sc	1 31.0	+30 24	5.79	720	2300	17	60	−18.9	−189	8×10^{9}

FIG. 32-24   Irregular cluster of galaxies in Hercules, photographed with the 200-inch telescope. *(Mount Wilson and Palomar Observatories.)*

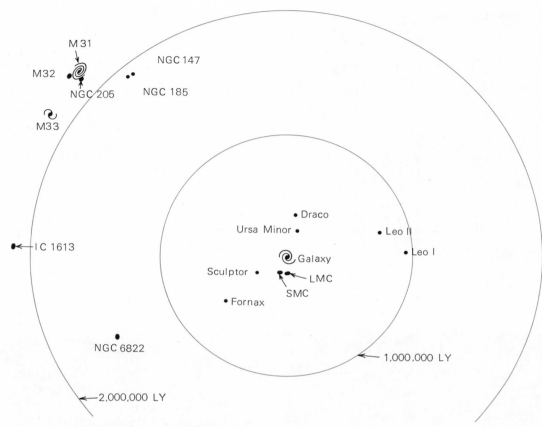

FIG. 32-25   Plot of the Local Group.

and that our Galaxy is moving at a speed of from 100 to 150 km/sec in the Local Group. The average of the motions of all the galaxies in the Group indicates that its total mass is near 5×10^{11} solar masses.

(c) The Neighboring Groups and Clusters and the Local Supercluster

Beyond the Local Group, at distances of a few times its diameter, we find other similar small

FIG. 32-26   NGC 185, an intermediate elliptical galaxy in the Local Group. *(Lick Observatory.)*

FIG. 32-27   Stephan's quintet, a small group of galaxies. *(Lick Observatory.)*

groups of galaxies. Distances to the nearer of these can be determined from cepheid variables that are observed in them. Examples are groups of galaxies centered about each of the bright spirals, M51 and M81 (the galaxies pictured in Figures 32-10 and 32-12, respectively).

The nearest rich cluster of galaxies is the Virgo cluster (so named because it is in the direction of the constellation Virgo). Its distance is probably between 30 and 50 million LY — too remote to observe the cepheid variables in it. The distance to that cluster is estimated from the apparent faintness of the O and B supergiant stars that are barely observable in some of its galaxies, and from the apparent brightness of globular clusters in M87, a bright elliptical galaxy in the cluster. The Virgo cluster appears as a great cloud covering a 10° by 12° region of the sky, within which at least 1000 galaxies can be seen. The brightest of these are giant elliptical galaxies and large spirals; the faintest (observable) are dwarf ellipticals more or less like the Sculptor and Fornax galaxies in the

Local Group (see Table 32.2). The linear diameter of the cluster is about 7 million LY.

The Virgo cloud is an example of a rich irregular cluster. Its structure is very complicated, there being several relatively dense subcondensations of galaxies within it, as well as many double and triple galaxies. It is possible that it may not even be a single cluster, but two clusters of slightly different distances, seen in projection.

Altogether, there are several thousand galaxies within about 70 million LY. Many of these galaxies are grouped into the Virgo cluster and a few other somewhat smaller irregular clusters, and most of the rest of them are in relatively small clusters or groups like the Local Group. Beyond 50 to 70 million LY, galaxies seem to thin out and are sparse in space until a much larger distance is reached. Many experts in this field of astronomy, therefore, are of the opinion that this large collection of clusters and smaller groups comprises a *second-order cluster* — that is, a *cluster of clusters.* G. de Vaucouleurs, who has studied the system extensively, has termed it the *local supergalaxy,* or *local supercluster.* The over-all diameter of the supercluster is believed to be between 100 and 150 million LY. Some evidence has been presented which suggests that the local supercluster is flattened and is rotating, but more observational data are needed to verify this suspicion. The total mass of the local supercluster is probably of the order 10^{15} solar masses.

(d) More Distant Systems of Galaxies

There must be small groups of galaxies that lie beyond the local supercluster, but which are not conspicuous to us. In a typical system like the Local Group, for example, we would see only the one or two brightest galaxies, and would not identify the group as a cluster. Rich clusters of galaxies, on the other hand, especially the *regular clusters,* stand out conspicuously and are recognized to very great distances.

The nearest rich regular cluster is the Coma cluster (in the constellation Coma Berenices), which lies at a distance that is probably about 300 million LY, and which has a linear diameter of at least 10 million LY. Despite its great distance, we can observe more than 1000 member galaxies. The brightest galaxies in the cluster are two giant ellipticals, whose absolute visual magnitudes are likely to be between −23 and −24. The Coma cluster contains more and more member galaxies at magnitudes that are successively fainter and fainter. There is every reason to expect there to be dwarf elliptical galaxies present; if so, the total number of galaxies in the cluster might be tens of thousands. Rather similar to the Coma cluster is the regular cluster of galaxies in Corona Borealis (Figure 32-23), whose distance may be as great as 1 billion LY.

From the Sky Survey photographs obtained with the 48-inch Schmidt telescope at Palomar, a fairly complete catalogue has been prepared of the very rich clusters of galaxies that are not hidden by the obscuring clouds of the Milky Way and that lie in the northern three quarters of the sky that is covered by the Palomar Sky Survey. The most remote clusters in the catalogue are at distances between 2 and 3 billion LY — about 10 times the distance of the Coma cluster. In all, 2712 rich clusters are listed. The total volume of space they occupy is about 10^{14} times that occupied by our own Galaxy — itself believed to be the entire "universe" less than half a century ago.

The distribution of these rich clusters (Figure 32-28) provides our best estimate of the general distribution of matter in the universe. We find that

FIG. 32-28   Distribution of rich clusters of galaxies. Each symbol on the plot (sky map) represents a cluster of galaxies; the larger the symbol, the closer the cluster. The most distant are probably more than 2 billion LY away. The empty region along the middle of the plot is the zone of avoidance, as in Figure 32-6. The large, oval, empty region on the left is part of the sky too far south to be surveyed from Palomar Mountain, where the study of rich clusters was carried out by the author.

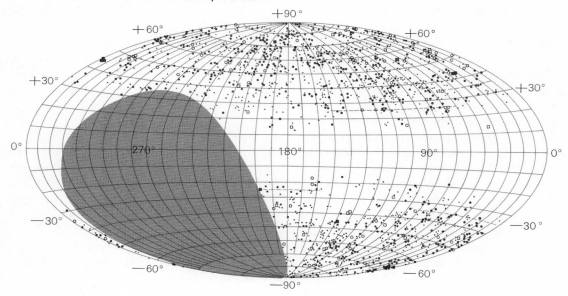

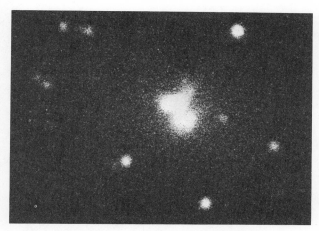

FIG. 32-29 The strong radio source Cygnus A has been identified with this remote pair of galaxies, photographed with the 200-inch telescope. *(Mount Wilson and Palomar Observatories.)*

on the large scale the clusters are spread uniformly throughout space in all directions, out to the greatest distance for which our survey of them is complete. The universe, in other words, seems to be homogeneous on the large scale.

On the small scale, however, the clusters are *not* distributed uniformly. Just as we find a tendency for galaxies to be found in clusters, so the clusters themselves sometimes are grouped into "clumps" or *second-order clusters*. Rigorous statistical tests appear to verify the reality of this clustering tendency of clusters of galaxies. The average size of the second-order clusters is similar to the size of the local supercluster, which suggests that they may be similar supersystems. Evidence is lacking for or against the existence of *third-order clusters*.

32.6 GALAXIES AS RADIO SOURCES

The investigations that followed Jansky's discovery of cosmic radio waves in 1931 (Section 11.8) revealed that continuous radio energy is emitted from the disk and corona of our Galaxy. In addition, since World War II, some thousands of discrete radio sources, each occupying a small region in the sky, have been discovered and catalogued. There is growing evidence that many or most of these discrete sources are extragalactic.

The first identification of a radio source with an extragalactic object was made in 1951. Astronomers at Palomar found that one of the strongest radio sources in the sky, discovered in 1948 and known as "Cygnus A," coincided in position with a remote cluster of galaxies in the constellation of Cygnus. In the cluster is observed what is either a very peculiar galaxy or two galaxies in contact or collision (Figure 32-29). There are now about 100 other radio sources that have been identified with individual galaxies.

FIG. 32-30   The elliptical galaxy NGC 4486 (M87). A 120-inch telescope photograph of very short exposure to show the nuclear "jet." *(Lick Observatory.)*

FIG. 32-31   The peculiar elliptical galaxy NGC 5128 in Centaurus;
a strong radio source. Photographed with the 200-inch telescope.
(Mount Wilson and Palomar Observatories.)

These "radio galaxies" fall into two groups: "normal" and "peculiar" galaxies. The normal radio galaxies are simply ordinary galaxies that, like our Galaxy, emit some of their radiation at radio wavelengths. Radio energy with a continuous spectrum, for example, has been observed to emanate from the disks and coronas of all nearby spirals. Their radiation at radio wavelengths amounts, on the average, to only about 10^{38} ergs/sec, or 10^{28} kilowatts, a tiny fraction of their energy output at visible wavelengths. In addition, radiation at the 21-cm line of neutral hydrogen (Section 26.2i) has been observed in some galaxies. The Doppler shift of this emission line indicates the same radial velocity for a galaxy as does the shift of a line in its visible spectrum.

The peculiar radio galaxies are galaxies which for some reason emit unusually large amounts of radio energy — from 10^{39} to 10^{44} ergs/sec. Some peculiar radio galaxies resemble ordinary galaxies in all their visual aspects.

Other peculiar galaxies have unique properties that may be associated with their unusual radio "luminosities." For example, the Seyfert galaxy NGC 1068, the spectrum of whose nucleus shows emission lines, emits 100 times as much radio energy as normal spirals. Another example is M87, a giant elliptical galaxy in the Virgo cluster, which emits thousands of times as much radio energy as is typical of bright galaxies. This galaxy looks normal on photographs of full exposure. On photographs of short exposure, however, where the outer parts of the galaxy are not "burned out," a luminous feature resembling a jet is seen emanating from the center of M87 (Figure 32-30). The visible light from this jet is highly polarized and is therefore believed to be synchrotron radiation, produced by electrons moving with speeds near that of light and spiraling around lines of force in a magnetic field. The radio radiation has a spectrum that shows it is also synchrotron radiation (compare with that from the Crab nebula — Section 25.3c). Whether some other peculiar galaxies may have similar jets is unknown; most of them are more distant than M87, and it would be difficult to detect such features in them.

Some radio galaxies are very complex. An example is the Cygnus A source, which is one of the strongest *observed* radio sources despite the fact that the cluster with which it is identified has an estimated distance of about 700 million LY. The power emitted by the Cygnus A source at radio frequencies is about 10^{44} ergs/sec — many times the visible light output of the galaxy or galaxies involved. Another complex and strong radio source is NGC 5128, a comparatively nearby elliptical galaxy that appears to have a dust lane running through it. A remarkable feature of these strong complex sources is that the radio energy from such a galaxy often appears to come from *two* regions on opposite sides of, and about 10^5 LY from, its center (Figure 32-32).

The continuous radio emission from extragalactic sources always increases in intensity with wavelength, which suggests that it is synchrontron radiation (Section 25.3c). The weak radio energy from normal galaxies can be accounted for (among other ways) by the electrons that would be ejected in collisions between cosmic rays and interstellar matter. These electrons would then emit energy as they spiral about the lines of force of intragalactic magnetic fields.

It is very difficult, however, to account for the radio emission from some peculiar radio galaxies. In the case of M87, the radio energy is probably associated with the presence of the jet. Some of the strong complex sources were once thought to be due to interactions between the interstellar media of two colliding galaxies. This hypothesis has now been rejected by most astronomers, however, because of the extremely high rate at which the kinetic energy of the galaxies would have to be converted into radio energy. Among the speculations on the origin of the radio energy emitted by these galaxies are hypotheses that involve chain reactions between supernovae, the fission of galactic nuclei, and potential energy released when a galaxy condenses from pregalaxian material. It is not clear, however, that any of the mechanisms suggested can produce the power required; if the radio energy is, indeed, synchrotron radiation, the amount of energy tied up in the magnetic fields and moving electrons in some galaxies must be near 10^{60} ergs — an appreciable fraction of the total nuclear energy available in a galaxy! It is difficult to understand what the source of so great an amount of energy could be. Still more remarkable are the *quasi-stellar sources*.

32.7 QUASI-STELLAR SOURCES

If the sun, as has been stated before, were typical among stars as a radio emitter, we would not expect

FIG. 32-32   Location of radio sources associated with a typical peculiar radio galaxy.

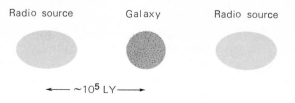

Radio source   Galaxy   Radio source

←—— $\sim 10^5$ LY ——→

to be able to observe a single other star at radio wavelengths; the radio emission from the stars would be too feeble to detect with existing instruments. It was with considerable surprise, therefore, that in 1960 two radio sources were identified with what appeared to be stars. There seemed to be no chance that the identifications were in error, because the precise positions of the radio sources were pinned down by noting the exact instants they were occulted by the moon. By 1963 the number of such "radio stars" had increased to four. They were especially perplexing objects because their optical spectra showed emission lines that at first could not be identified with known chemical elements.

The breakthrough came in 1963 when M. Schmidt, at the Mount Wilson and Palomar Observatories, recognized the emission lines in one of the objects to be the Balmer lines of hydrogen (Section 10.5b) shifted far to the red from their normal wavelengths. If the redshift is a Doppler shift, the object must be receding from us at about 15 percent the speed of light! With this hint, the emission lines in the other objects were reexamined to see if they too might be well-known lines with large redshifts. Such proved, indeed, to be the case, but the other objects were found to be receding from us at even greater speeds. Evidently, they could not be neighboring stars; their stellar appearance must be due to the fact that they are very distant. They are called, therefore, *quasi-stellar-radio sources*, or simply *quasi-stellar sources* (abbreviated QSS). In popular literature, the term is often shortened to *quasar*.

The discovery of these peculiar objects prompted a search for others. The procedure now followed is to look for stellar-appearing objects at the positions of all unidentified radio sources. By 1968, well over a hundred quasi-stellar sources had been discovered. It is estimated that about a third of all radio sources are QSSs. All have spectra that show enormous redshifts. The relative shifts of wavelength range up to $\Delta\lambda/\lambda = 2.3$, and for the

majority $\Delta\lambda/\lambda$ is greater than 1.0. If we apply the exact formula for the Doppler shift (from relativity theory — given in Section 10.3b) we find that $\Delta\lambda/\lambda = 2.2$ corresponds to a velocity of recession of 82 percent of the speed of light.

We shall see in Chapter 33 that remote galaxies are receding from us also, and that there is a correlation between their radial velocities and distances. Most investigators (at the time of writing) regard the redshifts of the QSSs as indicative that they are at very great extragalactic distances and that they conform to the same relation between radial velocity and distance that ordinary galaxies do; the vast majority of QSSs, however, have much

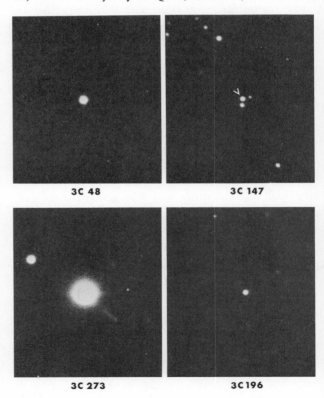

3C 48   3C 147

3C 273   3C 196

FIG. 32-33   Quasi-stellar radio sources photographed with the 200-inch telescope. *(Mount Wilson and Palomar Observatories.)*

higher speeds than any known galaxy, and must, therefore, be even more distant. In our description of QSSs, we assume that this interpretation of their red shifts is correct, but we shall also consider some alternatives.

(a) Characteristics of Quasi-Stellar Sources

The QSSs are all unresolved optically — that is, they appear stellar. A few, however, are associated with tiny wisps or filaments of nebulous-appearing matter. Some are resolved at radio wavelengths, which indicates that the radio energy (at least for some) comes from larger regions than the visible photographic images — perhaps not especially surprising, for radio radiation from galaxies, as we have seen, often originates from outside their optical images. At least some of the radio radiation appears to be synchrotron.

Although they differ considerably from each other in luminosity, the QSSs are nevertheless extremely luminous at all wavelengths. In radio energy, they are as bright as the brightest peculiar radio galaxies, and in visible light most are far more luminous than the brightest elliptical galaxies — their absolute magnitudes range down to -25 or -26. They are very blue in color — in fact, one of their recognizable characteristics is their excess amount of ultraviolet radiation, as compared to normal stars and galaxies. Most surprising of all is that almost all of them are variable, both in radio emission and visible light. Their variation is irregular, evidently at random, by a few tenths of a magnitude or so, but sometimes flareups of more than a magnitude are observed in an interval of a few weeks. Now QSSs are highly luminous, so a change in brightness by a magnitude (a factor of 2.5 in light) means an extremely great amount of energy is released rather suddenly. Moreover, since the fluctuations occur in such short times, the part of a QSS responsible for the light (and radio) variations, must be smaller than the distance light travels in a month or so; otherwise

light emitted at one time from different parts of the object would reach earth at different times (because of the range of distances light would have to travel to reach us), and we would see the increase spread over a longer time.

We have then the perplexing picture of a quasi-stellar source: an extremely luminous object of small size displaying enormous changes in energy output over intervals of months or less from regions less than a few light-months across; 100 times the luminosity of our entire Galaxy is released from a volume more than 10^{17} times smaller than the Galaxy.

(b) Quasi-Stellar Galaxies

The QSSs are stellar-appearing objects of extragalactic distance that are also radio sources. A natural question is whether there exist similar objects that are not radio sources or that radiate too weakly at radio wavelengths to be detected. In an effort to answer this question, A. R. Sandage, at the Mount Wilson and Palomar Observatories, initiated the investigation, in 1965, of faint, very blue objects far from the plane of the Milky Way that had been presumed to be stars in the corona of our Galaxy. Spectra revealed that most of such objects are, in fact, stars, but some showed large redshifts, as did the QSSs. Thus, radio-quiet quasi-stellar objects do exist. Sandage has dubbed them *quasi-stellar galaxies* (QSGs).

It now appears that the QSGs probably outnumber the QSSs by many times, although they are harder to find, because radio emission does not call attention to them. Many investigators believe that the radio-emitting QSSs are temporary evolutionary stages of longer-lived QSGs. In any event, another remarkable class of objects has been shown to exist.

(c) The Significance of the Redshifts

Until now we have assumed that the redshifts of QSSs and QSGs are Doppler shifts, and that their distances are those of hypothetical galaxies of similar redshifts. As we have seen, however, such enormous distances require that these objects have truly astounding luminosities, and, in particular, the variations involve fantastic changes in luminosity over short periods and in relatively small regions. Some investigators, therefore, look to other interpretations for the redshifts.

The only known cause of redshift, other than Doppler, is an exceedingly strong gravitational field (Sections 4.4a and 23.2d). It has not been absolutely ruled out that the redshifts of the QSSs are entirely or partially gravitational, but most experts regard it as very unlikely. It is difficult, even theoretically, to construct a configuration of matter that fits the general characteristics of a QSS and has so large a gravitational redshift. The weight of opinion, therefore, is that the observed redshifts imply that the QSSs do, in fact, have extremely large radial velocities.

There is less agreement, on the other hand, that the QSSs obey the same velocity-distance relation observed for ordinary galaxies. The luminosities need not be so extreme and the small variable regions are not so difficult to explain, if the QSSs are presumed to be relatively nearby. They still, however, must be extragalactic. Objects inside our Galaxy moving as fast as the QSSs would certainly have observable proper motions; the QSSs do not. We are left with the possibility, therefore, that they are at extragalactic distance, but not so extremely far away as galaxies of comparable radial velocities. But then where did they come from, and what gave them their great speeds? It has been suggested that they are objects that have been ejected from other nearby galaxies, or even from our own Galaxy. But if they are ejected from other galaxies, why are they all receding from us, rather than some approaching us? And a disturbance violent enough to eject them from our own Galaxy would be expected to produce other observed effects, for which no evidence exists. Moreover, to explain the energy required to eject

such objects from galaxies at such speeds is even harder than to explain the high luminosities required if they are at very great distances.

There are, finally, other reasons for believing the quasi-stellar sources to have the enormous distances indicated by their large radial velocities. First, if they all had the same intrinsic luminosity, their apparent magnitudes would be indicative of their distances, and there should thus be a relation between their redshifts and magnitudes. Actually, QSSs of the same redshift generally show a considerable range of magnitude, so they obviously do not all have the same luminosity. There is, however, a correlation between the redshifts and apparent magnitudes of the QSSs, in the sense that those of larger red shift are statistically fainter appearing at both optical and radio wavelengths. This is just what we would expect of objects having a range of luminosity that obey the general velocity-distance relation.

Second, there are other classes of objects that are certainly galaxies but that have some characteristics in common with the QSSs. The most important of these are the compact galaxies, the N galaxies, and the Seyfert galaxies. None of these has the great luminosity of a typical QSS, but all do, nevertheless, emit large amounts of energy from small regions. Many are radio sources, and some are variable in radio output. Finally, they tend to emit especially strongly in the ultraviolet. Many investigators suspect there to be a generic relation between these unusual galaxies and the QSSs, in which case the latter are rather extreme examples of peculiar galaxies.

(d) Theories of Quasi-Stellar Sources

It is, perhaps, unfortunate that we feel obliged to say anything about the various theories that have been advanced to explain the quasi-stellar sources, for they are all almost certain to be out of date, if not completely wrong, by the time this is read. We shall, in any case, be brief.

FIG. 32-34   The Andromeda galaxy, Messier 31. *(Lick Observatory.)*

One model is that QSSs are large masses of gas, probably hundreds or thousands of solar masses, in gravitational collapse, the material freely falling together under its mutual gravitation. Such bodies, too massive to ever reach hydrostatic equilibrium, would continue to collapse indefinitely. The energy they radiate comes, of course, from the gravitational potential energy they release as they contract (see Section 28.2). Such collapsing masses are presumed to be either isolated or perhaps within galaxies; the galaxies themselves (in the latter case) would be too faint, because of their distances, to be observed.

Another theory is that the QSSs are galaxies in the process of formation. Some elliptical galaxies are nearly spherical, and thus have relatively low angular momentum in relation to their masses. Galaxies presumably form initially as condensations from pregalaxian gas drawn together by its mutual gravitation, much as we imagine star formation to start. Stars within the galaxies are thus subcondensations in the main gas mass. According to this theory of QSSs, when the collapsing mass of gas achieves a critical density, the stellar subcondensations form within it. Now the stars occupy so little volume compared to the galaxy as a whole that they do not collide with each other but rather fall on through the new galaxy, interact gravitationally with each other, and finally, as a result of such gravitational interactions, reach the equilibrium configuration of the finished elliptical galaxy. At one point in its evolution, just after the star formation begins, the galaxy would be much smaller than when its equilibrium is reached. At the same time, there are many young stars of high luminosity — main-sequence O and B stars. It is mainly these young stars, according to the theory, that give rise to the great luminosity of the QSS, and coming from such a small volume of space. These high-mass stars evolve very rapidly (Section 30.1a) and in the course of their evolution produce supernova outbursts. Such supernovae, going off at random, are presumed (in this theory) to account for the light variations of the QSSs.

Still another suggestion is that the QSSs are old galaxies in which the central nuclear regions have evolved to very high stellar densities, as a result of stellar gravitational encounters. If the stars are crowded together close enough, they can actually collide with each other. The colliding stars would have very high relative speeds, and their collisions would rip them apart. The high-intensity, fluctuating luminosity of the QSSs in this model, then, is the energy emitted as stars collide in the nuclei of distant galaxies — the nuclei being the only parts of the galaxies bright enough to observe.

There are also other theories for quasi-stellar sources, many very esoteric. No model yet proposed, however, satisfactorily accounts for all observed phenomena associated with these mysterious objects, and probably none is correct.

32.8 INTERGALACTIC MATTER

Practically all of the observed matter in the universe seems to be concentrated into galaxies, which, in turn, are usually (if not always) associated with other galaxies in groups and clusters. We cannot rule out the possibility, however, that considerable matter exists in intergalactic space. Some of the reasons for expecting that such matter might exist are: (1) if galaxies have formed from gas and dust, some of this pregalaxian material could well have been left over; (2) individual, rapidly moving stars should, occasionally, escape from galaxies, just as stars sometimes escape from star clusters; (3) collisions or tidal interactions between galaxies could sweep interstellar matter from them — especially in clusters where collisions and close encounters between galaxies are common; and (4) a general feeble background of X-ray radiation has been observed that is believed to be of extragalactic origin — presumably coming from hot intergalactic gas of extremely low density.

There is no conclusive evidence of intergalactic matter outside of groups and clusters, but within clusters hints of intergalactic matter are present. Several observers have reported very faint luminosity in the central regions of some clusters which does not seem to be associated with the galaxies themselves. At least one investigator has reported evidence for absorption of light from remote galaxies behind certain nearer clusters, caused, presumably, by intergalactic dust within the foreground clusters. Finally, luminous "bridges" of matter are sometimes seen stretching between the members in a pair of galaxies that have, apparently, either collided or are interacting gravitationally.

32.9 EXTENT OF THE OBSERVABLE UNIVERSE

As far as we can see in all directions we find galaxies and clusters of galaxies. The more intrinsically luminous a galaxy is, the greater is the distance to which it can be observed. At the farthest depth of observable space, we see only the greatest giants among galaxies — that is, the brightest members of individual rich clusters — or perhaps quasi-stellar sources. If we knew the upper limit to the luminosities of galaxies or QSSs, we could calculate the distance of those most remote observed objects. At present, we can only guess that they are between 5 and 15 billion LY away. It is possible that some or many of the unidentified sources of radio waves may be galaxies, or clusters, or QSSs at still greater distances — beyond the range of our largest optical telescopes.

The size and extent of the universe, its motions, and its evolution, are the subjects of the next — and final — chapter.

Exercises

1. Why is the term "island universe" a misnomer?
2. In a hypothetical galaxy, a cepheid variable is observed, which, at median light, is at magnitude +15. From its period, the absolute magnitude of the cepheid is determined to be −5. What is the distance to the galaxy?

   Answer: 10^5 pc; 3.26×10^5 LY

3. In a very remote galaxy, a supernova is observed which reaches magnitude +17. Assume that the absolute magnitude of the supernova was −18, and calculate the distance to the galaxy.

   Answer: 10^8 pc; 3.26×10^8 LY

*4. Why, in 1920, were the Large and Small Clouds of Magellan thought to be only 75,000 LY away, while their true distance is about twice this figure?
5. Since globular clusters do not all lie in the plane of our Galaxy, how is it that they can be dimmed at all by interstellar absorption?
6. Make up a table of distance indicators for galaxies, and list the distances to which each can be used.
7. How can the visual light of an emission nebula in another galaxy exceed that from the star whose energy produces the nebula?
8. Starting with the determination of the size of the earth, outline all the steps one has to go through to obtain the distance to a remote cluster of galaxies.

9. Why do we use the *brightest* galaxies in a cluster as indicators of its distance, rather than average galaxies in it? (*Hint:* There are two reasons, one involving the definition of "average," the other involving the distances to typical clusters.)

10. The tenth brightest galaxy in cluster *A* is at apparent magnitude $+10$, while the tenth brightest galaxy in cluster *B* is at apparent magnitude $+15$. Which cluster is more distant, and by how many times?

11. How can we determine the inclination of M31 (the Andromeda galaxy) to our line of sight?

12. Where might the gas and dust (if any) in an elliptical galaxy come from?

13. If extragalatic globular clusters exist as "galaxies," what kind of galaxies would you class them? How about extragalactic stars?

14. Are there subunits or subcondensations in the Local Group? If so, discuss them.

°15. From the data given in the chapter, calculate the mean number of galaxies per unit volume of space and the mean separation of galaxies in the Virgo cluster of galaxies.

°16. From the data given in the chapter, try to make an estimate of the mean density of matter in the universe, in grams per cubic centimeter. If you cannot calculate the density unambiguously, see if you can place upper and lower limits on it. What assumptions must you make?

°17. Assume that its red shift is a Doppler shift, and verify that a quasi-stellar source with a red shift of $\Delta\lambda/\lambda = 2.2$ has a radial velocity of 82 percent the speed of light.

18. (a) Suppose a QSS has an absolute visual magnitude of -25. What is its visual luminosity in terms of the sun's? (Assume that for the sun, $M_v = +5$.)

    (b) Suppose that the QSS has a distance of 5×10^9 pc. Suppose, further, that the fact that it appears stellar implies that its angular diameter is less than 1 arc-second. What is the upper limit to its linear diameter in parsecs?

    (c) If it is variable, can you suggest a way of fixing a smaller upper limit to its linear size?

    Answers: (a) 10^{12}; (b) 2.5×10^4 pc

CHAPTER 33

Cosmology

In its modern usage by astronomers, the term *cosmology* embraces not only *cosmogony* — the origin and evolution of the universe — but also the content and organization of the universe. It is the problem of the modern cosmologist to piece together those properties of the universe that are revealed by observation into a self-consistent hypothesis that describes its structure and evolution.

In one respect the study of cosmology differs from research in other realms of physical science; the cosmological problem is unique. We can check the *generality of* gravitational theory, for example, by noting that it applies to a very great many different physical phenomena. By its very nature, however, a cosmological theory — a theory of the universe — can be applied only to one system — the universe itself. For this reason, cosmology is one of the most difficult, and most speculative, fields of science.

The observational basis for cosmology comes from two sources: the large-scale distribution of matter in space, which was discussed in Chapter 32, and the large-scale motions in the universe, which will be considered now.

33.1 THE "EXPANDING UNIVERSE"

Within the first decade after the true nature of the galaxies was established, it was found that the universe may be expanding. The evidence comes from the radial velocities of the galaxies.

(a) The Law of the Redshifts

The first radial velocities of galaxies were measured by V. M. Slipher at the Lowell Observatory. In

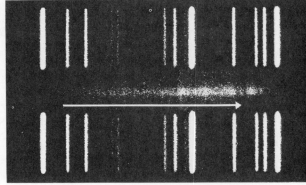

FIG. 33-1 Hydra cluster of galaxies. showing the galaxy observed for Doppler shift, and the spectrum showing the large redshift (indicated by the length of the arrow) equivalent to a radial velocity of about 60,000 km/sec. 200-inch telescope photographs. *(Mount Wilson and Palomar Observatories.)*

the years between 1912 and 1925 he obtained spectra from which he measured the velocities of more than 40 "nebulae." Slipher did not, of course, know the true nature of the objects he investigated, but he found very high velocities for those that were later shown to be galaxies. A very few of them showed velocities of approach (negative radial velocities) — these are some of the Local Group galaxies — but Slipher found the vast majority to be moving away from us, with speeds up to 1125 mi/sec.

In the mid-1920s there seemed to be some evidence that the velocities of the nebulae might be correlated with their distances, but the distances were too poorly known at that time to verify the suspicion. By 1929, however, Edwin Hubble, at the Mount Wilson Observatory, had determined new estimates for the distances to many of the galaxies whose velocities had been measured, and he found that the velocities of recession of those galaxies were, in fact, *proportional* to their distances.

Meanwhile, M. L. Humason, also at Mount Wilson, had begun to photograph spectra of fainter galaxies and galaxies in clusters with the 60- and 100-inch telescopes. He and Hubble collaborated, and had definitely established, by the early 1930s, that the more distant a galaxy, the greater, in direct proportion, is its speed of recession, as determined by the shift of its spectral lines to the longer (or red) wavelengths. This relation is now known as the *law of redshifts*, or sometimes the *Hubble law*.

The study of the radial velocities of distant galaxies and clusters of galaxies — that is, of the redshifts of their spectral lines — has been continued, most recently with the 200-inch telescope at Palomar. As the investigation progressed, more and more remote galaxies of greater and greater speed of recession have been discovered. The largest speed yet measured (at the time of writing) is of a very distant cluster (Figure 33-5) that is moving away from us at 41 percent of the speed of light, or about 76,000 mi/sec. The actual distances to the galaxies

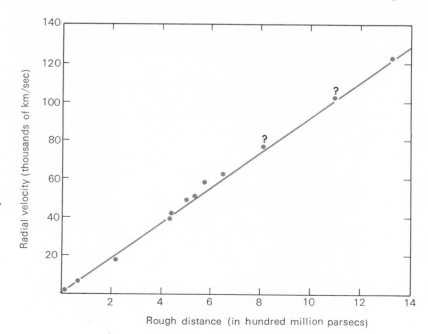

FIG. 33-2 Velocity-distance relation for clusters of galaxies.

whose velocities have been measured cannot be determined accurately, but the *relative* distances to clusters of known radial velocity (or redshift) are fairly well established (Section 32.3c). To the accuracy of the present observations, these clusters have radial velocities that are proportional to their distances. The velocity, V, of a cluster, in other words, is represented by the equation

$$V = Hr,$$

where r is the distance to the cluster, and H is a constant of proportionality, called the *Hubble constant,* that specifies the rate of recession of galaxies or clusters of various distances. The Hubble constant is now believed to lie in the range 50 to 150 km/sec per million parsecs, a recent determination giving 75 km/sec per million parsecs. In other words, a cluster moves away from us at a speed of 75 km/sec for every million parsecs of its distance. This velocity-distance relation for clusters of galaxies is shown in Figure 33-2.

(b) Interpretation of the Law of Redshifts

The fact that distant galaxies all seem to rush away from us may seem to imply that we must, somehow, be at the "center" of the universe, but such is not the case. A simple and familiar analogy is that of a balloon whose surface is covered with dots of ink. As the balloon is inflated, the dots all separate, and a tiny insect on any one dot would see all other dots moving away from it; yet none of the dots is at a "center."

For a three-dimensional analogy, consider the loaf of raisin bread shown in Figure 33-3. Before being placed in the oven, the loaf is 12 in. in diameter (a). The cook, however, mistakenly put far too much yeast in the dough, and within 1 hour the loaf has grown in size to a diameter of 24 in. (b). During the expansion of the bread, the raisins in it all separated from each other. Suppose, now, that we are microbes, sitting on raisin A (intentionally *not* at the center of the loaf), and that we measure the speeds with which the other raisins

FIG. 33-3 Separation of raisins in a hypothetical loaf of raisin bread that is 12 inches in diameter when it is placed in the oven (a), and grows to 24 inches in diameter at the end of 1 hour (b).

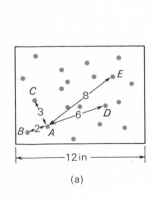

(a)

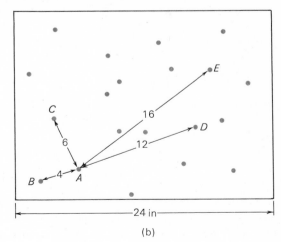

(b)

TABLE 33.1

RAISIN	PRESENT DISTANCE (IN.)	SPEED OF RECESSION (IN./HR)
B	4	2
C	6	3
D	12	6
E	16	8

recede from us. At the end of the hour, raisin B, originally 2 in. away, has increased its distance to 4 in.; evidently, its speed of recession has been 2 in./hour. Raisin C increased its distance from 3 to 6 in., and so moved at 3 in./hour, and so on for the other raisins; we obtain the data in Table 33.1. If we plot the data in the table, we will find a linear (straight-line) velocity-distance relation for the raisins analogous to that observed for galaxies. The same result (or plot) would be obtained irrespective of which raisin were chosen for "us." We would, in fact, observe an exactly linear velocity-distance relation for the raisins if we could measure their instantaneous velocities and distances, for *any* uniform expansion or contraction of the bread, whether or not it expands (or contracts) at a constant rate.

If the universe were now changing scale — either uniformly expanding or contracting — we should expect to observe a linear relation, or proportionality, between the speeds and distances of galaxies and clusters, just as in the raisin bread analogy. In an expanding universe, galaxies would move away from us at speeds proportional to their distances; in a contracting universe, they would *approach* us at speeds proportional to their distances. The law of redshifts, therefore, is generally interpreted as observational evidence that the universe is now expanding; in this expansion the galaxies and clusters are being carried farther and farther from each other.

It must be noted that we observe galaxies, not where they are "now," but at the distances they had when the light by which we see them left those objects. Because of the finite speed of light, those observed distances are millions to billions of years "out of date" (for galaxies millions to billions of light-years away). If light traveled with infinite speed, we would observe those objects at much greater distances (unless they have ceased to move away from us since light left them). Similarly, their radial velocities may have changed from the values indicated by the spectra of their observed light. The velocity-distance relation, therefore, should not necessarily be exactly linear for objects at great distances; we shall return to this point in Section 33.3.

The expansion of the universe does not necessarily imply that the galaxies and clusters of galaxies themselves are actually expanding. The raisins in our analogy need not grow in size as the cake expands. If the galaxies *were* growing at the same rate that the universe expands, their increase in size would be so gradual that it would be very difficult to observe. It is possible that everything in the universe — the galaxies, the stars, planets, even ourselves — takes part in the expansion. But atoms could not be expanding (or must at least maintain their spectral line wavelengths), else we could not measure redshifts, and the idea of expansion loses its meaning. We assume in the following discussion that the galaxies and clusters are constant in size but merely separate from each other as the universe grows.

Galaxies and clusters do, of course, have individual motions of their own, superimposed upon the general expansion. Galaxies in pairs, for example, revolve about each other, and those in clusters move about within the clusters. In fact, a few galaxies in nearby groups and clusters move fast enough within those systems so that they are actually *approaching* us even though the clusters of which they are a part are moving, as units, away.

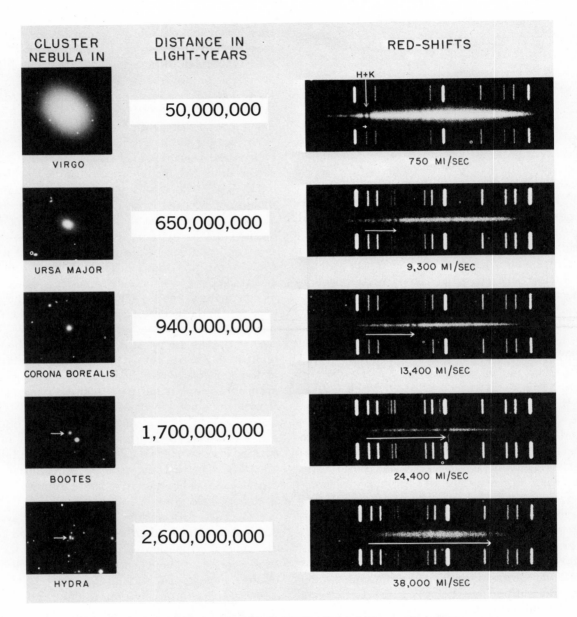

CLUSTER NEBULA IN	DISTANCE IN LIGHT-YEARS	RED-SHIFTS
VIRGO	50,000,000	750 MI/SEC
URSA MAJOR	650,000,000	9,300 MI/SEC
CORONA BOREALIS	940,000,000	13,400 MI/SEC
BOOTES	1,700,000,000	24,400 MI/SEC
HYDRA	2,600,000,000	38,000 MI/SEC

FIG. 33-4 *(Left)* Photographs of individual galaxies in successively more distant clusters; *(right)* the spectra of those galaxies, showing the Doppler shift of two strong absorption lines due to ionized calcium. The distances are estimated from the faintness of galaxies in each cluster, and are provisional. *(Mount Wilson and Palomar Observatories.)*

(c) Are the Redshifts Really Doppler Shifts?

Not all scientists accept the interpretation of the redshift law as an indication of an expanding universe. Some argue that the observed redshifts in the spectra of distant galaxies may not be Doppler shifts at all but may be caused by some unknown effect on light as it travels over large distances. It has been suggested, for example, that photons may lose energy, or "tire," as they traverse space; since the energy of a photon is inversely proportional to its wavelength, the "tired-light" hypothesis could explain the redshift of the lines in spectra of remote objects. However, there is not a shred of evidence, either observational or theoretical, to support the hypothesis.

The Doppler shift, on the other hand, is an effect of speed that is well confirmed in laboratory experiments, in observations in the solar system, and in observations of double stars. There we can see the light sources, or planets, or stars, moving. The Doppler effect is the only *known* cause that can account for the redshifts observed in the spectra of galaxies.

Many people, apparently, are troubled by the concept of the extremely high speeds with which galaxies are receding. On the other hand, it requires very special assumptions about the nature of natural laws to understand how the galaxies could be fixed in space in a *static* universe. Their mutual gravitation should pull them together, causing the universe to *contract*. It would be easy to understand their mutual separation, however, if an expansion were once started — say, by a primeval "explosion."

Einstein's general theory of relativity provides another basis for understanding an expanding universe. The solution of Einstein's equations, as they apply to a field of gravitating bodies, allows for the possibility of a *repulsion* between distant objects in the universe. The magnitude of this repulsion is represented in the equations by a term called the *cosmological constant*. Unfortunately, the theory of relativity does not specify what value the cosmo-

logical constant must have, but leaves it arbitrary. Einstein originally chose the constant so that the repulsion between distant masses would exactly balance their gravitational attraction, to maintain a static universe.

According to the theory, however, there is no a priori reason why the universe must be static. Larger or smaller values of the cosmological constant (including zero) are possible; an expanding, a static, or a contracting universe are all compatible with general relativity.

Most investigators, therefore, tentatively accept the expanding universe interpretation of the redshift law. They are ready, however, to revise any conclusions that are based upon this interpretation, when and if new evidence is brought to light that would support alternate explanations of the redshifts.

33.2 COSMOLOGICAL MODELS

Having examined the observational basis for cosmology, we may investigate some of the hypotheses that have been suggested to describe its structure and evolution. Such a hypothesis is called a *cosmological model.*

All cosmological models are based on certain assumptions, without which no hypothesis could be formed. For example, an assumption common to most models, and to all discussed here, is that the redshifts of galaxies are really Doppler shifts and that the universe is really expanding. Another even more basic postulate is that known as the cosmological principle.

(a) The Cosmological Principle

It would be impossible even to begin to formulate a theory of the universe if we did not assume that our observations give us information that applies to the whole universe, not just to our own part of it. In other words, we must assume that the part of the universe that we actually observe is representative

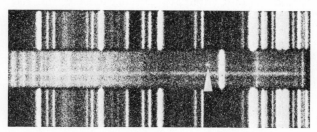

FIG. 33-5   The cosmic radio source 3C295 has been identified with the extremely remote cluster of galaxies marked with an arrow on the direct photograph *(top)*. The spectrum of the brightest galaxy in the cluster (nearest the arrow point) is shown below. A faint feature (marked) has been identified as an emission line due to ionized oxygen in interstellar space in that galaxy; its displacement from its nominal wavelength in the spectrum indicates that the galaxy is receding from us at nearly half the speed of light. *(Mount Wilson and Palomar Observatories.)*

of the entire cosmos, and that we are not located in some very unusual place, fundamentally different from the rest of the universe. Stated more generally, we must assume that whereas there will be local variations in the exact details of galaxies and clusters, all observers, everywhere in space, would view the same large-scale picture of the universe. This assumption is known as the *cosmological principle.*

The statement of the cosmological principle implies that it is possible to define some kind of universal or cosmic time, at some instant of which different parts of the universe can be compared with each other. It is beyond the scope of this discussion to define "cosmic time" rigorously. We can imagine it, however, by supposing that observers at different places in the universe could synchronize their clocks with each other my means of fictitious "light" signals that propagate with infinite speed.

(b) Evolutionary Cosmologies

As the universe expands, it thins out. If no new matter is being created — that is, if the total mass of the universe is constant — and if no previous contraction preceded the present expansion, it follows that all the matter of the universe must have once been close together. This suggests that some original "explosion" may have started galaxies moving away from each other. The fastest-moving galaxies would have gone the farthest since the explosion, and the slow-moving ones less far. From any one place in the exploding material, all other places would appear to move away at speeds proportional to their distances — that is, the redshift law would apply. The idea that such an explosion took place is called the "big-bang" theory.

The big-bang hypothesis implies that the universe has a finite age — at least since the explosion. Since no galaxy could have a greater age, all galaxies would be aging, and evolving, together. Models based on some version of the big-bang theory, therefore, are usually called *evolutionary models* or *evolutionary cosmologies.*

We can estimate an upper limit to the age of the universe since the explosion from the present rate of recession of the galaxies. If it is assumed that the speeds of galaxies have not changed since then, a galaxy at a distance *r,* moving away with a radial velocity V, must have been receding from us for a time r/V. But r/V is simply the reciprocal of the Hubble constant (Section 33.1a); the maximum "age" of the universe, depending upon the value of the Hubble constant, would lie in the range 7 to 20 billion years. Actually, if there is no force of repulsion present, the mutual gravitation of the galaxies would have slowed them down; in the past, in other words, they would have to have been moving faster than they now are and would have required a somewhat shorter time to reach their present distances.

The Belgian cosmologist Abbe Lemaitre proposed a version of the big-bang theory to explain cosmic rays in the 1920s which he called the theory of the "primeval atom," referring to an original single chunk of matter that exploded. A group of American physicists, led by George Gamow, have used the idea in an attempt to explain how the chemical elements were formed in the relative abundances we observe today. Gamow and his collaborators considered what nuclear reactions could go on in all the matter of the universe during the big bang. Their idea was that electrons, protons, and neutrons would combine to form hydrogen, helium, and heavier atoms. As the material expanded and the hot gases cooled, condensations formed that later became galaxies, and smaller condensations within these became stars.

There are difficulties in the calculations of element formation, and today there are alternative hypotheses that some of the heavy elements are formed within stars (Chapter 30). Otherwise, however, the general scheme of the big-bang theory provides a basis for understanding the universe as we see it today. The start of the explosion some billions of years ago takes on special significance; if all matter was then in the form of subatomic particles at a high temperature, we could never expect to discover by observation today what occurred before then. That is, the beginning of the explosion was a "beginning" in a broader sense; it can be considered that the universe was "created" then.

(c) General Relativity in Cosmology

If the mass of the universe is not infinite, the original exploding matter would have to have an "edge." This might seem to be a violation of the cosmological principle (for an observer at a boundary of the universe would have a different view of it than would one not near a boundary), but the unwanted edge is eliminated in certain kinds of geometry. In the "curved space" that is possible in non-Euclidean geometry, the universe can be finite, but unbounded.

Concepts like "straight" and "curved" lines can be defined only in terms of the paths followed by light. According to Einstein's general theory of relativity, the path of light, that is, the geometry of space, is affected by the presence of matter in it. The curvature is locally greater, for example, near a massive body than it is far from all objects. The general curvature of space depends on the average density of matter in the universe.

It is unlikely that any of us can visualize curved space, but an analogy may make it less mysterious. Consider the curved surface of the earth, which has a finite *area*, but is unbounded, and upon which the geometry is not Euclidean. In a very large triangle on the earth's surface, for example, the angles do not add up to 180°. There is no "center" to the surface area of the earth, just as there is no center to the finite volume of curved three-dimensional space. Moreover, there is a "radius" of curvature of space that is connected with its volume much in the way that the radius of the earth is connected with its surface area. An important difference between the curved surface of the earth and the curved volume of space, however, is that the space curvature may be changing — in fact, *must* be changing — if the density of matter in space is changing.

If we knew accurately the natural laws that describe the universe, we should, from observational data, be able to predict how it is expanding, thereby changing in scale and density. General relativity provides the best set of natural laws we know, but, as stated above, the theory includes a repulsive term of arbitrary magnitude. Depending on the value assigned to this cosmological constant, the universe can be depicted by an infinity of models.

Since we have no information that enables us to guess the magnitude of the cosmological constant, the simplest thing to do is to assume that it is zero — that is, that there is no repulsive force. There is no particular justification for this assumption other than its simplicity. Moreover, it leaves the cause of the original "big bang" unexplained. The assumption does, however, lead to definite cosmological models that can be checked (in principle) against observations and which have been given serious consideration.

Without repulsion the motions of galaxies after an initial explosion would be slowed down by gravitation in the same way that a rocket fired away from the earth is slowed down by the earth's gravity. If the rocket has an initial speed of about 11 km/sec (the "escape velocity"), it will move away from the earth forever. The galaxies, if they were started fast enough in the big bang, could likewise move away from each other forever; if they started slower than some critical escape velocity, however, they would eventually fall back together again. In the latter case, possibly a future explosion would send the galaxies out again; it is even conceivable that alternate expansions and contractions might lead to an "oscillating" universe.

The speeds with which galaxies must recede to escape from each other depends on the mean density of matter in the universe. The present mean density is estimated to lie in the range 10^{-29} to 10^{-31} gm/cm^3 but is not known accurately enough to predict the future motions of the galaxies. Thus the hypotheses that the universe will expand forever, and that it will eventually stop expanding and begin to contract, are both possible evolutionary cosmological models (for which the cosmological constant is assumed to be zero). These models require ages for the universe, however, that are probably less than 15 billion years, and may be less than 10 billion years. These ages are comparable to, or even less than, those we estimate for the oldest star clusters in our Galaxy. One of the difficulties with the models described (with the cosmological constant equal to zero) is that the age of the universe, since the "big bang," is not comfortably greater than the age of the oldest objects known. There is not quite a discrepancy here future observations and/or refinements in theory, of course, may completely remove it.

We emphasize that the models considered so far are among the simplest ones. We have assumed: (1) the cosmological principle; (2) that the mass and energy of the universe are jointly conserved; (3) that general relativity, the most accurate description of nature we know today, is, in fact, absolutely correct; (4) that the cosmological constant is zero; and (5) that there are no complications, such as rotation of the universe as a whole. It is appropriate that we try the simplest models first, and turn to more complex ones only as observations show that the simple ones must be rejected.

We now describe an even simpler model — the *steady-state theory*. Recent observations, as we shall see, have shown that the steady state is almost certainly wrong, but it is worth discussion anyway, because it has received a great deal of interest since the late 1940s, because it illustrates how an alternative cosmological model can be constructed, and because we shall be able to see how observational tests can be applied to force the rejection of a model.

(d) The Steady-State Theory

The problems of the finite age implied by the evolving cosmologies, and of beginnings and endings, led several astronomers (Bondi and Gold, and later Hoyle and others) to speculate on the possibilities of another cosmological hypothesis. Theirs was the simplest of all possible cosmological models, for it rested upon only a single assumption. This assumption, an extension of the cosmological principle, is called the *perfect cosmological principle;* it states that the universe is not only the same *everywhere* (except for local small-scale irregularities), but at *all times.* In other words, if we could return to life billions of years hence, we would find the universe, on the whole, as it is now. Since it was assumed that nothing changes in time, the hypothesis was known as the *steady-state cosmology.*

By starting with this premise, Bondi and Gold ruled out creation of the universe at a definite time in the past. Moreover, they explained how our view of the universe can remain the same when galaxies are moving apart — how the average density of matter can stay constant when galaxies are leaving our region of space at high speeds on all sides. They explained this by the bold assumption that matter is being *created continuously* in empty space, at a rate just sufficient to replace what is leaving due to the observed recession of galaxies. The theory, therefore, was sometimes called the "continuous-creation theory." Moreover, to satisfy the perfect cosmological principle, there must always be the same proportion of young and old galaxies in any given volume of space. This condition requires that the rate of creation of new matter be always proportional to the present amount of matter, and that the rate of increase of distance between any two galaxies (or clusters of galaxies) be proportional to their current separation. The new matter that comes into being (in this theory) was usually presumed to be in the form of gases composed of individual atoms (possibly pure hydrogen), which later condense into galaxies and stars.

In the steady-state cosmology, *individual* galaxies were assumed to form, age, grow old, and, as their stars evolve to the black dwarf state, finally die. Only when a large volume of space, containing many galaxies, is considered are things unchanging — there would always be young and old galaxies, side by side, and always in the same proportions. The formation of new matter would need to occur only at a very slow rate, and we would not have necessarily expected to find newly formed galaxies in our immediate vicinity in space. The universe itself, according to the theory, is infinitely large and infinitely old — with no beginning and no end.

To the objection that matter cannot be created spontaneously from nothing, the proponents of the theory answered that it is no more difficult to imagine matter being created gradually and steadily than all at once in or prior to some "big bang" in the past.

The steady-state cosmology was particularly appealing philosophically because of its symmetry and simplicity. It is contrasted with the evolutionary cosmologies in Figure 33-6.

(e) Summary of Cosmological Models

Two entirely different cosmological hypotheses have been described. These are:

(1) The evolutionary cosmologies, consistent with general relativity, and with the cosmological constant equal to zero. These models all predict a finite age for the universe, and that it is evolving in time — that is, that the galaxies are aging together. The universe, in these models, is now expanding, and its mean density is diminishing in time, but the expansion is slowing down. Whether or not it will continue until the galaxies are infinitely far apart, or will

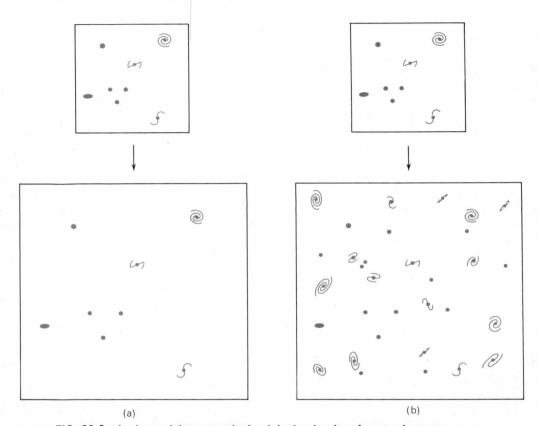

(a)   (b)

FIG. 33-6 In the evolving cosmologies (a), the density of matter in space
thins out as the universe expands; in the steady-state cosmology (b), new matter
is created spontaneously, and the density of the universe remains constant.

eventually stop and give way to contraction, depends upon the mean density of matter in space.

(2) The steady-state cosmology. This model assumes that the universe is infinite in age and extent. On the large scale, nothing changes with time. As the universe expands and galaxies separate, new matter is created, which keeps the mean density of the universe constant.

We shall see in Section 33.3 how we can subject these models to observational tests. At the time of writing (summer 1968) the observations appear to be inconsistent with the steady state. However, it must be emphasized that the two types of models described here are not *the* two opposing views of the universe, but are merely two of the simplest of an infinite number of possibilities. The rejection of one *does not* imply the correctness of the other.

33.3 TESTS FOR COSMOLOGICAL MODELS

There are three distinct stages in the progress of scientific research: (1) systematic observing, or collecting experimental results; (2) forming hypotheses to account for what is observed; and (3) testing these hypotheses by further observations or experiments. In the realm of cosmology we have described (1) the observations of galaxies, and (2) two hypotheses (evolutionary and steady-state cosmologies), which, without the testing of stage (3) would remain as separate, unproved speculations along with many others.

It is clear that evidence favoring one hypothesis over another will depend on *differences* between them. Since all the cosmological models described here refer to a universe expanding at the rate indicated by the Hubble-Humason (redshift) law, the differences between them must reflect the different ways in which the theories predict that the universe will change in time. In other words, for testing cosmological theories we need observa-

tions of the universe as it was long ago to compare with present conditions. Fortunately, we can and do look into the past. Light travels at a finite speed, so we see remote objects as they were when light left them — up to billions of years ago for galaxies at distances of billions of light-years.

One of the possibly observable differences between the steady-state and evolutionary cosmologies is in the predicted mixture of galaxies of different ages. According to the steady-state theory, individual galaxies age and die, but new ones are always forming, so that old and young galaxies have always existed side by side. No matter how far out in space we look, therefore, we should not find galaxies, on the average, any different from those nearby us. Evolutionary cosmology, on the other hand, assumes a distinct "beginning," so that

FIG. 33-7 Central part of the great cluster of galaxies in Coma Berenices, which has an estimated distance of about 300 million light years. Photographed with the 200-inch telescope. *(Mount Wilson and Palomar Observatories.)*

all galaxies have about the same age. Thus, if we look out to a great distance, and hence far back into time, we should find galaxies that are systematically younger than those near us — that is, remote galaxies should be at an earlier stage of evolution.

Another test for cosmology is the way the density of matter in space changes in time. Evolutionary cosmologies require that matter gradually thins out in space; remote galaxies, therefore, should appear relatively closer together (as they were in the past) than nearby ones. The steady-state theory, however, predicts that the average distance between galaxies never changes and so should be the same for galaxies at all distances.

A third, less obvious test involves the speeds of receding galaxies at great distances. The evolutionary models predict that for no repulsion (cosmological constant zero), the expansion should slow down, because of the gravitational forces between galaxies. In the steady-state theory, on the other hand, the rate of the separation of any two galaxies is ever increasing. Because of these changes in the rate of expansion of the universe, the radial velocities of remote galaxies should deviate from a relation exactly proportional to their observed distances; that is, a graph of radial velocity versus distance should not necessarily be a straight line for very distant galaxies (see Exercise 7). The exact form of the redshift distance relation is different as predicted by each cosmological theory.

The differences between the observable quantities predicted by various models are small until very large distances are reached — that is, until we look back through an appreciable interval of time. Unfortunately, precise distances of remote objects, as well as their luminosities and other characteristics, are very difficult to determine, and critically accurate observations are required. Therefore, the tests described above are not easy to apply. At present we may be on the verge of being able to observe some quantities with enough precision to

choose between some of the evolutionary models (or perhaps eliminate them all), but we have not yet succeeded. It is very difficult to ascertain systematic differences between galaxies that might be due to evolutionary effects, because the very distant galaxies (the ones that we see as they were billions of years ago) are very faint. For those very faint, distant galaxies, it is not yet possible to measure distances accurately enough to tell whether the density of the universe is changing in time. On the other hand, many of the sources of radio waves that have been catalogued (Section 32.6) are probably remote galaxies or clusters of galaxies, and there is hope that the numbers of weak cosmic radio sources may indicate that one or the other cosmological model is out of the question. Also, we are probably near to being able to apply a test for the third distinction we have mentioned between the different models. Red shifts have now been measured of galaxies that are moving away from us with speeds up to 41 percent of the speed of light. As soon as we have found a reliable procedure for measuring accurately the relative distances of these remote objects, we shall be able to determine whether the rate of expansion of the universe is decreasing, and at what rate.

Whereas we cannot yet discriminate between the various evolutionary models, there are at least two significant observations that do, if our interpretations of them are correct, appear to rule out the steady state. We shall describe them briefly next.

(a) Observational Evidence against the Steady State
One clue that the steady-state theory is probably wrong comes from the quasi-stellar sources (Section 32.7). If we assume that they obey the same velocity-distance relation that galaxies do (in the professional parlance, that their redshifts are "cosmological") the distribution of QSSs in space would have to be uniform to be consistent with

the steady state. M. Schmidt, of the Mount Wilson and Palomar Observatories, has recently studied a statistically homogeneous sample of about 30 quasi-stellar sources. He found that the overwhelming majority of them — far more than would be expected in a uniform distribution — are at very great distances. In other words, the QSSs lie at such distances (or at least, have such redshifts) that nearly all of them represent phenomena occurring long ago in the history of the universe. It is as if the QSSs are things that existed early in the evolution of the universe — perhaps things happening to or in galaxies far in the past — and that occur only rarely today. If this analysis is correct, it clearly violates the perfect cosmological principle, and hence the steady state cosmology.

The other evidence against the steady state is perhaps one of the most exciting discoveries in recent decades. In 1965, A. Penzias and R. Wilson, of the Bell Telephone Laboratories, attempted to calibrate the random noise in the Laboratories' very sensitive radio receiver. They found, however, that the antenna, operating at 7.35-cm wavelength, picked up a weak radio signal that they could not account for unless they assumed that it was extraterrestrial in origin. It seemed to be coming from all directions, as if the entire celestial sphere were emitting the radiation. The radio radiation was of about the intensity that would be received if the whole universe were emitting radiation like a very cold black body (perfect radiator) with a temperature of about 3°K (3 centigrade degrees above absolute zero).

Now both Gamow and the Princeton physicist R. H. Dicke had predicted that if there were a "big bang" to start the expansion of the universe, it should now be possible to observe some of the radiation originating in that explosion. Just after the "bang," according to the hypothesis, the matter in the universe was extremely dense, hot, and opaque. Very intense radiation flowed from atom to atom in this *primeval fireball*. As the universe rapidly expanded, however, the matter thinned out and soon became transparent. From that point, most of the radiation was no longer absorbed and flowed freely throughout the universe; only those photons that chanced to strike material bodies (such as stars, gas clouds, and planets) would be absorbed. Some of that radiation should strike the earth and we should be able to detect it. Soon after the fireball the radiation must have been extremely intense — mostly X-rays — and have been characteristic of the radiation from an extremely hot black body — 1 million degrees or more in temperature. However, we must look far back into time to see this early radiation, which means it is coming from very far away in space, and thus it should be greatly redshifted. Thus, Gamow and Dicke predicted that if we could observe the radiation escaped from the primeval fireball today, it should be like that from a very cold black body — only a few degrees absolute in temperature — and would be at radio wavelengths.

When Penzias and Wilson announced their discovery of radiation at 7.35 cm, it occurred to Dicke and his colleagues that we might actually be looking at the "glowing embers" of the "big bang." Hence they designed delicate equipment to see if radiation corresponding to a 3°K black body could be detected at other wavelengths as well. The observations that followed confirmed the Bell Laboratories' findings. By mid-1968, isotropic radiation (that is, radiation equally intense from all directions) has been observed at 0.86, 1.58, 3.2, 7.35, and 20.7 cm. The radiation at all these wavelengths fits very well that from a black body with a temperature of 2.7°K. An independent check comes from observations of interstellar absorption lines (Sections 10.5f and 26.2g) of cyanogen (CN) in the spectra of the stars Zeta Ophiuchi and Zeta Persei. The observed lines showed that a significant fraction of the CN molecules are in a level of excitation that would be expected if they were bathed in radio-wavelength radiation of 2.6 cm. The implied strength of this radiation is also compatible with a general background radiation like that from a black body of temperature between 2 and 3°K.

Note that in all directions in space, as we look out to great distances, we are also looking back in time to the same (hypothetical) "big bang." Thus the observed radiation comes equally from all directions and gives no direction to a "center" of the universe; the universe, its "center," and its origin are all around us.

Still another prediction comes from the theory. In the fireball, nuclear reactions should occur in the high-temperature matter. Heavy elements should have been broken down into lighter ones, but hydrogen and helium should have survived as the matter thinned out and cooled. If the present-day radiation corresponds to that from a 3° black body, the theory predicts that the over-all abundance of helium in the universe (not counting that formed by hydrogen burning in stars), by mass, should be about 25 percent — a value quite consistent with observations.

(b) Conclusion

Present-day evidence suggests that the universe is not a steady state but is evolving from a "big bang" billions of years ago. We should not,

however, take our ideas too seriously. Our thinking in the realm of cosmology may be as rudimentary as that of the Greek philosophers 2500 years ago, who believed that all heavenly motions must occur in perfect circles. It will be very surprising if, after the next century, scientists finally settle on one or another of the specific theories we have described here.

The following words, written by Edwin Hubble in 1936, are still appropriate†:

Thus the explorations of space end on a note of uncertainty. And necessarily so. We are, by definition, in the very center of the observable region. We know our immediate neighborhood rather intimately. With increasing distance, our knowledge fades, and fades rapidly. Eventually, we reach the dim boundary — the utmost limits of our telescopes. There, we measure shadows, and we search among ghostly errors of measurement for landmarks that are scarcely more substantial.

The search will continue. Not until the empirical resources are exhausted, need we pass on to the dreamy realms of speculation.

---

†*The Realm of the Nebulae*, Yale University Press, 1936, pp. 201-202. Quoted by permission of The Publisher.

Exercises

1. A cluster of galaxies is observed to have a radial velocity of 60,000 km/sec. Assuming that $H = 100$ km/sec per million parsecs, find the distance to the cluster.

2. Plot the "velocity-distance relation" for the raisins in the bread analogy from the data given in Table 33.1.

3. Repeat Exercise 2, but use some other raisin than A for a reference. Is your new plot the same as the last one?

4. Why can the red shift in the spectra of galaxies *not* be explained by the absorption of their light by intergalactic dust?

5. Show by numerical example that the spreading out of galaxies after a "big bang" would be consistent with the observed red-shift law.

6. There are about 3×10^{13} km/pc and about 3×10^7 sec/year. Compute the maximum age of the universe, according to the big-bang theory, for $H = 100$ km/sec per million parsecs.

*7. Assume that the radial velocities of galaxies have always been the same and are given *at this instant of time* by

$$V = Hr,$$

where $H = 100$ km/sec per million parsecs. Note, however, that we do not observe the *present* distances of galaxies, but the distances they had when light left them on its journey to us. Now plot the relation between velocity and distance that would be obtained directly from *observations* (that is, corresponding to *measured* distances, not present distance). Consider several distances, out to 2×10^9 pc. Discuss the shape of the curve. How would this curve differ if the expansion rate were decreasing (say, because of gravitational attraction between galaxies)? What if it were increasing?

*8. Suppose we were to count all galaxies out to a certain distance in space. If the universe were not expanding, the total number counted should be proportional to the *cube* of the limiting distance of our counts. (Why?) Taking account of the finite time required for light to reach us, describe the relation that would be *observed* between the total count and the limiting distance for

(a) The steady-state theory.

(b) The evolutionary theory of the universe.

Bibliography

(The more technical references are marked with an asterisk.)

General Texts

McLaughlin, D. B., *Introduction to Astronomy*. Boston: Houghton Mifflin Company, 1961.

Krogdahl, W., *The Astronomical Universe*. New York: The Macmillan Company, 1962.

Struve, O., Lynds, B., and Pillans, H., *Elementary Astronomy*. New York: Oxford University Press, 1959.

Russell, H. N., Dugan, R. S., and Stewart, J. Q., *Astronomy* (2 volumes). Boston: Ginn & Company, 1938, 1945.

Motz, L., and Duveen, A., *Essentials of Astronomy*. Belmont, Calif.: Wadsworth Publishing Co., Inc., 1966.

Wyatt, S. P., *Principles of Astronomy*. Boston: Allyn and Bacon, Inc., 1964.

Huffer, C. M., Trinklein, F. E., and Bunge, M., *An Introduction to Astronomy*. New York: Holt, Rinehart and Winston, Inc., 1967.

More Elementary Texts

Inglis, S. J., *Planets, Stars, and Galaxies* (2nd ed.). New York: John Wiley & Sons, Inc., 1967.

Baker, R. H., and Fredrick, L. W., *An Introduction to Astronomy*. Princeton, N. J.: D. Van Nostrand Company, Inc., 1968.

Alter, D., Cleminshaw, C. H., and Phillips, J., *Pictorial Astronomy*. New York: Thomas Y. Crowell Company, 1963.

Histories of Astronomy

Pannekoek, A., *A History of Astronomy*. New York: Interscience Publishers, 1961.

Berry, A., *A Short History of Astronomy*. New York: Dover Publications, Inc., 1960.

Hoyle, F., *Astronomy*. New York: Doubleday & Company, Inc., 1962.

Vaucouleurs, G. de., *Discovery of the Universe*. New York: The Macmillan Company, 1957.

Shapley, H., and Howarth, H. E., *Source Book in Astronomy*. New York: McGraw-Hill, Inc., 1929.

Struve, O., and Zebergs, V., *Astronomy of the Twentieth Century*. New York: The Macmillan Company, 1962.

King, H. C., *Exploration of the Universe*. New York: New American Library, Inc., 1964.

Koestler, A., *The Watershed* (A Biography of Johannes Kepler). New York: Doubleday & Company, Inc., 1960.

Koestler, A., *The Sleepwalkers*. New York: The Macmillan Company, 1959.

Celestial Mechanics

Ryabov, Y., *An Elementary Survey of Celestial Mechanics*. New York: Dover Publications, Inc., 1961.

Ahrendt, M. H., *The Mathematics of Space Exploration*. New York: Holt, Rinehart and Winston, 1965.

*Danby, J. M. A., *Fundamentals of Celestial Mechanics*. New York: The Macmillan Company, 1962.

*Baker, R., and Makemson, M., *Introduction to Astrodynamics*. New York: Academic Press, Inc., 1960.

*Moulton, F. R., *An Introduction to Celestial Mechanics*. New York: The Macmillan Company, 1923.

*Sterne, T., *An Introduction to Celestial Mechanics*. New York: Interscience Publishers, 1960.

Space Science

Newell, H. E., *Express to the Stars*. New York: McGraw-Hill, Inc., 1961.

*Blanco, V., and McCuskey, S., *Basic Physics of the Solar System*. Reading, Mass.: Addison-Wesley Publishing Company, 1961.

Lundquist, C. A., *Space Science*. New York: McGraw-Hill, Inc., 1966.

Telescopes and Optics

Miczaika, G., and Sinton, W., *Tools of the Astronomer.* Cambridge, Mass.: Harvard University Press, 1961.

Page, T., and Page, L. W., *Telescopes; How to Make Them and Use Them.* New York: The Macmillan Company, 1966.

*Kuiper, G., and Middlehurst, B. (editors), *Telescopes* (Volume I of *Compendium of Stars and Stellar Systems*). Chicago: University of Chicago Press, 1960.

*Sears, F. W., *Optics.* Reading, Mass.: Addison-Wesley Publishing Company, 1949.

*Jenkins, F. A., and White, H. E., *Fundamentals of Optics.* New York: McGraw-Hill, Inc., 1950.

Steinberg, J. L., and Lequeux, J., *Radio Astronomy.* New York: McGraw-Hill, Inc., 1963.

Earth and Solar System

Whipple, F., *Earth, Moon, and Planets.* Cambridge, Mass. Harvard University Press, 1963.

Watson, F., *Between the Planets.* Cambridge, Mass.: Harvard University Press, 1956.

*Kuiper, G. (editor), *The Earth as a Planet.* Chicago: University of Chicago Press, 1954.

*Kuiper, G. (editor), *The Atmospheres of the Earth and Planets.* Chicago: University of Chicago Press, 1957.

Menzel, D., *Our Sun.* Cambridge, Mass.: Harvard University Press, 1959.

Baldwin, R. B., *A Fundamental Survey of the Moon.* New York: McGraw-Hill, Inc., 1965.

Hawkins, G. S., *Meteors, Comets, and Meteorites.* New York: McGraw-Hill, Inc., 1964.

Stellar Astronomy

Brandt, J. C., *The Sun and Stars.* New York: McGraw-Hill, Inc., 1966.

Goldberg, L., and Aller, L. H., *Atoms, Stars, and Nebulae.* New York: McGraw-Hill, Inc., 1943.

Bok, B. J., and Bok, P. F., *The Milky Way.* 1957. Cambridge, Mass.: Harvard University Press,

Campbell, L., and Jacchia, L., *The Story of Variable Stars.* New York: McGraw-Hill, Inc., 1941.

*Hynek, J. A. (editor), *Astrophysics, a Topical Symposium.* New York: McGraw-Hill, Inc., 1951.

*Aller, L. H., *Astrophysics* (2 volumes). New York: The Ronald Press Company, 1954 (Vol. I), 1963 (Vol. II).

*Schwarzchild, M., *Structure and·Evolution of the Stars.* Princeton, N. J.: Princeton University Press, 1958.

*Mihalas, D., and Routly, P. M., *Galactic Structure.* San Francisco: W. H. Freeman, 1968.

Galaxies and Cosmology

Shapley, H., *Galaxies.* Cambridge, Mass.: Harvard University Press, 1961.

Hubble, E., *The Realm of the Nebulae.* New Haven, Conn.: Yale University Press, 1936.

Shapley, H., *The Inner Metagalaxy.* New Haven, Conn.: Yale University Press, 1957.

Sandage, A. R., *The Hubble Atlas of Galaxies.* Washington, D.C.: Carnegie Institution, 1961.

Baade, W., *Evolution of Stars and Galaxies.* Cambridge, Mass.: Harvard University Press, 1963.

Sciama, D. W., *The Unity of the Universe.* New York: Doubleday & Company, Inc., 1959.

*McVittie, G. C., *Fact and Theory in Cosmology.* New York: The Macmillan Company, 1961.

*Bondi, H., *Cosmology.* New York: Cambridge University Press, 1960.

*Couderc, P., *The Expansion of the Universe.* London: Faber & Faber, Ltd., 1952.

Hodge, P. W., *Galaxies and Cosmology.* New York: McGraw-Hill, Inc., 1966.

Journals and Periodicals

Sky and Telescope, published monthly by the Sky Publishing Corporation, Harvard College Observatory, Cambridge, Massachusetts.

The Griffith Observer, published monthly by the Griffith Observatory, Los Angeles, California.

The Review of Popular Astronomy, published bimonthly by the Sky Map Publications, St. Louis, Missouri.

Scientific American, published monthly by Scientific American, New York, New York.

Leaflets and *Publications of the Astronomical Society of the Pacific,* published monthly and bimonthly, respectively, by the Astronomical Society of the Pacific, care of the California Academy of Sciences, San Francisco, California.

Astrophysical Journal, published monthly by the University of Chicago Press, Chicago, Illinois.

Astronomical Journal, published by the Yale University Observatory, New· Haven, Connecticut.

Glossary

aberration (of starlight). Apparent displacement in the direction of a star due to the earth's orbital motion.

ablation. Fragmentation and vaporization of a meteorite upon entering the atmosphere.

absolute magnitude. Apparent magnitude a star would have at a distance of 10 pc.

absolute zero. A temperature of $-273°C$ (or $0°K$) where all molecular motion stops.

absorption spectrum. Dark lines superimposed on a continuous spectrum.

accelerate. To change velocity; either to speed up, slow down, or change direction.

acceleration of gravity. Numerical value of the acceleration produced by the gravitational attraction on an object at the surface of a planet, or star.

active sun. The sun during times of unusual solar activity — spots, flares, and associated phenomena.

age-zero main sequence. Main sequence for a system of stars that have completed their contraction from interstellar matter, are now deriving all their energy from nuclear reactions, but whose chemical composition has not yet been altered by nuclear reactions.

airglow. Fluorescence in the atmosphere.

albedo. The fraction of incident sunlight that a planet or minor planet reflects.

almanac. A book or table listing astronomical events.

alpha particle. The nucleus of a helium atom, consisting of two protons and two neutrons.

altitude. Angular distance above or below the horizon, measured along a vertical circle, to a celestial object.

amplitude. The range in variability, as in the light from a variable star.

angstrom (Å). A unit of length equal to 10^{-8} cm.

angular diameter. Angle subtended by the diameter of an object.

angular momentum. A measure of the momentum associated with motion about an axis or fixed point.

annular eclipse. An eclipse of the sun in which the moon is too distant to appear to cover the sun completely, so that a ring of sunlight shows around the moon.

anomalistic month. The period of revolution of the moon about the earth with respect to its line of apsides, or to the perigee point.

anomalistic year. The period of revolution of the earth about the sun with respect to its line of apsides, or to the perihelion point.

antarctic circle. Parallel of latitude $66\frac{1}{2}°$ S; at this latitude the noon altitude of the sun is $0°$ on the date of the summer solstice.

aphelion. Point in its orbit where a planet is farthest from the sun.

apogee. Point in its orbit where an earth satellite is farthest from the earth.

apparent magnitude. A measure of the observed light flux received from a star or other object at the earth.

apparent relative orbit. The projection onto a plane perpendicular to the line of sight of the relative orbit of the fainter of the two components of a visual binary star about the brighter.

apparent solar day. The interval between two successive transits of the sun's center across the meridian.

apparent solar time. The hour angle of the sun's center *plus* 12 hours.

arctic circle. Parallel of latitude $66\frac{1}{2}°$ N; at this latitude the noon altitude of the sun is $0°$ on the date of the winter solstice.

argument of perifocus (or of perihelion or perigee). Angle at the focus of the orbit of a body (or at the center of the sun or earth), measured in the orbital plane and in the direction of motion of the body, from the ascending node to the perifocus (or perihelion or perigee) point.

artificial satellite. A manmade object put into a

closed orbit about the earth.

ascendant. (astrological term). The point on the zodiac that is on the eastern horizon, just rising at the moment of birth.

ascending node. The point along the orbit of a body where it crosses from the south to the north of some reference plane, usually the plane of the celestial equator or of the ecliptic.

aspect. The situation of the sun, moon, or planets with respect to one another.

association. A loose cluster of stars whose spectral types, motions, or positions in the sky indicate that they have probably had a common origin.

asteroid. A synonym for "minor planet."

astigmatism. A defect in an optical system whereby pairs of light rays in different planes do not focus at the same place.

astrology. The pseudoscience that treats with supposed influences of the configurations and locations in the sky of the sun, moon, and planets on human destiny; a primitive religion having its origin in ancient Babylonia.

astrometric binary. A binary star in which one component is not observed, but whose presence is deduced from the orbital motion of the visible component.

astrometry. That branch of astronomy that deals with the determination of precise positions and motions of celestial bodies.

astronautics. The science of the laws and methods of space flight.

astronomical unit (AU). Originally meant to be the semimajor axis of the orbit of the earth; now defined as the semimajor axis of the orbit of a hypothetical body with the mass and period that Gauss assumed for the earth. The semimajor axis of the orbit of the earth is 1.000 000 230 AU.

astronomy. The branch of science that treats of the physics and morphology of that part of the universe which lies beyond the earth's atmosphere.

astrophysics. That part of astronomy which deals principally with the physics of stars, stellar systems, and interstellar material. Astrophysics also deals, however, with the structures and atmospheres of the sun and planets.

atmospheric refraction. The bending, or refraction, of light rays from celestial objects by the earth's atmosphere.

atom. The smallest particle of an element that retains the properties which characterize that element.

atomic mass unit. *Chemical*: one sixteenth of the mean mass of an oxygen atom. *Physical*: one twelfth of the mass of an atom of the most common isotope of carbon. The atomic mass unit is approximately the mass of a hydrogen atom, 1.67×10^{-24} gm.

atomic number. The number of protons in each atom of a particular element.

atomic transition. A change in the state of energy of an atom; the atom may gain or lose energy by collision with another particle or by the emission or absorption of a photon.

atomic weight. The mean mass of an atom of a particular element in atomic mass units.

aurora. Light radiated by atoms and ions in the ionosphere, mostly in the polar regions.

autumnal equinox. The intersection of the ecliptic and celestial equator where the sun crosses the equator from north to south.

azimuth. The angle along the celestial horizon, measured eastward from the north point, to the intersection of the horizon with the vertical circle passing through an object.

Baily's beads. Small "beads" of sunlight seen passing through valleys along the limb of the moon in the instant preceding totality in a solar eclipse.

ballistic missile. A missile or rocket that is given its entire thrust during a brief period at the beginning of its flight, and that subsequently "coasts" to its target along an orbit.

Balmer lines. Emission or absorption lines in the spectrum of hydrogen that arise from transitions between the second (or first excited) and higher energy states of the hydrogen atoms.

bands (in spectra). Emission or absorption lines, usually in the spectra of chemical compounds or radicals, so numerous and closely spaced that they coalesce into broad emission or absorption bands.

barred spiral galaxy. Spiral galaxy in which the spiral arms begin from the ends of a "bar" running through the nucleus rather than from the nucleus itself.

barycenter. The center of mass of two mutually revolving bodies.

base line. That side of a triangle used in triangulation or surveying whose length is known (or can be measured), and which is included between two angles that are known (or can be measured).

Be star. A spectral type B star with emission lines in its spectrum, which are presumed to arise from material ejected from or surrounding the star.

"big-bang" theory. A theory of cosmology in which the expansion of the universe is presumed to have begun with a primeval explosion.

binary star. A double star; two stars revolving about each other.

black body. A hypothetical perfect radiator, which absorbs and reemits all radiation incident upon it.

black dwarf. The presumed final state of evolution for a star, in which all of its energy sources are exhausted and it no longer emits radiation.

blink microscope. A microscope in which the user's view is shifted rapidly back and forth between the corresponding portions of two different photographs of the same region of the sky.

Bode's law. A scheme by which a sequence of numbers can be obtained that give the approximate distances of the planets from the sun in astronomical units.

Bohr atom. A particular model of an atom, invented by Niels Bohr, in which the electrons are described as revolving about the nucleus in circular orbits.

bolide. A very bright fireball or meteor; sometimes defined as a fireball accompanied by sound.

bolometer. An instrument for measuring radiation received from a luminous body over the entire electromagnetic spectrum.

bolometric correction. The difference between the visual (or photovisual) and bolometric magnitudes of a star.

bolometric magnitude. A measure of the flux of radiation from a star or other object received just outside the earth's atmosphere, as it would be detected by a device sensitive to *all* forms of electromagnetic energy.

bremsstrahlung. Radiation from free-free transitions, in which electrons gain or lose energy while being accelerated in the field of an atomic nucleus or ion.

bubble chamber. A chamber in which bubbles form along the electrically charged path of a high-energy charged particle, rendering the track of that particle visible.

burnout. The instant when a rocket stops firing.

calculus. A branch of mathematics that permits computations involving rates of change (*differential* calculus) or of the contribution of an infinite number of infinitesimal quantities (*integral* calculus).

carbon cycle. A series of nuclear reactions involving carbon as a catalyst, by which hydrogen is transformed to helium.

cardinal points. The four principal points of the compass: North, East, South, and West.

Cassegrain focus. An optical arrangement in a reflecting telescope in which light is reflected by a second mirror to a point behind the objective mirror.

cD galaxy. A supergiant elliptical galaxy of the type frequently found in the centers of clusters of galaxies.

celestial equator. A great circle on the celestial sphere 90° from the celestial poles; the circle of intersection of the celestial sphere with the plane of the earth's equator.

celestial mechanics. That branch of astronomy which deals with the motions and gravitational influences of the members of the solar system.

celestial navigation. The art of navigation at sea or in the air from sightings of the sun, moon, planets, and stars.

celestial poles. Points about which the celestial sphere appears to rotate; intersections of the celestial sphere with the earth's polar axis.

celestial sphere. Apparent sphere of the sky; a sphere of large radius centered on the observer. Directions of objects in the sky can be denoted by the positions of those objects on the celestial sphere.

center of gravity. Center of mass.

center of mass. The mean position of the various mass elements of a body or system, weighted according to their distances from that center of mass; that point in an isolated system which moves with constant velocity, according to Newton's first law of motion.

centrifugal force (or acceleration). An imaginary force (or acceleration) that is often introduced to account for the illusion that a body moving on a curved path tends to accelerate radially from the center of curvature. The actual force present is the one that diverts the body's motion from a straight line and is directed *toward* the center of curvature. It is, however, legitimate to introduce a fictitious centrifugal force field in a rotating (and hence noninertial) coordinate system.

centripetal force (or acceleration). The force required to divert a body from a straight path into a curved path (or the acceleration experienced by the body); it is directed toward the center of curvature.

Cepheid variable. A star that belongs to one of two classes (type I and type II) of yellow supergiant pulsating stars.

Ceres. Largest of the minor planets and first to be discovered.

chromatic aberration. A defect of optical systems whereby light of different colors is focused at different places.

chromosphere. That part of the solar atmosphere that lies immediately above the photospheric layers.

chronograph. A device for recording and measuring the times of events.

chronometer. An accurate clock.

circle of position. A small circle on the surface of the earth, centered at the substellar point of some star, and of radius equal to the coaltitude of that star as seen by some observer; the observer must, therefore, be somewhere on this circle of position.

circular velocity. The critical speed with which a revolving body can have a circular orbit.

circumpolar regions. Portions of the celestial sphere near the celestial poles that are either always above or always below the horizon.

cloud chamber. A chamber in which droplets of liquid condense along the electrically charged path of a high-energy charged particle, rendering the track of that particle visible.

Clouds of Magellan. Two neighboring galaxies visible to the naked eye from southern latitudes.

cluster of galaxies. A system of galaxies containing from several to thousands of member galaxies.

cluster variable (RR Lyrae variable). A member of a certain large class of pulsating variable stars, all with periods less than 1 day. These stars are often present in globular star clusters.

color excess. The amount by which the color index of a star is increased when its light is reddened in passing through interstellar absorbing material.

color index. Difference between the magnitudes of a star or other object measured in light of two different spectral regions, for example, photographic *minus* photovisual magnitudes.

color-magnitude diagram. Plot of the magnitudes (apparent or absolute) of the stars in a cluster against their color indices.

coma. A defect in an optical system in which off-axis rays of light striking different parts of the objective do not focus in the same place.

coma (of comet). The diffuse gaseous component of the head of a comet.

comet. A swarm of solid particles and gases, which revolves about the sun, usually in an orbit of high eccentricity.

commensurable. State of a quantity (such as the period of a minor planet) that can be reduced to another quantity (such as the period of another minor planet) by multiplication by a ratio of two small whole numbers.

compact galaxy. A galaxy of small size and high surface brightness.

comparison spectrum. The spectrum of a vaporized element (such as iron) photographed beside the image of a stellar spectrum, and with the same camera, for purposes of comparison of wavelengths.

compound. A substance composed of two or more chemical elements.

compound nucleus. An excited nucleus, usually temporary, formed by nuclei of two or more simpler atoms.

conduction. The transfer of energy by the direct passing of energy or electrons from atom to atom.

configuration. Any one of several particular orientations in the sky of the moon or a planet with respect to the sun.

conic section. The curve of intersection between a circular cone and a plane; these curves can be ellipses, circles, parabolas, or hyperbolas.

conjunction. The configuration of a planet when it has the same celestial longitude as the sun, or the configuration when any two celestial bodies have the same celestial longitude or right ascension.

conservation of angular momentum. The law that angular momentum is conserved in the absence of any force not directed toward or away from the point or axis about which the angular momentum is referred — that is, in the absence of a torque.

constellation. A configuration of stars named for a particular object, person, or animal; or the area of the sky assigned to a particular configuration.

contacts (of eclipses). The instants of certain stages of an eclipse.

continuous spectrum. A spectrum of light comprised of radiation of a continuous range of wavelengths or colors rather than only certain discrete wavelengths.

convection. The transfer of energy by moving currents of a fluid containing that energy.

core (of earth). The central part of the earth, believed to be a liquid of high density.

coriolis effect. The deflection (with respect to the ground) of projectiles moving across the surface of the rotating earth.

corona. Outer atmosphere of the sun.

corona (or halo) of Galaxy. The outer portions of the Galaxy, especially on either side of the plane of the Milky Way.

coronagraph. An instrument for photographing the chromosphere and corona of the sun outside of eclipse.

corpuscular radiation. Charged particles, mostly atomic nuclei and electrons, emitted into space by the sun and possibly other objects.

cosine. One of the trigonometric functions of an angle. If the angle is in a right triangle, the cosine is the ratio of the lengths of the shorter side adjacent to the angle to the hypotenuse.

cosmic rays. Atomic nuclei (mostly protons) that are observed to strike the earth's atmosphere with exceedingly high energies.

cosmogony. The study of the origin of the world or universe.

cosmological constant. A term that arises in the development of the field equations of general relativity, which represents a repulsive force in the universe. The cosmological constant is often assumed to be zero.

cosmological model. A specific model, or theory, of the organization and evolution of the universe.

cosmological principle. The assumption that, on the large scale, the universe at any given time is the same everywhere.

cosmology. The study of the organization and evolution of the universe.

Coudé focus. An optical arrangement in a reflecting telescope whereby light is reflected by two or more secondary mirrors down the polar axis of the telescope to a focus at a place separate from the moving parts of the telescope.

crater (lunar). A more or less circular depression in the surface of the moon.

crater (meteoritic). A crater on the earth caused by the collision of a meteoroid with the earth, and a subsequent explosion.

crescent moon. One of the phases of the moon when its elongation is less than 90° from the sun and it appears less than half full.

crust (of earth). The outer layer of the earth.

cyclonic motion. A counterclockwise circular circulation of winds (in the Northern Hemisphere) that results from the coriolis effect.

dark nebula. A cloud of interstellar dust that obscures the light of more distant stars, and appears as an opaque curtain.

daylight savings time. A time 1 hour more advanced than standard time, usually adopted in spring and summer to take advantage of long evening twilights.

declination. Angular distance north or south of the celestial equator to some object,

measured along an hour circle passing through that object.

deferent. A stationary circle in the Ptolemaic system along which moves the center of another circle (epicycle), along which moves an object or another epicycle.

degenerate gas. A gas in which the allowable states for the electrons have been filled; it behaves according to different laws from those that apply to "perfect" gases.

density. The ratio of the mass of an object to its volume.

descending node. The point along the orbit of a body where it crosses from the north to the south of some reference plane, usually the plane of the celestial equator or of the ecliptic.

deuterium. A "heavy" form of hydrogen, in which the nucleus of each atom consists of one proton and one neutron.

differential equation. An equation that involves derivatives or rates of change.

differential galactic rotation. The state of rotation of the galaxy whereby it does not rotate as a solid wheel, so that parts adjacent to each other do not always stay close together.

differential gravitational force. The difference between the respective gravitational forces exerted on two bodies near each other by a third, more distant body.

diffraction. The spreading out of light in passing the edge of an opaque body.

diffraction grating. A system of closely spaced equidistant slits or reflecting strips which, by diffraction and interference, produce a spectrum.

diffraction pattern. A pattern of bright and dark fringes produced by the interference of light rays, diffracted by different amounts, with each other.

diffuse nebula. A reflection or emission nebula produced by interstellar matter (not a planetary nebula).

disk (of planet or other object). The apparent disk that a planet (or the sun, or moon, or a star) displays when seen in the sky or viewed telescopically.

disk of Galaxy. The central disk or "wheel" of our Galaxy, superimposed on the spiral structure.

dispersion. Separation, from white light, of different wavelengths being refracted by different amounts.

distance modulus. Difference between the apparent and absolute magnitudes of an object.

diurnal. Daily.

diurnal circle. Apparent path of a star in the sky during a complete day due to the earth's rotation.

diurnal libration. The phenomenon whereby slightly different hemispheres of the moon can be observed during a day because of the motion of the observer caused by the earth's rotation.

diurnal motion. Motion during one day.

diurnal parallax. Apparent change in direction of an object caused by a displacement of the observer due to the earth's rotation.

Doppler shift. Apparent change in wavelength of the radiation from a source due to its relative motion in the line of sight.

draconic month. The period of revolution of the moon about the earth with respect to the nodes of the moon's orbit.

dwarf (star). A main-sequence star (as opposed to a giant or supergiant).

dynamical parallax. A distance (or parallax) for a binary star derived from the period of mutual revolution, the mass-luminosity relation, and the laws of mechanics.

dyne. The metric unit of force; the force required to accelerate a mass of 1 gram in the amount 1 centimeter per second per second.

east point. The point on the horizon 90° from the north point (measured clockwise as seen from the zenith).

eccentric. A point, about which an object revolves on a circular orbit, that is not at the center of the circle.

eccentricity (of ellipse). Ratio of the distance between the foci to the major axis.

eclipse. The cutting off of all or part of the light of one body by another passing in front of it.

eclipse path. The track along the earth's surface swept out by the tip of the shadow of the moon (or the extension of its shadow) during a total (or annular) solar eclipse.

eclipse season. A period during the year when an eclipse of the sun or moon is possible.

eclipsing binary star. A binary star in which the plane of revolution of the two stars is nearly edge on to our line of sight, so that the light of one star is periodically diminished by the other passing in front of it.

ecliptic. The apparent annual path of the sun on the celestial sphere.

ecliptic limit. The maximum angular distance from a node where the moon can be for an eclipse to take place.

electromagnetic radiation. Radiation consisting of waves propagated through the building up and breaking down of electric and magnetic fields; these include radio, infrared, light, ultraviolet, X-rays, and gamma rays.

electromagnetic spectrum. The whole array or family of electromagnetic waves.

electron. A negatively charged subatomic particle that normally moves about the nucleus of an atom.

electron volt. The kinetic energy acquired by an electron that is accelerated through an electric potential of 1 volt; 1 electron volt is 1.60207×10^{-12} erg.

electroscope. A device for measuring the amount of charge in the air.

element. A substance that cannot be decomposed, by chemical means, into simpler substances.

elements (of orbit). Any of several quantities that describe the size, shape, and orientation of the orbit of a body.

ellipse. A conic section: the curve of intersection of a circular cone and a plane cutting completely through the cone.

elliptical galaxy. A galaxy whose apparent photometric contours are ellipses and which contains no conspicuous interstellar material.

ellipticity. The ratio (in an ellipse) of the major axis *minus* the minor axis to the major axis.

elongation. The difference between the celestial longitudes of a planet and the sun.

emission line. A discrete bright spectral line.

emission nebula. A gaseous nebula that derives its visible light from the fluorescence of ultraviolet light from a star in or near the nebula.

emission spectrum. A spectrum consisting of emission lines.

emulsion stack. A number of layers of photographic emulsion piled on top of one another and used for the detection of higher energy particles.

encounter (gravitational). A near passing, on hyperbolic orbits, of two objects that influence each other gravitationally.

energy. The ability to do work.

energy equation. See *vis viva* equation.

energy level (in an atom or ion). A particular level, or amount, of energy possessed by an atom or ion above the energy it possesses in its least energetic state.

energy spectrum. A table or plot showing the relative numbers of particles (in cosmic rays or corpuscular radiation) of various energies.

ephemeris. A table that gives the positions of a celestial body at various times, or other astronomical data.

ephemeris time. A kind of time that passes at a strictly uniform rate; used to compute the instants of various astronomical events.

epicycle. A circular orbit of a body in the Ptolemaic system, the center of which revolves about another circle (the deferent).

equant. A stationary point in the Ptolemaic system not at the center of a circular orbit about which a body (or the center of an epicycle) revolves with uniform angular velocity.

equation of state. An equation relating the pressure, temperature, and density of a substance (usually a gas).

equation of time. The difference between apparent and mean solar time.

equator. A great circle on the earth, 90° from its poles.

equatorial mount. A mounting for a telescope, one axis of which is parallel to the earth's axis, so that a motion of the telescope about that axis can compensate for the earth's rotation.

equinox. One of the intersections of the ecliptic and celestial equator.

equivalent width. A measure of the strength of a spectral line; the width of an absorption line of rectangular profile and zero intensity at its center.

erg. The metric unit of energy; the work done by a force of 1 dyne moving through a distance of one centimeter.

eruptive variable. A variable star whose changes in light are erratic or explosive.

establishment of port. The time interval, at a particular port, between the meridian passage of the moon and high tide.

evolutionary cosmology. A theory of cosmology that assumes that all parts of the universe have a common age and evolve together.

eyepiece. A magnifying lens used to view the image produced by the objective of a telescope.

excitation. The process of imparting to an atom or an ion an amount of energy greater than that it has in its normal or least-energy state.

extinction. Attenuation of light from a celestial body produced by the earth's atmosphere, or by interstellar absorption.

extragalactic. Beyond the Galaxy.

faculus (p. faculae). Bright regions near the limb of the sun.

filar micrometer. A device used with a telescope to measure the angular separations of closely separated pairs of stars or the diameters of extended objects.

filtergram. A photograph of the sun (or part of it) taken through a special narrow-band-pass filter.

fireball. A spectacular meteor.

fission. The breakup of a heavy atomic nucleus into two or more lighter ones.

flare. A sudden and temporary outburst of light from an extended region of the solar surface.

flare star. A member of a class of stars that show occasional, sudden, unpredicted increases in light.

flash spectrum. The spectrum of the very limb of the sun obtained in the instant before totality in a solar eclipse.

flocculus (p. flocculi). Bright regions of the solar surface observed in the monochromatic light of some spectral line; flocculi are now usually called *plages*.

fluctions. Name given by Newton to the calculus.

fluorescence. The absorption of light of one wavelength and reemission of it at another wavelength; especially the conversion of ultraviolet into visible light.

focal length. The distance from a lens or mirror to the point where light converged by it comes to a focus.

focal ratio (speed). Ratio of the focal length of a lens or mirror to its aperture.

focus. Point where the rays of light converged by a mirror or lens meet.

focus of a conic section. Mathematical point associated with a conic section, whose distance to any point on the conic bears a constant ratio to the distance from that point to a straight line known as the *directrix*.

forbidden lines. Spectral lines that are not usually observed under laboratory conditions because they result from atomic transitions that are highly improbable.

force. That which can change the momentum of a body; numerically, the rate at which the body's momentum changes.

Foucault pendulum. An experiment first conducted by Jean Foucault in 1851 to demonstrate the rotation of the earth.

Fraunhofer line. An absorption line in the spectrum of the sun or of a star.

Fraunhofer spectrum. The array of absorption lines in the spectrum of the sun or of a star.

free-free transition. An atomic transition in which the energy associated with an atom or ion and passing electron changes during the encounter, but without capture of the electron by the atom or ion.

frequency. Number of vibrations per unit time;

number of waves that cross a given point per unit time (in radiation).

fringes (interference). Successive dark and light fringes, or lines, caused by interference of light waves with each other before they strike a screen or detecting device and are observed.

full moon. That phase of the moon when it is at opposition (180° from the sun) and its full daylight hemisphere is visible from the earth.

fusion. The building up of heavier atomic nuclei from lighter ones.

galactic cluster. An "open" cluster of stars located in the spiral arms or disk of the Galaxy.

galactic equator. Intersection of the principal plane of the Milky Way with the celestial sphere.

galactic latitude. Angular distance north or south of the galactic equator to an object, measured along a great circle passing through that object and the galactic poles.

galactic longitude. Angular distance, measured eastward along the galactic equator from the galactic center, to the intersection of the galactic equator with a great circle passing through the galactic poles and an object.

galactic poles. The poles of the galactic equator; the intersections with the celestial sphere of a line through the observer that is perpendicular to the plane of the galactic equator.

galactic rotation. Rotation of the Galaxy.

galaxy. A large assemblage of stars; a typical galaxy contains millions to hundreds of billions of stars.

Galaxy. The galaxy to which the sun and our neighboring stars belong; the Milky Way is light from remote stars in the Galaxy.

gamma rays. Photons (of electromagnetic radiation) of energy higher than those of X-rays; the most energetic form of electromagnetic radiation.

gauss. A unit of magnetic flux density.

gegenschein (counterglow). A very faint, diffuse glow of light opposite the sun in the sky, believed to be caused by sunlight reflected from interplanetary particles.

Geiger counter. A device for counting high-energy charged particles and hence for measuring the intensity of corpuscular radiation.

geod. The mean figure of the earth.

geomagnetic. Referring to the geometrical center of the earth's magnetic field.

geomagnetic poles. The poles of a hypothetical

bar magnet whose magnetic field most nearly matches that of the earth.

giant (star). A star of large luminosity and radius.

gibbous moon. One of the phases of the moon in which more than half, but not all, of the moon's daylight hemisphere is visible from the earth.

globular cluster. One of about 120 large star clusters that form a system of clusters centered on the center of the Galaxy.

globule. A small, dense, dark nebula; believed to be a possible protostar.

granulation. The "rice-grain"-like structure of the solar photosphere.

gravitation. The tendency of matter to attract itself together.

gravitational constant, G. The constant of proportionality in Newton's law of gravitation; in metric units G has the value 6.668×10^{-8} dyne $\cdot$ cm²/gm².

gravitational energy. Energy that can be released by the gravitational collapse, or partial collapse, of a system.

great circle. Circle on the surface of a sphere that is the curve of intersection of the sphere with a plane passing through its center.

greatest elongation (east or west). The largest separation in celestial longitude (to the east or west) that an inferior planet can have from the sun.

Greenwich meridian. The meridian of longitude passing through the site of the old Royal Greenwich Observatory, near London; origin of longitude on the earth.

Gregorian calendar. A calendar (now in common use) introduced by Pope Gregory XIII in 1582.

H I region. Region of neutral hydrogen in interstellar space.

H II region. Region of ionized hydrogen in interstellar space.

half-life. The time required for half of the radioactive atoms in a sample to disintegrate.

halo (around sun or moon). A ring of light around the sun or moon caused by refraction by the ice crystals of cirrus clouds.

halo (of galaxy). See corona.

harmonic law. Kepler's third law of planetary motion: the cubes of the semimajor axes of the planetary orbits are in proportion to the squares of the sidereal periods of the planets.

harvest moon. The full moon nearest the time of the autumnal equinox.

Hayaschi line. Track of evolution on the

Hertzsprung-Russell diagram of a completely convective star.

head (of comet). The main part of a comet, consisting of its nucleus and coma.

"heavy" elements. In astronomy, usually those elements of greater atomic number than helium.

heliacal rising. The rising of a star or planet the first time that it can be observed in the morning before sunrise.

helio-. Prefix referring to the sun.

heliocentric. Centered on the sun.

helium flash. The nearly explosive ignition of helium in the triple-alpha process in the dense core of a red giant star.

Hertzsprung gap. A V-shaped gap in the upper part of the Hertzsprung-Russell diagram where few stable stars are found.

Hertzsprung-Russell (H-R) diagram. A plot of absolute magnitude against temperature (or spectral class or color index) for a group of stars.

high-velocity star (or object). A star (or object) with high space motion; generally an object that does not share the high orbital velocity of the sun about the galactic nucleus.

homogeneous star (or stellar model). A star (or theoretical model of a star) whose chemical composition is the same throughout its interior.

horizon (astronomical). A great circle on the celestial sphere 90° from the zenith.

horizon system. A system of celestial coordinates (altitude and azimuth) based on the astronomical horizon and the north point.

horizontal branch. A sequence of stars on the Hertzsprung-Russell diagram of a typical globular cluster of approximately constant absolute magnitude (near $M_v = 0$).

horizontal parallax. The angle by which an object appears displaced (after correction for atmospheric refraction) when viewed on the horizon from a place on the earth's equator, as compared to its direction if it were viewed from the center of the earth.

horoscope (astrological term). A chart showing the positions along the zodiac and in the sky of the sun, moon, and planets at some given instant and place on earth — generally corresponding to the time and place of a person's birth.

hour angle. The angle measured westward along the celestial equator from the local meridian to the hour circle passing through an object.

hour circle. A great circle on the celestial sphere passing through the celestial poles.

Hubble constant. Constant of proportionality in the relation between the velocities of remote galaxies and their distances. The Hubble constant is approximately 75 km/sec · 10^6 pc.

hydrostatic equilibrium. A balance between the weights of various layers, as in a star or the earth's atmosphere, and the pressures that support them.

hyperbola. A conic section of eccentricity greater than 1.0; the curve of intersection between a circular cone and a plane that is at too small an angle with the axis of the cone to cut all of the way through it, and is not parallel to a line in the face of the cone.

hyperfine transition. A change in the energy state of an atom that involves a change in the spin of its nucleus.

hypothesis. A tentative theory or supposition, advanced to explain certain facts or phenomena, and which is subject to further tests and verification.

image. The optical representation of an object produced by light rays from the object being refracted or reflected by a lens or mirror.

image tube. A device in which electrons, emitted from a photocathode surface exposed to light, are focused electronically.

inclination (of an orbit). The angle between the orbital plane of a revolving body and some fundamental plane — usually the plane of the celestial equator or of the ecliptic.

Index Catalogue, IC. The supplement to Dreyer's *New General Catalogue* of star clusters and nebulae.

index of refraction. A measure of the refracting power of a transparent substance; specifically, the ratio of the speed of light in a vacuum to its speed in the substance.

inertia. The property of matter that requires a force to act on it to change its state of motion; momentum is a measure of inertia.

inferior conjunction. The configuration of an inferior planet when it has the same longitude as the sun, and is between the sun and earth.

inferior planet. A planet whose distance from the sun is less than the earth's.

infrared radiation. Electromagnetic radiation of wavelength longer than the longest (red) wavelengths that can be perceived by the eye, but shorter than radio wavelengths.

insolation. The rate at which all radiation from the sun is received per unit area on the ground.

intercalate. To insert, as a day in a calendar.

interferometer (stellar). An optical device, making use of the principle of interference of light waves, with which small angles can be measured.

iteration. The "closing in " on the solution to a mathematical problem by repetitive calculations.

International Date Line. An arbitrary line on the surface of the earth near longitude 180° across which the date changes by one day.

international magnitude system. The system of photographic and photovisual magnitudes, referring to the blue and yellow spectral regions, at one time adopted by international agreement, but now largely superseded by the *U, B, V* system.

interplanetary medium. The sparse distribution of gas and solid particles in the interplanetary space.

interstellar dust. Microscopic solid grains, believed to be mostly dielectric compounds of hydrogen and other common elements, in interstellar space.

interstellar gas. Sparse gas in interstellar space.

interstellar lines. Absorption lines superimposed on stellar spectra, produced by the interstellar gas.

interstellar matter. Interstellar gas and dust.

ion. An atom that has become electrically charged by the addition or loss of one or more electrons.

ionization. The process by which an atom gains or loses electrons.

ionization potential. The energy required to remove an electron from an atom.

ionosphere. The upper region of the earth's atmosphere in which many of the atoms are ionized.

irregular galaxy. A galaxy without rotational symmetry; neither a spiral or elliptical galaxy.

irregular variable. A variable star whose light variations do not repeat with a regular period.

island universe. Historical synonym for galaxy.

isotope. Any of two or more forms of the same element, whose atoms all have the same number but different masses.

isotropic. The same in all directions.

Jovian planet. Any of the planets Jupiter, Saturn, Uranus, and Neptune.

Julian calendar. A calendar introduced by Julius Caesar in 45 B.C.

Julian day. The number of the day in a running sequence beginning January 1, 4713 B.C.

Jupiter. The fifth planet from the sun in the solar system.

Kepler's laws. Three laws, discovered by J. Kepler, that describe the motions of the planets.

kinetic energy. Energy associated with motion; the kinetic energy of a body is one half the product of its mass and the square of its velocity.

kinetic theory (of gases). The science that treats the motions of the molecules that compose gases.

Kirkwood's gaps. Gaps in the spacing of the minor planets that arise from perturbations produced by the major planets.

Lagrangian points. Five points in the plane of revolution of two bodies, revolving mutually about each other in circular orbits, where a third body of negligible mass can remain in equilibrium with respect to the other two bodies.

latitude. A north-south coordinate on the surface of the earth; the angular distance north or south of the equator measured along a meridian passing through a place.

law. A statement of order or relation between phenomena that, under given conditions, is presumed to be invariable.

law of areas. Kepler's second law: the radius vector from the sun to any planet sweeps out equal areas in the planet's orbital plane in equal intervals of time.

law of the redshifts. The relation between the radial velocity and distance of a remote galaxy: the radial velocity is proportional to the distance of the galaxy.

lead sulfide cell. A device used to measure infrared radiation.

leap year. A calendar year with 366 days, intercalated approximately every 4 years to make the average length of the calendar year as nearly equal as possible to the tropical year.

libration. Any of several phenomena by which an observer on earth, over a period of time, can see more than one hemisphere of the moon.

libration in latitude. Libration caused by the fact that the moon's axis of rotation is not perpendicular to its plane of revolution.

libration in longitude. Libration caused by the regularity in the moon's rotation but irregularity in its orbital speed.

light. Electromagnetic radiation that is visible to the eye.

light curve. A graph that displays the variation in light or magnitude of a variable or eclipsing binary star.

light-year. The distance light travels in a

vacuum in 1 year; 1 LY = 9.46×10^{17} cm, or about 6×10^{12} mi.

limb (of sun or moon). Apparent edge of the sun or moon as seen in the sky.

limb darkening. The phenomenon whereby the sun is less bright near its limb than near the center of its disk.

limiting magnitude. The faintest magnitude that can be observed with a given instrument or under given conditions.

line broadening. The phenomenon by which spectral lines are not precisely sharp but have finite widths.

line of apsides. The line connecting the apses of an orbit (the perifocus and farthest from focus points); or the line along the major axis of the orbit.

line of nodes. The line connecting the nodes of an orbit.

line of position. A part of a circle of position, so small that it can be considered a straight line.

line profile. A plot of the intensity of light versus wavelength across a spectral line.

linear diameter. Actual diameter in units of length.

Local Group. The cluster of galaxies to which our Galaxy belongs.

local standard of rest. A coordinate system that shares the average motion of the sun and its neighboring stars about the galactic center.

local supercluster (or supergalaxy). A proposed cluster of clusters of galaxies, to which the Local Group belongs.

longitude. An east-west coordinate on the earth's surface; the angular distance, measured east or west along the equator from the Greenwich meridian, to the meridian passing through a place.

longitude of the ascending node. The angle measured eastward from a reference direction (usually the vernal equinox) in a fundamental plane (usually the plane of the celestial equator or of the ecliptic) to the ascending node of the orbit of a body.

low-velocity star (or object). A star (or object) that has low space velocity; generally an object that shares the sun's high orbital speed about the galactic center.

luminosity. The rate of radiation of electromagnetic energy into space by a star or other object.

luminosity class. A classification of a star according to its luminosity for a given spectral class.

luminosity function. The relative numbers of stars (or other objects) of various luminosities or absolute magnitudes.

luminous energy. Light.

lunar. Referring to the moon.

lunar eclipse. An eclipse of the moon.

Lyman lines. A series of absorption or emission lines in the spectrum of hydrogen that arise from transitions to and from the lowest energy states of the hydrogen atoms.

magnetic field. The region of space near a magnetized body within which magnetic forces can be detected.

magnetic pole. One of two points on a magnet (or the earth) at which the greatest density of lines of force emerge. A compass needle aligns itself along the local lines of force on the earth and points more or less toward the magnetic poles of the earth.

magnifying power. The number of times larger (in angular diameter) an object appears through a telescope than with the naked eye.

magnitude. A measure of the amount of light flux received from a star or other luminous object.

main sequence. A sequence of stars on the Hertzsprung-Russell diagram, containing the majority of stars, that runs diagonally from the upper left to the lower right.

major axis (of ellipse). The maximum diameter of an ellipse.

major planet. A Jovian planet.

mantle (of earth). The greatest part of the earth's interior, lying between the crust and the core.

mare. Latin for "sea"; name applied to many of the "sealike" features on the moon or Mars.

Mars. Fourth planet from the sun in the solar system.

mass. A measure of the total amount of material in a body; defined either by the inertial properties of the body or by its gravitational influence on other bodies.

mass defect. The amount by which the mass of an atomic nucleus is less than the sum of the masses of the individual nucleons that compose it.

mass function. A numerical relation between the masses of the components of a binary star and the inclination of their plane of mutual revolution to the plane of the sky; the mass function is determined from an analysis of the radial-velocity curve of a spectroscopic binary when the spectral lines of only one of the stars are visible.

mass-luminosity relation. An empirical relation

between the masses and luminosities of many (principally main-sequence) stars.

mass-radius relation (for white dwarfs). A theoretical relation between the masses and radii of white dwarf stars.

mean solar day. Interval between successive meridian passages of the mean sun; average length of the apparent solar day.

mean solar time. Local hour angle of the mean sun *plus* 12 hours.

mean sun. A fictitious body that moves eastward with uniform angular velocity along the celestial equator, completing one circuit of the sky with respect to the vernal equinox in a tropical year.

mechanics. That branch of physics which deals with the behavior of material bodies under the influence of, or in the absence of, forces.

Mercury. Nearest planet to the sun in the solar system.

meridian (celestial). The great circle on the celestial sphere that passes through an observer's zenith and the north (or south) celestial pole.

meridian (terrestrial). The great circle on the surface of the earth that passes through a particular place and the north and south poles of the earth.

meson. A subatomic particle of mass intermediate between that of a proton and that of an electron.

mesosphere. The layer of the ionosphere immediately above the stratosphere.

Messier catalogue. A catalogue of nonstellar objects compiled by Charles Messier in 1787.

metastable level. An energy level in an atom from which there is a low probability of an atomic transition accompanied by the radiation of a photon.

meteor. The luminous phenomenon observed when a meteoroid enters the earth's atmosphere and burns up; popularly called a "shooting star."

meteor shower. Many meteors appearing to radiate from a common point in the sky caused by the collision of the earth with a swarm of meteoritic particles.

meteorite. A portion of a meteoroid that survives passage through the atmosphere and strikes the ground.

meteorite fall. The occurrence of a meteorite striking the ground.

meteoroid. A meteoritic particle in space before any encounter with the earth.

micrometeorite. A meteoroid so small that, on entering the atmosphere of the earth, it is slowed quickly enough that it does not burn up or ablate but filters through the air to the ground.

microphotometer. A device for accurately recording variations in photographic density or transmission through the emulsion along a path in a photograph, especially a spectrogram.

Milky Way. The band of light encircling the sky, which is due to the many stars and diffuse nebulae lying near the plane of the Galaxy.

minor axis (of ellipse). The smallest or least diameter of an ellipse.

minor planet. One of several tens of thousands of small planets, ranging in size from a few hundred miles to less than 1 mile in diameter.

Mira Ceti–type variable star. Any of a large class of red-giant long-period or irregular pulsating variable stars, of which the star Mira is a prototype.

missile. A projectile, especially a rocket.

model atmosphere (or photosphere). The result of a theoretical calculation of the run of temperature, pressure, density, and so on, through the outer layers of the sun or a star.

molecule. A combination of two or more atoms bound together; the smallest particle of a chemical compound or substance that exhibits the chemical properties of that substance.

momentum. A measure of the inertia or state of motion of a body; the momentum of a body is the product of its mass and velocity. In the absence of a force, momentum is conserved.

monochromatic. Of one wavelength or color.

n-**body problem.** The problem of determining the positions and motions of more than two bodies in a system in which the bodies interact under the influence of their mutual gravitation.

N galaxy. A galaxy with a stellar-appearing nucleus with the remainder of the galaxy appearing as a surrounding faint haze.

nadir. The point on the celestial sphere 180° from the zenith.

nautical mile. The mean length of 1 minute of arc on the earth's surface along a meridian.

navigation. The art of finding one's position and course at sea or in the air.

neap tide. The lowest tides in the month which occur when the moon is near first or third quarter.

nebula. Cloud of interstellar gas or dust.

Neptune. Eighth planet from the sun in the solar system.

neutrino. A particle that has no mass or charge but that carries energy away in the course of certain nuclear transformations.

neutron. A subatomic particle with no charge and with mass approximately equal to that of the proton.

neutron star. A hypothetical star of extremely high density composed entirely of neutrons.

New General Catalogue (NGC). A catalogue of star clusters, nebulae, and galaxies compiled by J. L. E. Dreyer in 1888.

new moon. Phase of the moon when its longitude is the same as that of the sun.

Newtonian focus. An optical arrangement in a reflecting telescope where the light is reflected by a flat mirror to a focus at the side of the telescope tube just before it reaches the focus of the objective.

Newton's laws. The laws of mechanics and gravitation formulated by Isaac Newton.

night sky light. The faint illumination of the night sky; the main source is usually fluorescence by atoms high in the atmosphere.

node. The intersection of the orbit of a body with a fundamental plane — usually the plane of the celestial equator or of the ecliptic.

nodical month. The period of revolution of the moon about the earth with respect to the line of nodes of the moon's orbit.

nodical (eclipse) year. Period of revolution of the earth about the sun with respect to the line of nodes of the moon's orbit.

north point. That intersection of the celestial meridian and astronomical horizon lying nearest the north celestial sphere.

north polar sequence. A group of stars in the vicinity of the north celestial pole whose magnitudes serve as standards for the international magnitude system.

nova. A star that experiences a sudden outburst of radiant energy, temporarily increasing its luminosity by hundreds to thousands of times.

nuclear. Referring to the nucleus of the atom.

nuclear transformation. Transformation of one atomic nucleus into another.

nucleon. Any one of the subatomic particles that compose a nucleus.

nucleus (of atom). The heavy part of an atom, composed mostly of protons and neutrons, and about which the electrons revolve.

nucleus (of comet). A swarm of solid particles in the head of a comet.

nucleus (of galaxy). Central concentration of stars, and possibly gas, at the center of a galaxy.

nutation. A "nodding" of the earth's polar axis; a small periodic motion of the earth's axis superimposed on precession.

O-association. A stellar association in which the stars are predominately of types O and B.

objective. The principal image-forming component of a telescope or other optical instrument.

objective prism. A prismatic lens that can be placed in front of a telescope objective to transform each star image into an image of its spectrum.

oblate spheroid. A solid formed by rotating an ellipse about its minor axis.

oblateness. A measure of the "flattening" of an oblate spheroid; numerically, the ratio of the difference between the major and minor diameters (or axes) to the major diameter (or axis).

obliquity of the ecliptic. Angle between the planes of the celestial equator and the ecliptic; about $23\frac{1}{2}°$.

obscuration (interstellar). Absorption of starlight by interstellar dust.

occultation. An eclipse of a star or planet by the moon or another planet.

ocular. Eyepiece.

Oort's constants. Constants derived in the analysis of J. Oort that characterize the rotation of the Galaxy in the neighborhood of the sun.

opacity. Absorbing power; capacity to impede the passage of light.

open cluster. A comparatively loose or "open" cluster of stars, containing from a few dozen to a few thousand members, located in the spiral arms or disk of the Galaxy; galactic cluster.

opposition. Configuration of a planet when its elongation is 180°.

optical binary. Two stars at different distances nearly lined up in projection so that they appear close together, but which are not really dynamically associated.

optics. The branch of physics that deals with light and its properties.

orbit. The path of a body that is in revolution about another body or point.

osculating orbit. The strict two-body orbit of a body — a hypothetical planet or other object revolving about the sun (or other object) — that has the same position and velocity, at a given instant, as the actual body does. It is the orbit the body would follow if there were no other object to produce perturbations.

Pallas. Second minor planet to be discovered.

parabola. A conic section of eccentricity 1.0; the curve of intersection between a circular cone and a plane parallel to a straight line in the surface of the cone.

paraboloid. A parabola of revolution; a curved surface of parabolic cross section. Especially applied to the surface of the primary mirror in a standard reflecting telescope.

parallactic ellipse. A small ellipse that a comparatively nearby star appears to trace out in the sky, which results from the orbital motion of the earth about the sun.

parallax. An apparent displacement of an object due to a motion of the observer.

parallax (stellar). An apparent displacement of a nearby star that results from the motion of the earth around the sun; numerically, the angle subtended by 1 AU at the distance of a particular star.

parallelogram of forces. A geometrical construction that permits the determination of the resultant of two different forces.

parsec. The distance of an object that would have a stellar parallax of 1 second of arc; 1 parsec = 3.26 light-years.

partial eclipse. An eclipse of the sun or moon in which the eclipsed body does not appear completely obscured.

peculiar velocity. The velocity of a star with respect to the local standard of rest; that is, its space motion, corrected for the motion of the sun with respect to our neighboring stars.

penumbra. The portion of a shadow from which only part of the light source is occulted by an opaque body.

penumbral eclipse. A lunar eclipse in which the moon passes through the penumbra, but not the umbra, of the earth's shadow.

perfect cosmological principle. The assumption that, on the large scale, the universe appears the same from every place and at all times.

perfect gas. An "ideal" gas that obeys the perfect gas laws.

perfect gas laws. Certain laws that describe the behavior of an ideal gas; Charles' law, Boyle's law, and the equation of state for a perfect gas.

perfect radiator. Black body; a body that absorbs and subsequently reemits all radiation incident upon it.

periastron. The place in the orbit of a star in a binary star system where it is closest to its companion star.

perifocus. The place on an elliptical orbit that is closest to the focus occupied by the central force.

perigee. The place in the orbit of an earth satellite where it is closest to the center of the earth.

perihelion. The place in the orbit of an object revolving about the sun where it is closest to the center of the sun.

period. A time interval; for example, the time required for one complete revolution.

period-luminosity relation. An empirical relation between the periods and luminosities of cepheid-variable stars.

periodic comet. A comet whose orbit has been determined to have an eccentricity of less than 1.0.

perturbation. The disturbing effect, when small, on the motion of a body as predicted by a simple theory, produced by a third body or other external agent.

phases of the moon. The progression of changes in the moon's appearance during the month that results from the moon's turning different portions of its illuminated hemisphere to our view.

photocell (photoelectric cell). An electron tube in which electrons are dislodged from the cathode when it is exposed to light and are accelerated to the anode, thus producing a current in the tube, whose strength serves as a measure of the light striking the cathode.

photographic magnitude. The magnitude of an object, as measured on the traditional, blue- and violet-sensitive photographic emulsions.

photometry. The measurement of light intensities.

photomultiplier. A photoelectric cell in which the electric current generated is amplified at several stages within the tube.

photon. A discrete unit of electromagnetic energy.

photosphere. The region of the solar (or a stellar) atmosphere from which radiation escapes into space.

photovisual magnitude. A magnitude corresponding to the spectral region to which the human eye is most sensitive, but measured by photographic methods with suitable green- and yellow-sensitive emulsions and filters.

plage. A bright region of the solar surface observed in the monochromatic light of some spectral line; flocculus.

Planck's constant. The constant of proportionality relating the energy of a photon to its frequency.

Planck's radiation law. A formula from which can be calculated the intensity of radiation at various wavelengths emitted by a black body.

planet. Any of nine solid, nonluminous bodies revolving about the sun.

planetarium. An optical device for projecting on a screen or domed ceiling the stars and planets and their apparent motions as they appear in the sky.

planetary nebula. A shell of gas ejected from, and enlarging about, a certain kind of extremely hot star.

planetoid. Synonym for minor planet.

Pluto. Ninth planet from the sun in the solar system.

polar axis. The axis of rotation of the earth; also, an axis in the mounting of a telescope that is parallel to the earth's axis.

polarization. A condition in which the planes of vibration (or the *E* vectors) of the various rays in a light beam are at least partially aligned.

polarized light. Light in which polarization is present.

Polaroid. Trade name for a transparent substance that produces polarization in light.

Population I and II. Two classes of stars (and systems of stars), classified according to their spectral characteristics, chemical compositions, radial velocities, ages, and locations in the Galaxy.

position angle. Direction in the sky of one celestial object from another; for example, the angle, measured to the east from the north, of the fainter component of a visual binary star in relation to the brighter component.

positron. An electron with a positive rather than a negative charge.

postulate. An essential prerequisite to a hypothesis or theory.

potential energy. Stored energy that can be converted into other forms; especially gravitational energy.

Poynting-Robertson effect. An effect of the pressure of radiation from the sun on small particles that causes them to spiral slowly into the sun.

precession (of earth). A slow, conical motion of the earth's axis of rotation, caused principally by the gravitational torque of the moon and sun on the earth's equatorial bulge. *Luni-solar precession,* precession caused by the moon and sun only; *planetary precession,* a slow change in the orientation of the plane of the earth's orbit caused by planetary perturbations; *general precession,* the combination of these two effects on the motion of the earth's axis with respect to the stars.

precession of the equinoxes. Slow westward motion of the equinoxes along the ecliptic that results from precession.

primary cosmic rays. The cosmic-ray particles that arrive at the earth from beyond its atmosphere as opposed to the secondary particles that are produced by collisions between primary cosmic rays and air molecules.

primary minimum (in the light curve of an eclipsing binary). The middle of the eclipse during which the most light is lost.

prime focus. The point in a telescope where the objective focuses the light.

prime meridian. The terrestrial meridian passing through the site of the old Royal Greenwich Observatory; longitude 0°.

primeval atom. A single mass whose explosion (in some cosmological theories) has been postulated to have resulted in all the matter now present in the universe.

primeval fireball. The extremely hot opaque gas that is presumed to comprise the entire mass of the universe at the time of or immediately following the "big bang"; the exploding primeval atom.

Principia. Contraction of *Philosophiae Naturalis Principia Mathematica,* the great book by Newton in which he set forth his laws of motion and gravitation in 1687.

prism. A wedge-shaped piece of glass that is used to disperse white light into a spectrum.

prolate spheroid. The solid produced by the rotation of an ellipse about its major axis.

prominence. A phenomenon in the solar corona that commonly appears like a flame above the limb of the sun.

proper motion. The angular change in direction of a star per year.

proton. A heavy subatomic particle that carries a positive charge, and one of the two principal constituents of the atomic nucleus.

proton-proton chain. A chain of thermonuclear reactions by which nuclei of hydrogen are built up into nuclei of helium.

protoplanet (or star or galaxy). The original material from which a planet (or a star or galaxy) condensed.

pulsar. One of several pulsating radio sources of small angular size that emit radio pulses in very regular short periods.

pulsating variable. A variable star that physically pulsates in size and luminosity.

quadrature. A configuration of a planet in which its elongation is 90°.

quantum mechanics. The branch of physics that deals with the structure of atoms and their interactions with each other and with radiation.

quarter moon. Either of the two phases of the moon when its longitude differs by 90° from that of the sun; the moon appears half full at these phases.

quasar. Popular term for quasi-stellar source.

quasi-stellar galaxy (QSG). A stellar-appearing object of very large red shift presumed to be extragalactic and highly luminous.

quasi-stellar source (QSS). A stellar-appearing object of very large red shift that is a strong source of radio waves; presumed to be extragalactic and highly luminous.

R Coronae Borealis variables. Eruptive variable stars that show sudden and irregular drops in brightness; the class is named for the prototype, R Coronae Borealis.

RR Lyrae variable. One of a class of giant pulsating stars with periods less than 1 day; a cluster variable.

RW Aurigae stars. Variable stars, generally associated with interstellar matter, that show rapid and irregular light variations.

radar. A technique for observing the reflection of radio waves from a distant object.

radial velocity. The component of relative velocity that lies in the line of sight.

radial velocity curve. A plot of the variation of radial velocity with time for a binary or variable star.

radiant (of meteor shower). The point in the sky from which the meteors belonging to a shower seem to radiate.

radiation. A mode of energy transport whereby energy is transmitted through a vacuum; also the transmitted energy itself, either electromagnetic or corpuscular.

radiation pressure. The transfer of momentum carried by electromagnetic radiation to a body that the radiation impinges upon.

radical. A bond of two or more atoms that does not, in itself, comprise a molecule, but that has characteristics of its own and enters into chemical reactions as if it were a single atom.

radio astronomy. The technique of making astronomical observations in radio wavelengths.

radio telescope. A telescope designed to make observations in radio wavelengths.

radioactivity (radioactive decay). The process by which certain kinds of atomic nuclei naturally decompose with the spontaneous emission of other subatomic particles and gamma rays.

range (of a rocket or space vehicle). Distance away.

range rate. Rate at which range is changing; radial velocity.

ray (lunar). Any of a system of bright elongated streaks, sometimes associated with a crater on the moon.

Rayleigh scattering. Scattering of light (photons) by molecules of a gas.

recurrent nova. A nova that has been known to erupt more than once.

red giant. A large, cool star of high luminosity; a star occupying the upper right portion of the Hertzsprung-Russell diagram.

reddening (interstellar). The reddening of starlight passing through interstellar dust, caused by the dust scattering blue light more effectively than red.

redshift. A shift to longer wavelengths of the light from remote galaxies; presumed to be produced by a Doppler shift.

reflecting telescope. A telescope in which the principal optical component (objective) is a concave mirror.

reflection. The reflection of light rays by an optical surface.

reflection nebula. A relatively dense dust cloud in interstellar space that is illuminated by starlight.

refracting telescope. A telescope in which the principal optical component (objective) is a lens or system of lenses.

refraction. The bending of light rays passing from one transparent medium (or a vacuum) to another.

regression of nodes. A consequence of certain perturbations on the orbit of a revolving body whereby the nodes of the orbit slide westward in the fundamental plane (usually the plane of the ecliptic or of the celestial equator).

relative orbit. The orbit of one of two mutually revolving bodies referred to the other body as origin.

relativity. A theory formulated by Einstein that describes the relations between measurements of physical phenomena by two different observers who are in relative motion at constant velocity (the *special theory of relativity*), or at accelerated motion (the *general theory of relativity*).

resolution. The degree to which fine details in an image are separated or resolved.

resolving power. A measure of the ability of an optical system to resolve or separate fine details in the image it produces; in astronomy, the angle in the sky that can be resolved by a telescope.

restricted three-body problem. The study of the motion of a body of negligible mass in the gravitational field of two other bodies revolving about each other in circular orbits.

retrograde motion. An apparent westward motion of a planet on the celestial sphere or with respect to the stars.

revolution. The motion of one body around another.

right ascension. A coordinate for measuring the east-west positions of celestial bodies; the angle measured eastward along the celestial equator from the vernal equinox to the hour circle passing through a body.

rille (or rill). A crevasse or trenchlike depression in the moon's surface.

Roche's limit. The smallest distance from a planet or other body at which purely gravitational forces can hold together a satellite or secondary body of the same mean density as the primary; within this distance the tidal forces of the primary would break up the secondary.

rotation. Turning of a body about an axis running through it.

Russell-Vogt theorem. The theorem that the mass and chemical composition of a star determine its entire structure if it derives its energy entirely from thermonuclear reactions.

saros. A particular cycle of similar eclipses that recur at intervals of about 18 years.

satellite. A body that revolves about a larger one; for example, a moon of a planet.

Saturn. The sixth planet from the sun in the solar system.

scale (of telescope). The linear distance in the image corresponding to a particular angular distance in the sky; say, so many centimeters per degree.

Schmidt telescope. A type of reflecting telescope invented by B. Schmidt, in which certain aberrations produced by a spherical concave mirror are compensated for by a thin objective correcting lens.

science. The attempt to find order in nature or to find laws that describe natural phenomena.

scientific method. A specific procedure in science: (1) the observation of phenomena or the results of experiments; (2) the formulation of hypotheses that describe these phenomena, and that are consistent with the body of knowledge available; (3) the testing of these hypotheses by noting whether or not they adequately predict and describe new phenomena or the results of new experiments.

scintillation counter. Device for recording primary or secondary cosmic-ray particles from flashes of light produced when these particles strike fluorescent materials.

Sculptor-type system. A dwarf elliptical galaxy, of which the system in Sculptor is a typical example.

second-order cluster of galaxies. A cluster of clusters of galaxies.

secondary cosmic rays. Secondary particles produced by interactions between primary cosmic rays from space and the atomic nuclei in molecules of the earth's atmosphere.

secondary minimum (in an eclipsing binary light curve). The middle of the eclipse of the cooler star by the hotter, in which the light of the system diminishes less than during the eclipse of the hotter star by the cooler.

secular. Not periodic.

secular parallax. A mean parallax for a selection of stars, derived from the components of their proper motions that reflect the motion of the sun.

seeing. The unsteadiness of the earth's atmosphere, which blurs telescopic images.

seismic waves. Vibrations traveling through the earth's interior that result from earthquakes.

seismograph. An instrument used to record and measure seismic waves.

seismology. The study of earthquakes and the conditions that produce them, and of the internal structure of the earth as deduced from analyses of seismic waves.

selected areas. Small regions of the sky in which magnitudes of stars have been accurately measured and serve as standards for magnitude systems.

seleno-. Prefix referring to the moon.

semimajor axis. Half the major axis of a conic section.

semiregular variable. A variable star, usually a red giant or supergiant, whose period of pulsation is far from constant.

separation (in a visual binary). The angular separation of the two components of a visual binary star.

Seyfert galaxy. A spiral galaxy whose nucleus shows bright emission lines; one of a class of galaxies first described by C. Seyfert. (They are often radio sources.)

shadow cone. The umbra of the shadow of a

spherical body (such as the earth) in sunlight.

shell star. A type of star, usually of spectral type B to F, surrounded by a gaseous ring or shell.

shower (of cosmic rays). A large "rain" of secondary cosmic-ray particles produced by a very energetic primary particle impinging on the earth's atmosphere.

shower (meteor). Many meteors, all seeming to radiate from a common point in the sky, caused by the encounter by the earth of a swarm of meteoroids moving together through space.

sidereal day. The interval between two successive meridian passages of the vernal equinox.

sidereal month. The period of the moon's revolution about the earth with respect to the stars.

sidereal period. The period of revolution of one body about another with respect to the stars.

sidereal time. The local hour angle of the vernal equinox.

sidereal year. Period of the earth's revolution about the sun with respect to the stars.

sign (of zodiac) (Astrological term). Any of twelve equal sections along the ecliptic, each of length 30°. Starting at the vernal equinox, and commencing eastward, the signs are Aries, Taurus, Gemini, Cancer, Leo, Virgo, Libra, Scorpio, Sagittarius, Capricornus, Aquarius, and Pisces.

sine (of angle). One of the trigonometric functions; the sine of an angle (in a right triangle) is the ratio of the length of the side opposite the angle to that of the hypotenuse.

sine curve. A graph of the sine of an angle plotted against the angle.

small circle. Any circle on the surface of a sphere that is not a great circle.

solar activity. Phenomena of the solar atmosphere associated with sunspots, plages, and related phenomena.

solar antapex. Direction away from which the sun is moving with respect to the local standard of rest.

solar apex. The direction toward which the sun is moving with respect to the local standard of rest.

solar constant. Mean amount of solar radiation received per unit time, by a unit area, just outside the earth's atmosphere, and normal to the direction to the sun; the numerical value is 1.37×10^6 ergs/cm^2·sec.

solar motion. Motion of the sun, or the velocity of the sun, with respect to the local standard of rest.

solar parallax. Angle subtended by the equatorial radius of the earth at a distance of 1 AU.

solar system. The system of the sun and the planets, their satellites, the minor planets, comets, meteoroids, and other objects revolving around the sun.

solar time. A time based on the sun; usually the hour angle of the sun *plus* 12 hours.

solar wind. A radial flow of corpuscular radiation leaving the sun.

solstice. Either of two points on the celestial sphere where the sun reaches its maximum distances north and south of the celestial equator.

south point. Intersection of the celestial meridian and astronomical horizon 180° from the north point.

space motion. The velocity of a star with respect to the sun.

space probe. An unmanned interplanetary rocket carrying scientific instruments to obtain data on other planets or on the interplanetary environment.

space technology. The applied science of the immediate space environment of the earth.

specific gravity. The ratio of the density of a body or substance to that of water.

spectral class (or type). A classification of a star according to the characteristics of its spectrum.

spectral sequence. The sequence of spectral classes of stars arranged in order of decreasing temperatures of stars of those classes.

spectrogram. A photograph of a spectrum.

spectrograph. An instrument for photographing a spectrum; usually attached to a telescope to photograph the spectrum of a star.

spectroheliogram. A photograph of the sun obtained with a spectroheliograph.

spectroheliograph. An instrument for photographing the sun, or part of the sun, in the monochromatic light of a particular spectral line.

spectrophotometry. The measurement of the intensity of light from a star or other source at different wavelengths.

spectroscope. An instrument for directly viewing the spectrum of a light source.

spectroscopic binary star. A binary star in which the components are not resolved optically, but whose binary nature is indicated by periodic variations in radial velocity, indicating orbital moiton.

spectroscopic parallax. A parallax (or distance)

of a star that is derived by comparing the apparent magnitude of the star with its absolute magnitude as deduced from its spectral characteristics.

spectroscopy. The study of spectra.

spectrum. The array of colors or wavelengths obtained when light from a source is dispersed, as in passing it through a prism or grating.

spectrum analysis. The study and analysis of spectra, especially stellar spectra.

spectrum binary. A binary star whose binary nature is revealed by spectral characteristics that can only result from the composite of the spectra of two different stars.

spectrum (α^2 Canum Venaticorum) variable. Any of a class of main-sequence spectral type A stars that show anomalously strong lines of certain elements which vary periodically in intensity.

speed. The rate at which an object moves without regard to its direction of motion; the numerical or absolute value of velocity.

spherical aberration. A defect of optical systems whereby on-axis rays of light striking different parts of the objective do not focus at the same place.

spherical harmonics. A series of terms by which the shape of a body can be expressed mathematically to any desired degree of accuracy (by using enough terms in the series).

spicule. A narrow jet of rising material in the solar chromosphere.

spiral arms. Arms of interstellar material and young stars that wind out in a plane from the central nucleus of a spiral galaxy.

spiral galaxy. A flattened, rotating galaxy with pinwheel-like arms of interstellar material and young stars winding out from its nucleus.

sporadic meteor. A meteor that does not belong to a shower.

spring tide. The highest tide of the month, produced when the sun and moon have longitudes that differ from each other by nearly 0° or 180°.

Sputnik. Russian for "satellite," or "fellow traveler"; the name given to the first Soviet artificial satellite.

stadium. A Greek unit of length, based on the Olympic Stadium; roughly, $\frac{1}{10}$ mi.

standard time. The local mean solar time of a standard meridian, adopted over a large region to avoid the inconvenience of continuous time changes around the earth.

star. A self-luminous sphere of gas.

star cluster. An assemblage of stars held together by their mutual gravitation.

statistical equilibrium. A condition in a system of particles in which, statistically, energy is divided equally among particles of all types.

statistical parallax. The mean parallax for a selection of stars, derived from the radial velocities of the stars and the components of their proper motions that cannot be affected by the solar motion.

steady state (theory of cosmology). A theory of cosmology embracing the perfect cosmological principle, and involving the continuous creation of matter.

Stefan's law. A formula from which the rate at which a black body radiates energy can be computed; the total rate of energy emission from a unit area of a black body is proportional to the fourth power of its absolute temperature.

stellar evolution. The changes that take place in the sizes, luminosities, structures, and so on, of stars as they age.

stellar model. The result of a theoretical calculation of the run of physical conditions in a stellar interior.

stellar parallax. The angle subtended by 1 AU at the distance of a star; usually measured in seconds of arc.

stratosphere. The layer of the earth's atmosphere above the troposphere, where most weather takes place, and below the ionosphere.

Strömgren sphere. A region of ionized gas in interstellar space surrounding a hot star; an HII region.

Stonehenge. An assemblage of upright stones in Salisbury Plain, England, believed to have been constructed by early people for astronomical observations connected with timekeeping and the calendar.

subdwarf. A star of luminosity lower than that of main-sequence stars of the same spectral type.

subgiant. A star of luminosity intermediate between those of main-sequence stars and normal giants of the same spectral type.

summer solstice. The point on the celestial sphere where the sun reaches its greatest distance north of the celestial equator.

sun. The star about which the earth and other planets revolve.

sundial. A device for keeping time by the shadow a marker (gnomon) casts in sunlight.

sunspot. A temporary cool region in the solar photosphere that appears dark by contrast against the surrounding hotter photosphere.

sunspot cycle. The semiregular 11-year period

with which the frequency of sunspots fluctuates.

supergiant. A star of very high luminosity.

superior conjunction. The configuration of a planet in which it and the sun have the same longitude, with the planet being more distant than the sun.

superior planet. A planet more distant from the sun than the earth.

supernova. A stellar outburst or explosion in which a star suddenly increases its luminosity by from hundreds of thousands to hundreds of millions of times.

surface gravity. The weight of a unit mass at the surface of a body.

surveying. The technique of measuring distances and relative positions of places over the surface of the earth (or elsewhere); generally accomplished by triangulation.

synchrotron radiation. The radiation emitted by charged particles being accelerated in magnetic fields and moving at speeds near that of light.

synodic month. The period of revolution of the moon with respect to the sun, or its cycle of phases.

synodic period. The interval between successive occurrences of the same configuration of a planet; for example, between successive oppositions or successive superior conjunctions.

syzygy. A configuration of the moon in which its elongation is 0° or 180° (new or full).

T-association. A stellar association containing T Tauri stars.

T Tauri stars. Variable stars associated with interstellar matter that show rapid and erratic changes in light.

tail (of comet). Gases and solid particles ejected from the head of a comet and forced away from the sun by radiation pressure or corpuscular radiation.

tangent (of angle). One of the trigonometric functions; the tangent of an angle (in a right triangle) is the ratio of the length of the side opposite the angle to that of the shorter of the adjacent sides.

tangential (transverse) velocity. The component of a star's space velocity that lies in the plane of the sky.

tau component (of proper motion). The component of a star's proper motion that lies perpendicular to a great circle passing through the star and the solar apex.

tektites. Rounded glassy bodies that are suspected to be of meteoritic origin.

telescope. An optical instrument used to aid the eye in viewing or measuring, or to photograph distant objects.

telluric. Of terrestrial origin.

temperature (absolute). Temperature measured in centigrade degrees from absolute zero.

temperature (centigrade). Temperature measured on a scale where water freezes at 0° and boils at 100°.

temperature (color). The temperature of a star as estimated from the intensity of the stellar radiation at two or more colors or wavelengths.

temperature (effective). The temperature of a black body that would radiate the same total amount of energy that a particular body does.

temperature (excitation). The temperature of a star as estimated from the relative strengths of lines in its spectrum that originate from atoms in different stages of excitation.

temperature (Fahrenheit). Temperature measured on a scale where water freezes at 32° and boils at 212°.

temperature (ionization). The temperature of a star as estimated from the relative strengths of lines in its spectrum that originate from atoms in different stages of ionization.

temperature (Kelvin). Absolute temperature measured in centigrade degrees.

temperature (kinetic). A measure of the speeds or mean energy of the molecules in a substance.

temperature (radiation). The temperature of a black body that radiates the same amount of energy in a given spectral region as does a particular body.

terminator. The line of sunrise or sunset on the moon.

terrestrial planet. Any of the planets Mercury, Venus, Earth, Mars, and sometimes Pluto.

theory. A set of hypotheses and laws that have been well demonstrated as applying to a wide range of phenomena associated with a particular subject.

thermal energy. Energy associated with the motions of the molecules in a substance.

thermal equilibrium. A balance between the input and outflow of heat in a system.

thermocouple. A device for measuring the intensity of infrared radiation.

thermodynamics. The branch of physics that deals with heat and heat transfer among bodies.

thermonuclear energy. Energy associated with thermonuclear reactions or that can be released through thermonuclear reactions.

thermonuclear reaction. A nuclear reaction or

transformation that results from encounters between nuclear particles that are given high velocities (by heating them).

thermopile. A device consisting of a series or pile of thermocouples, which is used to measure the intensity of infrared radiation.

tidal force. A differential gravitational force that tends to deform a body.

tide. Deformation of a body by the differential gravitational force exerted on it by another body; in the earth, the deformation of the ocean surface by the differential gravitational forces exerted by the moon and sun.

ton (American short). 2000 lb.

ton (English long). 2240 lb.

ton (metric). One million grams.

topography. The configuration or relief of the surface of the earth, moon or a planet.

total eclipse. An eclipse of the sun in which the sun's photosphere is entirely hidden by the moon, or an eclipse of the moon in which it passes completely into the umbra of the earth's shadow.

train (of meteor). A temporarily luminous trail left in the wake of a meteor.

transit. An instrument for timing the exact instant a star or other object crosses the local meridian. Also, the passage of a celestial body across the meridian; or the passage of a small body (say, a planet) across the disk of a larger one (say, the sun).

triangulation. The operation of measuring some of the elements of a triangle so that other ones can be calculated by the methods of trigonometry, thus determining distances to remote places without having to span them directly.

triaxial ellipsoid. A solid figure whose cross sections along three planes at right angles to each other, and all passing through its center, are ellipses but of different sizes and eccentricities.

trigonometry. The branch of mathematics that deals with the analytical solutions of triangles.

trillion. In the United States — 1 thousand billions or million millions (10^{12}); in Great Britain — 1 million billions or million-million millions (10^{18}).

triple-alpha process. A series of two nuclear reactions by which three helium nuclei are built up into one carbon nucleus.

Trojan minor planet. One of several minor planets that share Jupiter's orbit around the sun, but located approximately 60° around the orbit from Jupiter.

Tropic of Cancer. Parallel of latitude 23½° N.

Tropic of Capricorn. Parallel of latitude 23½° S.

tropical year. Period of revolution of the earth about the sun with respect to the vernal equinox.

troposphere. Lowest level of the earth's atmosphere, where most weather takes place.

tsunami. A series of very fast waves of seismic origin traveling through the ocean; popularly called "tidal waves."

U,B,V system. A system of stellar magnitudes consisting of measures in the ultraviolet, blue, and green-yellow spectral regions.

UV Ceti stars. Main-sequence stars, mostly of spectral class M, which show occasional, unpredicted flare-ups in light; also called "flare stars."

ultraviolet radiation. Electromagnetic radiation of wavelengths shorter than the shortest (violet) wavelengths to which the eye is sensitive; radiation of wavelengths in the approximate range 100 to 4000 angstroms.

umbra. The central, completely dark part of a shadow.

universal time. The local mean time of the prime meridian.

universe. The totality of all matter and radiation and the space occupied by same.

upsilon component (of proper motion). The component of a star's proper motion that lies along a great circle passing through the star and the solar apex.

Uranus. Seventh planet from the sun in the solar system.

Van Allen layer. Doughnut-shaped region surrounding the earth where many rapidly moving charged particles are trapped in its magnetic field.

variable star. A star that varies in luminosity.

variation of latitude. A slight semiperiodic change in the latitudes of places on the earth that results from a slight shifting of the body of the earth with respect to its axis of rotation.

vector. A quantity that has both magnitude and direction.

velocity. A vector that denotes both the speed and direction a body is moving.

velocity of escape. The speed with which an object must move in order to enter a parabolic orbit about another body (such as the earth), and hence move permanently away from the vicinity of that body.

Venus. The second planet from the sun in the solar system.

vernal equinox. The point on the celestial sphere where the sun crosses the celestial equator passing from south to north.

vertical circle. Any great circle passing through the zenith.

virial theorem. A relation between the potential and kinetic energies of a system of mutually gravitating bodies in statistical equilibrium.

vis viva **(energy) equation.** An equation that expresses the conservation of energy for two mutually revolving bodies; it relates their relative speed to their separation and the semimajor axis of their relative orbit.

visual binary star. A binary star in which the two components are telescopically resolved.

visual photometer. An instrument used with a telescope for visually measuring the light flux from a star.

volume. A measure of the total space occupied by a body.

von Jolly balance. A balance invented by von Jolly in 1881 to measure the mass of the earth.

Vulcan. A hypothetical planet once believed to exist and have an orbit between that of Mercury and the sun; the existence of Vulcan is now generally discredited.

W Ursae Majoris star. Any of a class of eclipsing binaries whose components are nearly in contact and hence suffer tidal distortion and loss or transfer of matter.

W Virginis star (type II Cepheid). A variable star belonging to the relatively rare class of population II Cepheids.

walled plain. A large lunar crater.

wandering of the poles. A semiperiodic shift of the body of the earth relative to its axis of rotation; responsible for variation of latitude.

watt. A unit of power; 10 million ergs expended per second.

wavelength. The spacing of the crests or troughs in a wave train.

weight. A measure of the force due to gravitational attraction.

west point. The point on the horizon 270° around the horizon from the north point, measured in a clockwise direction as seen from the zenith.

whistles (of meteors). Brief interruptions in high-frequency radio broadcasting reception due to the ionization of the air by meteors.

white dwarf. A star that has exhausted most or all of its nuclear fuel and has collapsed to a very small size; believed to be a star near its final stage of evolution.

Widmanstätten figures. Crystalline structure that can be observed in cut and polished meteorites.

Wien's law. Formula that relates the temperature of a black body to the wavelength at which it emits the greatest intensity of radiation.

winter solstice. Point on the celestial sphere where the sun reaches its greatest distance south of the celestial equator.

Wolf-Rayet star. One of a class of very hot stars that eject shells of gas at very high velocity.

World Calendar. A suggested calendar reform in which each of the four quarters would be identical and the same date of the month would always fall on the same day of the week.

X-rays. Photons of wavelengths intermediate between those of ultraviolet radiation and gamma rays.

X-ray stars. Stars (other than the sun) that emit observable amounts of radiation at X-ray frequencies.

year. The period of revolution of the earth around the sun.

Zeeman effect. A splitting or broadening of spectral lines due to magnetic fields.

zenith. The point on the celestial sphere opposite to the direction of gravity; or the direction opposite to that indicated by a plumb bob.

zenith distance. Arc distance of a point on the celestial sphere from the zenith; 90° minus the altitude of the object.

zodiac. A belt around the sky centered on the ecliptic.

zodiacal light. A faint illumination along the zodiac, believed to be sunlight reflected and scattered by interplanetary dust.

zone of avoidance. A region near the Milky Way where obscuration by interstellar dust is so heavy that few or no exterior galaxies can be seen.

zone time. The time, kept in a zone 15° wide in longitude, that is the local mean time of the central meridian of that zone. Zone time is used at sea, but over land the boundaries are irregular to conform to political boundaries, and it is called *standard time.*

Some Principles of Arithmetic, Algebra, and Geometry

A3.1 POWERS AND ROOTS

If a number is raised to a power, it is multiplied by itself the number of times indicated by the power or exponent:

$$2^3 = 2 \times 2 \times 2 = 8;$$

$$a^5 = a \times a \times a \times a \times a.$$

Numbers can be raised to fractional as well as integral powers; thus $3^{2-1/2}$ and $7^{3.86}$ are numbers that exist. However, nonintegral powers cannot be evaluated with simple arithmetic. Any number raised to the zero power is unity:

$$a^0 = 1; \quad 2^0 = 1; \quad (8.7621)^0 = 1.$$

The nth root of a number is a quantity which, when raised to the nth power, is equal to the number. Examples:

square root of $64 = \sqrt{64} = 64^{1/2} = 8;$

4th root of $16 = \sqrt[4]{16} = 16^{1/4} = 2;$

nth root of $a = \sqrt[n]{a} = a^{1/n}.$

Usually roots are not integral numbers:

$$2^{1/2} = 1.414214 \ldots ;$$

$$17^{1/3} = 2.571282 \ldots .$$

Nonintegral roots, such as $\sqrt[3.47]{33}$ and $21^{1/1.73}$, exist but cannot be evaluated with simple arithmetic.

The nth power or root of a fraction is evaluated by calculating the ratio of the nth power or root of the numerator to the nth power or root of the denominator:

$$\left(\frac{a}{b}\right)^n = \frac{a^n}{b^n}; \qquad \left(\frac{a}{b}\right)^{1/n} = \frac{a^{1/n}}{b^{1/n}};$$

$$\left(\frac{4}{2}\right)^2 = \frac{16}{4} = 4; \qquad \left(\frac{8}{27}\right)^{1/3} = \frac{8^{1/3}}{27^{1/3}} = \frac{2}{3}.$$

The negative power or root of a number is the reciprocal of the same positive power or root of the number:

$$a^{-n} = \frac{1}{a^n}; \quad 2^{-2} = \frac{1}{2^2} = \frac{1}{4}; \quad 3^{-1/3} = \frac{1}{3^{1/3}}.$$

Further properties of powers and roots are:

$$(ab)^n = a^n b^n; \qquad (abc)^n = a^n b^n c^n;$$

$$(ab)^{1/n} = a^{1/n} b^{1/n}; \qquad (abc)^{1/n} = a^{1/n} b^{1/n} c^{1/n}.$$

Examples:

$$(2 \times 3)^3 = 2^3 \times 3^3 = 8 \times 27 = 216;$$

$$(4 \times 9)^{1/2} = 4^{1/2} \times 9^{1/2} = 2 \times 3 = 6.$$

$$a^n a^m = a^{n+m}; \qquad \frac{a^n}{a^m} = a^n a^{-m} = a^{n-m}.$$

Examples:

$$2^2 \times 2^3 = 2^5 = 32; \quad \frac{2^3}{2^2} = 2^{3-2} = 2.$$

$$(a^n)^m = a^{nm}; \qquad (a^n)^{1/m} = a^{n/m};$$

$$(a^{1/n})^m = a^{m/n}; \qquad (a^{1/n})^{1/m} = a^{1/nm}.$$

A3.2 POWERS-OF-10 NOTATION

It is often necessary to deal with very large or very small numbers. For example, the earth is 93,000,000 miles from the sun and the mass of the hydrogen atom is 0.000 000 000 000 000 000 000-000 06 oz. Instead of writing and carrying so many zeros, the numbers are usually written as figures between 1 and 10 multiplied by the appropriate power of 10. For example, 93,000,000 is $9.3 \times 10,000,000$, or 9.3×10^7. Similarly, 0.000-000 000 000 000 000 000 000 06 is 6/100 000 000-000 000 000 000 000 000 or $6/10^{26} = 6 \times 10^{-26}$. The rule in reading numbers written in this notation is that the exponent of 10 is the number of places the decimal point is to be moved to the right (if the exponent is positive) or to the left (if the exponent is negative).

Multiplication, division, and exponentiation of numbers are facilitated in powers-of-10 notation. Examples:

$$6,000,000 \times 400 = 6 \times 10^6 \times 4 \times 10^2$$
$$= (6 \times 4) \times (10^6 \times 10^2) = 24 \times 10^8$$
$$= 2.4 \times 10^9.$$

$$\frac{6 \times 10^{-26}}{9.3 \times 10^7} = \frac{6}{9.3} \times \frac{10^{-26}}{10^7} = \frac{6}{9.3} \times 10^{-26-7}$$
$$= 0.645 \times 10^{-33} = 6.45 \times 10^{-34}.$$

$$(4000)^3 = (4 \times 10^3)^3 = 4^3 \times (10^3)^3 = 64 \times 10^9$$
$$= 6.4 \times 10^{10}.$$

$$(64,000,000)^{1/2} = (64 \times 10^6)^{1/2} = 64^{1/2} \times (10^6)^{1/2}$$
$$= 8 \times 10^3.$$

A3.3 ANGULAR MEASURE

The most common units of angular measure used in astronomy are the following:
(1) arc measure:
   one circle contains 360 degrees $= 360°$;
   $1°$ contains 60 minutes of arc $= 60'$;
   $1'$ contains 60 seconds of arc $= 60''$.
(2) time measure:
   one circle contains 24 hours $= 24^h$;
   1^h contains 60 minutes of time $= 60^m$;
   1^m contains 60 seconds of time $= 60^s$.
(3) radian measure:
   one circle contains 2π radians.
A radian is the angle at the center of a circle subtended by a length along the circumference of the circle equal to its radius. Since the circumference of a circle is 2π times its radius, there are 2π radians in a circle.

Relations between these different units of angular measure are given in the following table.

ARC MEASURE	TIME MEASURE	RADIANS	SECONDS OF ARC
57°.2958	3^h.820	1.0	206,264.806
15°	1^h	0.2618	54,000
1°	4^m	1.745×10^{-2}	3,600
15′	1^m	4.363×10^{-3}	900
1′	4^s	2.909×10^{-4}	60
15″	1^s	7.27×10^{-5}	15
1″	0^s.0667	4.85×10^{-6}	1

A3.4 PROPERTIES OF CIRCLES AND SPHERES

The ratio of the circumference of a circle to its diameter is always the same, regardless of the size of the circle. This ratio is universally symbolized by the Greek letter pi (π). Because π is a type of number called an *irrational number*, it cannot be expressed as a simple ratio of two integers, and so its value can never be specified exactly. It can, however, be approximated to any desired degree of accuracy by methods of mathematical analysis. Even the Greeks had determined the value of π by geometrical means to considerable accuracy. For many purposes, a sufficient approximation to π is

$$\pi = 3\tfrac{1}{7} = \tfrac{22}{7}.$$

The value of π has been evaluated to hundreds of decimal places; however, it is seldom needed to greater accuracy than

$$\pi = 3.14159265.$$

In a circle of circumference C, diameter D, and radius R $(R = \tfrac{1}{2}D)$, we have, from the definition of π,

$$\pi = \frac{C}{D} \quad \text{or} \quad C = \pi D = 2\pi R.$$

The circumference of a sphere of radius R is the circumference of any circle on the surface of the sphere whose center coincides with the center of the sphere (a *great circle*), and hence which also has a radius R. Thus, also on a sphere

$$C = 2\pi R.$$

For example, the radius of the earth is about 4000 mi. Its circumference is thus

$$C = 2\pi(4000) = 25,000 \text{ mi.}$$

The *area* of a circle of radius R and diameter D is

$$A = \pi R^2 \quad \text{or} \quad A = \frac{\pi D^2}{4}.$$

The *surface area* of a sphere of radius R is

$$A = 4\pi R^2.$$

This is the total area over the outside surface and must not be confused with the volume. For example, the surface area of the earth is approximately:

$$A = 4\pi(4000)^2 = 2 \times 10^8 \text{ mi}^2.$$

The *volume* of a sphere of radius R is

$$V = \frac{4}{3}\pi R^3.$$

A3.5 PROPORTIONS

If a is to b as c is to d, the ratios a/b and c/d are equal, and the equation

$$\frac{a}{b} = \frac{c}{d}$$

expresses a proportion. This same equation, of course, could also be written

$$ad = bc; \qquad a = b\frac{c}{d}; \qquad c = d\frac{a}{b}.$$

If x and y are both variable quantities and a is a fixed quantity (a *constant*), x and y are said to be proportional to each other (for example, y is proportional to x) if

$$y = ax.$$

This equation means that if x is doubled, y is doubled; if x is increased by 10 times, so is y; if x is diminished to $\frac{1}{1000}$ its former value, so is y. The two always change proportionately. The constant a is thus called a *constant of proportionality*. The above equation can also be written

$$y \propto x,$$

where the symbol $\propto$ means "proportional to."

For example, if x is the length of a board in feet and y is its length in inches, y is proportional to x and the constant of proportionality is 12. If one board has a length of twice as many inches as another, it also has a length of twice as many feet. As a further example, the circumference of a circle is proportional to its radius, the constant of proportionality being 2π.

In an equation like $y = ax$, y is said to be *directly* proportional to x. If, on the other hand, $y = a/x$, which can also be written $y \propto 1/x$, y is said to be *inversely* proportional to x (it is pro-portional to the reciprocal of x). Again, a is the constant of proportionality. Here, if x is doubled, y is halved; if x is increased by 10 times, y is decreased by 10 times; if x is decreased by 10 times, y is increased by 10 times, and so on.

As an example, if one has a fixed amount of money and apples cost p cents each, the number of apples he can buy, N, is inversely proportional to their cost, p, for

$$N = \frac{A}{p} \qquad \text{or} \qquad N \propto \frac{1}{p};$$

here A, the constant of proportionality, is the total amount of money available in cents.

If $y = ax^2$, y is proportional to the *square* of x; a is still the constant of proportionality. For example, the area of a circle is proportional to the square of its radius (the constant of proportionality is π); a circle of 3 in. radius, in other words, has nine times the area of a circle of 1-in. radius.

In general, proportionalities hold regardless of the units used to express the values of the variables. Only the constant of proportionality changes with a change in units. Suppose the area, in square feet, is desired of a circle of radius R feet. Then $A = \pi R^2$; the constant of proportionality is π. Now, however, suppose R is still expressed in feet but that the area is desired in square inches. Then, since 1 square foot $= 12^2 = 144$ in.^{2}, $A = 144\pi R^2$; the constant of proportionality is now 144π. If R is given in feet but A is desired in square miles, since there are $(5280)^2$ square feet per square mile, we have

$$A = \frac{\pi}{(5280)^2}R^2 \text{ mi}^2.$$

Here the constant of proportionality is $\pi/(5280)^2$. In all cases, however, $A \propto R^2$; only the constant of proportionality changes as the units used to express R and A are changed.

If $y = ax^n$, y is proportional to the nth power of x; if $y = a/x^2$, y is inversely proportional to the square of x. In both cases a is the constant of proportionality.

Among objects of similar shape, no matter what that shape may be, linear distances on those objects (circumferences, diameters, and so on) are proportional to the linear dimensions of the objects. For example, the circumference of a cylinder is proportional to its diameter; the altitudes of cones, similar to each other, are in proportion to their bases, and so on.

Areas of similarly shaped objects, or parts thereof, are proportional to the *square* of their linear dimensions. The surface area of a sphere, for example, is proportional to the square of its radius, and the total outside area of a cube is proportional to the square of the length along one of its edges. The constants of proportionality are 4π and 6, respectively.

The *volumes* of similar objects are propor- tional to the *cubes* of their linear dimensions. For example, the volume of a sphere is proportional to the cube of its radius, and the volume of a cube is proportional to the cube of the length along one of its edges. The constants of proportionality here are $4\pi/3$ and 1, respectively.

APPENDIX 4

Metric and English Units

In the English system of measure the fundamental units of length, mass, and time are the yard, pound, and second, respectively. There are also, of course, larger and smaller units, which include the ton (2000 lb), the mile (1760 yd), the rod ($16\frac{1}{2}$ ft), the inch ($\frac{1}{36}$ yd), the ounce ($\frac{1}{16}$ lb), and so on. Such units are inconvenient for conversion and arithmetic computation.

In science, therefore, it is more usual to use the metric system, which has been adopted universally in nearly all except the English-speaking countries. The fundamental units of the metric system are:

length: 1 meter (m)
mass: 1 kilogram (kg)
time: 1 second (sec)

A meter was originally intended to be 1 ten-millionth of the distance from the equator to the North Pole along the surface of the earth. It is about 1.1 yd. A kilogram is about 2.2 lb. The second is the same in metric and English units. The most commonly used quantities of length and mass of the metric system are the following:

length

1 km	= 1 kilometer	= 1000 meters	= 0.6214 mile	
1 m	= 1 meter	= 1.094 yards	= 39.37 inches	
1 cm	= 1 centimeter	= 0.01 meter	= 0.3937 inch	
1 mm	= 1 millimeter	= 0.001 meter	= 0.1 cm	= 0.03937 inch
1μ	= 1 micron	= 0.000 001 meter	= 0.0001 cm	= 3.3937×10^{-5} inch

also: 1 mile = 1.6093 km
        1 inch = 2.5400 cm

mass

1 metric ton	= 10^6 grams	= 1000 kg	= 2.2046×10^3 lb
1 kg	= 1000 grams	= 2.2046 lb	
1 gm	= 1 gram	= 0.0022046 lb	= 0.0353 oz
1 mg	= 1 milligram	= 0.001 gm	= 2.2046×10^{-6} lb

also: 1 lb = 453.6 gm
        1 oz = 28.3495 gm

Temperature Scales

Three temperature scales are in general use:
(1) Fahrenheit (F); water freezes at 32°F and boils at 212°F.
(2) Centigrade (C); water freezes at 0°C and boils at 100°C.
(3) Kelvin or absolute (K); water freezes at 273°K and boils at 373°K.

All molecular motion ceases at $-459°F = -273°C = 0°K$. Thus Kelvin temperature is measured from this lowest possible temperature, called *absolute zero*. It is the temperature scale most often used in astronomy. Kelvin degrees have the same value as centigrade degrees, since the difference between the freezing and boiling points of water is 100 degrees in each.

On the Fahrenheit scale, water boils at 212 degrees and freezes at 32 degrees; the difference is 180 degrees. Thus to convert centigrade or Kelvin degrees to Fahrenheit it is necessary to multiply by $180/100 = 9/5$. To convert from Fahrenheit to centigrade or Kelvin degrees, it is necessary to multiply by $100/180 = 5/9$.

Example 1: What is 68°F in centigrade and in Kelvin?

$$68°F - 32°F = 36°F \text{ above freezing.}$$

$$\frac{5}{9} \times 36° = 20°;$$

thus,

$$68°F = 20°C = 293°K.$$

Example 2: What is 37°C in Fahrenheit and in Kelvin?

$$37°C = 273° + 37° = 310°K;$$

$$\frac{9}{5} \times 37° = 66.6°;$$

thus,

37°C is 66.6°F above freezing

or

$$37°C = 32° + 66.6° = 98.6°F.$$

Mathematical, Physical, and Astronomical Constants

MATHEMATICAL CONSTANTS:

$$\pi = 3.1415926536$$
$$1 \text{ radian} = 57°2957795$$
$$= 3437'.74677$$
$$= 206264''.806$$

Number of square degrees on a sphere = 41 252.96124

PHYSICAL CONSTANTS:

velocity of light	$c = 2.99793 \times 10^{10}$ cm/sec
constant of gravitation	$G = 6.668 \times 10^{-8}$ dyne·cm²/gm²
Planck's constant	$h = 6.624 \times 10^{-27}$ erg·sec
Boltzmann's constant	$k = 1.380 \times 10^{-16}$ erg/deg
mass of hydrogen atom	$m_H = 1.673 \times 10^{-24}$ gm
mass of electron	$m_e = 9.1085 \times 10^{-28}$ gm
charge on electron	$\epsilon = 4.803 \times 10^{-10}$ electrostatic units
Stefan-Boltzmann constant	$\sigma = 5.669 \times 10^{-5}$ erg/cm²·deg⁴·sec
constant in Wien's law	$\lambda_{max} T = 0.28979$ cm·deg
Rydberg's constant	$R = 1.09737 \times 10^5$ per cm
1 electron volt	$eV = 1.60207 \times 10^{-12}$ erg
1 angstrom	$\mathring{A} = 10^{-8}$ cm

ASTRONOMICAL CONSTANTS:

astronomical unit	$AU = 1.49597893 \times 10^{13}$ cm
parsec	$pc = 206265$ AU
	$= 3.262$ LY
	$= 3.086 \times 10^{18}$ cm
light-year	$LY = 9.4605 \times 10^{17}$ cm
	$= 6.324 \times 10^4$ AU
tropical year	$= 365.242199$ ephemeris days
sidereal year	$= 365.256366$ ephemeris days
	$= 3.155815 \times 10^7$ sec
mass of earth	$M_\oplus = 5.977 \times 10^{27}$ gm
mass of sun	$M_\odot = 1.991 \times 10^{33}$ gm
equatorial radius of earth	$R_\oplus = 6378.24$ km
radius of sun	$R_\odot = 6.960 \times 10^{10}$ cm
luminosity of sun	$L_\odot = 3.86 \times 10^{33}$ erg/sec
solar constant	$S = 1.36 \times 10^6$ erg/cm²·sec
	$= 1.97$ cal/m²·min
obliquity of ecliptic (1900)	$\epsilon = 23°27'8''.26$
direction of galactic center (1950)	$\alpha = 17^h42^m.4$
	$\delta = -28°55'$
direction of north galactic pole (1950)	$\alpha = 12^h49^m$
	$\delta = +27°.4$

Astronomical Coordinate Systems

There are several astronomical coordinate systems that are in common use. In each of these systems the position of an object in the sky, or on the celestial sphere, is denoted by two angles. These angles are referred to a *reference plane*, which contains the observer, and a *reference direction*, which is a direction from the observer to some arbitrary point lying in the reference plane. The intersection of the reference plane and the celestial sphere is a great circle, which defines the "equator" of the coordinate system. At two points, each 90° from this equator, are the "poles" of the coordinate system. Great circles passing through these poles intersect the equator of the system at right angles.

One of the two angular coordinates of each coordinate system is measured from the equator of the system to the object along the great circle passing through it and the poles. Angles on one side of the equator (or reference plane) are reckoned as positive; those on the opposite side are negative. The other angular coordinate is measured along the equator from the reference direction to the intersection of the equator with the great circle passing through the object and the poles.

The system of terrestrial latitude and longitude provides an excellent analogue. Here the plane of the terrestrial equator is the fundamental plane and the earth's equator is the equator of the system; the North and South terrestrial Poles are the poles of the system. One coordinate, the *latitude* of a place, is reckoned north (positive) or south (negative) of the equator along a meridian passing through the place. The other coordinate, *longitude*, is measured along the equator to the intersection of the equator and the meridian of the place from the intersection of the equator and the Greenwich meridian. The direction (from the center of the earth) to this latter intersection is the reference direction. Terrestrial longitude is either east or west (whichever is less), but the corresponding coordinate in celestial systems is generally reckoned in one direction from 0 to 360° (or, equivalently, from 0 to 24^h).

The following table lists the more important astronomical coordinate systems and defines how each of the angular coordinates is defined.

Astronomical Coordinate Systems

SYSTEM	REFERENCE PLANE	REFERENCE DIRECTION	"LATITUDE" COORDINATE	RANGE	"LONGITUDE" COORDINATE	RANGE
Horizon	Horizon plane	North point (formerly the south point was used by astronomers)	Altitude, h; toward the zenith $(+)$ toward the nadir $(-)$	$\pm 90°$	Azimuth, A; measured to the east along the horizon from the north point	0 to 360°
Equator	Plane of the celestial equator	Vernal equinox	Declination, δ; toward the north celestial pole $(+)$ toward the south celestial pole $(-)$	$\pm 90°$	Right ascension, α or R.A.; measured to the east along the celestial equator from the vernal equinox	0 to 24^h
Ecliptic	Plane of the earth's orbit (ecliptic)	Vernal equinox	Celestial latitude, β; toward the north ecliptic pole $(+)$ toward the south ecliptic pole $(-)$	$\pm 90°$	Celestial longitude, λ; measured to the east along the ecliptic from the vernal equinox	0 to 360°
Galactic	Mean plane of the Milky Way	Direction to the galactic center	Galactic latitude, b or b''; toward the north galactic pole $(+)$ toward the south galactic pole $(-)$	$\pm 90°$	Galactic longitude, l or l''; measured along the galactic equator to the east from the galactic center	0 to 360°

APPENDIX 8

Nuclear Reactions in Astronomy

Given here are the series of thermonuclear reactions that are most important in stellar interiors. The subscript to the left of a nuclear symbol is the atomic number; the superscript to the right is the atomic mass number. β^+ is the symbol for a positron, ν for neutrino, and γ for a photon (generally of gamma-ray energy).

1. THE PROTON-PROTON CHAIN
(Important below 15×10^6°K)
- (a) $_1H^1 + {}_1H^1 \rightarrow {}_1H^2 + \beta^+ + \nu$
- (b) $_1H^2 + {}_1H^1 \rightarrow {}_2He^3 + \gamma$
- (c) $_2He^3 + {}_2He^3 \rightarrow {}_2He^4 + 2{}_1H^1$

2. THE CARBON-NITROGEN CYCLE
(Important above 15×10^6°K)
- (a) $_6C^{12} + {}_1H^1 \rightarrow {}_7N^{13} + \gamma$
- (b) $\quad\quad {}_7N^{13} \rightarrow {}_6C^{13} + \beta^+ + \nu$
- (c) $_6C^{13} + {}_1H^1 \rightarrow {}_7N^{14} + \gamma$
- (d) $_7N^{14} + {}_1H^1 \rightarrow {}_8O^{15} + \gamma$
- $\quad\quad\quad {}_8O^{15} \rightarrow {}_7N^{15} + \beta^+ + \nu$
- (f) $_7N^{15} + {}_1H^1 \rightarrow {}_6C^{12} + {}_2He^4$

3. THE TRIPLE-ALPHA PROCESS
(Important above 10^8°K)
- (a) $_2He^4 + {}_2He^4 \rightarrow {}_4Be^8$
- (b) $_2He^4 + {}_4Be^8 \rightarrow {}_6C^{12}$

Orbital Data
for the Planets

| PLANET | SYMBOL | SEMIMAJOR AXIS | | SIDEREAL PERIOD | | SYNODIC PERIOD | MEAN ORBITAL SPEED | ORBITAL ECCEN- | INCLINA- TION OF ORBIT TO |
		AU	10^6 KM	TROPICAL YEARS	DAYS	(DAYS)	(KM/SEC)	TRICITY	ECLIPTIC
Mercury	☿	0.3871	57.9	0.24085	87.97	115.88	47.8	0.20563	7°004
Venus	♀	0.7233	108.1	0.61521	224.70	583.92	35.0	0.00679	3.394
Earth	⊕	1.0000	149.5	1.000039	365.26		29.8	0.01673	0.0
Mars	♂	1.5237	227.8	1.88089	686.98	779.94	24.2	0.09337	1.850
(Ceres)	①	2.7673	414	4.604		466.6	17.9	0.0765	10.615
Jupiter	♃	5.2028	778	11.86223		398.88	13.1	0.04844	1.305
Saturn	♄	9.5388	1426	29.4577		378.09	9.7	0.05568	2.490
Uranus	♅ or ♅	19.182	2868	84.013		369.66	6.8	0.04721	0.773
Neptune	♆	30.058	4494	164.793		367.48	5.4	0.00858	1.774
Pluto	♇	39.439	5896	247.686		366.72	4.7	0.25024	17.170

For the mean equator and equinox of 1960. Source: Supplement to *The American Ephemeris and Nautical Almanac*.

APPENDIX 10

Physical Data for the Planets

PLANET	DIAMETER KM	DIAMETER EARTH = 1	MASS (EARTH = 1)	MEAN DENSITY (GM/CM³)	PERIOD OF ROTATION	INCLINATION OF EQUATOR TO ORBIT	OBLATENESS	SURFACE GRAVITY (EARTH = 1)	ALBEDO	VISUAL MAGNITUDE AT MAXIMUM LIGHT	VELOCITY OF ESCAPE (KM/SEC)
Mercury	4,880	0.38	0.056	5.1	59 days	?	0	0.39	0.06	−1.9	4.3
Venus	12,112	0.95	0.82	5.3	242.9 days	23°	0	0.91	0.76	−4.4	10.3
Earth	12,742	1.00	1.00	5.52	$23^h56^m04^s$	23°27'	1/298.2	1.00	0.39	—	11.2
Mars	6,800	0.53	0.108	3.94	$24^h37^m23^s$	24°	1/192	0.38	0.15	−2.8	5.1
Jupiter	143,000	11.19	318.0	1.33	9^h50^m to 9^h55^m	3°	1/15	2.64	0.51	−2.5	57.5
Saturn	121,000	9.47	95.2	0.69	10^h14^m to 10^h38^m	27°	1/9.5	1.13	0.50	−0.4	35.4
Uranus	47,000	3.69	14.6	1.56	10^h45^m	98°	1/14	1.07	0.66	+5.6	21.9
Neptune	45,000	3.50	17.3	2.27	16^h?	29°	1/40	1.41	0.62	+7.9	24.4
Pluto	<6,000	<0.47	0.06?	?	6.387 days		0	?	?	+14.9	?

Satellites of Planets

PLANET	SATELLITE	DISCOVERED BY	MEAN DISTANCE FROM PLANET (KM)	SIDEREAL PERIOD (DAYS)	ORBITAL ECCENTRICITY	DIAMETER OF SATELLITE (KM)†	MASS (PLANET = 1)	APPROXIMATE MAGNITUDE AT OPPOSITION
Earth	Moon	—	384,405	27.322	0.055	3476	0.0123	−12.5
Mars	Phobos	A. Hall (1877)	9,380	0.319	0.021	(15)		+12
	Deimos	A. Hall (1877)	23,500	1.262	0.003	(8)		13
Jupiter	V	Barnard (1892)	180,500	0.498	0.003	(150)		13
	I Io	Galileo (1610)	421,600	1.769	0.000	3240	4×10^{-5}	5
	II Europa	Galileo (1610)	670,800	3.551	0.000	2830	2.5×10^{-5}	6
	III Ganymede	Galileo (1610)	1,070,000	7.155	0.002	4900	8×10^{-5}	5
	IV Callisto	Galileo (1610)	1,882,000	16.689	0.008	4570	5×10^{-5}	6
	VI	Perrine (1904)	11,470,000	250.57	0.158	(120)		14
	VII	Perrine (1905)	11,800,000	259.65	0.207	(50)		18
	X	Nicholson (1938)	11,850,000	263.55	0.130	(20)		19
	XII	Nicholson (1951)	21,200,000	631.1	0.169	(20)		18
	XI	Nicholson (1938)	22,600,000	692.5	0.207	(25)		19
	VIII	Melotte (1908)	23,500,000	738.9	0.378	(50)		17
	IX	Nicholson (1914)	23,700,000	758	0.275	(22)		19
Saturn	Janus	A. Dollfus (1966)	157,500	0.749	0.020	(350)	6.7×10^{-8}	14
	Mimas	W. Herschel (1789)	185,400	0.942	0.004	(500)	1.5×10^{-7}	12
	Enceladus	W. Herschel (1789)	237,900	1.370	0.000	(500)	1.1×10^{-6}	12
	Tethys	Cassini (1684)	294,500	1.888	0.000	(1000)	1.8×10^{-6}	11
	Dione	Cassini (1684)	377,200	2.737	0.002	(1000)	4×10^{-6}	11
	Rhea	Cassini (1672)	526,700	4.518	0.001	1350		10
	Titan	Huygens (1655)	1,221,000	15.945	0.029	4950	2.5×10^{-4}	8
	Hyperion	Bond (1848)	1,479,300	21.277	0.104	(400)		14
	Iapetus	Cassini (1671)	3,558,400	79.331	0.028	(1200)	2×10^{-7}	11
	Phoebe	W. Pickering (1898)	12,945,500	550.45	0.163	(300)	2.7×10^{-6}	14
Uranus	Miranda	Kuiper (1948)	123,000	1.414	0			17
	Ariel	Lassell (1851)	191,700	2.520	0.003	(600)		15
	Umbriel	Lassell (1851)	267,000	4.144	0.004	(400)		15
	Titania	W. Herschel (1787)	438,000	8.706	0.002	(1000)		14
	Oberon	W. Herschel (1787)	585,960	13.463	0.001	(900)		14
Neptune	Triton	Lassell (1846)	353,400	5.877	0.000	4000	1.3×10^{-3}	14
	Nereid	Kuiper (1949)	5,560,000	359.881	0.749	(300)		19

†A diameter of a satellite given in parentheses is estimated from the amount of sunlight it reflects.

The Nearest Stars

STAR	RIGHT ASCENSION (1950) h m	DECLINATION (1950) ° '	DISTANCE (pc)	PROPER MOTION "	RADIAL VELOCITY (km/sec)	SPECTRA OF COMPONENTS† A	B	C	VISUAL MAGNITUDES OF COMPONENTS A	B	C	ABSOLUTE VISUAL MAGNITUDES OF COMPONENTS A	B	C
α Centauri	14 36.2	−60 38	1.31	3.68	−23	G2V	K5V	M5eV	−0.01	+1.4	+10.7	+4.4	+5.8	+15
Barnard's Star	17 55.4	+4 33	1.83	10.30	−108	M5V	+		+9.54			+13.2		
Wolf 359	10 54.1	+7 19	2.35	4.84	+13	M6eV	+		+13.66			+16.8		
Lalande 21185	11 00.6	+36 18	2.49	4.78	−86	M2V	+		+7.47			+10.5		
Sirius	6 42.9	−16 39	2.67	1.32	−8	A1V	wd		−1.42	+8.7		+1.4	+11.5	
Luyten 726−8	1 36.4	−18 13	2.67	3.32	+29	M5.5eV	M6eV		+12.5	+12.9		+15.4	+15.8	
Ross 154	18 46.8	−23 54	2.94	0.67	−4	M4.5eV			+10.6			+13.3		
Ross 248	23 39.4	+43 55	3.16	1.58	−81	M5.5eV			+12.24			+14.7		
ε Eridani	3 30.6	−9 38	3.30	0.97	+15	K2V			+3.73			+6.1		
Ross 128	11 45.2	+1 06	3.37	1.36	−13	M5V			+11.13			+13.5		
Luyten 789−6	22 35.8	−15 36	3.37	3.27	−60	M5.5eV			+12.58			+14.9		
61 Cygni	21 04.7	+38 30	3.40	5.22	−64	K5V	K7V		+5.19	+6.02		+7.5	+8.3	
Procyon	7 36.7	+5 21	3.47	1.25	−3	F5 IV-V	wd		+0.38	+10.7		+2.7	+13.0	
ε Indi	21 59.6	−57 00	3.51	4.67	−40	K5V			+4.73			+7.0		
Σ 2398	18 42.2	+59 33	3.60	2.29	+8	M4V	M5V		+8.90	+9.69		+11.1	+11.9	
BD+43°44	0 15.5	+43 44	3.60	2.91	+18	M2.5eV	M4eV(?)		+8.07	+11.04		+10.3	+13.2	
τ Ceti	1 41.8	−16 12	3.64	1.92	−16	G8Vp			+3.50			+5.7		
CD−36°15693	23 02.6	−36 08	3.66	6.87	+10	M2V			+7.39			+9.6		
BD+5°1668	7 24.7	+5 23	3.76	3.73	+26	M4V			+9.82			+11.9		
CD−39°14192	21 14.3	−39 04	3.92	3.46	+21	M0V			+6.72			+8.7		
Kruger 60	22 26.2	+57 27	3.95	0.87	+26	M3V	M4.5eV		+9.77	+11.43		+11.8	+13.4	
Kapteyn's Star	5 09.7	−45 00	3.98	8.79	+242	M0			+8.81			+10.8		
Ross 614	6 26.8	−2 46	4.03	0.97	+24	M4.5eV(?)	?		+11.13	+14.8		+13.1	+16.8	
BD−12°4523	16 27.5	−12 32	4.10	1.24	−13	M4.5V			+10.13			+12.0		
v. Maanen's Star	0 46.5	+5 09	4.24	2.98	+30	wd			+12.36			+14.3		
Wolf 424	12 30.8	+9 18	4.45	1.87	−5	M5.5eV	M6eV		+12.7	+12.7		+14.4	+14.4	
BD+50°1725	10 08.3	+49 42	4.51	1.45	−27	M0V			+6.59			+8.3		
CD−37°15492	0 02.5	−37 36	4.57	6.09	+24	M3V			+8.59			+10.3		
CD−46°11540	17 24.9	−46 51	4.70	1.15	—	M4V			+9.34			+11.3		
BD+20°2465	10 16.9	+20 07	4.72	0.49	+10	M4eV	+		+9.5			+11.1		
CD−44°11909	17 33.5	−44 16	4.78	1.14	—	M5V			+11.2			+12.8		
CD−49°13515	21 30.2	−49 13	4.78	0.78	—	M3V			+9			+11		
AOe 17415−6	17 36.7	+68 23	4.84	1.31	−17	M3V			+9.1			+10.7		
Ross 780	22 50.5	−14 31	4.84	1.12	+9	M5V			+10.2			+11.8		
Lalande 25372	13 43.2	+15 10	4.87	2.30	+15	M2V			+8.6			+10.2		
CC 658	11 42.7	−64 33	4.90	2.69	—	wd			+11			+12.5		
40 Eridani	4 13.0	−7 44	5.00	4.08	−42	K0V	wd	M5eV	+4.5	+9.2	+11.0	+6.0	+10.7	+12.5

†A + symbol indicates that the star is an astrometric binary and has an unseen companion. See also the Notes to Appendix 13.

The Twenty Brightest Stars

STAR	RIGHT ASCENSION (1950) h m	DECLINATION (1950) ° '	DISTANCE (pc)†	PROPER MOTION "	SPECTRA OF COMPONENTS A	B	C	VISUAL MAGNITUDES OF COMPONENTS A	B	C	ABSOLUTE VISUAL MAGNITUDES OF COMPONENTS A	B	C
Sirius	6 42.9	−16 39	2.7	1.32	A1V	wd		−1.42	+8.7		+1.4	+11.5	
Canopus	6 22.8	−52 40	30	0.02	F0Ib-II			−0.72			−3.1		
α Centauri	14 36.2	−60 38	1.3	3.68	G2V	K5V	M5eV	−0.01	+1.4	+10.7	+4.4	+5.8	+15
Arcturus	14 13.4	+19 27	11	2.28	K2IIIp			−0.06			−0.3		
Vega	18 35.2	+38 44	8.0	0.34	A0V			+0.04			+0.5		
Capella	5 13.0	+45 57	14	0.44	GIII	M1V	M5V	+0.05	+10.2	+13.7	−0.7	+9.5	+13
Rigel	5 12.1	−8 15	250	0.00	B8 Ia	B9		+0.14	+6.6		−6.8	−0.4	
Procyon	7 36.7	+5 21	3.5	1.25	F5IV-V	wd		+0.38	+10.7		+2.7	+13.0	
Betelgeuse	5 52.5	+7 24	150	0.03	M2Iab			+0.41v			−5.5		
Achernar	1 35.9	−57 29	20	0.10	B5V			+0.51			−1.0		
β Centauri	14 00.3	−60 08	90	0.04	B1III	?		+0.63	+4		−4.1	−0.8	
Altair	19 48.3	+8 44	5.1	0.66	A7IV-V			+0.77			+2.2		
α Crucis	12 23.8	−62 49	120	0.04	B1IV	B3		+1.39	+1.9		−4.0	−3.5	
Aldebaran	4 33.0	+16 25	16	0.20	K5III	M2V		+0.86	+13		−0.2	+12	
Spica	13 22.6	−10 54	80	0.05	B1V			+0.91v			−3.6		
Antares	16 26.3	−26 19	120	0.03	M1Ib	B4eV		+0.92v	+5.1		−4.5	−0.3	
Pollux	7 42.3	+28 09	12	0.62	K0III			+1.16			+0.8		
Fomalhaut	22 54.9	−29 53	7.0	0.37	A3V	K4V		+1.19	+6.5		+2.0	+7.3	
Deneb	20 39.7	+45 06	430	0.00	A2Ia			+1.26			−6.9		
β Crucis	12 44.8	−59 24	150	0.05	B0.5IV			+1.28v			−4.6		

†Distances of the more remote stars have been estimated from their spectral types and apparent magnitudes, and are only approximate.

Notes: Several of the components listed are themselves spectroscopic binaries. A "v" after a magnitude denotes that the star is variable, in which case the magnitude at median light is given. A "p" after a spectral type indicates that the spectrum is peculiar. An "e" after a spectral type indicates that emission lines are present. When the luminosity classification is rather uncertain a range is given.

APPENDIX 14

The Messier Catalogue of Nebulae and Star Clusters

M	NGC OR (IC)	RIGHT ASCENSION (1950)		DECLI-NATION (1950)		APPARENT VISUAL MAGNITUDE	DESCRIPTION
		h	m	°	′		
1	1952	5	31.5	+22	00	11.3	"Crab" nebula in Taurus; remains of SN 1054
2	7089	21	30.9	−1	02	6.4	Globular cluster in Aquarius
3	5272	13	39.8	+28	38	6.3	Globular cluster in Canes Venatici
4	6121	16	20.6	−26	24	6.5	Globular cluster in Scorpio
5	5904	15	16.0	+2	17	6.1	Globular cluster in Serpens
6	6405	17	36.8	−32	10		Open cluster in Scorpio
7	6475	17	50.7	−34	48		Open cluster in Scorpio
8	6523	18	00.6	−24	23		"Lagoon" nebula in Sagittarius
9	6333	17	16.3	−18	28	8.0	Globular cluster in Ophiuchus
10	6254	16	54.5	−4	02	6.7	Globular cluster in Ophiuchus
11	6705	18	48.4	−6	20		Open cluster in Scutum Sobieskii
12	6218	16	44.7	−1	52	7.1	Globular cluster in Ophiuchus
13	6205	16	39.9	+36	33	5.9	Globular cluster in Hercules
14	6402	17	35.0	−3	13	8.5	Globular cluster in Ophiuchus
15	7078	21	27.5	+11	57	6.4	Globular cluster in Pegasus
16	6611	18	16.1	−13	48		Open cluster with nebulosity in Serpens
17	6618	18	17.9	−16	12		"Swan" or "Omega" nebula in Sagittarius
18	6613	18	17.0	−17	09		Open cluster in Sagittarius
19	6273	16	59.5	−26	11	7.4	Globular cluster in Ophiuchus
20	6514	17	59.4	−23	02		"Trifid" nebula in Sagittarius
21	6531	18	01.6	−22	30		Open cluster in Sagittarius
22	6656	18	33.4	−23	57	5.6	Globular cluster in Sagittarius
23	6494	17	54.0	−19	00		Open cluster in Sagittarius
24	6603	18	15.5	−18	27		Open cluster in Sagittarius
25	(4725)	18	28.7	−19	17		Open cluster in Sagittarius
26	6694	18	42.5	−9	27		Open cluster in Scutum Sobieskii
27	6853	19	57.5	+22	35	8.2	"Dumbbell" planetary nebula in Vulpecula
28	6626	18	21.4	−24	53	7.6	Globular cluster in Sagittarius
29	6913	20	22.2	+38	21		Open cluster in Cygnus
30	7099	21	37.5	−23	24	7.7	Globular cluster in Capricornus

M	NGC	RIGHT ASCENSION (1950)		DECLI-NATION (1950)		APPARENT VISUAL MAGNITUDE	DESCRIPTION
		h	m	°	′		
31	224	0	40.0	+41	00	3.5	Andromeda galaxy
32	221	0	40.0	+40	36	8.2	Elliptical galaxy; companion to M31
33	598	1	31.0	+30	24	5.8	Spiral galaxy in Triangulum
34	1039	2	38.8	+42	35		Open cluster in Perseus
35	2168	6	05.7	+24	21		Open cluster in Gemini
36	1960	5	33.0	+34	04		Open cluster in Auriga
37	2099	5	49.1	+32	33		Open cluster in Auriga
38	1912	5	25.3	+35	47		Open cluster in Auriga
39	7092	21	30.4	+48	13		Open cluster in Cygnus
40		12	20	+59			Close double star in Ursa Major
41	2287	6	44.9	−20	41		Loose open cluster in Canis Major
42	1976	5	32.9	−5	25		Orion nebula
43	1982	5	33.1	−5	19		Northeast portion of Orion nebula
44	2632	8	37	+20	10		Praesepe; open cluster in Cancer
45		3	44.5	+23	57		The Pleiades; open cluster in Taurus
46	2437	7	39.5	−14	42		Open cluster in Puppis
47	2478	7	52.4	−15	17		Loose group of stars in Puppis
48		8	11	−1	40		"Cluster of very small stars"; not identifiable
49	4472	12	27.3	+8	16	8.5	Elliptical galaxy in Virgo
50	2323	7	00.6	−8	16		Loose open cluster in Monoceros
51	5194	13	27.8	+47	27	8.4	"Whirlpool" spiral galaxy in Canes Venatici
52	7654	23	22.0	+61	20		Loose open cluster in Cassiopeia
53	5024	13	10.5	+18	26	7.8	Globular cluster in Coma Berenices
54	6715	18	51.9	−30	32	7.8	Globular cluster in Sagittarius
55	6809	19	36.8	−31	03	6.2	Globular cluster in Sagittarius
56	6779	19	14.6	+30	05	8.7	Globular cluster in Lyra
57	6720	18	51.7	+32	58	9.0	"Ring" nebula; planetary nebula in Lyra
58	4579	12	35.2	+12	05	9.6	Spiral galaxy in Virgo
59	4621	12	39.5	+11	56	10.0	Spiral galaxy in Virgo
60	4649	12	41.1	+11	50	9.0	Elliptical galaxy in Virgo
61	4303	12	19.3	+4	45	9.6	Spiral galaxy in Virgo
62	6266	16	58.0	−30	02	7.3	Globular cluster in Scorpio
63	5055	13	13.5	+42	17	8.6	Spiral galaxy in Canes Venatici
64	4826	12	54.2	+21	57	8.5	Spiral galaxy in Coma Berenices
65	3623	11	16.3	+13	22	9.4	Spiral galaxy in Leo
66	3627	11	17.6	+13	16	9.0	Spiral galaxy in Leo; companion to M65
67	2682	8	48.4	+12	00		Open cluster in Cancer
68	4590	12	36.8	−26	29	8.2	Globular cluster in Hydra
69	6637	18	28.1	−32	24	8.0	Globular cluster in Sagittarius
70	6681	18	40.0	−32	20	8.1	Globular cluster in Sagittarius
71	6838	19	51.5	+18	39		Globular cluster in Sagitta
72	6981	20	50.7	−12	45	9.3	Globular cluster in Aquarius
73	6994	20	56.2	−12	50		Open cluster in Aquarius
74	628	1	34.0	+15	32	9.3	Spiral galaxy in Pisces
75	6864	20	03.1	−22	04	8.6	Globular cluster in Sagittarius
76	650	1	39.1	+51	19	11.4	Planetary nebula in Perseus
77	1068	2	40.1	−0	12	8.9	Spiral galaxy in Cetus

M	NGC	RIGHT ASCENSION (1950) h m	DECLI-NATION (1950) ° ′	APPARENT VISUAL MAGNITUDE	DESCRIPTION
78	2068	5 44.2	+0 02		Small emission nebula in Orion
79	1904	5 22.1	−24 34	7.5	Globular cluster in Lepus
80	6093	16 14.0	−22 52	7.5	Globular cluster in Scorpio
81	3031	9 51.7	+69 18	7.0	Spiral galaxy in Ursa Major
82	3034	9 51.9	+69 56	8.4	Irregular galaxy in Ursa Major
83	5236	13 34.2	−29 37	8.3	Spiral galaxy in Hydra
84	4374	12 22.6	+13 10	9.4	Elliptical galaxy in Virgo
85	4382	12 22.8	+18 28	9.3	Elliptical galaxy in Coma Berenices
86	4406	12 23.6	+13 13	9.2	Elliptical galaxy in Virgo
87	4486	12 28.2	+12 40	8.7	Elliptical galaxy in Virgo
88	4501	12 29.4	+14 42	9.5	Spiral galaxy in Coma Berenices
89	4552	12 33.1	+12 50	10.3	Elliptical galaxy in Virgo
90	4569	12 34.3	+13 26	9.6	Spiral galaxy in Virgo
91		omitted			
92	6341	17 15.6	+43 12	6.4	Globular cluster in Hercules
93	2447	7 42.4	−23 45		Open cluster in Puppis
94	4736	12 48.6	+41 24	8.3	Spiral galaxy in Canes Venatici
95	3351	10 41.3	+11 58	9.8	Barred spiral galaxy in Leo
96	3368	10 44.1	+12 05	9.3	Spiral galaxy in Leo
97	3587	11 12.0	+55 17	11.1	"Owl" nebula; planetary nebula in Ursa Major
98	4192	12 11.2	+15 11	10.2	Spiral galaxy in Coma Berenices
99	4254	12 16.3	+14 42	9.9	Spiral galaxy in Coma Berenices
100	4321	12 20.4	+16 06	9.4	Spiral galaxy in Coma Berenices
101	5457	14 01.4	+54 36	7.9	Spiral galaxy in Ursa Major
102		omitted			
103	581	1 29.9	+60 26		Open cluster in Cassiopeia
104	4594	12 37.4	−11 21	8.3	Spiral galaxy in Virgo
105	3379	10 45.2	+13 01	9.7	Elliptical galaxy in Leo
106	4258	12 16.5	+47 35	8.4	Spiral galaxy in Canes Venatici
107	6171	16 29.7	−12 57	9.2	Globular cluster in Ophiuchus

The Chemical Elements

ELEMENT	SYMBOL	ATOMIC NUMBER	ATOMIC WEIGHT† (CHEMICAL SCALE)	NUMBER OF ATOMS PER TRILLION (10^{12}) HYDROGEN ATOMS‡
Hydrogen	H	1	1.0080	1.0×10^{12}
Helium	He	2	4.003	1.6×10^{11}
Lithium	Li	3	6.940	3.2×10^{3}
Beryllium	Be	4	9.013	6.3×10^{2}
Boron	B	5	10.82	7.6×10^{2}
Carbon	C	6	12.011	4.0×10^{8}
Nitrogen	N	7	14.008	1.1×10^{8}
Oxygen	O	8	16.0000	8.9×10^{8}
Fluorine	F	9	19.00	1.0×10^{6}
Neon	Ne	10	20.183	5.0×10^{8}
Sodium	Na	11	22.991	2.8×10^{5}
Magnesium	Mg	12	24.32	2.5×10^{7}
Aluminum	Al	13	26.98	1.7×10^{6}
Silicon	Si	14	28.09	3.2×10^{7}
Phosphorus	P	15	30.975	2.5×10^{5}
Sulfur	S	16	32.066	2.2×10^{7}
Chlorine	Cl	17	35.457	1.8×10^{6}
Argon	Ar(A)	18	39.944	7.6×10^{6}
Potassium	K	19	39.100	4.6×10^{4}
Calcium	Ca	20	40.08	1.4×10^{6}
Scandium	Sc	21	44.96	6.3×10^{2}
Titanium	Ti	22	47.90	6.0×10^{4}
Vanadium	V	23	50.95	1.3×10^{4}
Chromium	Cr	24	52.01	1.1×10^{5}
Manganese	Mn	25	54.94	7.4×10^{4}
Iron	Fe	26	55.85	5.4×10^{6}
Cobalt	Co	27	58.94	2.6×10^{4}
Nickel	Ni	28	58.71	8.9×10^{5}
Copper	Cu	29	63.54	3.2×10^{3}
Zinc	Zn	30	65.38	3.3×10^{3}
Gallium	Ga	31	69.72	3.2×10^{2}
Germanium	Ge	32	72.60	3.1×10^{2}
Arsenic	As	33	74.91	1.3×10^{2}
Selenium	Se	34	78.96	2.1×10^{3}
Bromine	Br	35	79.916	4.5×10^{2}
Krypton	Kr	36	83.80	1.6×10^{3}
Rubidium	Rb	37	85.48	2.2×10^{2}
Strontium	Sr	38	87.63	5.0×10^{2}

ELEMENT	SYMBOL	ATOMIC NUMBER	ATOMIC WEIGHT† (CHEMICAL SCALE)	NUMBER OF ATOMS PER TRILLION (10^{12}) HYDROGEN ATOMS‡
Yttrium	Y	39	88.92	1.6×10^3
Zirconium	Zr	40	91.22	4.5×10^2
Niobium (Columbium)	Nb(Cb)	41	92.91	2.0×10^2
Molybdenum	Mo	42	95.95	2.0×10^2
Technetium	Tc(Ma)	43	(99)	—
Ruthenium	Ru	44	101.1	66
Rhodium	Rh	45	102.91	23
Palladium	Pd	46	106.4	19
Silver	Ag	47	107.880	11
Cadmium	Cd	48	112.41	46
Indium	In	49	114.82	19
Tin	Sn	50	118.70	1.1×10^2
Antimony	Sb	51	121.76	2.6
Tellurium	Te	52	127.61	1.1×10^2
Iodine	I(J)	53	126.91	22
Xenon	Xe(X)	54	131.30	1.1×10^2
Cesium	Cs	55	132.91	14
Barium	Ba	56	137.36	3.2×10^2
Lanthanum	La	57	138.92	13
Cerium	Ce	58	140.13	20
Praseodymium	Pr	59	140.92	4.6
Neodymium	Nd	60	144.27	23
Promethium	Pm	61	(147)	—
Samarium	Sm(Sa)	62	150.35	7.8
Europium	Eu	63	152.0	3.0
Gadolinium	Gd	64	157.26	11
Terbium	Tb	65	158.93	1.7
Dysprosium	Dy(Ds)	66	162.51	12
Holmium	Ho	67	164.94	2.5
Erbium	Er	68	167.27	6.9
Thulium	Tm(Tu)	69	168.94	1.2
Ytterbium	Yb	70	173.04	6.0
Lutecium	Lu(Cp)	71	174.99	1.1
Hafnium	Hf	72	178.50	2.5
Tantalum	Ta	73	180.95	5.6
Tungsten	W	74	183.86	4.0
Rhenium	Re	75	186.22	8.0
Osmium	Os	76	190.2	25
Iridium	Ir	77	192.2	16
Platinum	Pt	78	195.09	50
Gold	Au	79	197.0	4.6
Mercury	Hg	80	200.61	5.6
Thallium	Tl	81	204.39	3.5
Lead	Pb	82	207.21	32
Bismuth	Bi	83	209.00	3.2
Polonium	Po	84	(209)	—
Astatine	At	85	(210)	—
Radon	Rn	86	(222)	—
Francium	Fr(Fa)	87	(223)	—
Radium	Ra	88	226.05	—
Actinium	Ac	89	(227)	—
Thorium	Th	90	232.12	1.0
Protoactinium	Pa	91	(231)	—

ELEMENT	SYMBOL	ATOMIC NUMBER	ATOMIC WEIGHT† (CHEMICAL SCALE)	NUMBER OF ATOMS PER TRILLION (10^{12}) HYDROGEN ATOMS‡
Uranium	U(Ur)	92	238.07	0.5
Neptunium	Np	93	(237)	—
Plutonium	Pu	94	(244)	—
Americium	Am	95	(243)	—
Curium	Cm	96	(248)	—
Berkelium	Bk	97	(247)	—
Californium	Cf	98	(251)	—
Einsteinium	E	99	(254)	—
Fermium	Fm	100	(253)	—
Mendeleevium	Mv	101	(256)	—
Nobelium	No	102	(253)	—

†Where mean atomic weights have not been well determined, the atomic mass numbers of the most stable isotopes are given in parentheses.
‡Adapted from a compilation by L. H. Aller.

APPENDIX 16

The Constellations

CONSTELLATION (LATIN NAME)	GENITIVE CASE ENDING	ENGLISH NAME OR DESCRIPTION	ABBRE-VIA-TION	APPROXIMATE POSITION α (h)	δ (°)
Andromeda	Andromedae	Princess of Ethiopia	And	1	+40
Antlia	Antliae	Air pump	Ant	10	−35
Apus	Apodis	Bird of Paradise	Aps	16	−75
Aquarius	Aquarii	Water bearer	Aqr	23	−15
Aquila	Aquilae	Eagle	Aql	20	+5
Ara	Arae	Altar	Ara	17	−55
Aries	Arietis	Ram	Ari	3	+20
Auriga	Aurigae	Charioteer	Aur	6	+40
Boötes	Boötis	Herdsman	Boo	15	+30
Caelum	Caeli	Graving tool	Cae	5	−40
Camelopardus	Camelopardis	Giraffe	Cam	6	−70
Cancer	Cancri	Crab	Cnc	9	+20
Canes Venatici	Canum Venaticorum	Hunting dogs	CVn	13	+40
Canis Major	Canis Majoris	Big dog	CMa	7	−20
Canis Minor	Canis Minoris	Little dog	CMi	8	+5
Capricornus	Capricorni	Sea goat	Cap	21	−20
†Carina	Carinae	Keel of Argonauts' ship	Car	9	−60
Cassiopeia	Cassiopeiae	Queen of Ethiopia	Cas	1	+60
Centaurus	Centauri	Centaur	Cen	13	−50
Cephus	Cephei	King of Ethiopia	Cep	22	+70
Cetus	Ceti	Sea monster (whale)	Cet	2	−10
Chamaeleon	Chamaeleontis	Chameleon	Cha	11	−80

CONSTELLATION (LATIN NAME)	GENITIVE CASE ENDING	ENGLISH NAME OR DESCRIPTION	ABBRE- VIA- TION	APPROXIMATE POSITION	
				α h	δ °
Circinus	Circini	Compasses	Cir	15	−60
Columba	Columbae	Dove	Col	6	−35
Coma Berenices	Comae Berenices	Berenice's hair	Com	13	+20
Corona Australis	Coronae Australis	Southern crown	CrA	19	−40
Corona Borealis	Coronae Borealis	Northern crown	CrB	16	+30
Corvus	Corvi	Crow	Crv	12	−20
Crater	Crateris	Cup	Crt	11	−15
Crux	Crucis	Cross (southern)	Cru	12	−60
Cygnus	Cygni	Swan	Cyg	21	+40
Delphinus	Delphini	Porpoise	Del	21	+10
Dorado	Doradus	Swordfish	Dor	5	−65
Draco	Draconis	Dragon	Dra	17	+65
Equuleus	Equulei	Little horse	Equ	21	+10
Eridanus	Eridani	River	Eri	3	−20
Fornax	Fornacis	Furnace	For	3	−30
Gemini	Geminorum	Twins	Gem	7	+20
Grus	Gruis	Crane	Gru	22	−45
Hercules	Herculis	Hercules, son of Zeus	Her	17	+30
Horologium	Horologii	Clock	Hor	3	−60
Hydra	Hydrae	Sea serpent	Hya	10	−20
Hydrus	Hydri	Water snake	Hyi	2	−75
Indus	Indi	Indian	Ind	21	−55
Lacerta	Lacertae	Lizard	Lac	22	+45
Leo	Leonis	Lion	Leo	11	+15
Leo Minor	Leonis Minoris	Little lion	LMi	10	+35
Lepus	Leporis	Hare	Lep	6	−20
Libra	Librae	Balance	Lib	15	−15
Lupus	Lupi	Wolf	Lup	15	−45
Lynx	Lyncis	Lynx	Lyn	8	+45
Lyra	Lyrae	Lyre or harp	Lyr	19	+40
Mensa	Mensae	Table Mountain	Men	5	−80
Microscopium	Microscopii	Microscope	Mic	21	−35
Monoceros	Monocerotis	Unicorn	Mon	7	−5
Musca	Muscae	Fly	Mus	12	−70
Norma	Normae	Carpenter's level	Nor	16	−50
Octans	Octantis	Octant	Oct	22	−85
Ophiuchus	Ophiuchi	Holder of serpent	Oph	17	0
Orion	Orionis	Orion, the hunter	Ori	5	+5
Pavo	Pavonis	Peacock	Pav	20	−65
Pegasus	Pegasi	Pegasus, the winged horse	Peg	22	+20
Perseus	Persei	Perseus, hero who saved Andromeda	Per	3	+45
Phoenix	Phoenicis	Phoenix	Phe	1	−50
Pictor	Pictoris	Easel	Pic	6	−55
Pisces	Piscium	Fishes	Psc	1	+15
Piscis Austrinus	Piscis Austrini	Southern fish	PsA	22	−30
†Puppis	Puppis	Stern of the Arognauts' ship	Pup	8	−40
†Pyxis (= Malus)	Pyxidis	Compass on the Argonauts' ship	Pyx	9	−30
Reticulum	Reticuli	Net	Ret	4	−60
Sagitta	Sagittae	Arrow	Sge	20	+10
Sagittarius	Sagittarii	Archer	Sgr	19	−25
Scorpius	Scorpii	Scorpion	Sco	17	−40

CONSTELLATION (LATIN NAME)	GENITIVE CASE ENDING	ENGLISH NAME OR DESCRIPTION	ABBRE-VIA-TION	APPROXIMATE POSITION	
				α (h)	δ (°)
Sculptor	Sculptoris	Sculptor's tools	Scl	0	−30
Scutum	Scuti	Shield	Sct	19	−10
Serpens	Serpentis	Serpent	Ser	17	0
Sextans	Sextantis	Sextant	Sex	10	0
Taurus	Tauri	Bull	Tau	4	+15
Telescopium	Telescopii	Telescope	Tel	19	−50
Triangulum	Trianguli	Triangle	Tri	2	+30
Triangulum Australe	Trianguli Australis	Southern triangle	TrA	16	−65
Tucana	Tucanae	Toucan	Tuc	0	−65
Ursa Major	Ursae Majoris	Big bear	UMa	11	+50
Ursa Minor	Ursae Minoris	Little bear	UMi	15	+70
†Vela	Velorum	Sail of the Argonauts' ship	Vel	9	−50
Virgo	Virginis	Virgin	Vir	13	0
Volans	Volantis	Flying fish	Vol	8	−70
Vulpecula	Vulpeculae	Fox	Vul	20	+25

†The four constellations Carina, Puppis, Pyxis, and Vela originally formed the single constellation, Argo Navis.

APPENDIX 17

Star Maps

The star maps, one for each month, are printed on cards inserted in a pocket inside the back cover of this book. To learn the stars and constellations, the card containing the map for the current month should be taken outdoors and compared directly with the sky. The maps were designed for a latitude of about 35° N but are useful anywhere in the continental United States. Each map shows the appearance of the sky at about 9:00 P.M. (Standard Time) near the middle of the month for which it is intended; near the beginning and end of the month, it shows the sky as it appears about 10:00 P.M. and 8:00 P.M., respectively. To use a map, hold the card vertically, and turn it so that the direction you are facing is shown at the bottom.

These star maps were prepared by C. H. Cleminshaw, director of the Griffith Observatory, for his *Monthly Star Maps* (published by the Griffith Observatory, Los Angeles, California, 1962), and are reproduced here by the very kind permission of Dr. Cleminshaw.

Index

Index

Page numbers in boldface refer to principal discussions. Page numbers in parentheses refer to figures.

D

E